动物传染病防控技术的研究与应用

Research and Application of Animal Infectious Diseases Prevention and Control Technology

谢芝勋　主　编

广西科学技术出版社

·南宁·

图书在版编目（CIP）数据

动物传染病防控技术的研究与应用 ：英文 / 谢芝勋主编. -- 南宁 ：广西科学技术出版社，2024. 11.

ISBN 978-7-5551-2261-6

Ⅰ. S855

中国国家版本馆 CIP 数据核字第 2024VU4275 号

Dongwu Chuanranbing Fangkong Jishu De Yanjiu Yu Yingyong

动物传染病防控技术的研究与应用

谢芝勋　主编

策　　划：黎志海　　封面设计：梁　良

责任编辑：覃　艳　　责任校对：吴书丽

责任印制：陆　弟

出 版 人：岑　刚

出版发行：广西科学技术出版社　　地　　址：广西南宁市东葛路 66 号

邮政编码：530023　　网　　址：http://www.gxkjs.com

经　　销：全国各地新华书店

印　　刷：广西民族印刷包装集团有限公司

开　　本：889mm × 1194mm　1/16

印　　张：71.75　　彩插 20　　字　　数：1062 千字

版　　次：2024 年 11 月第 1 版　　印　　次：2024 年 11 月第 1 次印刷

书　　号：ISBN 978-7-5551-2261-6

定　　价：598.00 元（上、下册）

Contents

Chapter Five
Studies on Rapid Diagnostic Technology

Part Ⅰ High-throughput GenomeLab Gene Expression Profiler (GeXP) Diagnostic Technology

Simultaneous typing of nine avian respiratory pathogens using a novel GeXP analyzer-based multiplex PCR assay

Xie Zhixun, Luo Sisi, Xie Liji, Liu Jiabo, Pang Yaoshan, Deng Xianwen, Xie Zhiqin, Fan Qing, and Khan Mazhar I

Abstract

A new, rapid, and high-throughput GenomeLab Gene Expression Profiler (GeXP) analyzer-based multiplex PCR method was developed for simultaneous detection and differentiation of nine avian respiratory pathogens. The respiratory pathogens included in this study were avian influenza subtypes H5, H7, and H9, infectious bronchitis virus (IBV), Newcastle disease virus (NDV), infectious laryngotracheitis virus (ILTV), *Mycoplasma gallisepticum* (MG), *Mycoplasma synoviae* (MS) and *Haemophilus paragallinarum* (HPG). Ten pairs of primers were designed using conserved and specific sequence genes of AIV subtypes and respiratory pathogens from GenBank. Single and mixed pathogen cDNA/DNA templates were used to evaluate the specificity of the GeXP-multiplex assay. The corresponding specific DNA products were amplified for each pathogen. The specific DNA product amplification peaks of nine respiratory pathogens were observed on the GeXP analyzer. Non-respiratory avian pathogens, including chicken infectious anemia virus, fowl adenovirus, avian reovirus and infectious bursal disease virus, did not produce DNA products. The detection limit for the GeXP-multiplex assay was determined to be 100 copies/μL using various pre-mixed plasmids/ssRNAs containing known target genes of the respiratory pathogens. Further, GeXP-multiplex PCR assay was 100% specific when 24 clinical samples with respiratory infections were tested in comparison with conventional PCR method. The GeXP-multiplex PCR assay provides a novel tool for simultaneous detection and differentiation of nine avian respiratory pathogens.

Keywords

avian respiratory pathogens, GenomeLab Gene Expression Profiler (GeXP), multiplex PCR

Introduction

Avian influenza virus (AIV) subtypes H5, H7, H9, Newcastle disease virus (NDV), infectious bronchitis virus (IBV), infectious laryngotracheitis virus (ILTV), *Mycoplasma gallisepticum* (MG), *Mycoplasma synoviae* (MS) and *Haemophilus paragallinarum* (HPG) are key avian respiratory pathogens. It can be difficult to alone definitively differentiate these respiratory pathogens by clinical symptoms and pathology.

Identification and differentiation of avian influenza subtypes and respiratory pathogens has traditionally involved time consuming and labor intensive procedures including isolation of the pathogenic agents and differentiation by serological tests[1-11]. Molecular typing methods for rapid detection and identification of avian respiratory pathogens have been developed and are used currently[12-22]. Although useful, most molecular methods are limited to the detection of a few pathogens in one reaction[13-15, 22]. Therefore, a rapid, cost effective, and high throughput method for detection and differentiation of pathogens in one test tube would be advantageous, enhancing rapid response for prompt treatment and control.

The GenomeLab Gene Expression Profiler (GeXP) analyzer is a multiplex gene expression analysis platform developed by Beckman Coulter Company. GeXP assays have been successfully used in the rapid identification of several inflammatory and cytokine gene targets in normal colon and polyps colon tumors[23, 24], as well as detection of 68 unique varicella zoster virus gene transcripts and genotyping of several human papilloma viruses[25, 26]. GeXP analyzer (Beckman) built-in software program determine PCR product according to sizes of amplicon and display differences of various specific genes under investigation and compare to the expected PCR product size to identify each reaction. The GeXP-multiplex PCR assay provide high sensitivity and specificity compared to other multiple detection methods[26].

In this study, a GeXP analyzer-based multiplex PCR (GeXP-multiplex PCR) assay was developed and optimized to simultaneously detect nine avian respiratory pathogens, including AIV subtypes (H5, H7, H9), NDV, IBV, ILTV, MG, MS and HPG. This assay can be implemented effectively in routine testing laboratories, allowing users to process more samples in less time than existing assays and platforms.

Materials and methods

Pathogens and DNA/RNA extractions

Avian respiratory pathogen reference strains, field isolates and other non-respiratory pathogens used in this study are described in Table 5-1-1. Viral DNA/RNAs were extracted from AIV, NDV, IBV, ILTV, chicken infectious anemia virus, fowl adenovirus, avian reovirus and infectious bursal disease viruses using the MiniBEST Viral RNA/DNA Extraction kit (TaKaRa, Dalian, China) according to the manual. The DNA/RNA was aliquoted and stored at −80 ℃ until used. The DNA was extracted from MG, MS, and HPG by using the TIANamp Bacteria DNA Kit (Tiangen, China). Genomic RNAs were used to generate cDNA by reverse transcription as described previously[27].

Table 5-1-1 Sources of pathogens used and GeXP assay results

Pathogens/field samples	Numbers of sample	Source	Results									
			M gene	H5 AIV	H7 AIV	H9 AIV	NDV	IBV	ILTV	MG	MS	HPG
Reference samples												
Inactivated H5N1 AIV Re-1	1	HVRI	+	+	−	−	−	−	−	−	−	−
cDNA of H5N3 AIV Duck/HK/313/78	1	HKU	+	+	−	−	−	−	−	−	−	−
cDNA of H5N2/chicken/QT35/87	1	PU	+	+	−	−	−	−	−	−	−	−
cDNA of H5N5/chicken/QT35/98	1	PU	+	+	−	−	−	−	−	−	−	−
cDNA of AIV H5N9/chicken/QT35/98	1	PU	+	+	−	−	−	−	−	−	−	−
H7N2 AIV Duck/HK/47/76	1	HKU	+	−	+	−	−	−	−	−	−	−
cDNA of AIV H7N2/chicken PA/3979/97	1	PU	+	−	+	−	−	−	−	−	−	−
H9 subtypes of AIV (Guangxi isolate)	9	GVRI	+	−	−	+	−	−	−	−	−	−
H9N6/Duck/HK/147/77	1	HKU	+	−	−	+	−	−	−	−	−	−
NDV F48E9	1	GVRI	−	−	−	−	+	−	−	−	−	−

continued

Pathogens/field samples	Numbers of sample	Source	Results									
			M gene	H5 AIV	H7 AIV	H9 AIV	NDV	IBV	ILTV	MG	MS	HPG
NDV La Sota	1	GVRI	−	−	−	−	+	−	−	−	−	−
NDV (GX1/00, GX6/02, GX7/02, GX8/03, GX9/03 and GX10/03)	6	GVRI	−	−	−	−	+	−	−	−	−	−
IBV Massachussetts 41	1	GVRI	−	−	−	−	−	+	−	−	−	−
IBV (4785/03, 5359/03, 5483/03, GXIB/02)	4	GVRI	−	−	−	−	−	+	−	−	−	−
IBV JMK	1	GVRI	−	−	−	−	−	+	−	−	−	−
ILTV (AV1231, AV22, AV23)	3	CIVDC	−	−	−	−	−	−	+	−	−	−
MG S6	3	GVRI	−	−	−	−	−	−	−	+	−	−
MS K1415	2	GVRI	−	−	−	−	−	−	−	−	+	−
HPG AV269	1	CIVDC	−	−	−	−	−	−	−	−	−	+
Other pathogens												
Chicken infectious anemia virus (GXC060821)	1	GVRI	−	−	−	−	−	−	−	−	−	−
Fowl adenovirus (Celo)	1	GVRI	−	−	−	−	−	−	−	−	−	−
Avian reovirus (S1133)	1	GVRI	−	−	−	−	−	−	−	−	−	−
Infectious bursal disease virus (AV6)	1	CIVDC	−	−	−	−	−	−	−	−	−	−
H1N3 AIV Duck/HK/717/79-d1	1	HKU	+	−	−	−	−	−	−	−	−	−
H1N1 AIV Human/NJ/8/76	1	HKU	+	−	−	−	−	−	−	−	−	−
H2N3 AIV Duck/HK/77/76	1	HKU	+	−	−	−	−	−	−	−	−	−
H3N6 AIV Duck/HK/526/79/2B	1	HKU	+	−	−	−	−	−	−	−	−	−
H4N5 AIV Duck/HK/668/79	1	HKU	+	−	−	−	−	−	−	−	−	−
H6N8 AIV Duck/HK/531/79	1	HKU	+	−	−	−	−	−	−	−	−	−
H8N4 AIV Turkey/ont/6118/68	1	HKU	+	−	−	−	−	−	−	−	−	−
H10N3 AIV Duck/HK/876/80	1	HKU	+	−	−	−	−	−	−	−	−	−
H11N3 AIV Duck/HK/661/79	1	HKU	+	−	−	−	−	−	−	−	−	−
H12N5 AIV Duck/HK/862/80	1	HKU	+	−	−	−	−	−	−	−	−	−
H13N5 AIV Gull/MD/704/77	1	HKU	+	−	−	−	−	−	−	−	−	−
Sample mixture												
AIV H5+AIV H9	1	GVRI	+	+	−	+	−	−	−	−	−	−
MG+IBV	1	GVRI	−	−	−	−	−	+	−	+	−	−
AIV H5+AIV H7+AIV H9+NDV+IBV+ILTV+MG+MS+HPG	1	GVRI	+	+	+	+	+	+	+	+	+	+
Clinical samples												
H9 subtypes of AIV	5	GVRI	+	−	−	+	−	−	−	−	−	−
NDV	6	GVRI	−	−	−	−	+	−	−	−	−	−
IBV	4	GVRI	−	−	−	−	−	+	−	−	−	−
ILTV	3	GVRI	−	−	−	−	−	−	+	−	−	−
MG	3	GVRI	−	−	−	−	−	−	−	+	−	−
MS	2	GVRI	−	−	−	−	−	−	−	−	+	−
HPG	1	GVRI	−	−	−	−	−	−	−	−	−	+

Note: HVRI, Harbin Veterinary Research Institute, China; HKU, The University of Hong Kong, China; GVRI, Guangxi Veterinary Research Institute, China; CIVDC, China Institute of Veterinary Drugs Control, China; PU, University of Pennsylvania.

Primers and plasmids

The GeXP-multiplex assay consisted of 10 pairs of primers including one pair of AIV universal primers and 9 pairs of avian respiratory pathogens specific primers (listed in Table 5-1-2). Each of the primers consisted of a gene-specific sequence for each pathogen fused at the 5' end to the universal sequence. The gene specific portions of the primers were designed based on the sequence information obtained from GenBank. This strategy of selecting primers was developed by Beckman Coulter Company and described previously[28]. The AIV universal primers were designed in a highly conserved region of the matrix (M) gene. The AIV primers for H5, H7 and H9 subtypes were designed in a specific region of the HA gene. The NDV primers were designed in a region of the fusion (F) gene. The IBV primers were designed in a specific region of the nucleocapsid protein (N) gene. The ILTV primers were designed in a highly conserved region of the thymidine kinase (TK) gene. The MG primers were designed in a region of the 16rRNA gene. The MS primers were designed in a conserved region of the phase-variable surface lipoprotein hemagglutinin (VLHA) gene. The HPG primers were designed in a specific region of the hemagglutinin antigen (hagA) gene. Pathogen-specific primers sequences were evaluated using the NCBI Primer-Blast and Primer Premier 5.0 software. All primers were synthesized and purified by polyacrylamide gel electrophoresis (Invitrogen, Guangzhou, China).

Table 5-1-2 Sequences of primers used for GeXP aasay

Pathogens	Forward primer sequence (5'-3')	Reverse primer sequence (5'-3')	Amplicon size / bp	Target region
AIV-type A	AGGTGACACTATAGAATAAGCCGAGAT CGCGCAGA	GTACGACTCACTATAGGGACGCTCACTG GGCACGGT	192	M
AIV-H5	AGGTGACACTATAGAATAGGAAAGTGT AAGAAACGGAACGTA	GTACGACTCACTATAGGGACACATCCAT AAAGAYAGACCAGC	223	HA
AIV-H7	AGGTGACACTATAGAATAAGTGGGGAC AAGTTGATAACAGT	GTACGACTCACTATAGGGAAGCCCCATT GAAGSTGAA	177	HA
AIV-H9	AGGTGACACTATAGAATAACCATTTATT CGACTGTCGCCT	GTACGACTCACTATAGGGACATTGGACA TGGCCCAGAA	118	HA
NDV	AGGTGACACTATAGAATAAGCATTGCTG CAACCAATGA	GTACGACTCACTATAGGGAACAAACTG CTGCATCTTCCC	128	F
IBV	AGGTGACACTATAGAATAGGAGGACCT GATGGTAATTTCC	GTACGACTCACTATAGGGATGCGAGAR CCCTTCTTCTGCT	240	N
ILTV	AGGTGACACTATAGAATAGTGCAGTTTT GCTCCGAGTT	GTACGACTCACTATAGGGAATAGACGG CAACCTCTCCAA	184	TK
MG	AGGTGACACTATAGAATACAAGGAAGC GCATGTCTAGG	GTACGACTCACTATAGGGACTGGGTTTT AATTGTTTCACCG	202	16rRNA
MS	AGGTGACACTATAGAATAGACCCTGTAG AGRCTGCTAAAA	GTACGACTCACTATAGGGAGCTTCAACT TGTCTTTTTAATGC	137	VLHA
HPG	AGGTGACACTATAGAATATCTGATTATA AACCAACTAAAAGAGCA	GTACGACTCACTATAGGGAGCAAGTTCT GGTAATGATGGTAAGTT	158	HA

Note: Universal tag sequences are underlined shows mutant sites. Abbreviations are as follows: Y= C or T; S= G or C.

Ten specific genes of the nine avian respiratory pathogens were amplified using the primers listed in Table 5-1-2, the specific amplified DNA (amplicon) for each pathogen was then cloned into the pEASY-T1 vector (TransGen Biotech, China) as per manufacturer's protocol. Constructed plasmids were transcribed in *E. coli* (TransGen Biotech, China) following the manufacturer's protocol. The isolation of plasmids was carried out using a plasmid extraction kit (TransGen Biotech, China). The recombinant plasmids obtained were sequenced. Sequence data were analyzed using DNASTAR software and were compared with corresponding sequence data in GenBank.

The copy number of the plasmid for ILTV (AV1231), MG (S6), MS (K1415) and HPG (AV269) was calculated according to the formula (copies/μL=6 × 10^{23} × plasmid DNA concentration (ng/μL) × 10^{-9}/molecular weight (g/mol))[29].

The plasmids for AIV (H5N1 AIV Re-1, H7N2 AIV Duck/HK/47/76, H9N6/Duck/HK/147/77), NDV (F48E9) and IBV (Massachusetts 41) were used to produce ssRNA via *in vitro* transcription using Thermo Scientific TranscriptAid T7 High Yield Transcription Kit (Thermo Scientific). The copy number of ssRNA for AIV, NDV and IBV were calculated according to the formula (copies/μL=6 × 10^{23} × ssRNA concentration (ng/μL) × 10^{-9}/molecular weight (g/mol))[29]. RNA concentration was determined measuring the ultraviolet light absorbance at 260 nm wavelength as described previously[29].

Specificity of the GeXP-multiplex assay

The mono-GeXP PCR assay was developed with DNA/cDNA from nine avian respiratory pathogens to evaluate the specificity of each pair of gene-specific primers and ascertain the actual amplicon size of each target region. Both the mono-GeXP PCR and the GeXP-multiplex PCR assays were performed with the Genome Lab GeXP Start kit (Beckman Coulter, USA) in a 10 μL volume containing 2 μL of 5 × PCR Buffer (including 0.25 μM concentration of Tag-F: 5'-AGGTGACACTATAGAATA-3'and 0.25 μM concentration of Tag-R: 5'-GTACGACTCACTATAGGGA-3'), 2 μL $MgCl_2$ (25 μM), 1 μL primer (pool of ten pairs of primers concentration of 0.2 μM each), 0.35 μL Thermo-Start DNA Polymerase, 1 μL cDNA/DNA, nuclease-free water was added to a volume of 10 μL. The mono-GeXP PCR assay was performed under the following conditions: 95 ℃ for 10 min followed by 30 cycles that each consisted of denaturing at 95 ℃ for 30 s, annealing at 50 ℃ for 30 s, and extension at 72 ℃ for 60 s. The sample was then heated to 72 ℃ for 10 min for final extension. Nucleic acids (DNA/cDNA) from chicken infectious anemia virus, fowl adenovirus, avian reovirus and infectious bursal disease virus were used as negative controls.

Separation by capillary electrophoresis and fragment analysis

PCR product separation and detection were performed on a GenomeLab GeXP genetic analysis system (Beckman Coulter) by capillary electrophoresis, following the protocols described previously[30]. After amplified DNA amplicons were separated, the peaks were initially analyzed using the GeXP system software and matched to the appropriate genes. The peak height for each gene was reported in the electropherograms.

Sensitivity of the GeXP-multiplex PCR assay

Absolute sensitivity was tested by using serial 10-fold dilutions of the plasmids (ILTV, MG, MS and HPG) and ssRNA *in vitro* transcription (AIV, NDV and IBV) (separately or 10 samples mixed together) containing the specific gene sequences of the nine avian respiratory pathogens. The concentrations of the specific primers were optimized according to the amplification efficiency of the GeXP-multiplex PCR assay using a single template. The sensitivity of the optimized GeXP-multiplex PCR assay was re-evaluated by using 10 premixed plasmids and ssRNA (*in vitro* transcription) ranging from 10^5 copies/μL to 10^1 copies/μL replicated three times.

Artificial mixture of respiratory pathogens

In order to emulate mixed infections, we randomly choose avian pathogens, and extracted their DNA/RNAs. RNA viruses were transcribed into cDNA as described above. DNA/cDNA were mixed together in equal concentrations for the templates and subjected to optimized GeXP-multiplex PCR assay.

Interference assay

Since the presence of other high quantity templates could alter the efficiency of GeXP-multiplex PCR amplification, different amount of templates (10^3 copies to 10^7 template copies) were mixed, and tested by the GeXP-multiplex PCR assay. Result was compared with those of single template GeXP-multiplex PCR assay.

Detection in clinical samples

Twenty four clinical samples, including five samples of H9 subtypes of AIV, six samples of NDV, four samples of IBV, three samples of ILTV, three samples of MG, two samples of MS and one sample of HPG were tested positive by routine PCR. These samples were then extracted by MiniBEST Viral RNA/DNA Extraction kit (TaKaRa, Dalian, China) and TIANamp Bacteria DNA Kit (Tiangen, China). Genomic RNA from AIV, NDV and IBV were transcribed into cDNA as described above and these DNA and cDNA were tested by the optimized GeXP-multiplex PCR assay.

Results

Specificity of the GeXP-multiplex PCR assay

The DNA/cDNA from nine avian respiratory pathogens was individually used as template to evaluate the specificity of each pair of gene-specific primers. In mono-GeXP PCR assays, the AIV universal primers could amplify the target M genes of all AIV serotypes, but each pair of pathogen specific primers generated corresponding genes only of the targeted pathogens without cross-amplification (Table 5-1-1). The amplicon sizes for the pathogen were as follows: AIV M gene, 190～192 bp; AIV-H5, 223～225 bp; AIV-H7, 175～177 bp; AIV-H9, 116～119 bp; NDV, 126～128 bp; IBV, 239～241 bp; ILTV, 183～185 bp; MG, 201～203 bp; MS, 137～140 bp; HPG, 157～159 bp (Table 5-1-2).

These nine avian respiratory pathogens were detected via the GeXP-multiplex PCR assay and the following specific amplification peaks were observed (Figure 5-1-1) : AIV-H5, 224.68 bp and 191.74 bp; AIV-H7, 176.02 bp and 191.30 bp; AIV-H9, 117.05 bp and 190.18 bp; NDV, 128 bp; IBV, 240 bp; ILTV, 184 bp; MG, 202 bp; MS, 139 bp and HPG, 156 bp. Specific amplification peaks of the M gene of AIV were observed when H1N3, H1N1, H2N3, H3N6, H4N5, H6N8, H8N4, H10N3, H11N3, H12N5 and H13N5 were tested. No specific amplification peaks were observed when infectious anemia virus, fowl adenovirus, avian reovirus, infectious bursal disease virus were tested.

Sensitivity of the GeXP multiplex PCR assay

The GeXP-multiplex PCR assay detected as little as 10 copies/μL of ILTV, MG, IBV and 10^2 copies/μL of AIV-type A, AIV-H5, AIV-H7, AIV-H9, NDV, MS and HPG when the GeXP-multiplex PCR assay used plasmids and ssRNA *in vitro* transcription individually (electropherograms not shown). The GeXP multiplex PCR assay detected as little as 10^2 copies/μL of AIV H5, AIV H7, AIV H9, ND, IBV, ILTV, MG, MS and HPG when the GeXP multiplex PCR assay used plasmids and ssRNA *in vitro* transcription mixed together (Figure 5-1-2).

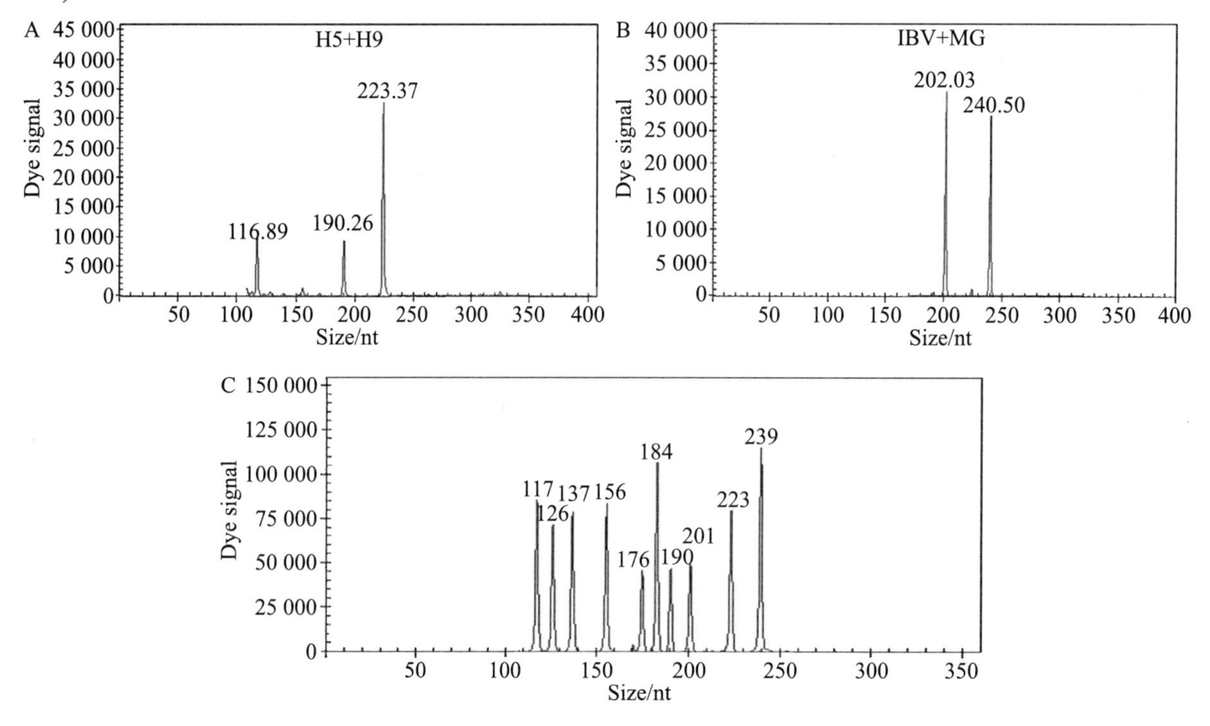

A: AIV-H5 and AIV-H9; B: IBV and MG; C: Nine pathogens.

Figure 5-1-1 Results of GeXP-multiplex PCR detection with mixed pathogen templates

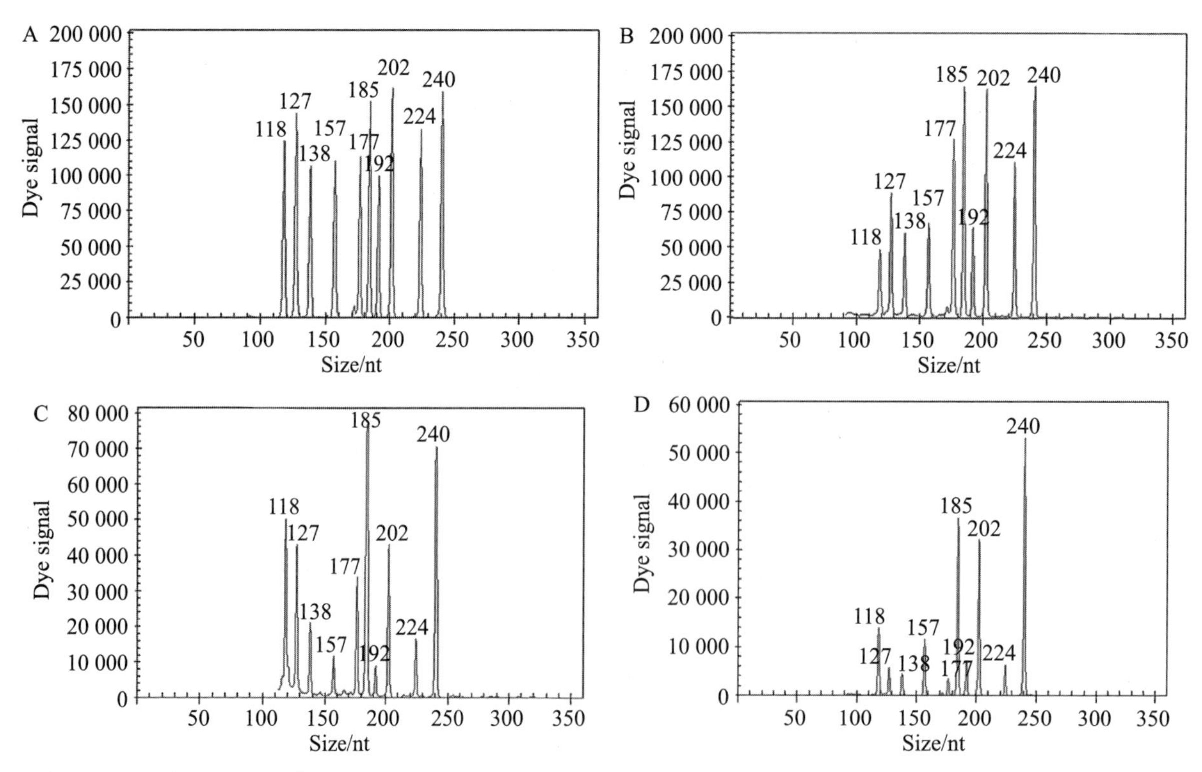

A: 10^5 copies; B: 10^4 copies; C: 10^3 copies; D: 10^2 copies. The GeXP-multiplex PCR assay was carried out using different concentrations of in plasmid. All 10 target genes were detected simultaneously at levels of 10^5, 10^4, 10^3 and 10^2 copies/μL.

Figure 5-1-2 The sensitivity results of GeXP-multiplex-PCR detection of nine pathogens templates

Artificial mixture

When cDNA/DNA from avian respiratory pathogens were mixed together to test the GeXP-multiplex PCR assay's ability to differentiate them, the appropriate specific amplification peaks and the universal amplification peaks were observed (Table 5-1-1, Figure 5-1-1 A and B). For example, when AIV H5 and AIV H9 were mixed together as the samples for the GeXP-multiplex PCR assay, 2 specific amplification peaks (AIV-H9, 116.89 bp; AIV-H5, 223.37 bp) and one universal amplification peaks (AIV M gene, 190.26 bp) were observed (Figure 5-1-1 A). When MG and IBV were mixed together as the samples for the GeXP-multiplex PCR assay, there are 2 specific amplification peaks (MG, 202.03 bp; IBV, 240.50 bp) (Figure 5-1-1 B). When nine pathogens RNA/DNA were mixed as the samples for the GeXP-multiplex PCR assay, 9 specific amplification peaks (AIV H5, 223 bp; AIV H7, 176 bp; AIV H9, 117 bp; IBV, 239 bp; NDV, 126 bp and ILTV, 184 bp; MG, 201 bp; MS, 137 bp; HPG, 156 bp) and one universal amplification peak (AIV M gene, 190 bp) were observed (Table 5-1-1, Figure 5-1-1C).

Interference assay

Two specific amplification peaks were observed when two different templates (one templates with 10^3 copies and another templates with 10^7 copies) were tested by the GeXP-multiplex PCR assay. The peak value of single template was similar to that of the mixed template. For example, two specific amplification peaks were observed when two different templates (AIV-H5 at 10^3 copies and IBV at 10^7 copies) were tested by the GeXP-multiplex PCR assay (Figure 5-1-3) and the peak values for AIV-H5 and for IBV were the same regardless of whether a single template (AIV-H5 or IBV) or a mixed template (AIV-H5+IBV) was utilized. The results of these experiments showed that this interference had a minimal effect on the mixed infection.

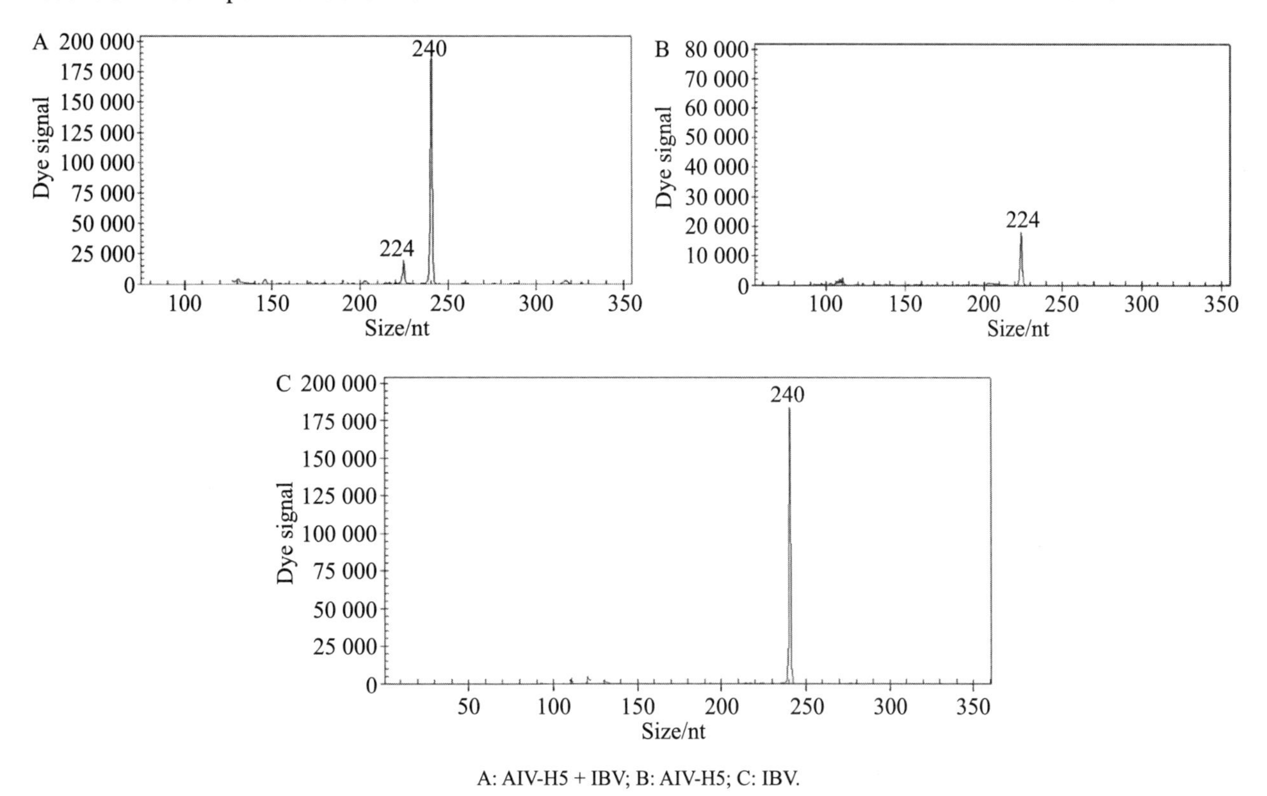

A: AIV-H5 + IBV; B: AIV-H5; C: IBV.

Figure 5-1-3 Results of interference assay for GeXP-multiplex PCR

GeXP-multiplex PCR assay detection in field samples

A total of 24 clinical samples were assayed by the optimized GeXP-multiplex PCR and the results showed that the assay could detect and differentiate the nine respiratory pathogens (Table 5-1-1, Figure 5-1-4, Figures 5-1-1 to 5-1-2).

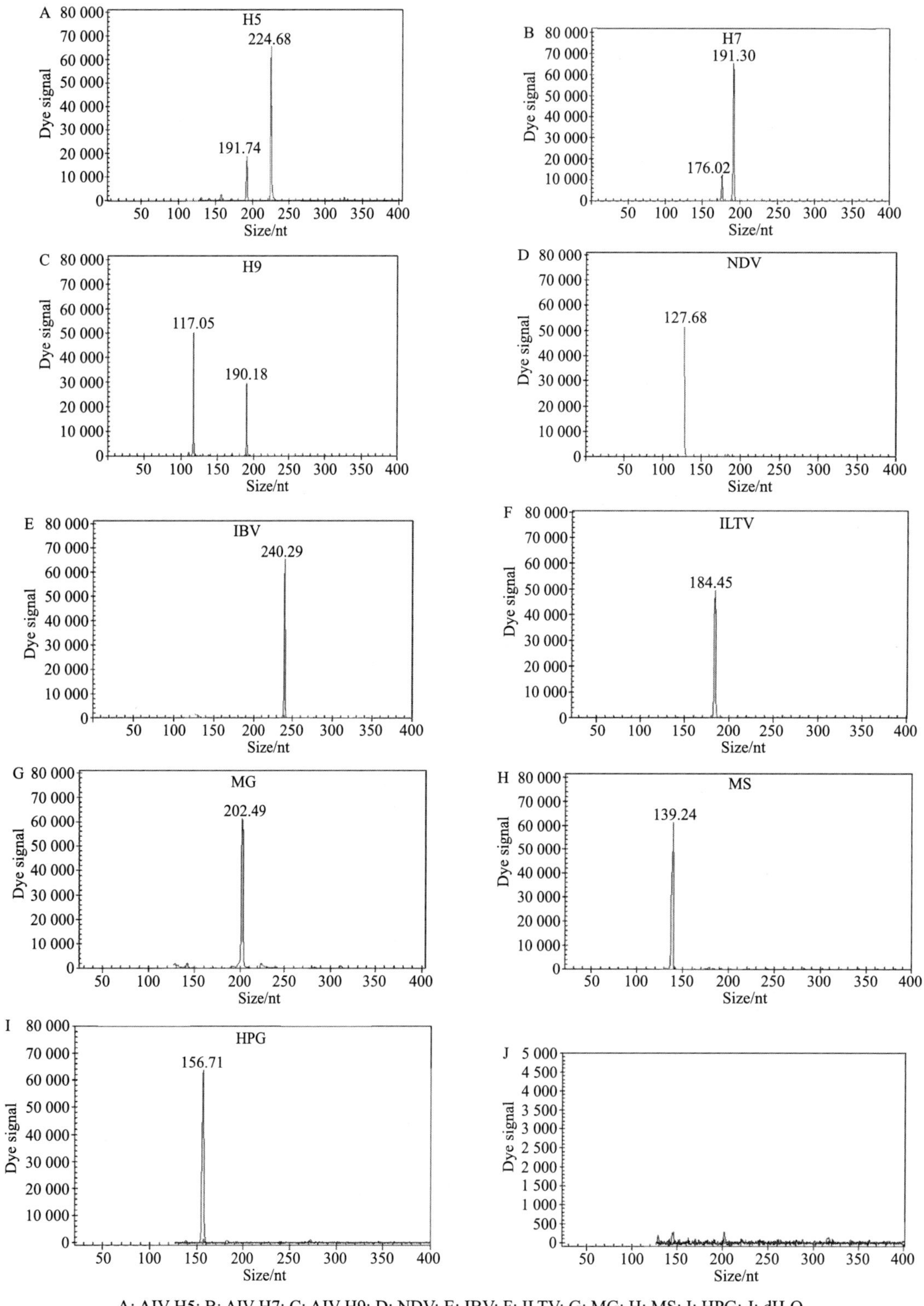

A: AIV H5; B: AIV H7; C: AIV H9; D: NDV; E: IBV; F: ILTV; G: MG; H: MS; I: HPG; J: dH_2O.

Figure 5-1-4 The specificity results of GeXP-multiplex PCR

Discussion

Avian influenza virus subtypes H5, H7, H9 and NDV, IBV, ILTV, MG, MS and HPG are the main nine avian respiratory pathogens in China and other parts of the world. These avian influenza subtypes and pathogens have similar clinical signs and lesions, and may present as multiple infections in chickens and in other avian species[13-15, 22]. Viral isolation and serological tests are laborious and require many days to identify the causative agents.

One main advantage of this GeXP-multiplex PCR assay compared to other available molecular tests is its ability to simultaneously detect and differentiate between these respiratory pathogens. By using this approach, it was possible to identify three subtypes of avian influenza and six respiratory pathogens in the same reaction vessel. Due to the use of second amplification by the Tag primer in the reaction, the presence of unrelated cDNA/DNA templates did not alter the efficiency of the GeXP-multiplex PCR amplification. In contrast, the conventional multiplex PCR and multiplex real time PCR[13-15, 22, 31] were only able to identify two to four pathogens in the same reaction vessel, and the presence of other templates could alter the efficiency of conventional multiplex PCR and multiplex real time PCR amplification[13, 15].

This GeXP-multiplex PCR assay is especially useful because these pathogens commonly cause infections in chickens and other avian species[12, 15, 22]. However, the multiplex feature of this assay is optional; it can also be utilized as single-target assay, or it can be combined into multiplex assays without impacting the quality of the results. This makes this assay adaptable to circumstances that may not require the simultaneous detection of three subtypes of avian influenza and six respiratory pathogens for diagnostic purposes.

Another important feature of this GeXP-multiplex PCR assay is better detection sensitivity than that noted with conventional detection methods[23, 24]. In this study, the absolute sensitivities were analyzed to evaluate the detection limit of the GeXP-multiplex PCR assay. The optimal GeXP-multiplex PCR assay detected as few as 10 copies/μL of AIV-H9, HPG, NDV and IBV, and as few as 10^2 copies/μL of AIV-type A, AIV-H5, AIV-H7, ILTV, MG and MS, which is comparable to the detection sensitivity of the real-time PCR assay[17-19]. In a test of 24 field samples, the specificity of the GeXP-multiplex PCR assay was 100% compared with the conventional methods.

Another distinct advantage of this GeXP-multiplex PCR assay is the time savings as described[32]. The whole reaction was completed in one tube within 2.5 h followed by capillary electrophoresis separation. In addition, two 96-well plates can be placed in parallel in a GeXP analyzer at the same time to further increase the throughput of the samples.

Conclusions

This study has demonstrated that the GeXP-multiplex PCR assay is a rapid method with high sensitivity and specificity for identifying very important avian influenza subtypes (H5, H7, H9) and six respiratory pathogens, which may be adopted for molecular epidemiologic survey of these respiratory pathogens.

References

[1] SAHU S P, OLSON N O. Evaluation of broiler breeder flocks for nonspecific *Mycoplasma synoviae* reaction. Avian Disease, 1976, 20(1): 49-64.

[2] CDC. Concepts and procedures for laboratory-based influenza surveillance. Washington DC: US Department of Health and Human Services, 1982.

[3] SNYDER D B, MARQUARDT W W, MALLINSON E T, et al. Rapid serological profiling by enzyme-linked immunosorbent assay. I. Measurement of antibody activity titer against Newcastle disease virus in a single serum dilution. Avian Disease, 1983, 27(1): 161-170.

[4] ROCKE T E, YUILL T M, AMUNDSON T E. Evaluation of serologic tests for *Mycoplasma gallisepticum* in wild turkeys. Journal Wildlife Diseases, 1985, 21(1): 58-61.

[5] OHKUBO Y, SHIBATA K, MIMURA T, et al. Labeled avidin-biotin enzyme-linked immunosorbent assay for detecting antibody to infectious laryngotracheitis virus in chickens. Avian Disease, 1988, 32(1): 24-31.

[6] KLEVEN S H, LEVISOHN S. Molecular and diagnostic procedures in mycoplasmology-2nd edition. Singapore: Elsevier Pte Ltd Press, 1996.

[7] SWAYNE D E, SENNE D A, BEARD C W. Isolation and identification of avian pathogens-4th edition. Singapore: Elsevier Pte Ltd Press, 1998.

[8] DE WIT J J. Detection of infectious bronchitis virus. Avian Pathology, 2000, 29(2): 71-93.

[9] WOOLCOCK P R, MCFARLAND M D, LAI S, et al. Enhanced recovery of avian influenza virus isolates by a combination of chicken embryo inoculation methods. Avian Disease, 2001, 45(4): 1030-1035.

[10] OFFICE INTERNATIONAL DES EPIZOOTIES. OIE Manual of standards for diagnostic tests and vaccines, 5 th edition. France: OIE Press, 2004.

[11] SELLERS H S, GARCIA M, GLISSON J R, et al. Mild infectious laryngotracheitis in broilers in the southeast. Avian Disease, 2004, 48(2): 430-436.

[12] KEELER C J, REED K L, NIX W A, et al. Serotype identification of avian infectious bronchitis virus by RT-PCR of the peplomer (S-1) gene. Avian Disease, 1998, 42(2): 275-284.

[13] PANG Y, WANG H, GIRSHICK T, et al. Development and application of a multiplex polymerase chain reaction for avian respiratory agents. Avian Disease, 2002, 46(3): 691-699.

[14] MARDASSI B B, MOHAMED R B, GUERIRI I, et al. Duplex PCR to differentiate between *Mycoplasma synoviae* and *Mycoplasma gallisepticum* on the basis of conserved species-specific sequences of their hemagglutinin genes. Journal of Clinical Microbiology, 2005, 43(2): 948-958.

[15] XIE Z, PANG Y S, LIU J, et al. A multiplex RT-PCR for detection of type A influenza virus and differentiation of avian H5, H7, and H9 hemagglutinin subtypes. Moleular and Cellular Probes, 2006, 20(3-4): 245-249.

[16] OLDONI I, RODRIGUEZ-AVILA A, RIBLET S, et al. Characterization of infectious laryngotracheitis virus (ILTV) isolates from commercial poultry by polymerase chain reaction and restriction fragment length polymorphism (PCR-RFLP). Avian Disease, 2008, 52(1): 59-63.

[17] FULLER C M, COLLINS M S, ALEXANDER D J. Development of a real-time reverse-transcription PCR for the detection and simultaneous pathotyping of Newcastle disease virus isolates using a novel probe. Archives of Virology, 2009, 154(6): 929-937.

[18] MEIR R, MAHARAT O, FARNUSHI Y, et al. Development of a real-time TaqMan RT-PCR assay for the detection of infectious bronchitis virus in chickens, and comparison of RT-PCR and virus isolation. Journal of Virological Methods, 2010, 163(2): 190-194.

[19] SPRYGIN A V, ANDREYCHUK D B, KOLOTILOV A N, et al. Development of a duplex real-time TaqMan PCR assay with an internal control for the detection of *Mycoplasma gallisepticum* and *Mycoplasma synoviae* in clinical samples from commercial and backyard poultry. Avian Pathology, 2010, 39(2): 99-109.

[20] ZHANG L, PAN Z, GENG S, et al. Sensitive, semi-nested RT-PCR amplification of fusion gene sequences for the rapid detection and differentiation of Newcastle disease virus. Research in Veterinary Science, 2010, 89(2): 282-289.

[21] MASE M, KANEHIRA K. Simple differentiation of avirulent and virulent strains of avian paramyxovirus serotype-1 (Newcastle disease virus) by PCR and restriction endonuclease analysis in Japan. Journal of Veterinary Medical Science, 2012, 74(12): 1661-1664.

[22] NGUYEN T T, KWON H J, KIM I H, et al. Multiplex nested RT-PCR for detecting avian influenza virus, infectious bronchitis virus and Newcastle disease virus. Journal of Virological Methods, 2013, 188(1-2): 41-46.

[23] DREW J E, MAYER C D, FARQUHARSON A J, et al. Custom design of a GeXP multiplexed assay used to assess expression profiles of inflammatory gene targets in normal colon, polyp, and tumor tissue. Journal of Molecular Diagnostics, 2011, 13(2): 233-242.

[24] FARQUHARSON A J, STEELE R J, CAREY F A, et al. Novel multiplex method to assess insulin, leptin and adiponectin regulation of inflammatory cytokines associated with colon cancer. Molecular Biology Reports, 2012, 39(5): 5727-5736.

[25] NAGEL M A, GILDEN D, SHADE T, et al. Rapid and sensitive detection of 68 unique varicella zoster virus gene transcripts in five multiplex reverse transcription-polymerase chain reactions. Journal of Virological Methods, 2009, 157(1): 62-68.

[26] YANG M J, LUO L, NIE K, et al. Genotyping of 11 human papillomaviruses by multiplex PCR with a GeXP analyzer. Journal of Medical Virology, 2012, 84(6): 957-963.

[27] XIE Z, XIE L, PANG Y, et al. Development of a real-time multiplex PCR assay for detection of viral pathogens of penaeid shrimp. Archives of Virology, 2008, 153(12): 2245-2251.

[28] QIN M, WANG D Y, HUANG F, et al. Detection of pandemic influenza A H1N1 virus by multiplex reverse transcription-PCR with a GeXP analyzer. Journal of Virological Methods, 2010, 168(1-2): 255-258.

[29] XIE Z, XIE L, FAN Q, et al. A duplex quantitative real-time PCR assay for the detection of *Haplosporidium* and *Perkinsus* species in shellfish. Parasitology Research, 2013, 112(4): 1597-1606.

[30] RAI A J, KAMATH R M, GERALD W, et al. Analytical validation of the GeXP analyzer and design of a workflow for cancer-biomarker discovery using multiplexed gene-expression profiling. Analytical and Bioanalytical Chemistry, 2009, 393(5): 1505-1511.

[31] XIE Z, DONG J, TANG X, et al. Sequence and phylogenetic analysis of three isolates of avian influenza H9N2 from chickens in southern China. Scholarly Research Exchange, 2008: 1-7.

[32] HU X, ZHANG Y, ZHOU X, et al. Simultaneously typing nine serotypes of enteroviruses associated with hand, foot, and mouth disease by a GeXP analyzer-based multiplex reverse transcription-PCR assay. Journal of Clinical Microbiology, 2012, 50(2): 288-293.

Simultaneous detection of eight avian influenza A virus subtypes by multiplex reverse transcription-PCR using a GeXP analyser

Li Meng, Xie Zhixun, Xie Zhiqin, Liu Jiabo, Xie Liji, Deng Xianwen, Luo Sisi, Fan Qing, Huang Li, Huang Jiaoling, Zhang Yanfang, Zeng Tingting, and Wang Sheng

Abstract

Recent studies have demonstrated that at least eight subtypes of avian influenza virus (AIV) can infect humans, including H1, H2, H3, H5, H6, H7, H9 and H10. A GeXP analyser-based multiplex reverse transcription (RT)-PCR (GeXP-multiplex RT-PCR) assay was developed in our recent studies to simultaneously detect these eight AIV subtypes using the haemagglutinin (HA) gene. The assay consists of chimeric primer-based PCR amplification with fluorescent labelling and capillary electrophoresis separation. RNA was extracted from chick embryo allantoic fluid or liquid cultures of viral isolates. In addition, RNA synthesised via *in vitro* transcription was used to determine the specificity and sensitivity of the assay. After selecting the primer pairs, their concentrations and GeXP-multiplex RT-PCR conditions were optimised. The established GeXP-multiplex RT-PCR assay can detect as few as 100 copies of premixed RNA templates. In the present study, 120 clinical specimens collected from domestic poultry at live bird markets and from wild birds were used to evaluate the performance of the assay. The GeXP-multiplex RT-PCR assay specificity was the same as that of conventional RT-PCR. Thus, the GeXP-multiplex RT-PCR assay is a rapid and relatively high-throughput method for detecting and identifying eight AIV subtypes that may infect humans.

Keywords

GeXP analyser, HA typing, eight subtypes, infect humans

Introduction

Influenza A viruses are important human and animal pathogens affecting human health, causing severe animal diseases and death. To date, 16 HA (haemagglutinin) types and nine NA (neuraminidase) influenza A viral types have been identified based on the combination of these two major antigens in avian populations. The 17 th and 18 th HA types and 10th and 11th NA types were recently discovered in bats in South America[1-4]. Avian influenza (AI) is an acute infectious disease caused by influenza A viruses or avian influenza viruses (AIVs) in domestic poultry and wild birds. Some AIV subtypes can break the species barrier and infect humans[5, 6]. Studies have shown that AIV subtypes H1, H2, H3, H5, H7, H9 and H10 can directly infect humans and could be potentially lethal pathogens that cause human influenza pandemics. Research has shown that most human influenza viruses originate from AIVs. Prior to 2013, human AIV infection cases included infections by H5N1, H5N2, H7N2, H7N3, H7N7, H9N2 and H10N7[7-10]. Since 2013, four additional AIV subtypes, H7N9, H6N1, H10N8 and H5N6, have been detected in humans, and new cases of these subtypes continue to appear in China. For example, H7N9 AIV can infect humans and poultry but has low pathogenicity in chickens. In 2017, H7N9 AIV mutated into a strain that is highly pathogenic against chickens and caused hundreds of cases of human infections in China[11-14]. To date, at least eight AIV subtypes, including H1, H2, H3, H5, H6, H7, H9 and H10, have been reported to infect humans[15].

Effective laboratory techniques are necessary to detect and identify AIV subtypes during outbreaks. Molecular biological diagnostic methods based on PCR technology, such as conventional PCR, reverse transcription PCR (RT-PCR), real-time RT-PCR (RRT-PCR) and quantitative RRT-PCR have been widely used for AIV detection and genotyping[16-19]. However, a method is needed that uses only one PCR reaction to simultaneously detect and differentiate all HA subtypes of AIV strains that can infect humans. In general, RRT-PCR and multiplex RT-PCR can detect multiple AIV pathogenic subtypes[20-22] but no more than four at once. The GeXP analyser is an instrument that can detect the expression of up to 35 genes simultaneously using a multiplex gene expression profling analysis platform (Beckman Coulter, Brea, CA, USA). Several human and animal pathogens, including those causing hand, foot and mouth disease, 16 human respiratory viral types or subtypes, and 11 human papilloma viruses have been successfully and rapidly detected and identified using the GeXP analyser[23-25]. Moreover, our laboratory also developed several procedures to simultaneously detect nine avian respiratory pathogens, eight swine reproductive and respiratory viruses, eleven duck viruses, six immunosuppressive chicken viruses and four different avian influenza A H5 NA viral types[26-30]. This report describes our recently developed multiplex RT-PCR assay using a GeXP analyser (GeXP-multiplex RT-PCR) to simultaneously detect eight AIV subtypes that can infect humans.

Methods

Ethics statement

The present study was approved and conducted in strict accordance with the recommendations in the guide for the care and use of routine sampled animals in LBMs by the Animal Ethics Committee of the Guangxi Veterinary Research Institute, which supervises all LBMs in Guangxi. Biological samples were gently collected from healthy chickens, ducks, birds and geese using aseptic cotton swabs. The birds were not anaesthetised before sampling, and poultry were observed for 30 min after sampling before being returned to their cages.

Sample collection and viral DNA/RNA nucleic acid extraction

The pathogens used in this study include different AIV subtype reference strains, AIV field isolates and other avian pathogens. All clinical swab samples were collected from the cloacae, larynges and tracheae of healthy chickens, geese and ducks. The sample treatment method was described previously[37]. A viral RNA/DNA Extraction Kit (TaKaRa, Dalian, China) was used to extract genomic RNA/DNA from samples (200 μL of chicken embryo allantoic fluid or liquid cultures) per the manufacturer's protocol. The extracted RNA was reverse transcribed to synthesise cDNA, and DNA was stored at −80 ℃ . All samples were manipulated inside a class- Ⅱ biosafety cabinet in a biosafety level-2 laboratory.

Primer design and plasmid preparation

The GeXP-multiplex assay consisted of nine gene-specific primer pairs, including one pair of AIV universal primers (AIV M) and eight HA AIV gene segment pairs. Sequence information obtained from the Influenza Sequence Database and the NCBI database were used to design the specific primers. A highly conserved region of the M gene was used to design the AIV universal primers; a specific region of the HA gene segment for all subtypes was used to design the Hl, H2, H3, H5, H6, H7, H9 and H10 primers. Premier 6.0, Oligo 7.0 and NCBI Primer BLAST were used for primer analysis and filtration. One universal primer pair was fused at the 5' end of each gene-specific primer as a universal sequence to generate nine chimeric primer pairs. In addition, we labelled one universal primer pair and one universal primer with Cy5 (Table 5-2-1). The primers were purified using high-performance liquid chromatography (HPLC) by Invitrogen (Guangzhou, China).

Table 5-2-1 GeXP primers designed to detect eight subtypes of the avian influenza A virus

Primer	Forward primer (5'→3')[a]	Reverse primer (5'→3')[a]	Size/bp
M	AGGTGACACTATAGAATAAGCCGAGATCGCGCAGA	GTACGACTCACTATAGGGACGCTCACTGGGCACGGT	192
Hl	AGGTGACACTATAGAATACCAGAAYGTGCATCCTATCACT	GTACGACTCACTATAGGGATATCATTCCTGTCCAWCCCCCT	198
H2	AGGTGACACTATAGAATATTCGAGAAAGTRAAGATTYTGCC	GTACGACTCACTATAGGGACCAGACCATGTTCCTGAAGAA	152
H3	AGGTGACACTATAGAATATTGCCATATCATGYTTTTTGCTTTG	GTACGACTCACTATAGGGAAATGCAAATGTTGCACCTAATGTTG	131
H5	AGGTGACACTATAGAATAGGAAAGTGTAAGAAACGGAACGTA	GTACGACTCACTATAGGGACACATCCATAAAGAYAGACCAGC	223
H6	AGGTGACACTATAGAATATCTCAAACAAGGCCCCTCTC	GTACGACTCACTATAGGGATCCCATTTCGGGCATTAGGC	173
H7	AGGTGACACTATAGAATAAGAATACAGATTGACCCAGTSAA	GTACGACTCACTATAGGGACCCATTGCAATGGCHAGAAG	142
H9	AGGTGACACTATAGAATAACCATTTATTCGACTGTCGCCT	GTACGACTCACTATAGGGACATTGGACATGGCCCAGAA	118
H10	AGGTGACACTATAGAATAAACACGGACACRGCTGA	GTACGACTCACTATAGGGAATTGTTCTGGTAWGTGGAAC	167
Cy5-Tag-F[b]	AGGTGACACTATAGAATA		
Tag-R	GTACGACTCACTATAGGGA		

Note: Underlined oligonucleotides are universal sequences. [a] Degenerate nucleotide abbreviations are as follows: R, A/G; W, A/T; Y, C/T. [b] The primer is contained in the PCR primer mix.

Plasmids carrying different HA genes from AIV (AIV H1N3 Duck/HK/717/79-d1, AIV H2N3 Duck/HK/77/76, AIV H3N2 A/Chicken/Guangxi/015C10/2009, H5N1 AIV Re-1, AIV H6N8 Duck/HK/531/79, AIV H7N2 AIV Duck/HK/47/76, AIV H9N2 A/chicken/Guangxi/NN1/2011, and AIV H10N3 Duck/HK/876/80) were used for ssRNA synthesis using an RNA production system T7 *in vitro* transcription kit (Promega, Madison, WI, USA). Published methods were used to calculate the ssRNA copy numbers for the target AIV genes (M, H1, H2, H3, H5, H6, H7, H9 and H10)[31].

Multiplex PCR reaction conditions for the GeXP analyser

The GeXP PCR reaction system contained a total volume of 25 μL, including 2.5 μL of 10 × PCR Buffer (Sigma, STL, MO, USA), 2.5 μL of $MgCl_2$ (25 μM, Sigma), 1.25 μL of universal primers (500 nmol/L), 1.25 μL of JumpStart Taq DNA polymerase (2.5 U/μL, Sigma), 1.25 μL of mixed primers (containing 50～150 nmol/L of nine gene-specific chimeric primer pairs), and 0.5 pg to 0.5 ng of template (cDNA or DNA). Next, nuclease-free water was used to bring the final volume of the PCR reaction to 25 μL. The PCR thermo cycling procedure was performed at 95 ℃ for 5 min, followed by 95 ℃ for 30 s, 55 ℃ for 30 s and 72 ℃ for 30 s (10 cycles). The second step was performed at 95 ℃ for 30 s, 65 ℃ for 30 s, and 72 ℃ for 30 s (10 cycles). The third step was performed at 95 ℃ for 30 s, 55 ℃ for 30 s, and 72 ℃ for 30 s (20 cycles). Detailed information regarding the primers is listed in Table 5-2-1; all primers were synthesised and purified by Invitrogen (Guangzhou, China).

The GenomeLab GeXP genetic analysis system (Brea, CA, USA) was used to separate and detect the PCR products by capillary electrophoresis following previous protocols[32]. After separating the products, the product peaks were analysed using the GeXP system software. The peak height for each gene is illustrated in an electrophoretogram.

Single-primer test for specificity

The assay specificity for all target genes was individually assessed using premixed cDNA/DNA in a multiplex PCR assay after optimisation. Other avian pathogens were tested, including all AIV subtypes (except H1, H2, H3, H5, H6, H7, H9 and H10, for which the reference strain HA genes (1-16) were confirmed by sequencing), infectious laryngotracheitis virus (ILTV), Newcastle disease virus (NDV), Abelson leukaemia virus (ALV), infectious bronchitis virus (IBV), *Mycoplasma gallinarum* (MG), avian reovirus (ARV), influenza B viruses, and reticuloendothelial hyperplasia (REV). Nuclease-free water was used as a negative control.

Evaluation of the sensitivity of the GeXP method and the interference assay

The sensitivity of the multiplex assay was evaluated using a GeXP analyser as previously described[32]. We diluted the same initial concentration for each target gene of eight premixed RNA templates to final concentrations of 10 to 1 copy/μL. Next, PCR products at each dilution were subjected to the multiplex assay using a GeXP system. Finally, specific primer concentrations and amplification systems were optimised based on the best response system. The sensitivity of the GeXP RT-PCR assay was tested three times on three different days in one month using the diluted sample as described previously. Because the simultaneous presence of different template concentrations may affect the amplification eficiency of multiple PCR, several template dilutions (10^2 to 10 copies/μL) were randomly mixed and tested in the GeXP multiplex RT-PCR assay. The results were also compared with those of the single-template multiplex PCR assay.

Applications for detecting clinical specimens

We randomly collected 120 clinical specimens from different poultry and wild bird species in LBMs, and all clinical specimens were subjected to virus isolation, GeXP-multiplex RT-PCR and RRT-PCR simultaneously. Using previously reported primers, HA genes from positive specimens in the methods described above were sequenced by BGl (Shenzhen, China) to confirm RT-PCR compliance[33].

Results

GeXP-multiplex RT-PCR specificity testing

Each pair of gene-specific primers was tested using cDNA samples from eight HA AIV subtypes (H1, H2, H3, H5, H6, H7, H9 and H10) in a mono-PCR-GeXP assay. The target matrix (M) gene was amplified from all AIV subtypes where each specific primer pair could not cross-amplify other HA AIV subtypes but could amplify only the corresponding target gene. All amplicon sizes are listed in Table 5-2-1. The nine target genes were amplified by the GeXP-multiplex RT-PCR assay, and each primer combination showed specific amplification peaks (Figure 5-2-1). The AIV M gene showed specific amplification peaks for H1N1, H2N3, H3N2, H3N6, H3N8, H4N5, H5N1, H6N1, H6N2, H6N6, H6N8, H7N2, H7N9, H8N4, H9N2, H10N3, H11N3, H12N5, H13N6, H14N5, H15N9 and H16N3.

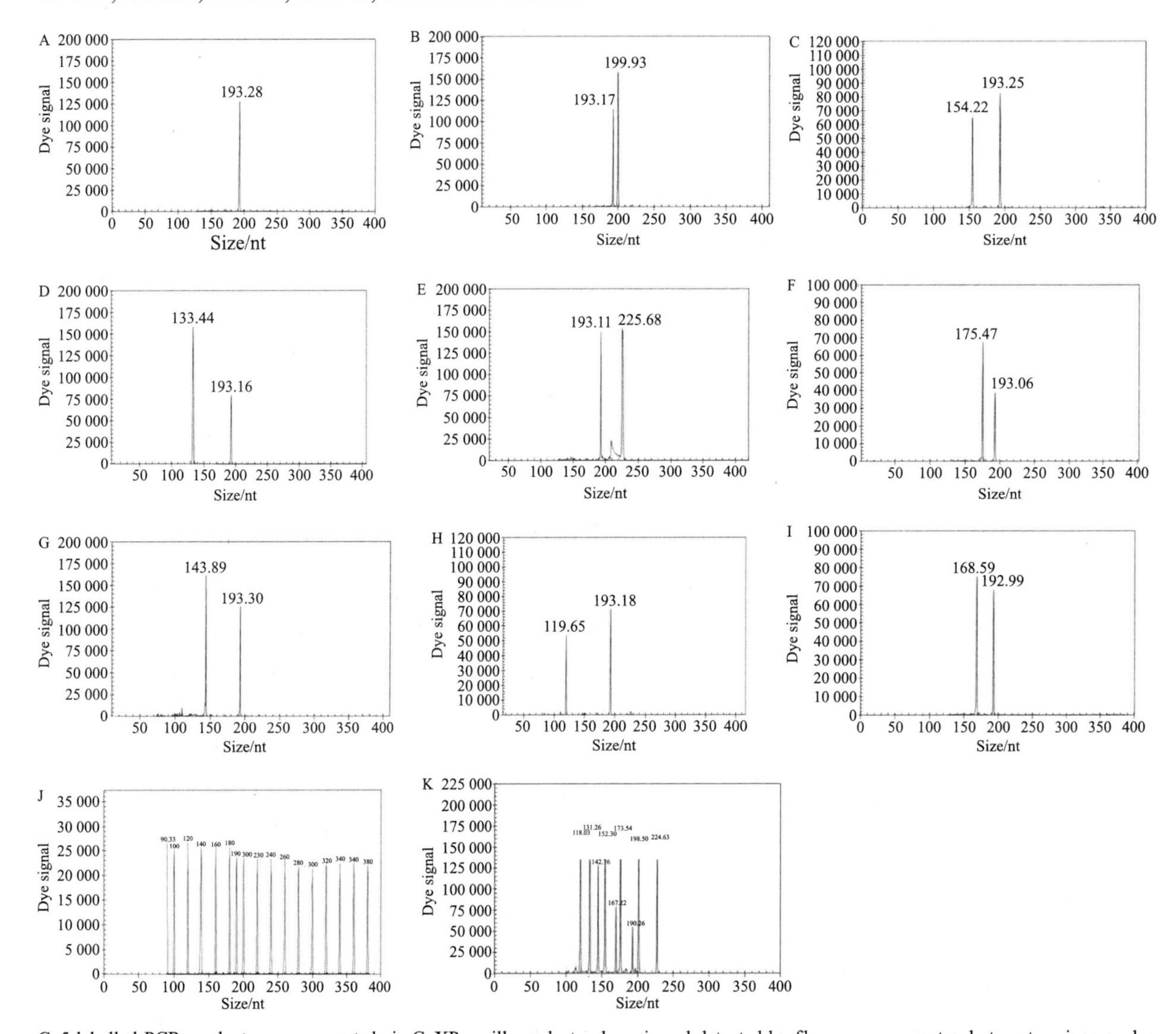

Cy5-labelled PCR products were separated via GeXP capillary electrophoresis and detected by fluorescence spectrophotometry given as dye signals in arbitrary units on the y-axis. Each peak was identified by comparing the expected to the actual PCR product size on the x-axis. A-I, K: The individual target gene amplification results for M, Hl, H2, H3, H5, H6, H7, H9, H10 and all amplified genes simultaneously, respectively; J: RNase-free water was used as the negative control. Peaks indicate the DNA size standard.

Figure 5-2-1 Specificity analyses of GeXP detection of eight avian influenza A viral subtypes with multiplex primers

GeXP multiplex RT-PCR assay sensitivity

For individual ssRNA transcribed *in vitro*, the sensitivity of this method for detecting the AIV type A, AIV-H1, AIV-H2, AIV-H3, AIV-H5, AIV-H6, AIV-H7, AIV-H9 and AIV-H10 templates by the GeXP-multiplex RT-PCR assay was as low as 10^2 copies/μL (Figure 5-2-2, selected electrophoresis results are shown). Each sample was assayed three times under the same conditions on different days, producing highly similar results.

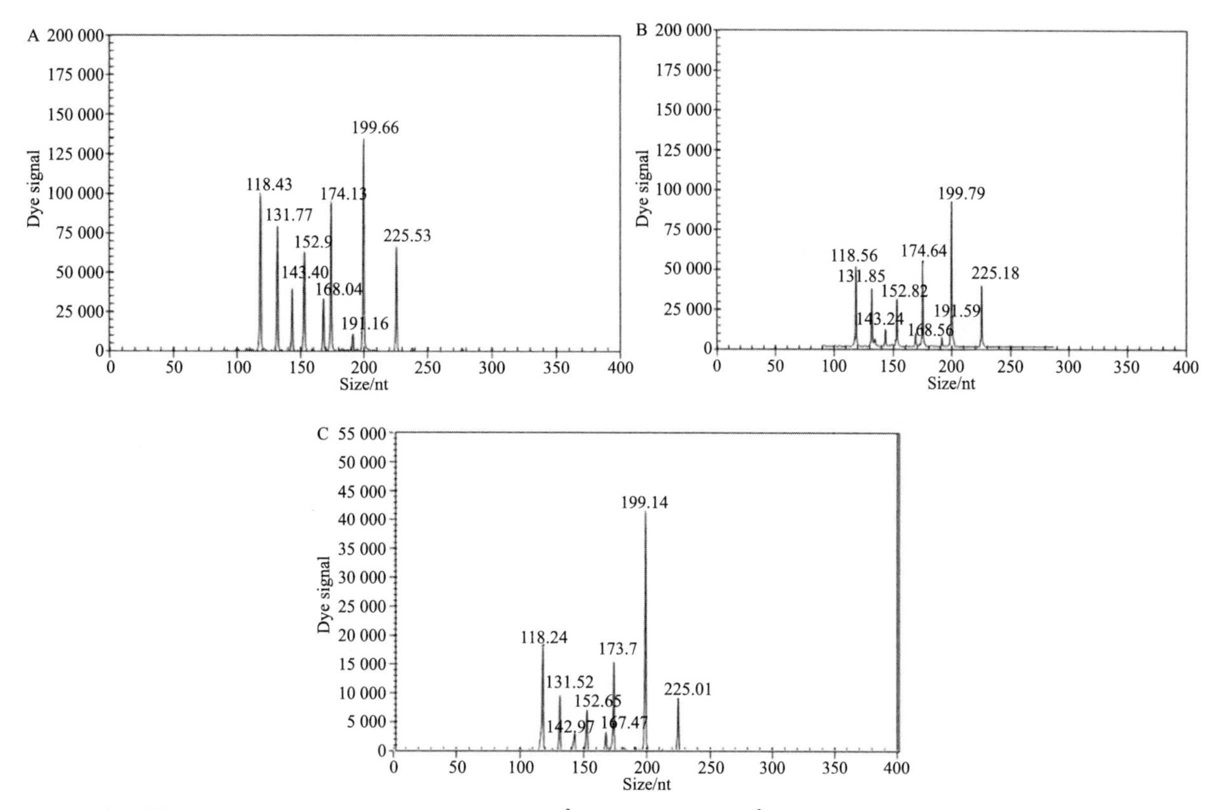

A and B: Nine different target genes were detected by GeXP at 10^3 copies/μL (A) and 10^2 copies/μL (B); C: Eight target genes were detected by GeXP at 10 copies/μL.

Figure 5-2-2 Sensitivity analyses of GeXP detection of nine premixed RNA templates of different concentrations with multiplex primers

Artificial mixture and interference assay

Two specific amplification peaks were observed when two different templates (10^2 copies of one template and 10^6 copies of the second template) were evaluated using the GeXP-multiplex RT-PCR assay, and the peak value observed using the single template was similar to that of the mixed template. For example, two specific amplification peaks were observed when different template quantities (from 10^2 to 10^6 copies) were tested, and the peak values for these target genes were identical, regardless of whether a single (AIV-H5 or H9) or mixed (AIV-H5+H9, AIV-H5+H7+H9, AIV-H1+H3+H6+H9) template was utilised (electrophoresis results are not shown). These results demonstrated that interference from different templates minimally impacted the detection of a mixed infection.

Detection of AIVs in clinical samples by the GeXP-multiplex RT-PCR assay

A total of 120 cloacal and oropharyngeal swab samples were randomly collected from poultry and wild birds at various live bird markets (LBMs) from January to May 2017 in Guangxi. All the swab samples were tested for HA AIV serotypes using the optimised GeXP-multiplex RT-PCR assay, and the samples underwent three confirmation RRT-PCR tests, virus isolation and sequencing. The positive and negative results obtained using the various methods are shown in Table 5-2-2. Agreement among the GeXP assay, RRT-PCR method and sequencing results is presented in Table 5-2-3. The H3, H6 and H9 AIV subtypes were the most common in the 66 positive samples. The GeXP-multiplex RT-PCR assay yielded 100% specificity compared with conventional approaches.

Table 5-2-2 Detection results for clinical specimens

Sample No.	Typing by RRT-PCR	Conformity	Numbers	GeXP assay								
				M	H1	H2	H3	H5	H6	H7	H9	H10
1-6	H1	+	2	+	+	−	−	−	−	−	−	−
7-21	H3	+	8	+	−	−	+	−	−	−	−	−
22-23	H5	+	2	+	−	−	−	+	−	−	−	−
24-43	H6	+	14	+	−	−	−	−	+	−	−	−
44-48	H7	+	5	+	−	−	−	−	−	+	−	−
49-63	H9	+	14	+	−	−	−	−	−	−	+	−
64	H10	+	1	+	−	−	−	−	−	−	−	+
65-66	H1+H3	+	2	+	+	−	+	−	−	−	−	−
67-68	H1+H3+H6	+	2	+	+	−	+	−	+	−	−	−
69-71	H3+H9	+	3	+	−	−	+	−	−	−	+	−
72-76	H6+H9	+	5	+	−	−	−	−	+	−	+	−
77-78	H5+H9	+	2	+	−	−	−	+	−	−	+	−
79-100	H2	+	1	+	−	+	−	−	−	−	−	−
101-120	H3+H6	+	5	+	−	−	−	−	+	−	−	−

Note: All the samples were collected from poultry and wild birds at live bird markets.

Table 5-2-3 Comparison of the 66 positive clinical samples using the GeXP assay, RRT-PCR methods and sequencing

Serotype	No. of samples testing positive via			
	GeXP assay	RRT-PCR	Sequencing	Coincidence rate
H1	2	2	2	100%
H2	1	−	1	100%
H3	8	8	8	100%
H5	2	2	2	100%
H6	14	14	14	100%
H7	5	5	5	100%
H9	14	14	14	100%
H10	1	1	1	100%
Mix	20	20	20	100%
Total	66	66	66	100%

Discussion

At present, the influenza virus is one of the primary pathogenic microorganisms threatening human health. This virus causes large economic losses and affects social stability. In addition to occasional large influenza outbreaks, certain AIV subtypes cause numerous human deaths each year. In particular, H7N9, which caused an outbreak in China in 2013, has infected more than 1500 humans to date[34]. Influenza viruses contain eight gene segments, and RNA segments of viral strains are prone to gene rearrangement when different influenza strains infect the same cell, resulting in the formation of new strains. Mixed infections with different

AIV subtypes are common among birds and appear to play a key role in the natural history of viruses with segmented genomes[35-37]. Multiple AIV subtypes were recently isolated from live poultry markets in southern China[38, 39].

Rapid and accurate identification of AIV HA subtypes is important for understanding AIV circulation in birds and provides useful epidemiological information for selecting appropriate control and elimination strategies. Routine serological tests and primary detection techniques for the influenza virus include RT-PCR, immunofluorescence, virus isolation, and culture methods. However, virus isolation is not always possible due to sample matrix conditions or low viral titres in clinical specimens. These conventional methods for differential diagnosis are complex and time consuming. Moreover, these tests are easily affected by factors such as antibodies. Although new methods for detecting AIV have been developed[40-42], PCR remains a highly specific and rapid method for accurately detecting pathogens. The GeXP multiplex RT-PCR assay is high-throughput, highly sensitive and specific, can analyse 192 samples in 6 h and can simultaneously detect as many as 35 genes in a single PCR reaction. To date, studies have demonstrated that at least eight AIV subtypes can infect humans[15]. These AIVs pose a serious threat to human health and are difficult to differentially and rapidly diagnose using conventional methods. To the best of our knowledge, this study is the first to successfully establish a GeXP multiplex RT-PCR method for rapidly and accurately detecting eight AIV subtypes (H1, H2, H3, H5, H6, H7, H9 and H10) within 4 h. This method effectively reduces the time required by the traditional method of identifying each HA subtype individually. The assay avoids cross-reactivity with other HA subtypes of AIV and human seasonal influenza viruses by employing a nucleic-acid specific validation test and has a detection limit of 10^2 copies/μL. This assay will greatly accelerates the differential diagnosis of AIV subtypes in infected patients with improved accuracy and thus will be important to develop for use in animal husbandry as well as for human public health.

Because the GeXP multiplex RT-PCR method is highly sensitive, and viral RNA degrades easily, samples and extracted RNA should be stored at −80 ℃ as quickly as possible to prevent RNA degradation. The high variability of the AIV genome, even within a single HA subtype, causes subtype-specific primer design to be complex; therefore, HA gene sequences were comprehensively selected to consider most circulating isolates. Nonetheless, certain degenerate positions had to be included in the designed primer sequences. More reference strains and clinical samples are needed to validate the reliability of these results due to variations among AIV subtypes. Two distinct advantages of this method are its low cost and short assay time when testing multiple mixed infection samples simultaneously[43, 44]. For example, the cost of this GeXP multiplex RT-PCR method for simultaneously detecting eight subtypes of avian influenza A viruses that infect humans is approximately $4 per test, versus $6 per test for each avian influenza A virus using a commercial RRT-PCR kit. The GeXP method cannot be widely used because the system is expensive. By using a one-step RT-PCR kit, the entire reaction can be completed in one tube within 2.5 h, followed by capillary electrophoresis separation. An automated workstation can be employed to reduce the number of steps requiring manual operation to further improve precision, reliability, and speed. Consequently, this method could be immensely helpful in surveillance studies targeting AIV subtypes.

References

[1] ZHU X, YU W, MECBRIDE R, et al. Hemagglutinin homologue from H17N10 bat influenza virus exhibits divergent receptor-binding and pH-dependent fusion activities. Proc Natl Acad Sci USA, 2013, 110: 1458-1463.

[2] TONG S, ZHU X, LI Y, et al. New world bats harbor diverse influenza A viruses. PLOS Pathog, 2013, 9: e1003657.

[3] WU Y, WU Y, TEFSEN B, et al. Bat-derived influenza-like viruses H17N10 and H18N11. Trends Microbiol, 2014, 22(4): 183-191.

[4] CHAN J F, TO K K, TSE H, et al. Interspecies transmission and emergence of novel viruses: lessons from bats and birds. Trends Microbiol, 2013, 21(10): 544-555.

[5] MEDINA RA, GARCIA-SASTRE A. Influenza A viruses: new research developments. Nat Rev Microbio, 2011, 9(8): 590-603.

[6] DE GRAAF M, FOUCHIER R A. Role of receptor binding specificity in influenza A virus transmission and pathogenesis. EMBO J, 2014, 33(8): 823-841.

[7] GUAN Y, SMITH G J. The emergence and diversification of panzootic H5N1 influenza viruses. Virus Res, 2013, 178(1): 35-43.

[8] ARZEY G G, KIRKLAND P D, ARZEY K E, et al. Influenza virus A (H10N7) in chickens and poultry abattoir workers, Australia. Emerg Infect Dis, 2012, 18(5): 814-816.

[9] CHENG V C, CHAN J F, WEN X, et al.Infection of immunocompromised patients by avian H9N2 influenza A virus. J Infect, 2011, 62(5): 394-399.

[10] FOUCHIER R A, SCHNEEBERGER P M, ROZENDAAL F W et al. Avian influenza A virus (H7N7) associated with human conjunctivitis and a fatal case of acute respiratory distress syndrome. Proc Natl Acad Sci U S A, 2004, 101(5): 1356-1361.

[11] YU H, COWLING B J, FENG L, et al. Human infection with avian influenza A H7N9 virus: an assessment of clinical severity. Lancet, 2013, 382(9887): 138-145.

[12] YUAN J, ZHANG L, KAN X et al. Origin and molecular characteristics of a novel 2013 avian influenza A(H6N1) virus causing human infection in Taiwan. Clin Infect Dis, 2013, 57(9): 1367-1368.

[13] XU Y, CAO H, LIU H, SUN H, et al. Identification of the source of A (H10N8) virus causing human infection. Infect Genet Evol, 2015, 30: 159-163.

[14] ZHOU L, TAN Y, KANG M, et al. Preliminary epidemiology of human infections with highly pathogenic avian influenza A(H7N9) virus, China, 2017. Emerg Infect Dis, 2017, 23(8): 1355-1359.

[15] ZHANG W, SHI Y. Molecular mechanisms on interspecies transmission of avian influenza viruses. Chin Bull Life Sci, 2015, 27: 539-548.

[16] TSUKAMOTO K, ASHIZAWA T, NAKANISHI K, et al. Use of reverse transcriptase PCR to subtype N1 to N9 neuraminidase genes of avian influenza viruses. J Clin Microbiol, 2009, 47(7): 2301-2303.

[17] TSUKAMOTO K, PANEI C J, SHISHIDO M, et al. SYBR green-based real-time reverse transcription-PCR for typing and subtyping of all hemagglutinin and neuraminidase genes of avian influenza viruses and comparison to standard serological subtyping tests. J Clin Microbio, 2012, 50(1): 37-45.

[18] ELIZALDE M, AGUERO M, BUITRAGO D, et al. Rapid molecular haemagglutinin subtyping of avian influenza isolates by specific real-time RT-PCR tests. J Virol Methods, 2014, 196: 71-81.

[19] HOFFMANN B, HOFFMANN D, HENRITIZI D et al. Riems influenza a typing array (RITA): An RT-qPCR-based low density array for subtyping avian and mammalian influenza a viruses. Sci Rep, 2016, 6: 27211.

[20] XIE Z, PANG YS, LIU J. et al. A multiplex RT-PCR for detection of type A influenza virus and differentiation of avian H5, H7, and H9 hemagglutinin subtypes. Mol Cell Probes, 2006, 20: 245-249.

[21] ZENG Z, LIU Z, WANG W, et al. Establishment and application of a multiplex PCR for rapid and simultaneous detection of six viruses in swine. J Virol Methods, 2014, 208: 102-106.

[22] HU X, ZHANG Y, ZHOU X, et al. Simultaneously typing nine serotypes of enteroviruses associated with hand, foot, and mouth disease by a GeXP analyzer-based multiplex reverse transcription-PCR assay. J Clin Microbiol, 2012, 50(2): 288-293.

[23] LI J, QI S, ZHANG C, HU X, et al. A two-tube multiplex reverse transcription PCR assay for simultaneous detection of sixteen human respiratory virus types/subtypes. Biomed Res Int, 2013, 2013: 327620.

[24] LIU Y, XU Z Q, ZHANG Q, et al. Simultaneous detection of seven enteric viruses associated with acute gastroenteritis by a multiplexed Luminex-based assay. J Clin Microbiol, 2012, 50(7): 2384-2389.

[25] YANG M J, LUO L, NIE K, et al. Genotyping of 11 human papillomaviruses by multiplex PCR with a GeXP analyzer. J Med Virol, 2012, 84(6): 957-963.

[26] XIE Z, LUO S, XIE L, et al. Simultaneous typing of nine avian respiratory pathogens using a novel GeXP analyzer-based multiplex PCR assay. J Virol Methods, 2014, 207: 188-195.

[27] ZHANG M, XIE Z, XIE L, et al. Simultaneous detection of eight swine reproductive and respiratory pathogens using a novel GeXP analyser-based multiplex PCR assay. J Virol Method, 2015, 224: 9-15.

[28] ZHANG Y F, XIE Z X, XIE L J, et al. GeXP analyzer-based multiplex reverse-transcription PCR assay for the simultaneous detection and differentiation of eleven duck viruses. BMC Microbiol, 2015, 15: 247.

[29] ZENG T, XIE Z, XIE L, et al. Simultaneous detection of eight immunosuppressive chicken viruses using a GeXP analyser-based multiplex PCR assay. Virol J, 2015, 12: 226.

[30] LI M, XIE Z, XIE Z, et al. Simultaneous detection of four different neuraminidase types of avian influenza A H5 viruses by multiplex reverse transcription PCR using a GeXP analyser. Influenza Other Respir Viruses, 2016, 10(2): 141-149.

[31] XIE Z, XIE L, PANG Y, et al. Development of a real-time multiplex PCR assay for detection of viral pathogens of penaeid shrimp. Arch Virol, 2008, 153: 2245-2251.

[32] RAI A J, KAMATH R M, GERALD W, et al. Analytical validation of the GeXP analyzer and design of a workflow for cancer-biomarker discovery using multiplexed gene-expression profiling. Anal Bioanal Chem, 2009, 393(5): 1505-1511.

[33] HOFFMANN E, STECH J, GUAN Y, et al. Universal primer set for the ful-length amplification of all influenza A viruses. Arch Virol, 2001, 146: 2275-2289.

[34] WHO. Influenza (avian and other zoonotic). [2017-06-19]. https: //www.who.int/china/health-topics/influenza-seasonal/Influenza-avian-and-other-zoonotic.

[35] LU L, LYCETT S J, LEIGH BROWN A J. Reassortment patterns of avian influenza virus internal segments among different subtypes. BMC Evol Biol, 2014, 14: 16.

[36] ZHANG Y, ZHANG Q, KONG H, et al. H5N1 hybrid viruses bearing 2009/H1N1 virus genes transmit in guinea pigs by respiratory droplet. Science, 2013, 340(6139): 1459-1463.

[37] PENG Y, XIE Z X, LIU J B, et al. Epidemiological surveillance of low pathogenic avian influenza virus (LPAIV) from poultry in Guangxi Province, southern China. PLOS ONE, 2013, 8(10): e77132.

[38] DENG G, TAN D, SHI J, et al. Complex reassortment of multiple subtypes of avian influenza viruses in domestic ducks at the Dongting Lake region of China. J Virol, 2013, 87(17): 9452-9462.

[39] HUANG K, ZHU H, FAN X, et al. Establishment and lineage replacement of H6 influenza viruses in domestic ducks in southern China. J Virol, 2012, 86(11): 6075-6083.

[40] XIE Z, HUANG J, LUO S, et al. Ultrasensitive electrochemical immunoassay for avian influenza subtype H5 using nanocomposite. PLOS ONE, 2014, 9(4): e94685.

[41] HUANG J, XIE Z, XIE Z, et al. Silver nanoparticles coated graphene electrochemical sensor for the ultrasensitive analysis of avian influenza virus H7. Anal Chim Acta, 2016, 913: 121-127.

[42] PENG Y, XIE Z, LIU J, et al. Visual detection of H3 subtype avian influenza viruses by reverse transcription loop-mediated isothermal amplification assay. Virol J, 2011, 8: 337.

[43] LI J, MAO N Y, ZHANG C, et al. The development of a GeXP-based multiplex reverse transcription-PCR assay for simultaneous detection of sixteen human respiratory virus types/subtypes. BMC Infect Dis, 2012, 12: 189.

[44] LUO S S, XIE Z X, XIE L J, et al. Development of a GeXP assay for simultaneous differentiation of six chicken respiratory viruses. Hinaese Journal of Virology, 2013, 29(3): 250-257.

Simultaneous detection of four different neuraminidase types of avian influenza A H5 viruses by multiplex reverse transcription PCR using a GeXP analyser

Li Meng, Xie Zhixun, Xie Zhiqin, Liu Jiabo, Xie Liji, Deng Xianwen, Luo Sisi, Fan Qing, Huang Li, Huang Jiaoling, Zhang Yanfang, Zeng Tingting, and Feng Jiaxun

Abstract

In order to develop a multiplex RT-PCR assay using the GeXP analyser for the simultaneous detection of four different NA serotypes of H5-subtype AIVs, effective to control and reduce H5 subtype of avian influenza outbreak. Six pairs of primers were designed using conserved and specific sequences of the AIV subtypes H5, N1, N2, N6 and N8 in GenBank. Each gene-specific primer was fused at the 5' end to a universal sequence to generate six pairs of chimeric primers, and one pair of universal primers was used for RT-PCR, and PCR product separation and detection were performed by capillary electrophoresis using the GenomeLab GeXP genetic analysis system. Single and mixed avian pathogen cDNA/DNA templates were employed to evaluate the specificity of a multiplex assay with a GeXP analyser. Corresponding specific DNA products were amplified for each gene, revealing amplification peaks for M, H5, N1, N2, N6 and N8 genes from four different NA subtypes of influenza A H5 virus. A total of 180 cloacal swabs were collected from poultry at live bird markets. The multiplex PCR assay demonstrated excellent specificity, with each pair of specific primers generating only products corresponding to the target genes and without cross-amplification with other NA-subtype influenza viruses or other avian pathogens. Using various premixed ssRNAs containing known AIV target genes, the detection limit for the multiplex assay was determined to be 10^2 copies/μL. The GeXP assay was further evaluated using 180 clinical specimens and compared with RRT-PCR (real-time reverse transcriptase PCR) and virus isolation. This GeXP analyser-based multiplex assay for four different NA subtypes of H5 HPAI viruses is both highly specific and sensitive and can be used as a rapid and direct diagnostic assay for testing clinical samples.

Keywords

differential diagnoses, GeXP analyser, H5 avian influenza viruses, HA typing, multiplex detection, NA typing

---

Introduction

Influenza virus serotypes are classified based on the combination of two major antigens on the virion, namely haemagglutinin (HA) and neuraminidase (NA). To date, 16 H types and 9 N types have been acknowledged, and a 17th and 18th HA type plus a 10 th and 11 th NA type have recently been discovered in bats[1, 2]. Avian influenza is an acute infectious disease caused by type A avian influenza viruses (AIVs), and certain subtypes have public health significance because they can also infect humans. In addition to causing considerable economic damage to the global poultry industry, highly pathogenic avian influenza viruses (HPAIVs) pose a major public health hazard, especially the H5 subtype[3]. Since 2003, H5N1 HPAI viruses have emerged in at least sixty countries worldwide[4, 5] and various NA subtypes of H5 HPAIVs (H5N2, H5N6 and H5N8) have recently been detected in different poultry and wild birds, with highly pathogenic AI epidemics. For example, Heilongjiang, Jiangsu and Hunan provinces, as well as other provinces in China, reported outbreaks of H5N6 on poultry farms in 2014 and 2015; in 2014, China, Japan, Germany, the Netherlands, the UK and Republic of Korea reported outbreaks of H5N8 on poultry farms and also viral presence in migratory birds. Furthermore, earlier this year, outbreaks of H5N8 and H5N2 were reported on poultry farms in Taiwan, China, and H5N2 HPAI viruses appeared in at least 14 states in the USA[6]. These AIV outbreaks resulted in extensive losses to the poultry industry worldwide[7].

Laboratory diagnosis is essential when AI is suspected in the field, and currently, RT-PCR tests are used routinely for AI diagnosis and genotyping. Although conventional and real-time RT-PCR protocols have been published for the detection of H5-subtype HPAIVs in poultry[8-10], none of these methods can be applied for simultaneous detection or for HA and NA subtyping of H5-subtype HPAI viruses such as H5N1, H5N2, H5N6 and H5N8 in a single assay. The GeXP analyser is a multiplex gene expression profiling analysis platform developed by Beckman Coulter Company (Brea, CA, USA) that was originally designed to allow for the high-throughput, robust and differential assessment of a multiplexed expression profile of up to 35 genes in a single RT-PCR[11]. This system has been successfully used for the rapid identification of several viral diseases of humans, such as nine serotypes of enteroviruses associated with hand, foot and mouth disease[12], pandemic influenza A H1N1 virus[13], sixteen human respiratory virus types/subtypes[14] and seven enteric viruses associated with acute gastroenteritis[15], as well as the genotyping of 11 human papillomaviruses[16]. Moreover, our laboratory has successfully developed three GeXP multiplex PCR assay for the simultaneous typing of nine different avian respiratory pathogens, eight swine reproductive and respiratory pathogens, as well as differentiation of eleven duck viruses[17-19]. In this study, a multiplex RT-PCR assay using the GeXP analyser was developed for the simultaneous detection of four different NA serotypes of H5-subtype AIVs.

Methods

Virus strains and DNA/RNA extraction

The AIV reference strains, field isolates and other avian pathogens used in this study are listed in Table 5-3-1. Genomic DNA/RNA was extracted from 200 μL of virus using MiniBEST Viral RNA/DNA Extraction Kit ver.5.0 (TaKaRa, Dalian, China) according to the manufacturer's protocol. DNA and RNA were eluted in 30 μL of nuclease-free water. The concentrations of total DNA and RNA were measured by UV spectrophotometry (Beckman Coulter). All DNA and RNA samples were stored immediately at −80 ℃ until used in experiments.

Table 5-3-1 Sources of pathogens used and GeXP assay results

Pathogens/field samples	Numbers of sample	Source	M	H5	N1	N2	N6	N8
Reference samples								
Inactivated H5N1 AIV Re-1	1	HVRI	+	+	+	−	−	−
cDNA of H5N3AIV Duck/HK/313/78	1	CU	+	+	−	−	−	−
cDNA of AIV H5N2/chicken/QT35/87	1	CU	+	+	−	+	−	−
cDNA of AIV H5N5/chicken/QT35/98	1	CU	+	+	−	−	−	−

continued

Pathogens/field samples	Numbers of sample	Source	M	H5	N1	N2	N6	N8
cDNA of AIV H5N7 A/waterfowl/G A/269452-56/03	1	CU	+	+	−	−	−	−
cDNA of AIV AIV H5N9/chicken/QT35/98	1	CU	+	+	−	−	−	−
cDNA of AIV H7N2AIV Duck/HK/47/76	1	HKU	+	−	−	+	−	−
cDNA of AIV H7N2/chickenPA/3979/97	1	PU	+	−	−	+	−	−
AIV H6N1 A/Duck/Guangxi/GXd-5/2010	1	GVRI	+	−	+	−	−	−
AIV H6N1 A/Duck/Guangxi/105/2011	1	GVRI	+	−	+	−	−	−
AIV H6N2 A/Goose/Guangxi/105/2011	1	GVRI	+	−	−	+	−	−
AIV H6N2 A/Goose/Guangxi/115/2012	1	GVRI	+	−	−	+	−	−
AIV H6N2 A/Duck/Guangxi/116/2012	1	GVRI	+	−	−	+	−	−
AIV H6N2 A/Chicken/Guangxi/121/2013	1	GVRI	+	−	−	+	−	−
AIV H6N2 A/Duck/Guangxi/121/2012	1	GVRI	+	−	−	+	−	−
AIV H9N2 A/Turtledove/Guangxi/49B6/2013	1	GVRI	+	−	−	+	−	−
AIV H9N2 A/Chicken/Guangxi/NN2/2011	1	GVRI	+	−	−	+	−	−
AIV H9N2 A/Chicken/Guangxi/NN1/2011	1	GVRI	+	−	−	+	−	−
AIV H9N2 A/Chicken/Guangxi/111C8/2012	1	GVRI	+	−	−	+	−	−
AIV H9N2 A/Chicken/Guangxi/116C4/2012	1	GVRI	+	−	−	+	−	−
AIV H9N2 A/Pheasant/Guangxi/49B2/2013	1	GVRI	+	−	−	+	−	−
AIV H9N2 A/Sparrow/Guangxi/35B15/2013	1	GVRI	+	−	−	+	−	−
AIV H9N2 A/Dove/Guangxi/31B6/2013	1	GVRI	+	−	−	+	−	−
AIV H9N2 A/Chicken/Guangxi/141C10/2013	1	GVRI	+	−	−	+	−	−
AIV H9N2 A/Chicken/Guangxi/CX/2013	1	GVRI	+	−	−	+	−	−
AIV H3N2 A/Chicken/Guangxi/015C10/2009	1	GVRI	+	−	−	+	−	−
AIV H3N2 A/Duck/Guangxi/015D2/2009	1	GVRI	+	−	−	+	−	−
AIV H3N6 A/Pigeon/Guangxi/020P/2009	1	GVRI	+	−	−	−	+	−
AIV H3N6 A/Duck/Guangxi/175D12/2014	1	GVRI	+	−	−	−	+	−
AIV H6N6 A/Duck/Guangxi/058/2010	1	GVRI	+	−	−	−	+	−
AIV H6N6 A/Chicken/Guangxi/129/2013	1	GVRI	+	−	−	−	+	−
AIV H6N6 A/Duck/Guangxi/131/2013	1	GVRI	+	−	−	−	+	−
AIV H6N6 A/Pigeon/Guangxi/161/2014	1	GVRI	+	−	−	−	+	−
AIV H6N6 A/Duck/Guangxi/GXd-7/2011	1	GVRI	+	−	−	−	+	−
AIV H6N8 A/Duck/Guangxi/GXd-6/2010	1	GVRI	+	−	−	−	−	+
AIV H6N8 A/Duck/Guangxi/113/2012	1	GVRI	+	−	−	−	−	+
AIV H3N8 A/Goose/Guangxi/020G/2009	1	GVRI	+	−	−	−	−	+
AIV H1N3 Duck/HK/717/79-d1	1	HKU	+	−	−	−	−	−
AIV H1N1 Human/NJ/8/76	1	HKU	+	−	+	−	−	−
AIV H2N3 Duck/HK/77/76	1	HKU	+	−	−	−	−	−

continued

Pathogens/field samples	Numbers of sample	Source	M	H5	N1	N2	N6	N8
AIV H3N6 AIV Duck/HK/526/79/2B	1	HKU	+	–	–	–	+	–
AIV H4N5 Duck/HK/668/79	1	HKU	+	–	–	–	–	–
AIV H6N8 Duck/HK/531/79	1	HKU	+	–	–	–	–	+
AIV H8N4 AIV Turkey/Ont/6118/68	1	HKU	+	–	–	–	–	–
AIV H10N3 Duck/HK/876/80	1	HKU	+	–	–	–	–	–
AIV H11N3 Duck/HK/661/79	1	HKU	+	–	–	–	–	–
AIV H12N5 Duck/HK/862/80	1	HKU	+	–	–	–	–	–
AIV H13N5AIV Gull/MD/704/77	1	HKU	+	–	–	–	–	–
AIV H13N6 A/Gull/Maryland/704/1977	1	PU	+	–	–	–	+	–
AIV H14N5 A/Mallard duck/Astrakhan/263/1982	1	PU	+	–	–	–	–	–
AIV H15N9 A/wedge-tailedshearwater/Western Australia/ 2576/1979	1	PU	+	–	–	–	–	–
AIV H16N3 A/shorebird/Delaware/168/06	1	PU	+	–	–	–	–	–
IAVH1N1 A/Guangxi/1415/15	1	GCDC	–	–	–	–	–	–
IAVH3N2 A/Guangxi/1632/15	1	GCDC	–	–	–	–	–	–
B/Guangxi/1470/15	1	GCDC	–	–	–	–	–	–
Other pathogens								
NDV La Sota	1	GVRI	–	–	–	–	–	–
IBV Massachussetts 41	1	GVRI	–	–	–	–	–	–
ILTV (AV1231)	1	GVRI	–	–	–	–	–	–
MGS6	1	GVRI	–	–	–	–	–	–
MSK1415	1	GVRI	–	–	–	–	–	–
HPGAV269	1	GVRI	–	–	–	–	–	–
Avian reovirus (S1133)	1	GVRI	–	–	–	–	–	–

Note: HVRI, Harbin Veterinary Research Institute, China; HKU, The University of Hong Kong, China; GVRI, Guangxi Veterinary Research Institute, China; CIVDC, China Institute of Veterinary Drugs Control, China; PU, University of Pennsylvania, USA; GCDC, Guangxi Center for Disease Control; CU, University of Connecticut, USA.

Primer design and plasmid preparation

Six pairs of gene-specific primers were designed based on sequence information obtained from Influenza Sequence Database. The designed primers were analysed and filtered using the Premier 5.0 (Primer, Montreal, Canada), NCBI Primer Blast (NCBI, Bethesda, MD, USA) and Oligo 7.0 (Biolytic, Fremont, CA, USA) tool. Each gene-specific primer was fused at the 5' end to a universal sequence to generate six pairs of chimeric primers, and one pair of universal primers was used for RT-PCR (Table 5-3-2). The AIV universal primers were designed to correspond to a highly conserved region of the matrix (M) gene, and the H5 primers were designed in a specific region of the HA gene segment for the H5 serotype. The other four pairs of primers were designed to correspond to the specific region of an NA gene segment for each of the NA types: N1, N2, N6 and N8. Primer synthesis and HPLC purification was performed by Invitrogen (Guangzhou, China).

Table 5-3-2 Sequences of primers used for GeXP assay

Primer	Forward primer (5'-3')[a]	Reverse primer (5'-3')[a]	Size/bp
M	AGGTGACACTATAGAATATCTTGCACTTGAYA TTGTGGATTC[b]	GTACGACTCACTATAGGGAACAAAATGACCA TCGTCAACATCC[b]	211
H5	AGGTGACACTATAGAATAGGAAAGTGTAAGA AACGGAACGTA[b]	GTACGACTCACTATAGGGACACATCCATAAA GAYAGACCAGC[b]	223
N1	AGGTGACACTATAGAATACTGTAATGACTGAY GGACCAAGTA[b]	GTACGACTCACTATAGGGACAGGAGCATTCC TCATAGTGGTAA[b]	162
N2	AGGTGACACTATAGAATAATGTTATCARTTTG CACTTGGGCAG[b]	GTACGACTCACTATAGGGACATGCTATGCACA CYTGTTTGGTTC[b]	188
N6	AGGTGACACTATAGAATACACTATAGATCCYG ARATGATGACC[b]	GTACGACTCACTATAGGGAGGAGTCTTTGCT AATWGTCCTTCCA[b]	240
N8	AGGTGACACTATAGAATAATGTGTACCAGGC AAGGTTTGA[b]	GTACGACTCACTATAGGGATTTGCTGGTCCAT CCGTCATTA[b]	280
Cy5-labelled Tag-F	AGGTGACACTATAGAATA[b]		
Tag-R	GTACGACTCACTATAGGGA[b]		

Note: Underlined oligonucleotides are universal sequences. [a] Degenerated primer abbreviations are as follows: R, A/G; W, A/T; Y, C/T. [b] Primer is in the PCR primer mix.

Plasmids harbouring genes from AIV (H5N1 AIV Re-1, H5N2 AIV chicken/QT35/87, H3N6 AIV Duck/HK/526/79/2B, H6N8 AIV Duck/HK/531/79) were used to produce ssRNA via in vitro transcription using a RiboMAX^{TM} Large Scale RNA Production System-T7 kit (Promega, Madison, wisconsin, USA). The copy numbers of the ssRNAs for the target genes of AIV (M, H5, N1, N2, N6 and N8) were calculated according to previous methods[20, 21].

Set-up of the multiplex PCR for the GeXP analyser

The reaction system was created using a total volume of 25 μL containing 2.5 μL of 10 × PCR buffer, 2.5 μL of $MgCl_2$ (25 μm), 1~25 μL of mixed primers (containing 20~100 nmol/L of 6 pairs of gene-specific chimeric primers), 1~25 μL of universal primers (100 nmol/L), 1~2 μL of JumpStart Taq DNA polymerase (2.5 U/μL) and 0.5 pg to 0.5 ng of cDNA or DNA template. Nuclease-free water was then added to the PCR to achieve a final volume of 25 μL. PCR was carried out using the GeXP system followed by three steps of amplification according to the temperature switch PCR (TSP) strategy[22]: step 1, 95 ℃ for 3 min, 10 cycles of 95 ℃ for 30 s, 55 ℃ for 30 s and 72 ℃ for 30 s; step 2, 10 cycles of 95 ℃ for 30 s, 65 ℃ for 30 s and 72 ℃ for 30 s; and step 3, 20 cycles of 95 ℃ for 30 s, 53 ℃ for 30 s and 72 ℃ for 30 s. The details of the primers are shown in Table 5-3-2, and all were purchased from Invitrogen. PCR product separation and detection were performed by capillary electrophoresis using the GenomeLab GeXP genetic analysis system (Beckman Coulter) following previously described protocols[23]. After the amplified fragments were separated, the peaks were initially analysed using the fragment analysis module of the GEXP SYSTEM software (Beckman Coulter, Brea, CA, USA) and matched to the appropriate genes. The peak height for each gene is illustrated in an electropherogram.

Single primer test for specificity of the GeXP multiplex assay

The assay specificities of all the targets were individually tested with pre-mixed cDNA/DNA in a multiplex PCR assay after optimisation. Other conventional chicken viruses, including all subtypes (except the H5 subtype, including HA (1-16) and NA (1-9); the HA and NA genes of the reference strains were confirmed by sequencing) of avian influenza virus (AIV), Newcastle disease virus (NDV), infectious bronchitis virus

(IBV), avian reovirus (ARV), infectious laryngotracheitis virus (ILTV), seasonal influenza A H1N1, H3N2 and influenza B viruses and nuclease-free water, were used as negative controls.

Evaluation of the sensitivity of the GeXP multiplex assay

The sensitivity of the multiplex assay using the GeXP analyser was evaluated using a previously described method[23]. The concentrated products for each target gene were diluted to final concentrations ranging from 10^5 copies to 1 copy per microlitre and then individually subjected to the multiplex assay. The concentrations of specific primers were optimised according to the amplification efficiency of the assay using a single template. The sensitivity of the optimised multiplex GeXP PCR assay was reevaluated three times on three different days using 6 premixed RNA templates ranging from 10^5 copies to 1 copy per microlitre.

Interference assay

Because high quantities of different templates could alter the efficiency of multiplex PCR amplification, different amounts of template (10^2 to 10^6 copies) were selected at random, mixed and tested in the multiplex PCR assay. The results were then compared with those of the single template multiplex PCR assay.

Application to clinical specimens

A total of 180 cloacal swabs were collected from poultry at live bird markets (LBMs), and the swab samples were injected into 9- to 11-day-old embryonated specific pathogen-free (SPF) chicken eggs, as previously described[24]. At 48～96 h, all allantoid fluids were recovered incubation for virus detection and titration, as described[25]. The 180 clinical specimens were randomly divided into five groups and tested by RRT-PCR, GeXP multiplex PCR and virus isolation, respectively (Table 5-3-3), and the HA and NA genes of positive samples were sequenced for demonstration of RT-PCR compliance using previously reported primers[26].

Table 5-3-3 Comparative detection of cloacal swab samples using virus isolation, RRT-PCR and GeXP

Sample number	Virus isolate	Samples negative for avian influenza virus		Samples positive for avian influenza virus						
		RRT-PCR	GeXP assay	RRT-PCR	GeXP assay					
					M	H5	N1	N2	N6	N8
1～30	H5N1	29/30	29/30	1/30	1/30	1/30	1/30	0/30	0/30	0/30
31～60	H5N2	29/30	28/30	1/30	2/30	2/30	0/30	2/30	0/30	0/30
61～102	H9N2	39/42	38/42	3/42	4/42	0/42	0/42	4/42	0/42	0/42
103～137	H6N2	34/35	34/35	1/35	1/35	0/35	0/35	1/35	0/35	0/35
138～180	H6N6	41/43	41/43	2/43	2/43	0/43	0/43	0/43	32/43	0/43

Results

Specificity of the GeXP multiplex PCR assay

cDNA samples from four different NA subtypes of influenza A H5 virus (H5N1, H5N2, H5N6 and H5N8) were individually used as template to evaluate the specificity of each pair of gene-specific primers. In mono-PCR GeXP assays, AIV universal primers were able to amplify the target M gene of all AIV serotypes, although each pair of specific primers generated only the corresponding targeted gene, without cross-amplification (Table 5-3-1). The amplicon sizes for the viruses were as follows: AIV-H5, 223～226 bp; AIV-N1, 160～163 bp; AIV-N2, 188～191 bp; AIV-N6, 240～243 bp; AIV-N8, 280～283 bp; and AIV M gene, 210～213 bp (Table 5-3-2). These six targeted genes were detected via the multiplex GeXP PCR assay, and specific amplifcation peaks were observed (Figure 5-3-1). For example, specific amplification peaks of the AIV M gene were observed for HIN1, H2N3, H3N6, H4N5, H5N2, H5N3, H5N7, H6N8, H7N2, H8N4, H9N2, H10N3, H11N3, H12N5, H13N6, H14 N5, H15N9 and H16N3.

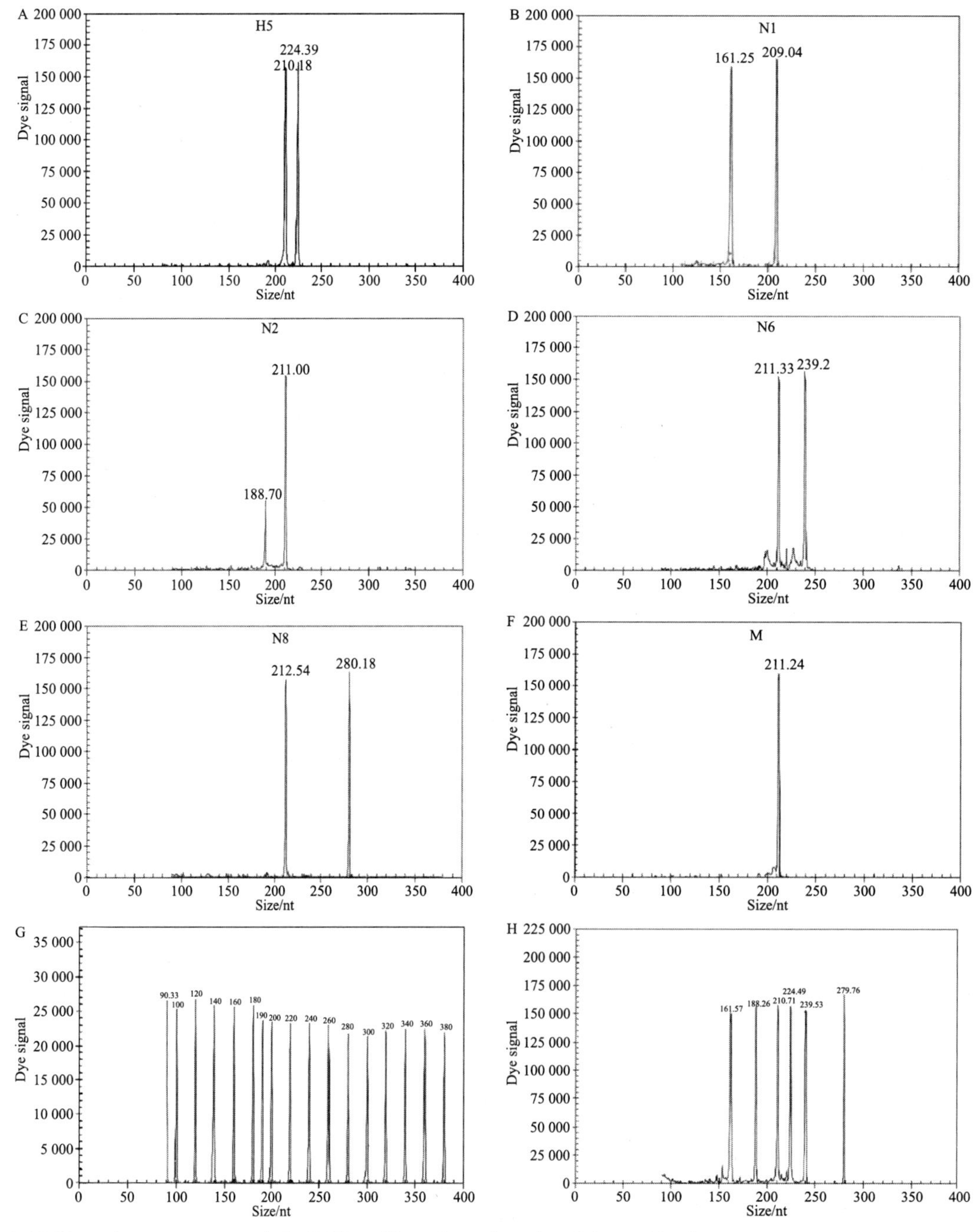

Cy5-labeled PCR products were separated via GeXP capillary electrophoresis and detected by fluorescence spectrophotometry, given as dye signals in arbitrary units on the y axis. Each peak was identified by comparing the expected to the actual PCR product size on the x axis. A-F, H: The results of amplification of target genes H5, N1, N2, N6, N8, M, and all, respectively; G: Nuclease-free water was used as the negative control. The red peaks indicate the DNA size standard.

Figure 5-3-1 Specificity of the multiplex RT-PCR assay

Sensitivity of the GeXP multiplex PCR assay

When using ssRNA in vitro transcription individually, the GeXP multiplex PCR assay detected as little as 10^2 copies/μL of AIV type A, AIV-H5, AIV-N1, AIV-N2, AIV-N6 and AIV-N8 (Figure 5-3-2, all electropherograms are not shown). The reactions were repeated three times for each template concentration on different days, and similar results were obtained.

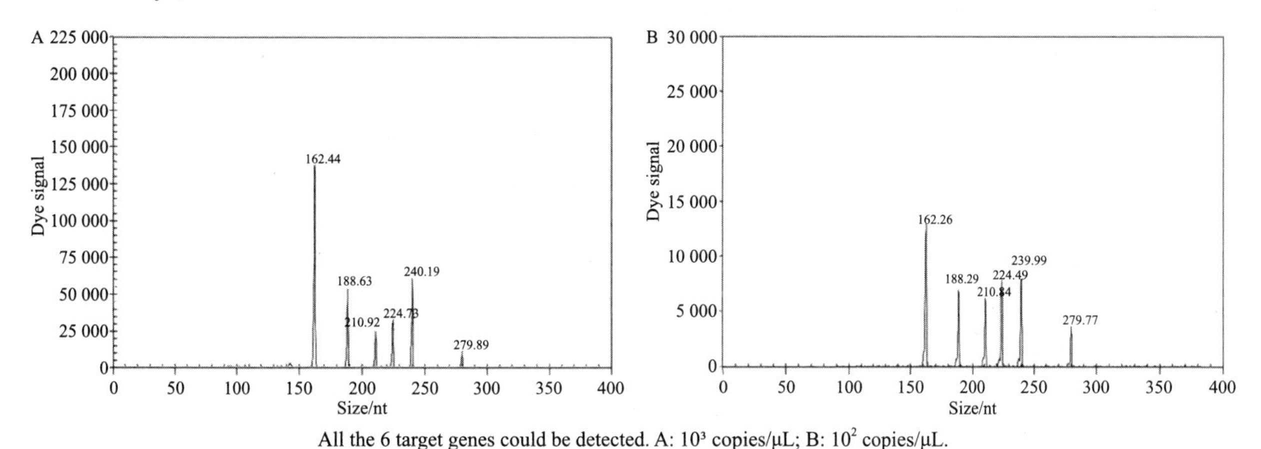

All the 6 target genes could be detected. A: 10^3 copies/μL; B: 10^2 copies/μL.

Figure 5-3-2　Sensitivity of GeXP detection of 6 premixed RNA templates with multiplex primers

Artificial mixture and interference assay

Two specific amplification peaks were observed when two different templates (one template with 10^2 copies and another with 10^6 copies) were tested by the GeXP multiplex PCR assay, and the peak value of the single template was similar to that of the mixed template. For example, two specific amplification peaks were observed when two different templates (AIV-H5N1 at 10^2 copies and H5N2 at 10^6 copies) were tested, and the peak values for AIV-H5N1 and H5N2 were the same, regardless of whether a single (AIV-H5N1 or H5N2) or a mixed (AIV-H5N1+H5N2) template was utilised (electropherograms not shown). The results of these experiments demonstrate that this interference had a minimal effect on the detection of mixed infection.

Detection of clinical samples by the GeXP multiplex PCR assay

A total of 180 random cloacal swab samples were collected from poultry at various LBMs during May 2015; the samples were assayed using optimised GeXP multiplex PCR (Table 5-3-3) and confirmed by DNA sequencing of the HA and NA genes. Virus isolation revealed ten different genotypes among the clinical samples, and three positive samples of H5-subtype AIVs were isolated among 60 cloacal swab samples. Of the ten positive samples according to the GeXP PCR assay, one sample was negative by RRT-PCR but positive when retested by virus isolation. Moreover, the HA and NA genes of the positive samples were sequenced, showing complete agreement for 100% of the GeXP PCR assay results (Table 5-3-3). We also found some nucleotide variation, which resulted in the failure of the real-time PCR detection of an H9N2 isolate.

Discussion

Although the vaccination strategy against H5N1 influenza has been effective at reducing the incidence of H5N1 infection in poultry during the past decade, recent studies have highlighted the continued presence of H5N1, H5N2, H5N6 and H5N8, posing a threat to both poultry and human health[27, 28]. Outbreaks of various

H5 HPAIV NA subtypes are characterised by their sudden occurrence, dissemination speed, similar clinical symptoms and highly infectious nature with common characteristics. As such factors are unfavourable for the implementation of rational measures to prevent and control disease, the rapid and accurate identification of different NA subtypes circulating in different avian species within a geographical region can provide important epidemiological information and is crucial for the selection of appropriate control and eradication strategies.

Although HI and NI tests have long been used as the Office International Des Epizooties (OIE) standard for AIV subtyping, these methods often have low levels of specificity and sensitivity; in addition, standardisation and reference antisera are difficult to obtain[29]. Accordingly, laboratory diagnosis of AIV, including conventional and real-time RT-PCR protocols, has been recently used for AIV detection and subtyping[9, 30-32]. Indeed, multiplex PCR and multiplex fluorescence real-time quantitative PCR techniques are widely used for the detection and subtyping of mixed infections with multiple pathogens[33, 34]. However, conventional multiple PCR is vulnerable to variations in template and primer concentration as well as concentrations of reagents that interfere with each other during amplification; such factors will cause different PCR efficiencies, decreasing the sensitivity of the results.

In this study, we successfully established a GeXP method that can simultaneously detect and identify the H5N1, H5N2, H5N6 and H5N8 subtypes of AIV in a single reaction within 4 hours; this GeXP method utilises six different genes primers for M, H5, N1, N2, N6 and N8. Additionally, the specificity of the test was examined using different HA (1-16) and NA (1-9) subtypes of avian influenza virus and nucleic acid of seasonal influenza virus and influenza B virus infecting humans (Table 5-3-1). The failure of the M gene primer set in detecting human H1N1 and H3N2 is because the primer set is based on GenBank data of avian influenza virus, with variation in a few nucleotides between human H1N1 and H3N2. We also found no cross-reaction between different pair primers, with specificity for detecting AIV H5, N1, N2, N6 and N8 subtypes and a sensitivity of 10^2 copies/μL. When 180 specimens were analysed, the results for the GeXP PCR assay and the reference method (Virus isolation and sequencing) were in complete agreement for 100% of the specimens, as shown in Table 5-3-3. However, one H9N2 isolate was not detected by RRT-PCR using the N2 primer set (Table 5-3-3). The comparison of the results for NA sequences revealed that the new H9N2 isolates exhibited variation in the target gene of the N2 primer designed for RRT-PCR. Considering that the genetic diversity of AIVs in the natural ecosystem may be higher than the current understanding, it may be necessary to develop primers for all subtypes.

Two distinct advantages of the GeXP assay compared with other nucleic acid-based tests are the reduced time required and cost-effectiveness. The cost of the multiplex GeXP assay for simultaneous detection of the four different NA subtypes of H5 AIVs is approximately $6 per test, including primers representing the HA and NA subtypes, when the entire reaction is performed in a single tube. Utilising two 96-well plates at the same time in the GeXP analyser will further increase the throughput. This method will save much time and cost compared with classical methods of AIV HA and NA subtyping in surveillance programmes.

References

[1] ZHU X, YU W, MECBRIDE R, et al. Hemagglutinin homologue from H17N10 bat influenza virus exhibits divergent receptor-binding and pH-dependent fusion activities. Proc Natl Acad Sci USA, 2013, 110: 1458-1463.

[2] TONG S, ZHU X, LI Y, et al. New world bats harbor diverse influenza A viruses. PLOS Pathog, 2013, 9: e1003657.

[3] SWAYNE D E, SUAREZ D L. Highly pathogenic avian influenza. Rev Sai Tech, 2000, 19: 463-482.

[4] SUH H, LIU J. Epidemicity and pathogenicity of avian influenza A H5N1 virus. Chin Bull Life Sci, 2015, 27: 525-530.

[5] JEONG J, KANG H M, LEE E K, et al. Highly pathogenic avian influenza virus (H5N8) in domestic poultry and its relationship with migratory birds in South Korea during 2014. Vet Microbiol, 2014, 173: 249-257.

[6] OIE. Avian influenza portal2015.(2015-08-03)[2015-12-02]. http: //www.oie.int/ animal-health-in-the-world/update-on-avian-influenza/2015/.

[7] CHEN H. H5N1 avian influenza in China. Sci China Ser C Life Sci, 2009, 52: 419-427.

[8] XIE Z, PANG Y S, LIU J, et al. A multiplex RT-PCR for detection of type A influenza virus and differentiation of avian H5, H7, and H9 hemagglutinin subtypes. Mol Cell Probes, 2006, 20: 245-249.

[9] TSUKAMOTO K, PANEI C J, SHISHIDO M, et al. SYBR green-based real-time reverse transcription-PCR for typing and subtyping of all hemagglutinin and neuraminidase genes of avian influenza viruses and comparison to standard serological subtyping tests. J Clin Microbiol, 2012, 50: 37-45.

[10] HEINE H G, FOORD A J, WANG J, et al. Detection of highly pathogenic Zoonotic Influenza Virus H5N6 chain reaction. Virol J, 2015, 8: 12-18.

[11] LIU Y, XU Z Q, LI J S, et al. A novel method for multiplex detection of gastroenteritis-associated viruses. Chinese Journal of Virdogy, 2011, 27: 288-293.

[12] HU X, ZHANG Y, ZHOU X, et al. Simultaneously typing nine serotypes of enteroviruses associated with hand, foot, and mouth disease by a GeXP analyzer-based multiplex reverse transcription-PCR assay. J Clin Microbiol, 2012, 50: 288-293.

[13] QIN M, WANG D Y, HUANG F, et al. Detection of pandemic influenza AH1N1 virus by multiplex reverse transcription-PCR with a GeXP analyser. Virol Methods, 2010, 168: 255-258.

[14] LI J, QI S, ZHANG C, et al. A two-tube multiplex reverse transcription PCR assay for simultaneous detection of sixteen human respiratory virus types/subtypes. Biomed Res Int, 2013, 2013: 327620.

[15] LIU Y, XU Z, ZHANG Q, et al. Simultaneous detection of seven enteric viruses associated with acute gastroenteritis by a multiplexed Luminex-based assay. J Clin Microbiol, 2012, 50: 2384-2389.

[16] YANG M, LUO L, NIE K, et al. Genotyping of 11 human papillomaviruses by multiplex PCR with a GeXP analyzer. J Med Virol, 2012, 84: 957-963.

[17] XIE Z, LUO S, XIE L, et al. Simultaneous typing of nine avian respiratory pathogens using a novel GeXP analyzer-based multiplex PCR assay. J Virol Methods, 2014, 207: 188-195.

[18] ZHANG M, XIE Z, XIE L, et al. Simultaneous detection of eight swine reproductive and respiratory pathogens using a novel GeXP analyser-based multiplex PCR assay. J Virol Methods, 2015, 224: 9-15.

[19] ZHANG Y F, XIE Z X, XIE U, et al. GeXP analyzer-based multiplex reverse-transcription PCR assay for the simultaneous detection and differentiation of eleven duck viruses. BMC Microbiol, 2015, 15: 247.

[20] XIE Z, XIE L, PANG Y, et al. Development of a real-time multiplex PCR assay for detection of viral pathogens of penaeid shrimp. Arch Virol, 2008, 153: 2245-2251.

[21] XIE Z, XIE L, FAN Q, et al. A duplex quantitative real-time PCR assay for the detection of *Haplosporidium* and *Perkinsus* species in shellfish. Parasitol Res, 2013, 112: 1597-1606.

[22] TABON T, MATHER D E, HAYDEN M. Temperature switch PCR (TSP) : robust assay design for reliable amplification and genotyping of SNPs. BMC Genom, 2009, 510: 580.

[23] RAI A J, KAMATH R M, GERALD W, et al. Analytical validation of the GeXP analyzer and design of a work flow for cancer-bio marker discovery using multiplexed gene-expression profiling. Anal Bioanal Chem, 2009, 393: 1505-1511.

[24] PENG Y, XIE Z, LIU J, et al. Epidemiological surveillance of low pathgenic avian influenza virus (LPAIV) from poultry in Guangxi Province, southern China. PLOS ONE, 2013, 8: e77132.

[25] XIE Z, DONG J, TANG X, et al. Sequence and phylogenetic analysis of three isolates of avian influenza H9N2 from chickens in southern China. Sch Res Exch, 2008: 1-7.

[26] HOFFMANN E, STECH J, GUAN Y, et al. Universal primer set for the ful-length amplification of all influenza A viruses. Arch Virol 2001; 146: 2275-2289.

[27] CHEN H. Avian influenza vaccination: the experience in China. Rev Sci Tech, 2009, 28: 267-274.

[28] WANG M, DI B, ZHOU D, et al. Food markets with live birds as source of avian influenza. Emerg Infect Dis, 2006, 12: 1773-1775.

[29] FOUCHIER R A M, MUNSTERR V, WALLENSTEN A, et al. Characterization of a novel influenza A virus hemagglutinin subtype (H16) obtained from black-headed gulls. J Virol, 2005, 79: 2814-2822.

[30] HUANG Y, KHAN M I, MANDOIU I. Neuraminidase subtyping of avian influenza viruses with primerhunter-designed primers and quadruplicate primer pools. PLOS ONE, 2013, 8: e81842.

[31] QIU B, LIU W, PENG D, et al. A reverse transcription-PCR for subtyping of the neuraminidase of avian influenza viruses. J Virol Methods, 2009, 155: 193-198.

[32] ELIZALDE M, AG € UERO M, BUITRAGO D, et al. Rapid molecular haemagglutinin subtyping of avian influenza isolates by specific real-time RT-PCR tests. J Virol Methods, 2014, 196: 71-81.

[33] PANG Y, WANG H, GIRSHICK T, et al. Development and application of a multiplex polymerase chain reaction for avian respiratory agents. Avian Dis, 2002, 46: 691-699.

[34] KURIAKOSE T, HILT D A, JACKWOOD M W. Detection of avian influenza viruses and differentiation of H5, H7, N1, and N2 subtypes using a multiplex microsphere assay. Avian Dis, 2012, 56: 90-96.

Simultaneous differentiation of the N1 to N9 neuraminidase subtypes of avian influenza virus by a GeXP analyzer-based multiplex reverse transcription PCR assay

Luo Sisi, Xie Zhixun, Huang Jiaoling, Xie Zhiqin, Xie Liji, Zhang Minxiu, Li Meng, Wang Sheng, Li Dan, Zeng Tingting, Zhang Yanfang, Fan Qing, and Deng Xianwen

Abstract

To date, nine neuraminidase (NA) subtypes of avian influenza virus (AIV) have been identified in poultry and wild birds. Rapid and effective methods for differentiating these nine NA subtypes are needed. We developed and validated a rapid, sensitive, and robust method utilizing a GeXP analyzer-based multiplex RT-PCR assay and capillary electrophoresis for the simultaneous differentiation of the N1 to N9 subtypes in a single-tube reaction. Ten pairs of primers-nine subtype-specific pairs and one pan-AIV pair-were screened and used to establish the GeXP multiplex RT-PCR assay. A single subtype was detected using the developed GeXP assay; the N1 to N9 AIV subtypes individually generated two target peaks: the NA subtype-specific peak and the general AIV peak. Different concentrations of multiplexed subtypes were tested with this GeXP assay and the peaks of the corresponding NA subtypes were generated, suggesting that this GeXP assay is useful for identifying NA subtypes in mixed samples. Moreover, no peaks were generated for other important avian viruses, indicating negative results and validating the lack of cross-reactions between AIV subtypes and other avian pathogens. RNA templates synthesized through *in vitro* transcription were used to analyze the sensitivity of the assay; the limit of detection was 100 copies per reaction mixture. The results obtained from clinical samples using this GeXP method were consistent with the results of the neuraminidase inhibition (NI) test, and the ability of the GeXP assay to identify mixed infections was superior to amplicon sequencing of isolated viruses. In conclusion, this GeXP assay is proposed as a specific, sensitive, rapid, high-throughput, and versatile diagnostic tool for nine NA subtypes of AIV.

Keywords

avian influenza virus, neuraminidase, GeXP analyzer, multiplex PCR, differentiation diagnosis

---

Introduction

Four types of influenza viruses, designated *Influenza A* virus, *Influenza B* virus, *Influenza C* virus, *Influenza D* virus, have been identified[1]. Influenza A virus infects birds, humans, swine, and many other mammalian species. Avian influenza virus (AIV) is a negative-sense, segmented, single-stranded, enveloped RNA virus belonging to influenza A virus in the Orthomyxoviridae family. Two major antigens of AIV, hemagglutinin (HA) and neuraminidase (NA), allow subtype classification based on antigenic diversity. Sixteen HA and nine NA subtypes have traditionally been identified in poultry and wild birds, and the 17th and 18th HA subtypes and 10th and 11th NA subtypes were recently discovered in bats[2, 3]. Four influenza-driven pandemics occurred in the 20th century (H2N2 in 1957, H3N2 in 1968, and H1N1 in 1918 and

2009), in which millions of people perished. China is recognized as a geographical area with an environment suitable for the emergence of novel influenza viruses[4]. New strains of AIV, including at least six subtypes of AIV (H5N1, H5N6, H7N9, H10N8, H9N2, and H6N1), are currently circulating widely in China or have resulted in occasional human infections in recent decades. The first confirmed human case of infection with the highly pathogenic H5N1 strain of AIV was reported in Hong Kong in 1997, the first recognized case of virus transmission directly from poultry to humans; a second outbreak of H5N1 viruses occurred in 2003, and continuing occurrences have been reported[5]. H5N1 viruses have currently spread to avian species in many countries and threaten animal and human health[6]. In May 2014, China formally confirmed the first human infection with a novel H5N6 strain of AIV in Sichuan Province, and H5N6 has replaced H5N1 as the dominant epidemic strain in China. Currently, H5N6 strains of AIV are the most prevalent, widespread and harmful subtypes observed in Asia and Southeast Asia (World Organization for Animal Health (OIE))[7]. Up to March 31, 2013, the H7N9 subtype of AIV was not known to infect humans. Since the first outbreak of human infection with the H7N9 subtype of AIV was identified, five waves of human infection have occurred in China[8]. Originally, H7N9 viruses were non-pathogenic in chickens, but mutated to a highly pathogenic strain in early 2017 and caused severe disease outbreaks in chickens. A bivalent inactivated H5/H7 vaccine for chicken was introduced in September 2017, and the rate of H7N9 virus isolation in poultry decreased substantially after vaccination. More importantly, only three H7N9 cases were reported in humans between October 1, 2017 and September 30, 2018, indicating that the vaccination of poultry successfully eliminated human infection with the H7N9 virus[9]. In December 2013, China formally confirmed the first human infection with an avian influenza A (H10N8) virus in Jiangxi Province[10]. The H9N2 virus acts as a gene donor for H7N9 and H10N8 viruses, and the role of the H9N2 virus in enabling human infection by wild bird influenza viruses deserves further study[11]. A case of human infection with an H6N1 virus was reported for the first time in Taiwan, China in June 2013[12]. Based on current evidence, H10N8 and H6N1 AIV infection patterns indicate that this case was most likely sporadic[13]. However, AIV has pandemic potential and is a global concern. The development of appropriate and effective methods for the accurate diagnosis and tracking of AIV is required to control economic losses due to AIV.

Currently, surveillance to prepare for influenza A virus pandemics is conducted worldwide. Oral and cloacal swabs are collected and inoculated into the allantoic cavity of specific pathogen-free (SPF) chickens for the propagation of AIV. Allantoic fluid is then harvested and examined for HA activity with an HA assay. HA-positive samples are examined further using a hemagglutination inhibition (HI) assay to determine HA subtypes. The HI assay is the most commonly used method for the initial identification of HA subtypes in epidemiological surveys of AIVs. Many tests have been reported to identify HA subtypes, but few are able to identify NA subtypes. After the HA subtype is identified, the NA subtype is usually determined by amplicon sequencing, but an effective, convenient and simple method for identifying the nine NA subtypes is unavailable. The traditional method for NA subtyping is a neuraminidase inhibition (NI) assay with cultured AIV; indeed, this method is the gold standard suggested by the OIE. However, NI assay are more complex and time-consuming than the HI test and require an array of high-quality monospecific antisera and reference antigens. Thus, this method is used less frequently in many labs. Rapid and convenient molecular diagnostic assays for the detection of the nine NA subtypes have been developed. Currently, several reports have described the subtyping of the nine NAs using conventional RT-PCR, all of which used nine pairs of NA-specific primers to amplify NA genes in separate reactions, with nine reactions and nine tubes subsequently subjected to agarose gel electrophoresis[14-16]. Some studies improved and simplified the procedure for subtyping the nine

NAs using real-time fluorescent RT-PCR with different fluorophores, which decreased the number of reactions from 9 to 3 or 4[17-18]. However, few methods enable the simultaneous detection of all nine NA subtypes in a single tube. In this study, we developed a method for simultaneously differentiating the N1 to N9 subtypes in a single reaction, thus decreasing the number of reactions from 9 to 1.

GenomeLab Gene Expression Profler (GeXP) technology represents a novel platform for high-throughput nucleic acid detection based on multiplex RT-PCR assays and an analysis of amplicons size using capillary electrophoresis, a method that is capable of differentially assessing the expression profile of up to 30 genes in one tube by converting amplification with multiple primers to amplification with a pair of universal primers. The GeXP analyzer has been widely adopted in veterinary diagnostics and medical examinations, and displays high sensitivity and specificity[19-23]. Currently, a method for the rapid and high-throughput diagnosis of nine NA subtypes is urgently needed. This study aimed to develop a rapid, sensitive, and robust method for differentiating the nine NA subtypes of AIV.

Materials and methods

Primer design

The sequences of the nine NA genes and the M gene were obtained from the Influenza Sequence Database. Analyses of sequences from representative strains of nine NA subtypes of AIV were conducted and conserved regions were identified using Lasergene 8.0 software. The primers were designed in relatively conserved regions, and primer parameters, such as the annealing temperature, mismatch and dimerization, were evaluated using Primer Premier 5.0 software. The primer sequences were subjected to an in silico BLAST analysis in the nucleotide database (NCBI) to evaluate specificity. Chimeric primers consisted of two parts: a designed gene-specific primer and a universal tag primer; the universal forward and reverse tag sequences were attached to the 5' end of the designed specific primers. Generally, the size of the designed amplicons for GeXP was 105～350 bp without the universal tags and 142～387 bp with the universal tags. Ten pairs of primers with universal tags were finally chosen (Table 5-4-1) from the initial evaluation panel of 30 primer pairs, including nine pairs of subtype-specific primers targeting the AIV NA genes and one pair of pan-AIV primers targeting the M gene, a gene conserved across all AIV subtypes. All primers were synthesized and purified by Invitrogen (Guangzhou, China).

Table 5-4-1 Primer information for the GeXP assay

Gene#	Gene name	Forward primer (5'-3')	Reverse primer (5'-3')	GenBank accession number	Position	Size/bp
1	N1	AGGTGACACTATAGAATA GGTGTTTGGATCGG**R**AGAAC	GTACGACTCACTATAGGGA TCAACCCAGAA**R**CAAGGTC	KJ907641.1	1 006-1 025 1 214-1 196	246
2	N2	AGGTGACACTATAGAATA TTGGGTGTTCCGTTTCA	GTACGACTCACTATAGGGA CCATCCGTCATTACTAC	KF768231.1	506-522 750-734	282
3	N3	AGGTGACACTATAGAATA TTCCCAATAGGAACAGC**Y**CCAGT	GTACGACTCACTATAGGGA TTCTCCATGATTT**R**ATGGAGTC	CY129336.1	487-509 667-646	218
4	N4	AGGTGACACTATAGAATA CAGA**Y**AAGGA**Y**TCAAATGGTGT	GTACGACTCACTATAGGGA CATGGTACAGTGCAATTCCT	CY133359.1	1 151-1 172 1 265-1 246	152
5	N5	AGGTGACACTATAGAATA GTGAGGTCATGGAGAAAGCA	GTACGACTCACTATAGGGA TGG**Y**CTATTCATTCC**R**TTCCA	CY196019.1	646-665 906-886	298

continued

Gene#	Gene name	Forward primer (5'-3')	Reverse primer (5'-3')	GenBank accession number	Position	Size/bp
6	N6	AGGTGACACTATAGAATA CACTATAGATCC**Y**GA**R**ATGATGACC	GTACGACTCACTATAGGGA GGAGTCTTTGCTAAT**W**GTCCTTCCA	KR919741.1	879-903 1 080-1 056	239
7	N7	AGGTGACACTATAGAATA GACAG**R**AC**W**GCTTTCAGAGG	GTACGACTCACTATAGGGA GTTGCGTTGTCATTATTTCC	CY167234.1	455-474 612-593	195
8	N8	AGGTGACACTATAGAATA AGGGAATACAATGAAACAGT	GTACGACTCACTATAGGGA TGCAAAACCCTTAGCATCACA	JX304764.1	171-190 288-308	175
9	N9	AGGTGACACTATAGAATA CGCCCTGATAAGCTGGCCACT	GTACGACTCACTATAGGGA ACAGGCCTTCTGTTGTACCA	CY129263.1	472-492 642-623	208
10	AIV-M	AGGTGACACTATAGAATA AGGCTCTCATGGAGTGGCTA	GTACGACTCACTATAGGGA TGGACAAAGCGTCTACGCTG	MH341597.1	119-138 242-223	161

Note: Universal tag sequences are underlined. Bold type shows degenerate sites. R, A/G; Y, C/T; W, A/T.

Preparation of viruses and cDNAs

The reference isolates of the N1 to N9 subtypes of AIV and several important avian pathogens (Newcastle disease virus (NDV), infectious bronchitis virus (IBV), infectious laryngotracheitis virus (ILTV), avian reovirus (ARV), and fowl adenovirus 4 (FAdV-4)) were used to assess the specificity of the GeXP assay (Table 5-4-2). Viral RNA/DNA was extracted from 200 μL of allantoic fluid of various virus stocks using an EasyPure Viral DNA/RNA kit (TransGen) according to the manual and eluted in 35 μL of nuclease-free water. Reverse transcription was performed in 50 μL reaction mixtures containing 10 μL of 5 × M-MLV buffer, 0.6 μL of 40 U/μL RNase inhibitor, 2 μL of 10 mM dNTPs, 1 μL of 200 U/μL M-MLV reverse transcriptase (TaKaRa), 1.5 μL of a 25 μM stock of the 12 bp primer (5'-AGCGAAAGCAGG-3'), and 35 μL of extracted RNA and were then incubated at 42 ℃ for 1 h and stored at −20 ℃ until needed.

Table 5-4-2 Reference strains used in the GeXP assay

Number	Strain designation	Source	GeXP assay	Identification of isolate subtype
1	A/Duck/Guangxi/030D/2009 (H1N1)	GVRI	N1	N1
2	Duck/Guangxi/1/04 (H5N1)	GVRI	N1	N1
3	Chicken/Guangxi/1/04 (H5N1)	GVRI	N1	N1
4	A/Duck/Guangxi/GXd-5/2010 (H6N1)	GVRI	N1	N1
5	A/Sparrow/Guangxi/GXs-1/2012 (H1N2)	GVRI	N2	N2
6	A/Duck/Guangxi/GXd-2/2012 (H1N2)	GVRI	N2	N2
7	A/Duck/Guangxi/M20/2009 (H3N2)	GVRI	N2	N2
8	A/Duck/Guangxi/N42/2009 (H3N2)	GVRI	N2	N2
9	A/Chicken/Guangxi/045C2/2009 (H4N2)	GVRI	N2	N2
10	A/Duck/Guangxi/125D17/2012 (H4N2)	GVRI	N2	N2
11	A/Duck/Guangxi/GXd-2/2009 (H6N2)	GVRI	N2	N2
12	A/Chicken/Guangxi/DX/2008 (H9N2)	GVRI	N2	N2
13	A/Chinese Francolin/Guangxi/020B7/2010 (H9N2)	GVRI	N2	N2
14	A/Duck/HK/77/76 (H2N3)	UHK	N3	N3

continued

Number	Strain designation	Source	GeXP assay	Identification of isolate subtype
15	A/Duck/HK/876/80 (H10N3)	UHK	N3	N3
16	A/Shorebird/Delaware/168/06 (H16N3)	UC	N3	N3
17	A/Turkey/Ontario/6118/68 (H8N4)	UHK	N4	N4
18	A/Duck/Guangxi/GXd-1/2009 (H6N5)	GVRI	N5	N5
19	A/Duck/HK/862/80 (H12N5)	UHK	N5	N5
20	A/Gull/Md/704/77 (H13N5)	UHK	N5	N5
21	A/Mallard/Astrakhan/263/82 (H14N5)	UC	N5	N5
22	A/Pigeon/Guangxi/020P/2009 (H3N6)	GVRI	N6	N6
23	A/Duck/Guangxi/175D12/2014 (H3N6)	GVRI	N6	N6
24	A/Duck/Guangxi/070D/2010 (H4N6)	GVRI	N6	N6
25	A/Duck/Guangxi/101D18/2011 (H4N6)	GVRI	N6	N6
26	A/Duck/Guangxi/149D24/2013 (H4N6)	GVRI	N6	N6
27	A/Duck/Guangxi/GXd-4/2009 (H6N6)	GVRI	N6	N6
28	A/Duck/Guangxi/GXd-7/2011 (H6N6)	GVRI	N6	N6
29	A/Chicken Guangxi/129/2013 (H6N6)	GVRI	N6	N6
30	A/Duck/42846/07 (H7N7) RNA	UP	N7	N7
31	A/Goose/Guangxi/020G/2009 (H3N8)	GVRI	N8	N8
32	A/Duck/Guangxi/GXd-6/2010 (H6N8)	GVRI	N8	N8
33	A/Duck/Guangxi/113/2012 (H6N8)	GVRI	N8	N8
34	A/Chicken/QT35/98 (H5N9) RNA	UP	N9	N9
35	A/Duck/PA/2099/12 (H11N9)	UP	N9	N9
36	A/Shearwater/Western Australia/2576/79 (H15N9)	UC	N9	N9
37	B/Guangxi/1418/15	GXCDC		Influenza B virus
38	B/Guangxi/1470/15	GXCDC		Influenza B virus
39	NDV F48	CIVDC		NDV
40	IBV M41	CIVDC		IBV
41	ILTV Beijing strain	CIVDC		ILTV
42	ARV S1133	CIVDC		ARV
43	FAdV-4-GX001	GVRI		FAdV-4

Note: GVRI, Guangxi Veterinary Research Institute, China; UHK, University of Hong Kong, China; UP, University of Pennsylvania, United States; UC, University of Connecticut, United States; CIVDC, China Institute of Veterinary Drug Control; GXCDC, Guangxi Provincial Center for Disease Control and Prevention, China.

Evaluation of the designed primers using the GeXP mono-RT-PCR assay

The N1 to N9 subtypes of AIV and other important avian viruses (NDV, IBV, ILTV, ARV, and FAdV-4) were detected using the GeXP mono-RT-PCR assay to validate the ability of each primer pair to identify the NA subtype. The PCR master mix was prepared in a 20 μL volume containing 4 μL of 5 × GeXP PCR buffer (containing the fluorophore-labeled forward universal primer and the unlabeled reverse universal primer, AB Sciex), 4 μL of 25 mM $MgCl_2$, 1 μL of 2.5 U/μL JumpStart Taq DNA polymerase (Sigma), 2 μL each of

200 nM solutions of the forward and reverse primers, 2 μL of transcribed cDNA and 7 μL of nuclease-free water. Since the fluorescent dye is light-sensitive, each PCR should be prepared in dimly lit conditions. PCR was performed with the following cycling conditions: initial denaturation at 95 ℃ for 5 min; followed by 30 cycles at 95 ℃ for 30 s, 50 ℃ for 30 s, and 72 ℃ for 40 s; and a final extension at 72 ℃ for 5 min.

Separation using capillary electrophoresis and fragment analysis

PCR products were separated and detected using a GenomeLab GeXP genetic analysis system (Beckman Coulter) through capillary electrophoresis, according to previously described protocols[24]. After the amplified fragments were separated, the peaks were initially analyzed using the fragment analysis module of the GeXP system software and matched to the appropriate genes. The peak height for each gene was reported in the electropherogram.

Evaluation of the ability of multiplex primers to amplify single and multiple templates using the GeXP multiplex RT-PCR assay

We next selected the optimal primers validated in the GeXP mono-RT-PCR assay for each subtype as candidates for GeXP multiplex RT-PCR. Nine pairs of subtype-specific primers and one pan-AIV primer were combined in a reaction mixture (2 μL each of 200 nM primer stocks), the other reagents and protocols were the same as those used in the GeXP mono-RT-PCR assays. If certain primers did not amplify their target NA subtype after combination, other primers from the candidate primer pools were substituted until the optimal combination of ten primer pairs was identified, ensuring that each primer pair amplified a specific product without cross-reactions. Ten pairs of primers were ultimately selected for the GeXP multiplex RT-PCR assay (Table 5-4-1). A single AIV subtype (N1 to N9 of the reference strains, Table 5-4-2), random mixtures of multiple AIV subtypes (such as N1+N2, N6+N8+N9, and N3+N4+N5+N7) and other important avian viruses (NDV, IBV, ILTV, ARV, and FAdV-4) were examined using the GeXP multiplex RT-PCR assay.

Evaluation of the sensitivity of the GeXP multiplex RT-PCR assay

The sensitivity of the GeXP assay was evaluated using *in vitro* transcribed RNAs for the N1 to N9 and M genes. Briefly, the nine NA genes and one M gene were amplified using the primers listed in Table 5-4-1, PCR amplicons were ligated into the pGEM-T vector, and expanded in competent DH5α cells to produce ten recombinant plasmids. The ten plasmids were purified, sequenced, linearized, and subjected to *in vitro* transcription according to the instructions of T7 RiboMAX™ Express Large Scale RNA Production System kit (Promega, Madison, WI, United States). The *in vitro* transcribed RNAs were quantified using a NanoDrop 2 000 (Thermo Fisher Scientific); then, serial 10-fold dilutions were prepared. Ten premixed RNA templates at the same concentrations were prepared.

Detection in clinical samples

Three hundred fifty swab samples (the oral pharyngeal and cloacal swabs from the same bird were pooled as a single sample) were obtained as part of AIV surveillance programs in live bird markets (LBMs) in Nanning, the capital of Guangxi, from 2016-2017. RNA was extracted from the washing solution of the swabs and analyzed according to the protocol established for the GeXP multiplex RT-PCR assay. All samples were inoculated in parallel in the 9-day-old SPF embryonated chicken eggs for virus isolation. Isolated allantoic fluids were identified using the neuraminidase (NA) assay and neuraminidase inhibition (NI) test, according to the World Health Organization (WHO) protocol. The allantoic fluids were also amplified with conventional RT-PCR followed by NA amplicon sequencing.

Results

Evaluation of the single primers with the GeXP mono-RT-PCR assay

RNA samples extracted from nine NA subtypes of AIV were used as individual templates for GeXP mono-RT-PCR in separate reactions to evaluate the specificity of each pair of gene-specific primers. In the mono-RT-PCR assays, the pan-AIV primers amplified all AIV subtypes, and each pair of subtype-specific primers only generated a product for the NA gene corresponding to the target subtype.

Screening of the optimal multiplex primers

Amplicons were designed to ensure that each fragment was no less than 5 nucleotides away from its nearest neighbor and allowed for variation in peak migration to meet the minimum peak separation distance of 3 nucleotides. In this study, fragments of the expected sizes were amplified for the nine NA subtypes: N1, 244 to 249 bp; N2, 279 to 285 bp; N3, 215 to 221 bp; N4, 149 to 155 bp; N5, 295 to 301 bp; N6, 236 to 241 bp; N7, 192 to 198 bp; N8, 173 to 178 bp; N9, 205 to 211 bp; and AIV-M, 158 to 164 bp.

Evaluation of the multiplex primers using the GeXP multiplex RT-PCR assay with templates for single and multiple subtypes

Forty-three reference isolates (Table 5-4-2), including N1 to N9 subtypes of the influenza A virus originating from different host species, such as duck, chicken, and pigeon, were subjected to GeXP multiplex RT-PCR. Two specific amplification peaks were observed, representing the subtype-specific target amplicon and the AIV M gene amplicon (Figure 5-4-1). No cross-reactivity among these nine subtypes was observed. Genomic RNA and DNA extracted from other important avian viruses, such as NDV, IBV, ILTV, ARV, and FAdV-4, tested negative in the GeXP multiplex RT-PCR (data not shown). The high specificity of the GeXP multiplex RT-PCR assay described here was also confirmed by the investigation of well-characterized AIV reference viruses and non-AIV avian viruses. Each sample was assayed three times under the same conditions on different days and produced very similar results.

We randomly mixed RNAs from two different subtypes at different concentrations (10^2 copies of one template and 10^6 copies of the other template) as templates to assess the performance characteristics of GeXP for the diagnosis of a mixed infection with multiple NA subtypes. Regardless of whether a high or low concentration was used, the two subtype-specific peaks were both observed, with values similar to each template alone. Furthermore, templates with random mixtures of three or four subtypes (from 10^2 to 10^6 copies) were evaluated using the GeXP assay, and all target subtypes were accurately amplified. The amplification results for the multiplex templates were similar to the single-template amplifications, suggesting that this GeXP assay is useful for identifying different NA subtypes in a mixed sample. Each group was assayed three times under the same conditions on different days and produced very similar results.

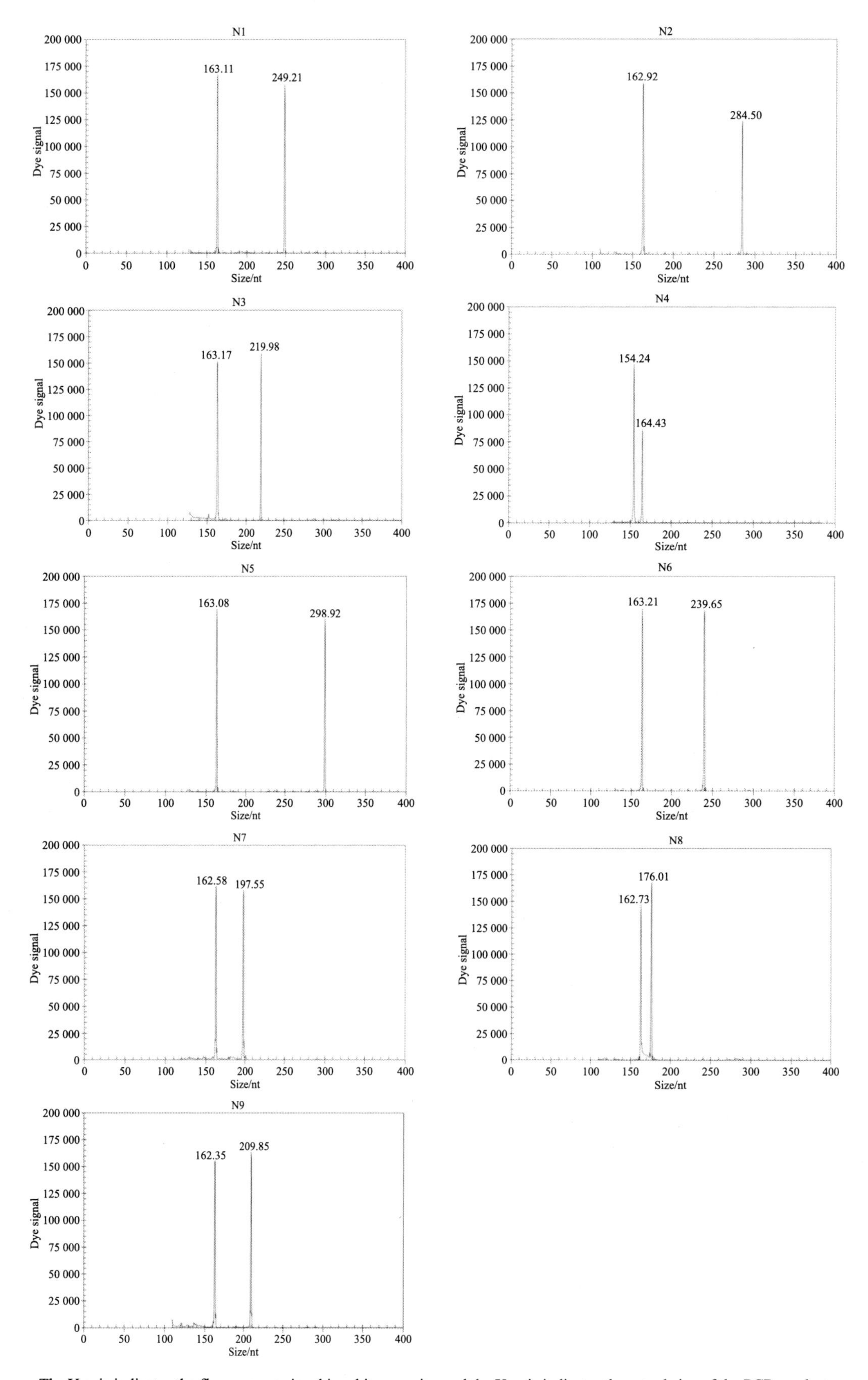

The Y-axis indicates the fluorescence signal in arbitrary units, and the X-axis indicates the actual size of the PCR product.

Figure 5-4-1 Detection of single AIV subtypes using GeXP multiplex RT-PCR

Sensitivity of the GeXP multiplex RT-PCR assay

The sensitivity of the GeXP assay was measured using serial 10-fold dilutions of ten premixed quantitative *in vitro* transcribed RNAs from the viral targets. The limit of simultaneous detection of the ten templates of N1, N2, N3, N4, N5, N6, N7, N8, N9, and AIV using the GeXP multiplex RT-PCR assay was as low as 100 copies per reaction mixture (Figure 5-4-2). Typically, the cut-off CT value for positive and negative results was determined as 2 000 A. U. (absorbance units) by default. Each sample was assayed three times under the same conditions on different days and produced very similar results.

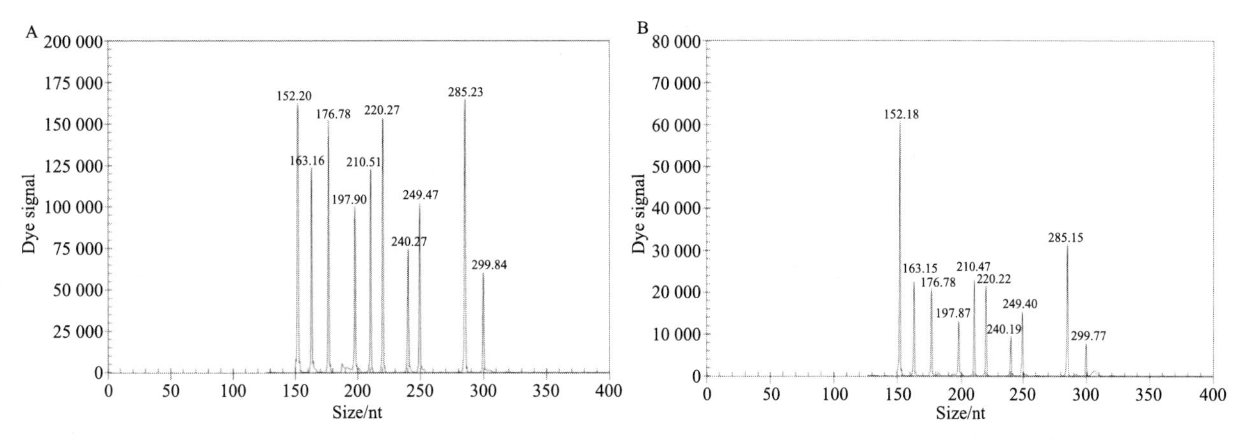

All 10 premixed viral targets were detected. A: 10^3 copies per reaction mixture; B: 100 copies per reaction mixture.

Figure 5-4-2 The sensitivity of the GeXP assay for the simultaneous detection of nine NA subtypes of AIV

Detection in clinical samples

Among three hundred fifty samples, twenty-seven samples were positive and the remaining were negative for AIV and NA subtypes using three methods: GeXP, NA amplicon sequencing, and the NI test. Among the twenty-seven positive cases, two were identified as two subtypes of mixed infection using GeXP and the NI test, but amplicon sequencing only identified one subtype. Consistent results for the remaining twenty-five cases were obtained with these three methods. The results of NA subtyping are shown in Table 5-4-3.

Table 5-4-3 Results obtained from clinical samples using GeXP multiplex RT-PCR, the NI test, and amplicon sequencing method

Number	Host	GeXP	NI assay	Amplicon sequencing
1	Chicken	N2	N2	N2
2	Chicken	N2	N2	N2
3	Chicken	N2	N2	N2
4	Chicken	N2	N2	N2
5	Chicken	N2	N2	N2
6	Chicken	N6	N6	N6
7	Chicken	N6	N6	N6
8	Chicken	N6	N6	N6
9	Chicken	N9	N9	N9
10	Duck	N1	N1	N1
11	Duck	N1	N1	N1
12	Duck	N2	N2	N2

continued

Number	Host	GeXP	NI assay	Amplicon sequencing
13	Duck	N2	N2	N2
14	Duck	N2	N2	N2
15	Duck	N2/N6	N2/N6	N2
16	Duck	N6/N2	N6/N2	N6
17	Duck	N6	N6	N6
18	Duck	N8	N8	N8
19	Duck	N8	N8	N8
20	Duck	N8	N8	N8
21	Duck	N9	N9	N9
22	Duck	N2	N2	N2
23	Duck	N2	N2	N2
24	Duck	N2	N2	N2
25	Duck	N6	N6	N6
26	Duck	N6	N6	N6
27	Duck	N6	N6	N6

Discussion

Recently, the reassortment of HA subtypes with multiple NA subtypes has increased; for example, H5N1, H5N2, and H5N3 were previously detected, and H5N6 and H5N8 are now emerging. H7 assorts with different NA subtypes (N2, N3, N4, N7, and N9 have been found in cases of human infection). H6N1, H6N2, H6N5, H6N6, and H6N8 were identified during the surveillance of live poultry markets in China[25]. H4 and H3 reassort with N2, N6, and N8. H9 usually assorts with N2[26]. Epidemiological surveys of some viruses use only real-time PCR to analyze DNA/RNA from swab samples, with no further isolation. However, AIV is a segmented virus, and rearrangement of its eight gene segments during coinfection with two or more subtypes produces new recombinant strains. For an early warning of emerging flu outbreaks, we must understand the ecological evolution of AIV isolates. Therefore, surveys of AIV usually require the isolation of the virus followed by the selection of representative strains for gene sequencing and a subsequent genetic evolution analysis. Swab-inoculated SPF chick embryos are used for virus propagation, allantoic fluid is harvested and screened with an HA assay, and HA subtypes are identified with an HI test. Here, the confirmation of the NA subtype is important and a key step in the process. In addition, the routine monitoring of AIV frequently involves a large number of samples. Therefore, the development of this high-throughput GeXP assay for the simultaneous and rapid detection of these nine NA subtypes is important for the effective prevention and control of AIV.

In clinical samples, sequencing identified only a single NA subtype. This result was expected, because sequencing is performed on samples with positive single colony in the panel and is preceded by conventional RT-PCR to amplify a high concentration of NA subtypes. Thus, in the panel of clones and transformants, most single colonies should possess the dominant NA genes. Therefore, the determination of the existence of a mixed infection with different NA subtypes difficult using amplicon sequencing, which guarantees only the detection of the dominant NA subtype. Certainly, amplicon sequencing occasionally, but not always, identifies different NA subtypes in different colonies. This uncertainty is a limitation of sequence identification based on amplicon cloning. However, GeXP, which simultaneously detects nine NA subtypes in a single tube and rapidly identifies any number of coinfecting NA subtypes, addresses this limitation. The ability of the GeXP assay to identify mixed infections was superior to clone sequencing.

In the traditional multiplex RT-PCR assay, multiplex primers compete for amplification, and the primers with a high amplification efficiency or smaller fragments are more likely to dominate and be preferred. As the number of primers increased, the probability of forming complex primer dimers also increases, leading to a decrease in sensitivity. In addition, the results of multiple PCR products must be observed using agarose electrophoresis, and bands less than 100 bp are difficult to distinguish. In the real-time multiplex RT-PCR assay, multiple probes are labeled with fluorescent dyes that are excited at different wavelengths. A greater number of fluorescent tags will increase the mutual interference and decrease the sensitivity. The number of target genes that can be detected using multiplex PCR and multiplex real-time PCR is limited, as only 2 to 4 pathogens are generally able to be detected, and neither assay achieves the purpose of rapid and high-throughput detection and analysis of multiple pathogens. The GeXP technology has recently overcome these shortcomings. In the present study, nine pairs of NA subtype-specific primers and one pair of pan-AIV primers were designed to develop a GeXP assay for the simultaneous identification of nine NA subtypes of AIV, including N1, N2, N3, N4, N5, N6, N7, N8, and N9. The pan-AIV primer was added to the set of primers to provide evidence that the samples were indeed AIV. The subtype-specific primers were designed to target the NA gene of each NA subtype, and the pan-AIV primer was designed to target the M gene, which is relatively conserved among subtypes. In the GeXP assay, which was based on both subtype-specific and universal primers, the forward and reverse universal primers were constant, and the universal primer was attached to the 5' end of each specific primer. Another pair of universal primers was mixed in 5 × GeXP PCR buffer, and the 5' end of this universal forward primer was labeled with a fluorescent dye. At the beginning of the multiplex PCR process, specific primer sequences were combined with different NA subtype templates, and amplicons with universal sequence tags at both the 5' and 3' ends were generated. The amplification was quickly controlled by the universal primers, since the universal primers were present in the PCRs at significantly higher concentrations than the gene-specific primers. Amplification by multiplex primers was converted into amplification by a pair of universal primers, and thus the amplification efficiency of each template tended to be concordant, without effects on the amplification efficiency of each pair of primers. This method results in the highly sensitive and specific amplification of different genes in a single multiplex RT-PCR assay, preventing preferred and inferior amplification, and minimizing non-specific reactions. PCR products were separated using capillary gel electrophoresis, and the separated fragments were initially analyzed using the fragment analysis module of the GeXP system software. The amplified fragments were visualized as fluorescence signal peaks and different gene fragments were identified by size. Despite the limited number of positive samples for the other four subtypes of AIV (N3, N4, N5, and N7), we propose that the methods described here could be extended to the routine diagnosis and epidemiological detection of AIV infections.

The GeXP assay developed here simultaneously differentiated nine NA subtypes in a single reaction tube with a cost of approximately $8 per comprehensive test compared with $8 per test for each subtype using a commercial real-time RT-PCR kit. The whole reaction, including RNA extraction, was completed in one tube in a multiplex PCR system within 3 h, followed by a capillary electrophoresis separation step lasting approximately 1 h. In addition, two 96-well plates can be simultaneously placed in parallel in a GeXP machine to further increase sample throughput. Moreover, the procedure is easily automated and requires minimal hands-on time by technical staff. In conclusion, this GeXP assay is proposed as a specific, sensitive, rapid, high-throughput, and versatile diagnostic tool for nine NA subtypes of AIV. In the future, we will develop additional methods to simultaneously detect important HA and NA subtypes of AIV.

References

[1] YU J, HIKA B, LIU R, et al. The hemagglutinin-esterase fusion glycoprotein is a primary determinant of the exceptional thermal and acid stability of influenza D virus. mSphere, 2017, 2(4): e00254-17.

[2] TONG S, LI Y, RIVAILLER P, et al. A distinct lineage of influenza A virus from bats. Proceedings of the National Academy of Sciences of the United States of America, 2012, 109(11): 4269-4274.

[3] TONG S, ZHU X, LI Y, et al. New world bats harbor diverse influenza A viruses. PLOS Pathogens, 2013, 9(10): e1003657.

[4] SU S, BI Y, WONG G, et al. Epidemiology, evolution, and recent outbreaks of avian influenza virus in China. Journal of Virology, 2015, 89(17): 8671-8676.

[5] YAMAJI R, YAMADA S, Le M Q, et al. Identification of PB2 mutations responsible for the efficient replication of H5N1 influenza viruses in human lung epithelial cells. Journal of Virology, 2015, 89(7): 3947-3956.

[6] LI K S, GUAN Y, WANG J, et al. Genesis of a highly pathogenic and potentially pandemic H5N1 influenza virus in eastern Asia. Nature, 2004, 430(6996): 209-213.

[7] SUN W, LI J, HU J, et al. Genetic analysis and biological characteristics of different internal gene origin H5N6 reassortment avian influenza virus in China in 2016. Veterinary Microbiology, 2018, 219: 200-211.

[8] SHEN Y, LU H. Global concern regarding the fifth epidemic of human infection with avian influenza A (H7N9) virus in China. Bioscience Trends, 2017, 11(1): 120-121.

[9] ZENG X, TIAN G, SHI J, et al. Vaccination of poultry successfully eliminated human infection with H7N9 virus in China. Science China-Life Sciences, 2018, 61(12): 1465-1473.

[10] TO K K, TSANG A K, CHAN J F, et al. Emergence in China of human disease due to avian influenza A (H10N8)-cause for concern?. The Journal of Infection, 2014, 68(3): 205-215.

[11] CHEN H, YUAN H, GAO R, et al. Clinical and epidemiological characteristics of a fatal case of avian influenza A H10N8 virus infection: a descriptive study. Lancet, 2014, 383(9918): 714-721.

[12] WANG F, QI J, BI Y, et al. Adaptation of avian influenza A (H6N1) virus from avian to human receptor-binding preference. Embo Journal, 2015, 34(12): 1661-1673.

[13] QI W, ZHOU X, SHI W, et al. Genesis of the novel human-infecting influenza A (H10N8) virus and potential genetic diversity of the virus in poultry, China. Euro Surveillance, 2014, 19(25): 20841.

[14] FEREIDOUNI S R, STARICK E, GRUND C, et al. Rapid molecular subtyping by reverse transcription polymerase chain reaction of the neuraminidase gene of avian influenza A viruses. Veterinary Microbiology, 2009, 135(3-4): 253-260.

[15] QIU B F, LIU W J, PENG D X, et al. A reverse transcription-PCR for subtyping of the neuraminidase of avian influenza viruses. Journal of Virological Methods, 2009, 155(2): 193-198.

[16] TSUKAMOTO K, ASHIZAWA T, NAKANISHI K, et al. Use of reverse transcriptase PCR to subtype N1 to N9 neuraminidase genes of avian influenza viruses. Journal of Clinical Microbiology, 2009, 47(7): 2301-2303.

[17] HUANG Y, KHAN M I, MANDOIU I. Neuraminidase subtyping of avian influenza viruses with primerhunter-designed primers and quadruplicate primer pools. PLOS ONE, 2013, 8(11): e81842.

[18] SUN Z, QIN T, MENG F, et al. Development of a multiplex probe combination-based one-step real-time reverse transcription-PCR for NA subtype typing of avian influenza virus. Scientific Reports, 2017, 7(1): 13455.

[19] HU X, ZHANG Y, ZHOU X, et al. Simultaneously typing nine serotypes of enteroviruses associated with hand, foot, and mouth disease by a GeXP analyzer-based multiplex reverse transcription-PCR assay. Journal of Clinical Microbiology, 2012, 50(2): 288-293.

[20] LI J, MAO N Y, ZHANG C, et al. The development of a GeXP-based multiplex reverse transcription-PCR assay for simultaneous detection of sixteen human respiratory virus types/subtypes. BMC Infectious Diseases, 2012, 12: 189.

[21] ZENG T, XIE Z, XIE L, et al. Simultaneous detection of eight immunosuppressive chicken viruses using a GeXP analyser-based multiplex PCR assay. Virology Journal, 2015, 12: 226.

[22] ZHANG M, XIE Z, XIE L, et al. Simultaneous detection of eight swine reproductive and respiratory pathogens using a novel GeXP analyser-based multiplex PCR assay. Journal of Virological Methods, 2015, 224: 9-15.

[23] FAN Q, XIE Z, XIE Z, et al. Development of a GeXP-multiplex PCR assay for the simultaneous detection and differentiation of six cattle viruses. PLOS ONE, 2017, 12(2): e171287.

[24] RAI A J, KAMATH R M, GERALD W, et al. Analytical validation of the GeXP analyzer and design of a workflow for cancer-biomarker discovery using multiplexed gene-expression profiling. Analytical and Bioanalytical Chemistry, 2009, 393(5): 1505-1511.

[25] PENG Y, XIE Z X, LIU J B, et al. Epidemiological surveillance of low pathogenic avian influenza virus (LPAIV) from poultry in Guangxi Province, southern China. PLOS ONE, 2013, 8(10): e77132.

[26] LUO S, XIE Z, XIE Z, et al. Surveillance of live poultry markets for low pathogenic avian influenza viruses in Guangxi Province, southern China, from 2012-2015. Scientific Reports, 2017, 7(1): 17577.

Simultaneous detection of eight immunosuppressive chicken viruses using a GeXP analyser-based multiplex PCR assay

Zeng Tingting, Xie Zhixun, Xie Liji, Deng Xianwen, Xie Zhiqin, Luo Sisi, Huang Li, and Huang Jiaoling

Abstract

Immunosuppressive viruses are frequently found as co-infections in the chicken industry, potentially causing serious economic losses. Because traditional molecular biology methods have limited detection ability, a rapid, high-throughput method for the differential diagnosis of these viruses is needed. The objective of this study is to develop a GenomeLab Gene Expression Profiler Analyser-based multiplex PCR method (GeXP-multiplex PCR) for simultaneous detection of eight immunosuppressive chicken viruses. Using chimeric primers, eight such viruses, including Marek's disease virus (MDV), three subgroups of avian leucosis virus (ALV-A/B/J), reticuloendotheliosis virus (REV), infectious bursal disease virus (IBDV), chicken infectious anaemia virus (CIAV) and avian reovirus (ARV), were amplified and identified by their respective amplicon sizes. The specificity and sensitivity of the optimised GeXP-multiplex PCR assay were evaluated, and the data demonstrated that this technique could selectively amplify these eight viruses at a sensitivity of 100 copies/20 μL when all eight viruses were present. Among 300 examined clinical specimens, 190 were found to be positive for immunosuppressive viruses according to this novel assay. The GeXP-multiplex PCR assay is a high-throughput, sensitive and specific method for the detection of eight immunosuppressive viruses and can be used for differential diagnosis and molecular epidemiological surveys.

Keywords

GeXP analyser, multiplex PCR, immunosuppressive viruses

Background

Immunosuppression causes major economic losses in poultry farming because immunosuppressed chickens are more susceptible to viral and bacterial pathogens, respond poorly to vaccination, and display lower feed conversion efficiency as well as growth retardation. Immunosuppressive chicken viruses include Marek's disease virus (MDV), avian leucosis virus (ALV), reticuloendotheliosis virus (REV), infectious bursal disease virus (IBDV), chicken infectious anaemia virus (CIAV) and avian reovirus (ARV), all of which affect immune function in chickens and lead to immunosuppression[1-7]. The typical symptoms elicited by these viruses differ, and some symptoms are not readily observable. Immunosuppression in chickens infected with MDV, ALV or REV occurs much earlier than does tumour development and death[1, 2]. Although early infection with IBDV in chicks less than 3 weeks old may not result in the typical symptoms of IBD, this infection none-theless causes serious immunosuppression[8]. Chickens older than 3 weeks of age are resistant to anaemia after infection with CIAV yet remain susceptible to immunosuppression[9]. Furthermore, the possibility of co-infection makes it difficult to differentiate among these immunosuppressive viruses[10-13].

The detection and differential diagnosis of immunosuppressive viruses are important for the poultry industry. However, conventional methods, such as virus isolation and serum neutralisation tests, are typically time-consuming and labour-intensive procedures[14]. Molecular methods have been used to rapidly detect immunosuppression in chickens, but they are limited by their ability to detect only a few pathogens per reaction[15-20]. Therefore, a rapid, cost-effective, and high-throughput detection technique is needed for the clinical diagnosis of immunosuppressive viral infection in chickens.

The GenomeLab Gene Expression Profiler genetic analysis system (GeXP) is a new multi-target, high-throughput detection platform that integrates RT-PCR or PCR with a labelled, amplified product in a multiplex RT-PCR/PCR assay followed by fluorescence capillary electrophoresis separation based on the sizes of the amplified products[21]. The GeXP profiler utilises gene-specific primers containing 5'-universal adaptor sequences[21]: the chimeric primers consist of a universal sequence fused to the 5'-end of a gene-specific sequence. The forward primer consists of a universal dye-labelled sequence fused to the 5'-end of a gene-specific sequence, whereas the reverse primer consists of a universal unlabelled sequence fused to the 5'-end of a gene-specific sequence. Products differing by 7～10 bp in size are separated by capillary electrophoresis. This technique has been used to identify various diseases in humans, including 11 genotypes of HPV[22]; 9 serotypes of hand, foot, and mouth disease[23]; cancer[21, 24]; 7 enteric viruses[25]; and H1N1[26]. The GeXP genetic analysis system has also been successfully utilised to simultaneously detect 9 avian respiratory pathogens in clinical samples, 8 swine reproductive and respiratory pathogens and 11 duck viruses[27-29].

In this study, a GeXP-multiplex PCR assay was developed and optimised to simultaneously detect eight immunosuppressive chicken viruses: MDV, ALV (three subgroups of ALV, ALV-A/B/J), REV, IBDV, CIAV and ARV.

Methods

Viruses and clinical specimens

MDV, ALV (three subgroups of ALV, A/B/J), REV, IBDV, CIAV, ARV and the other viruses used in this study are listed in Table 5-5-1. MDV, ALV-A/B/J, and REV were propagated in chicken embryo fibroblasts (CEFs), and IBDV and ARV were propagated in SPF chicken embryos. CIAV was propagated using a homogenate of positive samples of chicken liver and bone marrow to infect SPF 1-day-old broilers, from which the bone marrow was collected at 10 days after infection. Clinical tissue specimens of the thymus, bursa, spleen, bone marrow, blood and liver were collected from diseased chickens at poultry farms.

Table 5-5-1 Sources of pathogens

Pathogen reference viruses	Source	Pathogen other viruses	Source
Marek's disease virus KC453972, KC453973, GX130112, GX140301, 050118, 070123, 090201, 100428	GVRI	Inactivated H5N1 avian influenza virus Re-1	HVRI
Avian leucosis virus subgroup A isolate RSV-1	CVCC	Avian influenza virus H9N6/Duck/HK/147/77	HKU
Avian leucosis virus subgroup A isolate GX110521, GX110522, ALVA01, ALVA02, ALVA03	GVRI	Avian influenza virus H7N2/chicken PA/3979/97	PU
Avian leucosis virus subgroup B isolate RSV-2	CVCC	Newcastle disease virus F48E9	GVRI
Avian leucosis virus subgroup B isolate GX111230, GX130401, ALVB15, ALVB23, ALVB28	GVRI	Newcastle disease virus GX6/02	GVRI

continued

Pathogen reference viruses	Source	Pathogen other viruses	Source
Avian leucosis virus subgroup J isolate KC453974, KC453975, GX090201, GX090521, GX110110, GX120081, GX130018, GX140010	GVRI	Infectious bronchitis virus Massachusetts 41	GVRI
Reticuloendotheliosis virus AV235	CVCC	Infectious laryngotracheitis virus AV1231	CIVDC
Reticuloendotheliosis virus KC453976, KC453977, GX120825, GX131118	GVRI	*Mycoplasma synoviae* CAU0748	CVCC
Avian reovirus S1133, 1733, 526, C78, GuangxiR1, GuangxiR2, GX110058	GVRI		
Infectious bursal disease virus CA, AV162, AV144	CVCC		
Infectious bursal disease virus AV6	CIVDC		
Infectious bursal disease virus 070124, 080113, 090053, 100008, 110110, 130223	GVRI		
Chicken infectious anaemia virus CAU0728, CAU0729, CAU0730, CAU0731, CAU0732	CVCC		
Chicken infectious anaemia virus GXC060821	GVRI		

Note: HVRI, Harbin Veterinary Research Institute, China; HKU, The University of Hong Kong, China; GVRI, Guangxi Veterinary Research Institute, China; CIVDC, China Institute of Veterinary Drug Control, China; PU, University of Pennsylvania; CVCC, China Veterinary Culture Collection Centre.

RNA/DNA extraction and RNA reverse transcription

MDV DNA from infected CEFs, CIAV DNA from the bone marrow of affected chickens, REV and ALV-A/B/J genomic proviral DNA from infected CEFs, and IBDV and ARV RNA from allantoic fluid were extracted using the E. Z. N. A.® Total DNA/RNA Isolation Kit (OMEGA, Norcross, GA, USA). The extracted DNA/RNA was eluted in DNase and RNase-free dH_2O and stored at −80 ℃ . The RNA was used to generate cDNA via reverse transcription, as described previously[30].

GeXP-multiplex PCR primer design

Gene-specific primers were designed based on sequence information obtained from GenBank using Primer premier 5.0 software (Premier, Palo Alto, USA) and NCBI Primer-Blast. ALV-A/B/J primers were designed to correspond to a specific region of the envelope gene gp85, MDV primers to a specific region of the gene encoding Marek's EcoRI-Q protein (meq), REV primers to a specific region of the envelope gene gp90, IBDV primers to a specific region of the capsid protein gene VP2, CIAV primers to a specific region of the capsid protein gene VP1, and ARV primers to a specific region of the S1 gene. The primer sequences, the sizes of the resulting amplicons, and the target regions are listed in Table 5-5-1. The chimeric primers consisted of a universal sequence fused to the 5'-end of a gene-specific sequence. The forward universal primer was labelled at the 5'-end with the fluorescent dye Cy5. All chimeric primers and universal primers were synthesised and purified by polyacrylamide gel electrophoresis (Invitrogen, Shanghai, China).

GeXP-multiplex PCR assay

GeXP-multiplex PCR assays were performed using the Genome Lab GeXP Starter Kit (Beckman Coulter,

Brea, USA) in a 20-μL volume containing 4 μL of 5 × buffer, 0.25 μM (final concentration) universal forward and reverse primers, 2 μL of $MgCl_2$ (25 μM), 1 μL of chimeric primer mixture, 0.35 μL of Thermo-Start DNA polymerase, 1 μL of cDNA/DNA, and nuclease-free H_2O. The concentration of the chimeric primers was optimised according to the amplification efficiency of the GeXP-multiplex PCR assay.

The GeXP-multiplex PCR assay was performed via a three-step amplification procedure after a 5-min incubation at 95 ℃: 10 cycles of 30 s at 94 ℃, 30 s at 55 ℃, and 30 s at 72 ℃; 10 cycles of 30 s at 94 ℃, 30 s at 68 ℃, and 30 s at 72 ℃; 20 cycles of 30 s at 94 ℃, 30 s at 50 ℃, and 30 s at 72 ℃; and 5 min at 72 ℃. The reactions were then held at 4 ℃ in the thermal cycler (Thermo, Milford, USA).

Separation by capillary electrophoresis and fragment analysis

After amplification, 1 μL of PCR product was added to 38.75 μL of sample loading solution along with 0.25 μL of DNA size standard 400 (Beckman Coulter, Brea, USA). The fluorescently labelled amplicons were separated into distinct peaks on a electropherogram via GeXP high-resolution capillary electrophoresis and then identified by their respective sizes. The dye signal strength of each peak was measured in A. U. of optical fluorescence and was defined as the fluorescence signal minus the background signal. The data were imported into the analysis module of express Profiler software (Beckman Coulter, Brea, USA) as a tab-delimited file for subsequent analyses.

Specificity and sensitivity of the GeXP-multiplex PCR assay

The assay specificity for each immunosuppressive viral target was individually tested with a mixture of 8 sets of chimeric primers in a multiplex PCR assay after optimisation. Other conventional chicken viruses, including the H5/H7/H9 serotypes of avian influenza virus (AIV), Newcastle disease virus (NDV), infectious bronchitis virus (IBV) and infectious laryngotracheitis virus (ILTV), were used as negative controls. DNA from the thymus, spleen and bursa of SPF chickens was also used as a negative control.

Specific PCR amplicons for each virus were individually cloned into the pGEM-T vector (Promega, Madison, USA), and the plasmids were purified and sequenced. The sequence data were analysed and compared with the corresponding sequence data in GenBank. The IBDV and ARV plasmids were linearised with *SpeI* (TaKaRa, Dalian, China) and then in vitro transcribed into ssRNA using the RiboMAX Large Scale RNA Production Systems SP6/T7 Kit (Promega, Madison, USA). After DNase I digestion, ssRNA was purified with TRIzol (Invitrogen, Shanghai, China) and chloroform. The concentration of plasmid DNA and transcribed ssRNA was measured at 260 nm using a NanoDrop 2000 (Thermo Fisher Scientific, Waltham, USA), and the copy number was calculated[31]. Plasmid DNA and transcribed ssRNA were diluted to a final concentration ranging from 10^5 copies/μL to 1 copy/μL and then subjected to the GeXP-multiplex PCR assay with 8 sets of chimeric primers, both individually and in premixed solutions. The sensitivity of the GeXP-multiplex PCR assay was reevaluated three times on three different days.

Artificial mixture and interference assays

To simulate co-infection, thymus, spleen, bursa, bone marrow, blood and liver samples from chickens previously diagnosed with the immunosuppressive viruses were randomly chosen and mixed together in various amounts; DNA/RNA was then extracted. cDNA was generated from RNA as described above. GeXP-multiplex PCR assays were performed using the mixed DNA/cDNA.

Different concentrations of specific DNA-containing plasmids and in vitro ssRNA transcripts (10^7 copies/μL to 10^2 copies/μL) were used as templates for GeXP-multiplex PCR assays to evaluate interference, and the results were compared with those of single-template GeXP-multiplex PCR assays.

Evaluation of the GeXP-multiplex PCR assay with clinical specimens

A total of 300 clinical specimens, including the thymus gland, spleen, bursa, bone marrow, blood and liver of diseased chickens, were collected from farms. These diseased chickens showed various symptoms, with most exhibiting depression and anepithymia and some showing extreme emaciation. One-third of the diseased chickens were diarrheic, and approximately half showed respiratory symptoms; some chickens displayed both diarrhea and respiratory symptoms. Approximately one-fourth of the diseased chickens exhibited a poor reaction to vaccination or was prone to bacterial infection. Mortality ranged from 2% to 30%. Tumours in the liver, heart, spleen or skin were observed in only approximately 20 chickens and not always in those showing extreme emaciation.

DNA/RNA was extracted, and RNA was reverse transcribed as described above. The GeXP-multiplex PCR assay was performed using DNA/cDNA as the template, and the results were confirmed using independent realtime PCR/RT-PCR and sequencing to determine true positives. Independent real-time PCR/RT-PCR was performed using the primers described above but without the universal sequence at the 5'-end. The results of GeXP-multiplex PCR, real-time PCR/RT-PCR and sequencing were analysed by the Kappa statistical method using SPSS software (IBM, New York, USA).

Results

Specificity of the GeXP-multiplex PCR assay

The concentrations of the GeXP-multiplex PCR-specific primers (Table 5-5-2) were optimised, and DNA/cDNA from the immunosuppressive viruses described in Table 5-5-1 was used individually as a template for evaluating the specificity of the GeXP-multiplex PCR assay. The expected size of each immunosuppressive viral amplicon was determined. A single peak for the complex PCR was detected using the GeXP analyser system (Figure 5-5-1 A-I), and no mispriming with or nonspecific amplification of other avian pathogens or the chicken genome was observed.

Table 5-5-2 Primer sequences and PCR product sizes

Virus	Forward primer sequence (5'-3')	Reverse primer sequence (5'-3')	Amplicon size/bp	Target region	Primer concentration/μM
MDV	AGGTGACACTATAGAATAAGGGAGCAGACGTACTATGTAGACAA	GTACGACTCACTATAGGGATGGTAAGCAGTCCAAGGGTCA	227	meq	0.16
ALV-A	AGGTGACACTATAGAATACAAGGGGTTCCTTGGTATCT	GTACGACTCACTATAGGGATGTGCCTATCCGCTGTCA	155	gp85	0.2
ALV-B	AGGTGACACTATAGAATATCAATCACGATTCTCCCACC	GTACGACTCACTATAGGGATGTGACGCTTCGTTTACGTCTT	285	gp85	0.2
ALV-J	AGGTGACACTATAGAATACTGATGCAACAACCAGGAAA	GTACGACTCACTATAGGGAGCAGTAACATTAGTGACATACCC	204	gp85	0.2
REV	AGGTGACACTATAGAATAGACCAGGCGAGCAAAATC	GTACGACTCACTATAGGGAGGTGTAATAGGTAGGTATGGAGGA	182	gp90	0.2

continued

Virus	Forward primer sequence (5'-3')	Reverse primer sequence (5'-3')	Amplicon size/bp	Target region	Primer concentration/μM
IBDV	AGGTGACACTATAGAATAGGGTCAG GGCTAATTGTCTT	GTACGACTCACTATAGGGATCTGTCA GTTCACTCAGGCTTC	294	VP2	0.2
CIAV	AGGTGACACTATAGAATAAAAGGCG AACAACCGATGA	GTACGACTCACTATAGGGATGCCCTG GAGGAAAAGACC	269	VP1	0.2
ARV	AGGTGACACTATAGAATAGGACCCCT ACTTCTGTTCTCA	GTACGACTCACTATAGGGAATTTCCC GTGGACGACAT	215	S1	0.16
Universal primers	AGGTGACACTATAGAATA	GTACGACTCACTATAGGGA			0.25

Note: Universal primers sequences are underlined. Chimeric primers were synthesised using universal primers and specific primers.

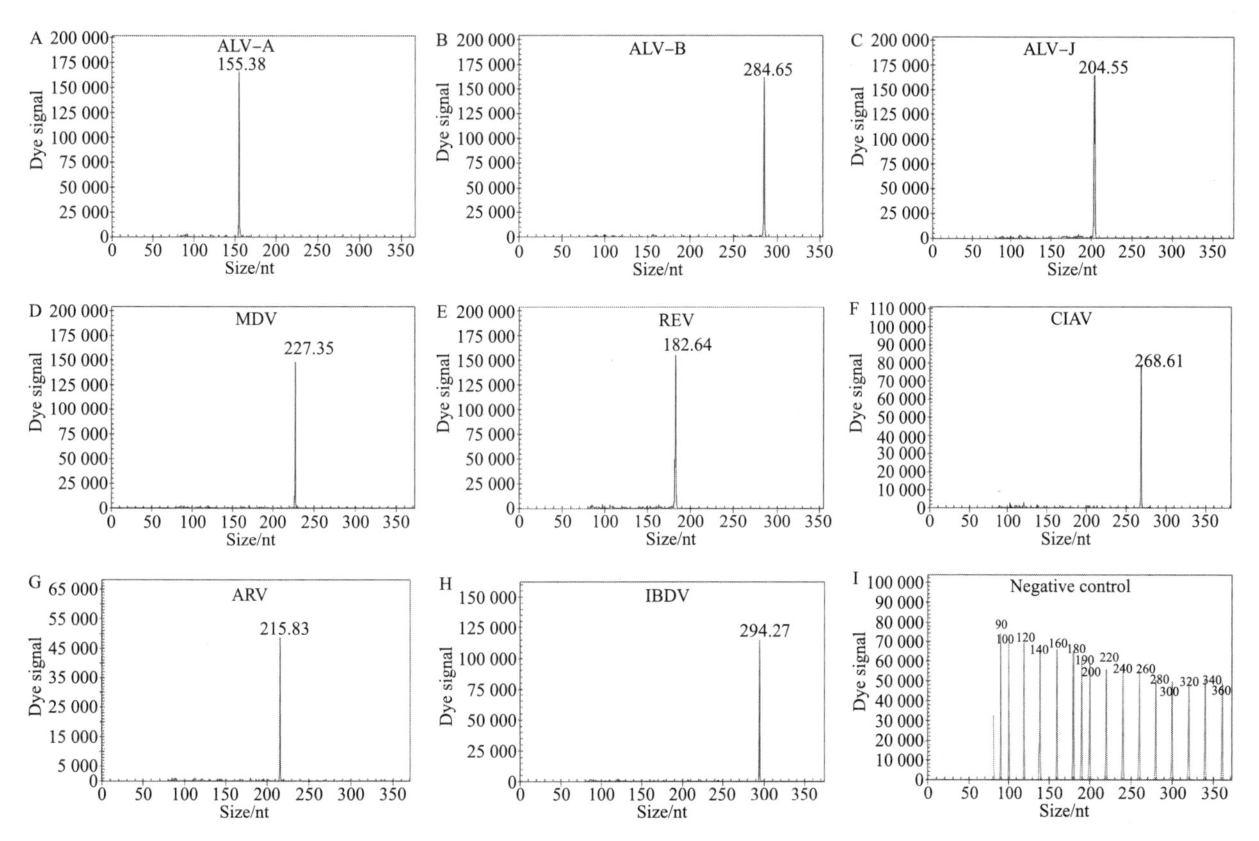

A: ALV-A, 155.38 bp; B: ALV-B, 284.65 bp; C: ALV-J, 204.55 bp; D: MDV, 227.35 bp; E: REV, 182.64 bp; F: CIAV, 268.61 bp; G: ARV, 215.83 bp; H: IBDV, 294.27 bp; I: DNA from the thymus, spleen and bursa of SPF chickens was used as a negative control. The x axes represent the sizes of PCR products in bp, and the y axes represent the dye signal in absorbance units (A.U.). Blue peaks denote specific amplification peaks, and red peaks denote marker peaks.

Figure 5-5-1　GeXP-multiplex PCR assay specificity

Sensitivity of the GeXP-PCR assay

The GeXP-multiplex PCR assay achieved the following minimum sensitivity levels in the detection of each of the eight detectable immunosuppressive viruses using eight sets of primers and either plasmid or in vitro ssRNA transcripts: 10 copies for the ALV-A, ALV-J, REV, CIAV and ARV viruses; and 100 copies for the MDV, ALV-B and IBDV viruses (electropherograms not shown). Tenfold serial dilutions of specific

DNA-containing plasmids and *in vitro* ssRNA transcripts from the eight immunosuppressive viruses, i. e., 10^5 (Figure 5-5-2 A), 10^4 (Figure 5-5-2 B), 10^3 (Figure 5-5-2 C) and 10^2 (Figure 5-5-2 D) copies per reaction, were prepared and amplified using an equal amount of template. When all of the premixed, specific DNA-containing plasmids and *in vitro* ssRNA transcripts corresponding to the eight immunosuppressive viruses were present, the GeXP-multiplex PCR assay achieved a minimum sensitivity of 100 copies.

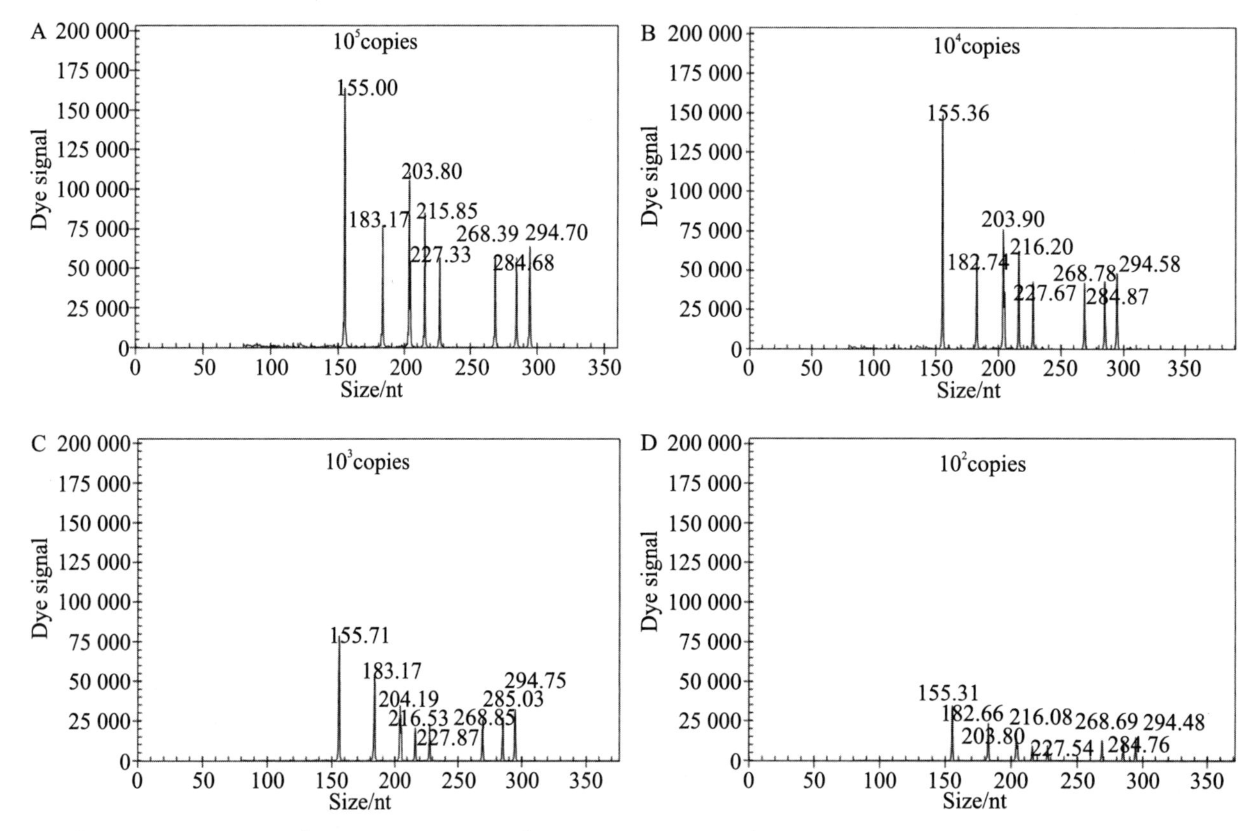

A: 10^5 copies per reaction; B: 10^4 copies per reaction; C: 10^3 copies per reaction; D: 10^2 copies per reaction. The GeXP-multiplex PCR assay was performed with mixed primers using equal amounts of specific DNA-containing plasmid template and in vitro-transcribed ssRNA corresponding to eight immunosuppressive viruses at concentrations of 10^5, 10^4, 10^3 or 10^2 copies per reaction. The x axes represent the sizes of PCR products in bp, and the y axes represent the dye signal in absorbance units.

Figure 5-5-2 GeXP-multiplex PCR assay sensitivity

Artificial mixtures and interference assays

To test the differentiation ability of the GeXP-multiplex PCR assay, samples previously deemed positive for avian immunosuppressive viruses were randomly mixed; DNA/RNA was then extracted, and the appropriate specific amplification peaks were observed (Figure 5-5-3). When either ARV and IBDV or ALV-J and ALV-B were mixed, two specific amplification peaks were observed (ARV, 216.16 bp; IBDV, 294.55 bp; ALV-J, 203.93 bp; and ALV-B, 284. 83 bp), and when DNA/cDNA from all eight immunosuppressive viruses was mixed, eight specific amplification peaks were observed (ALV-A, 155.37 bp; REV, 182.69 bp; ALV-J, 203.71 bp; ARV, 216.01 bp; MDV, 227.49 bp; CIAV, 268.60 bp; ALV-B, 284.65 bp; and IBDV, 294.55 bp).

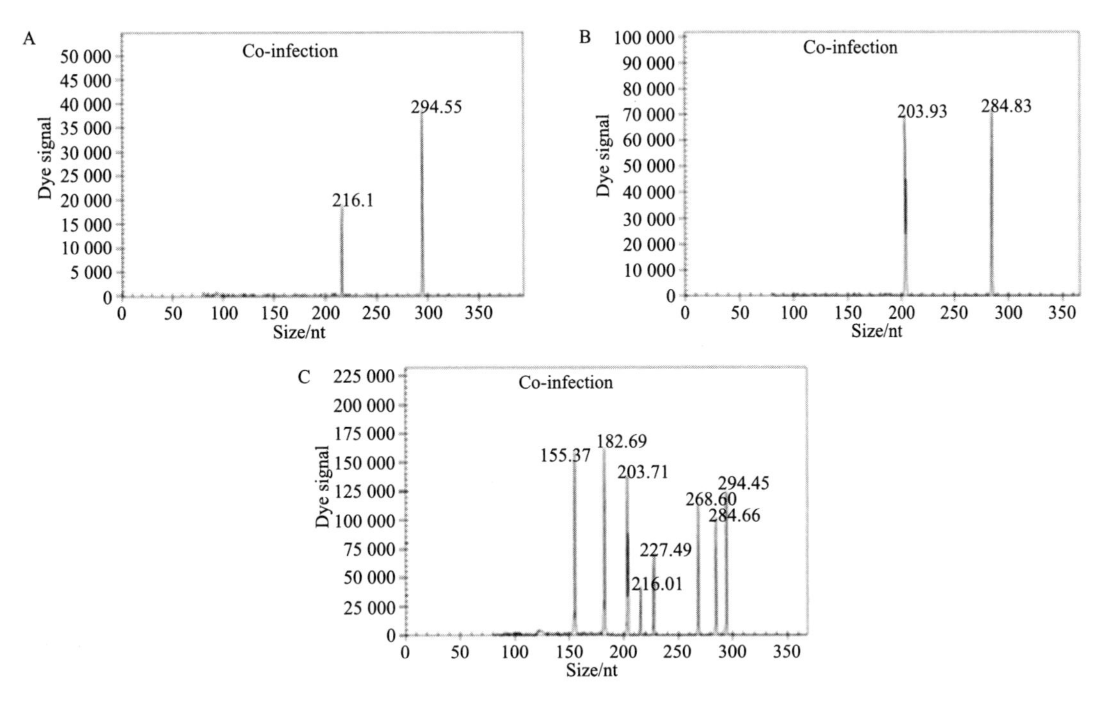

A: Using artificially mixed templates and mixed primers for ARV and IBDV; B: Using artificially mixed templates and mixed primers for ALV-J and ALV-B; C: Using artificially mixed templates and mixed primers for eight immunosuppressive viruses. The x axes represent the sizes of PCR products in bp, and the y axes represent the dye signal in absorbance units.

Figure 5-5-3 Artificially mixed templates for the GeXP-multiplex PCR assay

Specific amplification peaks were also observed when 10^7 copies of ALV-J, 10^2 copies of MDV and 10^3 copies of CIAV were mixed in a single reaction, and similar amplification peaks were observed when 10^7 copies of ALV-J or 10^2 copies of MDV were individually tested (Figure 5-5-4). These results demonstrate minimal to no interference by mixed infections in the GeXP-multiplex PCR assay.

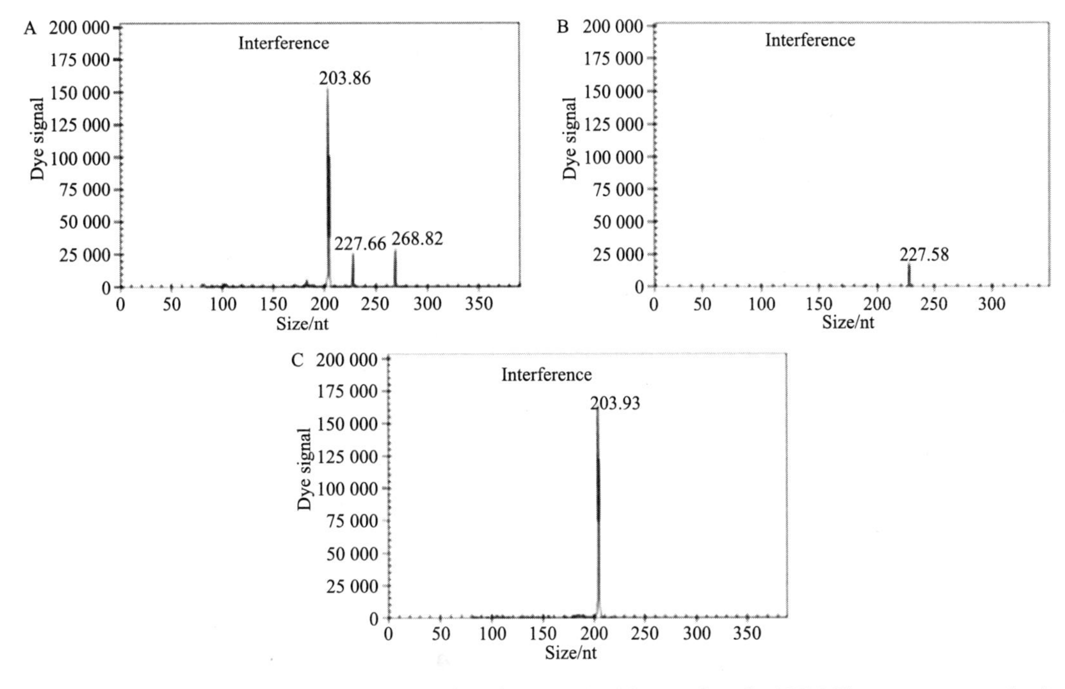

A: Using templates for ALV-J, MDV and CIAV; B: Using templates for MDV; C: Using templates for ALV-J. The x axes represent the sizes of PCR products in bp, and the y axes represent the dye signal in absorbance units.

Figure 5-5-4 GeXP-multiplex PCR interference assays

Evaluation of the GeXP-PCR assay using clinical specimens

A total of 300 clinical specimens were collected from diseased farm chickens and tested using the GeXP-multiplex PCR assay. The chickens ranged from 50 to 135 days old, and the cohort included layer, broiler and breeding birds. The results of the GeXP-multiplex PCR assay are provided in Table 5-5-3. Independent real-time PCR/RT-PCR and sequencing to identify true positives and negatives confirmed the GeXP-multiplex PCR results, with the positive results being 100% comparable to those for realtime PCR/RT-PCR and sequencing. Among a total of 190 positive results, 119 specimens displayed a single infection with an immunosuppressive virus. A total of 110 specimens were negative for the eight immunosuppressive viruses. A Kappa value of 1 was found with regard to consistency among GeXP-multiplex PCR, real-time PCR/RT-PCR and sequencing results.

Table 5-5-3 Detection results for clinical specimens

Single infection						Co-infection					
	Number	Rate /%	Real-time PCR/ RT-PCR results	Sequencing results	Measures of agreement Kappa values		Number	Rate /%	Real-time PCR/ RT-PCR results	Sequencing results	Measures of agreement Kappa values
MDV	26	8.7	26	26	1 ($P<0.001$)	MDV+ALV-A	4	1.3	4	4	1 ($P<0.001$)
ALV-A	6	2.0	6	6	1 ($P<0.001$)	MDV+ALV-J	25	8.3	25	25	1 ($P<0.001$)
ALV-B	2	0.7	2	2	1 ($P<0.001$)	ALV-B+ALV-J	1	0.4	1	1	1 ($P<0.001$)
ALV-J	21	7.0	21	21	1 ($P<0.001$)	MDV+ALV-J+REV	6	2	6	6	1 ($P<0.001$)
REV	15	5	15	15	1 ($P<0.001$)	MDV+CIAV	5	1.6	5	5	1 ($P<0.001$)
IBDV	17	5.7	17	17	1 ($P<0.001$)	MDV+REV	6	2	6	6	1 ($P<0.001$)
CIAV	9	3	9	9	1 ($P<0.001$)	ALV-J+REV	3	1.5	3	3	1 ($P<0.001$)
ARV	23	7.7	23	23	1 ($P<0.001$)	IBDV+ALV-J	5	1.6	5	5	1 ($P<0.001$)
						MDV+CIAV+ALV-J	4	1.3	4	4	1 ($P<0.001$)
						REV+ARV	3	1	3	3	1 ($P<0.001$)
						ALV-J+ARV	6	2	6	6	1 ($P<0.001$)
						ALV-J+IBDV	3	1	3	3	1 ($P<0.001$)

Discussion

MDV, ALV (three subgroups found predominantly in China, ALV-A/B/J), REV, IBDV, CIAV and ARV are the major immunosuppressive viruses causing economic losses to the chicken industry. Furthermore, the possibility of co-infection increases the difficulty of differentially diagnosing individual viral infections[16, 32-35]. However, conventional diagnostic methods are time consuming, and molecular methods are limited by their ability to detect only a few pathogens per reaction.

The advantages of the GeXP-multiplex PCR assay include its specificity and its high-throughput ability to differentiate eight immunosuppressive viruses. These advantages stem from the use of chimeric and universal primers in a 3-step PCR procedure with different annealing temperatures: the first step amplifies gene-specific sequences within specific regions of the chimeric primers; the second step utilises the entire chimeric primer; and the last step uses universal primers for amplification. As non-specific amplification is minimised by using chimeric primers in the second step at a temperature 13 ℃ higher than the temperature in step one, only specific amplicons are produced. Furthermore, false-positive reactions are minimised via capillary electrophoresis separation to confirm the identity of the bands[21]. Based on electropherograms, amplicon sizes were found to deviate from their theoretical size by approximately+/−1 bp; for example, amplicons of 154～156 bp indicated a positive result for ALV-A. In contrast, conventional multiplex PCR and multiplex realtime PCR are able to identify only two to four pathogens in one reaction[16, 34-36].

Another advantage of this GeXP-multiplex PCR assay is improved detection sensitivity[27]. In this study, the minimum detection limit of the GeXP-multiplex PCR assay was 100 copies (DNA plasmids or in vitro ssRNA transcripts) for all mixed templates, and the minimum absorbance unit (A. U.) was 9 000; by default, a reaction was considered positive when A. U.>2 000. Further evaluation of clinical specimens confirmed that the positive results were 100% comparable to the results of sequencing.

The third advantage of this GeXP-multiplex PCR assay is the ability to avoid interference due to the use of universal primers in the last step, ensuring equal amplification efficiency of each target gene, regardless of differences in pathogen concentration. Of the viruses evaluated in this study, MDV and CIAV are DNA viruses, and IBDV, ALV, REV and ARV are RNA viruses. ALV and REV are retroviruses, which insert their proviral DNA into the host genome for viral replication[38]. In these cases, we recommend using a DNA/RNA kit to extract DNA/RNA together, followed by the procedure described above. All of these immunosuppressive viruses affect the immune organs of chickens, including the thymus, spleen and bursa, and these organs should be collected from diseased chickens for diagnosis.

An analysis of 300 specimens using the GeXP-multiplex PCR assay revealed MDV as the most prevalent single infection, followed by ARV. MDV+ALV-J was the most prevalent coinfection, yet different co-infections were also present. Our data demonstrate that MDV and ALV-J are the main pathogens that co-infect with other immunosuppressive viruses and that various coinfections are common in chickens in south China. Chickens infected with one immunosuppressive virus are more susceptible to attack by another immunosuppressive virus, and vaccine contamination might be a cause of common co-infections[39, 40].

As no assay can be 100% accurate, there are some limitations to this assay. Virus genomes vary; thus, some primers will likely not be suitable for future use, and additional primers may be required for different varia-tions. Furthermore, the GeXP instrument is expensive for clinical applications; therefore, the availability of less expensive machines will be important.

Conclusions

In conclusion, the GeXP-multiplex PCR assay is a high-throughput, sensitive and specific method for detecting eight immunosuppressive viruses in chickens. Accordingly, this assay is a potentially useful tool for the detection and differentiation of immunosuppressive viruses and for molecular epidemiologic testing, especially in situations in which chicken flocks are not performing well and the underlying cause may be immunosuppressive viruses. Identifying the immunosuppressive viruses responsible for these infections will be helpful for designing better disease-control programmes.

References

[1] CALNEK B W. Pathogenesis of Marek's disease virus infection. Current Topics in Microbiology and Immunology, 2001, 255: 25-55.

[2] ENRIETTO P J, WYKE J A. The pathogenesis of oncogenic avian retroviruses. Advances in Cancer Research, 1983, 39: 269-314.

[3] CUI Z, SUN S, WANG J. Reduced serologic response to Newcastle disease virus in broiler chickens exposed to a Chinese field strain of subgroup J avian leukosis virus. Avian Dis, 2006, 50(2): 191-195.

[4] FADLY A M, WITTER R L, LEE L F. Effects of chemically or virus-induced immunodepression on response of chickens to avian leukosis virus. Avian Dis, 1985, 29(1): 12-25.

[5] SHARMA J M, KIM I J, RAUTENSCHLEIN S, et al. Infectious bursal disease virus of chickens: pathogenesis and immunosuppression. Developmental and Comparative Immunology, 2000, 24(2-3): 223-235.

[6] SMYTH J A, MOFFETT D A, CONNOR T J, et al. Chicken anaemia virus inoculated by the oral route causes lymphocyte depletion in the thymus in 3-week-old and 6-week-old chickens. Avian Pathol, 2006, 35(3): 254-259.

[7] ROESSLER D E. Studies on the pathogenicity and persistence of avian reovirus pathotypes in relation to age resistance and immunosuppression. Newark: University of Delaware Press, 1987.

[8] FARAGHER J T, ALLAN W H, WYETH P J. Immunosuppressive effect of infectious bursal agent on vaccination against Newcastle disease. The Veterinary Record, 1974, 95(17): 385-388.

[9] MCCONNELL C D, ADAIR B M, MCNULTY M S. Effects of chicken anemia virus on cell-mediated immune function in chickens exposed to the virus by a natural route. Avian Dis, 1993, 37(2): 366-374.

[10] DONG X, JU S, ZHAO P, et al. Synergetic effects of subgroup J avian leukosis virus and reticuloendotheliosis virus co-infection on growth retardation and immunosuppression in SPF chickens. Veterinary Microbiology, 2014, 172(3-4): 425-431.

[11] DIAO X G, ZHANG L, CHENG Z Q, et al. Dynamic pathology and antigen location study on broiler breeders with coinfection of MDV and REV. Scientia Agricultura Sinica, 2008, 41 (6): 1838-1844.

[12] CLOUD S S, LILLEHOJ H S, ROSENBERGER J K. Immune dysfunction following infection with chicken anemia agent and infectious bursal disease virus. I. Kinetic alterations of avian lymphocyte subpopulations. Veterinary Immunology and Immunopathology, 1992, 34(3-4): 337-352.

[13] MCNEILLY F, SMYTH J A, ADAIR B M, et al. Synergism between chicken anemia virus (CAV) and avian reovirus following dual infection of 1-day-old chicks by a natural route. Avian Dis, 1995, 39(3): 532-537.

[14] ZUO T, ZHAO Z, WEI P, et al. Isolation and identification of a field isolate of Marek's disease virus with acute oncogenicity. Chinese Journal of Virology, 2007, 23(03): 218-223.

[15] DRÉN C, FARKAS T, NÉMETH I. Serological survey on the prevalence of chicken anaemia virus infection in Hungarian chicken flocks. Veterinary Microbiology, 1996, 50(1-2): 7-16.

[16] GOPAL S, MANOHARAN P, KATHAPERUMAL K, et al. Differential detection of avian oncogenic viruses in poultry layer farms and Turkeys by use of multiplex PCR. Journal of Clinical Microbiology, 2012, 50(8): 2668-2673.

[17] JACKWOOD D J, NIELSEN C K. Detection of infectious bursal disease viruses in commercially reared chickens using the reverse transcriptase/polymerase chain reaction-restriction endonuclease assay. Avian Dis, 1997, 41(1): 137-143.

[18] KE G M, CHENG H L, KE L Y, et al. Development of a quantitative Light Cycler real-time RT-PCR for detection of avian reovirus. Journal of Virological Methods, 2006, 133(1): 6-13.

[19] KIM H R, KWON Y K, BAE Y C, et al. Molecular characterization of chicken infectious anemia viruses detected from breeder and broiler chickens in South Korea. Poultry Science, 2010, 89(11): 2426-2431.

[20] CATERINA K M, FRASCA S, J R., GIRSHICK T, et al. Development of a multiplex PCR for detection of avian adenovirus, avian reovirus, infectious bursal disease virus, and chicken anemia virus. Molecular and Cellular Probes, 2004, 18(5): 293-298.

[21] RAI A J, KAMATH R M, GERALD W, et al. Analytical validation of the GeXP analyzer and design of a workflow for cancer-biomarker discovery using multiplexed gene-expression profiling. Analytical and Bioanalytical Chemistry, 2009, 393(5): 1505-1511.

[22] YANG M J, LUO L, NIE K, et al. Genotyping of 11 human papillomaviruses by multiplex PCR with a GeXP analyzer. Journal of Medical Virology, 2012, 84(6): 957-963.

[23] HU X, ZHANG Y, ZHOU X, et al. Simultaneously typing nine serotypes of enteroviruses associated with hand, foot, and mouth disease by a GeXP analyzer-based multiplex reverse transcription-PCR assay. Journal of Clinical Microbiology, 2012,

50(2): 288-293.

[24] DREW J E, MAYER C D, FARQUHARSON A J, et al. Custom design of a GeXP multiplexed assay used to assess expression profiles of inflammatory gene targets in normal colon, polyp, and tumor tissue. The Journal of Molecular Diagnostics, 2011, 13(2): 233-242.

[25] LIU Y, XU Z Q, ZHANG Q, et al. Simultaneous detection of seven enteric viruses associated with acute gastroenteritis by a multiplexed Luminex-based assay. Journal of Clinical Microbiology, 2012, 50(7): 2384-2389.

[26] QIN M, WANG D Y, HUANG F, et al. Detection of pandemic influenza A H1N1 virus by multiplex reverse transcription-PCR with a GeXP analyzer. Journal of Virological Methods, 2010, 168(1-2): 255-258.

[27] XIE Z X, LUO S S, XIE L J, et al. Simultaneous typing of nine avian respiratory pathogens using a novel GeXP analyzer-based multiplex PCR assay. Journal of Virological Methods, 2014, 207(0): 188-195.

[28] ZHANG Y F, XIE Z X, XIE L J, et al. GeXP analyzer-based multiplex reverse-transcription PCR assay for the simultaneous detection and differentiation of eleven duck viruses. BMC Microbiology, 2015, 15(1): 247.

[29] ZHANG M X, XIE Z X, XIE L J, et al. Simultaneous detection of eight swine reproductive and respiratory pathogens using a novel GeXP analyser-based multiplex PCR assay. Journal of Virological Methods, 2015, 224: 9-15.

[30] XIE Z X, XIE L J, PANG Y S, et al. Development of a real-time multiplex PCR assay for detection of viral pathogens of penaeid shrimp. Archives of Virology, 2008, 153(12): 2245-2251.

[31] CHEN N H, CHEN X Z, HU D M, et al. Rapid differential detection of classical and highly pathogenic north American porcine reproductive and respiratory syndrome virus in China by a duplex real-time RT-PCR. Journal of Virological Methods, 2009, 161 (2): 192-198.

[32] ZHANG H H, LIU Q, QIU B, et al. Mixed infection of ALV-J and MDV in a flock of Shandong free range Chickens. Acta Veterinaria et Zootechnica Sinica, 2009, 40(08): 1215-1221.

[33] QING L T, GAO Y L, PAN W, et al. Investigation of co-infection of ALV-J with REV, MDV, CAV in layer chicken f locks in some regions of China. Chinese Journal of Preventive Veterinary Medicine, 2010, 32(2): 90-96.

[34] CUI Z Z, MENG S S, JIANG S J. Serological surveys of chicken anemia virus, avian reticuloendotheliosis virus and avian reovirus infections in white meat-type chickens in China. Acta Veterinaria et Zootechnica Sinica, 2006, 37 (2): 152.

[35] DUCATEZ M F, OWOADE A A, ABIOLA J O, et al. Molecular epidemiology of chicken anemia virus in Nigeria. Archives of virology, 2006, 151(1): 97-111.

[36] XIE Z X, PANG Y S, LIU J B, et al. A multiplex RT-PCR for detection of type A influenza virus and differentiation of avian H5, H7, and H9 hemagglutinin subtypes. Molecular and cellular probes, 2006, 20(3-4): 245-249.

[37] XIE Z X, XIE L J, FAN Q, et al. A duplex quantitative real-time PCR assay for the detection of *Haplosporidium* and *Perkinsus* species in shellfish. Parasitology Research, 2013, 112(4): 1597-1606.

[38] TELESNITSKY A, GOFF S P. Reverse transcriptase and the generation of retroviral DNA//COFFIN J M, HUGHES S H, VARMUS H E. Retroviruses. Cold Spring Harbor: Cold Spring Harbor Laboratory Press,1997.

[39] ZAVALA G, CHENG S. Detection and characterization of avian leukosis virus in Marek's disease vaccines. Avian Dis, 2006, 50(2): 209-215.

[40] SHARMA J M. Effect of infectious bursal disease virus on protection against Marek's disease by turkey herpesvirus vaccine. Avian Dis, 1984, 28(3): 629-640.

GeXP analyzer-based multiplex reverse-transcription PCR assay for the simultaneous detection and differentiation of eleven duck viruses

Zhang Yanfang, Xie Zhixun, Xie Liji, Deng Xianwen, Xie Zhiqin, Luo Sisi, Huang Li, Huang Jiaoling, and Zeng Tingting

Abstract

Background: Duck viral pathogens primarily include the avian influenza virus (AIV) subtypes H5, H7, and H9; duck hepatitis virus (DHV); duck Tembusu virus (DTMUV); egg drop syndrome virus (EDSV); duck enteritis virus (DEV); Newcastle disease virus (NDV); duck circovirus (DuCV); muscovy duck reovirus (MDRV); and muscovy duck parvovirus (MDPV). These pathogens cause great economic losses to China's duck breeding industry. Result: A rapid, specific, sensitive and high-throughput GeXP-based multiplex PCR assay consisting of chimeric primer-based PCR amplification with fluorescent labeling and capillary electrophoresis separation was developed and optimized to simultaneously detect these eleven viral pathogens. Single and mixed pathogen cDNA/DNA templates were used to evaluate the specificity of the GeXP-multiplex assay. Corresponding specific DNA products were amplified from each pathogen. Other pathogens, including duck *Escherichia coli*, duck *Salmonella*, duck *Staphylococcus aureus, Pasteurella multocida,* infectious bronchitis virus, and *Mycoplasma gallisepticum,* did not result in amplification products. The detection limit of GeXP was 10^3 copies when all twelve pre-mixed plasmids containing the target genes of eleven types of duck viruses were present. To further evaluate the reliability of GeXP, 150 clinical field samples were evaluated. Comparison with the results of conventional PCR methods for the field samples, the GeXP-multiplex PCR method was more sensitive and accurate. Conclusions: This GeXP-based multiplex PCR method can be utilized for the rapid differential diagnosis of clinical samples as an effective tool to prevent and control duck viruses with similar clinical symptoms.

Keywords

duck virus, Genome Lab Gene Expression Profiler (GeXP), multiplex PCR, separation identification

Background

According to the Food and Agriculture Organization (FAO) statistics, the 771 million ducks annually raised for meat purposes in China accounted for 65.73% of the world's stock in 2009 (FAO, 2009). There are eleven important viral pathogens that cause infections in ducks in China, including the avian influenza virus (AIV) subtypes H5, H7, H9; duck hepatitis virus (DHV); duck Tembusu virus (DTMUV); egg drop syndrome virus (EDSV); duck enteritis virus (DEV); Newcastle disease virus (NDV); duck circovirus (DuCV); muscovy duck reovirus (MDRV); and muscovy duck parvovirus (MDPV)[1-4]. With the development of the commercial duck industry in China and in other parts of the world in recent years, the incidence of duck diseases has increased[5].

The increase in duck disease has become an important factor that restricts the growth and further development of the duck industry. Differential diagnosis of infectious duck diseases using traditional methods requires isolation of the pathogens and identification using serological techniques[6-9]. The accuracy of these methods is often affected by the freshness of the clinical material, contamination with bacteria and the length of time between the collection of the material and its analysis. Thus conventional methods of diagnosis are very tedious and time-consuming and are further complicated by the differential diagnosis of mixed infections[10, 11]. Molecular typing methods for the rapid detection and identification of pathogens have been developed and are used currently. Although useful, most molecular methods are limited to the detection of a few pathogens in one reaction[12-16]. Therefore, a rapid, cost-effective and high-throughput method for the detection and differentiation of viral pathogens in one test tube would be advantageous and would ensure prompt treatment and infection control.

The GenomeLab Gene Expression Profiler (GeXP) analyzer is a multiplex gene expression analysis platform developed by Beckman Coulter (Brea, CA, USA). GeXP assays have been successfully used for rapid identification in human medicine including assays to detect several inflammatory and cytokine gene targets in normal colon tissue as well as in colon polyps and tumors[17, 18], prostate cancer biomarker gene expression signatures in biological samples[19], pandemic influenza A H1N1 virus[20]; nine types of enteroviruses associated with hand, foot, and mouth disease[21]; and eleven human papillomaviruses in a rapid and sensitive manner[22]. Recently, a GeXP assay was used for the detection of nine avian respiratory disease agents[23]. The GeXP analyzer has a built-in software program that can be used to evaluate PCR products based on amplicon size. The software displays the differences between the various specific genes under investigation and compares the expected PCR product sizes to identify each product. The GeXP-multiplex PCR assay provide high sensitivity and specificity compared to other multiple-detection methods[24].

In this study, a GeXP analyzer-based multiplex RT-PCR assay (GeXP-multiplex PCR) was developed to simultaneously detect eleven common duck viral diseases in China, including AIV-H5, AIV-H7, AIV-H9, DHV, DTMUV, EDSV, DEV, NDV, DuCV, MDRV, and MDPV. The diagnostic sensitivities and specificities of this assay were evaluated with 150 clinical specimens.

Materials and methods

Extraction of DNA/RNA from pathogens and sample preparation

The avian pathogen reference strains, field isolates and other duck pathogens used in this study are described in Table 5-6-1. Viral RNA and DNA were extracted using the MiniBEST Viral RNA/DNA Extraction Kit Ver 5.0 (TaKaRa, Dalian, China) according to the manufacturer's protocol. The DNA and RNA samples were aliquoted and stored at −30 ℃ until use.

Table 5-6-1 Sources of pathogens used and GeXP assay results

Pathogen/field samples	Number of samples	Source	Results											
			AIV-M	AIV-H5	AIV-H7	AIV-H9	DHV	DEV	DTMUV	NDV	EDSV	MDRV	MDPV	DuCV
Reference sample														
[a] cDNA of H5N3 AIV Duck/HK 313/78	1	HKU	+	+	−	−	−	−	−	−	−	−	−	−
[a] H7N2 AIV Duck/HK/47/76	1	HKU	+	−	+	−	−	−	−	−	−	−	−	−
[a] H9N6 AIV Duck/HK/147/77	1	HKU	+	−	−	+	−	−	−	−	−	−	−	−
[b] DHV (AV2111)	1	CIVDC	−	−	−	−	+	−	−	−	−	−	−	−

continued

Pathogen/field samples	Number of samples	Source	Results											
			AIV-M	AIV-H5	AIV-H7	AIV-H9	DHV	DEV	DTMUV	NDV	EDSV	MDRV	MDPV	DuCV
[c] DEV (AV1221)	1	CIVDC	−	−	−	−	−	+	−	−	−	−	−	−
[d] DTMUV (GX201301, GX201302)	2	GVRI	−	−	−	−	−	−	+	−	−	−	−	−
[e] NDV (GX1/00, GX6/02)	2	GVRI	−	−	−	−	−	−	−	+	−	−	−	−
[f] EDSV (GEV)	1	GVRI	−	−	−	−	−	−	−	−	+	−	−	−
[g] MDRV (NM1, NM2)	2	GVRI	−	−	−	−	−	−	−	−	−	+	−	−
[h] MDPV (GX-5, GX-6)	2	GVRI	−	−	−	−	−	−	−	−	−	−	+	
[i] DuCV (GX1006, GX1008)	2	GVRI	−	−	−	−	−	−	−	−	−	−	−	+
Other pathogens														
H1N3 AIV Duck/HK/717/79-d1	1	HKU	+	−	−	−	−	−	−	−	−	−	−	−
H1N1 AIV Human/NJ/8/76	1	HKU	+	−	−	−	−	−	−	−	−	−	−	−
H2N3 AIV Duck/HK/77/76	1	HKU	+	−	−	−	−	−	−	−	−	−	−	−
H3N6 AIV Duck/HK/526/79/2B	1	HKU	+	−	−	−	−	−	−	−	−	−	−	−
H4N5 AIV Duck/HK/668/79	1	HKU	+	−	−	−	−	−	−	−	−	−	−	−
H6N8 AIV Duck/HK/531/79	1	HKU	+	−	−	−	−	−	−	−	−	−	−	−
H8N4AIV Turkey/ont/6118/68	1	HKU	+	−	−	−	−	−	−	−	−	−	−	−
H10N3 AIV Duck/HK/876/80	1	HKU	+	−	−	−	−	−	−	−	−	−	−	−
H11N3 AIV Duck/HK/661/79	1	HKU	+	−	−	−	−	−	−	−	−	−	−	−
H12N5 AIV Duck/HK/862/80	1	HKU	+	−	−	−	−	−	−	−	−	−	−	−
H6N8 AIV Duck/HK/531/79	1	HKU	+	−	−	−	−	−	−	−	−	−	−	−
duck *Escherichia coli*	1	GVRI	−	−	−	−	−	−	−	−	−	−	−	−
duck *Salmonella*	1	GVRI	−	−	−	−	−	−	−	−	−	−	−	−
duck *Staphylococcus aureus*	1	GVRI	−	−	−	−	−	−	−	−	−	−	−	−
Pasteurella multocida	1	GVRI	−	−	−	−	−	−	−	−	−	−	−	−
Infectious bronchitis virus	1	GVRI	−	−	−	−	−	−	−	−	−	−	−	−
Mycoplasma gallisepticum	1	GVRI	−	−	−	−	−	−	−	−	−	−	−	−
Sample mixture														
AIV-H5+AIV-H7+AIV-H9	1	GVRI	+	+	+	+	−	−	−	−	−	−	−	−
AIV-H5+DEV+DTMUV+NDV+EDSV	1	GVRI	+	+	−	−	−	+	+	+	+	−	−	−
AIV-H5+AIV-H7+AIV-H9+DHV+DEV+DTMUV+NDV+EDSV+MDRV+MDPV+DuCV	1	GVRI	+	+	+	+	+	+	+	+	+	+	+	+

Note: HVRI, Harbin Veterinary Research Institute, China; HKU, the University of Hong Kong, China; GVRI, Guangxi Veterinary Research Institute, China; CIVDC, China Institute of Veterinary Drug Control, China; PU, University of Pennsylvania.
[a] References [23], [b] GenBank accession number: EF442073.1, [c] GenBank accession number: EU315247, [d] GenBank accession number: KJ700462.1, [e] GenBank accession number: JX193083.1, [f] References [34], [g] References [34], [h] GenBank accession number: KM093740.1, [i] GenBank accession number: JX241046.1.

Primer design and plasmid preparation

The GeXP-multiplex assay consists of twelve pairs of chimeric primers including one pair of AIV universal primers (AIV-M) and eleven pairs of duck virus primers. These twelve pairs of duck virus primers were designed based on the sequences of the type A AIV-M gene; the HA genes from the H5, H7 and H9 subtypes of AIV; the 5'UTR region of the DHV gene; the DTMUV E gene; the DEV UL6 gene; the NDV L gene; the EDSV Penton gene; the MDRV S1 gene; the MDPV VP1 gene and the DuCV Red gene. All sequences were obtained from GenBank. The primers were designed by selecting the highly conserved regions using DNAstar software (DNASTAR Inc., Madison, WI, USA) and Primer Premier 5.0 software (Premier Biosoft International, Palo Alto, CA, USA). AIV universal primers were utilized to amplify all 16 subtypes. The 5' end of the forward universal primer (Tag-F: AGGTGACACTATAGAATA) was labeled and purified with high-pressure liquid chromatography. All chimeric primers and the reverse universal primer (Tag-R: GTACGACTCACTATAGGGA) were synthesized and purified by polyacrylamide gel electrophoresis (BGI, China). The primer sequences, the size of the resulting amplicons, and the target regions are listed in Table 5-6-2.

Table 5-6-2 Primers used in this study

Virus	Forward primer sequence (5'-3')	Reverse primer sequence (5'-3')	Amplicon size/bp	Target region
AIV-M	AGGTGACACTATAGAATACAGAAACGGATGGGAGTGC	GTACGACTCACTATAGGGATATCAAGTGCAAGATCCCAATGAT	122	M
AIV-H5	AGGTGACACTATAGAATACTTCAGGCATCAAAATGCACA	GTACGACTCACTATAGGGATAGTTTGTTCATTTCTGAGTCGGTC	285	HA
AIV-H7	AGGTGACACTATAGAATAAATGGGGCHTTCATAGCTCC	GTACGACTCACTATAGGGATGATAGCARTCRCCTTCACAA	144	HA
AIV-H9	AGGTGACACTATAGAATAACAACAAGTGTGACAACAGAAGA	GTACGACTCACTATAGGGATCTTCCGTGGCTCTCTCC	237	HA
DHV	AGGTGACACTATAGAATATCTTCGTTGTGAAACGGATTACC	GTACGACTCACTATAGGGATGCCTGGACAGATDTGTGCCTACT	133	5' UTR
DTMUV	AGGTGACACTATAGAATAATGGACAGGGTCATCAGCGG	GTACGACTCACTATAGGGAGAATRGCTCCYGCCAATGCT	176	E
EDSV	AGGTGACACTATAGAATAAATCGGCAACTCAAGACATC	GTACGACTCACTATAGGGACCCATTCATAAACAGGATTC	208	Penton
DEV	AGGTGACACTATAGAATAGGGAGGAGCAAACAAAGA	GTACGACTCACTATAGGGAATCGCAAATTCCATCACATA	150	UL6
NDV	AGGTGACACTATAGAATAGTRGCAGCAAGRACAAGG	GTACGACTCACTATAGGGACATATCYGCATACATCAA	196	L
DuCV	AGGTGACACTATAGAATATGCKCCAAAGAGTCGACATA	GTACGACTCACTATAGGGACAAAYGCATAACGGCTCTTTCC	300	Red
MDRV	AGGTGACACTATAGAATACAGTTGAGCCGGAYGGTAATT	GTACGACTCACTATAGGGAACTCGGTTGGTGTTAGTVGCVTAGAA	219	S1
MDPV	AGGTGACACTATAGAATACTTTCAGGCTACATCTTCAA	GTACGACTCACTATAGGGAAATTCTCTTTTCACCCATCC	253	VP1
Universal tag sequences	Tag-F: AGGTGACACTATAGAATA	Tag-R: GTACGACTCACTATAGGGA		

Note: Universal tag sequences are underlined. Abbreviations: M, A or C; R, A or G; W, A or T; S, G or C; Y, C or T; K, G or T; V, A, G, or C; H, A, C, or T; D, A, G, or T; B, G, C, or T.

Twelve specific genes from the eleven duck pathogens were amplified using the primers listed in Table 5-6-2. Plasmid-encoding genes from DEV (AV1221), EDSV (GEV), MDPV (GX-5) and DuCV (GX1006) were prepared and quantified according to Xie[23, 25]. Plasmids encoding genes from AIV (H5N1 AIV Re-1, H7N2 AIV Duck/HK/47/76, and H9N6/Duck/HK/147/77), DHV (AV2111), DTMUV (GX201301), NDV (GX1/00) and MDRV (NM1) were used to produce ssRNA via in vitro transcription using a High-Yield

Transcription Kit (Thermo Scientific, USA). The copy numbers of the ssRNAs for AIV (H5N1 AIV Re-1, H7N2 AIV Duck/HK/47/76, H9N6/Duck/HK/147/77), DHV (AV2111), DTMUV (GX201301), NDV (GX1/00) and MDRV (NM1) were calculated as described previously[23, 25].

Setup of the GeXP assay and the reaction procedure

The reaction system was created using the GeXP Start-up Kit in a total volume of 20 μL containing 4 μL of GenomeLab™ GeXP Start Kit 5 × PCR buffer, 4 μL of $MgCl_2$ (25 μM), 2 μL of mixed primers (containing 20～100 nmol/L of 12 pairs of gene-specific chimeric primers), 1.4 μL of JumpStart Taq DNA polymerase, and 0.5 pg to 0.5 ng of cDNA or DNA template. Nuclease-free water was then added to the PCR reaction to achieve a final volume of 20 μL. Three optimized PCR amplification steps were performed according to the temperature-switch PCR (TSP) strategy[22]: step 1 was carried out with gene-specific sequences of chimeric forward and reverse primers (10 cycles of 30 s at 95 ℃ , 30 s at 55 ℃ , and 30 s at 72 ℃); step 2 was carried out predominantly with chimeric forward and reverse primers (10 cycles of 30 s at 95 ℃ , 30 s at 63 ℃ , and 30 s at 72 ℃); and step 3 was carried out predominantly with universal forward and reverse primers (20 cycles of 30 s at 95 ℃ , 30 s at 50 ℃ , and 30 s at 72 ℃). The reactions were held at 4 ℃ after the amplification cycles.

Separation by capillary electrophoresis and fragment analysis

After amplification, 1 μL of the PCR product was added to 20 μL of sample loading solution along with 0.16 μL of DNA Size Standard-400 (GenomeLab GeXP Start Kit Beckman Coulter), following protocols described previously[19]. After amplified DNA amplicons were separated, the data were imported into the analysis module of ExpressProfiler software as a tab-delimited file for subsequent analyses.

Evaluating the specificity of the GeXP assay

cDNA and DNA templates were used for the GeXP-mono PCR assay and the GeXP-multiplex PCR assay. The GeXP-mono PCR assay was performed using a single template along with each pair of chimeric primers to determine the size of the amplification products for each target gene. The specificity of the GeXP-multiplex PCR assay was tested for each individual viral target gene, and the assay was performed using a single template. The GeXP-mono PCR assay and GeXP-multiplex PCR assay were developed using the reaction system and procedure described in Material and Methods Section *Primer design and plasmid preparation.*

Evaluating the accuracy of the GeXP assay

The reaction system and procedures described above were used for the validation of GeXP-multiplex PCR for the detection of the twelve viruses. cDNA/DNA hybrids of all twelve duck viral disease agents were used as templates along with a primer mixture (0.2 μL). PCR product separation and detection were performed on a Genome Lab GeXP Genetic Analysis System (Beckman Coulter) via capillary electrophoresis, following the protocols described previously in Section *Setup of the GeXP assay and the reaction prcedure.*

After the amplified fragments were separated, the peaks were initially analyzed using the fragment analysis module of the GeXP system software and matched to the appropriate genes. The peak height for each gene was reported in the electropherogram.

To emulate mixed infections, we randomly chose duck pathogens and extracted their DNA and RNA.

RNA viruses were transcribed into cDNA as described above. DNA and cDNA were mixed together in equal concentrations to serve as templates for the optimized GeXP-multiplex PCR assay. To simulate the detection of a clinically mixed infection by GeXP, random cDNAs from eleven viruses or DNA samples from three to five viruses were mixed, and two groups were established. Group 1 consisted of a mixture of cDNAs from three avian influenza viruses (AIV-H5, AIV-H7 and AIV-H9) per template, while group 2 consisted of cDNA/DNA mixtures of DEV, DTMUV, NDV, EDSV and AIV-H5 viruses per template. A mixture of primers (0.2 μL) was added, and the remainder of the procedure was performed as described in Materials and Methods.

Evaluating the sensitivity of the GeXP assay

The sensitivity of the GeXP-multiplex assay for each type of duck virus was examined using serial 10-fold dilutions of the plasmids (DEV, EDSV, MDPV, and DuCV) and ssRNAs obtained via in vitro transcription (AIV-5, AIV-7, AIV-9, DHV, DTMUV, NDV, and MDRV), either separately or with the twelve samples mixed together (10^6 to 10^1 copies/μL). The assays were performed in triplicate. Plasmid or ssRNA mixtures were used to test the detection limit when all twelve duck virus genotypes were present. PCR was performed using the same experimental conditions described above for the GeXP-multiplex PCR assay, and the detection limit of the GeXP-PCR was determined based on the most dilute template that yielded a positive result. After amplification, 2 μL of each Cy5-labeled PCR product was separated via GeXP capillary electrophoresis and detected by fluorescence spectrophotometry.

Interference assay

Because the presence of other templates in high quantities could alter the efficiency of GeXP-multiplex PCR amplification, different amounts of the templates (10^3 to 10^7 copies) were selected at random, mixed and tested in the GeXP-multiplex PCR assay. The results were compared with those of the single-template GeXP-multiplex PCR assay.

Evaluation of the GeXP-PCR assay using clinical samples

A total of 150 archived clinical specimens were collected from ducks in Guangxi, China, between July, 2012 and November 2013. RNA and DNA from the 150 specimens were then extracted using the MiniBEST Viral RNA/DNA Extraction Kit (TaKaRa, Dalian, China) and a TIANamp Bacteria DNA Kit (Tiangen, Beijing, China).

Genomic RNA from all samples was transcribed into cDNA as described above, and cDNA samples were analyzed using the optimized GeXP-multiplex PCR assay in addition to conventional simplex PCR methods with same primers as the GeXP assay. The obtained positive samples were sent to a company (BGI, China) for sequencing.

Results

Specificity of the GeXP-multiplex PCR assay

The DNA and cDNA from eleven duck pathogens were used as templates to evaluate the specificity of each pair of gene-specific primers. In mono-GeXP PCR assays, the AIV universal primers that we designed could amplify the target M genes of all AIV serotypes; however, each pair of pathogen-specific primers amplified only the corresponding genes from the targeted pathogens without cross-amplification. The expected amplification peaks were observed in the GeXP-multiplex PCR assay (Table 5-6-1). No specific amplification peaks were observed when duck *Escherichia coli*, duck *Salmonella*, duck *Staphylococcus aureus, Pasteurella multocida*, infectious bronchitis virus, or *Mycoplasma gallisepticum* were tested (Table 5-6-1).

The GeXP-multiplex PCR assay used twelve pairs of primers to detect eleven duck viruses, and specific amplification peaks were observed (Figure 5-6-1). Specific amplification peaks corresponding to the M gene of AIV were observed upon testing H1N3, H1N1, H2N3, H3N6, H4N5, H6N8, H8N4, H10N3, H11N3, H12N5 and H13N5.

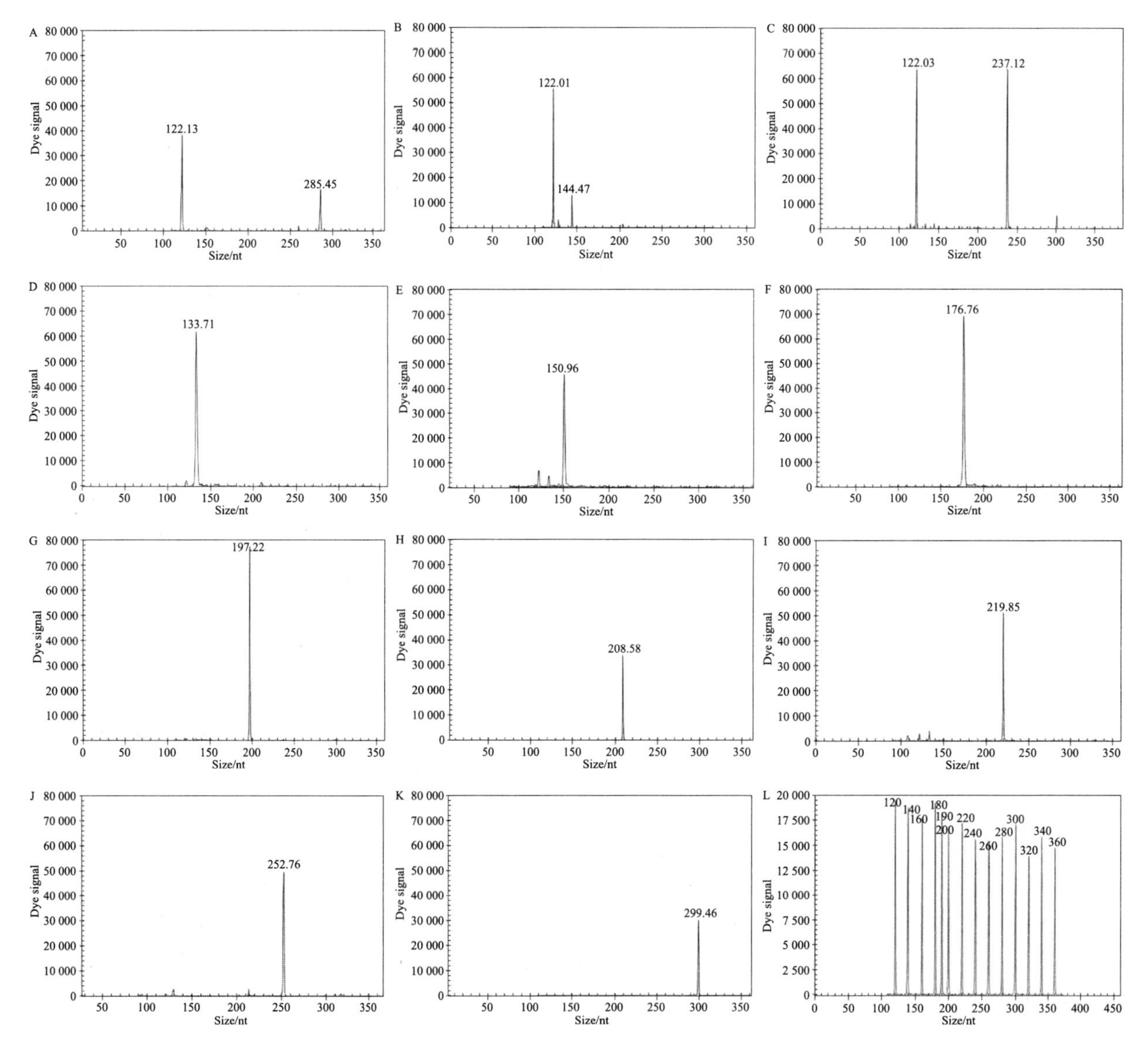

A-K: The results of the amplification of AIV-H5, AIV-H7, AIV-H9, DHV, DEV, DTMUV, NDV, EDSV, MDRV, MDPV, and DuCV, respectively; L: Nuclease-free water was used as the negative control. The Y-axis indicates the dye signal in A.U, and the X-axis indicates the actual PCR product size. The peaks indicate the target genes and the DNA size standard.

Figure 5-6-1 Specificity analyses of the GeXP-PCR assay

Sensitivity of the GeXP assay

The GeXP-PCR assay was capable of detecting as few as 10～100 copies of each of the twelve recombinant plasmids of duck viruses (data not shown) and as few as 10^3 copies when all of the twelve premixed duck virus targets were present in a mixture (Figure 5-6-2). Typically, a reaction is considered positive when the A. U. value is over 2 000 by default[22], and 10 duck viruses were detected at a concentration of 10^2 copies per reaction (DuCV being the exception). The reactions were repeated three times at each template concentration, and similar results were obtained.

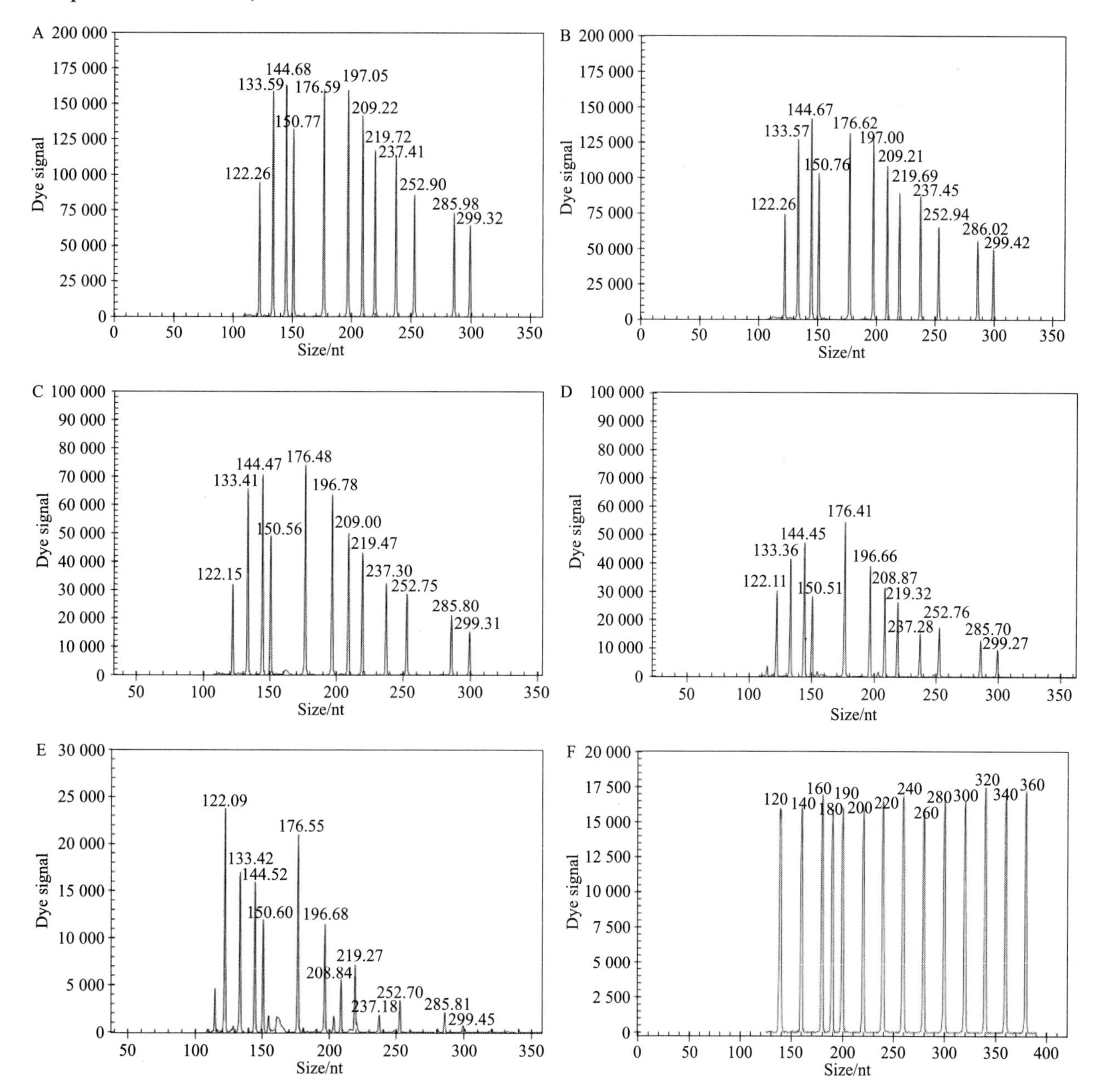

A-E: 10^6, 10^5, 10^4, 10^3 and 10^2 copies of plasmids of per reaction in the GeXP-PCR assay; F: Nuclease-free water was used as the negative control. The viral targets from left to right are as follow: AIV-M, DHV, AIV-H7, DEV, DTMUV, NDV, EDSV, MDRV, AIV-H9, MDPV, AIV-H5, and DuCV.

Figure 5-6-2 Sensitivity of the GeXP-PCR assay

Artificial mixture

The cDNA and DNA from the duck pathogens were mixed together to test the ability of the GeXP-multiplex PCR assay to differentiate among them, and the appropriate specific amplification peaks and universal amplification peaks were observed (Table 5-6-1; Figure 5-6-3 A, B). When AIV-H5, AIV-H7 and AIV-H9 were mixed together and used in the GeXP-multiplex PCR assay, 3 specific amplification peaks (AIV-H7, 144.92 bp; AIV-H9, 237.21 bp; AIV-H5, 285.73 bp) and one universal amplification peak (AIV M gene, 122.14 bp) were observed (Figure 5-6-3 A). When AIV-H5, DEV, DTMUV, NDV and EDSV were mixed together and used in the GeXP-multiplex PCR assay, there were six specific amplification peaks (Figure 5-6-3 B). When the cDNA and DNA of all eleven pathogens were mixed and used in the GeXP-multiplex PCR assay, eleven specific amplification peaks were observed (Table 5-6-1; Figure 5-6-3 C).

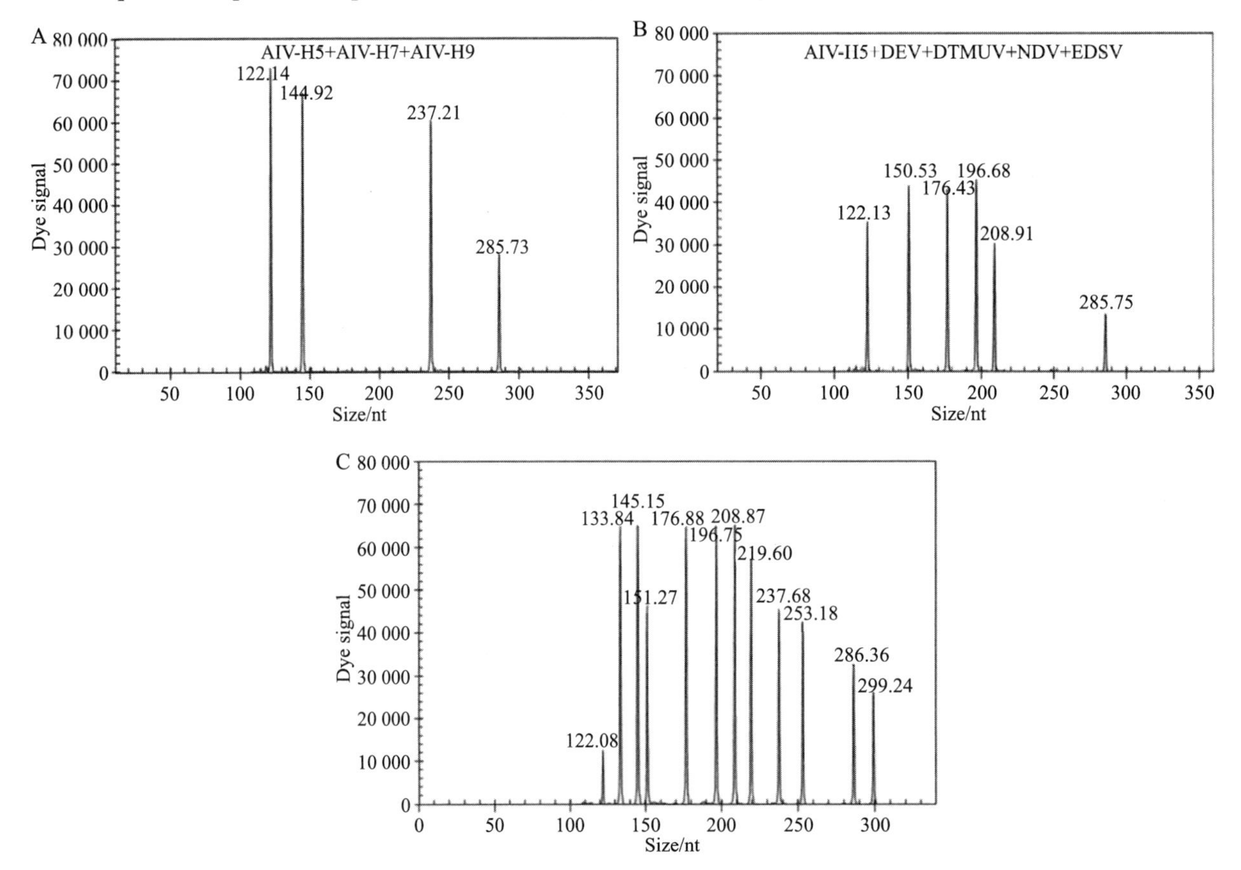

A: The result of mixed templates for AIV-H5, AIV-H7 and AIV-H9 by GeXP-multiplex assay; B: The result of mixed templates for AIV-H5, DEV, DTUMV, NDV and EDSV by GeXP-multiplex assay; C: The result of mixed templates for all eleven viruses by GeXP-multiplex assay.

Figure 5-6-3 GeXP-multiplex PCR detection of mixed pathogen templates

Interference assay

Three specific amplification peaks were observed when three different templates (one template with 10^3 copies and two templates with 10^7 copies) were tested using the GeXP-multiplex PCR assay. Additionally, the peak value of a single template was similar to that of a mixed template. For example, three specific amplification peaks were observed when three different templates (10^3 copies of DuCV and 10^7 copies of AIV-M and AIV-H9) were tested using the GeXP-multiplex PCR assay (Figure 5-6-4), and the peak values for AIV-M, AIV-H9 and for DuCV were the same regardless of whether a single template (AIV-M and AIV-H9 or DuCV) or a mixed template (AIV-M+AIV-H9+DuCV) was utilized. The results of these experiments showed that mixed infections can be detected by GeXP-multiplex PCR with only minimal interference.

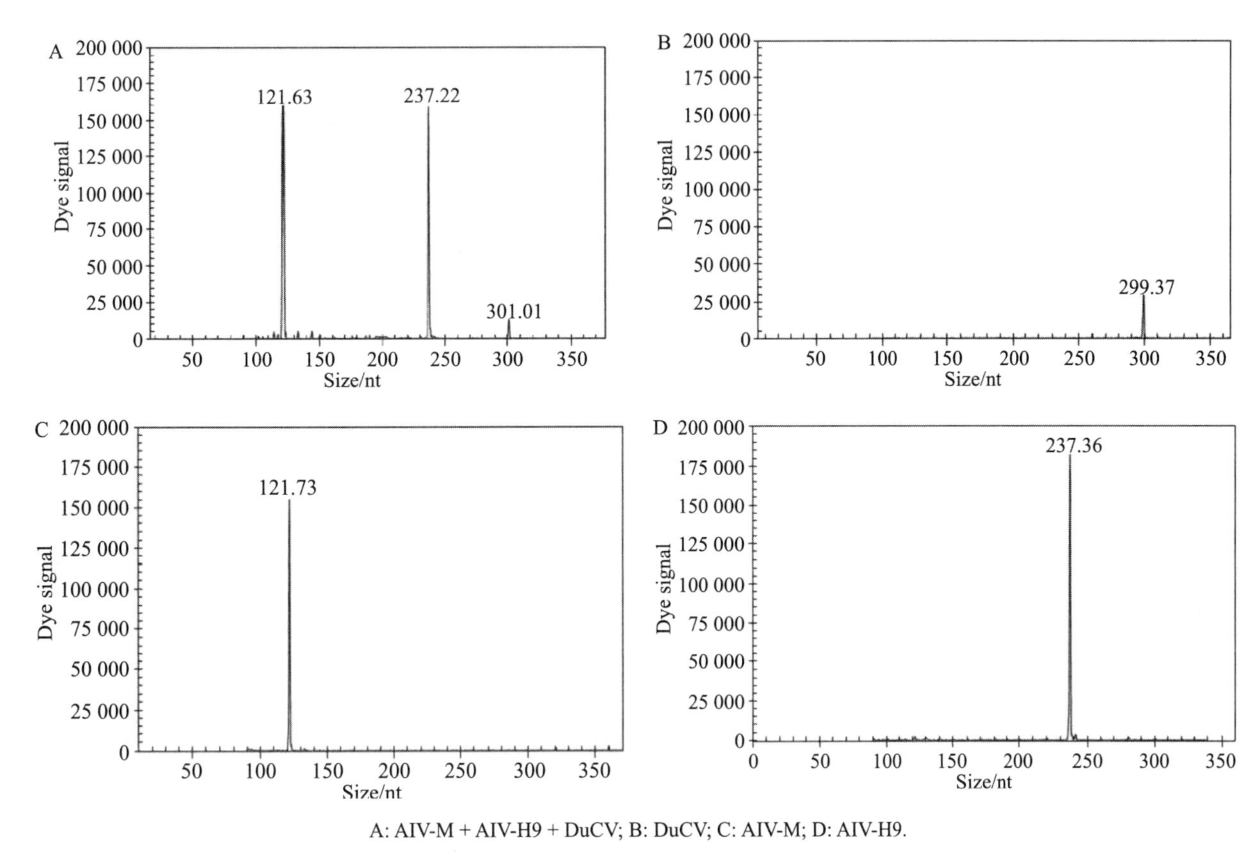

A: AIV-M + AIV-H9 + DuCV; B: DuCV; C: AIV-M; D: AIV-H9.

Figure 5-6-4 GeXP-multiplex PCR interference assay

Assay of field samples using GeXP-multiplex PCR

A total of 150 clinical samples were assayed using the optimized GeXP-multiplex PCR method in addition to the conventional simplex PCR method. The positive and negative results obtained with the different methods are shown in Table 5-6-3. The degree of consistency between the GeXP assay and conventional PCR methods is shown in Table 5-6-4. The kappa values as a measure of agreement for the GeXP assay and conventional PCR methods were as follow: K=0.908, $K>0$; u=28.375, $u>1.96$, $P<0.05$. The two experimental results appear to have a high degree of consistency. All positive specimens in the GeXP assay and conventional methods were identified via sequencing (BGI, China) to be true positive samples.

Table 5-6-3 Analysis of clinical samples using the GeXP assay and conventional PCR methods

Clinical samples	No. of GeXP assay results		No. of conventional PCR method results	
	Positive	Negative	Positive	Negative
H9 subtypes of AIV	20	130	18	132
DHV	5	145	5	145
DEV	1	149	1	149
DTMUV	5	145	5	145
NDV	10	140	9	141
EDSV	3	142	3	147
MDRV	1	149	1	149

continued

Clinical samples	No. of GeXP assay results		No. of conventional PCR method results	
	Positive	Negative	Positive	Negative
MDPV	4	146	4	146
DuCV	40	110	35	115
Total	89	61	81	69

Table 5-6-4 Comparison of clinical samples using the GeXP assay and conventional PCR methods

GeXP assay results	Conventional PCR methods results		Total
	+	–	
+	81	8	89
–	0	61	61
Total	81	69	150

Discussion

AIV-H5, AIV-H7, AIV-H9, DHV, DEV, DTMUV, NDV, EDSV, MDRV, MDPV, and DuCV are the eleven types of viruses that cause infectious diseases in duck. In this study, we developed a GeXP method for the simultaneous detection of these eleven duck viruses. Twelve pairs of specific primers were designed according to the conserved sequences of the genes from each pathogen; these sequences were obtained from GenBank. The GeXP genetic analysis system is a multitarget, high-throughput detection platform, and its application in the differential detection of nine avian respiratory pathogens was reported recently by our laboratory[23].

The amplification specificity and accuracy of each primer set in GeXP-multiplex PCR was verified by the inspection of samples containing a single virus or a mixed population and positive clinical samples (including mixed infections, single infections and infections with more than two types of pathogen) were used to further verify the method. The evaluation of sensitivity using recombinant plasmids for each duck virus revealed that all twelve of the hybrid templates were detected simultaneously when they were present at a concentration of 10^3 copies per reaction, indicating the high sensitivity of the GeXP-PCR assay. One hundred and fifty specimens were analyzed by the GeXP-PCR assay (89 were positive) and conventional PCR methods (81 were positive). Additionally, 8 clinical samples that were negative based on conventional single PCR but positive in the GeXP assay were confirmed later by sequencing to be true positives (data not shown). Compared with the results of conventional PCR method, the GeXP-multiplex PCR method was more sensitive and accurate.

In recent years, multiplex PCR and multiplex fluorescence real-time quantitative PCR techniques have been widely used for the detection of mixed infections with multiple pathogens[26]. Multiple PCR involves the use of several primer pairs in one reaction system at the same time. Thus competition occurs during amplification as the primer pairs interfere with each other. The probability of forming complex primer dimers increases, and the sensitivity decrease. Additionally, multiplex PCR products are typically observed by agarose gel electrophoresis, which is not an optimal method for distinguishing bands that are less than 100 bp in length. Furthermore, the probes used in multiplex fluorescence real-time quantitative PCR are conjugated to fluorophores that emit light at different wavelengths. General PCR methods, such as conventional multiplex PCR and multiplex real-time PCR[27-31], can only detect 2 to 6 types of pathogens; thus, these methods are not ideal for the rapid high-throughput detection of gene expression or analysis of multiple pathogens. Quantitative GeXP expression analysis combines multiplex PCR and capillary electrophoresis, employing fluorescently labeled universal primers and specific primer combinations to trigger multiple amplifications. Using this technique, multiple genes can be amplified using 1 pair of universal primers. The amplification efficiency of each template is consistent, and the amplification efficiency of each primer pair is not affected. Thus, high-throughput detection and identification of multiple pathogens can be achieved using this method.

Two other distinct advantages of this GeXP-multiplex PCR method are the short assay time and the low cost[32-34]. The entire reaction can be completed in one tube within 2.5 h followed by capillary electrophoresis separation. In addition, two 96-well plates can be placed in parallel in a GeXP analyzer at the same time to further increase the throughput of this method.

The findings described here may lead to increased utilization of these gene-based tests for the routine diagnosis of viral infection. In addition, some of the viruses we detected in the assay may not be pathogenic in animals or could be live virus vaccine strains found in the animal herds. However, the data may still be valuable for epidemiological studies. To make a definite diagnosis, we must consider the clinical symptoms and whether live vaccine was recently injected. Furthermore, we have found that the GeXP analyzer is an efficient and easy-to-use tool; thus it may be a useful new technology for the identification of new biomarkers for diagnosing viral diseases in ducks.

Conclusions

This study has demonstrated that the GeXP-multiplex PCR assay is a rapid method with high sensitivity and specificity for the identification of three very important avian influenza subtypes (H5, H7, H9) in addition to eight other duck pathogens. This method may therefore be adopted for the molecular epidemiologic surveillance of these duck pathogens.

References

[1] ZHANG D. The current epidemic status of duck disease and its prevention and control strategies. China Poultry, 2012, 34: 38-40.

[2] BUTTERFIELD W K, ATA F A, DARDIRI A H. Duck plague virus distribution in embryonating chicken and duck eggs. Avian Dis, 1969, 13 (1): 198-202.

[3] LIU P, LU H, LI S, et al. Duck egg drop syndrome virus: an emerging Tembusu-related flavivirus in China. Sci China Life Sci, 2013, 56(8): 701-710.

[4] REN L, LI J, BI Y, et al. Overview on duck virus hepatitis A. Chinese Journal of Biotechnology, 2012, 28(7): 789-799.

[5] HOU S. The waterfowl industry faced with the challenge in China. China Poultry, 2013, 35(10): 36-67.

[6] PLUMMER P J, ALEFANTIS T, KAPLAN S, et al. Detection of duck enteritis virus by polymerase chain reaction. Avian Dis, 1998, 42(3): 554-564.

[7] HOTTA K, TAKAKUWA H, YABUTA T, et al. Antibody survey on avian influenza viruses using egg yolks of ducks in Hanoi between 2010 and 2012. Vet Microbiol, 2013, 166(1-2): 179-183.

[8] SCHMITZ A, LE BRAS M O, GUILLEMOTO C, et al. Evaluation of a commercial ELISA for H5 low pathogenic avian influenza virus antibody detection in duck sera using Bayesian methods. J Virol Methods, 2013, 193(1): 197-204.

[9] WANG G, QU Y, WANG F, et al. The comprehensive diagnosis and prevention of duck plague in northwest Shandong province of China. Poult Sci, 2013, 92(11): 2892-2898.

[10] MASON R A, TAURASO N M, GINN R K. Growth of duck hepatitis virus in chicken embryos and in cell cultures derived from infected embryos. Avian Dis, 1972, 16(5): 973-979.

[11] ZHANG Y, LI Y, LIU M, et al. Development and evaluation of a VP3-ELISA for the detection of goose and Muscovy duck parvovirus antibodies. J Virol Methods, 2010, 163(2): 405-409.

[12] JI J, XIE Q M, CHEN C Y, et al. Molecular detection of Muscovy duck parvovirus by loop-mediated isothermal amplification assay. Poult Sci, 2010, 89(3): 477-483.

[13] LIU S N, ZHANG X X, ZOU J F, et al. Development of an indirect ELISA for the detection of duck circovirus infection in duck flocks. Vet Microbiol, 2010, 145(1-2): 41-46.

[14] WOŹNIAKOWSKI G, SAMOREK-SALAMONOWICZ E, KOZDRUŃ W. Quantitative analysis of waterfowl parvoviruses in geese and Muscovy ducks by real-time polymerase chain reaction: correlation between age, clinical symptoms and DNA copy number of waterfowl parvoviruses. BMC Vet Res, 2012, 8: 29.

[15] XIE Z, XIE L, XU Z, et al. Identification of a genotype Ⅸ Newcastle disease virus in a Guangxi white duck. Genome Announc, 2013, 1(5): e00836.

[16] XIE L, XIE Z, ZHAO G, et al. A loop-mediated isothermal amplification assay for the visual detection of duck circovirus. Virol J, 2014, 11: 76.

[17] DREW J E, MAYER C D, FARQUHARSON A J, et al. Custom design of a GeXP multiplexed assay used to assess expression profiles of inflammatory gene targets in normal colon, polyp, and tumor tissue. J Mol Diagn, 2011, 13(2): 233-242.

[18] FARQUHARSON A J, STEELE R J, CAREY F A, et al. Novel multiplex method to assess insulin, leptin and adiponectin regulation of inflammatory cytokines associated with colon cancer. Mol Biol Rep, 2012, 39(5): 5727-5736.

[19] RAI A J, KAMATH R M, GERALD W, et al. Analytical validation of the GeXP analyzer and design of a workflow for cancer-biomarker discovery using multiplexed gene-expression profiling. Anal Bioanal Chem, 2009, 393(5): 1505-1511.

[20] QIN M, WANG D Y, HUANG F, et al. Detection of pandemic influenza A H1N1 virus by multiplex reverse transcription-PCR with a GeXP analyzer. J Virol Methods, 2010, 168(1-2): 255-258.

[21] HU X, ZHANG Y, ZHOU X, et al. Simultaneously typing nine serotypes of enteroviruses associated with hand, foot, and mouth disease by a GeXP analyzer-based multiplex reverse transcription-PCR assay. J Clin Microbiol, 2012, 50(2): 288-293.

[22] YANG M J, LUO L, NIE K, et al. Genotyping of 11 human papillomaviruses by multiplex PCR with a GeXP analyzer. J Med Virol, 2012, 84(6): 957-963.

[23] XIE Z, LUO S, XIE L, et al. Simultaneous typing of nine avian respiratory pathogens using a novel GeXP analyzer-based multiplex PCR assay. J Virol Methods, 2014, 207: 188-195.

[24] HU X, XU B, YANG Y, et al. A high throughput multiplex PCR assay for simultaneous detection of seven aminoglycoside-resistance genes in Enterobacteriaceae. BMC Microbiol, 2013, 13: 58.

[25] XIE Z, XIE L, FAN Q, et al. A duplex quantitative real-time PCR assay for the detection of *Haplosporidium* and *Perkinsus* species in shellfish. Parasitol Res, 2013, 112(4): 1597-1606.

[26] PANG Y, WANG H, GIRSHICK T, et al. Development and application of a multiplex polymerase chain reaction for avian respiratory agents. Avian Dis, 2002, 46(3): 691-699.

[27] XIE Z, PANG Y S, LIU J, et al. A multiplex RT-PCR for detection of type A influenza virus and differentiation of avian H5, H7, and H9 hemagglutinin subtypes. Mol Cell Probes, 2006, 20(3-4): 245-249.

[28] HUANG Q, YUE H, ZHANG B, et al. Development of a real-time quantitative PCR for detecting duck hepatitis a virus genotype C. J Clin Microbiol, 2012, 50(10): 3318-3323.

[29] PARK S H, RICKE S C. Development of multiplex PCR assay for simultaneous detection of Salmonella genus, Salmonella subspecies I, Salm. Enteritidis, Salm. Heidelberg and Salm. Typhimurium. J Appl Microbiol, 2015, 118(1): 152-160.

[30] ZENG Z, LIU Z, WANG W, et al. Establishment and application of a multiplex PCR for rapid and simultaneous detection of six viruses in swine. J Virol Methods, 2014, 208: 102-106.

[31] HYMAS W C, MILLS A, FERGUSON S, et al. Development of a multiplex real-time RT-PCR assay for detection of

influenza A, influenza B, RSV and typing of the 2009-H1N1 influenza virus. J Virol Methods, 2010, 167(2): 113-118.

[32] LI J, MAO N Y, ZHANG C, et al. The development of a GeXP-based multiplex reverse transcription-PCR assay for simultaneous detection of sixteen human respiratory virus types/subtypes. BMC Infect Dis, 2012, 12: 189.

[33] LUO S S, XIE Z X, XIE L J, et al. Development of a GeXP assay for simultaneous differentiation of six chicken respiratory viruses. Chinese Journal of Virology, 2013, 29(3): 250-257.

[34] ZHANG Y F, Xie Z X, Xie L J, et al. Development of a duplex RT-PCR assay for detection of duck Tembusu virus and duck plague virus. Journal of Southern Agriculture, 2014, 45: 314-317.

Simultaneous detection of eight swine reproductive and respiratory pathogens using a novel GeXP analyser-based multiplex PCR assay

Zhang Minxiu, Xie Zhixun, Xie Liji, Deng Xianwen, Xie Zhiqin, Luo Sisi, Liu Jiabo, Pang Yaoshan, Khan Mazhar

Abstract

A new high-throughput GenomeLab Gene Expression Profiler (GeXP) analyser-based multiplex PCR assay was developed for the detection of eight reproductive and respiratory pathogens in swine. The reproductive and respiratory pathogens include North American porcine reproductive and respiratory syndrome virus (PRRSV-NA), classical swine fever virus (CSFV), porcine circovirus 2 (PCV-2), swine influenza virus (SIV) (including H1 and H3 subtypes), porcine parvovirus (PPV), pseudorabies virus (PRV) and Japanese encephalitis virus (JEV). Nine pairs of specific chimeric primers were designed and used to initiate PCRs, and one pair of universal primers was used for subsequent PCR cycles. The specificity of the GeXP assay was examined using positive controls for each virus. The sensitivity was evaluated using serial ten-fold dilutions of in vitro-transcribed RNA from all of the RNA viruses and plasmids from DNA viruses. The GeXP assay was further evaluated using 114 clinical specimens and was compared with real-time PCR/single RT-PCR methods. The specificity of the GeXP assay for each pathogen was examined using single cDNA/DNA template. Specific amplification peaks of the reproductive and respiratory pathogens were observed on the GeXP analyser. The minimum copies per reaction detected for each virus by the GeXP assay were as follows: 1 000 copies/μL for PRV; 100 copies/μL for CSFV, JEV, PCV-2 and PPV; and 10 copies/μL for SIV-H1, SIV-H3 and PRRSV-NA. Analysis of 114 clinical samples using the GeXP assay demonstrated that the GeXP assay had comparable detection to real-time PCR/single RT-PCR. This study demonstrated that the GeXP assay is a new method with high sensitivity and specificity for the identification of these swine reproductive and respiratory pathogens. The GeXP assay may be adopted for molecular epidemiological surveys of these reproductive and respiratory pathogens in swine populations.

Keywords

swine reproductive and respiratory, pathogens, GenomeLab Gene Expression Profiler (GeXP), multiplex PCR

Introduction

Reproductive and respiratory viruses are widespread and endemic in swine populations, and they seriously affect the development of the swine industry worldwide. Common epidemic swine reproductive and respiratory pathogens include North American porcine reproductive and respiratory syndrome virus (PRRSV-NA), classical swine fever virus (CSFV), porcine circovirus 2 (PCV-2), swine influenza virus (SIV, including H1 and H3 subtypes), porcine parvovirus (PPV), pseudorabies virus (PRV) and Japanese encephali-tisvirus (JEV). The accurate laboratory diagnosis of these pathogens is essential for disease management and epidemiological surveillance[1]. These reproductive and respiratory viruses cause identical reproductive and/

or respiratory symptoms in swine. For example, PRRSV infection is characterised by abortion and/or stillbirth, and it can be easily confused with CSFV, PPV, PRV and JEV infections, which makes etiological diagnoses difficult. Additionally, CSFV, PRRSV, PCV-2 and PPV infections can damage porcine immune systems to different degrees and can decrease immunity[2], thus causing swine to be easily infected with other pathogens, which in turn makes the clinical signs more severe. Clinical signs are variable and can be complicated by multiple infections, thereby leading to difficulty in determining the causal agent based on clinical signs. Thus, in some cases, a method to differentiate these reproductive and respiratory pathogens is important.

Conventional methods to diagnose and differentiate these viruses include viral isolation, immunofluorescence staining and PCR methods. Viral culture is the gold standard for the detection of infectious viral diseases in swine, but this process is time-consuming and has a low sensitivity for the detection of these viruses[3]. Immunofluorescence staining is a rapid testing method. However, the availability and sensitivity of antisera for immunofluorescence staining are limiting factors[4]. Molecular assays have been developed for the rapid identification of reproductive and respiratory viruses, including reverse transcriptase (RT)-PCR, real-time RT-PCR, nested-PCR, duplex PCR and triplex PCR[5-8]. These molecular methods are sensitive and rapidly identify reproductive and respiratory viruses. However, real-time RT-PCR, nested-PCR, duplex PCR and triplex PCR are used for the detection of only a limited number of viruses. Gene microarrays can simultaneously detect several viruses, but they are expensive and time-consuming. Therefore, the development of a rapid, high-throughput and cost-effective method for the simultaneous detection of reproductive and respiratory viruses is necessary.

The Beckman Coulter Company developed the GenomeLabGene Expression Profiler (GeXP) as a multiplex gene expression profiling analysis platform. The GeXP-multiplex PCR assay uses two sets of primers for multiple PCR reactions, including universal primers and specific chimeric primers, and the PCR products are separated using capillary electrophoresis technology. GeXP-multiplex PCR assays have been used for the rapid identification of human and animal diseases[9, 10]. The GeXP analyser-based multiplex PCR assay (GeXP-multiplex PCR assay) was developed and applied in this study for the simultaneous detection of eight reproductive and respiratory pathogens, including PRRSV-NA, CSFV, PCV-2, SIV subtype H1, SIV subtype H3, PPV, JEV and PRV. The specificity and sensitivity of the GeXP assay were evaluated, and 114 clinical samples collected from pig farms were assayed for virological detection using the GeXP assay. The results using this method were then compared with results using the single RT-PCR/real-time PCR assay.

Material and methods

Virus isolates and clinical samples

The eight swine reproductive and respiratory pathogens used in this study are listed in Table 5-7-1. During the period from April 2012 to October 2014, 114 clinical specimens (lungs, lymph nodes, spleens, kidneys and aborted foetuses) were collected with the consent of swine farmers from 21 farms in 5 cities of the Guangxi.

Table 5-7-1 Sources of the pathogens used and the viruses detected using the GeXP-multiplex PCR assay

Pathogens/field samples	Numbers of sample	Source	M gene	H1 SIV	H3 SIV	PRRSV-NA	CSFV	JEV	PCV-2	PPV	PRV
H1N1 swine/GX/BB1/2013	1	GVRI	+	+	−	−	−	−	−	−	−
cDNA of H1N2 SIV	1	GU	+	+	−	−	−	−	−	−	−
cDNA of H3N2 SIV	1	GU	+	−	+	−	−	−	−	−	−

continued

Pathogens/field samples	Numbers of sample	Source	M gene	H1 SIV	H3 SIV	PRRSV-NA	CSFV	JEV	PCV-2	PPV	PRV
PRRSV-NA	1	CIVDC	–	–	–	+	–	–	–	–	–
cDNA of PRRSV-NA	1	GU	–	–	–	+	–	–	–	–	–
CSFV (AV281)	1	CIVDC	–	–	–	–	+	–	–	–	–
JEV (20060416)	1	CEC	–	–	–	–	–	+	–	–	–
JEV (GX/FC792)	1	GVRI	–	–	–	–	–	+	–	–	–
PCV-2 (Guangxi isolates)	5	GVRI	–	–	–	–	–	–	+	–	–
PPV (N strain)	1	GVRI	–	–	–	–	–	–	–	+	–
PRV (Guangxi isolates)	1	GVRI	–	–	–	–	–	–	–	–	+
PRV (AV24, AV25)	2	CIVDC	–	–	–	–	–	–	–	–	+

Note: GVRI, Guangxi Veterinary Research Institute, China; CIVDC, China Institute of Veterinary Drugs Control, China; GU, Guangxi University. China; CEC, China Animal Health & Epidemiology Center.

Primers

Nine pairs of gene-specific primers were designed using Primer Premier 5.0 software and the Primer designing tool (Table 5-7-2). Each gene-specific primer was fused at the 5' end to a universal sequence to generate 9 pairs of chimeric primers, and one pair of universal primers was used for RT-PCR (Table 5-7-2). The forward universal primer was Cy5-labeled at the 5' end of the sequence. SIV universal primers were designed in the conserved region of the matrix (M) gene. SIV primers for the H1 and H3 subtypes were designed in a specific region of the HA gene. PRRSV-NA, CSFV, JEV, PCV, PPV and PRV primers were designed in a region of the Nsp2, E2, PrM, ORF1, VP2 and gD genes, respectively. All primers were synthesised and purified using polyacrylamide gel electrophoresis (Invitrogen, Guangzhou, China).

Table 5-7-2 Primer information and tissue tropisms of swine reproductive and a respiratory pathogens

Pathogens	Tissue tropism	Primer name	Primer sequence (5'-3')	Gene	Size/bp
SIV	Tracheal epithelium and lungs	SIV-M-F	AGGTGACACTATAGAATATCAAAGCCGAGATCGCACAG	M	181
		SIV-M-R	GTACGACTCACTATAGGGACGGTGAGCGTGAACACAA		
SIV-H1	Tracheal epithelium and lungs	SIV-H1-F	AGGTGACACTATAGAATAGGCTTTATTGAAGGRGGDTGGAC	HA	121
		SIV-H1-R	GTACGACTCACTATAGGGAGCATATCCTGATCCCTGYTCATT		
SIV-H3	Tracheal epithelium and lungs	SIV-H3-F	AGGTGACACTATAGAATACACTCAAGCAGCCATCRACC	HA	325
		SIV-H3-R	GTACGACTCACTATAGGGACCATTGCCCATGTCCTCAGC		
PRRSV-NA	Lungs, lymph nodes, blood and tonsils	PRRSV-NA-F	AGGTGACACTATAGAATAAGCGGAGGCTGCAAGTTAAT	Nsp2	260
		PRRSV-NA-R	GTACGACTCACTATAGGGACACTTGTGACTGCCAAACCG		
CSFV	Lungs, lymph nodes, spleen, blood, kidneys and tonsil	CSFV-F	AGGTGACACTATAGAATACCCAGGGACAGCTATTTCCA	E2	159
		CSFV-R	GTACGACTCACTATAGGGAGCCACTACCACCAAGACAACAA		
JEV	Boar seminal fluid and abortion fetus	JEV-F	AGGTGACACTATAGAATAGGATGTGGATTGCTGGTGTG	PrM	303
		JEV-R	GTACGACTCACTATAGGGACGTTGACCGTTGTTACTGCC		
PCV-2	Lungs, lymph nodes, blood and spleens	PCV-F	AGGTGACACTATAGAATAGACGAACACCTCACCTCCA	ORF1	274
		PCV-R	GTACGACTCACTATAGGGAGACTCCCGCTCTCCAACAA		
PPV	Lungs, kidneys, lymph nodes, blood and spleens	PPV-F	AGGTGACACTATAGAATAAGCAACCTCACCACCAACCAA	VP2	139
		PPV-R	GTACGACTCACTATAGGGAGGTGCTGCTGGTGTGTATGGAA		

continued

Pathogens	Tissue tropism	Primer name	Primer sequence (5'-3')	Gene	Size/ bp
PRV	Brains, lungs, kidneys and tonsils	PRV-F	AGGTGACACTATAGAATAGAATAAACATCCTCACCGACTTC	gD	227
		PRV-R	GTACGACTCACTATAGGGAAGTAGTTCACCACCTCCC		
Universal-primers		Cy5-Tag-F	AGGTGACACTATAGAATA		
		Tag-R	GTACGACTCACTATAGGGA		

Note: Universal tag sequences are underlined. Abbreviations are as follows: Y=C or T; R=A or G; D=A or G or T.

Evaluation of GeXP assay specificity

RNA and DNA were extracted from 200 μL of cell-cultured isolates and clinical samples using a TranseGen Biotech EasyPure Viral DNA/RNA kit (TransGen, Beijing, China) according to the manufacturer's instructions. RNA and DNA were stored at 80 ℃ until use.

The extracted RNAs from SIV-H1, SIV-H3, PRRSV-NA, CSFV and JEV were reverse transcribed into cDNA as described previously[11]. The GeXP-mono PCR assay and the GeXP-multiplex PCR assay were developed using the Sigma PCR kit (Sigma, Shanghai, China). The GeXP-mono PCR assay was performed using a single template and each pair of chimeric primers to determine the size of the amplification product for each target gene. The specificity of the GeXP assay was tested individually in the GeXP-multiplex PCR assay for all viral targets. The GeXP-mono PCR assay and the GeXP-multiplex PCR assay were performed in a 25 μL volume containing 2.5 μL of 10 × PCR buffer, 2.5 μL of $MgCl_2$ (25 mmol/L), 2 μL of dNTP mix (2.5 mmol/L), 1.2 μL of JumpStart Taq DNA Polymerase (2.5 units/μL), 10 μL of cDNA or DNA template and 6. 8 μL of distilled water. The GeXP-mono PCR contained a 50 nM final concentration of each pair of chimeric primers, and the GeXP-multiplex PCR contained 50 nM final concentrations of 9 pairs of chimeric primers. The GeXP-mono PCR and GeXP-multiplex PCR both contained 500 nM final concentrations of the universal tag primers. The GeXP-mono PCR and the GeXP-multiplex PCR mixture were amplified in the following three steps: step (1) 10 cycles of 95 ℃ for 30 s, 55 ℃ for 30 s, and 72 ℃ for 30 s; step (2) 10 cycles of 95 ℃ for 30 s, 63 ℃ for 30 s, and 72 ℃ for 30 s; and step (3) 20 cycles of 95 ℃ for 30 s, 48 ℃ for 30 s, and 72 ℃ for 30 s.

Separation using capillary electrophoresis (CE) and fragment analysis

PCR products were analysed in a GenomeLab GeXP genetic analysis system (Brea, CA, USA) using capillary electrophoresis according to previously reported protocols[12, 13]. All of the amplified products were ascertained to be positive according to the fluorescence values following the positive cut-off previously described[14].

Evaluation of GeXP assay sensitivity

The recombinant plasmids of nine target genes were constructed, and six recombinant plasmids containing the M gene of SIV, HA genes (H1 and H3) of SIV, Nsp2 gene of PRRSV, E2 gene of CSFV, and PrM gene of JEV were transcribed into synthetic RNAs in vitro as described previously[10]. An additional three recombinant plasmids, including PCV, PPV and PRV (which are DNA viruses and do not need to be transcribed to synthetic ssRNAs in vitro), were purified using a TransGen Plasmid Miniprep Kit (TransGen, Beijing, China) to directly evaluate sensitivity. The six ssRNAs transcribed in vitro were purified using a Promega Mag-Bind RNA cleanup kit (Promega, Madison, Wisconsin, USA). The sensitivity of GeXP was evaluated with a One Step RT-PCR kit (TaKaRa, Dalian, China) in a 25 μL volume containing 12.5 μL of 2 × 1 step buffer; 2 μL of RT-PCR enzyme mix; the final concentration of the specific chimeric primers and universal tag primers as used in the GeXP-multiplex PCR assay (as described Section *Evaluation of GeXP assay specificity*); and 2 μL of template and RNase-free water. The sensitivity of the optimised GeXP-multiplex PCR assay for the simultaneous detection of nine genes was reevaluated using three premixed recombinant plasmids and six RNA templates ranging from 10^4 copies/μL to 10 copies/μL after quantitation using a NanoDrop 2 000 (Thermo Fisher Scientific, Shanghai, China). The RT-PCR was tested under the following the conditions: 50 ℃ for 30 and 95 ℃ for 3 min followed by the three steps as described Section *Evaluation of GeXP assay specificity*. Each template concentration was tested in triplicate.

Application to clinical samples

A total of 114 clinical tissue samples (spleens, lungs, kidneys, lymph nodes and aborted foetuses) were collected from 21 swine farms in 5 cities of the Guangxi Province. Individual samples (500 mg) were taken from the lung, lymph node, spleen and kidney, and these samples were mixed and ground into homogenates. Nucleic acids were extracted from the mixed tissue homogenate samples and tested using both the GeXP-multiplex PCR and single RT-PCR/real-time PCR assays. These PCR assays included real-time PCR for the detection of CSFV, PCV-2, PPV, PRRSV, PRV, swine influenza virus subtype H1 and swine influenza subtype H3[15-22] and a single RT-PCR for the detection of JEV[23]. All of the positive clinical samples detected by the GeXP assay were confirmed by DNA sequencing of single RT-PCR/PCR products using the same specific chimeric primer of the GeXP assay. The detection of clinical tissue samples in this study is described in Table 5-7-3.

Table 5-7-3 Performance of the GeXP-multiplex PCR assay compared to the real-time PCR/single RT-PCR assay *

Virus	GeXP assay	Real-time PCR/single RT-PCR methods		Performance of the GeXP assay		Measures of agreement Kappa values
		Positive	Negative	Sensitivity/%	Specificity/%	
SIV[a]	Positive	25	0	100	100	1 (P<0.001)
	Negative	0	89			
SIV-H1[a]	Positive	23	0	95. 8	100	0.97 (P<0.001)
	Negative	1	90			
SIV-H3[a]	Positive	1	0	100	100	1 (P<0.001)
	Negative	0	113			
PRRSV-NA[b]	Positive	58	0	100	100	1 (P<0.001)
	Negative	0	56			
CSFV[b]	Positive	34	0	100	100	1 (P<0.001)
	Negative	0	80			
PCV-2[b]	Positive	76	0	100	100	1 (P<0.001)
	Negative	0	38			
PPV[b]	Positive	23	0	100	100	1 (P<0.001)
	Negative	0	91			
PRV[b]	Positive	5	0	100	100	1 (P<0.001)
	Negative	0	109			
JEV[c]	Positive	3	0	100	100	1 (P<0.001)
	Negative	0	111			

Note: [a] Confirmed by real-time PCR detection of swine influenza virus H1 and H3 subtypes[15, 19, 20]; [b] Confirmed by real-time PCR detection of PCV-2, PPV, CSFV, PRRSV and PRV[16-18, 21, 22]; [c] Confirmed by a single RT-PCR detection of JEV[23]; * The numbers of positive and negative samples detected by either testing method are shown; sensitivity and specificity indicate true positive and true negative results as determined by real-time PCR/single RT-PCR assay.

Results

Evaluation of GeXP assay specificity

DNA/cDNA samples from eight reproductive and respiratory viruses were used as templates to evaluate the specificity of each pair of gene-specific primers. The SIV universal primers amplified the target M genes of all SIV subtypes (H1 and H3 subtypes) in the GeXP-mono PCR assays, but each pair of the other specific primers only generated the corresponding genes of the targeted pathogens without cross-amplification. The amplicon sizes for the pathogens conformed with the target sequences (Table 5-7-2). The specificity of the GeXP assay was evaluated using the GeXP-multiplex PCR assay. All viral targets were tested individually in the GeXP-multiplex PCR assay, and specific amplification peaks were observed when each pathogen was tested as follows (Figure 5-7-1 A-H): SIV-M and SIV-H1, 183.04 bp and 120.91 bp, respectively; SIV-M and SIV-H3, 183.01 bp and 324.45 bp, respectively; PRRSV-NA, 262.22 bp; CSFV, 161.59 bp; JEV, 303. 85 bp; PCV-2 275.17 bp; PPV, 138.49 bp; and PRV, 226.75 bp. SIV universal primers were designed for the detection of the SIV H1 and H3 subtypes, and specific amplification peaks of the SIV M gene were observed when SIV H1N1, H1N2, and H3N2 were tested by GeXP-multiplex PCR (Figure 5-7-1 A and B).

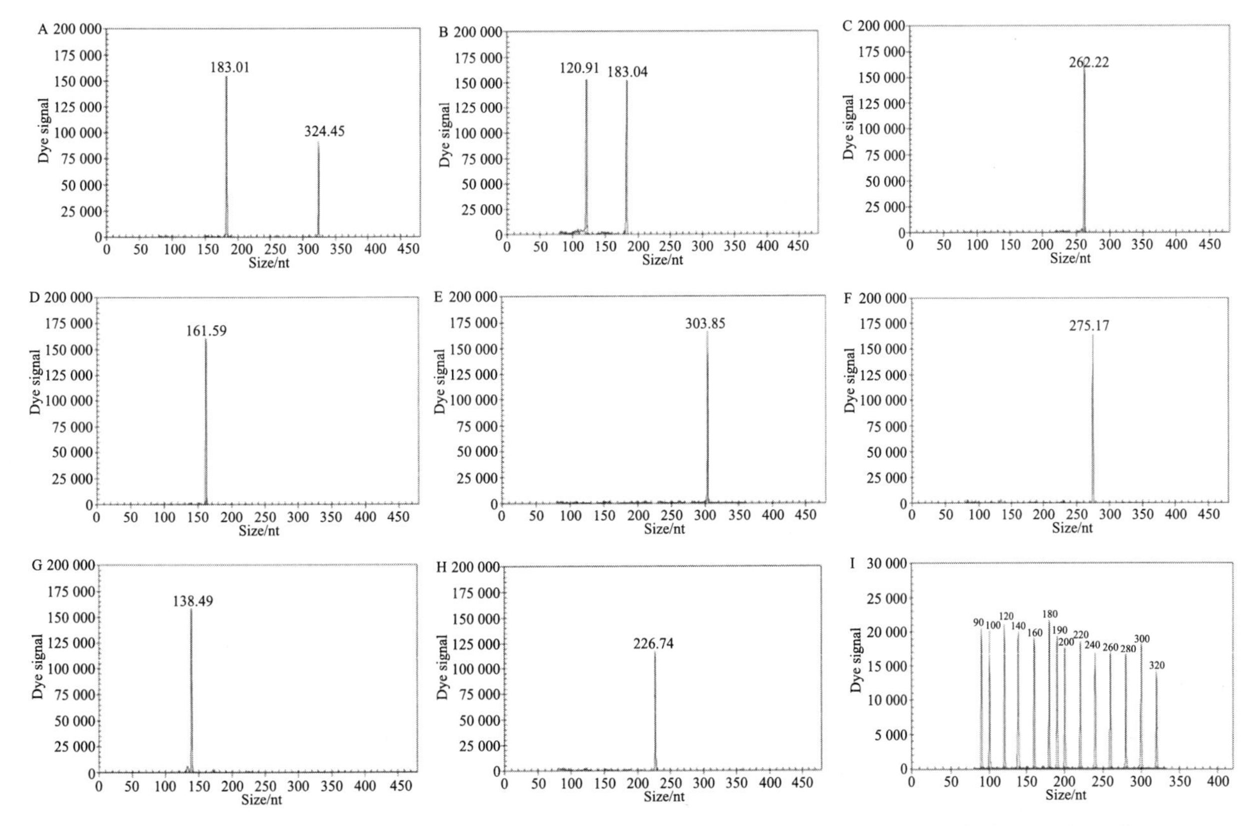

A-H: The results of the amplification of SIV (H3), SIV (H1), PRRSV, CSFV, JEV, PCV-2, PPV and PRV, respectively; I: Nuclease-free water was used as the negative control. The peaks indicate the target genes and the DNA size standard.

Figure 5-7-1 Specificity results of the GeXP-multiplex PCR assay and GeXP-multiplex PCR detection for a mixture of eight pathogen templates

Evaluation of GeXP assay sensitivity

The sensitivity of the GeXP-multiplex PCR assay for each reproductive and respiratory virus was examined individually using serial 10-fold dilutions from 10^3 to 10 copies/μL of RNA transcribed in vitro from all RNA viruses and plasmids encoding DNA viruses. The detection limit of each virus was 10 copies/μL per reaction in the multiplex assay. The sensitivity of the GeXP-multiplex PCR assay was also examined using nine premixed samples with the same copies of each virus. The detection limits in the GeXP-multiplex assay were as follows: 1 000 copies/μL for PRV; 100 copies/μL for CSFV, JEV, PCV-2 and PPV; and 10 copies/μL for SIV-H1, SIV-H3 and PRRSV-NA (Figure 5-7-2). The coefficients of variation (CVs) for each concentration of the template were less than 10%.

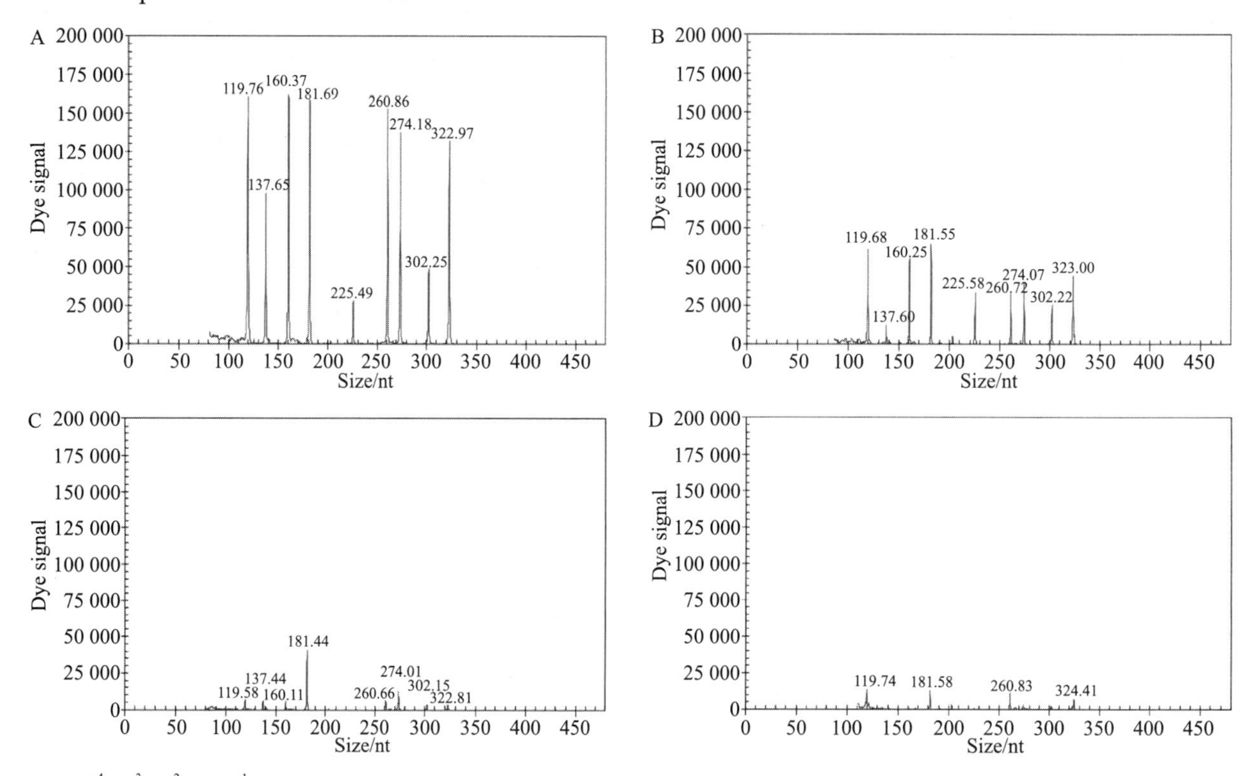

A-D: 10^4, 10^3, 10^2 and 10^1 copies of synthetic RNAs and plasmids of per reaction in the GeXP-multiplex assay. All 9 target genes were detected simultaneously at levels of 10^4 and 10^3 copies/μL. SIV-H1, SIV-H3, PRRSV-NA, CSFV, JEV, PCV-2 and PPV could be detected at a level of 100 copies/μL. SIV-H1, SIV-H3 and PRRSV-NA could be detected at a level of 10 copies/μL, the peaks denote specific amplification peaks.

Figure 5-7-2 Sensitivity results of the GeXP-multiplex PCR detection of the eight pathogen templates

Application to clinical samples

The isolates used in this study were detected using the GeXP assay, and 114 clinical samples were assayed using the optimised GeXP-multiplex PCR (Tables 5-7-1 and 5-7-3) and confirmed by DNA sequencing of single RT-PCR/PCR products using the same 9 sets of specific primers employed in the GeXP assay. The assay successfully detected and differentiated the 8 reproductive and respiratory pathogens. Compared with the results of the real-time PCR/single RT-PCR assay, the sensitivities of the GeXP assay for the detection of SIV, SIV-H3, PRRSV-VA, CSFV, PCV-2, PPV, PRV and JEV were all 100%, and the sensitivity for the detection of SIV-H1 was 95. 8%. The specificities were all 100% (Table 5-7-3).

Discussion

Reproductive and respiratory diseases cause significant economic losses in the swine industry, and the aetiologies of reproductive and respiratory diseases are complex[1]. It is common for swine to be simultaneously infected with two or more viral pathogens under typical conditions of intensive swine production, which can make illnesses more complicated and serious[6]. Furthermore, multiple infections are often difficult to diagnose, especially for PRRSV and CFSV, which have some clinical signs in common. Therefore, a high-throughput, reliable and rapid detection method for these viruses is essential for molecular epidemiological surveillance and disease management.

Recent reports have demonstrated the application of single PCR, single real-time PCR, multiplex real-time PCR and multiplex conventional PCR for the detection of SIV, PRRSV, CSFV, and PPV in one reaction[1, 5, 6, 11, 19, 24]. A maximum of six pathogens can be tested in a multiple setting using multiplex PCR[25]. The GeXP genetic analysis system used in this study is a new multitarget, high-throughput detection platform that has been successfully applied for the detection of tumour markers in the mechanism of drug toxicology and the expression of various varicella zoster virus infection genes[9, 26, 27]. In this study, 9 pairs of specific primers were designed to develop a GeXP-multiplex PCR assay based on multiplex PCR for simultaneous detection of SIV (including H1 and H3), PRRSV-NA, CSFV, JEV, PCV-2, PPV and PRV. The results of the specificity of the GeXP-multiplex PCR assay and GeXP-multiplex PCR detection with mixed pathogen templates suggested that eight reproductive and respiratory viruses were simultaneously detected using the GeXP-multiplex PCR assay without cross-amplification.

In general, multiplex PCR assays need to be optimised. The most common problem is that the primers used in the same reaction may interact with each other and block the amplification by different primers, which can significantly alter the sensitivity[6, 28]. There are three steps in the PCR amplification process in the GeXP-multiplex PCR assay. Two steps are needed for the formation of specific chimeric primers, which include forward and reverse primers, and one step involves universal forward and reverse tag primers[12, 29]. Therefore, the GeXP-multiplex PCR assay was primarily guided by fluorescent dye-labelled universal tag primer amplification to reduce the interference among the primers, inferior amplification and non-specific reactions[12]. The minimum copies per reaction detected by the GeXP-multiplex PCR assay in this study ranged from 10 to 100 copies/μL for CSFV, JEV, PCV-2, PPV, SIV-H1, SIV-H3 and PRRSV-NA, thereby indicating the high sensitivity of the GeXP-multiplex PCR assay and suggesting that the GeXP-multiplex PCR and real-time PCR/single RT-PCR assays exhibited comparable detection sensitivities and minimised interference among primers. When 114 specimens were analysed using GeXP-multiplex PCR, the sensitivities of the GeXP assay for the detection of SIV, SIV-H3, PRRSV-VA, CSFV, PCV-2, PPV, PRV and JEV were all 100%, and the sensitivity for SIV-H1 was 95. 8%. The specificities were all 100%, thus revealing the high sensitivity and specificity in the detection of these viruses. The kappa correlation between the GeXP-multiplex PCR and real-time PCR/single RT-PCR was >0.75, thereby revealing a comparable detection of these viruses with both assays.

Additionally, the detection rate for SIV-H1 using the GeXP-multiplex PCR assay exceeded the detection rate by real-time PCR, and the true positives confirmed by independent RT-PCR and sequencing suggested that the GeXP-multiplex PCR assay was more sensitive than the real-time PCR assay for detection. Because of the limited number of positive samples for the other three pathogens (SIV-H3, PRV and JEV), the GeXP-multiplex PCR assay needs to be validated with a larger number of clinical samples for these viruses.

With the development of the swine industry and the increasing complexity of the infectious disease profile in the swine population, infections with PCV-2, PPV, PRRSV-NA, CSFV, PRV, SIV and JEV have caused large economic losses. In the present study, the developed PCV-2, PPV, PRRSV-NA, CSFV, PRV, SIV and JEV GeXP-multiplex PCR diagnostic method can accurately detect these viruses, including live virus vaccine strains recently used in the animal herds, so the detection of clinical samples using this method must be diagnosed accurately through comprehensive analysis. In addition, to avoid bias, diagnosis should be made by evaluating results of molecular and/or serological tests together with clinical symptoms. This approach also has a few disadvantages. First, most swine respiratory and reproductive pathogens can replicate in more than one organ and this study used pooled tissue samples in the GeXP assay with the purpose of detecting pathogens infecting different tissues and organs, but PRV, JEV and SIV cannot be detected in blood samples using this assay, which could easily result in false negatives; therefore, clinical samples from pigs should be collected according to clinical symptoms and pathological changes. Second, the amplicon sizes for the pathogens are strictly limited to 100～400 bp, and a difference of at least 7 bp between amplicons is necessary to allow the GenomeLab GeXP genetic analysis system to distinguish the target sequences. Third, the GenomeLab GeXP genetic analysis system contains multiple steps, and artificial errors are easily produced. Therefore, modification of the procedures to reduce the number of steps is required.

Conclusions

This study has demonstrated that the GeXP assay is a sensitive and high-throughput method for the detection of multiple reproductive and respiratory virus infections, which may be adopted for molecular diagnosis of these reproductive and respiratory viruses.

References

[1] JIANG Y, SHANG H, XU H, et al. Simultaneous detection of porcine circovirus type 2, classical swine fever virus, porcine parvovirus and porcine reproductive and respiratory syndrome virus in pigs by multiplex polymerase chain reaction. Vet J, 2010, 183(2): 172-175.

[2] CHEN H Y, WEI Z Y, ZHANG H Y, et al. Use of a Multiplex RT-PCR assay for simultaneous detection of the North American genotype porcine reproductive and respiratory syndrome virus, swine influenza virus and Japanese encephalitis virus. Agric Sci China, 2010, 9(7): 1050-1057.

[3] LOPEZ ROA P, CATALAN P, GIANNELLA M, et al. Comparison of real-time RT-PCR, shell vial culture, and conventional cell culture for the detection of the pandemic influenza A (H1N1) in hospitalized patients. Diagn Microbiol Infect Dis, 2011, 69(4): 428-431.

[4] HENRICKSON K J. Advances in the laboratory diagnosis of viral respiratory disease. Pediatr Infect Dis J, 2004, 23(1 Suppl): S6-10.

[5] PANG Y, WANG H, GIRSHICK T, et al. Development and application of a multiplex polymerase chain reaction for avian respiratory agents. Avian Dis, 2002, 46(3): 691-699.

[6] GIAMMARIOLI M, PELLEGRINI C, CASCIARI C, et al. Development of a novel hot-start multiplex PCR for simultaneous detection of classical swine fever virus, African swine fever virus, porcine circovirus type 2, porcine reproductive and respiratory syndrome virus and porcine parvovirus. Vet Res Commun, 2008, 32(3): 255-262.

[7] MCKILLEN J, HJERTNER B, MILLAR A, et al. Molecular beacon real-time PCR detection of swine viruses. J Virol Methods, 2007, 140(1-2): 155-165.

[8] XIE Z, XIE L, FAN Q, et al. A duplex quantitative real-time PCR assay for the detection of *Haplosporidium* and *Perkinsus* species in shellfish. Parasitol Res, 2013, 112(4): 1597-1606.

[9] QIN M, WANG D Y, HUANG F, et al. Detection of pandemic influenza A H1N1 virus by multiplex reverse transcription-PCR with a GeXP analyzer. J Virol Methods, 2010, 168(1-2): 255-258.

[10] XIE Z, LUO S, XIE L, et al. Simultaneous typing of nine avian respiratory pathogens using a novel GeXP analyzer-based multiplex PCR assay. J Virol Methods, 2014, 207: 188-195.

[11] XIE Z, XIE L, PANG Y, et al. Development of a real-time multiplex PCR assay for detection of viral pathogens of penaeid shrimp. Arch Virol, 2008, 153(12): 2245-2251.
[12] HU X, ZHANG Y, ZHOU X, et al. Simultaneously typing nine serotypes of enteroviruses associated with hand, foot, and mouth disease by a GeXP analyzer-based multiplex reverse transcription-PCR assay. J Clin Microbiol, 2012, 50(2): 288-293.
[13] YANG M J, LUO L, NIE K, et al. Genotyping of 11 human papillomaviruses by multiplex PCR with a GeXP analyzer. J Med Virol, 2012, 84 (6): 957-963.
[14] LI J, MAO N Y, ZHANG C, et al. The development of a GeXP-based multiplex reverse transcription-PCR assay for simultaneous detection of sixteen human respiratory virus types/subtypes. BMC Infect Dis, 2012, 12: 189.
[15] RICHT J A, LAGER K M, CLOUSER D F, et al. Real-time reverse transcription-polymerase chain reaction assays for the detection and differentiation of North American swine influenza viruses. J Vet Diagn Invest, 2004, 16(5): 367-373.
[16] YOON H A, EO S K, ALEYAS A G, et al. Molecular survey of latent pseudorabies virus infection in nervous tissues of slaughtered pigs by nested and real-time PCR. J Microbiol, 2005, 43(5): 430-436.
[17] RISATTI G, HOLINKA L, LU Z, et al. Diagnostic evaluation of a real-time reverse transcriptase PCR assay for detection of classical swine fever virus. J Clin Microbiol, 2005, 43(1): 468-471.
[18] YANG Z Z, HABIB M, SHUAI J B, et al. Detection of PCV2 DNA by SYBR Green I-based quantitative PCR. J Zhejiang Univ Sci B, 2007, 8(3): 162-169.
[19] CARR M J, GUNSON R, MACLEAN A, et al. Development of a real-time RT-PCR for the detection of swine-lineage influenza A (H1N1) virus infections. J Clin Virol, 2009, 45(3): 196-199.
[20] LIU S, HOU G, ZHUANG Q, et al. A SYBR Green I real-time RT-PCR assay for detection and differentiation of influenza A (H1N1) virus in swine populations. J Virol Methods, 2009, 162(1-2): 184-187.
[21] LI Y D, YU Z D, BAI C X, et al. Development of a SYBR Green I real-time PCR assay for detection of novel porcine parvovirus 7. Pol J Vet Sci, 2021, 24(1): 43-49.
[22] XIAO S, CHEN Y, WANG L, et al. Simultaneous detection and differentiation of highly virulent and classical Chinese-type isolation of PRRSV by real-time RT-PCR. J Immunol Res, 2014, 2014: 809656.
[23] SAPKAL G N, WAIRAGKAR N S, AYACHIT V M, et al. Detection and isolation of Japanese encephalitis virus from blood clots collected during the acute phase of infection. Am J Trop Med Hyg, 2007, 77(6): 1139-1145.
[24] XIE Z, PANG Y S, LIU J, et al. A multiplex RT-PCR for detection of type A influenza virus and differentiation of avian H5, H7, and H9 hemagglutinin subtypes. Mol Cell Probes, 2006, 20(3-4): 245-249.
[25] ZENG Z, LIU Z, WANG W, et al. Establishment and application of a multiplex PCR for rapid and simultaneous detection of six viruses in swine. J Virol Methods, 2014, 208: 102-106.
[26] NAGEL M A, GILDEN D, SHADE T, et al. Rapid and sensitive detection of 68 unique varicella zoster virus gene transcripts in five multiplex reverse transcription-polymerase chain reactions. J Virol Methods, 2009, 157(1): 62-68.
[27] RAI A J, KAMATH R M, GERALD W, et al. Analytical validation of the GeXP analyzer and design of a workflow for cancer-biomarker discovery using multiplexed gene-expression profiling. Anal Bioanal Chem, 2009, 393(5): 1505-1511.
[28] ELNIFRO E M, ASHSHI A M, COOPER R J, et al. Multiplex PCR: optimization and application in diagnostic virology. Clin Microbiol Rev, 2000, 13(4): 559-570.
[29] TABONE T, MATHER D E, HAYDEN M J. Temperature switch PCR (TSP): Robust assay design for reliable amplification and genotyping of SNPs. BMC Genomics, 2009, 10: 580.

Development of a GeXP-multiplex PCR assay for the simultaneous detection and differentiation of six cattle viruses

Fan Qing, Xie Zhixun, Xie Zhiqin, Deng Xianwen, Xie Liji, Huang Li, Luo Sisi, Huang Jiaoling, Zhang Yanfang, Zeng Tingting, Wang Sheng, Liu Jiabo, and Pang Yaoshan

Abstract

Foot-and-mouth disease virus (FMDV), bluetongue virus (BTV), vesicular stomatitis virus (VSV), bovine viral diarrheal (BVDV), bovine rotavirus (BRV), and bovine herpesvirus 1 (IBRV) are common cattle infectious viruses that cause a great economic loss every year in many parts of the world. A rapid and high-throughput GenomeLab Gene Expression Profiler (GeXP) analyzer-based multiplex PCR assay was developed for the simultaneous detection and differentiation of these six cattle viruses. Six pairs of chimeric primers consisting of both the gene-specific primer and a universal primer were designed and used for amplification. Then capillary electrophoresis was used to separate the fluorescent labeled PCR products according to the amplicons size. The specificity of GeXP-multiplex PCR assay was examined with samples of the single template and mixed template of six viruses. The sensitivity was evaluated using the GeXP-multiplex PCR assay on serial 10-fold dilutions of ssRNAs obtained via in vitro transcription. To further evaluate the reliability, 305 clinical samples were tested by the GeXP-multiplex PCR assay. The results showed that the corresponding virus specific fragments of genes were amplified. The detection limit of the GeXP-multiplex PCR assay was 100 copies/μL in a mixed sample of ssRNAs containing target genes of six different cattle viruses, whereas the detection limit for the Gexp-mono PCR assay for a single target gene was 10 copies/μL. In detection of viruses in 305 clinical samples, the results of GeXP were consistent with simplex real-time PCR. Analysis of positive samples by sequencing demonstrated that the GeXP-multiplex PCR assay had no false positive samples of nonspecific amplification. In conclusion, this GeXP-multiplex PCR assay is a high throughput, specific, sensitive, rapid and simple method for the detection and differentiation of six cattle viruses. It is an effective tool that can be applied for the rapid differential diagnosis of clinical samples and for epidemiological investigation.

Keywords

Gexp-multiplex PCR, FMDV, BTV, VSV, BVDV, BRV, IBRV

Introduction

Foot-and-mouth disease virus (FMDV), bluetongue virus (BTV), vesicular stomatitis virus (VSV), bovine viral diarrheal virus (BVDV), bovine rotavirus (BRV), and bovine herpesvirus 1 (IBRV) are common cattle infectious viruses[1, 2]. It was reported recently that these infectious diseases increased the beef cattle mortality to 5% and lead to economic losses estimated at $200 billion during 2015 in China[3]. These viral diseases in cattle exhibit similar clinical symptoms and are difficult to differentiate from each other. FMDV, VSV, BTV, BVDV and IBRV infections display skin lesions of various degrees, including vesicular lesions, erythema, skin cracking, and necrosis on the mouth, feet, noses, cunnus and teats etc[4-10]. Moreover, FMDV, VSV, and BTV are listed on The World Organization for Animal Health (OIE) Terres-trial Animal Health Code and countries are obligated to report these diseases to OIE[11]. Bovine rotavirus infection in young cattle less than 6 months old may not show typical symptoms as of BVDV. Both BRV and BVDV are frequently exhibits acute watery diarrhea and emaciation[12, 13]. Therefore, it is important to diagnose and differentiate these infectious viruses accurately for the rapid control and prevention strategies[14].

Currently, OIE recommends antigen capture ELISA, virus isolation, and PCR, including real-time PCR, for the laboratory diagnosis of these viruses. However, PCR with low sensitivity and real-time PCR with limited plexity. The GenomeLab Gene Expression Profiler (GeXP) analyzer is a multiplex gene expression analysis platform that integrates PCR with capillary electrophoresis separation based on the size of the amplified products, and was designed to allow for the high-throughput, robust and differential assessment of multiplexed expression profile of up to 30 genes in one tube[15-18]. GeXP multiplex PCR assay has been successfully used for the rapid identification and differentiation of several animal infectious diseases[19-23]. In this study, a GeXP analyzer-based multiplex PCR assay was developed for the specific detection of six cattle infectious viruses: FMDV, BTV, VSV, BVDV, BRV and IBRV so that the assay can be applied for rapid differential diagnosis of these viral agents from clinical samples to adopt preventive and control measures against the cattle infectious diseases.

Materials and methods

Pathogens and DNA/RNA extraction

The pathogens used in this study were listed in Table 5-8-1. The genomic DNA of bacteria and mycoplasma strains were extracted from culture by using MiniBEST Universal Genomic DNA Extraction Kit (TaKaRa, Dalian, China) according to the manufacturer's protocol. Each virus's genomic RNA was extracted from 200 μL of virus suspension or clinical samples using MiniBEST Universal RNA Extraction Kit (TaKaRa, Dalian, China) according to the manual. The extracted DNA/RNA were eluted in 30 μL of distilled water. The RNA was synthesized to cDNA via reverse transcription using the PrimerScript™ cDNA Synthesis Kit (TaKaRa, Dalian, China) with random primers (Nona-deoxyribonucleotide mixture) according to the manual, then quantified at 260 nm using a Nano Drop 2 000 (Thermo Fisher Scientific, Waltham, USA). All the DNA/RNA were stored at −70 ℃ until used.

Table 5-8-1 Pathogens used and GeXP assay results

Pathogen	Source	GeXP Results					
		FMDV	BTV	VSV	BVDV	BRV	IBRV
FMDV							
FMDV serotype A inactivated virus	YNCIQ	+	−	−	−	−	−
FMDV serotype O inactivated virus	YNCIQ	+	−	−	−	−	−
FMDV serotype Asial inactivated virus	YNCIQ	+	−	−	−	−	−
FMDV serotype A inactivated vaccine	LVRI	+	−	−	−	−	−
FMDV serotype O inactivated vaccine	LVRI	+	−	−	−	−	−
FMDV serotype Asial inactivated vaccine	LVRI	+	−	−	−	−	−
BTV							
BTV serotype 4 inactivated virus	YNCIQ	−	+	−	−	−	−

continued

Pathogen	Source	GeXP Results					
		FMDV	BTV	VSV	BVDV	BRV	IBRV
BTV serotype8 inactivated virus	YNCIQ	−	+	−	−	−	−
BTV serotype 9 inactivated virus	YNCIQ	−	+	−	−	−	−
BTV serotype 15 inactivated virus	YNCIQ	−	+	−	−	−	−
BTV serotype 17 inactivated virus	YNCIQ	−	+	−	−	−	−
BTV serotype 18 inactivated virus	YNCIQ	−	+	−	−	−	−
VSV							
VSV serotype New Jersey inactivated virus	YNCIQ	−	−	+	−	−	−
VSV serotype Indiana inactivated virus	YNCIQ	−	−	+	−	−	−
BVDV							
Oregon CV24 (BVDV-1)	CVCC	−	−	−	+	−	−
NADL (BVDV-1)	CVCC	−	−	−	+	−	−
AV68 (BVDV-1)	CVCC	−	−	−	+	−	−
GX-BVDV1 (BVDV-1)	GVRI	−	−	−	+	−	−
GX-BVDV2 (BVDV-1)	GVRI	−	−	−	+	−	−
GX-BVDV3 (BVDV-1	GVRI	−	−	−	+	−	−
GX-BVDV4 (BVDV-1)	GVRI	−	−	−	+	−	−
GX-BVDV5 (BVDV-1	GVRI	−	−	−	+	−	−
GX-BVDV6 (BVDV-1)	GVRI	−	−	−	+	−	−
GX-BVDV7 (BVDV-1)	GVRI	−	−	−	+	−	−
GX-BVDV8 (BVDV-1)	GVRI	−	−	−	+	−	−
GX-BVDV9 (BVDV-1)	GVRI	−	−	−	+	−	−
GX-BVDV10 (BVDV-1)	GVRI	−	−	−	+	−	−
GX-BVDV11 (BVDV-1)	GVRI	−	−	−	+	−	−
GX-BVDV12 (BVDV-1)	GVRI	−	−	−	+	−	−
GX-BVDV13 (BVDV-1)	GVRI	−	−	−	+	−	−
GX-041 (BVDV-2)	GVRI	−	−	−	+	−	−
BRV							
NCDV	CVCC	−	−	−	−	+	−
BRV014	CVCC	−	−	−	−	+	−
GX-BRV1	GVRI	−	−	−	−	+	−
GX-BRV2	GVRI	−	−	−	−	+	−
GX-BRV3	GVRI	−	−	−	−	+	−
GX-BRV4	GVRI	−	−	−	−	+	−
GX-BRV5	GVRI	−	−	−	−	+	−
GX-BRV6	GVRI	−	−	−	−	+	−
GX-BRV7	GVRI	−	−	−	−	+	−
GX-BRV8	GVRI	−	−	−	−	+	−
IBRV							
AV20/Barta Nu/67	CVCC	−	−	−	−	−	+
AV21/BK125	CVCC	−	−	−	−	−	+

continued

Pathogen	Source	GeXP Results					
		FMDV	BTV	VSV	BVDV	BRV	IBRV
Reference strain							
PPRV inactivated virus	YNCIQ	–	–	–	–	–	–
ETEC							
GX-ETEC1	GVRI	–	–	–	–	–	–
GX-ETEC2	GVRI	–	–	–	–	–	–
GX-ETEC3	GVRI	–	–	–	–	–	–
Escherichia coli							
C83919/1676	CVCC	–	–	–	–	–	–
C83924/x114/83	CVCC	–	–	–	–	–	–
C83922/b41	CVCC	–	–	–	–	–	–
Mycoplasma bovis							
GX/MB1	GVRI	–	–	–	–	–	–
GX/MB2	GVRI	–	–	–	–	–	–
Mycobacterium bovis							
GXmt304	GVRI	–	–	–	–	–	–
GXmt397	GVRI	–	–	–	–	–	–
C680001	CVCC	–	–	–	–	–	–
Salmonellosis/GXsal71	GVRI	–	–	–	–	–	–

Note: GVRI, Guangxi Veterinary Research Institute; YNCIQ, Yunnan Entry-Exit Inspection and Quarantine Bureau; LVRI, Lanzhou Veterinary Research Institute; CVCC, Chinese Veterinary Culture Collection Center.

Primers design

The GeXP-multiplex PCR assay included six pairs of chimeric primers, and each of chimeric primers consisted of a gene-specific primer for each virus's conserved sequence fused at 5' end to a universal primer. The conserved nucleotide sequences of six cattle infectious disease viruses from GenBank were aligned using MegAlign 7.0 software (DNAStar, USA). Gene-specific primers were designed using the "Primer premier 5.0" (PRMIER Biosoft international, Canada) according to the restrict design rules of GeXP-multiplex PCR primer. A BLAST search program of GenBank website was performed to verify oligonucleotide specificity. All primers were synthesized and HPIC purified by the Invitrogen Inc (Guangzhou, China). The details of the oligonucleotides for primers were listed in Table 5-8-2.

Table 5-8-2 Primer information

Primer	Forward primer sequence (5'-3')	Reverse primer sequence (5'-3')	Amplicon size/bp	Target region	Primer concentration/ (μmol/L)
BTV	AGGTGACACTATAGAATAAGGGTAACTCACAGCAAACTCAA	GTACGACTCACTATAGGGAGAGCAGCCTGTCCATCCC	136	VP7	0.2
FMDV	AGGTGACACTATAGAATAGCCGTGGGACCATACAGG	GTACGACTCACTATAGGGAAAGTGATCTGTAGCTTGGAATCTC	166	3D	0.2
BRV	AGGTGACACTATAGAATAGCGTCATTTACAAGGAGAACATC	GTACGACTCACTATAGGGAATCTCGCCCATGCCCAC	188	gB	0.2

continued

Primer	Forward primer sequence (5'-3')	Reverse primer sequence (5'-3')	Amplicon size/bp	Target region	Primer concentration/ (μmol/L)
BRV	AGGTGACACTATAGAATACAGTGGC TTCCATTAGAAGCAT	GTACGACTCACTATAGGGAGGTCAC ATCCTCTCACTA	211	VP6	0.2
VSV	AGGTGACACTATAGAATAAAACTAC TGGACGGGCTTGA	GTACGACTCACTATAGGGATGAGAT GCCCAAATGTTGC	278	N	0.2
BVDV	AGGTGACACTATAGAATAGTGAGTT CGTTGGATGGC	GTACGACTCACTATAGGGATATGTTT TGTATAAGAGTTCATTTG	308	5'-UTR	2

Note: Universal tag sequences were underlined. Chimeric primers were synthesized using universal primers and gene-specific primers.

GeXP-multiplex PCR assay

The reaction system was created using the GeXP Start-up Kit (Beckman Coulter, Brea, USA) in a total volume of 20 μL containing 4 μL of GenomeLab™ GeXP Start Kit 5 × PCR Buffer (containing 0.25 μM concentration of each universal tag primer: Tag-F: 5'-AGGTGACACTATAGAATA-3' and Tag-R: 5'-GTACGACTCACTATAGGGA-3', the 5' end of forward universal primer was labeled with Cy5 fluorophore), 4 μL of $MgCl_2$ (25 μM), 2 μL of mixed primers (the concentration of each primer was listed in Table 5-8-2), 10 U JumpStart Tag DNA polymerase (Sigma-Aldrich, USA), and 1 μL of cDNA (0.5 pg~0.5 ng). Nuclease-free water was then added to the PCR reaction to achieve a final volume of 20 μL.

GeXP-multiplex PCR was performed using the thermal cycler (Thermo, Milford, USA). The optimized GeXP-multiplex PCR amplification condition as followed: 95 ℃ for 3 min; 10 cycles of 9 ℃ for 30 s, 55 ℃ for 30 s and 72 ℃ for 30 s; then 10 cycles of 95 ℃ for 30 s, 65 ℃ for 30 s and 72 ℃ for 30 s; and 20 cycles of 95 ℃ for 30 s, 53 ℃ for 30 s and 72 ℃ for 30 s; held at 4 ℃ for conservation.

PCR product separation and analysis were performed by capillary electrophoresis using GenomeLab GeXP Genetic Analysis System (Beckman Coulter, Brea, USA) following previously described[20]. The fluorescently labeled amplicons were separated into distinct peaks on a electropherogram via GeXP high-resolution capillary electrophoresis and then identified by their respective sizes. The peaks were initially analyzed by fragment analysis module of the GeXP system 10.2 software (Beckman Coulter, Brea, USA).

Standards preparation

The specific genes of six cattle viruses were amplified by using the primers listed in Table 5-8-2. The specific PCR amplicons for each virus were cloned into the pEASY-T1vector (TransGen Biotech, China) for sequencing. Sequence data were analyzed and blasted in GenBank. The six recombinant plasmids carrying the partial gene from each virus (VP7 gene of BTV, 3D gene of FMDV, gB gene of IBRV, VP6 gene of BRV, N gene of VSV, 5-UTR of BVDV) were linearized with restriction enzyme Spe Ⅰ (TaKaRa, Dalian, China) and then in vitro transcribed into ssRNA using a T7 RiboMAX™ Express Large Scale RNA production system kit (Promega, Madison, WI, USA). The DNA template was removed by digestion with DNase following the transcription reaction, and then removed unincorporated nucleotides by chromatography. The concentration of transcribed ssRNAs were measured at 260 nm using a NanoDrop 2000 (ThermoFisher Scientific, Waltham, USA), and copy number of transcribed ssRNAs were calculate according to previously described[22, 23]. Serial 10-fold dilutions, containing each transcribed ssRNA ranging from 10^8 copies/μL to 1 copies/μL, were stored at −70 ℃ until used.

Specificity and sensitivity of GeXP-multiplex PCR assay

The GeXP-mono PCR assay and GeXP-multiplex PCR assay were used to evaluate its specificity. The GeXP-mono PCR assay was performed using a single template (cDNA extracted from each virus listed in Table 5-8-1) along with a primer mixture of six sets of chimeric primers to determine the size of the amplification products for each virus. GeXP-multiplex PCR assay was performed using a mixture template containing cDNA of six viruses and a primer mixture to evaluate its cross-amplification in GeXP-multiplex PCR system. The other references strains of bacterial or viruses commonly found in cattle (listed in Table 5-8-1) were tested by the GeXP-multiplex PCR to confirm its specificity.

The sensitivity of the GeXP-mono PCR assay for single target gene was examined by serial 10-fold dilutions of each transcribed ssRNA ranging from 10^8 to 10^0 copies/μL. The sensitivity of the GeXP-multiplex PCR assay was also examined by serial 10-fold dilutions of premixed templates, containing same copies of each transcribed ssRNA (FMDV, BTV, VSV, BVDV, BRV, and IBRV), which contained the specific gene sequences of the six cattle infectious viruses. The standards of the mixed template used ranging from 10^8 to 10^0 copies/μL were prepared from stock using serial 10-fold dilutions in RNase-free H_2O and were used as templates to test the sensitivity of the new assay. 1 μL standard cDNA (0.5 pg～0.5 ng) was used in the reaction system. Profile of the reaction was described in GeXP-multiplex PCR assay section.

Interference assay

The presence of other templates in high quantities could suppress the amplification of other low concentration templates and alter the amplification efficiency of GeXP-multiplex PCR. Two artificial samples containing various concentration of transcribed ssRNAs were prepared, mixed and detected by the GeXP-multiplex PCR assay to assess the interference between high concentration and low concentration nucleic acid templates. The results were compared with those of a single-template GeXP-multiplex PCR assay.

Application to field samples

Three hundred and five field samples, including 156 fecal swabs, 30 conjunctival swabs, 30 nasal mucus swabs, 70 blood samples, 2 oesophageal-pharyngeal fluid, 2 vesicular fluid and tissue (10 mucous membranes, 2 vesicular skins, 3 lymph nodes) were collected from the various cattle farms in Guangxi, China during 2012 to 2014. More than three quarters of samples collected from cattle that did not have any typical clinical and pathological symptoms. A quarter of the samples were collected from diseased cattle showing different symptoms of diseases including metal lassitude, rhinorrhea, dysphagia, high fever, oral erosion, blisters and foaming at the mouth. The swab samples were placed into 1 mL sterilized water. Then supernatant was used to extract RNA after centrifugation. The liquid samples were used for the extraction of RNA as described previously. The tissue samples were ground into homogenates for RNA extraction. RNA was reverse-transcribed as described previously. The cDNA were assayed by both the optimized GeXP-multiplex PCR assay and simplex real-time PCR assays using previously published primers[24-29]. These simplex real-time PCR assays included five OIE recommended real-time PCR assays for detection of BTV, FMDV, IBRV, VSV, BVDV and one simplex real-time PCR for detection of BRV. All the positive field samples detected by the Gexp-multiplex PCR products were confirmed by DNA sequencing using conventional simplex PCR assays with same primers as the GeXP-multiplex PCR assay (Huada, Guangzhou, China).

Results

Specificity results

The cDNA samples from six cattle infectious viruses listed in Table 5-8-1 were individually used as a template to evaluate the specificity of gene-specific primers. In GeXP-mono PCR assay, each of the corresponding genes from the target viruses was amplified as expected (Table 5-8-1 and Figure 5-8-1), BTV: 135～137 bp, FMDV: 165～167 bp, IBRV: 187～189 bp, BRV: 211～213 bp, VSV: 277～279 bp, BVDV: 308～310 bp. In GeXP-multiplex PCR assay, six specific amplification peaks generated by each target virus were detected simultaneously (Figure 5-8-2): BTV: 136.23 bp, FMDV: 165.78 bp, IBRV: 188.42 bp, BRV: 212.37 bp, VSV: 278.54 bp, BVDV: 308. 86 bp. No cross amplification peak was observed in GeXP-mono PCR assay and GeXP-multiplex PCR assay. The GeXP-multiplex PCR assay specifically amplified six cattle infectious viruses, and exhibits no cross-reactivity with other cattle pathogens (Table 5-8-1). The results indicated that GeXP-multiplex PCR has a high specificity to detect six cattle infectious agents without any nonspecific amplification.

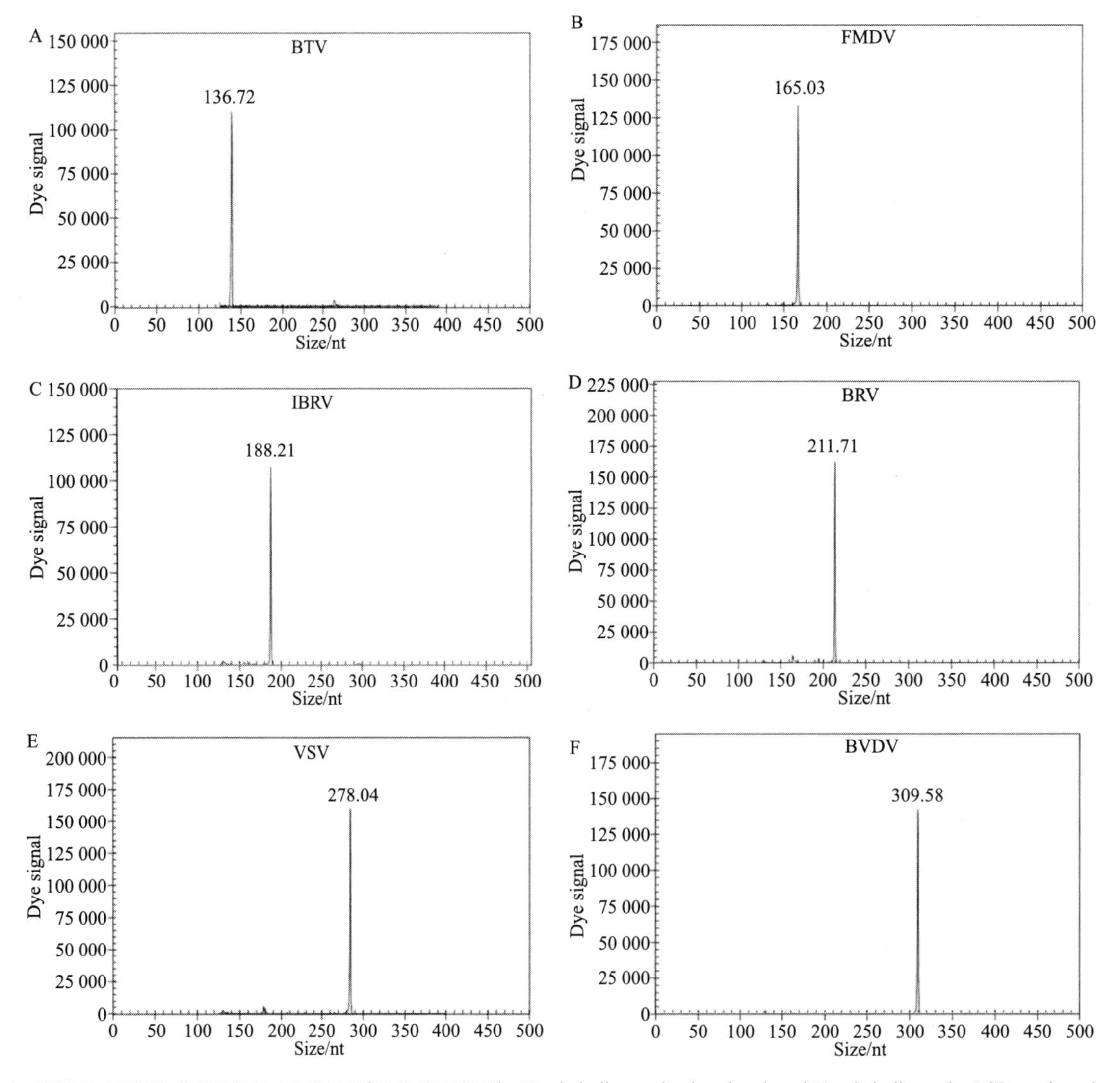

A: BTV; B: FMDV; C: IBRV; D: BRV; E: VSV; F: BVDV. The Y-axis indicates the dye signal, and X-axis indicate the PCR product size.

Figure 5-8-1 Specificity results of the GeXP-mono PCR assay

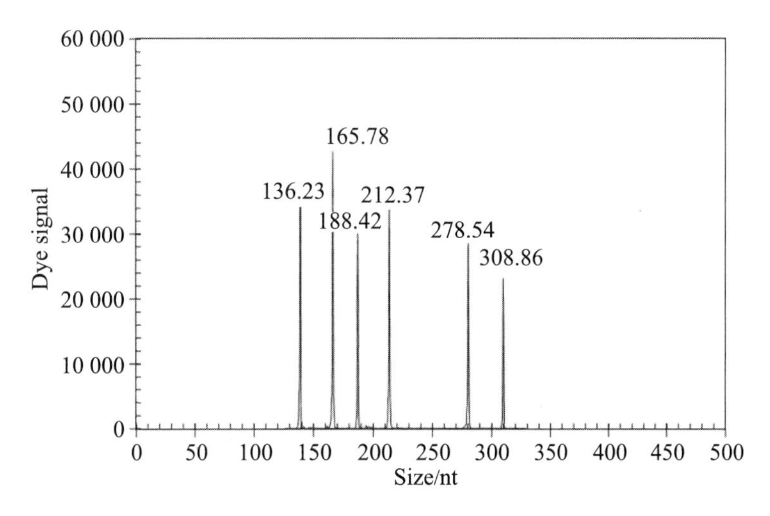

Figure 5-8-2 Specificity results of the GeXP-multiplex PCR assay with mixed template of six cattle infectious viruses

Sensitivity results

The detection limit for the Gexp-mono PCR assay for a single target gene was 10 copies/μL of each transcribed ssRNA (data not shown). The sensitivity of the GeXP-multiplex PCR assay was examined using premixed ssRNAs mixtures with adjusted equal copies of each virus. The detection limit of the GeXP-multiplex PCR assay was 100 copies/μL when all of six premixed ssRNAs containing target genes of 6 cattle viruses were tested (Figure 5-8-3). Each tests were repeated three times at each template concentration and similar results were obtained. Typically the cut-off CT value for positive and negative results was determined as 2 000 A. U. value (absorbance unite) by default. The results indicated that the GeXP-multiplex PCR assay has a good sensitivity to detect six cattle infectious viruses at the same time.

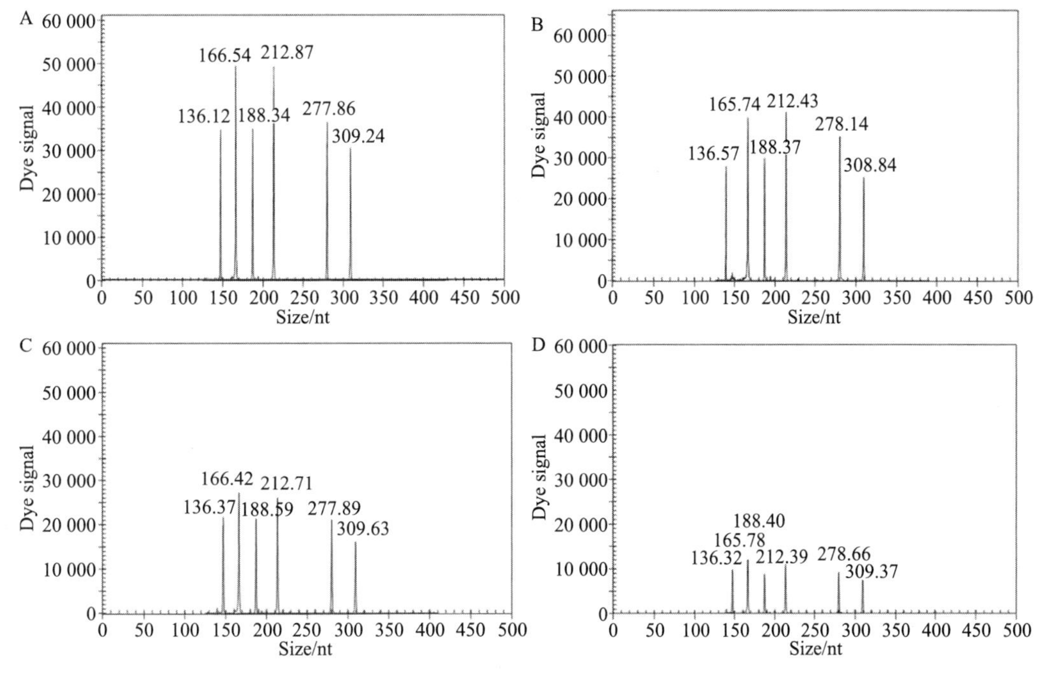

A: 10^5 copies/reactions; B: 10^4 copies/reactions; C: 10^3 copies/reactions; D: 10^2 copies/reactions in the GeXP-multiplex PCR assay. GeXP-muliplex PCR assay was performed using serial 10-fold dilutions of premixed transcribed ssRNAs containing specific gene sequences of the 6 cattle viruses. The viruses form left to right were as follow: BTV, FMDV, IBRV, BRV, VSV and BVDV.

Figure 5-8-3 Sensitivity results of GeXP-multiplex PCR assay

Interference results

Two artificial samples: sample A: FMDV (10^6copies/μL)+IBRV (10^3copies/μL)+BRV (10^3 copies/μL)+ VSV (10^8 copies/μL), sample B: BTV (10^7 copies/μL)+IBRV (10^3 copies/μL)+BRV (10^3 copies/μL)+BVDV (10^5 copies/μL), were prepared and tested by GeXP-multiplex PCR assay. The corresponding amplification peaks were observed in electrophoretogram (Figure 5-8-4). Additionally, the peaks of A. U. values of a simple template were similar to that of mixed templates (Table 5-8-3). Although there were some systematic deviations in the A. U. values, when comparing the mixed template with the single template, it did not affect the detection level. No differences in amplification efficiency were observed between the simple template and mixed template formats. The results suggest that variable viral concentration did not result in significant differences in amplification performance.

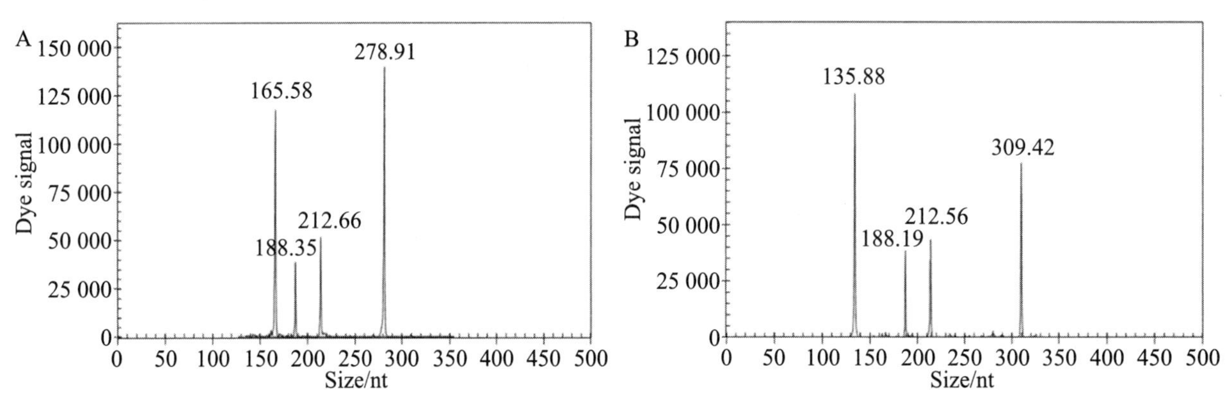

A: The artificial mixture sample, FMDV (10^6 copies/μL) +IBRV (10^3 copies/μL) +BRV (10^3 copies/μL) +VSV (10^8 copies/μL); B: The artificial mixture sample, BTV (10^7 copies/μL)+IBRV (10^3 copies/μL)+BRV (10^3 copies/μL)+BVDV (10^5 copies/μL).

Figure 5-8-4 Interference results of GeXP-multiplex PCR assay

Table 5-8-3 Results of comparing the artificial mixed template with the single template by GeXP-multiplex PCR assay

Template	A. U. value of GeXP-multiplex PCR assay					
	BTV	FMDV	IBRV	BRV	VSV	BVDV
Sample A		165.58	188.35	212.66	278.91	
FMDV (10 copies/μL)		165.07				
IBRV (10^3 copies/μL)			188.24			
BRV (10^3 copies/μL)				212.15		
VSV (10^8 copies/μL)					278.59	
Sample B	135. 88		188.19	212.56		309.42
BTV (10^7 copies/μL)	135.93					
IBRV (10^3 copies/μL)			188.75			
BRV (10^3 copies/μL)				212.21		
BVDV (10^5 copies/μL)						309.57

Detection in field samples

A total of 305 clinical samples were tested by the optimized GeXP-multiplex PCR assay and simplex real-time PCR assay to assess the reliability for the rapid detection of clinical samples. The positive and negative results obtained with the two different methods are shown in Table 5-8-4. The detection rates for each virus were 10.5% (BTV), 2.0% (FMDV), 1.3% (IBRV), 2.6% (BRV), 0 (VSV), 13.4% (BVDV), respectively. The results of the GeXP-multiplex PCR assay has 100% agreement with simplex real-time PCR assays without any inconsistent results. Moreover, all positive samples in the GeXP-multiplex PCR and simplex real-time PCR were confirmed via sequencing to rule out false positive samples. This GeXP-multiplex PCR assay could detect and differentiate the six cattle viruses.

Table 5-8-4　Analysis of clinical samples using GeXP-multiplex PCR assay and simplex real-time PCR methods

Background of clinical samples	Clinical sample	Number	Positive results (GeXP-multiplex PCR/simplex real-time PCR/ sequencing					
			BTV[a]	FMDV[a]	IBRV[a]	BRV[b]	VSV[a]	BVDV[a]
Cattle without any morbid symptoms and signs	fecal swab	141				3/3/3		18/18/18
	blood sample	70	32/32/32					
	conjunctival swab	30						
	nasal mucus swab	22						
Cattle showed typical symptoms of disease	oesophageal-pharyngeal fluid	2		2/2/2				
	vesicular skins	2		2/2/2				
	vesicular fluid	2		2/2/2				
	mucous membrane	10						10/10/10
	Fecal sample	15				5/5/5/		10/10/10
	lymph node	3						3/3/3
	nasal mucus swab	8			4/4/4			

Note: [a]Confirmed by OIE recommended real-time PCR detection of BTV, FMDV, IBRV, VSV, BVDV[24-28]; [b]confirmed by simplex real-time PCR detection of BRV[29].

Discussion

FMDV, BTV, VSV, BVDV, BRV, and IBRV are the six main cattle infectious viruses with a high infection rate and prevalent worldwide. Several global outbreaks have occurred in history, resulting in severe economic loss of stockbreeding and damage to international trade of animal products[14]. These diseases are potential threat to cattle industry. Therefore, a rapid, high-throughput and effective detection and differentiation technique is needed for the clinical diagnosis of these cattle viruses.

Although multiplex conventional PCR and multiplex fluorescence real-time quantitative PCR have been used for the detection of multiple viruses, they are limited by their high interference and fail to detect multiple target genes in one tube[30-32]. The GenomeLab Gene Expression Profiler (GeXP) analyzer is a novel multi-target, high-throughput detection technique that is capable of differentially assessing the expression profile of up to 30 genes in one tube based on analysis of amplicons size by capillary electrophoresis. The analytical procedure includes modified reverse transcription and PCR amplification, followed by capillary electrophoretic separation. Two-stage amplification using fluorescent dye-labeled universal tag primers reduces the interference among the primers, and inferior amplification and non-specific reaction. The GeXP-multiplex PCR assay has highly specificity and sensitivity. By far, the GeXP-multiplex PCR assay has been widely used in veterinary diagnostics and medical examination[19-23]. For example: simultaneous detection of sixteen human respiratory virus types/subtypes, 11 human papilloma viruses, nine serotypes of enteroviruses associated with hand, foot, and mouth disease, influenza A H1N1 virus has been reported[33-36].

Therefore, high-throughput detection and accurate identification of multiple viruses can be achieved by using this technique in large numbers of samples with limited amounts of starting material.

In this study, we have successfully established a GeXP-multiplex PCR assays that can simultaneously identify the FMDV, BTV, VSV, BVDV, BRV, and IBRV in a single reaction. The optimal detection limit of GeXP-multiplex PCR assay was 100 copies/μL when all of six premixed transcribed ssRNAs containing target genes of 6 bovine viruses. In detection of 305 clinical samples, the results of GeXP-multiplex PCR were consistent with that of simplex real-time PCR recommended by OIE. The subsequent analysis of positive samples by sequencing demonstrated that the GeXP-multiplex PCR assay had no false positive samples of non-specific amplification. Although two hundred sixty three samples were collected from cattle without any morbid symptoms and signs, 32 blood samples were positive for BTV, 3 fecal swabs were positive for BRV, and 18 fecal swabs were positive for BVDV by GeXP-multiplex PCR detection. This necessitates the epidemiological surveillance for BTV, BRV and BVDV in clinically normal cattle. Accurate diagnosis of BVDV positive cattle and timely elimination of them can be incorporated in the disease control of cattle herds programs to purify herd.

In practice, it only needs one single RNA extraction, one PCR, and one capillary electrophoresis, which will obtain detection results of six cattle viruses. Single capillary electrophoresis can analyze 96 samples at a time. This high-throughput advantage can meet the demand for a large scale of epidemiological investigation.

Conclusions

The GeXP-multiplex PCR assay described in the present study will provide a high throughput diagnostic method with high specificity and sensitivity for the simultaneous identification of the six very important cattle viruses. GeXP-multiplex PCR assay may therefore be adopted for the molecular epidemiologic surveillance of cattle infectious diseases for designing effective disease-control programs.

References

[1] TOMAS J D, SIMON F P. Rebhun's diseases of dairy cattle-2nd edition. Singapore: Elsevier Pte Ltd Press, 2007.

[2] CERNICCHIARO N, WHITE B J, RENTER D G, et al. Evaluation of economic and performance outcomes associated with the number of treatments after an initial diagnosis of bovine respiratory disease in commercial feeder cattle. Am J Vet Res, 2013, 74(2): 300-309.

[3] WEN W, HUANG Z, YE J. The current analysis of the status and prospect of cattle industry in China. China Animal Husbandry and Veterinary, 2016, 32(1): 45-46.

[4] WERNERY U, KINNE J. Foot and mouth disease and similar virus infections in camelids: a review. Rev Sci Tech, 2012, 31(3): 907-918.

[5] SIERRA S, DAVILA M, LOWENSTEIN P R, et al. Response of foot-and-mouth disease virus to increased mutagenesis: influence of viral load and fitness in loss of infectivity. J Virol, 2000, 74(18): 8316-8323.

[6] BRITO B P, RODRIGUEZ L L, HAMMOND J M, et al. Review of the Global Distribution of Foot-and-Mouth Disease Virus from 2007 to 2014. Transbound Emerg Dis, 2017, 64(2): 316-332.

[7] MACLACHLAN N J. Bluetongue: history, global epidemiology, and pathogenesis. Prev Vet Med, 2011, 102(2): 107-111.

[8] SMITH P F, HOWERTH E W, CARTER D, et al. Host predilection and transmissibility of vesicular stomatitis New Jersey virus strains in domestic cattle (*Bos taurus*) and swine (*Sus scrofa*). BMC Vet Res, 2012, 8: 183.

[9] WALDNER C L, KENNEDY R I. Associations between health and productivity in cow-calf beef herds and persistent infection with bovine viral diarrhea virus, antibodies against bovine viral diarrhea virus, or antibodies against infectious bovine rhinotracheitis virus in calves. Am J Vet Res, 2008, 69(7): 916-927.

[10] SANTMAN-BERENDS I M, MARS M H, VAN DUIJN L, et al. Evaluation of the epidemiological and economic consequences of control scenarios for bovine viral diarrhea virus in dairy herds. J Dairy Sci, 2015, 98(11): 7699-7716.

[11] WORD ORGANIZATION FOR ANIMAL HEALTH (OIE). The OIE list of notifiable terrestrial and aquatic animal diseases 2016. (2016-01-10) [2016-03-05]. http://www.oie.int//intenational-standard-setting/terrestrial-manual/access-online/.

[12] XIE J X, DUAN Z J, LI D D, et al. Detection of bovine rotavirus G10P[11] in a diary farm in Daqing, China. Chinese Journal of Virology, 2010, 26(5): 407-409.

[13] HASHISH E A, ZHANG C, RUAN X, et al. A multiepitope fusion antigen elicits neutralizing antibodies against enterotoxigenic Escherichia coli and homologous bovine viral diarrhea virus in vitro. Clin Vaccine Immunol, 2013, 20(7): 1076-1083.

[14] KARREMAN H J. Disease control on organic and natural cattle operations. Anim Health Res Rev, 2009, 10(2): 121-124.

[15] DREW J E, MAYER C D, FARQUHARSON A J, et al. Custom design of a GeXP multiplexed assay used to assess expression profiles of inflammatory gene targets in normal colon, polyp, and tumor tissue. J Mol Diagn, 2011, 13(2): 233-242.

[16] YANG M J, LUO L, NIE K, et al. Genotyping of 11 human papillomaviruses by multiplex PCR with a GeXP analyzer. J Med Virol, 2012, 84(6): 957-963.

[17] HU X, ZHANG Y, ZHOU X, et al. Simultaneously typing nine serotypes of enteroviruses associated with hand, foot, and mouth disease by a GeXP analyzer-based multiplex reverse transcription-PCR assay. J Clin Microbiol, 2012, 50(2): 288-293.

[18] RAI A J, KAMATH R M, GERALD W, et al. Analytical validation of the GeXP analyzer and design of a workflow for cancer-biomarker discovery using multiplexed gene-expression profiling. Anal Bioanal Chem, 2009, 393(5): 1505-1511.

[19] XIE Z, LUO S, XIE L, et al. Simultaneous typing of nine avian respiratory pathogens using a novel GeXP analyzer-based multiplex PCR assay. J Virol Methods, 2014, 207: 188-195.

[20] ZHANG Y F, XIE Z X, XIE L J, et al. GeXP analyzer-based multiplex reverse-transcription PCR assay for the simultaneous detection and differentiation of eleven duck viruses. BMC Microbiol. 2015, 15: 247.

[21] ZHANG M, XIE Z, XIE L, et al. Simultaneous detection of eight swine reproductive and respiratory pathogens using a novel GeXP analyser-based multiplex PCR assay. J Virol Methods, 2015, 224: 9-15.

[22] ZENG T, XIE Z, XIE L, et al. Simultaneous detection of eight immunosuppressive chicken viruses using a GeXP analyser-based multiplex PCR assay. Virol J, 2015, 12: 226-231.

[23] LI M, XIE Z, XIE Z, et al. Simultaneous detection of four different neuraminidase types of avian influenza A H5 viruses by multiplex reverse transcription PCR using a GeXP analyser. Influenza Other Respir Viruses, 2016, 10(2): 141-149.

[24] HOFMANN M, GRIOT C, CHAIGNAT V, et al. Bluetongue disease reaches Switzerland. Schweiz Arch Tierheilkd, 2008, 150(2): 49-56.

[25] SHAW A E, REID S M, EBERT K, et al. Implementation of a one-step real-time RT-PCR protocol for diagnosis of foot-and-mouth disease. J Virol Methods, 2007, 143(1): 81-85.

[26] WILSON W C, LETCHWORTH G J, JIMENEZ C, et al. Field evaluation of a multiplex real-time reverse transcription polymerase chain reaction assay for detection of Vesicular stomatitis virus. J Vet Diagn Invest, 2009, 21(2): 179-186.

[27] HOFFMANN B, DEPNER K, SCHIRRMEIER H, et al. A universal heterologous internal control system for duplex real-time RT-PCR assays used in a detection system for pestiviruses. J Virol Methods, 2006, 136(1-2): 200-209.

[28] WANG J, O'KEEFE J, ORR D, et al. An international inter-laboratory ring trial to evaluate a real-time PCR assay for the detection of bovine herpesvirus 1 in extended bovine semen. Vet Microbiol, 2008, 126(1-3): 11-19.

[29] OTTO P H, ROSENHAIN S, ELSCHNER M C, et al. Detection of rotavirus species A, B and C in domestic mammalian animals with diarrhoea and genotyping of bovine species A rotavirus strains. Vet Microbiol, 2015, 179(3-4): 168-176.

[30] ZENG Z, LIU Z, WANG W, et al. Establishment and application of a multiplex PCR for rapid and simultaneous detection of six viruses in swine. J Virol Methods, 2014, 208: 102-106.

[31] YEH J Y, LEE J H, SEO H J, et al. Simultaneous detection of Rift Valley Fever, bluetongue, rinderpest, and Peste des petits ruminants viruses by a single-tube multiplex reverse transcriptase-PCR assay using a dual-priming oligonucleotide system. J Clin Microbiol, 2011, 49(4): 1389-1394.

[32] FERNANDEZ J, AGUERO M, ROMERO L, et al. Rapid and differential diagnosis of foot-and-mouth disease, swine vesicular disease, and vesicular stomatitis by a new multiplex RT-PCR assay. J Virol Methods, 2008, 147(2): 301-311.

[33] LI J, MAO N Y, ZHANG C, et al. The development of a GeXP-based multiplex reverse transcription-PCR assay for simultaneous detection of sixteen human respiratory virus types/subtypes. BMC Infect Dis, 2012, 12: 189-194.

[34] YANG M J, LUO L, NIE K, et al. Genotyping of 11 human papillomaviruses by multiplex PCR with a GeXP analyzer. J Med Virol. 2012, 84(6): 957-963.

[35] HU X, ZHANG Y, ZHOU X, et al. Simultaneously typing nine serotypes of enteroviruses associated with hand, foot, and mouth disease by a GeXP analyzer-based multiplex reverse transcription-PCR assay. J Clin Microbiol. 2012, 50(2): 288-293.

[36] QIN M, WANG D Y, HUANG F, et al. Detection of pandemic influenza A H1N1 virus by multiplex reverse transcription-PCR with a GeXP analyzer. J Virol Methods, 2010, 168(1-2): 255-288.

Simultaneous differential detection of H5, H7, H9 and nine NA subtypes of avian influenza viruses via a GeXP assay

Luo Sisi, Xie Zhixun, Li Meng, Li Dan, Zhang Minxiu, Ruan Zhihua, Xie Liji, Wang Sheng, Fan Qing, Zhang Yanfang, Huang Jiaoling, and Zeng Tingting

Abstract

H5, H7 and H9 are the most important subtypes of avian influenza viruses (AIVs), and nine neuraminidase (NA) subtypes (N1-N9) of AIVs have been identified in poultry. A method that can simultaneously detect H5, H7, H9 and the nine NA subtypes of AIVs would save time and effort. In this study, 13 pairs of primers, including 12 pairs of subtype-specific primers for detecting particular subtypes (H5, H7, H9 and N1-N9) and one pair of universal primers for detecting all subtypes of AIVs, were designed and screened. The 13 pairs of primers were mixed in the same reaction, and the 13 target genes were simultaneously detected. A GeXP assay using all 13 pairs of primers to simultaneously detect H5, H7, H9 and the nine NA subtypes of AIVs was developed. The GeXP assay showed specific binding to the corresponding target genes for singlet and multiplex templates, and no cross-reactivity was observed between AIV subtypes and other related avian pathogens. Detection was observed even when only 10^2 copies of the 13 target genes were present. This study provides a high-throughput, rapid and labor-saving GeXP assay for the simultaneous rapid identification of three HA subtypes (H5, H7 and H9) and nine NA subtypes (N1-N9) of AIVs.

Keywords

H5 subtype, H7 subtype, H9 subtype, nine NA subtypes, avian influenza, GeXP assay

Introduction

Influenza can occur in pandemics and localized outbreaks. Avian influenza viruses (AIVs) belong to the influenza A type in the family Orthomyxoviridae. The hemagglutinin (HA) and neuraminidase (NA) proteins are often considered the most important viral antigens, and eighteen HA subtypes (H1-H18) and eleven NA subtypes (N1-N11) have been identified based on differences in the HA and NA antigens; sixteen HA subtypes (H1-H16) and nine NA subtypes (N1-N9) have been recognized and found in poultry and wild birds, while two additional HA and NA subtypes, H17N10 and H18N11, have been identified in bats[1, 2]. AIVs, especially subtypes H5, H7 and H9, have contributed to enormous economic losses and pose a potential threat to global human public health. H5 and H7 are highly pathogenic and cause serious illness and death in domestic poultry and humans[3, 4]. Infection with influenza A virus subtypes H5, H7 and H9 causes several respiratory diseases in humans. H5, H7 and H9 are the most important subtypes in poultry. In 1997, a human fatality occurred due to H5N1-subtype AIV infection in Hong Kong, and an outbreak of the disease caused panic and serious economic losses in the poultry industry[5]. In recent years, new H5 subtype combinations, H5N6 and H5N8, have emerged and caused a pandemic in poultry, even causing human infections[6, 7]. In 2013, a fatal case of H7N9 infection occurred in the Yangtze River Delta region of China, and this virus spread to most parts of the country, causing five waves of infection[8]. Currently, H7N9 is controlled by vaccination, but it still poses a serious threat to public health. H9 is commonly isolated and identified in poultry and provides internal genes to H5 and H7 viruses that can infect humans, such as H7N9 and H5N6[9, 10]. The H9 subtype has been detected in domestic fowl and wild birds in various regions and is considered one of the most likely causes of new influenza pandemics in humans. In recent years, viruses in which H5 and H7 are combined with different NA subtypes, such as H5N2, H5N6, H5N8, H7N2, H7N7 and H7N9, have emerged continuously[11]. The H9 subtype of AIV has been associated with every one of the known nine NA subtypes described and mainly combines with N2[12]. Rapid differential diagnosis of avian influenza viruses is of paramount importance to the poultry industry and public health.

Viral culture is still used as a standard reference method for routine surveillance of influenza viruses. However, the sample must first be cultured by inoculating cells or chicken embryos, and 3～5 days are needed for virus proliferation; then, the viral culture medium is collected and tested by hemagglutination inhibition and neuraminidase inhibition to determine the HA and NA subtypes, which requires a series of specific positive serums. This process is time-consuming and laborious. The use of viral culture, an essential tool in the epidemiological surveillance and diagnosis of influenza viruses, has been relegated by the use of more sensitive and affordable molecular techniques[13]. For example, in molecular assays, especially in multiplex formats, a multiplex reverse transcriptase polymerase chain reaction (conventional RT-PCR) was developed to simultaneously differentiate the avian H5, H7 and H9 subtypes[14]. Accurate, rapid and triplex real-time fluorescent quantitative RT-PCR assays were developed for the simultaneous detection of the AIV subtypes H5, H7 and H9[15]. Three loop-mediated isothermal amplification (LAMP) methods were developed to detect the presence of AIVs and discriminate between the H5 and H9 subtypes, which need three different reaction systems[16]. However, all of these methods can simultaneously detect only two to four pathogens (target genes) and cannot realize high-throughput detection. Simultaneous detection of HA and NA subtypes of AIVs has rarely been reported, and no previous studies have conducted simultaneous detection of H5, H7, H9 and the nine NA subtypes.

Rapid differentiation of the important AIV subtypes H5, H7 and H9 and the nine NA subtypes from other pathogens may require many reactions. There are many subtypes of AIVs, and rapid differential diagnosis is difficult. Detection methods for the simultaneous identification of the H5, H7 and H9 subtypes and nine NA subtypes are very important and rare. The GenomeLab Gene Expression Profiler genetic analysis system (GeXP) is a new high-throughput detection platform that uses a novel, modified RT-PCR process that converts multiplexed RT-PCR with several pairs of primers to a primer-pair process using universal primers, followed by fluorescence capillary electrophoresis separation based on the size of the amplified products. The amplified fragments are separated by capillary electrophoresis to generate visual fluorescence signal peaks to simultaneously detect and identify multiplex pathogens or genes. Our study aimed to establish a GeXP assay for the simultaneous identification of the H5, H7 and H9 subtypes and nine NA subtypes of AIVs by using GeXP technology; we simultaneously detected 13 target genes in one tube, indicating that this is a time-saving, simple, rapid and high-throughput process. We previously obtained partial data from a GeXP assay for simultaneous differential detection of N1-N9-subtype AIVs, and that assay was used for differential diagnosis of NA-subtype AIVs[17]. This study focused on the simultaneous detection of H5-, H7-, H9- and nine candidate NA-subtype AIVs and decreased the number of reactions from 13 to 1. This method could be used not only for the simultaneous detection of the 12 specific subtypes but also for determining the combinations of H5, H7 and

H9 with the NA subtypes in a single reaction tube, which is especially important for the prevention and control of AIV infection.

Materials and methods

Design of high-performance, gene-specific primers for a GeXP multiplex

To design the gene-specific primers, which included H5, H7, H9 and N1-N9 AIV subtype-specific primers and AIV subtype-universal primers, the HA gene sequences of the H5, H7 and H9 subtypes and other HA subtypes, the NA gene sequences of the N1-N9 subtypes and the M gene sequences of all subtypes were obtained from the Influenza Virus database at the National Center for Biotechnology Information (NCBI). First, the conserved regions and sequences of the HA and NA subtypes of AIVs were identified via MegAlign in DNASTAR-Lasergene 8.0 software. Second, specific primers were designed for the conserved sequences using Primer Premier 5.0 software, in which the annealing temperature, mismatches and dimers of primers were evaluated, and some candidate primers were preliminarily selected. Finally, the primers used were examined in silico via BLAST comparisons online to determine the specificity of crossing with other subtypes of AIVs and pathogens *in vitro*. After comparison and verification of the primers, two to three pairs of candidate primers were screened for each target gene. Primers' amplicons 100~350 bp in length were designed (without universal tags). The GeXP universal tag (the underlined sequence in Table 5-9-1) was added to the 5' end of the gene-specific forward and reverse primers, and the complete primers obtained were referred to as chimeric primers. The universal tag sequence was added to each of the gene-specific primer sequences to obtain the final primer sequence. The designed fragment size was the specific gene size plus 37 bp for the universal tags; thus, amplicons between 137 and 387 bp in size were designed with universal tags. Amplicons were designed such that each fragment was no less than 5 nucleotides away from its nearest neighbor, which allowed for variation in migration to meet the minimum peak separation distance of 3 nucleotides. All primers were synthesized and purified by Invitrogen (Guangzhou, China).

Viral DNA/RNA nucleic acid extraction

Reference strains and field isolates of the H5, H7, H9 and N1-N9 AIV subtypes, other avian pathogens (including infectious Newcastle disease virus (NDV), infectious bronchitis virus (IBV), laryngotracheitis virus (ILTV), avian reovirus (ARV) and fowl adenovirus virus serotype four (FAdV-4) and human influenza B virus) were used in this study. Viral RNA/DNA was extracted from 200 μL of virus stock using a Viral DNA/RNA Extraction Kit (TransGen, Beijing, China) according to the manufacturer's instructions. The extracted RNA was reverse transcribed to synthesize cDNA using the cDNA Synthesis Kit (TaKaRa, Dalian, China), AIVs were reverse transcribed using a 12 bp primer (5'-agcgaaagcagg-3'), non-AIV viruses were reverse transcribed using random primers in the kit, and the cDNA and DNA were stored at −20 ℃ .

Reaction procedures and conditions of the GeXP assay

The GeXP assay reaction system had a total volume of 20 μL, including 4 μL of 5 × PCR buffer (GenomeLab GeXP Start Kit, Beckman Coulter, Brea, CA, USA; universal forward and reverse primers were formulated in 5 × PCR buffer; fluorescently labeled universal forward primer: 5'-Cy5-AGGTGACACTATAGAATA-3'; universal reverse primer: 5'-GTACGACTCACTATAGGGA-3'), 4 μL of 25 mM $MgCl_2$, 1 μL of 2.5 U/μL JumpStart Taq DNA polymerase (Sigma, St. Louis, MO, USA), 2 μL of the 13-primer-pair mixture (200 nM each primer, Table 5-9-1) and 1～2 μL of cDNA, with nuclease-free water added to 20 μL. The reaction system was placed in a PCR amplification instrument (Bio-Rad, Hercules, CA, USA). The tubes were incubated at 95 ℃ for 5 min, followed by three steps of amplification according to the temperature switch PCR (TSP) strategy[18]: step 1, 10 cycles of 30 s at 94 ℃, 30 s at 55 ℃ and 30 s at 72 ℃; step 2, 10 cycles of 30 s at 94 ℃, 30 s at 62 ℃ and 30 s at 72 ℃; step 3, 20 cycles of 30 s at 94 ℃, 30 s at 50 ℃ and 30 s at 72 ℃; 5 min at 72 ℃ and held at 12 ℃ in a thermal cycler (Bio-Rad). After amplification, 2 μL of PCR product was added to 37.75 μL of sample loading solution along with 0.25 μL of DNA standard-400 (including in the GenomeLab GeXP Start Kit). The PCR products (the fluorescence-labeled amplicons) were separated and analyzed with a GeXP instrument (Beckman Coulter). After approximately 50 min, the PCR products were separated through capillary electrophoresis, and the results were presented as separated peaks on the electropherogram, which were identified according to the respective sizes. The dye signal strength of each peak was measured as the A. U. of optical fluorescence and defined as the fluorescence signal minus the background. The data were analyzed using GeXP system. The horizontal coordinate is the size of the amplified fragment, and the vertical coordinate is the fluorescence signal value.

Table 5-9-1 Information for the 13 pairs of primers used in the GeXP assay

Virus	Forward primer (5'-3')	Reverse primer (5'-3')	Gene	GenBank accession number
AIV	AGGTGACACTATAGAATA AGGCTCTCATGGAGTGGCTA	GTACGACTCACTATAGGGA TGGACAAAGCGTCTACGCTG	M	OQ830456.1
AIV-H5	AGGTGACACTATAGAATA ATACACCCTCTCACCATCGG	GTACGACTCACTATAGGGA TTGCTGTGGTGGTACCCATA	HA	OQ546777.1
AIV-H7	AGGTGACACTATAGAATA AGAATACAGATTGACCCAGT**S**AA	GTACGACTCACTATAGGGA CCCATTGCAATGGC**H**AGAAG	HA	MK453329.1
AIV-H9	AGGTGACACTATAGAATA ATGGCAA**Y**CCTTC**Y**TGTGA	GTACGACTCACTATAGGGA TTGTGTATTGGGCGTC**Y**TG	HA	OR528241.1
AIV-N1	AGGTGACACTATAGAATA GGTGTTTGGATCGG**R**AGAAC	GTACGACTCACTATAGGGA TCAACCCAGAA**R**CAAGGTC	NA	OP373692.1
AIV-N2	AGGTGACACTATAGAATA TTGGGTGTTCCGTTTCA	GTACGACTCACTATAGGGA CCATCCGTCATTACTAC	NA	OQ954751.1
AIV-N3	AGGTGACACTATAGAATA TTCCCAATAGGAACAGC**Y**CCAGT	GTACGACTCACTATAGGGA TTCTCCATGATTT**R**ATGGAGTC	NA	MK978938.1
AIV-N4	AGGTGACACTATAGAATA CAGA**Y**AAGGA**Y**TCAAATGGTGT	GTACGACTCACTATAGGGA CATGGTACAGTGCAATTCCT	NA	MT421389.1
AIV-N5	AGGTGACACTATAGAATA GTGAGGTCATGGAGAAAGCA	GTACGACTCACTATAGGGA TGG**Y**CTATTCATTCC**R**TTCCA	NA	LC339605.1
AIV-N6	AGGTGACACTATAGAATA CACTATAGATCC**Y**GA**R**ATGATGACC	GTACGACTCACTATAGGGA GGAGTCTTTGCTAAT**W**GTCCTTCCA	NA	MT375537.1
AIV-N7	AGGTGACACTATAGAATA GACAG**R**AC**W**GCTTTCAGAGG	GTACGACTCACTATAGGGA GTTGCGTTGTCATTATTTCC	NA	MN253549.1
AIV-N8	AGGTGACACTATAGAATA AGGGAATACAATGAAACAGT	GTACGACTCACTATAGGGA TGCAAAACCCTTAGCATCACA	NA	MT421085.1
AIV-N9	AGGTGACACTATAGAATA CGCCCTGATAAGCTGGCCACT	GTACGACTCACTATAGGGA ACAGGCCTTCTGTTGTACCA	NA	KP418553.1
GeXP universal primer	AGGTGACACTATAGAATA	GTACGACTCACTATAGGGA		

Note: The universal primer tag sequences are underlined. The bold type shows degenerate sites. R: A/G; Y: C/T; W: A/T; S: G/C; H: A/T/C.

Validation of a single primer and preliminary evaluation of multiple primers

The primers used for each target gene were first validated with a single pair of primers and cDNA to validate whether they could be amplified efficiently, and DNA/cDNA from other related pathogens was detected as a template to validate primer specificity according to the GeXP procedures. Only primers with good amplification efficiency and strong specificity can be candidate multiplex GeXP primers. Thirteen pairs of primers were selected and mixed as multiplex primers for the GeXP assay to determine whether the corresponding target gene could be amplified and whether cross-reaction occurred between primers.

Evaluation of the specificity of the GeXP assay

The GeXP assay was established using 13 screened and verified pairs of primers. To assess the specificity of the GeXP assay and determine whether the GeXP assay could correctly identify H5-, H7-, H9- and nine NA-subtype AIVs and whether any cross-reactivity occurred with other related pathogens, reference strains and field isolates of the H5N1, H5N6, H7N2, H7N7, H9N2 and N1-N9 subtypes (including H3N2, H4N2, H6N2, H2N3, H10N3, H8N4, H12N5, H3N6, H4N6, H6N6, H3N8, H4N8, H5N9 and H7N9) and other AIVs; human influenza B viruses; and other related avian pathogens (NDV, IBV, ILTV, ARV and FAdV-4) listed in Table 5-9-2 were individually detected using cDNA/DNA from the GeXP assay. Multiplex templates were also detected using the GeXP assay.

Evaluation of the sensitivity of the GeXP assay

The target fragments of the 13 pairs of primers (Table 5-9-1) were amplified, cloned and inserted into the pGEM-T vector to obtain 13 recombinant plasmids, which were verified by sequencing. The 13 recombinant plasmids were mixed in equal amounts and serially diluted with 10^6～10 copies. Based on the sensitivity of GeXP, we expected to detect at least 10^3 copies. A total of 10^3～10 copies of these recombinant plasmids, including the 13 target fragments, were evaluated via the GeXP assay for sensitivity. The concentrations of the 13 recombinant plasmids were adjusted according to the test results, and sensitivity evaluation was performed via the GeXP assay.

Application for detecting clinical samples

A total of 150 swab samples were randomly collected from the cloacae and larynges of healthy chickens, geese and ducks from five live bird markets (LBMs) in 2022 in Guangxi, China. The swabs were then suspended in 1 mL of storage medium containing antibiotics at 4 ℃ until arrival at the laboratory. The swab sample treatment method was previously described[19]. Viral RNA was extracted from swab samples, and cDNA was reverse transcribed according to previously described protocols (Section *Viral DNA/RNA nucleic acid extraction* in this paper). The cDNA of different tissue samples of three H5N2-infected SPF chickens was donated by the University of Connecticut, and the cDNA of different tissue samples of five H7N9-infected SPF chickens was donated by China Agricultural University. Swab and tissue samples were simultaneously examined via the GeXP assay and real-time PCR (using the same primers as the GeXP assay and TB Green Premix Ex Taq from TaKaRa). The HA and NA genes of the positive samples were amplified, cloned using previously described primers[20] and sequenced by Invitrogen (Guangzhou, China).

Results

Screening results for the 13 pairs of primers

First, to evaluate the single pair of primers for each target, each pair of primers was tested based on its corresponding singlet template; other subtypes of AIVs, other related avian pathogens and human influenza B virus were subsequently used to test the specificity of the single pair of primers. Then, each target was screened with a pair of candidate primers, which were combined without overlapping amplification fragment sizes. Second, to evaluate the multiplex primers used for singlet and multiplex templates, 13 pairs of target primers were mixed together, and singlet and multiplex templates of H5, H7, H9 and N1-N9 were detected to determine whether the primer combinations could amplify the corresponding targets. Other subtypes of AIVs, other related avian pathogens and human influenza B virus were subsequently used to test the specificity of the multiplex primers. If the amplification efficiency of the primers was low or if no target peak was observed, the primers were replaced until the optimal 13 pairs of primers were identified. The 13 pairs of primers that were ultimately selected are shown in Table 5-9-1, and these were used to develop the GeXP assay. In this study, fragments of the expected sizes were amplified for the H5, H7, H9 and N1-N9 subtypes AIVs: H5, 227 to 231 bp; H7, 140 to 145 bp; H9, 329 to 334 bp; N1, 244 to 249 bp; N2, 279 to 285 bp; N3, 215 to 221 bp; N4, 149 to 155 bp; N5, 295 to 301 bp; N6, 236 to 241 bp; N7, 192 to 198 bp; N8, 173 to 178 bp; N9, 205 to 211 bp; and AIV, 158 to 164 bp.

Evaluation of the specificity of the GeXP assay

Specific results were obtained for different subtypes of AIVs and other pathogens using the GeXP assay (Table 5-9-2). Single templates of the H5, H7, H9 and nine NA subtypes were detected using the GeXP assay, and the corresponding target peaks were amplified (Figure 5-9-1). For the H5, H7 and H9 subtypes, three target peaks were generated; for example, H5N1 produced H5- and N1-subtype-specific peaks and the universal AIV detection peak; H7N7 produced H7- and N7-subtype-specific peaks and the universal AIV detection peak; and H9N2 produced H9- and N2-subtype-specific peaks and the universal AIV detection peak. The other subtypes exhibited two peaks: an NA-subtype-specific peak and a universal detection peak. When multiple templates were mixed, the corresponding target peaks could be detected. Each primer showed specific amplification peaks, and in the detection of other related pathogens, no amplification peaks were observed, demonstrating the superior specificity of the GeXP assay.

Table 5-9-2 Information and detection result of strains used in the GeXP assay

Number	Pathogens/field samples	Source	AIV	H5	H7	H9	N1	N2	N3	N4	N5	N6	N7	N8	N9
1	A/Duck/Guangxi/1/04 (H5N1) AIV cDNA	GVRI	+	+	–	–	+	–	–	–	–	–	–	–	–
2	A/Chicken/Guangxi/1/04 (H5N1) AIV cDNA	GVRI	+	+	–	–	+	–	–	–	–	–	–	–	–
3	A/Goose/Guangxi/2/04 (H5N1) AIV cDNA	GVRI	+	+	–	–	+	–	–	–	–	–	–	–	–
4	A/Chicken/QT35/98 (H5N9) AIV cDNA	UP	+	+	–	–	–	–	–	–	–	–	–	–	+
5	A/Chicken PA/3979/97 (H7N2) AIV cDNA	UP	+	–	+	–	–	+	–	–	–	–	–	–	–
6	A/Chicken/NY/273874/03 (H7N2) AIV cDNA	UC	+	–	+	–	–	+	–	–	–	–	–	–	–

continued

Number	Pathogens/field samples	Source	AIV	H5	H7	H9	N1	N2	N3	N4	N5	N6	N7	N8	N9
7	A/Duck/42846/07 (H7N7) AIV cDNA	UP	+	−	+	−	−	−	−	−	−	−	+	−	−
8	A/Chicken/Guangxi/NN1/2017 (H7N9) AIV cDNA	GVRI	+	−	+	−	−	−	−	−	−	−	−	−	+
9	A/Chicken/Guangxi/YL1/2017 (H7N9) AIV cDNA	GVRI	+	−	+	−	−	−	−	−	−	−	−	−	+
10	A/Dove/Guangxi/151B24/2016 (H9N2) AIV	GVRI	+	−	−	+	−	+	−	−	−	−	−	−	−
11	A/Chicken/Guangxi/178C38/2014 (H9N2) AIV	GVRI	+	−	−	+	−	+	−	−	−	−	−	−	−
12	A/Dove/Guangxi/408P55/2020 (H9N2) AIV	GVRI	+	−	−	+	−	+	−	−	−	−	−	−	−
13	A/Chicken/Guangxi/449C11/2021 (H9N2) AIV	GVRI	+	−	−	+	−	+	−	−	−	−	−	−	−
14	A/Duck/Guangxi/413D40/2020 (H9N2) AIV	GVRI	+	−	−	+	−	+	−	−	−	−	−	−	−
15	A/Duck/HK/77/76 (H2N3) AIV	UHK	+	−	−	−	−	−	+	−	−	−	−	−	−
16	A/Duck/HK/876/80 (H10N3) AIV	UHK	+	−	−	−	−	−	+	−	−	−	−	−	−
17	A/Turkey/Ontario/6118/68 (H8N4) AIV	UHK	+	−	−	−	−	−	−	+	−	−	−	−	−
18	A/Duck/HK/862/80 (H12N5) AIV	UHK	+	−	−	−	−	−	−	−	+	−	−	−	−
19	A/Dove/Guangxi/288P43/2017 (H3N6) AIV	GVRI	+	−	−	−	−	−	−	−	−	+	−	−	−
20	A/Duck/Guangxi/LZ038E117/2019 (H4N6) AIV	GVRI	+	−	−	−	−	−	−	−	−	+	−	−	−
21	A/Duck/Guangxi/330D18/2018 (H6N6) AIV	GVRI	+	−	−	−	−	−	−	−	−	+	−	−	−
22	A/Dove/Guangxi/295D25/2017 (H4N8) AIV	GVRI	+	−	−	−	−	−	−	−	−	−	−	+	−
23	A/Duck/Guangxi/423D20/2020 (H4N8) AIV	GVRI	+	−	−	−	−	−	−	−	−	−	−	+	−
24	A/Goose/Guangxi/292G39/2017 (H3N8) AIV	GVRI	+	−	−	−	−	−	−	−	−	−	−	+	−
25	A/Duck/PA/2099/12 (H11N9) AIV	UP	+	−	−	−	−	−	−	−	−	−	−	−	+
26	B/Guangxi/1418/15	GXCDC	−	−	−	−	−	−	−	−	−	−	−	−	−
27	B/Guangxi/1470/15	GXCDC	−	−	−	−	−	−	−	−	−	−	−	−	−
28	NDV F48	CIVDC	−	−	−	−	−	−	−	−	−	−	−	−	−
29	IBV M41	CIVDC	−	−	−	−	−	−	−	−	−	−	−	−	−
30	ILTV Beijing strain	CIVDC	−	−	−	−	−	−	−	−	−	−	−	−	−
31	ARV S1133	CIVDC	−	−	−	−	−	−	−	−	−	−	−	−	−
32	FAdV-4-GX005	GVRI	−	−	−	−	−	−	−	−	−	−	−	−	−

Note: GVRI, Guangxi Veterinary Research Institute, China; UHK, University of Hong Kong, China; UP, University of Pennsylvania, USA; UC, University of Connecticut, USA; CIVDC, China Institute of Veterinary Drug Control; GXCDC, Guangxi Provincial Center for Disease Control and Prevention, China.

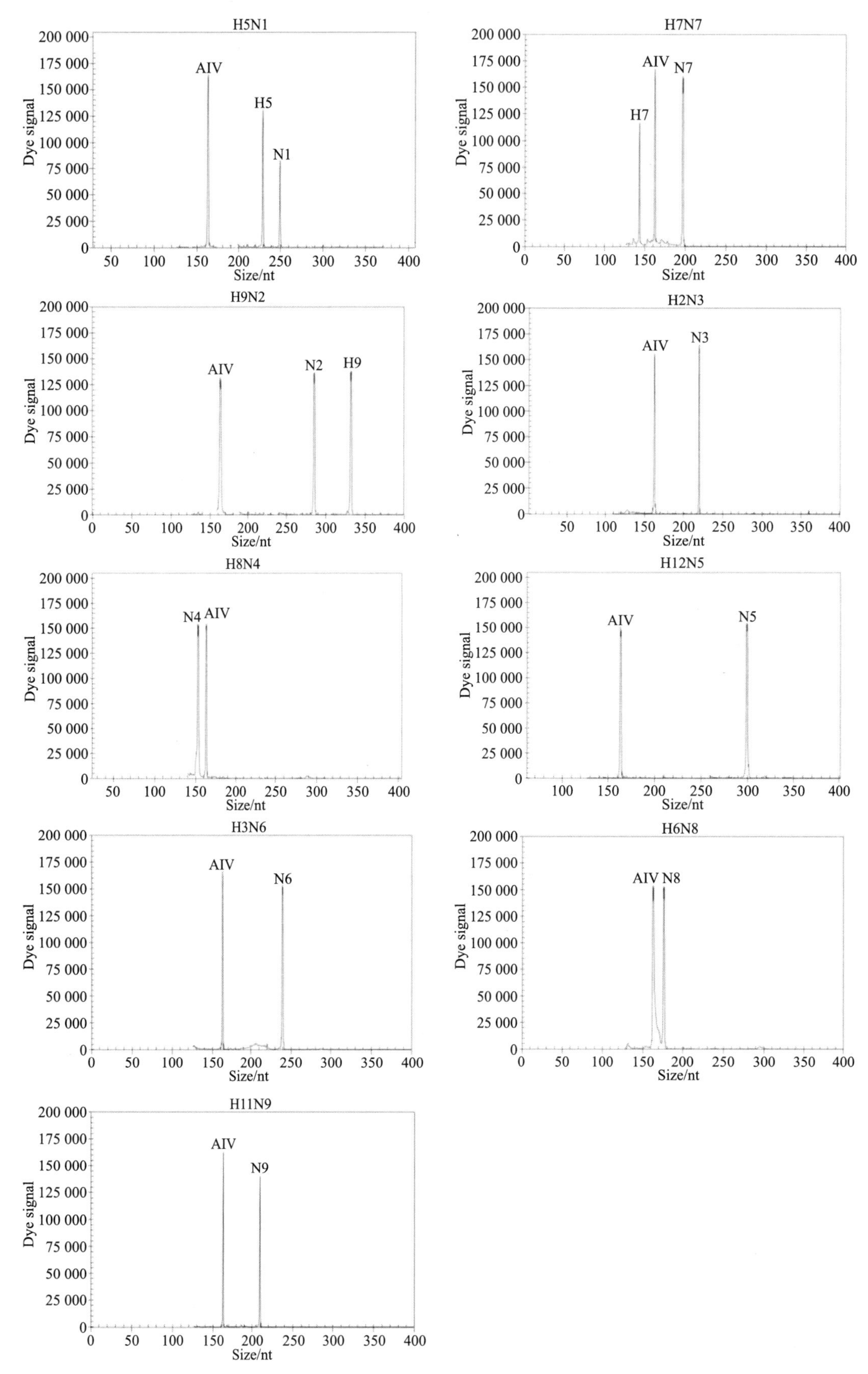

The Y-axis indicates the dye signal in arbitrary units, and the X-axis indicates the actual PCR product size.

Figure 5-9-1 Specificity of the GeXP assay for different AIV subtype templates

Evaluation of the sensitivity of the GeXP assay

To evaluate the sensitivity of the GeXP assay, 13 recombinant plasmids, each containing a target fragment, were mixed, diluted and detected using the GeXP assay, wherein 13 target genes were simultaneously detected in one tube. The results showed that when all 13 recombinant plasmids were present at 10^3 copies, the target peaks were all observed (Figure 5-9-2A). When the concentration of the 13 plasmids was diluted so that 10^2 copies of each plasmid were present, all the plasmids were still detected via the GeXP assay (Figure 5-9-2B). The reaction at each template concentration was repeated three times, and similar results were obtained (CV≤7.23% for each concentration).

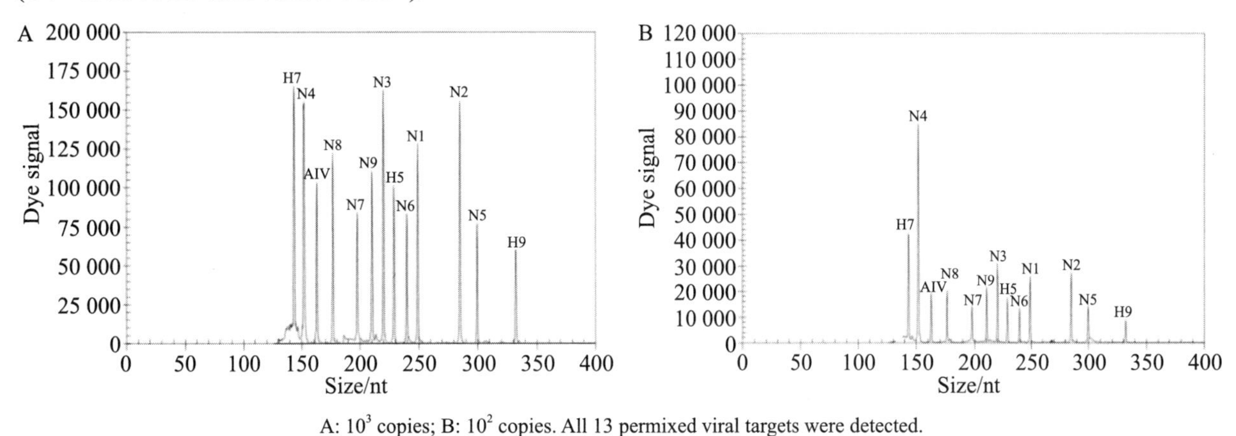

A: 10^3 copies; B: 10^2 copies. All 13 permixed viral targets were detected.

Figure 5-9-2 Sensitivity of the GeXP assay for the simultaneous detection of 13 target genes

Detection of AIVs in clinical samples using the GeXP assay

All the swab samples were tested using the GeXP assay, and the samples were confirmed by real-time PCR and sequencing. The positive and negative results obtained using the various methods and the agreement among the GeXP assay, real-time RT-PCR method and sequencing results are presented in Table 5-9-3. The H9N2, HxN2 and HxN6 subtypes of AIV were the most common in the positive samples. The GeXP assay yielded 100% specificity, in contrast to the conventional approaches, and GeXP was superior to the simultaneous detection of multiple pathogens. The cDNA of different tissue samples of three H5N2-infected SPF chickens donated by the University of Connecticut and the cDNA of different tissue samples of five H7N9-infected SPF chickens donated by China Agricultural University were detected using the GeXP assay, real-time PCR and sequencing, and the results indicated that the GeXP assay could detect H5, H7 and the corresponding NA subtypes (Table 5-9-3).

Table 5-9-3 Detection results for clinical samples using the GeXP assay, real-time RT-PCR and sequencing

Type of chicken samples	Number of positive samples	GeXP assay	Real-time RT-PCR		Sequencing	
			Three methods for H5, H7 and H9 real-time RT-PCR	Nine methods for N1-N9 real-time RT-PCR	HA gene	NA gene
Swab samples from LBMs	8	H9, N2, AIV	H9	N2	H9	N2
	2	N1, AIV	None[a]	N1	–[b]	N1
	9	N2, AIV	None[a]	N2	–[b]	N2
	5	N6, AIV	None[a]	N6	–[b]	N6
	3	N8, AIV	None[a]	N8	–[b]	N8

continued

Type of chicken samples	Number of positive samples	GeXP assay	Real-time RT-PCR		Sequencing	
			Three methods for H5, H7 and H9 real-time RT-PCR	Nine methods for N1-N9 real-time RT-PCR	HA gene	NA gene
Tissue samples from challenged SPF chickens	3[c]	H5, N2, AIV	H5	N2	H5	N2
	3[c]	H5, N2, AIV	H5	N2	H5	N2
	3[c]	H5, N2, AIV	H5	N2	H5	N2
	3[d]	H7, N9, AIV	H7	N9	H7	N9
	3[d]	H7, N9, AIV	H7	N9	H7	N9
	3[d]	H7, N9, AIV	H7	N9	H7	N9

Note: [a] No positive results according to the three methods for H5, H7 and H9 real-time RT-PCR. [b] The non-H5, H7 and H9 subtypes were AIVs, which was not relevant to the comparative detection of the developed GeXP method; therefore, the results are not shown here. [c] Heart, spleen and lung samples from the same H5N2-infected SPF chicken. [d] Indicates heart, spleen and lung samples from the same H7N9-infected SPF chicken.

Discussion

Globalization and industrialization over the past few decades have contributed to the emergence of novel influenza viruses that threaten animal and human health. Avian influenza is an important pathogen that continually threatens both human and animal health. H5- and H7-subtype AIVs have caused severe problems in the global poultry industry and pose severe threats to public health[21]. Highly pathogenic H5N1 AIVs continue to circulate in avian populations, resulting in sporadic infections in humans with a high mortality rate[22]. The H7-subtype AIV HA gene has been found in combination with all nine NA-subtype genes[23]. During the past few years, infections in poultry and humans with H7 subtypes have increased markedly. H9-subtype AIV, which exhibits low pathogenicity, has been reported to be capable of infecting humans, and the World Health Organization has warned that the H9N2 subtype could trigger a global influenza outbreak in humans[24, 25]. H9 is found mainly in N2, and H7N9 and H9N2 coinfection has been shown to produce H9N9[26]. The detection of H5, H7 and H9 is essential. In this study, using the developed GeXP assay, H5-, H7- and H9-subtype AIVs were simultaneously identified, and nine NA subtypes were also identified; one pair of primers was used for the detection and validation of all AIV subtypes.

Subtyping by conventional RT-PCR or real-time RT-PCR is a common method that can be performed directly on nucleic acids from clinical specimens. However, these approaches are often used to detect single pathogens, and even the detection of multiple pathogens is often limited to 2～3 species, as the detection results are disrupted by competitive amplification between primers or interference by different fluorophores. Conventional RT-PCR and real-time RT-PCR can thus be difficult to develop as high-throughput tests and have limited benefits in terms of the number of pathogens that can be simultaneously detected. GeXP can integrate reverse transcription PCR (RT-PCR) and labeled amplified products in multiplex PCR assays. GeXP technology can enable more effective detection of multiple pathogens, wherein 30 target genes can be detected simultaneously, saving reaction time and accelerating detection. The unique feature of GeXP-based multiplex PCR is that amplification by multiplex primers was converted into amplification by a pair of universal primers. At the beginning of the reaction, chimeric primers were first used to amplify the target gene to produce PCR products with GeXP universal primers. Amplification by the universal primers is rapid since universal primers are present at significantly higher concentrations than gene-specific primers within the PCRs; this gradually led to the transformation of a procedure based on multiple chimeric primers into a procedure involving a single pair of GeXP universal primers. Thus, we used a temperature switch PCR (TSP) procedure referring to a previous study[18]: step 1 was carried out using gene-specific sequences of chimeric forward and reverse primers, step 2 was carried out mainly using chimeric forward and reverse primers and step 3 was predominantly carried out using universal forward and reverse primers. Compared with conventional PCR, real-time PCR and LAMP, the GeXP assay can detect many more target genes simultaneously, and competitive amplification and interference between primers are greatly reduced via this approach. A major advantage of GeXP is that this technique can provide a broad detection range for the identification of pathogens that previously needed multiple laboratory subtype assays. Moreover, GeXP provides more intuitive and detailed data than conventional molecular subtyping tests.

The primers used were designed based on the HA and NA gene sequences of different AIV subtypes, allowing differentiation of the HA and NA subtypes of AIVs, respectively. The M gene is the conserved gene of all AIV subtypes and was used to detect all AIV subtypes. Thirteen pairs of gene-specific primers were selected for analysis as follows: (1) After the 13 pairs of primers were mixed, each pair of primers in the mixture could amplify its own target with the corresponding template, and cross-amplification of the remaining 12 target genes and other related pathogens did not occur. (2) Primers had little influence on each other's amplification efficiency, and the amplification efficiency was similar to that of a single template with a single pair of primers. (3) The amplicon lengths obtained with the 13 pairs of primers ranged from 137 to 387 bp, and the amplification size of each target fragment was greater than 5 bp. The whole GeXP process for eight samples was completed in approximately 5 h (1.5 h for nucleic acid extraction and reverse transcription, 2.5 h for PCR, and eight samples/1 h for electrophoresis). The time used was the same as that used for conventional RT-PCR; real-time RT-PCR may take 2.5 h to detect the same number of samples, but the GeXP assay can be used to test more targets in a short period. If real-time RT-PCR is used to detect the same targets, additional time maybe needed. The cost of the GeXP assay for the simultaneous detection of 12 subtypes is approximately USD 7 per test, and the cost of detection is comparable to that of real-time RT-PCR. In addition, two 96-well plates can be placed in parallel in a GeXP machine at the same time to further increase sample throughput.

Among the clinical samples, the N3-, N4-, N5-, N7- and N9-subtype AIVs were not detected. These NA subtypes were less prevalent. This GeXP method has been developed based on local AIV strains, allowing simultaneous and specific detection of H5, H7, H9 and nine NA subtypes of AIVs, and can be used as an efficient and novel method for the prevention and control of AIVs. In conclusion, in this study, we developed a GeXP assay that serves as a specific, sensitive, rapid, high-throughput tool for the simultaneous detection of the H5, H7, H9 and N1-N9 subtypes of AIVs.

References

[1] YOON S W, WEBBY R J, WEBSTER R G. Evolution and ecology of influenza A viruses. Current Topics in Microbiology and Immunology, 2014, 385: 359-375.

[2] TONG S, LI Y, RIVAILLER P, et al. A distinct lineage of influenza A virus from bats. Proceedings of the National Academy of Sciences of the United States of America, 2012, 109(11): 4269-4274.

[3] NAGUIB M M, VERHAGEN J H, MOSTAFA A, et al. Global patterns of avian influenza A (H7): virus evolution and zoonotic threats. FEMS Microbiology Reviews, 2019, 43(6): 608-621.

[4] YAMAJI R, SAAD M D, DAVIS C T, et al. Pandemic potential of highly pathogenic avian influenza clade 2.3.4.4 A(H5) viruses. Reviews in Medical Virology, 2020, 30(3): e2099.

[5] SUBBARAO K, KLIMOV A, KATZ J, et al. Characterization of an avian influenza A (H5N1) virus isolated from a child with

a fatal respiratory illness. Science, 1998, 279(5349): 393-396.

[6] BUI C, KUOK D, YEUNG H W, et al. Risk assessment for highly pathogenic avian influenza A (H5N6/H5N8) Clade 2.3.4.4 viruses. Emerging Infectious Diseases, 2021, 27(10): 2619-2627.

[7] HARFOOT R, WEBBY R J. H5 influenza, a global update. Journal of Microbiology, 2017, 55(3): 196-203.

[8] YIN X, DENG G, ZENG X, et al. Genetic and biological properties of H7N9 avian influenza viruses detected after application of the H7N9 poultry vaccine in China. PLOS Pathogens, 2021, 17(4): e1009561.

[9] ZHOU Y, GAO W, SUN Y, et al. Effect of the interaction between viral PB2 and host SphK1 on H9N2 AIV replication in mammals. Viruses, 2022, 14(7): 1585.

[10] KAGEYAMA T, FUJISAKI S, TAKASHITA E, et al. Genetic analysis of novel avian A (H7N9) influenza viruses isolated from patients in China, February to April 2013. Euro Surveillance, 2013, 18(15): 20453.

[11] SHI J, ZENG X, CUI P, et al. Alarming situation of emerging H5 and H7 avian influenza and effective control strategies. Emerging Microbes Infections, 2023, 12(1): 2155072.

[12] CARNACCINI S, PEREZ D R. H9 influenza viruses: an emerging challenge. Cold Spring Harbor Perspectives in Medicine, 2020, 10(6): a038588.

[13] ORTIZ D L L R, ROJO R S, SANZ M I. Diagnostic challenges in influenza. Enfermedades Infecciosas Microbiologia Clinica, 2019, 37(1): 47-55.

[14] XIE Z, PANG Y S, LIU J, et al. A multiplex RT-PCR for detection of type A influenza virus and differentiation of avian H5, H7, and H9 hemagglutinin subtypes. Molecular and Cellular Probes, 2006, 20(3-4): 245-249.

[15] LIU J, YAO L, ZHAI F, et al. Development and application of a triplex real-time PCR assay for the simultaneous detection of avian influenza virus subtype H5, H7 and H9. Journal of Virological Methods, 2018, 252: 49-56.

[16] ZHANG S, SHIN J, SHIN S, et al. Development of reverse transcription loop-mediated isothermal amplification assays for point-of-care testing of avian influenza virus subtype H5 and H9. Genomics Informatics, 2020, 18(4): e40.

[17] LUO S, XIE Z, HUANG J, et al. Simultaneous differentiation of the N1 to N9 neuraminidase subtypes of avian influenza virus by a GeXP analyzer-based multiplex reverse transcription PCR assay. Frontiers in Microbiology, 2019, 10: 1271.

[18] YANG M J, LUO L, NIE K, et al. Genotyping of 11 human papillomaviruses by multiplex PCR with a GeXP analyzer. Journal of Medical Virology, 2012, 84(6): 957-963.

[19] PENG Y, XIE Z X, LIU J B, et al. Epidemiological surveillance of low pathogenic avian influenza virus (LPAIV) from poultry in Guangxi Province, southern China. PLOS ONE, 2013, 8(10): e77132.

[20] HOFFMANN E, STECH J, GUAN Y, et al. Universal primer set for the full-length amplification of all influenza A viruses. Archives of Virology, 2001, 146(12): 2275-2289.

[21] SUTTON T C. The pandemic threat of emerging H5 and H7 avian influenza viruses. Viruses, 2018, 10(9): 461.

[22] CHAROSTAD J, REZAEI Z R M, MAHMOUDVAND S, et al. A comprehensive review of highly pathogenic avian influenza (HPAI) H5N1: An imminent threat at doorstep. Travel Medicine and Infectious Disease, 2023, 55: 102638.

[23] ABDELWHAB E M, VEITS J, METTENLEITER T C. Prevalence and control of H7 avian influenza viruses in birds and humans. Epidemiology and Infection, 2014, 142(5): 896-920.

[24] SONG W, QIN K. Human-infecting influenza A (H9N2) virus: A forgotten potential pandemic strain?. Zoonoses Public Health, 2020, 67(3): 203-212.

[25] LI S, ZHOU Y, SONG W, et al. Avian influenza virus H9N2 seroprevalence and risk factors for infection in occupational poultry-exposed workers in Tai'an of China. Journal of Medical Virology, 2016, 88(8): 1453-1456.

[26] BHAT S, JAMES J, SADEYEN J R, et al. Coinfection of chickens with H9N2 and H7N9 avian influenza viruses leads to emergence of reassortant H9N9 virus with increased fitness for poultry and a zoonotic potential. Journal of Virology, 2022, 96(5), e0185621.

Part Ⅱ Multiplex (fluorescent) PCR/RT-PCR/LAMP detection technology

Simultaneous differential detection of H5, H7 and H9 subtypes of avian influenza viruses by a triplex fluorescence LAMP assay

Fan Qing, Xie Zhixun, Zhao Junke, Hua Jun, Wei You, Li Xiaofeng, Li Dan, Luo Sisi, Li Meng, Xie Liji, Zhang Yanfang, Zhang Minxiu, Wang Sheng, Ren Hongyu, and Wan Lijun

Abstract

H5, H7, and H9 are pivotal avian influenza virus (AIV) subtypes that cause substantial economic losses and pose potential threats to public health worldwide. In this study, a novel triplex fluorescence reverse transcription-loop-mediated isothermal amplification (TLAMP) assay was developed in which traditional LAMP techniques were combined probes for detection. Through this innovative approach, H5, H7, and H9 AIV subtypes can be simultaneously identified and differentiated, thereby offering crucial technical support for prevention and control efforts. Three primer sets and composite probes were designed based on conserved regions of the haemagglutinin gene for each subtype. The probes were labelled with distinct fluorophores at their 3' ends, which were detached to release the fluorescence signal during the amplification process. The detection results were interpreted based on the colour of the TLAMP products. Then, the reaction conditions were optimized, and three primer sets and probes were combined in the same reaction system, resulting in a TLAMP detection assay for the differential diagnosis of AIV subtypes. Sensitivity testing with *in vitro*-transcribed RNA revealed that the detection limit of TLAMP assay was 205 copies per reaction for H5, 360 copies for H7, and 545 copies for H9. The TLAMP assay demonstrated excellent specificity and no cross-reactivity with related avian viruses, and 100% consistency with a previously published quantitative polymerase chain reaction (qPCR) assay. Therefore, due to its simplicity, rapidity, sensitivity, and specificity, the TLAMP assay is suitable for epidemiological investigations and is a valuable tool for detecting and distinguishing H5, H7, and H9 subtypes of AIV in clinical samples.

Keywords

AIV, H5, H7, H9, TLAMP, probe, differential diagnosis

Introduction

Avian influenza (AI) is a zoonotic disease caused by the avian influenza virus (AIV) that results in human infections and economic losses each year[1]. AIV is a type A influenza virus in the family Orthomyxoviridae with a negative-sense segmented RNA genome. Based on differences in haemagglutinin (HA) and neuraminidase (NA) antigens, 18 HA (H1-H18) and 11 NA (N1-N11) subtypes have been identified[2, 3]. These strains are classified based on their pathogenicity as either high-pathogenicity AIV (HPAIV) or a low-pathogenicity AIV (LPAIV). Strains H5 and H7 (particularly the H5N1 and H7N9 subtypes) are HPAIVs that can cause severe illness and and mortality in domestic poultry and humans[4, 5]. In 1997, the H5N1 virus infected 18 people in Hong Kong, resulting in 6 fatalities[6]. The emergence of H7N9 influenza viruses in China in early 2013 caused five waves of human infection from 2013-2017, with a total of 1 568 cases and 615 fatalities[7]. LPAIVs typically show mild or no symptoms in birds but have the potential to cause pathogenic avian influenza through antigenic drift or shift and pose the risk of human infection[8]. The H9 subtype of AIV, which has been classified as a LPAIV, is reportedly capable of infecting humans[9]. This strain contributes internal genes to H5 and H7 viruses with human infectivity, such as H7N9 and H5N6, making it an important candidate that could cause new influenza pandemics in humans[10-12]. The H5, H7, and H9 subtypes of AIV are crucial in poultry[13-15], emphasizing the urgent need to develop quick and sensitive diagnostic tools for the early detection of AIVs, especially the H5, H7, and H9 subtypes.

Molecular biology-based diagnostic methods using polymerase chain reaction (PCR) technology, such as reverse transcription PCR (RT-PCR), quantitative PCR (qPCR), and GeXP, have become commonplace for AIV detection and genotyping[16-19]. However, these methods are expensive and rely on sophisticated laboratory instruments, including real-time fluorescence PCR and GeXP instruments; therefore, these methods are impractical for use in rural areas. To prevent and control infectious diseases, swift and accurate diagnostics must be performed immediately in endemic areas, which underscores the need for timely point-of-care (POC) testing platforms to meet the evolving challenges of disease detection. Amongst numerous nucleic acid amplification assays, loop-mediated isothermal amplification (LAMP) stands out in terms of its sample-to-answer time, sensitivity, specificity, cost, robustness, and accessibility, making it ideal for field-deployable diagnostics in resource-limited regions[20]. One of the greatest opportunities for LAMP assays as a POC tool is their ability to eliminate nucleic acid purification or extraction steps and perform direct amplification from pretreated crude samples. Reverse-transcription-LAMP (RT-LAMP) sample pretreatment methods, such as thermal lysis, proteinase K and RNAsecure, can inactivate or inhibit RNase in the sample and lyse the virus particles[21]. Another advantage of using the LAMP assay as a POC test is the ability to visually detect amplicons. Techniques with a visual endpoint can allow the direct detection of amplicons using probes for a downstream immunoassay or allow the measurement of amplification indirectly by detecting the formation of pyrophosphate precipitates[22, 23].

The sensitivity and simplicity of LAMP methods generally fall between those of qPCR and rapid antigen tests. At present, LAMP technology is widely used to diagnose a variety of diseases. RT-LAMP assays that can detect SARS-CoV-2, the virus that causes COVID-19, have already penetrated commercial markets and were authorized by the US Food and Drug Administration (FDA) for emergency clinical POC diagnosis[21]. Moreover, these assays have found applications in three settings: highly complex laboratories, POC testing, and at home. However, all commercially available LAMP kits or settings can detect only a single target and fail to achieve the differential diagnosis of multiple targets. In this study, the conventional LAMP assay was innovatively improved by incorporating a quencher-fluorophore composite probe labelled with different fluorophores that display distinct colours at the corresponding wavelengths. Such modification led to the establishment of a groundbreaking triplex fluorescence LAMP (TLAMP) assay capable of simultaneously identifying H5, H7, and H9 subtypes in a single tube. The TLAMP results are directly interpretable by the naked eye and more accurate and intuitive than those of traditional methods. This study offers technical support for advancing AIV POC diagnostic devices as well as innovative strategies for utilizing LAMP technology to achieve multitarget detection.

Materials and methods

Strain source and nucleic acid extraction

The viruses used in this study are outlined in Table 5-10-1. RNA was extracted from four specific pathogen-free (SPF) chicken swab samples and served as the negative control. Nucleic acids from the entities listed in Table 5-10-1 were extracted with the *EasyPure* Viral DNA/RNA Kit (TransGen Biotech, Beijing, China). The DNA and cDNA were stored at −30 ℃, and the RNA was stored at −70 ℃ until use. Two microlitres of DNA/cDNA/RNA was used as a template to evaluate the specificity of TLAMP.

Table 5-10-1 The viruses used along with the corresponding TLAMP results.

Strains/sample type	Source	TLAMP results		
		H5	H7	H9
A/Turkey/GA/209092/02 (H5N2) AIV cDNA	UC, USA	+	−	−
A/Chicken/QT35/98 (H5N9) AIV cDNA	UP, USA	+	−	−
A/Chicken/NY/273874/03 (H7N2) AIV cDNA	UC, USA	−	+	−
A/Duck/HK/47/76 (H7N2) AIV cDNA	UHK, China	−	+	−
A/Duck/42848/07 (H7N7) AIV cDNA	UP, USA	−	+	−
A/Dove/Guangxi/408P55/2020 (H9N2) AIV	GVRI, China	−	−	+
A/Chicken/Guangxi/449C11/2021(H9N2) AIV	GVRI, China	−	−	+
A/Duck/Guangxi/413D40/2020 (H9N2) AIV	GVRI, China	−	−	+
A/Duck/Guangxi/291D16/2017 (H1N6) AIV	GVRI, China	−	−	−
A/Broiler/PA/117/04 (H2N2) AIV	UP, USA	−	−	−
A/Goose/Guangxi/318G39/2018(H3N2) AIV	GVRI, China	−	−	−
A/Duck/Guangxi/201D19/2016 (H4N8) AIV	GVRI, China	−	−	−
A/Duck/Guangxi/330D18/2018 (H6N6) AIV	GVRI, China	−	−	−
A/Turkey/ont/6118/68 (H8N4) AIV	UHK, China	−	−	−
A/Duck/HK/876/80 (H10N3) AIV	UHK, China	−	−	−
A/Duck/HK/661/79 (H11N3) AIV	UHK, China	−	−	−
A/Duck/HK/862/80 (H12N5) AIV	UHK, China	−	−	−
A/Gull/Md/704/77 (H13N5) AIV	UHK, China	−	−	−
A/Mallard Duck/Astrakhan/263/82 (H14N5) AIV	UC, USA	−	−	−
A/Shearwater/Australia/2576/79 (H15N9) AIV	UC, USA	−	−	−
A/Shorebird/Delaware/168/06 (H16N3) AIV	UC, USA	−	−	−
Newcastle disease virus (NDV), F48	CIVDC, China	−	−	−
Infectious bronchitis virus (IBV), M 41	CIVDC, China	−	−	−
Avian reovirus (ARV), S1133	CIVDC, China	−	−	−
Avian infectious laryngotracheitis virus (ILTV), Beijing	CIVDC, China	−	−	−
Chicken infectious anaemia virus (CIAV), GX1804	GVRI, China	−	−	−
Marek's disease virus (MDV) live-attenuated vaccine, CVI988	Boehringer Ingelheim, China	−	−	−
Fowl aviadenovirus serotype 4 (FAdV-4), GX005	GVRI, China	−	−	
Chicken parvovirus (ChPV), GX-ChPV-1, DNA	GVRI, China	−	−	−
Negative control, SPF chickens	Boehringer Ingelheim, China	−	−	−

Note: GVRI, Guangxi Veterinary Research Institute; UHK, University of Hong Kong; UP, University of Pennsylvania; UC, University of Connecticut; CIVDC, China Institute of Veterinary Drug Control; and CIVDC, China Institute of Veterinary Drug Control.

Primer and probe design and preparation of the composite probe (FIP-FD)

Thousands of AIV H5, H7, and H9 sequences were procured from the influenza virus database housed at the National Center for Biotechnology Information (NCBI). Using MegAlign within DNASTAR-Lasergene 8.0 software,the conserved regions and sequences of HA subtypes in AIVs were meticulously identified. Subsequently, Primer Premier 5 and PrimerExplorer V5 online software were used to designed three sets of TLAMP primers and corresponding FD probes, with each set was specifically tailored to target the identified subtype.

The inner primer contained two primers in series, FIP (F1c+F2) and BIP (B1c+B2). The probe FD is a complementary sequence to the F1c fragment of FIP. The 3' end of the FD probe of each subtype was labelled with a fluorescent group, each with a different emission wavelength (FAM, Cy5, or Cy3), while the 5' end of the inner primer FIP was labelled with the corresponding quenching group from the BHQ series. Before the reaction, FIP-quenched (FIP-Q) and FD-fluorescence (FD-F) were annealed, forming a "fluorescencequenched" composite probe (hereafter referred to as FIP-FD) that did not fluoresce prior to the reaction. Throughout the TLAMP assay, FIP maintained its role as an internal primer to guide amplification. However, during synthesis from the reverse direction directed by BIP, the FD probe detached from FIP-FD, releasing a discernible fluorescence signal (Figure 5-10-1). After the reaction, the detection results were interpreted based on the colour of the TLAMP product, highlighting the innovative nature of TLAMP.

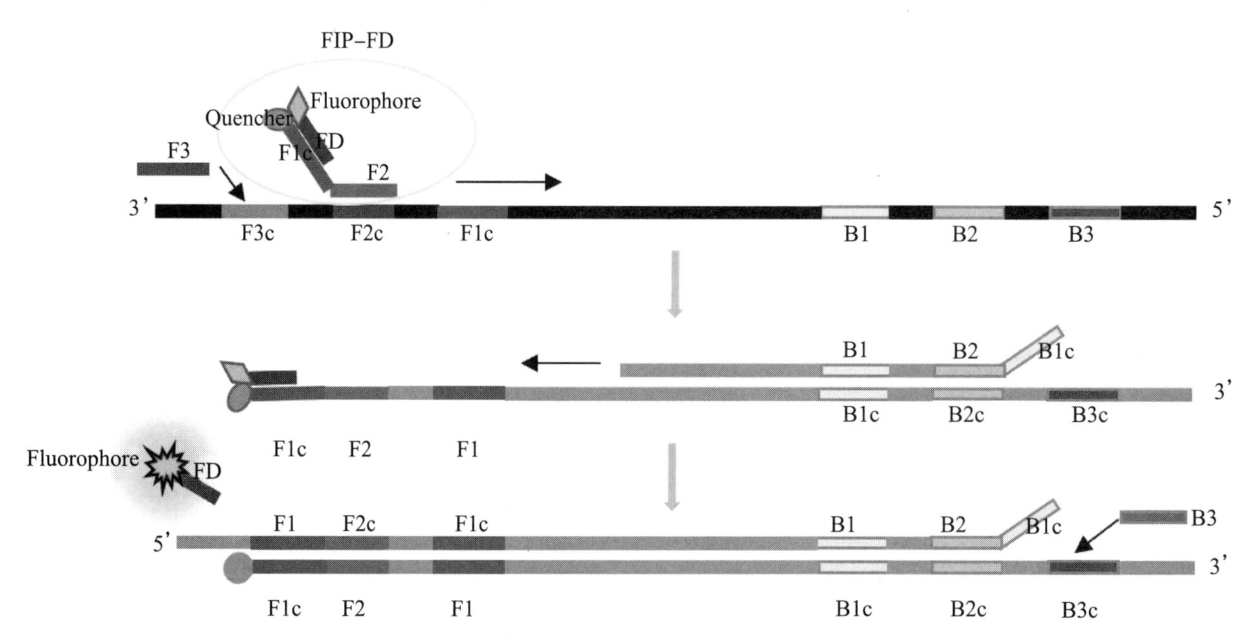

Figure 5-10-1 Schematic diagram of the TLAMP mechanism

Before the reaction, the fluorescence-quenched annealed composite probes were prepared by mixing 50 μmol/L FIP-Q and 50 μmol/L FD-F and heating at 90 ℃ for 5 min, after which the mixture was gradually cooled to room temperature to yield the annealed FIP-FD composite probe, which was stored at –20 ℃ until use. The primers and probes used were synthesized by TaKaRa Biomedical Technology (Dalian, China) Co., Ltd., and the sequences are provided in Table 5-10-2.

Table 5-10-2 Sequences of primers and probes used in the TLAMP assay

Primers and probes	sequence (5'-3')	TM/℃
H5-F3	TTCCATGACTCAAATGTCAAG	57.6
H5-B3	GCTAGGGAACCCGCCACT	59.7
H5-FIP-Q	BHQ1-CATACATTCATTATCACATTTGTGGACTACAGCTTAGGGA**Y**AATGCAAA[b]	63.2/59.6[a]
H5-BIP	GAAAGTGTGAGAAATGGGACGTTCTTCCCTTTTTA**R**TCTTGCTT[b]	60.5/55.7[a]
H5-Floop	CGAAACAACCATTACCCAGCTC	61.3
H5-Bloop	ATGACTACCCCCAGTATTCAGAAG	58.7
H5-FD	ATGTGATAATGAATGTATG-FAM	39.2
H7-F3	TCACATACAATGGAATAAGAAC	55.2
H7-B3	CCCATATAGCTTGGTTTGCT	57.2
H7-FIP-Q	BHQ3-CTGTGTTTGACAGGAGCCATTTCGTGACCAGTGCATGTAGG	56.5/61.9 a
H7-BIP	TTCCCGCAGATGACTAA**R**TCATAAGTTGAAACGGAATG**R**TGGA[b]	56.7/61.3 a
H7-Floop	TCTGCATAGAATGAAGATCCTGATC	59.7
H7-Bloop	CCAGCTATAGTAGTATGGGG	57.8
H7-FD	CTCCTGTCAAACACAG-Cy5	41.5
H9-F3	ACAAAATGAACAAGCAGTATGAAAT	57.6
H9-B3	TCCATGCATTGGTCATCACATT	60.8
H9-FIP-Q	BHQ2-GATCGTCAATCTTATTGTTAATCATGATCATGAATTCAGCGAGG	55.0/55.0[a]
H9-BIP	TATGGGCATATAATGCAGA**R**TTGCTTTTGCATCATGCTCATCGA[b]	56.8/60.9[a]
H9-Floop	TTGTTAATCATGTTAAGTCTAGTTT	57.1
H9-Bloop	AGAATTGCTAGTTCTGCTTGAAAA	58.0
H9-FD	AACAATAAGATTGACGATC-Cy3	44.0

Note: [a] The inner primer consists of two primers joined together, such as FIP=F1c+F2 and BIP=B1c+B2, so there are two annealing temperatures. [b] The bold type shows degenerate sites. R: A/G; Y: C/T.

Preparation of standards

The HA genes of H5, H7 and H9 were individually cloned and inserted into the pEASY T-18T vector (TransGen Biotech, Beijing, China) to generate three recombinant plasmids, each of which was verified via sequencing. Subsequently, the recombinant plasmid that contained the target gene was isolated using the EasyPure® HiPure Plasmid MiniPrep Kit (TransGen Biotech, Beijing, China) and transcribed into purified RNA using the In vitro Transcription T7 Kit (TaKaRa, Dalian, China). The concentrations of the obtained purified RNA transcripts were determined using a NanoDrop 2 000 (Thermo Fisher Scientific, Waltham, MA, United States). The number of RNA transcripts was calculated following a previously described formula[17]. Equal amounts of the three RNA transcripts were then mixed and serially diluted twofold with TE buffer to prepare the RNA standards, the concentrations of which are listed in Table 5-10-3. This twofold series of diluted RNA standards comprising eight concentrations of RNA was used to evaluate the sensitivity of the TLAMP assay.

Table 5-10-3 The trials for the TLAMP assay were repeated 10 times for each standard

RNA standard	H5 RNA/ (copies/μL)	Number of H5-positive samples	H7 RNA/ (copies/μL)	Number of H7-positive samples	H9 RNA/ (copies/μL)	Number of H9-positive samples	Number of repeat tests
1	1 600	10	2 820	10	4 130	10	10
2	800	10	1 410	10	2 065	10	10
3	400	10	705	10	1 033	10	10
4	200	10	353	10	516	10	10
5	100	9	176	10	258	9	10
6	50	6	88	2	129	5	10
7	25	0	44	0	65	0	10
8	12.5	0	22	0	32	0	10

Clinical sample collection

From January to October 2023, 240 broilers, including Ma chicken, Sanhuang chicken, and Fufeng native chicken breeds, were randomly selected from live bird markets (LBMs) in Guangxi, China. The sampled broilers exhibited good health without any signs of disease, and swab samples from the cloacae and larynges were collected. The swab specimens were directly eluted with 1mL of PBS, and RNA was extracted from the centrifuged supernatant using the MagicPure® Simple Viral DNA/RNA Kit (TransGen Biotech, Beijing, China). cDNA from different tissue samples of eight H5N2-infected SPF chickens was donated by the University of Connecticut, and cDNA from different tissue samples of three H7N9-infected SPF chickens was donated by China Agricultural University[19]. The swab and tissue samples were evaluated by performing the TLAMP assay and previously published H5, H7, and H9 qPCR assays[16]. qPCR-positive products were subsequently forwarded to BGI Company (Shenzhen, China) for sequencing to rule out false-positive results. A comparative analysis of the TLAMP results, qPCR results, and sequencing data was conducted to assess the clinical applicability of the TLAMP method.

Results

TLAMP reaction system

The TLAMP reaction mixture was established with a volume of 25 μL after each reaction condition was optimized. The optimal ratio of outer primer to inner primer to loop primer for the H5 and H7 subtypes was determined to be 1:8:2, while for H9, this ratio was 1:16:4. The concentrations of each component are presented in Table 5-10-4, which highlights the adjustments made to increase the efficacy of the TLAMP reaction system. Negative and positive controls were also used for each reaction.

Table 5-10-4 TLAMP reaction system

Reagent	Concentration in a 25 μL system/(μmol/L)		
WarmStart Multi-Purpose LAMP/RT-LAMP 2×Premix	12.5 μL		
Bst 2.0 WarmStart DNA polymerase	2 U		
Subtypes	H5	H7	H9
FIP-FD	0.133	0.067	0.067
Unlabelled FIP		0.1	0.267
FIP-Q	0.134	0.1	0.2

continued

Reagent	Concentration in a 25 μL system/(μmol/L)		
BIP	0.266	0.266	0.534
F3	0.033	0.033	0.033
B3	0.033	0.033	0.033
Floop	0.066	0.066	0. 132
Bloop	0.066	0.066	0.132
ddH_2O	Add to a total volume of 25 μL		

The TLAMP reaction was performed with amplification at 63 ℃ for 75 min followed by termination at 80 ℃ for 5 min. A real-time turbidimeter (LA-320; Eiken Chemical Co., Ltd., Tokyo, Japan) or a water bath was used.

To interpret the results, an image analyser (Universal Hood III, 731BR01622, Bio-Rad, Hercules, CA, United States) was used after the reaction. The reaction products were compared with the negative control. The results were determined based on the colour of the fluorophore in the reaction tube at the corresponding channels. The positive results were determined as follows: green fluorescence in the 520 nm channel indicated H5-positive samples, red fluorescence in the 670 nm channel indicated H7-positive samples, and blue fluorescence in the 570nm channel indicated H9-positive samples; overlapping fluorescence in multiple channels was considered indicative of coinfected samples. Notably, in the visible spectrum, Cy3 appears orange, and to distinguish Cy3 from the overlapping colours of FAM (green) and Cy5 (red), Cy3 is shown in blue in the output image (Figure 5-10-2).

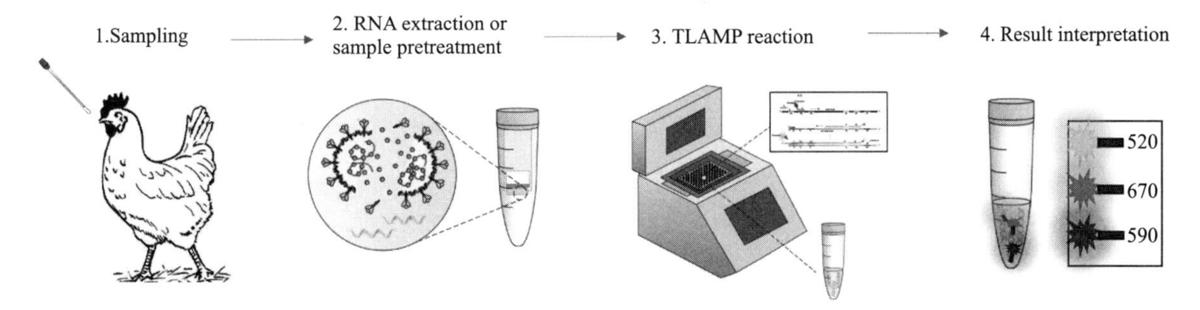

Figure 5-10-2 Flow chart of the TLAMP assay

Identifying the optimal proportions of the components in the composite probe FIP-FD among the total FIP primers

The composite probe FIP-FD was created by annealing the inner primer FIP with the probe FD. The ability of FIP-FD to guide new chain synthesis was attenuated by the fluorophore and quenching group, and the use of FIP-FD alone in the reaction system inhibited amplification. Previous studies have indicated that FIP-FD must be combined with a specific proportion of the unlabelled quenched group FIP (hereinafter referred to as unlabelled FIP) to effectively mitigate this inhibition[24-26]. In the single-template system, the proportions of FIP-FD among the total FIP primer were set to 0, 25%, 50%, 75%, and 100%, as outlined in Table 5-10-5. A real-time turbidimeter was used to generate a turbidity curve, with the x-axis denoting the reaction time and the y-axis indicating the turbidity intensity, and the curve reflected the quantity of the white precipitate byproduct from the TLAMP assay, magnesium pyrophosphate. The parameters that gave the most significant negative to positive contrast in fluorescence after the reaction and the shortest initiation time were determined to be the optimal reaction conditions. Specifically, the optimal proportion of the H5 subtype was 50%, while the optimal proportion of both the H7 and H9 subtypes was 25%. The results of this optimization process are illustrated in Figure 5-10-3.

Table 5-10-5 Proportion of FIP-FD composite probes among the total FIP primers

Proportion	0		25%		50%		75%		100%	
FIP-FD/(μmol/L)	0		0.067		0.133		0.200		0.267	
Unlabelled FIP/(μmol/L)	0.267		0.200		0.134		0.067		0	
Sample number in Figure 5-10-3	+	−	+	−	+	−	+	−	+	−
	1	6	2	7	3	8	4	9	5	10

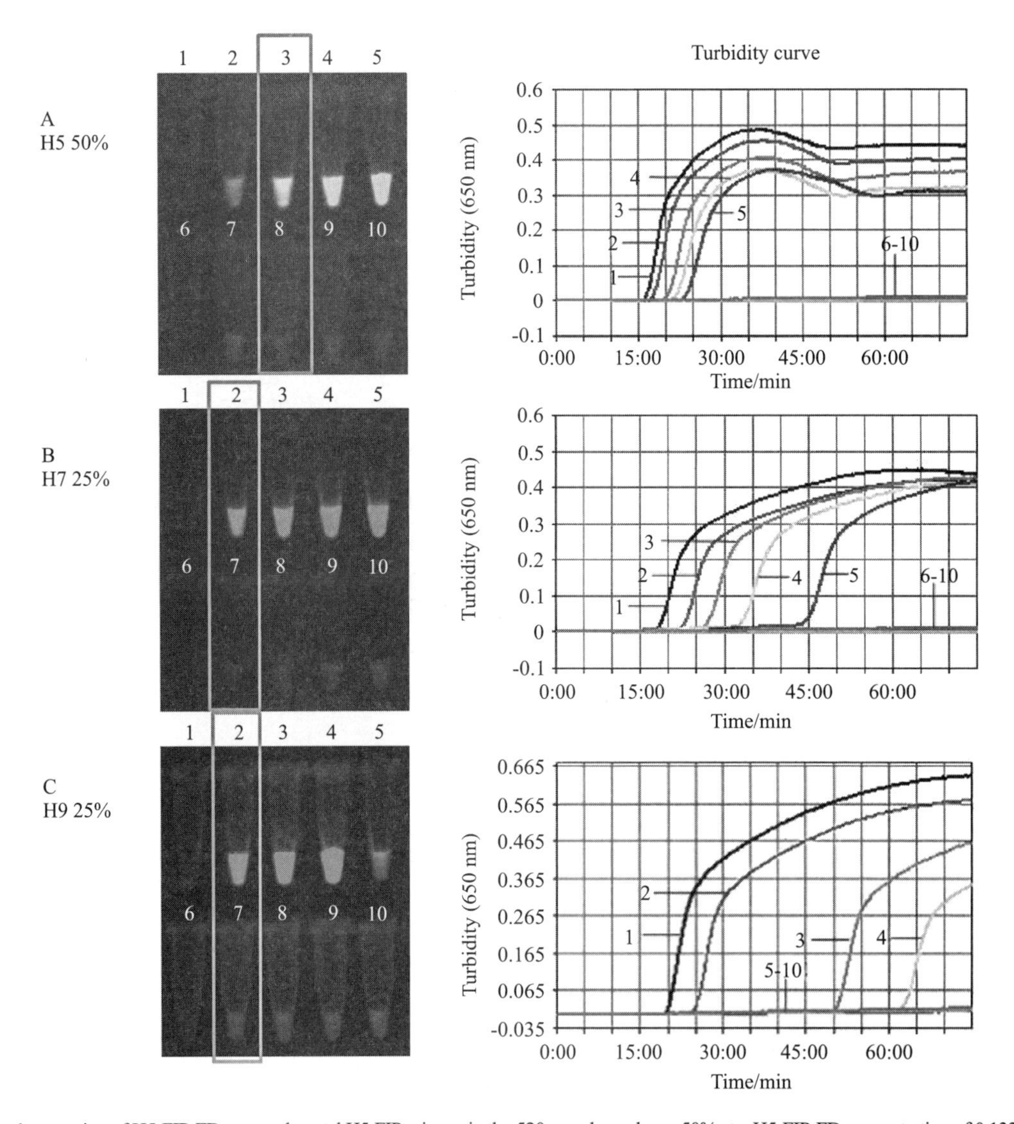

A: The optimal proportion of H5-FIP-FD among the total H5-FIP primers in the 520 nm channel was 50% at a H5-FIP-FD concentration of 0.133 μmol/L and an unlabelled H5-FIP concentration of 0.134 μmol/L and an initiation time of 20 min; B: The optimal proportion of H7 in the 670 nm channel was 25% with 0.067 μmol/L H7-FIP-FD, 0.2 μmol/L unlabelled H7-FIP, and an initiation time of 23 min; C: The optimal proportion of H9 in the 570 nm channel was 25% with 0.067 μmol/L H9-FIP-FD, 0.2 μmol/L unlabelled H9-FIP, and an initiation time of 25 min. When the H9 proportion was 100%, only 0.267 μmol/L H9-FIP-FD was used with no H9-FIP, and the reaction was completely suppressed. 1: FD-FIP composite probe among the total FIP=0; 2: 25%; 3: 50%; 4: 75%; and 5: 100%; 6, 7, 8, 9, and 10 are the negative controls for the corresponding proportions (color figure in appendix).

Figure 5-10-3 Optimal proportion of FIP-FD among the total FIP primers for each subtype

Determination of the optimal working concentration of FIP-Q and unlabelled FIP for each subtype

In this study, we found that adding an appropriate amount of FIP-Q effectively decreased the background fluorescence, which led to the positive results were more precisely identified. Nevertheless, since FIP-Q carries quenching groups, an excessively high concentration could impede the amplification process. Therefore, optimizing the FIP-Q concentration is crucial. Building upon the optimal proportion of the composite probe FIP-FD among the total FIP primers determined above, the concentrations of FIP-Q and unlabelled FIP were further fine-tuned, as outlined in Table 5-10-6. Based on the fluorescence intensity postreaction and the initiation time, the optimal working concentrations for each subtype of FIP-Q and unlabelled FIP were determined. The results of this optimization process are visually depicted in Figure 5-10-4.

Table 5-10-6 Working concentration of FIP-Q to unlabelled FIP

Subtypes	H5 (50%)						H7 (25%)						H9 (25%)					
FIP-FD/(μmol/L)	0.133						0.067						0.067					
Unlabelled FIP /(μmol/L)	0.134		0.067		0		0.2		0.1		0		0.2		0.1		0	
FIP-Q	0		0.067		0.134		0		0.1		0.2		0		0.1		0.2	
Sample number in Figure 5-10-4	+	−	+	−	+	−	+	−	+	−	+	−	+	−	+	−	+	−
	1	2	3	4	5	6	1	2	3	4	5	6	1	2	3	4	5	6

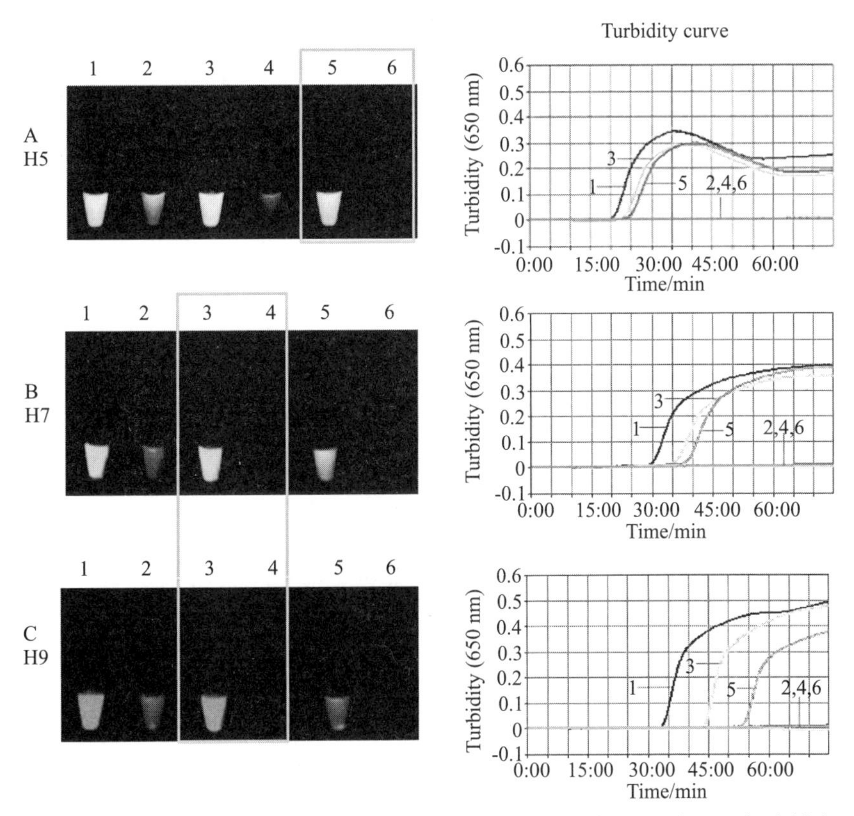

A: When only 0.134 μmol/L H5-FIP-Q was used instead of unlabelled H5-FIP in the H5 primer set, the reaction initiation time was 24 min, the background fluorescence was the lowest, and the difference between the negative and positive samples was the most pronounced; B: In the optimal reaction, the concentrations of the H7 primer group were 0.1 μmol/L unlabelled H7-FIP and 0.1 μmol/L H7-FIP-Q, the initiation time was 35 min, and the difference between the negative and positive samples was the most pronounced; C: In the optima reaction, the concentration of the H9 primer set was the same as that of the H7 primer set with a 43 min initiation time, and the difference between the negative and positive samples was the most pronounced. 1: Unlabelled FIP only (no FIP-Q was present); 2: The negative control of 1; 3: FIP-Q: unlabelled FIP=1:1; 4: The negative control of 3; 5: FIP-Q only (no unlabelled FIP was present); and 6: The negative control of 5 (color figure in appendix).

Figure 5-10-4 Optimal concentrations of FIP-Q and unlabelled FIP for each subtype

Determining the optimal ratio of the H9 primers for the TLAMP reaction

Following the combination of the three primers, the amplification efficiency of the H9 primer set was lower than those of H5 and H7, and H9 was inhibited via multiple amplifications. Within the triple reaction system, the quantities of the H5 and H7 primer sets were held constant while the amount of the H9 primer set was optimized by adjusting the quantities of H9-FIP-Q, unlabelled H9-FIP, H9-BIP, H9-Floop, and H9-Bloop used in the reaction. This process involved increasing the outer primer to inner primer to loop primer ratio in the reaction system. The specific sequences of the H9 primer sets are outlined in Table 5-10-7. The aim of this procedure was to determine the optimal amount of the H9 primer set, ensuring that the three subtypes are amplified in a steady and independent manner in the triple reaction.

Table 5-10-7 Amounts of H9 primers used in TLAMP

H9 primers	Concentration of H9 primers in 25 μL system/(μmol/L)					
Outer primer：Inner primer: Loop primer	1：8：2		1：16：4		1：24：6	
H9-FIP-FD	0.067					
Unlabelled H9-FIP	0.1		0.267		0.434	
H9-FIP-Q	0.1		0.2		0.3	
H9-BIP	0.267		0.534		0.801	
H9-F3	0.033		0.033		0.033	
H9-B3	0.033		0.033		0.033	
H9-Floop	0.066		0. 132		0.198	
H9-Bloop	0.066		0.132		0.198	
Sample number in Figure 5-10-5	+	−	+	−	+	−
	1	2	3	4	5	6

As illustrated in Figure 5-10-5, the optimal working concentration of H9-FIP-FD was 0.067 μmol/L. Additionally, the optimal ratio for the H9 primer set was determined to be 1:16:4, and the initiation time was 27min. Further increasing the amount of H9 primer did not improve H9 amplification during the reaction. After the primer combinations were optimized, each of the three subtypes could be consistently and independently amplified, demonstrating the robust performance of the TLAMP assay. After this adjustment was made, the TLAMP assay exhibited better performance and amplification.

Determining the optimal amount of Bst for the TLAMP reaction

The WarmStart Multi-Purpose LAMP/RT-LAMP 2 × Premix (including UDG) from New England Biolabs contained 8 mmol/L Mg^{2+}, Bst 2.0 WarmStart DNA polymerase, and WarmStart RT reverse transcriptase. Although the premix performs well for single-target amplification, its efficacy is diminished during tripletarget amplification due to sluggish amplification kinetics. This phenomenon arises from the coexistence of three sets of primers and probes within a single reaction tube, resulting in competitive reaction systems. Moreover, the presence of fluorophores and quenching groups inhibits amplification. Adding BST WarmStart DNA polymerase is essential for eliminating this issue because the polymerase facilitates smooth amplification of all three targets. The gradual inclusion of Bst 2.0 WarmStart DNA polymerase into the triple reaction system yielded the optimization results illustrated in Figure 5-10-6. In the absence of Bst 2.0, the TLAMP reaction exhibited slow kinetics, with an initiation time of 33 min. However, as the amount of Bst 2.0 increased, the amplification efficiency notably improved. The time needed to initiate the reaction and reach peak turbidity gradually decreased, which was accompanied by enhanced postreaction fluorescence. After 2 units of Bst 2.0 were integrated into the TLAMP system, the initiation time could be reduced to 16 min, and the postreaction fluorescence was sufficiently intense to distinguish between positive and negative samples. Further increases in the BST polymerase concentration had negligible effects on the reaction results; therefore, 2 units was established as the optimal amount of BST (2.0). This dose significantly enhanced amplification efficiency while expediting the TLAMP reaction..

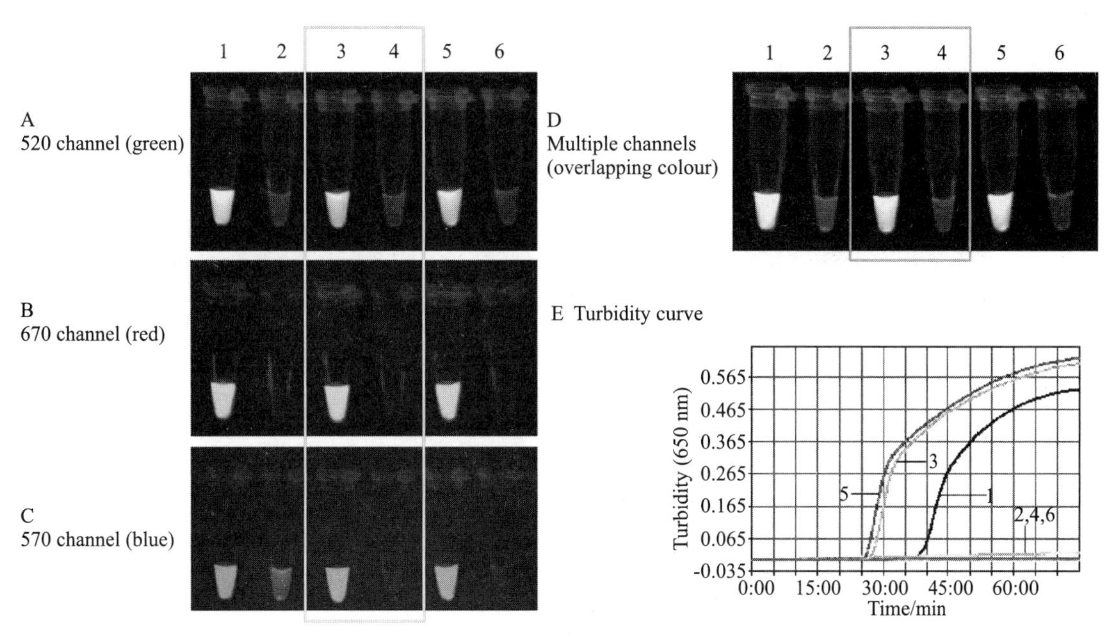

A: 520 nm channel; B: 670 nm channel; C: 570 nm channel; D: Multiple channels; E: Turbidity curve. 1: The ratio of H9 outer primer: inner primer: loop primer=1 : 8 : 2; 2: The negative control of 1; 3: H9 primer set with a ratio of 1:16:4; 4: The negative control of 3; 5: H9 primer set with a ratio of 1 : 24 : 6; and 6: The negative control of 5 (color figure in appendix).

Figure 5-10-5 Optimal amount of the H9 primers for the TLAMP reaction

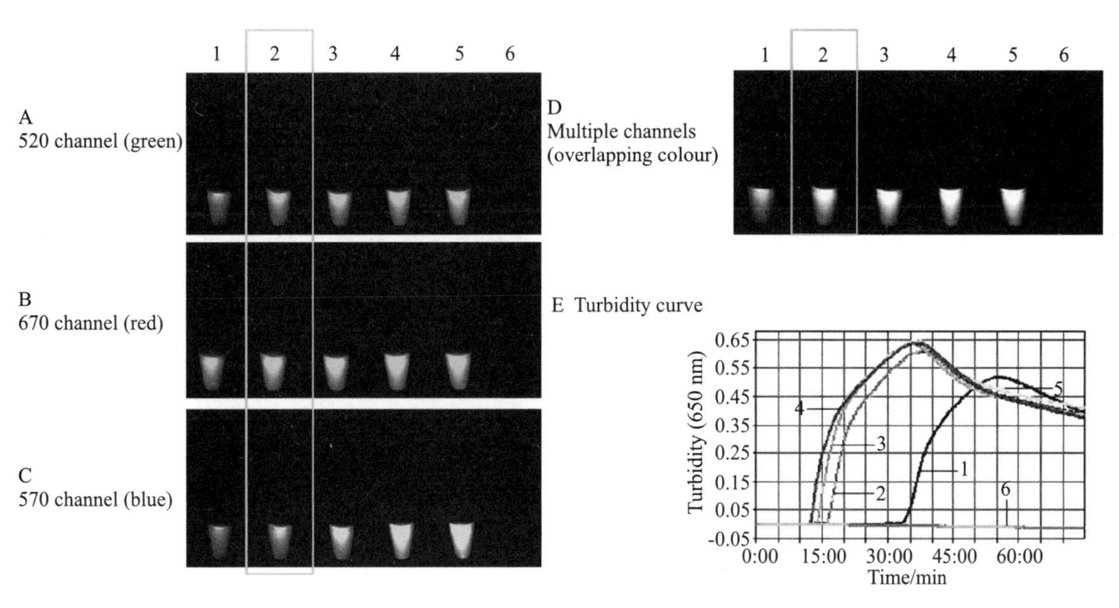

A: 520 nm channel; B: 670 nm channel; C: 570 nm channel; D: Multiple channels; E: Turbidity curve. 1: When no DNA polymerase was added, the reaction was slow, and 33 minutes was needed to initiate the reaction; 2: Each reaction with 2 U Bst 2.0 had an initial reaction time of 16 minutes; 3: 4 U Bst 2.0 was added over 14 minutes; 4: 6 U Bst 2.0 was added over 12 minutes; 5: 8 U Bst 2.0 was added over 12 minutes, and the effect of adding 8 U Bst 2.0 and 6 U; 6: The negative control (color figure in appendix).

Figure 5-10-6 Optimal amount of Bst 2.0 WarmStart DNA polymerase in the TLAMP reaction

Specificity of the TLAMP assay

To assess the specificity of the TLAMP assay, AIV H1-H16 subtypes and samples artificially coinfected with H5, H7, H9, and other avian control virus nucleic acids, as listed in Table 5-10-1, were analysed using the TLAMP assay. The specificity test results are presented in Table 5-10-1 and Figure 5-10-7. H5-positive samples exhibited green fluorescence (FAM) in the 520nm channel, H7-positive samples displayed red fluorescence (Cy5) in the 670nm channel, and H9-positive samples showed blue fluorescence (Cy3) in the 570nm channel. The coinfected samples exhibited overlapping colours across multiple channels. In contrast, other control viruses, including NDV, IBV, ARV, ILTV, CIAV, FAdV-4, ChPV and MDV, did not produce fluorescence or amplification across multiple channels, and the detection results were negative. These findings confirm the robust specificity of the TLAMP assay for identifying these three AIV subtypes.

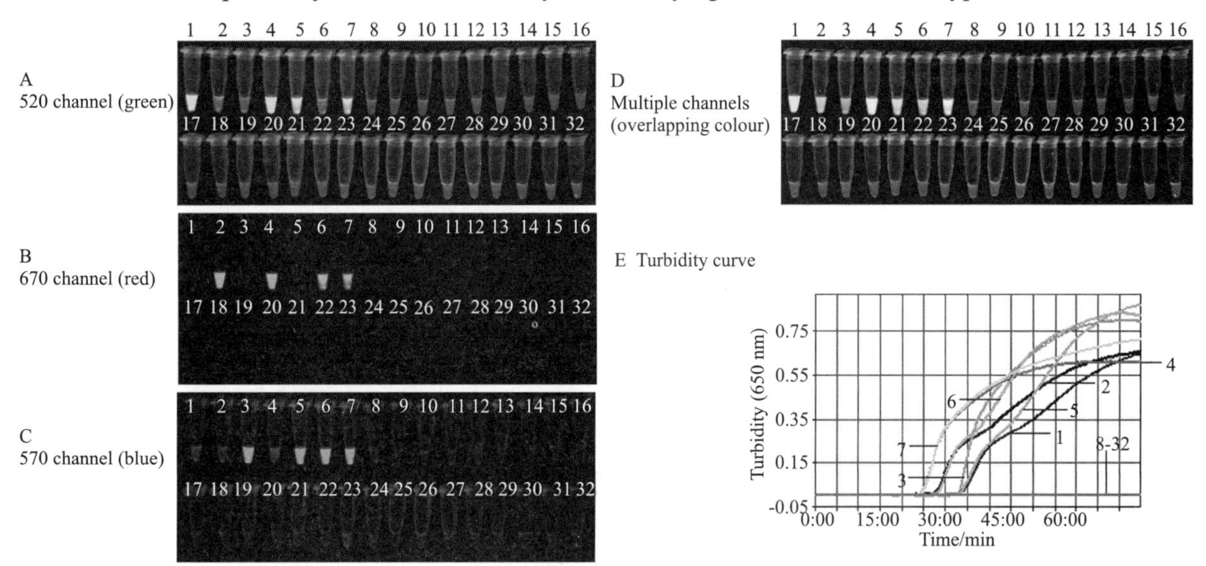

A: 520 nm channel; B: 670 nm channel; C: 570 nm channel; D: Multiple channels; E: Turbidity curve. 1: H5; 2: H7; 3: H9; 4: H5+H7; 5: H5+H9; 6: H7+H9; 7: H5+H7+H9; 8: H1; 9: H2; 10: H3; 11: H4; 12: H6; 13: H8; 14: H10; 15: H11; 16: H12; 17: H13: 18: H14; 19: H15; 20: H16; 21: NDV; 22: IBV; 23: ARV; 24: ILTV; 25: CIAV; 26: FAdV-4; 27: ChPV; 28: MDV; and 29-32: Negative control (color figure in appendix).

Figure 5-10-7 Specificity of the TLAMP assay

Sensitivity of the TLAMP assay

To assess the sensitivity of the TLAMP assay, eight concentrations of the RNA standard were analysed. Two microlitres of each RNA standard served as the template for testing, and 10 replicates were conducted for each standard. Table 5-10-3 summarizes the test results. Probit analysis was performed with SPSS software (SPSS, Inc., Chicago) based on the results shown in Table 5-10-3; the TLAMP detection limits for the three subtypes, along with the 95% confidence intervals (CIs), were also computed, as presented in Table 5-10-8 [27]. The TLAMP assay could detect a minimum of 205 copies of H5 RNA, 360 copies of H7 RNA, and 545 copies of H9 RNA per reaction.

Table 5-10-8 The TLAMP detection limit obtained by probit analysis

Probit[a]	Template concentration/(copies/μL), 95% CI		
	H5	H7	H9
0.95	205 (150～481)	360 (272～1 062)	545 (402～1 299)

Note: [a]Proportion of replication predicted by SPSS software based on the data from the 8 RNA standards listed in Table 5-10-3 analysed with 10 replicates.

Interference in the TLAMP assay

In clinical scenarios, the concentration of each subtype template in the sample under scrutiny remains unknown. Templates with higher concentrations can swiftly deplete the reaction reagents, impeding the amplification of templates present in lower concentrations. To address this issue, seven artificially coinfected samples featuring varying concentrations of H5, H7, or H9 *in vitro*-transcribed RNA were prepared and analysed using the TLAMP assay. The interference test results, as depicted in Figure 5-10-8, revealed that all the subtypes were independently amplified even though the concentrations of the H5, H7, and H9 templates differed within a single coinfection sample. After the reaction, the corresponding fluorescence was observed in the respective channel, rendering the detection results accurate and decipherable. These results underscore the minimal interference observed in the TLAMP assay, indicating that high-concentration templates within the same sample do not impede the amplification of low-concentration templates.

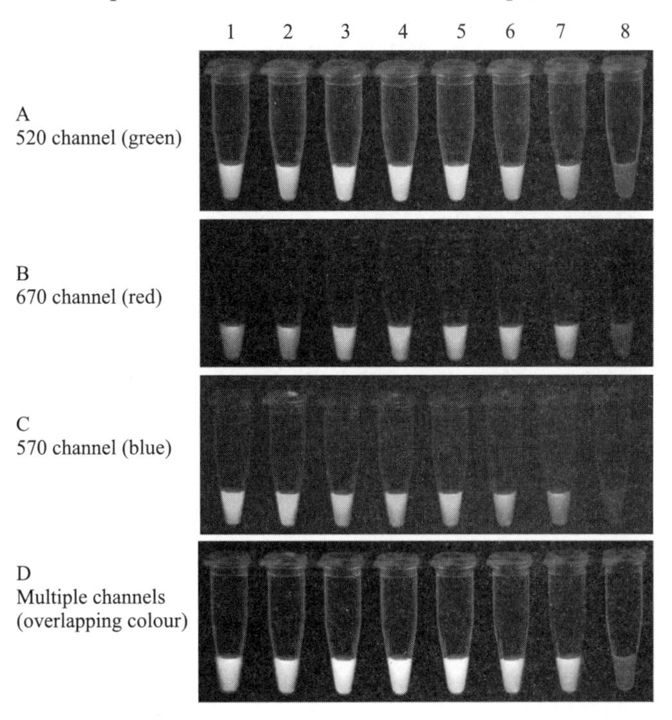

A: 520 nm channel; B: 670 nm channel; C: 570 nm channel; D: Multiple channels. 1: Sample 1, H5 (10^2 copies/μL) +H7 (10^2 copies/μL) + H9 (10^8 copies/μL); 2: Sample 2, H5 (10^4 copies/μL) +H7 (10^4 copies/μL) +H9 (10^7 copies/μL); 3: Sample 3, H5 (10^4 copies/μL) +H7 (10^5 copies/μL) + H9 (10^6 copies/μL); 4: Sample 4, H5 (10^5 copies/μL) +H7 (10^3 copies/μL) +H9 (10^5 copies/μL); 5: Sample 5, H5 (10^6 copies/μL) +H7 (10^8 copies/μL) + H9 (10^4 copies/μL); 6: Sample 6, H5 (10^4 copies/μL) +H7 (10^7 copies/μL) +H9 (10^3 copies/μL); 7: Sample 7, H5 (10^2 copies/μL) +H7 (10^8 copies/μL) + H9 (10^2 copies/μL); 8: Sample 8, negative control (color figure in appendix).

Figure 5-10-8 The results of the TLAMP assay

Clinical application of the TLAMP assay

From 240 clinical swab samples, 26 samples were identified as only H9-positive by both the TLAMP assay and qPCR, for an infection rate of 10.8%. However, no coinfected samples were detected. The clinical results obtained both the TLAMP assay and qPCR are shown in Table 5-10-9. Notably, the results of the TLAMP assay were 100% consistent with those of a previously published qPCR assay. The cDNA of different tissue samples from eight H5N2-infected SPF chickens donated by the University of Connecticut and the cDNA of different tissue samples from three H7N9-infected SPF chickens donated by China Agricultural

University were analysed using the TLAMP assay, qPCR and sequencing, and the results indicated that the TLAMP assay could detect H5 and H7. These sequencing results corroborated the authenticity of all the positive identifications. Additionally, a cycle threshold (CT) >30 was found in 4 of the 37 positive samples. The TLAMP assay reliably detected viruses in samples amplified by qPCR at a CT<35. Notably, this TLAMP assay offers several advantages, including low cost, convenient operation, and wide applicability.

Table 5-10-9 TLAMP and qPCR results for 37 positive samples

Type of chicken samples	Positive sample number	TLAMP (col or)	qPCR (CT value)	Sequencing
Tissue samples from challenged SPF chickens	1	H5+ (green)	H5+ (13.63)	H5 +
	2	H5+ (green)	H5+ (17.51)	H5 +
	3	H5+ (green)	H5+ (19.62)	H5+
	4	H5+ (green)	H5+ (18.73)	H5 +
	5	H5+ (green)	H5+ (14.23)	H5 +
	6	H5+ (green)	H5+ (15.03)	H5 +
	7	H5+ (green)	H5+ (20.07)	H5 +
	8	H5+ (green)	H5+ (22.75)	H5 +
	9	H7+ (red)	H7+ (14.77)	H7 +
	10	H7+ (red)	H7+ (18.26)	H7 +
	11	H7+ (red)	H7+ (19.67)	H7 +
Swab samples from LBMs	12	H9+ (blue)	H9+ (28.41)	H9 +
	13	H9+ (blue)	H9+ (23.69)	H9+
	14	H9 (blue)	H9+ (22.72)	H9 +
	15	H9+ (blue)	H9+ (25.82)	H9 +
	16	H9+ (blue)	H9+ (**31.80**)	H9 +
	17	H9+ (blue)	H9+ (21.92)	H9 +
	18	H9+ (blue)	H9+ (29.61)	H9 +
	19	H9+ (blue)	H9+ (28.40)	H9 +
	20	H9+ (blue)	H9+ (**33.26**)	H9 +
	21	H9+ (blue)	H9+ (26.85)	H9 +
	22	H9+ (blue)	H9+ (22.72)	H9 +
	23	H9+ (blue)	H9+ (20.55)	H9 +
	24	H9+ (blue)	H9+ (22.57)	H9 +
	25	H9+ (blue)	H9+ (27.31)	H9 +
	26	H9+ (blue)	H9+ (26.42)	H9 +
	27	H9+ (blue)	H9+ (28.56)	H9 +
	28	H9+ (blue)	H9+ (22.77)	H9 +
	29	H9+ (blue)	H9+ (19.36)	H9 +
	30	H9+ (blue)	H9+ (27.17)	H9 +
	31	H9+ (blue)	H9+ (**34.05**)	H9 +
	32	H9+ (blue)	H9+ (19.15)	H9 +
	33	H9+ (blue)	H9+ (28.62)	H9 +
	34	H9+ (blue)	H9+ (15.24)	H9 +
	35	H9+ (blue)	H9+ (23.63)	H9 +
	36	H9+ (blue)	H9+ (26.92)	H9 +
	37	H9+ (blue)	H9+ (**34.85**)	H9 +

Discussion

Due to intensive poultry farming, wild bird migration, and live poultry trading in markets, humans are in close proximity to many birds, creating favourable conditions for the increased transmission of AIVs and the emergence of new subtypes. The H5, H7, and H9 subtypes of avian influenza viruses have become endemic to domestic poultry in China and are persistent threats to public health and the poultry industry[28].

Although there have been many studies on LAMP assays for AIV detection, these methods are unable to discriminate among different subtypes. Golabi et al.[29] introduced a universal LAMP method targeting matrix gene sequences that can detect all AIV subtypes but cannot differentiate between individual subtypes. Moreover, Zhang et al.[30] developed a LAMP assay to specifically detect AIV subtypes H5 and H9 with a detection limit of 100～1 000 copies per reaction. However, this method requires fluorescence real-time PCR equipment and can only ascertain whether a sample is positive without identifying the specific subtype responsible for the positive result.

This study introduces an innovative approach involving the integration of LAMP technology with a probe. For each AIV subtype, an FD probe complementary to the F1c segment of the inner primer was designed. The FD probe incorporates distinct fluorophores, which become detached during amplification and release fluorophores with varying colours. The TLAMP results are interpreted based on both emission in a fluorescence channel and the colour and can be directly observed by the naked eye. Compared to conventional LAMP methods, the TLAMP method allows the high-throughput discrimination of multiple targets in the same reaction tube; in addition, the output is simplified and easier to interpret. The TLAMP uses fluorophores with three different emission wavelengths, presenting varied colours under different fluorescence channels. Compared to conventional methods that rely on colour change or the formation of white precipitates, the TLAMP assay produces results that can be more intuitively and accurately assessed due to the vivid contrast of the colours green, red, and blue. Furthermore, the TLAMP reaction eliminates several steps, including opening the lid, adding dye, and performing electrophoresis, which substantially reduces the risk of laboratory contamination. In future endeavours, this method could be applied in conjunction with a portable multiplex fluorescence channel analyser to develop a POC detection instrument, enabling real-time field detection at the grassroots level.

When developing a diagnostic tool, one always desires the best possible sensitivity and specificity. It is a tall order for any method, including LAMP, to outperform qPCR in this respect, since qPCR can routinely attain a sensitivity of only a few copies of the target. To date, the disadvantages of qPCR have been described in many publications, but in terms of absolute sensitivity, other methods can only approach that of qPCR. The high sensitivity of qPCR is attributed to its simple reaction system with 2 primers and one probe, complete denaturation at high temperatures, precise signal detection by a fluorescence detector, and skilled operation by laboratory personnel. However, due to the numerous primers used in LAMP assays, the samples require only a pretreatment without nucleic acid extraction, and the interpretation of the results depends on subjective judgement. Particularly in a multireaction system, competition among multiple primers and probes decreases the sensitivity, so the sensitivity of the TLAMP assay is inevitably slightly less than that of qPCR[31]. In this study, TLAMP reliably detected three AIV subtypes with CT values as low as 35. Nevertheless, the reliability of CT values ranging from 35 to 38 could not be assessed in this study. Moreover, H9 sensitivity was slightly lower than that of the other two subtypes. This may have occurred because Cy3 was used as the fluorophore on H9, which exhibits lower fluorescence intensity than FAM and Cy5[32]. Consequently, in the triple reaction system, the fluorescence signal of H9 was somewhat inferior amidst the backdrop of strong fluorescence from FAM and Cy5 in multiple channels.

To successfully establish a multi-LAMP approach, the optimal quantities of primers labelled with fluorophores or quenched groups, FIP-FD and FIP-Q, must be determined. While more fluorophores lead to an increase in visual fluorescence, the background fluorescence increases concurrently, which directly impairs the interpretation of the results. Adding an appropriate amount of FIP-Q to the reaction system effectively mitigated the background fluorescence; as a result, the positive and negative reaction results were clearer, and the results could be more accurately interpreted. However, regardless of whether fluorophores or quenched groups are added to label the primers, the primers must guide the synthesis of new chains during amplification, and the labelling or quenching of primers diminishes their ability to synthesize new chains. Therefore, unlabelled FIP was added to the reaction mixture to promote the reaction. Furthermore, the components of the TLAMP reaction system are intricate, and optimizing the ratio of each primer set is critical. In conventional LAMP assays, six primers are needed to amplify a target. In the TLAMP system, amplifying three targets requires 18 primers and three probes in a single reaction system, which leads to potential competition among different subtype primer sets. If the amplification efficiency of one target primer is low, the other primers may consume the reaction components first, resulting in amplification failure. In practice, the amplification efficiency varies among primer sets. Therefore, after combining the three primer sets, it is imperative to adjust the quantity of the set with lower amplification efficiency to ensure that each target is independently and robustly amplified without influencing the other sets. In this study, we meticulously optimized the quantities of the FIP-FD, FIP-Q, and H9 primer sets and successfully identified H5, H7, and H9 subtypes of AIV.

Conclusions

In summary, we successfully developed a TLAMP assay, which is a distinctive, sensitive, rapid, and high-throughput tool for the concurrent detection of H5, H7, and H9 subtypes of AIVs. Through this innovative approach, comprehensive prevention and control measures can be efficiently implemented, thereby reducing the incidence of avian influenza.

References

[1] JAVANIAN M, BARARY M, GHEBREHEWET S, et al. A brief review of influenza virus infection. J Med Virol, 2021, 93(8):4638-4646.

[2] SUN Y, ZHANG T, ZHAO X, et al. High activity levels of avian influenza upwards 2018–2022: A global epidemiological overview of fowl and human infections. One Health, 2023, 16: 100511.

[3] FEREIDOUNI S, STARICK E, KARAMENDIN K, et al. Genetic characterization of a new candidate hemagglutinin subtype of influenza a viruses. Emerg Microbes Infec, 2023, 12(2): 2225645.

[4] NAGUIB M M, VERHAGEN J H, MOSTAFA A, et al. Global patterns of avian influenza a (H7): Virus evolution and zoonotic threats. Fems Microbiol Rev, 2019, 43(6):608-621.

[5] YAMAJI R, SAAD M D, DAVIS C T, et al. Pandemic potential of highly pathogenic avian influenza clade 2.3.4.4 A(H5) viruses. Rev Med Virol, 2020, 30(3): e2099.

[6] SUBBARAO K, KLIMOV A, KATZ J, et al. Characterization of an avian influenza a (H5N1) virus isolated from a child with a fatal respiratory illness. Science, 1998, 279(5349): 393-396.

[7] YIN X, DENG G, ZENG X, et al. Genetic and biological properties of H7N9 avian influenza viruses detected after application of the H7N9 poultry vaccine in China. PLOS Pathog, 2021, 17(4): e1009561.

[8] LIU W J, XIAO H, DAI L, et al. Avian influenza a (H7N9) virus: From low pathogenic to highly pathogenic. Front Med-Prc, 2021, 15(4): 507-527.

[9] YU X, JIN T, CUI Y, et al. Influenza H7N9 and H9N2 viruses: Coexistence in poultry linked to human H7N9 infection and genome characteristics. J Virol, 2014, 88(6): 3423-3431.

[10] ZHOU Y, GAO W, SUN Y, et al. Effect of the Interaction between Viral PB2 and Host SphK1 on H9N2 AIV Replication in Mammals. Viruses, 2022, 14(7):1585.

[11] GOHRBANDT S, VEITS J, BREITHAUPT A, et al. H9 avian influenza reassortant with engineered polybasic cleavage site displays a highly pathogenic phenotype in chicken. The Journal of General Virology, 2011, 92(Pt 8):1843-1853.

[12] CARNACCINI S, PEREZ D R. H9 influenza viruses: an emerging challenge. Cold Spring Harb Perspect Med, 2020, 10(6):a038588.

[13] CHEN T, TAN Y, SONG Y, et al. Enhanced environmental surveillance for avian influenza A/H5, H7 and H9 viruses in Guangxi, China, 2017-2019. Biosafety and Health, 2023, 5: 30-36.

[14] CAPUA I. Three open issues on Avian Influenza-H5, H7, H9 against all odds. Brit Poultry Sci, 2013, 54(1): 1-4.

[15] SHI J, ZENG X, CUI P, et al. Alarming situation of emerging H5 and H7 avian influenza and effective control strategies. Emerg Microbes Infect, 2023, 12(1): 2155072.

[16] YANG F, DONG D, WU D, et al. A multiplex real-time RT-PCR method for detecting H5, H7 and H9 subtype avian influenza viruses in field and clinical samples. Virus Res, 2022, 309: 198669.

[17] XIE Z, LUO S, XIE L, et al. Simultaneous typing of nine avian respiratory pathogens using a novel GeXP analyzer-based multiplex PCR assay. Journal of Virol Methods, 2014, 207: 185-189.

[18] XIE Z, PANG Y S, LIU J, et al. A multiplex RT-PCR for detection of type a influenza virus and differentiation of avian H5, H7, and H9 hemagglutinin subtypes. Mol Cell Probes, 2006, 20(3-4): 245-249.

[19] LUO S, XIE Z, LI M, et al. Simultaneous differential detection of H5, H7, H9 and nine NA subtypes of avian influenza viruses via a GeXP assay. Microorganisms, 2024, 12(1):143.

[20] MOEHLING T J, CHOI G, DUGAN L C, et al. LAMP diagnostics at the Point-of-Care: Emerging trends and perspectives for the developer community. Expert Rev Mol Diagn, 2021, 21(1): 43-61.

[21] CHOI G, MOEHLING T J, MEAGHER R J. Advances in RT-LAMP for COVID-19 testing and diagnosis. Expert Rev Mol Diagn, 2023, 23(1): 9-28.

[22] DENG H, JAYAWARDENA A, CHAN J, et al. An ultra-portable, self-contained point-of-care nucleic acid amplification test for diagnosis of active COVID-19 infection. Sci Rep, 2021, 11(1): 15176.

[23] GANGULI A, MOSTAFA A, BERGER J, et al. Rapid isothermal amplification and portable detection system for SARS-CoV-2. Proc Natl Acad Sci U S A, 2020, 117: 22727-22735.

[24] FAN Q, XIE Z, WEI Y, et al. Development of a visual multiplex fluorescent LAMP assay for the detection of foot-and-mouth disease, vesicular stomatitis and bluetongue viruses. PLOS ONE, 2022, 17(12): e278451.

[25] TANNER N A, ZHANG Y, EVANS T J. Simultaneous multiple target detection in real-time loop-mediated isothermal amplification. Biotechniques, 2012, 53(2): 81-89.

[26] FAN Q, XIE Z, ZHANG Y, et al. A multiplex fluorescence-based loop-mediated isothermal amplification assay for identifying chicken parvovirus, chicken infectious anaemia virus, and fowl aviadenovirus serotype 4. Avian Pathol, 2023, 52(2): 128-136.

[27] SMIEJA M, MAHONY J B, GOLDSMITH C H, et al. Replicate PCR testing and probit analysis for detection and quantitation of Chlamydia pneumoniae in clinical specimens. J Clin Microbiol, 2001, 39(5): 1796-801.

[28] CHEN X, LI C, SUN H T, et al. Prevalence of avian influenza viruses and their associated antibodies in wild birds in China: A systematic review and meta-analysis. Microb Pathogenesis, 2019, 135: 103613.

[29] GOLABI M, FLODROPS M, GRASLAND B, et al. Development of reverse transcription Loop-Mediated isothermal amplification assay for rapid and On-Site detection of avian influenza virus. Front Cell Infect Microbiol, 2021, 11: 652048.

[30] ZHANG S, SHIN J, SHIN S, et al. Development of reverse transcription loop-mediated isothermal amplification assays for point-of-care testing of avian influenza virus subtype H5 and H9. Genomics Inform, 2020, 18(4): e40.

[31] DAO T V, HERBST K, BOERNER K, et al. A colorimetric RT-LAMP assay and LAMP-sequencing for detecting SARS-CoV-2 RNA in clinical samples. Sci Transl Med, 2020, 12(556): eabc7075.

[32] MAO H, LUO G, ZHANG J, et al. Detection of simultaneous multi-mutations using base-quenched probe. Anal Biochem, 2018, 543: 79-81.

A multiplex RT-PCR for detection of type A influenza virus and differentiation of avian H5, H7, and H9 hemagglutinin subtypes

Xie Zhixun, Pang Yaoshan, Liu Jiabo, Deng Xianwen, Tang Xiaofei, Sun Jianhua, and Khan Mazhar I

Abstract

A multiplex reverse transcriptase-polymerase chain reaction (mRT-PCR) was developed and optimized for the detection of type A influenza virus; the assay simultaneously differentiates avian H5, H7 and H9 hemagglutinin subtypes. Four sets of specific oligonucleotide primers were used in this test for type A influenza virus, H5, H7 and H9 hemagglutinin subtypes. The mRT-PCR DNA products were visualized by gel electrophoresis and consisted of fragments of 860 bp for H5, 634 bp for H7, 488 bp for H9 hemagglutinin subtypes, and 244 bp for type A influenza virus. The common set primers for type A influenza virus were able to amplify a 244 bp DNA band for any of the other subtypes of AIV. The mRT-PCR assay developed in this study was found to be sensitive and specific. Detection limit for PCR-amplified DNA products was 100 pg for the subtypes H5, H7, and H9 and 10 pg for type A influenza virus in all subtypes. No specific amplification bands of the same sizes (860, 634 and 488 bp) could be amplified for RNA of other influenza hemagglutinin subtypes, nor specific amplification bands of type A influenza (244 bp) for other viral or bacterial pathogens.

Keywords

multiplex, polymerase chain reaction, avian influenza virus, H5 subtype, H7 subtype, H9 subtype, hemagglutinin

Introduction

Influenza is a zoonotic disease, infecting a wide variety of warm-blooded animals, including birds and mammals. Influenza viruses are classified into types A, B, and C. Influenza A viruses are responsible for major disease problem in birds, as well as in humans[1-10]. Infections among domestic or confined birds have been associated with a variety of disease syndromes ranging from sub-clinical to mild upper respiratory disease, to loss of egg production, to acute generalized fatal disease. In domestic avian species, influenza viruses have caused considerable economic losses[1, 2].

Influenza A are enveloped, negative-sense RNA viruses. Influenza A viruses are further classified into subtypes on the basis of the antigenic properties of their two surface glycoproteins, hemagglutinin (HA) and neuraminidase (NA). To date, 15 HA and 9 NA subtypes have been identified. All influenza A virus subtypes have been found in aquatic and domestic birds, but only a few subtypes have been recovered from mammals and humans.

Among 15 HA subtypes, only H5 and H7 are highly virulent in poultry[1]. Historically, highly pathogenic avian influenza viruses (AIV) of poultry only belonged to the H5 and H7 hemagglutinin (HA) subtypes.

Therefore, because there is a greater risk for these subtypes to become highly pathogenic, it is important to identify them specifically in surveillance programs[10, 16]. Recently, H9 subtypes have been seen to cause infections in poultry[3, 6]. Separate from the H5N1 influenza virus, another subtype of influenza virus, H9N2, has become panzootic in the last decade and has been isolated from different types of terrestrial poultry worldwide[3, 6, 9].

Diagnosis of influenza A virus infection is routinely done by the isolation and identification of the virus. Serotyping is required to differentiate the subtypes of the AI viruses and is laborious and time-consuming. Furthermore, other tests required to determine the HA cleavage site sequence must be done to determine its potential virulence[12, 13]. Although single band PCR has been used to detect and differentiate subtypes, it only recognizes one specific subtype at a time[14]. Standard RT-PCR has been previously applied to the detection of avian influenza virus[10, 15, 16] and each of the 15 HA subtypes[14]. In addition, real time-RT-PCR assays for influenza virus have been developed for the detection of influenza virus types A and B[17] and differentiation of two subtypes H5, and H7[11]. Equipment costs and specific technical training requirements limit usefulness of these assays as routine laboratory tests. In our study, we have developed a specific and sensitive multiplex RT-PCR that can simultaneously detect and differentiate the three most important subtypes of avian influenza viruses.

Materials and methods

Avian pathogens and culture conditions

The avian pathogens used in this study are listed in Table 5-11-1. All avian influenza virus (AIV) subtypes, Newcastle disease virus (NDV), and infectious bronchitis virus (IBV) were propagated in the allantoic cavity of 10-day-old specific-pathogen-free (SPF) embryonated chicken eggs, whereas infectious laryngotracheitis virus (ILT) was propagated on the chorioallantoic membrane in 10-day-old SPF embryonated chicken eggs as described[18]; the allantoic fluids from embryonated eggs infected with AI, NDV, ILT and IBV were harvested after 36 h of incubation at 37 ℃ [18]. *Mycoplasma gallisepticum* (MG) was propagated in Frey's broth and incubated at 37 ℃ as described[19].

Table 5-11-1 Avian pathogens used in multiplex PCR

Avian pathogen	Subtype	Source	Results of mRT-PCR			
			Type A	H5	H7	H9
Duck/HK/717/79-d1	H1N3	Uni. Of HK	+	−	−	−
Duck/HK/717/79-d7	H1N3	Uni. of HK	+	−	−	−
Human/NJ/8/76	H1N1	Uni. of HK	+	−	−	−
Duck/HK/77/76 d77/3	H2N3	Uni. of HK	+	−	−	−
Duck/HK/77/76	H2N3	Uni. of HK	+	−	−	−
Duck/HK/526/79/2B	H3N6	Uni. of HK	+	−	−	−
Duck/HK/526/79/2B-1	H3N6	Uni. of HK	+	−	−	−
Duck/HK/668/79	H4N5	Uni. of HK	+	−	−	−
Duck/HK/668/79-1	H4N5	Uni. of HK	+	−	−	−
Duck/HK/313/78	H5N3	Uni. of HK	+	+	−	−

continued

Avian pathogen	Subtype	Source	Results of mRT-PCR			
			Type A	H5	H7	H9
Duck/HK/313/78-1	H5N3	Uni. of HK	+	+	−	−
Duck/Guangxi/1/04	H5N1	GVRI	+	+	−	−
Duck/Guangxi/2/04	H5N1	GVRI	+	+	−	−
Duck/Guangxi/3/04	H5N1	GVRI	+	+	−	−
Chicken/Guangxi/1/04	H5N1	GVRI	+	+	−	−
Chicken/Guangxi/2/04	H5N1	GVRI	+	+	−	−
Goose/Guangxi/1/04	H5N1	GVRI	+	+	−	−
Goose/Guangxi/2/04	H5N1	GVRI	+	+	−	−
Duck/HK/531/79-1	H6N8	Uni. of HK	+	−	−	−
Duck/HK/531/79	H6N8	Uni. of HK	+	−	−	−
Duck/HK/47/76	H7N2	Uni. of HK	+	−	+	−
Duck/HK/47/76-1	H7N2	Uni. of HK	+	−	+	−
Turkey/ont/6118/68	H8N4	Uni. of HK	+	−	−	−
Turkey/ont/6118/68-1	H8N4	Uni. of HK	+	−	−	−
Duck/HK/147/77	H9N6	Uni. of HK	+	−	−	+
Duck/HK/147/77-1	H9N6	Uni. of HK	+	−	−	+
Duck/Guangxi/1/00	H9N2	GVRI	+	−	−	+
Duck/Guangxi/2/00	H9N2	GVRI	+	−	−	+
Duck/Guangxi/3/00	H9N2	GVRI	+	−	−	+
Duck/HK/876/80	H10N3	Uni. of HK	+	−	−	−
Duck/HK/876/80-1	H10N3	Uni. of HK	+	−	−	−
Duck/HK/661/79	H11N3	Uni. of HK	+	−	−	−
Duck/HK/661/79-1	H11N3	Uni. of HK	+	−	−	−
Duck/HK/862/80	H12N5	Uni. of HK	+	−	−	−
Duck/HK/862/80-1	H12N5	Uni. of HK	+	−	−	−
Gull/MD/704/77	H13N5	Uni. of HK	+	−	−	−
Gull/MD/704/77-1	H13N5	Uni. of HK	+	−	−	−
NDV	F68-E9	CIVDC, Beijing	−	−	−	−
ILTV	Field	CIVDC, Beijing	−	−	−	−
IBV	M41	CIVDC, Beijing	−	−	−	−
MG	S6	UCDavis, Calif	−	−	−	−
Liver	SPF chicken		−		−	−
Lung	SPF chicken		−		−	−
Small intestine	SPF chicken		−		−	−

Note: Uni. of HK, University of Hong Kong, China; GVRI, Guangxi Veterinary Research Institute; CIVDC, China Institute of Veterinary Drug Control; UCDavis, University of California, Davis.

Extraction of RNA and DNA

Allantoic fluids from embryonated eggs infected with AI, NDV, IBV and ILT were first clarified by centrifugation at 500 g for 15 min. The supernatants were transferred to new tubes and then centrifuged at 45 000 g for 30 min at 4 ℃ . The supernatants were discarded and the pellets treated with 25 mL Rnase-free TE buffer (10 mM Tris-HCl and 1 mM ethylendiaminetetraacetic acid (EDTA)). The RNA extraction from AI, NDV, and IBV was carried out according to the Trizol LS manufacturer's protocol (Trizol, Invitrogen, Carlsbad, CA, USA). DNA from ILT and MG was extracted using the phenol: chloroform: isoamyl alcohol (25: 24: 1 v/v) (Amersham Life Science, Cleaveland, OH, USA) method as described by Pang et al. [20]. The concentrations of the RNA or DNA were determined by spectrophotometry using the UV2501PC (Shimadzu Corporation, Tokyo, Japan) and stored at –20 ℃ . An aliquot of 250 μL each was used for RNA extraction from the AI virus isolates listed in Table 5-11-1 according to the Trizol LS manufacturer's protocol (Trizol, Invitrogen, Carlsbad, CA, USA). Four-week-old SPF chicken (Jinan City, Shandong, China) was euthanized and 30 mg tissue samples from lung, liver, and small intestine were minced in TE buffer and RNA extraction was carried out according to the Trizol LS manufacturer's protocol for tissue samples (Trizol, Invitrogen, Carlsbad, CA, USA).

Primers designs and selection

Four sets of primers that specifically amplify type A influenza virus and simultaneously detect and differentiate H5, H7 and H9 subtypes of hemagglutinins are listed in Table 5-11-2. The primers for type A avian influenza were designed after reviewing the matrix gene sequences of over 10 different subtypes of AIV from chickens, ducks, geese, swine and humans; whereas primers for H5 subtypes were designed from hemagglutinin genes of avian origin. Primers for H7 and H9 subtypes were used from published data[14]. All four sets of oligonucleotide primers were synthesized at the TaKaRa Shuzo Co., Ltd (Dalian, Shandong, China). The primers were aliquoted to a final concentration of 100 pmol/mL and stored at –20 ℃ .

Table 5-11-2 Multiplex RT-PCR primers

Primer's name	Primer's oligonucleotide sequence (5'-3')[a]	Product/bp
Type A influenza virus		
XZ145-2	CTTCTAACCGAGGTCGAAAC	244
XZ146	AGGGCATTTTGGACAAAKCGTCTA	
Subtype H5		
XZ H5-1	ACACATGCYCARGACATACT	860
XZ H5-5	CAGGAACGYTCWCCTGAKTCT	
Subtype H7		
XZ H7-1	GGGATACAAAATGAAYACTC	634
XZ H7-2	CCATABARYYTRGTCTGYTC	
Subtype H9		
XZ H9-1	CTYCACACAGARCACAATGG	488
XZ H9-2	GTCACACTTGTTGTTGTRTC	

Note: [a]Codes for mixed bases position: Y, C/T; R, A/G; W, A/T; B, G/C/T; K, G/T.

Optimization of Reverse transcription and multiplex PCR reaction

The mRT-PCR consists of a two-step procedure, which includes reverse transcription (RT) and PCR amplification. An RT-PCR kit (TaKaRa Shuzo Co., Ltd, Dalian, Shandong, China) was used for the reverse transcription reaction. RT was performed in 20 μL volumes, in which the reaction mixture contained 2 μL RNA in different concentrations and 5 mM $MgCl_2$, in PCR buffer Buffer (500 mM KCl, 100 mM Tris-HCl, pH 8.3), 1 mM of each dinucleoside triphosphate (dNTP), 2 units RNase inhibitor, 0.25 units avian myeloblastosis virus (AMV) reverse transcriptase, 1.25 pmol upstream primers of XZ145-2, XZ H5-1, XZ H7-1 and XZ H9-1. DEPC treated water was added to bring the final volume to 20 mL. RT was performed in a thermal cycler (Model 9600, Perkin Elmer Cetus, Norwalk, CT) for one cycle at 42 ℃ for 25 min, 99 ℃ for 3 min and 4 ℃ for 5 min.

The multiplex PCR was performed in a 50 μL volume using a PCR kit (TaKaRa Shuzo Co., Ltd, Dalian, Shandong, China). The reaction contained 5 mM $MgCl_2$, 1 × PCR buffer, 10 mM of each dNTP, 0.5 pmol of each down stream primer XZ146, XZ H5-5, XZ H7-2 and XZ H9-2, and 1 unit TaKaRa LA Taq™ (TaKaRa Shuzo Co., Ltd, Dalian, Shandong, China). This mixture was added to the RT reaction tubes. Sterile deionized water was added to the mixture to bring the total volume to 50 μL. The mPCR was carried out in the same thermal cycler used for RT. The cycling protocol consisted of an initial denaturing at 94 ℃ for 5 min, then 35 cycles that each consisted of denaturing at 94 ℃ for 45 s, annealing at 55 ℃ for 45 s, and extension at 72 ℃ for 105 s. The sample was then heated at 72 ℃ for 10 min for a final extension. A negative control did not contain template cDNA and consisted of PCR master mix, all four sets of primers and deionized water.

Detection of amplified nucleic acid products

Agarose gel electrophoresis was used to detect mRT-PCR nucleic acid products. A volume of 10 mL of amplified PCR nucleic acid products was subjected to electrophoresis at 80 V in horizontal gels containing 1% agarose with Tris-borate buffer (45 mM Tris-borate, 1 mM EDTA) as described[21]. The Gels were stained with ethidium bromide (0.5 mg/mL or 0.5 μg/mL), and exposed to UV light to visualize the amplified nucleic acid products, and photographed using a Bio-vision Post-electrophoresis instrument (Vilber Lourmat, Paris, France).

Multiplex-RT-PCR sensitivity and specificity

Specificity of the mRT-PCR was determined by examining the ability of the test to detect type A influenza viruses and simultaneously differentiate H5, H7, and H9 subtypes. The mRT-PCR was tested using other avian pathogens that produce similar clinical signs or that can be present in mixed infections with AI subtypes. These pathogens and AIV subtypes are listed in Table 5-11-1. Random samples from the amplified mRT-PCR DNA bands were isolated from the gel and purified using a DNA Glass-milk purification kit (BioVed, Beijing, China). The purified DNA products were sent out to TaKaRa Shuzo Co., Ltd, Dalian, Shandong, China for DNA sequencing. DNA sequences of the amplified products were than analyzed using the DNAStar software to confirm the amplified DNA sequence in the products. To determine the ability of the multiplex PCR assay to detect type A influenza virus and differentiate three subtypes in the same reaction, we used a mixture of RNA concentrations ranging from 200 ng to 10 fg RNA in various combinations of all subtypes of avian influenza virus. Sensitivity of the mRT-PCR was determined by making 10-fold serial dilutions of a mixture containing 10 ng of the RNA templates of each of the three viruses of AIV.

Results

Multiplex RT-PCR was developed to detect group A avian influenza viruses and simultaneously differentiate three hemagglutinin subtypes H5, H7, and H9 in a single reaction through 35 cycles of PCR. The mRT-PCR products were 244 bp for type A avian influenza viruses (group specific), 860 bp for H5, 634 bp for H7, 488 bp for H9. These products were visualized by electrophoresis (Figure 5-11-1). This mRT-PCR was found to be a specific assay for type A avian influenza and hemagglutinin subtypes H5, H7, and H9, with no amplification of nucleic acid from NDV-F48, ILTV, IBV-M41, MG-S6 and SPF chicken tissues. The mRT-PCR specifically detected the H5, H7, and H9 hemagglutinin subtypes along with their group specific type A specification. Thirty-seven avian influenza type A, including nine H5 subtypes, two H7 subtypes, and five H9 subtypes (listed in Table 5-11-1) were detected by the mPCR with the amplification of 244 bp DNA products. Limit of detection by visualization of PCR-amplified DNA products was 100 pg for the subtypes hemagglutinins, and 10 pg for type A avian influenza viruses (Figure 5-11-2). Throughout the development of mRT-PCR, various modifications were made to the annealing temperature, extension time, cycle quantity, and primer concentrations in order to obtain the optimal conditions for this mRT-PCR. No spurious PCR amplification reactions among all influenza subtypes and other pathogens were noticed with various amounts of DNA and RNA mixtures. All the negative controls including RNA samples from the SPF chickens tissues were negative. DNAStar software analysis indicated that the mRT-PCR amplified DNA products were similar to the group A, subtypes H5, H7, and H9 genes sequences of avian influenza viruses.

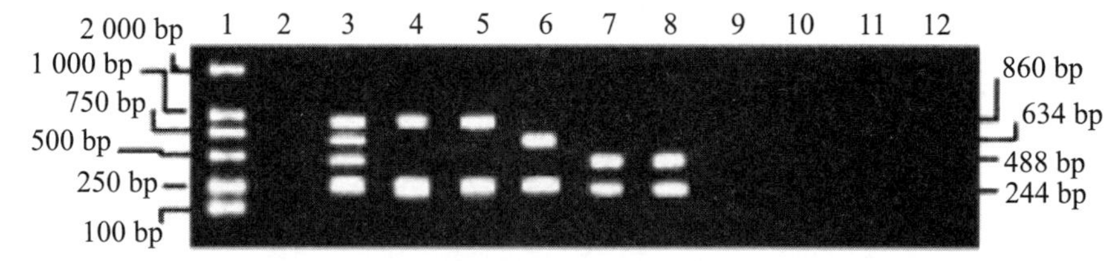

Lane 1: Molecular size marker; lane 2: PCR reagent buffer as a negative control; lane 3: H5N1 (Duck/Guangxi/2/04), H7N2 (Duck/HK/47/76-1) and H9N2(Duck/Guangxi/3/00) subtypes of AIV; lane 4: H5N3 (Duck/HK/313/78); lane 5: H5N1(Duck/Guangxi/2/04); lane 6: H7N2 (Duck/HK/47/76); lane 7: H9N2(Duck/Guangxi/2/00); lane 8: H9N6 (Duck/HK/147/77); lane 9: NDVF68-E9; lane 10: IBV-M41; lane 11: ILTV; lane 12: MG-S6.

Figure 5-11-1 Agarose gel-electrophoresis of multiplex RT-PCR amplified products from purified DNA and RNAs of known avian influenza subtypes and other avian pathogens

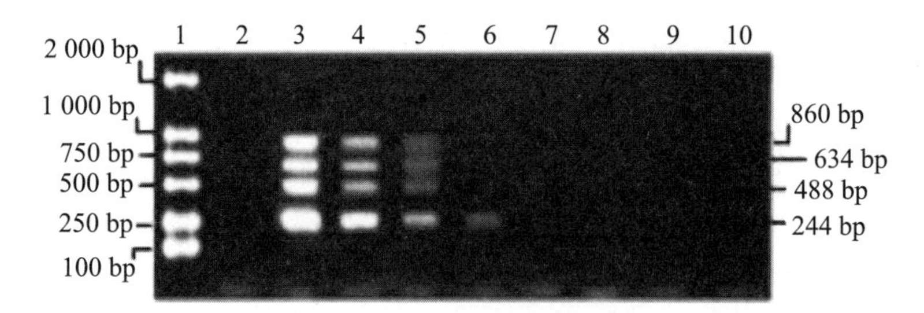

Lane 1: Molecular size marker; lane 2: PCR reagent buffer as a negative control; lane 3: 10 ng; lane 4: 1 ng; lane 5: 100 pg; lane 6: 10 pg; lane 7: 1 pg; lane 8: 100 fg; lane 9: 10 fg; lane10: 1 fg. Agarose gel electrophoresis of multiplex RT-PCR amplified products from purified RNAs from avian influenza subtypes H5N1 (Duck/Guangxi/2/04), H7N2 (Duck/HK/47/76-1) and H9N2 (Duck/Guangxi/3/00) subtype of AIV.

Figure 5-11-2 Sensitivity of multiplex RT-PCR

Discussion and conclusions

The mRT-PCR developed here was able to detect type A influenza virus hemagglutin subtypes from H1 to H13 as tested (Table 5-11-1) and simultaneously detected and differentiated the very important hemagglutinin subtypes H5, H7 and H9 in one single reaction (Figure 5-11-1 and Figure 5-11-2). These hemagglutin subtypes recently have been a cause of human infections[2, 4-7]. Therefore, a mRT-PCR which can rapidly identify type A influenza as well as subtypes H5, H7 and H9, will be very important for the control of disease transmission in poultry and in humans. Use of this assay will also help reduce the economic losses in poultry associated with an AIV outbreak. This mRT-PCR is sensitive, specific, cost effective and it may be useful in diagnosis, screening and surveillance of poultry, including live bird market population. This mRT-PCR has the added benefits of being time saving, and using fewer reagents. Studies will be carried out to further test the specificity and sensitivity of this mRT-PCR on avian influenza subtypes from the various diagnostic and research laboratories as well as on clinical samples originated in the USA and southeastern China.

References

[1] ALEXANDER D J. A review of avian influenza in different bird species. Vet Microbiol, 2000, 74: 3-13.

[2] NORMILE D. North Korea* collaborates to fight bird flu. Science, 2005, 308: 175.

[3] CAMERON K R, GREGORY V, BANKS J, et al. H9N2 subtype influenza viruses in poultry in Pakistan are closely related to the H9N2 viruses responsible for human infection in Hong Kong. Virology, 2000, 278: 36-41.

[4] FOUCHIER R A, SCHNEEBERGER P M, ROZENDAAL F W, et al. Avian influenza A virus (H7N7) associated with human conjunctivitis and a fatal case of acute respiratory distress syndrome. Proc Natl Acad Sci USA, 2004, 101: 1356-1361.

[5] NORMILE D. Avian flu. First human case in Cambodia highlights surveillance shortcomings. Science, 2005, 307: 1027.

[6] GUO Y, LI J, CHEN X, et al. Discovery of men infected by avian influenza A (H9N2) virus. Chin J Exp Clin Virol, 1999, 13: 105-108.

[7] TRAN T H, NGUYEN T L, NGUYEN T D, et al. Avian influenza A (H5N1) in 10 patients in Vietnam. N Engl J Med, 2004, 350: 1179-1188.

[8] KOOPMANS M, WILBRINK B, CONYN M, et al. Transmission of H7N7 avian influenza A virus to human being during a large outbreak in commercial poultry farm in the Netherlands. Lancet, 2004, 363: 587-593.

[9] LI K S, XU K M, PEIRIS J S, et al. Characterization of H9 subtype influenza viruses from the ducks of southern China: a candidate for the next influenza pandemic in humans?. J Virol, 2003, 77: 6988-6994.

[10] WEBSTER R G, KAWAOKA Y. Avian influenza. Rev Poult Biol, 1987, 1: 212-246.

[11] SPACKMAN E, SENNE D A, MYERS T J, et al. Development of a real-time reverse transcriptase PCR assay for type a influenza virus and the avian H5 and H7 hemagglutinin subtypes. J Clin Microbiol, 2002, 40: 3256-3260.

[12] SENNE D A, PANIGRAHY B, KAWAOKA Y, et al. Survey of the hemagglutinin (HA) cleavage site sequence of H5 and H7 avian influenza viruses: amino acid sequence at the HA cleavage site as a marker of pathogenicity potential. Avian Dis, 1996, 40: 425-437.

[13] HORIMOTO T, KAWAOKA Y. Direct reverse transcriptase PCR to determine virulence potential of influenza A viruses in birds. J Clin Microbiol, 1995, 33: 748-751.

[14] LEE S, CHANG P, SHIEN J, et al. Identification and subtyping of avian influenza viruses by reverse transcription-PCR. J Virol Methods, 2001, 97: 13-22.

[15] MUNCH M, NIELSEN L, HANDBERG K, et al. Detection and sub typing (H5 and H7) of avian type A influenza virus by reverse transcription-PCR and PCR-ELISA. Arch Virol, 2001, 146: 87-97.

[16] STARICK E, ROMER-OBERDORFER A, WERNER O. Type-and subtype-specific RT-PCR assays for avian influenza A viruses (AIV). J Vet Med B Infect Dis Vet Public Health, 2000, 47: 295-301.

[17] VAN ELDEN L, NIJHUIS M, SCHIPPER P, et al. Simultaneous detection of influenza viruses A and B using real-time quantitative PCR. J Clin Microbiol, 2001, 39: 196-200.

* "North Korea" refers to the Democratic People's Republic of Korea. The same applies to other occurrences in the book.

[18] SWAYNE D, SENNE D A, BEARD C. Avian influenza// SWAYNE D E, GLISSON J R, JACKSON M W, et al. A laboratory manual for the isolation and identification of avian pathogens, American association of avian pathologists, 4th ed. Pennsylvania: Kennett Square, 1998.

[19] FREY M I, HANSON R P, ANDERSON D P. A medium for the isolation of avian mycoplasma. Am J Vet Res, 1968, 29: 2163-2171.

[20] PANG Y S, WANG H, GIRSHICK T, et al. Development and application of a multiplex polymerase chain reaction for avian respiratory agents. Avian Dis, 2002, 46: 691-699.

[21] SAMBROOK J, FRITSCH EF, MANIATIS T. Molecular cloning A laboratory manual, 2nd ed. New York: Cold Spring Harbor Laboratory Press, 1989.

Development and application of a multiplex polymerase chain reaction for avian respiratory agents

Pang Yaoshan, Wang Han, Girshick Thcodore, Xie Zhixun, and Khan Mazhar I

Abstract

A multiplex polymerase chain reaction (PCR) was developed and optimized to simultaneously detect 6 avian respiratory pathogens. Six sets of specific oligonucdeotide primers for infectious bronchitis virus (IBV), avian influenza virus (AIV), infectious laryngotracheitis virus (ILTV), Newcastle disease virus (NDV), *Mycoplasma gallisepticum* (MG), and *Mycoplasma synoviae* (MS) were used respectively in the test. With the use of agarose gel electrophoresis for detection of the PCR-amplified DNA products, the sensitivity of detection was between 10 pg for IBV, AIV, MG, and ILTV and 100 pg for NDV and MS after 35 cycles of PCR. Similar sensitivity of these primers was achieved with chickens experimentally infected with respiratory pathogens. In experimental infections, the multiplex PCR was able to detect all the infected chickens in each group at 1 and 2 wk postinfection as compared with serologic tests at 2 wk postinfection that confirmed the presence of specific antibodies. The multiplex PCR was also able to detect and differentiate coinfections with two or more pathogens. No specific DNA amplification for respiratory avian pathogens was observed among noninoculated birds kept separately as a negative control group.

Keywords

avian, respiratory pathogens, multiplex, RT, PCR

Introduction

Infectious bronchitis virus (IBV), avian influenza virus (AIV), infectious laryngotracheitis virus (ILTV), Newcastle disease virus (NDV), *Mycoplasma galliepticum* (MG), and *Mycoplasma synoviae* (MS) are 6 major avian respiratory pathogens[1, 4, 8, 11, 18, 22]. They can cause high mortality in avian species. Infected birds present respiratory syndromes and other lesions, such as cough, respiratory distress, poor growth, or production leading to economic losses[4, 8, 10, 11, 18, 22]. Mixed infections of different respiratory agents may occur because of extensive use of multiple live vaccines, high geographic populations, and housing densities. Mixed infections involving mycoplasmas and respiratory viral pathogens are recognized in chickens[6, 15]. On the other hand, multiple infections of avian pathogenic mycoplasma are not uncommon in chicken and turkey flocks[7, 9].

Multiple diagnostic methods such as isolation and serology are required for detecting and differentiating bacterial and viral respiratory infections[14, 16, 29]. However, culture and virus isolation methods are time consuming and labor intensive. Furthermore, nonspecific reactions or cross-reactions[18, 19] often hamper serologic tests. Rapid and sensitive methods to detect and differentiate respiratory disease pathogens in poultry are critical for adopting proper preventive and control measures to reduce economic losses. Molecular methods, such as DNA probes and polymerase chain reaction (PCR) techniques, have been used for rapid and sensitive detection of avian pathogens[8]. Various avian respiratory disease-specific PCR assays have been reported in the literature to detect IBV[12, 20, 23], AIV[3, 26], ILTV[2], NDV[16, 28, 31], MG[24, 25], and MS[21, 35]. Traditionally, this kind of PCR has always been developed to detect a specific nucleic acid of one pathogen. The multiplex PCR, a relatively new technique, has the ability to amplify, detect, and differentiate multiple specific nucleic acids simultaneously[5, 17, 33, 34]. We describe here a multiplex PCR for the identification and differentiation of 6 important avian respiratory pathogens simultaneously.

Materials and methods

Avian pathogens

Table 5-12-1 shows a list of the avian respiratory pathogens studied here and other avian pathogens and their sources. IBV, AIV, NDV, and adenovirus were propagated in the allantoic cavity of 10-day-old-specifc-pathogen-free (SPF) embryonated eggs, whereas ILTV was propagated on the chorioallantoic membrane in 10-day-old SPF embryonated eggs. The allantoic fluids from embryonated eggs infected with IBV, AIV, NDV, ILTV, and adenovirus were harvested after 36 hr of incubation at 37 ℃. Avian reovirus was propagated in chicken embryo fibroblast monolayers as described previously[32]. MG, *Mycoplasma meleagridis* (MM), and *Mycoplasma iowae* (MI) were grown in Frey medium[13]. MS isolates were cultured in Frey medium supplemented with nicotinamide adenine dinucleotide. *Salmonella enteritidis, Pasteurella multocida*, and *Escherichia coli* cultures were grown in Luria-Bertani (LB) broth media[27]. All organisms were incubated at 37 ℃.

Table 5-12-1 Avian pathogens used for multiplex PCR

Species	Isolate/strain/serotype	Origin
Infectious bronchitis	Mass41	University of Connecticut
	Conn	SPAFAS, Inc., Storrs, CT
	Ark 99	University of Connecticut
Avian influenza	T/W/66	SPAFAS, Inc., Storrs, CT
Infectious laryngotracheitis	950802	University of Connecticut
	9415962	University of Connecticut
	932612	University of Connecticut
	993822	University of Connecticut
Newcastle disease	B1 LaSota	Washington State University
	89-720	Washington State University
	90-1062	Washington State University
	K-NDV	University of Indiana
M. gallisepticum	F2F10	University of California
	S6 (208)	University of California
	96-3179	Auburn University, Alabama
	K810	PDRCA Georgia
	A5969	University of Massachusetts
M. synoviae	K3146	PDRC, Georgia
	K3505	PDRC, Georgia
	K3181	PDRC, Georgia
	Alabama	PDRC, Georgia
	1853	University of California
M. meleagridis	RY 39	University of California
M. iowae	695	University of Connecticut

continued

Species	Isolate/strain/serotype	Origin
Reovirus	S1133	University of Connecticut
Adenovirus	Group I, type	SPAFAS, Inc. Storrs, CT
S. enteritidis	Field isolate	University of Maine
P. multocida	Field isolate	University of Connecticut
E. coli	Field isolate	University of Connecticut

Note: PDRC, Poultry Diagnostic and Research Center.

Extraction of RNA and DNA

Allantoic fluids from embryonated eggs infected with IBV, AIV, ILTV, NDV, and adenovirus and cell culture infected with reovirus were first clarified by centrifugation at 500 × g for 15 min. The supernatants were transferred to new tubes and then centrifuged at 15 000 × g for 30 min at 4 ℃. The supernatants were discarded and the pellets treated with 25 μL RNase-free TE buffer (10 mM Tris HCl and 1 mM ethylene-diaminetetraacetic acid (EDTA)). The RNA from IBV, AIV, NDV, and reovirus was extracted and purified with the RNA/DNA isolation kit from QIA-GEN Inc. (Valencia, CA), following the manufacturers protocol. Similarly, DNA from ILTV, MG, MS, MM, MI, adenovirus, *S. enteritidis, P. multocida,* and *E. coli* was extracted and purified with the QIAGEN RNA/DNA isolation kit. The concentration of the DNA and RNA was determined by spectrophotometry[27] and stored at −20 ℃.

Primer selection

Six sets of primers that specifically amplify with IBV, AIV, ILTV, NDV, MG, or MS originated from the published data listed in Table 5-12-2. All 6 sets of oligonudeotide primers were synthesized on a model 380B DNA synthesizer (Applied Biosystem Inc., Foster City, CA) with the assistance of the University of Connecticut Biotechnology Center. The primers were desalted through a Sephadex G-25 column (Pharmacia, Inc., Piscataway, Nj). The concentration of the primers was determined by spectrophotometry[27], and the primers were divided into 25 μL volumes and stored at −20 ℃.

Table 5-12-2 PCR primers used for respiratory pathogens

Respiratory pathogen	Primer designation	Primer sequence (5’-3’)	Expected PCR product
IBV[a]	Upstream	CATAACTAACATAAGGGCA	1 720 bp
	Downstream	TGAAAACTGAACAAAAGACA	
AIV[b]	Upstream	AGCAAAAGCAGGGGATAC	1 050 bp
	Downstream	GTCTGAAACCATACCATCC	
ILTV[c]	Upstream	ACGATGACTCCGACTTTC	647 bp
	Downstream	CGTTGGAGGTAGGTGGTA	
NDV[d]	Upstream	GGAGGATGTTGGCAGCATT	320 bp
	Downstream	GTCAACATATACACCTCATC	
MG[e]	Upstream	GGATCCCATCTCGACCACGAGAAAA	732 bp
	Downstream	CCTTCAATCAGTGAGTAACTGATGA	
MS[f]	Upstream	GAAGCAAATAGTGATATCA	207 bp
	Downstream	GTCGTCTCGAAGTTAACAA	

Note: [a] Reference [20]; [b] Mike Perdue, USDA, ARS, Athens, GA; [c] Fehhat Abbas, VRI, Quetta, Pakistan; [d] Reference [31]; [e] Reference [24]; [f] Reference [21].

Reverse transcription (RT) and multiplex PCR reaction

The multiplex PCR system is a two-step procedure that includes RT and PCR amplification. RT is performed in 20-μL volumes, each RT mixture containing 5 mM of $MgCl_2$; 500 mM KCl; 150 mM Tris-HCl, pH 8.0; 1 mM (each) deoxyadenosine triphosphate (dATP), deoxythymidine triphosphate (dTTP), deoxycytidine triphosphate (dCTP), and deoxyguanosine triphosphate (dGTP); 50 units of Moloney murine leukemia virus reverse transcriptase, and 0.5 mM of random hexamers. For IBV, 25 pM of IBV upstream primer MK38 was added. For AIV, 25 pM of AI upstream primer MK119 was added. For NDV, 25 pM of NDV upstream primer MK57 was added. Different concentrations of respiratory pathogen DNA or RNA in 4 μL volumes were then added to the mixture. The RT was carried out with thermal cycler settings of 42 ℃ for 60 min, 99 ℃ for 5 min, and 4 ℃ for 5 min for one cycle.

For the multiplex PCR reaction, 6.5 mM $MgCl_2$; 500 mM KCl; 150 mM Tris-HCL, pH 8.0; 1.5 mM (each) dATP, dTTP, dCTP, and dGTP; 25 pM of each primer (IBV downstream primer MK39; AIV downstream primer MK120; NDV downstream primer MK 58; ILT primers MK55, MK56; MG primers MK49, MK50; MS primers MK51, MK52); and 10 units Gold AmpliTaq DNA polymerase were added in the above RT reaction tubes, and 50 μL of total volume was obtained by adding diethylpyrocarbonate-treated distilled water. Various controls were included such as template DNAs and RNAs without any primers, primers without any DNA or RNA templates, and primers and templates with and without Gold AmpliTaq DNA polymerase. The mixtures were overlaid with 50 μL of mineral oil. Amplifications were performed in a thermal cycler (Model 480; Perkin-Elmer Cetus Corporation, Norwalk, CT). After extensive preliminary trials with different annealing temperatures and times and with various concentrations of DNA and RNA, the thermal cycler was programmed for optimum conditions. Initially, the reaction mixture was denatured at 96 ℃ for 10 min. Then the PCR was run for 35 cycles at a melting temperature of 95 ℃ for 1 min and an annealing and extension temperature of 60 ℃ for 3 min. The sample was then heated at 60 ℃ for 10 min for the final extension reaction.

Detection of amplified multiplex PCR products

Agarose gel electrophoresis was used to detect amplified DNA products. A volume of 20 μL amplified DNA PCR products was subjected to electrophoresis at 50 V in horizontal gels containing 1.5% agarose (Ultrapure; Bethesda Research Laboratories, Bethesda, MD) with Tris-borate buffer (45 mM Tris-borate, 1 mM EDTA). The gel was stained with ethidium bromide (0.5 μg/mL), exposed to ultraviolet light to visualize the amplified products, and photographed.

Multiplex PCR sensitivity and specificity

To determine the specifcity of the technique, we examined 22 different strains/isolates/serotypes of 6 respiratory agents, MM, MI, adenovirus, reovirus, *S. enteritidis, P. multocida,* and *E. coli* as listed in Table 5-12-1. To determine the ability of the multiplex PCR technique to detect and differentiate six respiratory pathogens in the same reaction, we used a mixture of DNA/RNA concentrations ranging from 500 ng to 10 fg DNA/RNA in various combinations of all six respiratory pathogens. To determine the sensitivity of the multiplex PCR, we used serial 10-fold dilutions of the mixture of 100 ng of each respiratory pathogen as template RNA/DNA.

Detection of respiratory pathogens in tracheal swabs from experimentally infected chickens

Twenty-one 4-wk-old SPF white leghorn chickens (SPAFAS, Norwich, CT) were separated into 7 groups. Six groups were infected separately with one of the six aforementioned avian respiratory pathogens. We inoculated 10^7 plaque-forming units (PFU) inoculum for IBV (Mass41) and NDV (B1 LaSota) and 10^5 PFU inoculum for AIV (T/W/66) and ILTV (950802) by intraocular route per group and 10^6 color-forming units inoculum for MG (K810) and MS (1853) by intraperitoneal injections as described[19]. One group of chickens was given 0.5 mL phosphate-buffered saline (PBS), pH 7.2, as a negative control and kept in a separate room. All infected groups were housed in separate cages in one room. Feed and water were provided ad libitum. Prior to infection and twice a week after 10 days postinoculation, tracheal/cloacal swab samples were collected from all the groups with sterile, premoistened, cotton-tipped applicators. Every swab was swirled in 1mL of PBS, pH 7.2, and squeezed completely. The PBS-containing swab samples were centrifuged at 1 000 × g for 5 min. The supernatants were transferred to new tubes and then centrifuged at 15 000 × g for 30 min at 4 C. The supernatants were discarded and the pellets were treated with 25 μL RNase-free TE buffer containing 1 mg/mL (final concentration) of lysozyme. The RNA/DNA was extracted from these pellets with the QIAamp RNA/DNA isolation kit (QIAGEN) following the manufacturer's protocol. The extracted RNA/DNAs were tested by multiplex RT-PCR. In order to confirm the respective pathogens from the experimental chickens, pathogens were isolated from tracheal swabs and identified by methods as described previously[12, 16]. Furthermore, at days 7 and 14 postinoculation, chickens were bled for serum. Every serum sample was tested by serologic methods. For IBV antibody detection, we used enzyme-linked immunosorbent assay (ELISA); for AIV and ILTV, agar gel precipitation (AGP); for NDV, MG, and Ms, hemagglutination inhibition (HI). Serologic tests for the identification of the respective infections were done at the Connecticut Veterinary Diagnostic Laboratory, Department of Pathobiology, Storrs, CT.

Results

We have optimized a multiplex PCR amplification technique that can identify 6 respiratory pathogens in a single PCR reaction of 35 cycles. The multiplex PCR products consisted of 1 720 bp for IBV, 1 050 bp for AIV, 647 bp for ILTV, 320 bp for NDV, 732 bp for MG, and 207 bp for MS and were visualized by gel electrophoresis (Figure 5-12-1 and Figure 5-12-2). The multiplex PCR assay developed and evaluated in this study was a specific assay for IBV, AIV, ILTV, NDV, MG, and MS with no amplification of nucleic acids from MM, MI, S. enteritidis, P multocida, E. coli, adenovirus, or reovirus. The multiplex PCR was able to detect nucleic acid for IBV, AIV, ILTV, and MG at a level as low as 10 pg and NDV and MS as low as 100 pg (Figure 5-12-3). Various annealing and extension temperatures as well as different time periods were carried out to obtain the best and optimal annealing and extension temperatures and the time period required for each thermal cycle. No spurious PCR amplification reactions among all six pathogens were noticed with various amounts of DNA and RNA mixtures. All the negative controls were negative.

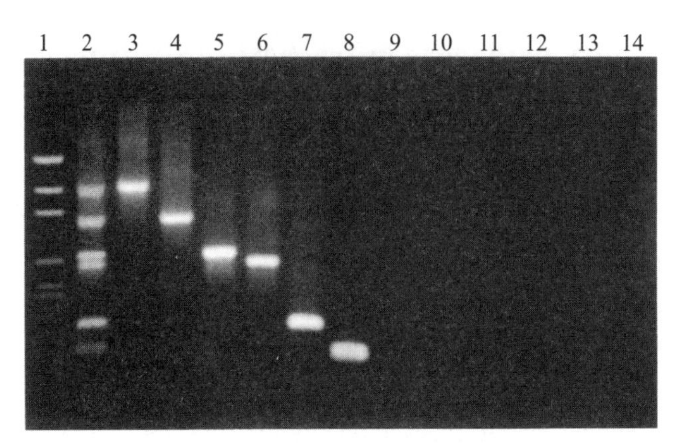

Lane 1: DNA marker; lane 2: IBV (Mass41); AIV (T/W/66); MG (S6-208); ILTV (950802); NDV (B, LaSota); MS (1853); lane 3: IBV (Mass41); lane 4: MG (S6-208); lane 5: ILTV (950802); lane 6: NDV (B, LaSota); lane 7: MS (1853); lane 8: reovirus (S1133); lane 9: Adenovirus (group I, type 1); lane 10: *Salmonella* sp. (field isolate); lane 11: *E. coli* (field isolate); lane 12: *P. multocida* (field isolate); lane 13: MI (675); lane 14: MM (RY39).

Figure 5-12-1 Agarose gel elecrophoresis of multiplex PCR-amplified products from purified DNAs and RNAs of known avian pathogens

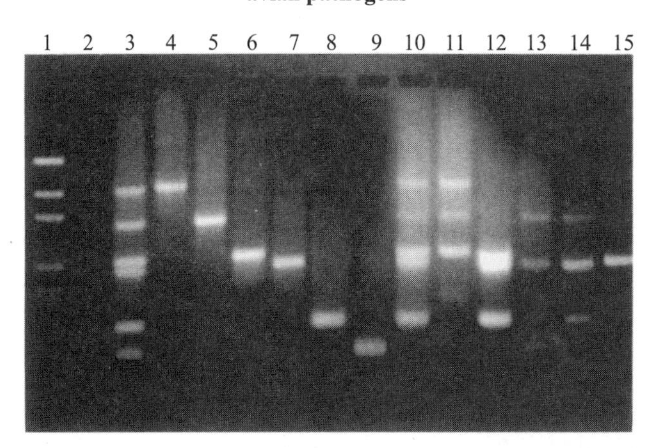

Lane 1: DNA marker; lane 2: Negative control (PCR buffer); lane 3: Positive control DNA and RNAs from IBV (Mass41), AIV (T/WI 66), MG(S6-208), ILTV(950802), NDV(B, LaSota), MS(1853); lane 3: IBV(Mass41) in a sample; lanes 4-9: Individual positive DNA and RNA samples from IBV, AIV, MG, ILTV, NDV, and MS as in lane 3; lane 10: Tracheal swab sample from chickens exposed/infected with IBV, AIV, MG, ILTV, and NDV; lane 11: Tracheal swab sample from chicken exposed/infected with IBV, AIV and MG; lane 12: Tracheal swab sample from chicken exposed/infected with MG, ILTV and NDV; lane 13: Tracheal swab sample from chicken exposed/infected with AIV, ILTV, and MS; lane 14: Tracheal swab sample from chicken exposed/infected with AIV, ILTV and NDV; lane 15: Tracheal swab sample from chicken exposed/infected with ILTV only.

Figure 5-12-2 Agarose gel electrophoresis of multiplex PCR-amplified products from the tracheal swab samples from experimentally infected chickens with avian respiratory pathogens

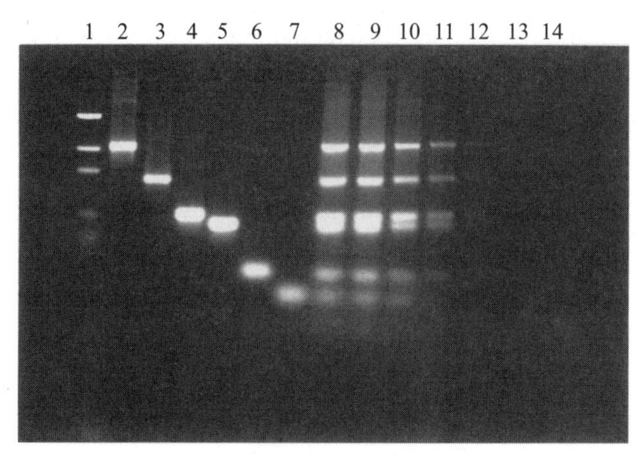

Lane 1: Molecular size marker; lane 2: IBV (Mass41); lane 3: AIV; lane 4: MG (K810); lane 5: ILTV (950802); lane 6: NDV (B, LaSota); lane 7: MS (1853); lane 8: 100 ng of DNA or RNA each of IBV, AIV, MG, ILTV, NDV, and MS; lane 9: 10 ng; lane 10: 1 ng; lane 11: 100 pg; lane 12: 10 pg; lane 13: 1 pg; lane 14: 100 fg.

Figure 5-12-3 Sensitivity of multiplex PCR

Multiplex PCR for chickens experimentally coinfected with respiratory pathogens

The results indicated that the multiplex PCR was able to detect and differentiate the presence of each pathogen in the inoculated birds, as well as in the birds that were also infected with two or more avian pathogens (Figure 5-12-2). The results of multiplex PCR and serologic tests at 7 days postinoculation indicated that all swab samples were PCR positive for specific pathogens from each group, but serum samples were negative for specific antibodies (data not shown). The results of the multiplex PCR and serologic tests at 14 days postinoculation are shown in Table 5-12-3. The results clearly showed detection of multiple infections by serologic as well as multiplex PCR tests. In each group, the multiplex PCR was positive for all individuals in the homologous group. IBV, AIV, ILTV, NDV, and MG all spread readily to infect neighboring uninfected birds, as demonstrated by at least one individual in each group being PCR positive for these organisms. Seropositive birds were also seen in each group for the abovementioned organisms. In contrast, neither PCR nor serology detected MS in any group other than the MS group. All the six respiratory pathogens were recovered at 7 days postinoculation from each group (homologous) as well as from the other groups (Table 5-12-4). We found no indication that any of the negative control chickens were infected by any of the six inoculated pathogens either by multiplex PCR or by traditional serologic methods.

Table 5-12-3 Results of multiplex PCR and serologic tests at 14 days postinoculation

Group	MG		MS		IBV		ILT		NDV		AIV	
	PCR	HI	PCR	HI	PCR	ELISA	PCR	AGP	PCR	HI	PCR	AGP
MG	**3/3**	1/3	0/3	0/3	1/3	3/3	2/3	1/3	3/3	2/3	1/3	2/3
MS	1/3	0/3	**3/3**	3/3	1/3	0/3	1/3	0/3	1/3	1/3	1/3	0/3
IBV	1/3	1/3	0/3	0/3	**3/3**	3/3	3/3	1/3	2/3	1/3	2/3	1/3
ILT	1/3	0/3	0/3	0/3	3/3	3/3	**3/3**	3/3	2/3	2/3	2/3	2/3
NDV	1/3	1/3	0/3	0/3	2/3	1/3	2/3	1/3	**3/3**	3/3	3/3	3/3
AIV	0/3	0/3	0/3	0/3	1/3	3/3	3/3	2/3	1/3	1/3	**3/3**	3/3
Total	7/18	3/18	3/18	3/18	11/18	13/18	13/18	8/18	12/18	10/18	12/18	11/18
Control	0/18	0/18	0/18	0/18	0/18	0/18	0/18	0/18	0/18	0/18	0/18	0/18

Note: Bold indicates homologous injected groups.

Table 5-12-4 Results of reisolations of respiratory pathogens at 7 days postinoculation

Group	Isolation					
	MG	MS	AIV	NDV	IBV	ILTV
MG	**2/3**	1/3	2/3	2/3	1/3	2/3
MS	1/3	**2/3**	2/3	1/3	1/3	1/3
AIV	0/3	0/3	**3/3**	2/3	1/3	3/3
NDV	2/3	0/3	3/3	**3/3**	3/3	2/3
IBV	2/3	0/3	3/3	2/3	**3/3**	2/3
ILT	0/3	0/3	2/3	2/3	1/3	**3/3**
Total	7/18	4/18	15/18	12/18	10/18	13/18

Note: Bold indicates homologous injected groups.

Discussion

The multiplex PCR described here was able to detect RNA from IBV, AIV, and NDV, as well as DNA from ILTV, MG, and MS, in one single reaction. Although several studies describing individual PCR methods can be found for each avian respiratory agent[3, 12, 20, 21, 23, 24, 26, 28, 30, 31, 35], no one has been able to detect all these simultaneously. The multiplex PCR assay developed and evaluated in this study was specific and sensitive (Figure 5-12-3). The results of multiplex PCR for chickens experimentally infected with the six respiratory pathogens suggested that this optimized multiplex PCR was able to detect and differentiate the presence of these respiratory pathogens in infected chickens as well as chickens exposed by lateral transmission. A multiplex PCR that simultancously detects and differentiates the 6 major pathogens of poultry will be highly advantageous to the poultry industry. Rapid, specific detection, without the need for subculture in host systems, would greatly aid diagnosis and control of outbreaks. A multiplex PCR system will be more economical and will require less time than a single PCR for each of these 6 avian respiratory pathogens. Further studies are in progress to test this multiplex PCR on clinical samples.

References

[1] ALEXANDER D J. Newcastle diseases and other avian Paramyxoviridae infections//CALNEK B W, BARNES H J, BEARD C W, et al, Diseases of poultry, 10th ed. Ames, IA: Iowa State University Press, 1977.

[2] ALEXANDER H S, NAGY E. Polymerase chain reaction to detect infectious laryngotracheitis virus in conjunctival swabs from experimentally in-fected chickens. Avian Dis, 1997, 41: 646-653.

[3] ATMAR R L, BAXTER B D, DOMINGUEZ E A, et al. Comparison of reverse transcription-PCR with tissue culture and other rapid diagnostic assays for detection of type A influenza virus. J Clin Microbiol, 1996, 34: 2604-2606.

[4] BAGUST T J, GUY J S. Laryngotracheitis// CALNEK B W, BARNES H J, BEARD C W, et al. Diseases of poultry, 10th ed. Ames, IA: Iowa State University Press, 1997.

[5] BEJ A K, MAHBUBANI M H, DICESARE R, et al. Multiplex PCR amplify-cation and immobilized capture probes for detection of bacterial pathogens and indicators in water. Mol Cell Probes, 1990, 4: 353-365.

[6] BRADBURY J M. Avian mycoplasma infections: prototypes of mixed infections with mycoplasmas, bacteria and viruses. Ann. Microbiol, 1984, 135: 83-89.

[7] BRADBURY J M, MCCLENAGHAN M. Detection of mixed mycoplasma species. J Clin. Microbiol, 1982,16: 314-318.

[8] CAVANAGH D, NAQI S A. Infectious bronchitis// CALNEK B W, BARNES H J, BEARD C W, et al. Diseases of poultry, 10th ed. Ames, IA: Iowa State University Press, 1997.

[9] CHARLTON KG. Antibodies to selected disease agents in translocated wild turkeys in California. J Wildl Dis, 2001, 36: 161-164.

[10] DUEWER L A, KRAUSE K R, NELSON K E U S. poultry and red meat consumption, prices, spreads and margins. United States Department of Agriculture, Economic Research Service Information Bulletin, 1993, 684.

[11] EASTERDAY B C, HINSHAW V S, HALVORSON D A. Influenza // CALNEK B W, BARNES H J, BEARD C W, et al. Diseases of poultry, 10th ed. Ames, IA: Iowa State University Press, 1997.

[12] FALCONE E, AMORE E D, DITRANI L, et al. Rapid diagnosis of avian bronchitis virus by the polymerase chain reaction. J Virol Methods, 1997, 64: 125-130.

[13] FREY M I, HANSON R P, ANDERSON D P. A medium for the isolation of avian mycoplasmas. Am J of Vet Res, 1968, 29: 2163-2171.

[14] GELB J JR, JACKWOOD M W. Infectious bronchitis// SWAYNE D, GLISSON J R, JACKWOOD M W, et al. A laboratory manual for the isolation and identification of avian pathogens, 4th ed. New Bolton Center, Kennett Square, PA: American Association of Avian Pathologists, 1998.

[15] GEORGIADES G, LORDANIDIS P, KOUMBATI M. Cases of swollen head syndrome in broiler chickens in Greece. Avian Dis, 2001, 45: 745-750.

[16] JESTIN V, JESTIN A. Detection of Newcastle disease virus RNA in infected allantoic fluids by in vitro enzymatic amplification. Arch. Virol, 1991, 118: 151-161.

[17] KARLSE E, KALANTAN M, JENKINS A, et al. Use of multiplex PCR primer sets for optimal detection of human papillomavirus. J Clin. Microbiol, 1996, 34: 2095-2100.

[18] KLEVEN S H. *Mycoplasma synoviae* infection// CALNEK B W, BARNES H J, BEARD C W, et al. Diseases of poultry, 10th ed. Ames, IA: Iowa State University Press, 1997.

[19] KLEVEN S H. Mycoplasmosis// SWAYNE D, GLISSON J R, JACKWOOD M W, et al. A laboratory manual for the isolation and identification of avian pathogens, 4th ed. New Bolton Center, Kennett Square, PA: American Association of Avian Pathologists, 1998.

[20] KWON H M, JACKWOOD M W, GELB J. Differentiation of infectious bronchitis virus serotypes using polymerase chain reaction and restriction fragment length polymorphism analysis. Avian Dis.37: 194-202.1993.

[21] LAUERMAN L H, HOERR E J, SHARPTON A R, et al. Development and application of polymerase chain reaction assay for *Mycoplasma synoviae*. Avian Dis, 1993, 37: 832-834.

[22] LEY D H, YODER J R H W, *Mycoplasma gallisepticum* infection// CALNEK B W, BARNES H J, BEARD C W, et al. Diseases of poultry, 10th ed. Ames, IA: Iowa State University Press, 1997.

[23] LIN Z, KATO A, KODOU Y, et al. A new typing method for the avian infectious bronchitis virus using polymerase chain reaction and restriction enzyme fragment length polymorphism. Arch. Virol, 1991, 116: 19-31.

[24] NISCIMENTO E R, YAMAMOTO R, HERRICK K R, et al. Polymerase chain reaction for detection of *Mycoplasma gallisepticum*. Avian Dis, 1991, 35: 62-69.

[25] NISCIMENTO E R, YAMAMOTO R, KHAN M I. *Mycoplasma gallisepticum* F-vaccine strain-specific polymerase chain reaction. Avian Dis, 1993, 37: 203-211.

[26] PISEREVA M, BECHTEREVA T, PLYUSNIN A, et al. PCR-amplifcation of influenza A virus specific sequences. Arch. Virol, 1992, 125: 313-318.

[27] SAMBROOK J, FRISCH E T, MANIATIS T. Molecular cloning: a laboratory manual. Cold Spring Harbor, NY: Cold Spring Harbor Laboratory Press, 1989.

[28] SEAL B S, KING D J, BENNET J D. Characterization of Newcastle disease virus isolates by reverse transcription PCR coupled to direct nucleotide sequencing and development of sequence database for pathotype prediction and molecular epidemiological analysis. J Clin Microbiol, 1995, 33: 2624-2630.

[29] SENNE D A. Virus propagation in embryonating eggs// SWAYNE D, GLISSON J R, JACKWOOD M W, et al. A laboratory manual for the isolation and identification of avian pathogens, 4th ed. New Bolton Center, Kennett Square, PA: American Association of Avian Pathologists, 1998.

[30] SLAVIK M E, WANG R F, CAO W W. Development and evaluation of the polymerase chain reaction method for diagnosis of *Mycoplasma gallisepticum* infection in chickens. Mol Cell Probes, 1993, 7: 459-463.

[31] STAUBER N, BRECHTBUHL K, BRUEKNER L, et al. Detection of Newcastle disease virus in poultry vaccines using the polymerase chain reaction and direct sequencing of amplified cDNA. Vaccine, 1995, 13: 360-364.

[32] VAN DER HEIDE L, KALBAC M, BRUSTOLON M, et al. Pathogenicity for chickens of a reovirus isolated from turkeys. Avian Dis, 1980, 24: 989-997.

[33] WANG H, FADL A A, KHAN M I. Multiplex PCR for avian mycoplasmas. Mol Cell Probes, 1997, 11: 211-216.

[34] WANG X, KHAN M I. A multiplex PCR for Massachusetts and Arkansas serotypes of infectious bronchitis virus. Mol Cell Probes, 1999, 13: 1-7.

[35] ZHAO S, YAMAMOTO R. Detection of *Mycoplasma synoviae* by polymerase chain reaction. Avian Pathol, 1993, 22: 533-542.

A multiplex RT-PCR for simultaneous differentiation of three viral pathogens of penaeid shrimp

Xie Zhixun, Pang Yaoshan, Deng Xianwen, Tang Xiaofei, Liu Jiabo, Lu Zhaofa, and Khan Mazhar I

Abstract

A multiplex reverse transcription polymerase chain reaction (mRT-PCR) was developed and optimized to simultaneously detect 3 viral pathogens of shrimp. Three sets of specific oligonucleotide primers for Taura syndrome virus (TSV); white spot syndrome virus (WSSV) and infectious hypodermal and hematopoietic necrosis virus (IHHNV) were used in the assay. The mRT-PCR DNA products were visualized by gel electrophoresis and consisted of fragments of 231 bp for TSV, 593 bp for WSSV and 356 bp for IHHNV. No specific bands of the same size were amplified from other penaeid shrimp pathogenic viruses or bacteria. As little as 10 pg of TSV RNA and 100 pg of WSSV DNA and IHHNV DNA could be detected using gel electrophoresis. Studies are in progress to further test the specificity and sensitivity of this mRT-PCR method on viral isolates, as well as on clinical samples.

Keywords

infectious hypodermal and hematopoietic necrosis virus, multiplex, polymerase chain reaction, Taura syndrome virus, white spot syndrome virus

Introduction

Infectious hypodermal and hematopoietic necrosis virus (IHHNV); Taura syndrome virus (TSV) and white spot syndrome virus (WSSV) are 3 major viral pathogens that infect penaeid shrimp[1-5]. Mixed infections involving these viruses have been described in penaeid shrimps[6] and can cause high mortality, leading to economic losses that are detrimental to the shrimp farming industry[7-9].

Multiple diagnostic methods, such as histological examination, electron microscopy and histological studies using in situ hybridization, are required to detect and differentiate these viral pathogens[4, 7, 8, 10]. However, these methods are time consuming and labor intensive. Molecular assays, such as DNA probes[11, 12] and PCR methods, have been used for rapid and sensitive detection of these viruses[13-17]. Traditionally, specific probes and PCR have been developed to detect a specific nucleic acid of one pathogen. Recently, real-time PCR assays have been developed to detect WSSV, TSV and IHHNV[18, 19] and to differentiate 2 viruses at a time[20, 21]. Conventional singleplex PCR is potentially expensive and resource intensive, whereas the cost of real-time PCR equipment and the specific technical training required limit the usefulness of such assays as routine laboratory tests. In our study, we have developed a specific and sensitive multiplex PCR that can simultaneously detect and differentiate these 3 important viral agents infecting shrimp.

Materials and methods

Viral isolates and clinical samples

TSV, WSSV and IHHNV isolates are listed in Table 5-13-1.

Table 5-13-1 Shrimp pathogens and field samples used in this experiment

Shrimp pathogens	Source
TSV (GXPRC/1/02)	Beihai, Guangxi, China
TSV (GXPRC/1/03)	Hepu, Guangxi, China
TSV (GXPRC/2/03)	Qinzhou, Guangxi, China
WSSV (GXPRC/1/02)	Beihai, Guangxi, China
WSSV (GXPRC/2/02)	Hepu, Guangxi, China
WSSV (GXPRC/1/03)	Qinzhou, Guangxi, China
IHHNV (GXPRC/1/03)	Beihai, Guangxi, China
IHHNV (GXPRC/2/03)	Hepu, Guangxi, China
IHHNV (GXPRC/1/04)	Qinzhou, Guangxi, China
Vibrio spp. (extracted DNA)	CIVDC, Beijing, China
Streptococcus spp. (extracted DNA)	CIVDC, Beijing, China
Yellow head virus (extracted RNA)	CIVDC, Beijing, China
Tissue sample	Disease-free Penaeus vannamei

Note: TSV, Taura syndrome virus; WSSV, white spot syndrome virus; IHHNV, infectious hypodermal and hematopoietic necrosis virus; GXPRC, Guangxi, China; CIVDC, China Institute of Veterinary Drug Control.

Isolation of nucleic acids from viruses

RNA and DNA extractions from TSV, IHHNV and WSV isolates were carried out using Trizol according to the manufacturer's protocol (Invitrogen).

Extraction of nucleic acids from tissue samples.

Total nucleic acid extracts from tissue samples from disease-free Penaeus vannamei were extracted by homogenizing 10 mg frozen infected tissues (stomach tissue homogenate) in 700 μL lysis buffer containing 2% hexadecyl-trimethy-ammonium-bromide, 1.4 mM NaCl, 20 mM EDTA and 20 mM Tris-HCl (pH 7.5), adding isoamyl alcohol to the final concentration of 2.5%. Samples were incubated at room temperature for 1 h. After incubation, 500 μL phenol/chloroform (1 : 1) was added to each sample and mixed vigorously by vortexing and incubated for 15 min at room temperature. The mixtures were centrifuged at 12 000 r/min (14 000 × g) for 15 min at 4 ℃, and supernatants were transferred to 1.5 mL tubes. Equal volumes of isopropanol were added to precipitate the nucleic acid, and the mixtures were inverted several times during the incubation at room temperature for 15 min. The mixtures were centrifuged at 12 000 r/min (14 000 × g) for 15 min at 4 ℃. Pellets were rinsed with 500 μL 75% ethanol, then air dried for about 15 min.

Pellets containing nucleic acid were resuspended in 20 μL of RNase-free distilled water. The purity and concentration of DNA and RNA were determined spectrophotometrically by 260 : 280 nm ratios and 260 nm readings, respectively, using a spectrophotometer (Shimadzu UV-1 200); and the nucleic acids were stored at −20 ℃ until use. Extracted DNA of *Vibro* spp. and *Streptococcus* spp. and RNA of yellow head virus were kindly provided by the China Institute of Veterinary Drug Control, Beijing.

Oligonucleotide primers

Three sets of primers that specifically amplify WSSV, IHHNV and TSV were used. WSSV primers GATGAGACAGCCAAGTTGTTAAAC and GCATCAACTTCCACAGCTTTATC amplify 593 bp DNA product, IHHNV primers ATCGGTGCACTACTCGGA and TCGTACTGGCTGTTCATC amplify 356 bp DNA product and TSV primers TCAATGAGAGCTTGGTCC and AAGTAGACAGCCGCGCTT amplify 231 bp DNA product. The primers for WSSV were designed using DNASTAR software against a conserved region of the WSSV genomic sequence (GenBank No. AF 369029) that encodes for a non-structural protein. IHHNV primers were from a region in between the non-structural and the structural protein-coding regions of the genome (GenBank No. AF 218266). TSV-specific primers had been reported previously in the literature[15]. All 3 sets of primers were synthesized at TaKaRa Shuzo Stet. Primers were diluted to a final concentration of 100 pmol/μL using RNase-free distilled water and stored at −20 ℃ .

Multiplex reverse transcription polymerase reaction (mRT-PCR)

The mRT-PCR consisted of a 2-step procedure as described[22], which includes reverse transcription (RT) and PCR amplification. The RT-PCR kit (TaKaRa Shuzo Co.) was used for all RT and PCR amplifications. RT is performed in 20 μL volumes, each RT mixture containing 5 mM $MgCl_2$; 500 mM KCL; 100 mM Tris HCL, pH 8.3; 1 mM of each deoxyadenosine triphosphate (dATP); deoxythymidine triphosphate (dTTP); deoxycytidine triphosphate (dCTP) and deoxyguanosine (dGTP); 2 U RNase inhibitor; 0.25 U avian myeloblastosis virus (AMV) reverse transcriptase; and 1.25 pmol of TSV forward primer. Different concentrations of DNA or RNA of WSSV, IHHNV and TSV in 4 μL volumes were then added to the mixture. Diethylpyrocarbonate (DEPC)-treated distilled H_2O was added to bring the final volume to 20 μL. RT was performed in a thermal cycler (Model 9 600, Perkin Elmer Cetus) for 1 cycle at 42 ℃ for 25 min, 99 ℃ for 3 min and 4 ℃ for 5 min.

For the multiplex PCR reaction, 5 mM $MgCl_2$; 500 mM KCl; 100 mM Tris HCl, pH 8.3; 10 mM each dATP, dTTP, dCTP and dGTP; 0.5 pmol of each WSSV and IHHNV forward and reverse primers with TSV reverse primer; and 1.25 U TaKaRa LA Taq (TaKaRa Shuzo Co.) polymerase were added in the above RT reaction tubes, and 50 μL of the total volume was obtained by adding DEPC-treated distilled water. The mPCR was carried out in the same thermal cycler used for RT. After extensive preliminary trials with different annealing temperatures and times and with various concentrations of DNA and RNA, the thermal cycler was programmed for optimum conditions. The optimized cycling protocol consisted of an initial denaturing at 94 ℃ for 5 min, then 35 cycles that each consisted of denaturing at 94 ℃ for 45 s and annealing and extension at 68 ℃ for 2 min. The sample was then heated to 68 ℃ for 10 min for a final extension. The negative control did not contain template cDNA/DNA; it consisted of PCR master mix, all 3 sets of primers and deionized water.

Detection of amplified mRT-PCR products

Agarose gel electrophoresis was used to detect mRT-PCR nucleic acid products. A volume of 10 μL of amplified product was subjected to 1% agarose horizontal gel electrophoresis with 0.5 μg ethidium bromide/mL at 5 v/m V using Tris-borate buffer (45 mM Tris-borate, 1 mM EDTA)[23]. Gels were visualized and photographed using the Bio-Vision Post-Electrophoresis Instrument (Vilber Lourmat).

Specificity and sensitivity of the mRT-PCR

To determine the specificity of the mRT-PCR, the amplified DNA fragments from TSV, WSSV and IHHNV were cloned into the pMD18-T cloning vector (TaKaRa Shuzo Co.). The recombinant plasmid DNA was sequenced in an automated DNA sequencer (TaKaRa Shuzo Co.), and the sequence data were analyzed by DNASTAR software and compared with the corresponding sequences in GenBank. For further confirmation of specificity, 200 ng of RNA from yellow head disease virus and 200 ng of DNA from both *Vibrio* spp. and *Streptococcus* spp., as well as same amount of nucleic acid isolated from specific-pathogen-free (SPF) *Penaeus vannamei* were also tested by mRT-PCR. Sensitivity of this mPCR was determined by making 10-fold dilutions of a mixture containing 100 ng of template of each TSV RNA, WSSV DNA and IHHNV DNA.

Results and discussion

A method of mRT-PCR was developed and optimized to detect and simultaneously differentiate 3 viral pathogens in a single tube reaction through 35 cycles of PCR. The optimal mRT-PCR yielded 3 amplified fragments, i.e. 231 bp for TSV, 593 bp for WSSV and 365 bp for IHHNV (Figure 5-13-1 and Figure 5-13-2). The identity of each mRT-PCR product was further confirmed by DNA sequencing. The sequences of the mRT-PCR products were matched to those of TSV, WSSV and IHHNV based on sequence data in GenBank using DNASTAR software. The mRT-PCR method was found to be specific and to be able to detect and differentiate TSV, WSSV and IHHNV. No amplifications were observed when nucleic acid from *Vibrio* spp. and *Streptococcus* spp., as well as SPF *Penaeus vannamei* were used as mRT-PCR templates (Figure 5-13-1). The detectable limit of the mRT-PCR was 10 pg for TSV and 100 pg for both WSSV and IHHNV (Figure 5-13-2).

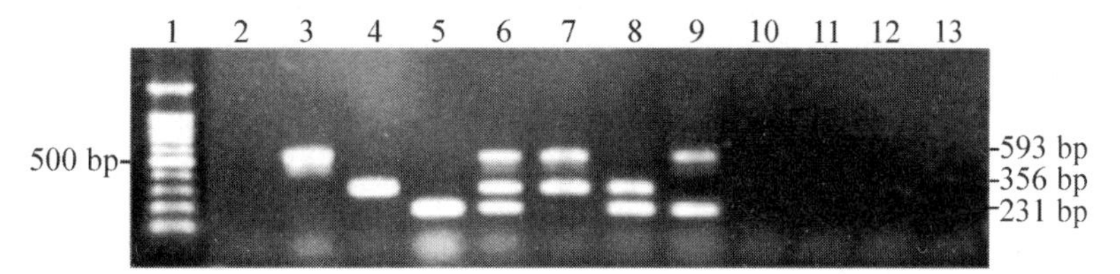

Lane 1: 100 bp DNA size marker; lane 2: Specific pathogen-free (SPF) Penaeus vannamei; lane 3: WSSV (GXPRC/1/02); lane 4: IHHNV (GXPRC/1/03); lane 5: TSV (GXPRC/1/02); lane 6: WSSV (GXPRC/1/02)+IHHNV (GXPRC/1/03)+TSV (GXPRC/1/02); lane 7: WSSV (GXP-RC/1/02)+IHHNV (GXPRC/1/03); lane 8: IHHNV (GXPRC/1/03)+TSV (GXPRC/1/02); lane 9: WSSV (GXPRC/1/02)+TSV (GXPRC/1/02); lane 10: Yellow head disease virus; lane 11: *Vibrio* spp.; lane 12: *Streptococcus* spp.; lane 13: Buffer control. For pathogen abbreviations, see Table 5-13-1.

Figure 5-13-1 Specificity of multiplex RT-PCR

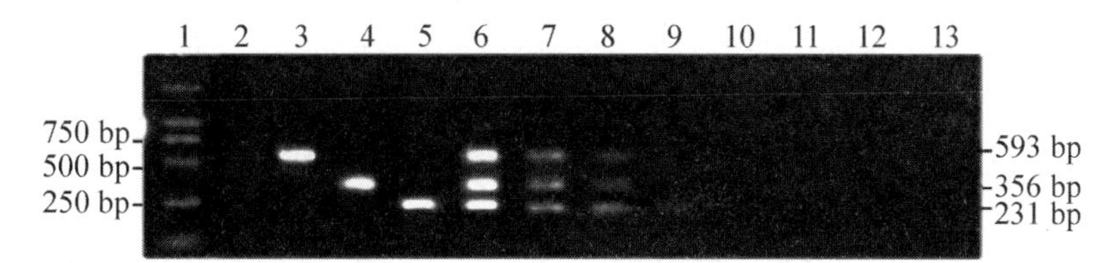

Lane 1: 200 bp DNA size marker; lane 2: SPF Penaeus vannamei; lane 3: WSSV (GXPRC/2/02); lane 4: IHHNV (GXPRC/1/04); lane 5: TSV (GXPRC/2/03). Lanes 6 to 13: Different amounts of DNA or RNA from each of WSSV, IHHNV and TSV; lane 6: 100 ng; lane 7: 10 ng; lane 8: 100 pg; lane 9: 10 pg; lane 10: 1 pg; lane 11: 100 fg; lane 12: 10 fg; lane 13: 1 fg. For pathogen abbreviations, see Table 5-13-1.

Figure 5-13-2 Sensitivity of multiplex RT-PCR

Throughout development of the mRT-PCR method, various modifications were made to the annealing temperature, extension time, cycle quantity and primer concentrations in order to obtain optimal conditions.

No spurious PCR amplification reactions among any shrimp or other pathogens were noted with various amounts of RNA and DNA mixtures. All negative controls included RNA/DNA samples from disease-free

Penaeus vannamei. DNASTAR software analysis indicated that the mRT-PCR-amplified DNA products were similar to the TSV-, WSSV- and IHHNV-specific gene sequences.

An mRT-PCR that can rapidly identify and differentiate these 3 viral infections, and possibly detect multiple infections, will be useful for the control of viral diseases in shrimp. Further studies are in progress to test the specificity and sensitivity of this mRT-PCR method on viral isolates of TSV, WSSV and IHHNV from various diagnostic and research laboratories, as well as on clinical samples originating from shrimp farms in south China.

Reference

[1] BONAMI J R, TRUMPER B, MARI J, et al. Purification and characterization of the infectious hypodermal and haematopoietic necrosis virus of penaeid shrimps. J Gen Virol, 1990, 71: 2657-2664.

[2] BONAMI J R., HASSON K, MARI J, et al. Taura syndrome of marine penaeid shrimp: characterization of the viral agent. J Gen Virol, 1997, 78: 313-319.

[3] BROCK J A. Special topic review: Taura syndrome, a disease important to shrimp farms in the Americas. World J Microbiol Biotechnol, 1997, 13: 415-418.

[4] LIGHTNER D V, REDMAN R M. Shrimp diseases and current diagnostic methods. Aquaculture, 1998, 164: 201-220.

[5] ERICKSON H S, ZARAIN-HERZBERG M, LIGHTNER D V. Detection of Taura syndrome virus (TSV) strain differences using selected diagnostic methods: diagnostic implications in penaeid shrimp. Dis Aquat Org, 2002, 52: 1-10.

[6] MANIVANNAN S, OTTA S K, KARUNASAGAR I, et al. Multiple viral infection in Penaeus monodon shrimp postlarvae in an Indian hatchery. Dis Aquat Org, 2002, 48: 233-236.

[7] LIGHTNER D V. The penaeid shrimp viruses IHHNV and TSV: epizootiology, production impacts and role of international trade in their distribution in the Americas. Rev Sci Tech Off Int Epizoot, 1996, 15: 579-601.

[8] PLUMB J A. Trends in freshwater fish disease research. FAO Aquacult News, 1997, 16: 35-47.

[9] WANG Y C, LO C F, CHANG P S, et al. White spot syndrome associated virus (WSSV) infection in cultured and wild decapods in Taiwan. Aquaculture, 1998, 164: 221-231.

[10] TAKAHASHI Y, ITAMI T, KONDO M, et al. Electron microscopic evidence of bacilliform virus infection in Kuruma shrimp (*Penaeus japonicus*). Fish Pathol, 1994, 29: 121-125.

[11] NUNAN LM, LIGHTNER D V. Development of a non-radioactive gene probe by PCR for detection of white spot syndrome virus (WSSV). J Virol Methods, 1997, 63: 193-201.

[12] MARI J, BONAMI J R, LIGHTNER D V. Taura syndrome of penaeid shrimp: cloning of viral genome fragments and development of specific gene probes. Dis Aquat Org, 1998, 33: 11-17.

[13] LO C F, LEU J H, HO C H, et al. Detection of baculovirus associated with white spot syndrome (WSBV) in penaeid shrimps using polymerase chain reaction. Dis Aquat Org, 1996, 25: 133-141.

[14] WANG S Y, HONG C, LOTZ J M. Development of a PCR procedure for the detection of Baculovirus penaei in shrimp. Dis Aquat Org, 1996, 25: 123-131.

[15] NUNAN L M, POULOS B T, LIGHTER D V. Reverse transcription polymerase chain reaction (RT-PCR) used for the detection of Taura syndrome virus (TSV) in experimentally infected shrimp. Dis Aquat Org, 1998, 34: 87-91.

[16] KIATPATHOMCHAI W, BOONSAENG V, TASSANAKAJON A, et al. A non-stop, singletube, seminested PCR technique for grading the severity of white spot syndrome virus infections in Penaeus monodon. Dis Aquat Org, 2001, 47: 235-239.

[17] TANG K F J, WANG J, LIGHTNER D V. Quantitation of Taura syndrome virus by realtime RT-PCR with TaqMan assay. J Virol Methods, 2004, 115: 109-114.

[18] DHAR A K, ROUX M M, KLIMPEL K R. Detection and quantification of infectious hypodermal and hematopoietic necrosis virus and white spot virus in shrimp using real-time quantitative PCR and SYBR green chemistry. J Clin Microbiol, 2001, 39: 2835-2845.

[19] YUE Z Q, LIU H, WANG W, et al. Development of real-time polymerase chain reaction assay with TaqMan probe for the quantitative detection of infectious hypodermal and hematopoietic necrosis virus from shrimp. AOAC Int J, 2006, 89: 240-244.

[20] TSAI J M, SHIAU L J, LEE H H, et al. Simulta-neous detection of white spot syndrome virus (WSSV) and Taura syndrome virus (TSV) by multiplex reverse transcription-polymerase chain reaction (RT-PCR) in Pacific white Penaeus vannamei. Dis Aquat Org, 2002, 50: 9-12.

[21] YANG B, SONG X L, HUANG J, et al. A single step multiplex PCR for simultaneous detection of white spot syndrome virus and infectious hypodermal and haematopoietic necrosis virus in penaeid shrimp. J Fish Dis, 2006, 29: 301-305.

[22] PANG Y, WANG H, GIRSHICK T, et al. Development and application of a multiplex polymerase chain for avian respiratory agents. Avian Dis, 2002, 46: 691-699.

[23] SAMBROOK J, FRITSCH E F, MANIATIS T. Molecular cloning: a laboratory manual, 2nd ed. Cold Spring Harbor, NY: Cold Spring Harbor Laboratory Press, 1989.

Development of a real-time multiplex PCR assay for detection of viral pathogens of penaeid shrimp

Xie Zhixun, Xie Liji, Pang Yaoshan, Lu Zhaofa, Xie Zhiqin, Sun Jianhua, Deng Xianwen, Liu Jiabo, Tang Xiaofei, and Khan Mazhar I

Abstract

A real-time multiplex polymerase chain reaction (rtm-PCR) assay was developed and optimized to simultaneously detect three viral pathogens of shrimp in one reaction. Three sets of specific oligonucleotide primers for white spot syndrome virus (WSSV), infectious hypodermal and haematopoietic necrosis virus (IHHNV) and Taura syndrome virus (TSV), along with three TaqMan probes specific for each virus were used in the assay. The rtm-PCR results were detected and analyzed using the Light Cycler 2.0 system. Forty-five PCR-positive samples and four negative samples were used to confirm the sensitivity and specificity of the rtm-PCR. The rtm-PCR identified and differentiated the three pathogens. With one viral infection of shrimp, a specific amplified standard curve was displayed. When samples from shrimp infected with two or three pathogens were analyzed, two or three specific standard curves were displayed. The sensitivity of the rtm-PCR assay was 2 000, 20, and 2 000 template copies for WSSV, IHHNV and TSV, respectively. No positive results (standard curves) were displayed when nucleic acid from *Vibro* spp., and *Streptococcus* spp. DNA were used as PCR templates. The results indicate that real-time multiplex PCR is able to detect the presence of and differentiate each pathogen in infected shrimp. This real-time multiplex PCR assay is a quick, sensitive, and specific test for detection of WSSV, IHHNV and TSV and will be useful for the control of these viruses in shrimp.

Keywords

real-time multiplex PCR, infectious hypodermal and haematopoietic necrosis virus, white spot syndrome virus, Taura syndrome virus

Introduction

White spot syndrome virus (WSSV), infectious hypodermal and haematopoietic necrosis virus (IHHNV) and Taura syndrome virus (TSV) are three major viral pathogens that infect penaeid shrimp [2-4, 6, 8]. Mixed infections with these viruses have been described[12] and can cause high mortality, leading to economic losses that are detrimental to the shrimp farming industry[9, 16, 21].

Multiple diagnostic methods such as histologic examination, electron microscopy and in situ hybridization are required to detect and differentiate these viral pathogens[8, 9, 16, 18]. However, these methods are time consuming and labor intensive. Molecular assays, such as DNA probes[13, 15] and PCR methods, have been used for rapid and sensitive detection of these viruses[7, 10, 14, 19, 20]. Recently we developed multiplex reverse transcription-PCR for the simultaneous differentiation of three viral pathogens of penaeid shrimp[22]. Real-time PCR is preferred over conventional PCR in clinical laboratories because there is no need for post-amplification handling, leading to faster analysis and reduced risk of amplicon contamination [11, 17]. Real-time

PCR can also provide an estimate of pathogen titer [1]. Recently, real-time PCR assays have been developed for the separate detection of WSSV, TSV and IHHNV[5, 23]. This study describes a real-time multiplex PCR assay for simultaneous detection of these three viruses.

Materials and methods

Clinical samples and plasmids

WSSV-, IHHNV- and TSV-infected clinical tissue samples and positive controls consisted of recombinant plasmids containing specific and conserved genes of the three viruses (TSV-pMD18-T, WSSV-pMD18-T, IHHNV-pMD18-T). These three plasmids were cloned using the specific primers described previously[22] and are listed in Table 5-14-3.

Isolation of nucleic acids from clinical tissues samples

RNA and DNA extractions from WSSV, IHHNV, and TVS isolates were carried out using Trizol according to the manufacturer's protocol (Invitrogen, Carlsbad, CA, USA). Total nucleic acid was extracted from clinical tissue samples from shrimp infected with WSSV, IHHNV, and TVS as well as from disease-free white shrimp and *Penaeus orientalis* Kishinouye according to a method described previously [22]. Extracted DNA of *Vibro* spp. and *Streptococcus* spp. were kindly provided by the China Institute of Veterinary Drug Control, Beijing.

Oligonucleotide primers and DNA probes

Three sets of primers and DNA probes, listed in Table 5-14-1, were designed to amplify highly conserved gene sequences of WSSV (AF369029), IHHNV (AF218226) and TSV (NC003005) (GenBank sequence data). All three sets of primers and probes were synthesized at TaKaRa, Dalian, China.

Table 5-14-1　Sequences of primers and TaqMan probes

Primer	Sequence(5'-3')	Size
Ihhnv24	AAACTGAACACTGGCCTAGTAACAA	77 bp
Ihhnv100	TAGGACTTTCCGATGAGGTTTTG	
Thhny50T	FAM-AACAGGAGACTCAAACACCTTCCATCT-ECLIPSE	
Wssv270	ACCATGGAGAAGATATGTACAAGCA	76 bp
Wssy345	GGCATGGACAGTCAGGTCTTT	
Wssv296T	ROX-TTACAGTGATGGAATTTCGTTTATC-ECLIPSE	
Tsy20F	GCTTGCGTGGTGGGACTAAAT	76 bp
Tsv95R	CCTCCACTGGTTGTTGTATCAAAA	
Tsv42T	HEX-AATGCCTGCTAACCCAGTCGAAATT-ECLIPSE	

Real-time multiplex PCR assay

Amplification reactions were performed in volumes of 20 μL with TaKaRa premix, 40 μM MLV reverse transcriptase, 16 U RNase inhibitor, 0.6 μM primer for WSSV and TSV or 0.4 μM primer for IHHNV, 0.4 μM probe for WSSV and TSV or 0.2 μM probe for IHHNV, and 2 μL of the DNA/RNA sample. The PCR amplification consisted of 2 min at 95 ℃, followed by 40 cycles of 10 s at 95 ℃ and 30 s at 60 ℃. Amplification, detection, and data analysis were performed with the Light Cycler 2.0 system (Roche Molecular Biochemicals, Mannheim, Germany).

Sensitivity and specificity of rtm-PCR

The sensitivity of the rtm-PCR was determined using tenfold dilutions of template of each specific plasmid containing specific genes of viruses (WSSV-pMD18-T, IHHNV-pMD18-T and TSV-pMD18-T). The results of the rtm-PCR assay were compared to those of our previously developed PCR assay[22]. To determine the specificity of the rtm-PCR assay, specific DNA fragments from WSSV, IHHNV and TSV were amplified and cloned in pMD18-T cloning vector according to the manufacturer's protocol (TaKaRa Dalian, China). These three recombinant plasmids were sequenced, and sequence data were analyzed using Dnastar software and compared with the corresponding sequence data in GenBank. For negative controls, DNA from *Vibrio* and *Streptococcus* and distilled water were included.

Interference assay and reproducibility

Various concentrations of plasmid containing WSSV, IHHNV and TSV genes (10^4, 10^8 and 10^8 copies of each; or 10^8, 10^1 and 10^8 copies of each; or 10^8, 10^8 and 10^4 copies of each) were mixed together and subjected to rtm-PCR. The copy numbers of the genes were calculated according to the following formula: copies/μL= (The plasmid's concentration × 6 × 10^{14})/(The plasmid's size in base pairs × 324.5).

Detection of clinical samples

DNA from 15 clinical samples each from WSSV, IHHNV and TSV that were known to be positive by routine PCR were also subjected to real-time multiplex PCR.

Results

The real-time multiplex PCR assay for detection of three viral pathogens of penaeid shrimp was designed as a multiplex assay for simultaneous detection of WSSV, IHHNV and TSV. Since the presence of other oligonucleotides and fluorescent probe could alter the efficiency of PCR amplification, each set of primers and probe was tested in an individual format as well as in a multiplex format, using different concentrations of the three viruses. The result of these experiments indicates that there was no systematic deviation in the amplification curves when comparing the multiplex assay with the single-target assays. No difference in amplification efficiency was observed between the singleplex and multiplex formats, as measured by the slopes of amplification curves during the exponential phase and the cycle threshold (CT) values obtained with individual samples. Furthermore, the detection limits for the multiplex and individual assay formats were nearly identical, since even the most diluted samples were detected in both types of assay.

Detection limit

The limit of detection for the real-time multiplex assay was determined with TSV-pMD18-T, WSSV-pMD18-T and IHHNV-pMD18-T that were serially diluted tenfold. The sensitivity of the real-time multiplex PCR assay was 20 000 for WSSV, 20 for IHHNV and 20 000 for TSV template copies, respectively (Figure 5-14-1 A-C), and its sensitivity was 10, 1 000 and ten times higher than that of the routine PCR. Standard curves are shown in Figure 5-14-2 A-C. Different concentrations of WSSV, IHHNV and TSV, when mixed together, still could be identified by this assay, which implies that the rtm-PCR assay can be used for simultaneous detection of infection with the three viruses.

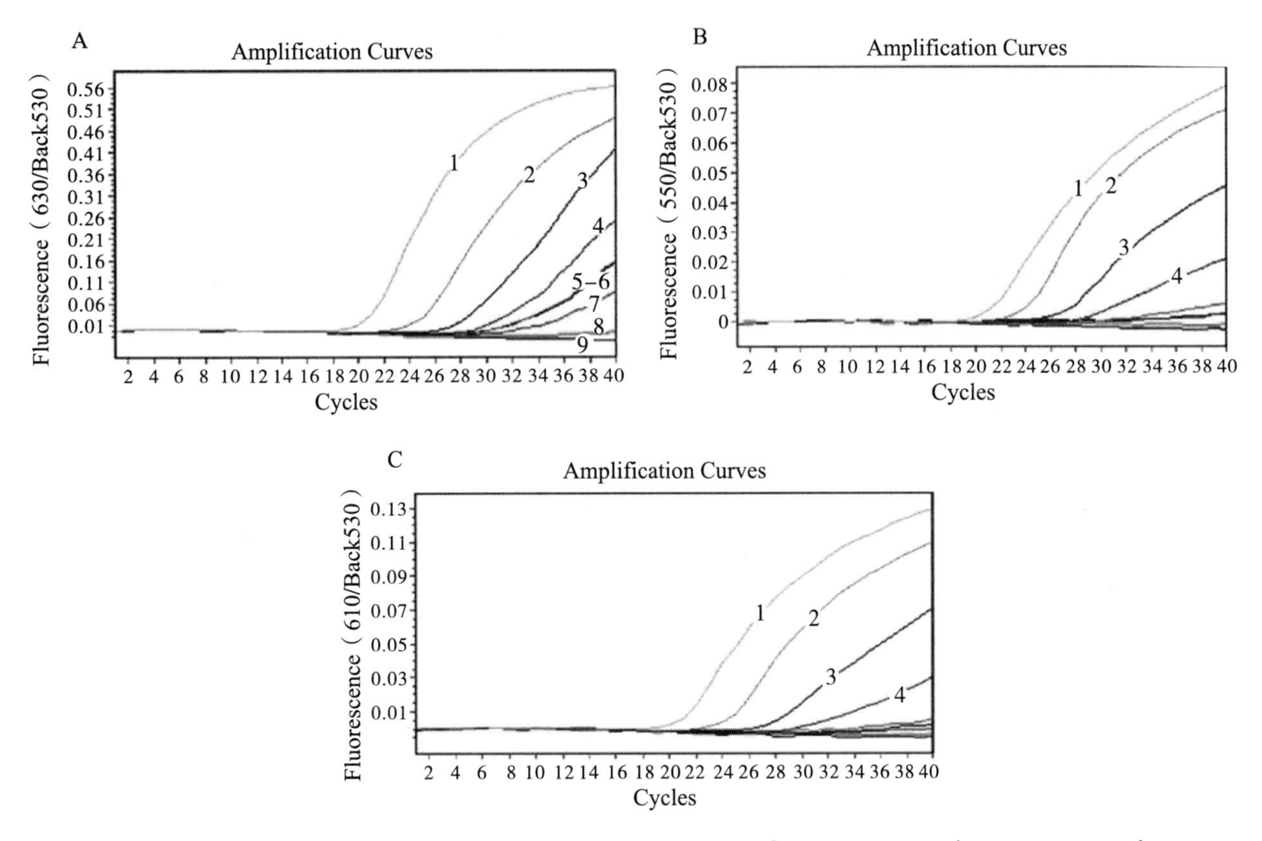

A: Sensitivity for IHHNV; B: Sensitivity for TSV; C: Sensitivity for WSSV. 1: 2×10^7 copies/μL; 2: 2×10^6 copies/μL; 3: 2×10^5 copies/μL; 4: 2×10^4 copies/μL; 5: 2×10^3 copies/μL; 6: 2×10^2 copies/μL; 7: 2×10^1 copies/μL; 8: 2×10^0 copies/μL; 9: Negative control.

Figure 5-14-1 Sensitivity of the real-time multiplex PCR

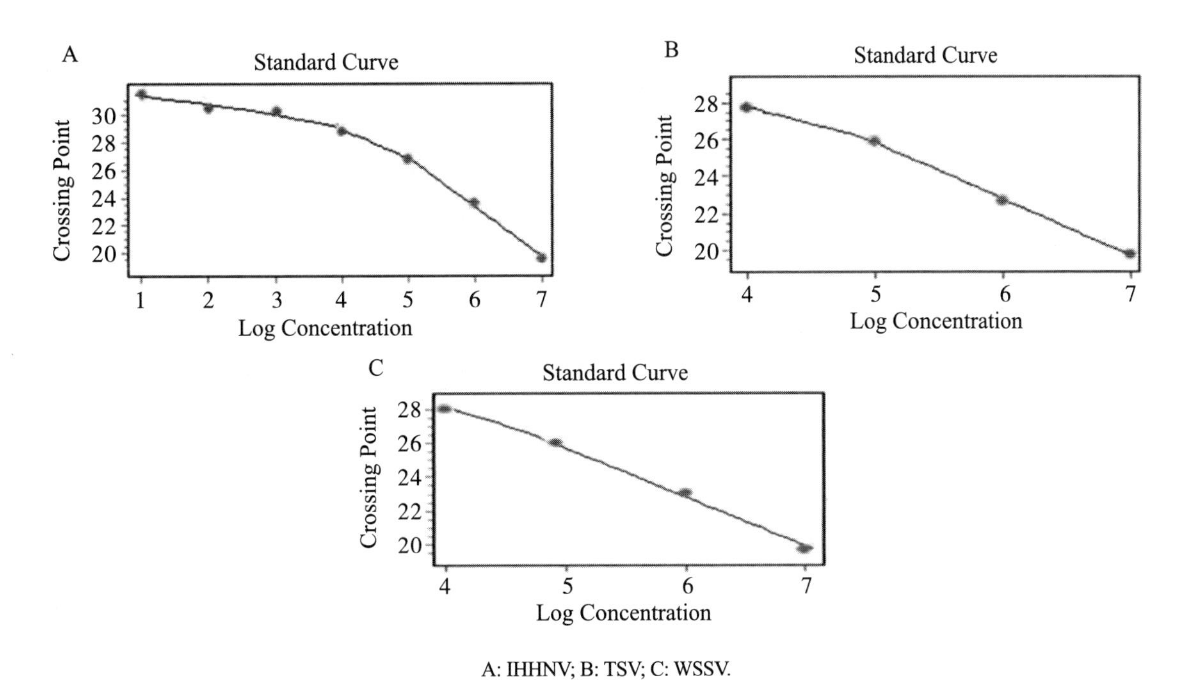

A: IHHNV; B: TSV; C: WSSV.

Figure 5-14-2 Standard curve of IHHNV, TSV, WSSV

Reproducibility and specificity

The samples were examined repeatedly using the rtm-PCR (Table 5-14-2), and the results indicated that the rtm-PCR was reproducible. The rtm-PCR results of different samples showed that one specific amplification curve was displayed when shrimp were infected by only one of these three viral pathogens, whereas two or three specific amplification curves were displayed when shrimp were infected by two or three viral pathogens, and no amplification curves were displayed for samples containing *Streptococcus*, *Vibrio* and water (Figure 5-14-2, Figure 5-14-3 A-C). The results indicate that rtm-PCR was able to detect and differentiate the presence of each pathogen in clinically infected shrimp.

Table 5-14-2 Real-time multiplex PCR results from three repeated detections

Sample	Ct value/copy number of the same sample at different times				
	First day	Fourth day	Seventh day	SD	CV/%
WSSV	$25.66/1\times10^6$	$25.98/1\times10^6$	$26.01/1\times10^6$	0.158	0.61
IHHNV	$25.32/1\times10^6$	$25.62/1\times10^6$	$25.99/1\times10^6$	0.336	1.31
TSV	$25.01/1\times10^6$	$26.01/1\times10^6$	$26.32/1\times10^6$	0.658	2.65

Note: SD, standard deviation; Ct, cycle threshold; CV, coefficient of variation.

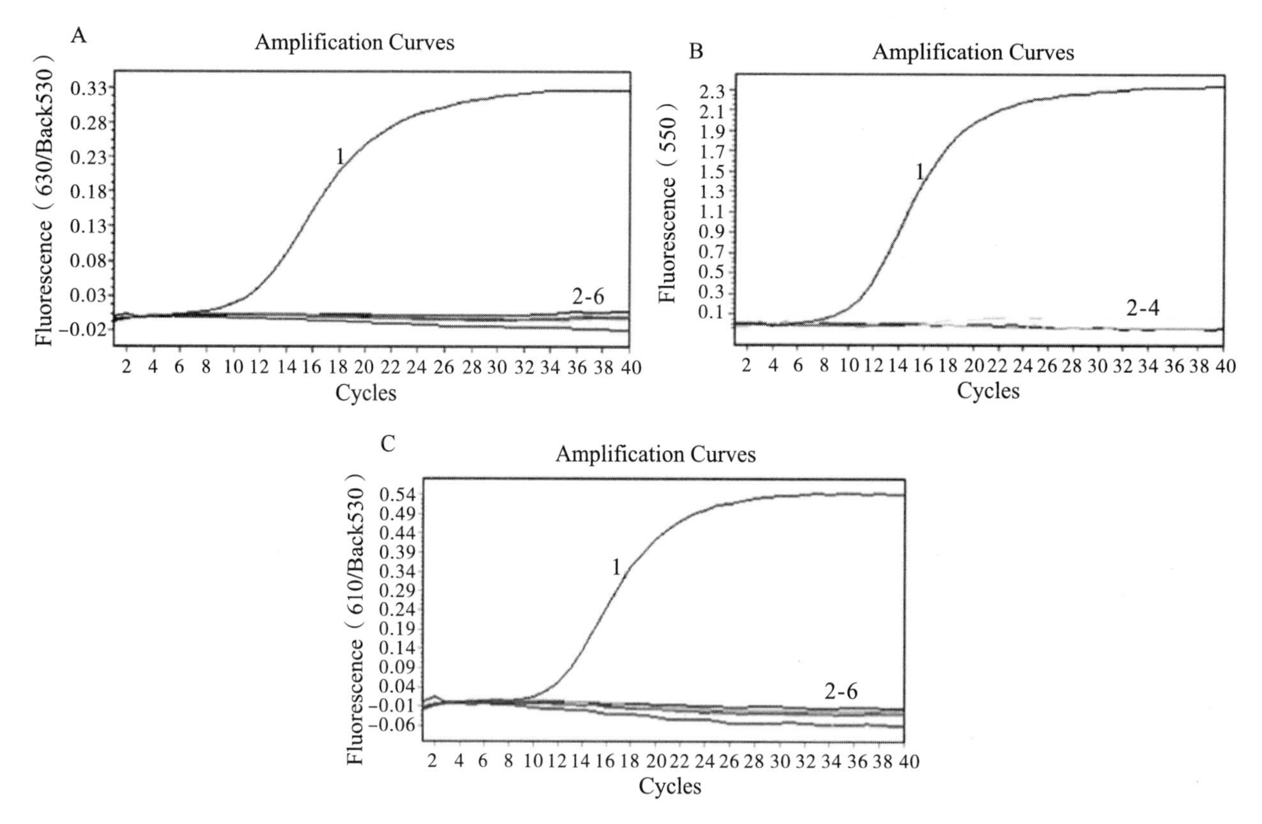

A: The specificity of IHHNV; 1: IHHNV; 2: WSSV; 3: TSV; 4: Streptococcus; 5: Vibrio; 6: Negative. B: The specificity of TSV; 1: TSV; 2: WSSV; 3: IHHNV; 4: Negative. C: The specificity of WSSV; 1: WSSV; 2: IHHNV; 3: TSV; 4: Streptococcus; 5: Vibrio; 6: Negative.

Figure 5-14-3 Specificity results of the real-time multiplex PCR

Clinical samples

Forty-five samples that were positive foreach single virus by routine PCR were detected by real-time multiplex PCR (Table 5-14-3). The results showed that $1.82 \times 10^7 \sim 3.34 \times 10^5$ copies/μL of WSSV were detected from 15 WSSV samples, and one sample was WSSV and IHHNV positive (7.72×10^6 and 2.10×10^2 copies/μL); $2.13 \times 10^7 \sim 8.49 \times 10^3$ copies/μL of IHHNV were detected from 15 IHHNV samples, and one sample was both IHHNV and WSSV positive (6.78×10^6 and 1.12×10^4 copies/μL); $1.16 \times 10^7 \sim 5.49 \times 10^5$ copies/μL of TSV were detected from 15 TSV samples, and none were IHHNV or WSSV positive.

Table 5-14-3 Results of real-time multiplex PCR of clinical samples

Species of shrimp	Shrimp pathogens/ field samples	Origin/source	Results					
			Real-time multiplex PCR			Routine PCR		
			WSSV	IHHNV	TSV	WSSV	IHHNV	TSV
POK+white shrimp	WSSV1+IHHNV	Guangxi, China	+	+	–	+	–	–
POK	WSSV (GXPRC/1/06)	Beihai, Guangxi, China	+	–	–	+	–	–
	WSSV (GXPRC/2/06)	Beihai, Guangxi, China	+	–	–	+	–	–
	WSSV (GXPRC/3/06)	Beihai, Guangxi, China	+	–	–	+	–	–
	WSSV (GXPRC/4/06)	Beihai, Guangxi, China	+	–	–	+	–	–
	WSSV (GXPRC/5/06)	Beihai, Guangxi, China	+	–	–	+	–	–
White shrimp	WSSV (GXPRC/7/06)	Beihai, Guangxi, China	+	–	–	+	–	–
	WSSV (GXPRC/8/06)	Beihai, Guangxi, China	+	–	–	+	–	–
POK	WSSV (GXPRC/1/06)	Hepu, Guangxi, China	+	–	–	+	–	–
	WSSV (GXPRC/2/06)	Hepu, Guangxi, China	+	–	–	+	–	–
	WSSV (GXPRC/3/06)	Hepu, Guangxi, China	+	–	–	+	–	–
	WSSV (GXPRC/4/06)	Hepu, Guangxi, China	+	–	–	+	–	–
White shrimp	WSSV (GXPRC/5/06)	Hepu, Guangxi, China	+	–	–	+	–	–
POK	WSSV (GXPRC/1/06)	Qinzhou, Guangxi, China	+	–	–	+	–	–
	WSSV (GXPRC/2/06)	Qinzhou, Guangxi, China	+	–	–	+	–	–
	WSSV-pMD18-T	GVRI	+	–	–	+	–	–
White shrimp+POK	IHHNV1+WSSV	Guangxi, China	+	+	–	–	+	–
POK	IHHNV (GXPRC/1/06)	Beihai, Guangxi, China	–	+	–	–	+	–
	IHHNV (GXPRC/2/06)	Beihai, Guangxi, China	–	+	–	–	+	–
	IHHNV (GXPRC/3/06)	Beihai, Guangxi, China	–	+	–	–	+	–
	IHHNV (GXPRC/4/06)	Beihai, Guangxi, China	–	+	–	–	+	–
White shrimp	IHHNV (GXPRC/5/06)	Beihai, Guangxi, China	–	+	–	–	+	–
POK	IHHNV (GXPRC/1/06)	Hepu, Guangxi, China	–	+	–	–	+	–
	IHHNV (GXPRC/2/06)	Hepu, Guangxi, China	–	+	–	–	+	–
	IHHNV (GXPRC/3/06)	Hepu, Guangxi, China	–	+	–	–	+	–
	IHHNV (GXPRC/4/06)	Hepu, Guangxi, China	–	+	–	–	+	–
	IHHNV (GXPRC/5/06)	Hepu, Guangxi, China	–	+	–	–	+	–
	IHHNV (GXPRC/6/06)	Hepu, Guangxi, China	–	+	–	–	+	–

continued

Species of shrimp	Shrimp pathogens/ field samples	Origin/source	Results					
			Real-time multiplex PCR			Routine PCR		
			WSSV	IHHNV	TSV	WSSV	IHHNV	TSV
POK	IHHNV (GXPRC/7/06)	Hepu, Guangxi, China	–	+	–	–	+	–
	IHHNV (GXPRC/1/06)	Qinzhou, Guangxi, China	–	+	–	–	+	–
	IHHNV (GXPRC/2/06)	Qinzhou, Guangxi, China	–	+	–	–	+	–
	IHHNV–pMD18–T	GVRI	–	+	–	–	+	–
	TSV (GXPRC/1/06)	Beihai, Guangxi, China	–	–	+	–	–	+
	TSV (GXPRC/2/06)	Beihai, Guangxi, China	–	–	+	–	–	+
	TSV (GXPRC/3/06)	Beihai, Guangxi, China	–	–	+	–	–	+
	TSV (GXPRC/4/06)	Beihai, Guangxi, China	–	–	+	–	–	+
	TSV (GXPRC/5/06)	Beihai, Guangxi, China	–	–	+	–	–	+
	TSV (GXPRC/6/06)	Beihai, Guangxi, China	–	–	+	–	–	+
	TSV (GXPRC/1/06)	Hepu, Guangxi, China	–	–	+	–	–	+
	TSV (GXPRC/2/06)	Hepu, Guangxi, China	–	–	+	–	–	+
	TSV (GXPRC/3/06)	Hepu, Guangxi, China	–	–	+	–	–	+
	TSV (GXPRC/4/06)	Hepu, Guangxi, China	–	–	+	–	–	+
	TSV (GXPRC/1/06)	Qinzhou, Guangxi, China	–	–	+	–	–	+
	TSV (GXPRC/2/06)	Qinzhou, Guangxi, China	–	–	+	–	–	+
	TSV (GXPRC/3/06)	Qinzhou, Guangxi, China	–	–	+	–	–	+
	TSV (GXPRC/4/06)	Qinzhou, Guangxi, China	–	–	+	–	–	+
White shrimp	TSV (GXPRC/5/06)	Qinzhou, Guangxi, China	–	–	+	–	–	+
	TSV–pMD18–T	GVRI	–	–	+	–	–	+
	pWSSV+pIHHNV+ pTSV	GVRI	+	+	+	+	+	+
	Vibrio (Extracted DNA)	CIVDC, Beijing, China	–	–	–	–	–	–
	Streptococcus (Extracted DNA)	CIVDC, Beijing, China	–	–	–	–	–	–
	Distilled water	Qinzhou, Guangxi, China	–	–	–	–	–	–
	White shrimp (Nucleic acid)	Qinzhou, Guangxi, China	–	–	–	–	–	–
	POK (Nucleic acid)		–	–	–	–	–	–

Note: CIVDC, China Institute of Veterinary Drug Control; GVRI, Guangxi Veterinary Research Institute; POK, Penaeus Orientalis Kishinouye.

Discussion

The rtm-PCR assay described here uses PCR primers and TaqMan probes targeting conserved regions of WSSV, IHHNV and TSV genes. One main advantage of this assay compared to other available tests is that it is multiplex. By using this approach, it was possible to identify all three pathogens in the same reaction vessel. The simultaneous detection of WSSV, IHHNV and TSV is especially useful, because these viruses commonly cause mixed infection in shrimp[9, 12, 16, 21].

In addition to its use in clinical diagnostics, the multiplex assay may be of value for detection of pathogen-free shrimp in environmental samples. However, the multiplex feature of this assay is optional; if so preferred, the three components can be utilized as single-targeting assays or combined into duplex assays without impacting the quality of the results. This makes this assay adaptable to circumstances that may not require the simultaneous detection of all three for a diagnostic decision.

Another important aspect of this real-time PCR approach is the short turnaround time. The confirmatory result of a suspected WSSV, IHHNV and TSV infection was obtained within 5 h of receiving the sample in the laboratory. This included time for DNA extraction and multiplex real-time PCR assay. The current methods for laboratory diagnosis of these three viruses are labor intensive and may lack the sensitivity and speed required to reveal the cause of infection before it is too late to take the appropriate measure. The real-time multiplex PCR assay presented here can therefore be extremely useful as a fast and sensitive complement to existing diagnostic methods. In addition, this real-time multiplex PCR does not require the unique expertise involved in morphology-based tests but can be performed in any laboratory with adequate infrastructure for real-time PCR testing.

References

[1] BELL A S, RANFORD-CARTWRIGHT L C. Real-time quantitative PCR in parasitology. Trends Parasitol, 2002, 18: 337-342.

[2] BONAMI J R, HASSON K, MARI J, et al. Taura syndrome of marine penaeid shrimp: characterization of the viral agent. J Gene Virol, 1997, 78: 313-319.

[3] BONAMI J R, TRUMPER B, MARI J, et al. Purification and characterization of the infectious hypodermal and haematopoietic necrosis virus of penaeid shrimps. J Gen Virol, 1990, 71: 2657-2664.

[4] BROCK J A. Special topic review: Taura syndrome. A disease important to shrimp farms: in the Americas. World J Microb Biotech, 1997, 13: 415-418.

[5] DHAR A K, ROUX M M, KLIMPEL K R. Detection and quantification of Infectious hypodermal and hematopoietic necrosis virus and white spot virus in shrimp using real-time quantitative PCR and SYBR green chemistry. J Clin Microb, 2001, 39: 2835-2845.

[6] ERICKSON H S, ZARAIN-HERZBERG M, LIGHTNER D V. Detection of Taura syndrome virus (TSV) strain differences using selected diagnostic methods: diagnostic implications in penaeid shrimp. Dis Aquat Org, 2002, 52: 1-10.

[7] KIATPATHOMCHAI W, BOONSAENG V, TASSANAKAJON A, et al. A non-stop, singletube, semi-nested PCR technique for grading the severity of white spot syndrome virus infections in Penaeus monodon. Dis Aquat Org, 2001, 47: 235-239.

[8] LIGHTNER D V, REDMAN R M. Shrimp diseases and current diagnostic methods. Aquaculture, 1998, 164: 201-220.

[9] LIGHTNER D V. The penaeid shrimp viruses IHHNV and TSV: epizootiology, production impacts and role of international trade in their distribution in the Americas. Revues Scientifique Et Technique Office International Des Epizooties, 1996, 15: 579-601.

[10] LO C F, LEU J H, HO C H, et al. Detection of baculovirus associated with white spot syndrome (WSBV) in penaeid shrimps using polymerase chain reaction. Dis Aquat Org, 1996, 25: 133-141.

[11] MACKAY I M. Real-time PCR in the microbiology laboratory. Clin Microbiol Infect, 2004, 10: 190-212.

[12] MANIVANNAN S, OTTA S K, KARUNASAGAR I, et al. Multiple viral infections in Penaeus monodon shrimp postlarvae in an Indian hatchery. Dis Aquat Org, 2002, 48: 233-236.

[13] MARI J, BONAMI J R, LIGHTNER D V. Taura syndrome of penaeid shrimp: cloning of viral genome fragments and development of specific gene probes. Dis Aquat Org, 1998, 33: 11-17.

[14] NUNAN L M, POULOS B T, LIGHTER D V. Reverse transcription polymerase chain reaction (RT-PCR) used for the detection of Taura syndrome virus (TSV) in experimentally infected shrimp. Dis Aquat Org, 1998, 34: 87-91.

[15] NUNAN L M, LIGHTNER D V. Development of a non-radio-active gene probe by PCR for detection of white spot syndrome virus (WSSV). J Virol Meth, 1997, 63: 193-201.

[16] PLUMB J A. Trends in freshwater fish disease research//FLEGEL T W, MACRAE I H. Diseases in Asian aquaculture Ⅲ. Fish health section, Asian fish soc, Manila. Rome: FAO Aquaculture Newsletter no. 16.FAO, 1997.

[17] SILVA A J, PIENIAZEK N J. Latest advances and trends in PCR-based diagnostic methods//DIONISIO D. Textbook-atlas of intestinal infections in AIDS. Milan: Springer-Verlag Italia, 2003.

[18] TAKAHASHI Y, ITAMI T, KONDO M, et al. Electron microscopic evidence of bacilliform virus infection in Kuruma shrimp (*Penaeus japonicus*). Fish Pathol, 1994, 29: 121-125.

[19] TANG K F J, WANG J, LIGHTNER D V. Quantitation of Taura syndrome virus by real-time RT-PCR with TaqMan assay. J Virol Meth, 2004, 115: 109-114.

[20] WANG S Y, HONG C, LOTZ J M. Development of a PCR procedure for the detection of Baculovirus penaei in shrimp. Dis Aquat Org, 1996, 25: 123-131.

[21] WANG Y C, LO C F, CHANG P S, et al. White spot syndrome associated virus (WSSV) infection in cultured and wild decapods in Taiwan. Aquaculture, 1998, 164: 221-231.

[22] XIE Z, PANG Y, DENG X, et al. A multiplex RT-PCR for simultaneous differentiation of three viral pathogens of penaeid shrimp. Dis Aquat Organ, 2007, 76: 77-80.

[23] YUE Z Q, LIU H, WANG W, et al. Development of real-time polymerase chain reaction assay with TaqMan probe for the quantitative detection of infectious hypodermal and hematopoietic necrosis virus from shrimp. J AOAC Int, 2006, 89: 240-244.

A duplex real-time PCR assay for the detection and quantification of avian reovirus and *Mycoplasma synoviae*

Huang Li, Xie Zhixun, Xie Liji, Deng Xianwen, Xie Zhiqin, Luo Sisi, Huang Jiaoling, Zeng Tingting, and Feng Jiaxun

Abstract

Background: infectious arthritis in broilers represents an economic and health problem, resulting in severe losses due to retarded growth and downgrading at the slaughterhouse. The most common agents associated with cases of infectious arthritis in poultry are avian reovirus (ARV) and *Mycoplasma synoviae* (MS). The accurate differentiation and rapid diagnosis of ARV and MS are essential prerequisites for the effective control and prevention of these avian pathogens in poultry flocks. This study thus aimed to develop and validate a duplex real-time PCR assay for the simultaneous detection and quantification of ARV and MS. Methods: specific primers and probes for each pathogen were designed to target the special sequence of the ARV σC gene or the MS phase-variable surface lipoprotein hemagglutinin (vlhA) gene. A duplex real-time PCR assay was developed, and the reaction conditions were optimized for the rapid detection and quantification of ARV and MS. Results: the duplex real-time PCR assay was capable of ARV- and MS-specific detection without cross-reaction with other non-targeted avian pathogens. The sensitivity of this assay was 2×10^1 copies for a recombinant plasmid containing ARV σC or MS vlhA gene, and 100 times higher than that of conventional PCR. This newly developed PCR assay was also reproducible and stable. All tested field samples of ARV and/or MS were detectable with this duplex real-time PCR assay compared with pathogen isolation and identification as well as serological tests. Conclusions: this duplex real-time PCR assay is highly specific, sensitive and reproducible and thus could provide a rapid, specific and sensitive diagnostic tool for the simultaneous detection of ARV and MS in poultry flocks. The assay will be useful not only for clinical diagnostics and disease surveillance but also for the efficient control and prevention of ARV and MS infections.

Keywords

duplex real-time PCR assay, avian reovirus, *mycoplasma synoviae*

Background

Infectious arthritis in broilers represents an economic and health problem, resulting in severe losses due to retarded growth and downgrading at the slaughterhouse. The most common agents associated with cases of infectious arthritis in poultry are avian reovirus (ARV) and *Mycoplasma synoviae* (MS). ARV belongs to the *Orthoreovirus* genus, one of nine genera of the Reoviridae family[1, 2]. ARV infection is associated with several disease syndromes and especially viral arthritis/tenosynovitis in chickens[3, 4]. Meanwhile, MS is a common pathogen found in turkeys and chickens that causes diseases of the respiratory tract, urogenital tract and joints and impairs growth[5, 6]. Mixed infections of ARV and MS have occurred in poultry flocks worldwide and have similar clinical signs, including severe immunosuppression, arthritis, depression, retarded growth, weight loss and decreased egg production. Bradbury[7] and Reck[8] also found that in chickens, a synergistic relationship exists between ARV and MS, which causes much more severe clinical signs and pathological lesions than the additive effects of these two pathogens alone do. The main feature of possible economic importance in ARV and MS infection is the incidence of decreased egg production and fertility, sternal bursitis leading to carcass downgrading and leg abnormalities related to condemnation of broilers. As the elimination of lesioned carcasses at the slaughterhouse is important[3, 9], the rapid and efficient detection and diagnosis of ARV and MS are essential prerequisites for the effective control and prevention of these avian pathogens in poultry flocks.

The current methods for ARV and MS detection include serological assays; pathogen isolation and identification; and molecular detection methods, such as single PCR and multiplex PCR[10-13]. However, these assays are laborious and time consuming, have limited specificity and sensitivity, and require post-amplification procedures. Real-time PCR assays for the specific identification of a target sequence by fluorescent probes can overcome these limitations and provide distinct advantages, such as a shorter detection time, improved sensitivity and specificity, simplified closed-tube procedures and the potential for pathogen screening and surveillance in commercial poultry flocks[14-16].

Therefore, the present study developed and validated a duplex real-time PCR assay for the differential diagnosis and quantitative detection of ARV and MS.

Materials and methods

Pathogens and construction of recombinant plasmids

DNA was first extracted from MS samples as described previously[17], and total RNA was extracted from ARV samples using TRIzol reagent (Life Technologies, Carlsbad, CA, USA) following the manufacturer's instructions. Next, cDNA was generated as described previously[18] and used as a template for a duplex real-time PCR assay. Target gene fragments from the ARV σC gene or the MS phase-variable surface lipoprotein hemagglutinin (vlhA) gene were then amplified with the primers listed in Table 5-15-1 and inserted into the pMD18-T vector (TaKaRa, Dalian, China). Subsequently, the constructed plasmids were transformed into DH5 α *Escherichia coli*. The recombinant plasmids carrying each target gene were confirmed by sequencing and were used as positive standards for ARV and MS. The copy number of each positive-standard plasmid was calculated as described previously[19].

Table 5-15-1 Specific primers used to clone ARV and MS specific genes

Primer name	Primer sequence(5'-3')	Amplicon length	Target gene
ARV C703F	TGTGGATCCATGGCGGGTCTCAAT	981 bp	σC gene
ARV C703R	CCGGAATTCTAAGGTGTCGATGCC		
MS vlhA F2	CTGTTATAGCAATTTCATGTGGTG	283 bp	phase-variable surface lipoprotein hemagglutinin (vlhA)
MS vlhA R2	TGTTGTAGTTGCTTCAACTTGTCT		

Oligonucleotide primers and DNA probes for duplex real-time PCR

DNASTAR software (DNASTAR Inc., Madison, WI, USA) was used to confirm the highly conserved regions in the ARV and MS genomes, and Primer Express 3.0 software (Applied Biosystems, Foster City, CA, USA) was used to design the primers and probes listed in Table 5-15-2 for ARV and MS, based on their highly conserved regions. The cross-reactivity of the oligonucleotides was assessed by BLAST analysis. Both sets of primers and probes were synthesized by TaKaRa (Dalian, China).

Table 5-15-2 Primers and probes used for the duplex real-time PCR assay

Primer/probe name	Primer/probe sequence (5'-3')	Target gene	Amplicon length
ARV F	CGTTCCCTGTGGACGTATCA	σC	69 bp
ARV R	GAGTACACCCCATACGCTTGGT		
ARV P	(FAM) TCACCCGCGATTCTGCGACTCAT (Eclipse)		
MS-F	ATAGCAATTTCATGTGGTGATCAA	vlhA	143 bp
MS-R	TGGATTTGGGTTTTGAGGATTA		
MS-P	(ROX) CAGCACCTGAACCAACACCTGGAA (Eclipse)		

Duplex real-time PCR assay for simultaneous MS and ARV detection

The duplex real-time PCR was performed in a 20 μL volume. The reaction mixture included 1 × real-time PCR Premix (Perfect Real Time PCR Kit, TaKaRa, Dalian, China); 0.3 μM ARV F, ARV R and ARV P primers; 0.3 μM MS F, MS R and MS P primers; and 2 μL of positive-plasmid template. Sterilized H_2O was added to bring the final volume to 20 μL. The protocol for the reaction was 95 ℃ for 30 s; 45 cycles of 95 ℃ for 10 s and 60 ℃ for 30 s; and, finally, 40 ℃ for 5 s. The fluorescence was measured at the end of each 60 ℃ incubation. The data analysis was performed using Light Cycler 2.0 system software (Roche, Molecular Biochemical, Mannheim, Germany).

Conventional RT-PCR and PCR

Conventional RT-PCR for ARV amplification and conventional PCR for MS amplification were performed. The PCR mixture contained 2 × Premix Taq (TaKaRa, Dalian, China), 0.4 μM forward primer or reverse primer, 2 μL of template and sterilized H_2O to bring the final reaction volume to 25 μL. The conditions for PCR were 95 ℃ for 5 min; 72 ℃ for 7 min; and three-step cycling 35 times at 95 ℃ for 30 s, 60 ℃ for 30 s and 72 ℃ for 30 s. The PCR product was run on a 2% agarose gel at 80 V for 45 min and visualized on a molecular imager Gel Doc XR+ imaging system with Image Lab software (Bio-Rad, Life Science Research, Hercules, CA, USA).

Specificity and sensitivity of the duplex real-time PCR assay

To assess the specificity of the assay, DNA from *Mycoplasma gallisepticum* (MG), *Mycoplasma iowae* (MI) and *Mycoplasma meleagridis* (MM) were extracted as described previously[17]. Additionally, cDNA was generated from total RNA that was extracted from cases of Newcastle disease virus (NDV), infectious bursal disease virus (IBDV), avian infectious bronchitis virus (AIBV), the H9 subtype of the avian influenza virus (AIV), Marek's disease virus (MDV), reticuloendotheliosis virus (REV), and avian leukosis virus (ALV) using TRIzol reagent (Life Technologies, Carlsbad, CA, USA) following the manufacturer's instructions. The DNA and cDNA were mixed together in equal concentrations as the templates and were subjected to the optimized duplex real-time PCR assay to detect ARV and MS. The sensitivity of the duplex real-time PCR assay was determined using serial 10-fold dilutions ($10^1 \sim 10^8$ copies/μL) of positive-plasmid combinations carrying the MS and ARV target genes as templates. These results were compared with the results of conventional PCR. To generate a standard curve for ARV and MS, the threshold cycle (Ct) of these standard dilutions was plotted against the log value of the copy number of the corresponding standard plasmid.

Reproducibility and interference tests of the duplex real-time PCR assay

To assess the intra- and inter-assay reproducibility, three samples with the same concentration (10^8 copies/μL) of the MS or ARV target gene were assessed using the duplex real-time PCR assay. The same experiments were repeated in triplicate every two days for seven days. The reproducibility was then analyzed based on the standard deviation (SD) and the coefficient of variability (CV) of the Ct average. To determine the reaction efficiency interference, different concentrations of positive plasmids carrying the ARV or MS target gene (10^8 and 10^1 copies/μL, respectively) were analyzed using the duplex real-time PCR assay.

Duplex real-time PCR analysis of field samples

All field samples, such as joints and joint contents, were collected from chicks and broilers exhibiting clinical signs of MS or ARV infections and were used to validate the duplex real-time PCR assay. The results were compared with those of traditional diagnostic methods, such as pathogen isolation and identification and serological tests.

Results

Specificity test

The specificity of the duplex real-time PCR assay was verified by examining DNA/cDNA from different samples infected with different pathogens. As shown in Figure 5-15-1 and Table 5-15-3, the duplex real-time PCR assay was able to detect and differentiate ARV and MS independently and simultaneously. In contrast, the other avian pathogens (NDV, IBDV, AIBV, AIV, MDV, REV, ALV, MG, MI and MM) were not detected using the duplex real-time PCR assay. When samples were coinfected with both ARV and MS, unique amplification curves were simultaneously produced in the 530 nm and 610 nm channels, whereas a single amplification curve was observed in the 530 nm or 610 nm channel when samples were infected with only ARV or MS, respectively. Thus, the specificity of the duplex real-time PCR assay was 100%, with no detectable fluorescent signals for other avian pathogens or negative controls.

Table 5-15-3 Pathogens used and Ct values of the duplex real-time PCR assay's specificity

Pathogen	Number of samples	Ct values of duplex real-time PCR assay	
		MS	ARV
NDV	1	Negative	Negative
H9 subtype of AIV	1	Negative	Negative
AIBV	1	Negative	Negative
IBDV	1	Negative	Negative
MG	1	Negative	Negative
MM	1	Negative	Negative
MI	1	Negative	Negative
MDV	1	Negative	Negative
pMD18-T-ARV	1	Negative	15.58
pMD18-T-MS	1	15.31	Negative
REV	1	Negative	Negative
ALV	1	Negative	Negative

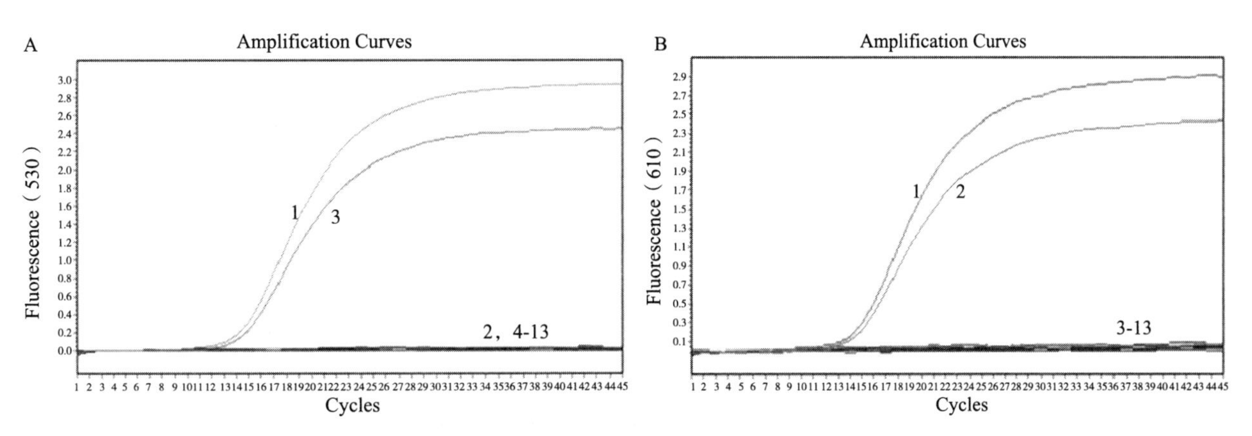

A: Specificity of the duplex real-time PCR assay for ARV; B: Specificity of the duplex real-time PCR assay for MS. 1: MS+ARV; 2: MS; 3: ARV; 4: IBDV; 5: NDV; 6: AIBV; 7: MDV; 8: H9 subtype of AIV; 9: REV; 10: ALV; 11: MG; 12: MI; 13: MM; 14: Negative control.

Figure 5-15-1 Specificity of the duplex real-time PCR assay for ARV and MS

Sensitivity test

The sensitivity of the duplex real-time PCR assay was verified by testing 10^1~10^8 copies/μL of recombinant plasmids carrying the ARV or MS target gene. In Figure 5-15-2, the ARV amplification curves are shown in the 530 nm channel (Figure 5-15-2 A), and the MS amplification curves are shown in the 610 nm channel (Figure 5-15-2 B). Moreover, the standard curves for ARV and MS are shown in Figure 5-15-3, and the Ct values are listed in Table 5-15-4.The results revealed that even with a template amount as low as 2×10^1 copies, the ARV or MS target gene was still detectable. In contrast, the detection limit of the conventional PCR template was 2×10^3 copies for ARV and MS (Figure 5-15-4), which is 100 times lower than that of the duplex real-time PCR assay. Thus, the duplex real-time PCR assay is highly sensitive.

Table 5-15-4 Ct values from the serial dilution of positive plasmids

Tenfold dilution	2×10^8 copies	2×10^7 copies	2×10^6 copies	2×10^5 copies	2×10^4 copies	2×10^3 copies	2×10^2 copies	2×10^1 copies
ARV (Ct)	14.86	17.21	20.20	22.49	24.90	28.36	30.45	33.18
MS (Ct)	14.79	16.83	19.77	22.18	24.49	28.24	29.72	32.41

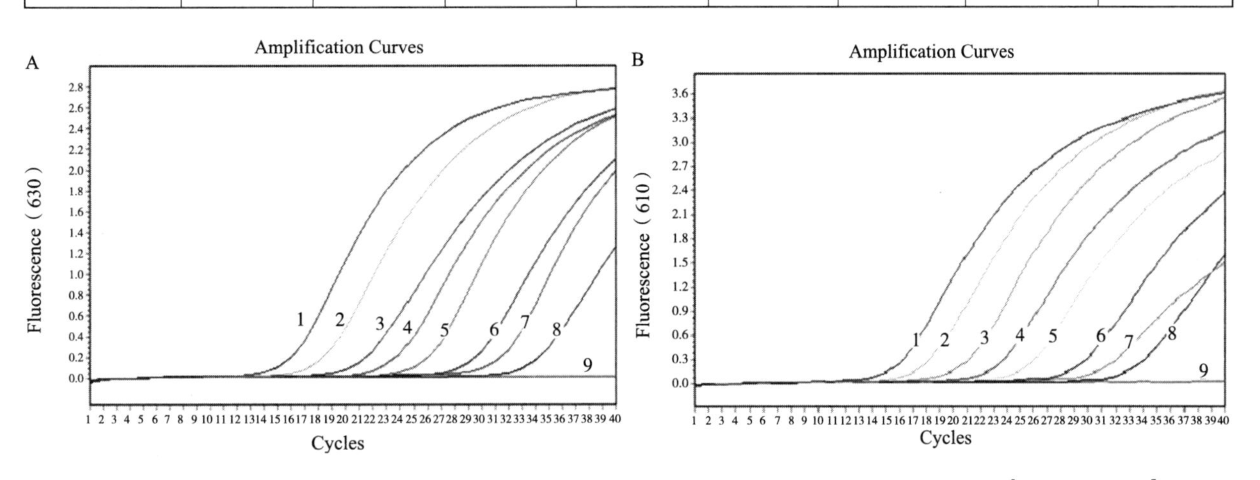

A: Sensitivity of the duplex real-time PCR assay for ARV; B: Sensitivity of the duplex real-time PCR assay for MS. 1: 2×10^8 copies; 2: 2×10^7 copies; 3: 2×10^6 copies; 4: 2×10^5 copies; 5: 2×10^4 copies; 6: 2×10^3 copies; 7: 2×10^2 copies; 8: 2×10^1 copies; 9: Negative control.

Figure 5-15-2 Sensitivity of the duplex real-time PCR assay for ARV and MS

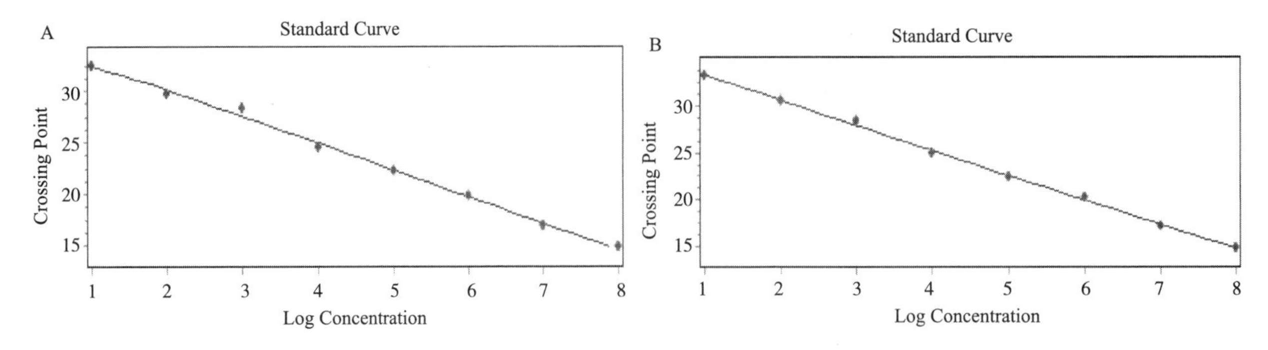

A: ARV standard curve; B: MS standard curve.

Figure 5-14-3 ARV and MS standard curves

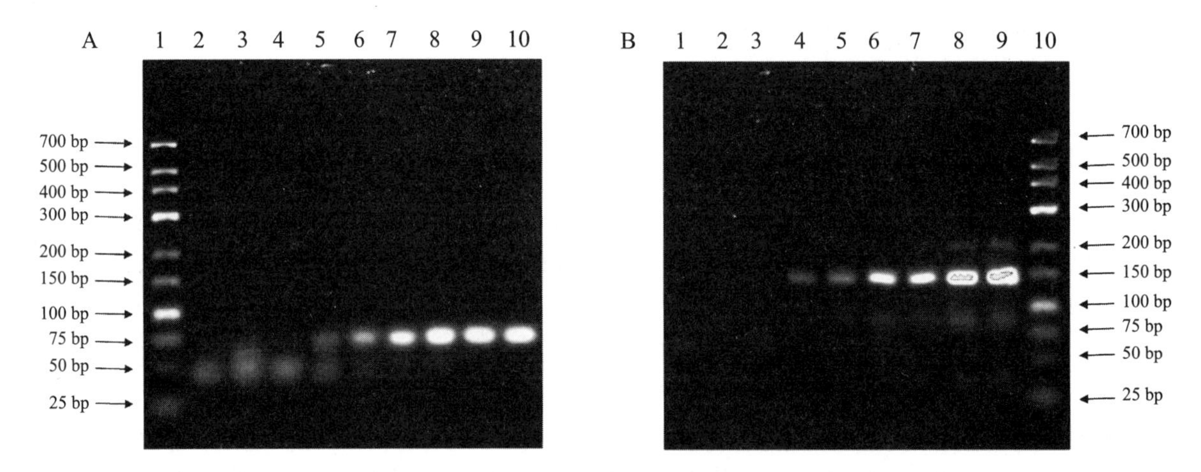

A: Sensitivity of conventional PCR for ARV; 1: Low ladder, 2: Negative control, 3: 2×10^1 copies, 4: 2×10^2 copies; 5: 2×10^3 copies; 6: 2×10^4 copies; 7: 2×10^5 copies 8: 2×10^6 copies, 9: 2×10^7 copies, 10: 2×10^8 copies. B: Sensitivity of conventional PCR for MS; 1: Negative control; 2: 2×10^1 copies; 3: 2×10^2 copies; 4: 2×10^3 copies, 5: 2×10^4 copies; 6: 2×10^5 copies; 7: 2×10^6 copies; 8: 2×10^7 copies; 9: 2×10^8 copies; 10: Low ladder.

Figure 5-15-4 Sensitivity of conventional PCR for ARV and MS

Reproducibility and interference tests

The reaction of reproducibility was determined by testing three samples of the same concentration at the same time points and was assessed using the SD and CV of the Ct values for each sample. The intra-assay reproducibility results are shown in Figure 5-15-5, and the inter-assay reproducibility results are listed in Table 5-15-5.The CV values were 1.61% for ARV and 1.89% for MS. These data indicate that the findings produced by the duplex real-time PCR assay are reproducible.

Table 5-15-5 Reproducibility of the duplex real-time PCR assay for ARV and MS

Pathogen	Ct values of same samples at different time points				
	Day 1	Day 4	Day 7	SD	CV
ARV	$15.23/1\times10^8$	$15.4/1\times10^8$	$15.36/1\times10^8$	0.24	1.61%
MS	$14.96/1\times10^8$	$14.7/1\times10^8$	$15.09/1\times10^8$	0.28	1.89%

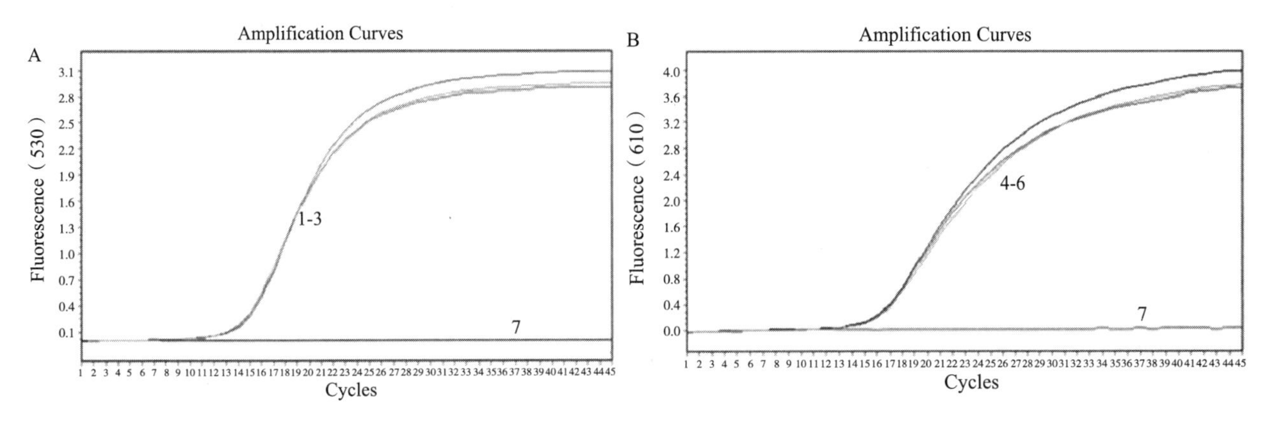

A: ARV; B: MS; 1-3: ARV; 4-6: MS; 7: Negative control.

Figure 5-15-5 Reproducibility of the duplex real-time PCR assay

Because the presence of other templates may affect the amplification eficiency of a PCR assay, we tested the influence of mixtures of different concentrations on reaction sensitivity. In particular, a combination of a high concentration (10^8 copies/μL) and a low concentration (10^2 copies/μL) of positive plasmids carrying the ARV or MS target gene was used in an interference test (Table 5-15-6). The results indicated that there were no systematic deviations in the amplification curves of the mixed templates compared with those of the single template; moreover, the CV value was less than 3% (data not shown). The results suggest that the newly developed duplex real-time PCR assay is stable.

Table 5-15-6 Samples used in the interference test

Pathogen	Sample 1	Sample 2	Sample 3	Sample 4	Sample 5	Sample 6
ARV	2×10^8 copies	2×10^2 copies	2×10^8 copies		2×10^2 copies	
MS	2×10^2 copies	2×10^8 copies		2×10^8 copies		2×10^2 copies

Field samples

The detection results of the duplex real-time PCR assay for 40 field samples are listed in Table 5-15-7 and the results were confirmed by pathogen isolation and identification or serological tests. The ARV detection rate was 7.5%, and the MS detection rate was 5%. Additionally, the range of Ct values for ARV was 15.29～34.42, and the range of Ct values for MS was 13.53～30.68.Thus, the results of this new assay were comparable with the results of other detection approaches.

Table 5-15-7 Detection of field samples using the duplex real-time PCR assay

	ARV	MS
Positive samples / total samples	3/40	2/40
Positive rate	7.50%	5.00%

Discussion

Both MS and ARV can cause similar clinical signs and lesions and may be present as co-infections in chickens and other avian species, which can lead to huge economic losses[20]. In this paper, we developed a duplex real-time PCR assay and described its use for the rapid, sensitive and accurate quantitative detection of ARV and MS.

The primary advantage of this duplex real-time PCR assay is the simultaneous detection and differentiation of ARV and MS. By using unique primer and probe sets within the highly conserved gene regions of ARV and MS, this duplex real-time PCR assay is readily able to detect and differentiate these pathogens via one reaction. Furthermore, this assay is optional and can be utilized as a single-target assay or combined into duplex assays, without impacting the quality of the results. Specifically, duplexing reduces the expense of reagents and the required time for analysis, and the single-target assay makes this assay adaptable to circumstances that may not require the simultaneous detection of these two pathogens for diagnostic purposes. These advantages greatly facilitate clinical application, which is an important criterion for the usefulness of a diagnostic assay for the early surveillance and prevention of diseases[21].

For a method of pathogen detection to be used as a clinical diagnostic tool, sensitivity is a key criterion [22, 23]. Using the newly developed assay, as few as 2×10^1 copies could be detected for both ARV and MS, which was more sensitive than the results of a duplex real-time PCR assay reported by Sprygin[24] and the results of the multiplex PCR performed by Reck[11]. Moreover, during detection with mixed samples (other non-targeted pathogens) and field samples, the specificity of this new assay was comparable with that of traditional methods, such as pathogen isolation and identification and serological tests. Therefore, this duplex real-time PCR assay with higher sensitivity rates could be promising as a tool for rapid clinical differentiation and diagnosis at the early stage of ARV and/or MS infection.

Another distinct feature of this duplex real-time PCR assay is the short turn-around time for the results. In the present study, the results for ARV and MS infections were obtained within 2 h with this duplex real-time PCR assay, which is very important for rapid diagnosis, especially during emergent disease outbreaks. Furthermore, the obtained results could be directly visualized on a computer connected to the real-time PCR station. Compared with the conventional diagnostic approaches for ARV and MS infections (and even single and multiplex PCRs[25, 26]), this assay does not require additional unique equipment or specialized labor. This method also minimizes post-amplification procedures, such as electrophoresis and UV visualization, which are time consuming. As compared to recently developed isothermal methods for ARV or MS detection, including loop-mediated isothermal amplification[27, 28] and cross-priming amplification[29], for which there is no need for expensive equipment except a water bath, the main drawback of the duplex real-time PCR assay is the absolute need for the thermal cycle. However, the method capability of simultaneous detection for ARV and MS highlights its importance and great value for the rapid detection of ARV and MS infections in the laboratory.

Considering the high cost of probe synthesis and the possibility of different genotypes as well as variant or vaccine strains of ARV or MS, the development of new technology or novel reagents for probe synthesis and the design of more primers based on more highly conserved regions of the ARV and MS genomes would be necessary to investigate further modification and optimization of this new assay.

Conclusions

In this study, we developed a rapid, specific and sensitive duplex real-time PCR assay for the simultaneous detection of ARV and MS. Based on its speed and sensitivity, this newly developed assay could be useful not only for the clinical diagnosis of ARV and MS infections but also for the control and prevention of these infections.

References

[1] TENG L, XIE Z, XIE L, et al. Complete genome sequences of an avian orthoreovirus isolated from Guangxi, China. Genome

Announc, 2013, 1 (4): e00495-13.

[2] TENG L, XIE Z, XIE L, et al. Sequencing and phylogenetic analysis of an avian reovirus genome. Virus Genes, 2014, 48 (2): 381-386.

[3] JONES R C. Avian reovirus infections. Rev Sci Tech, 2000, 19 (2): 614-625.

[4] VAN DER HEIDE L. The history of avian reovirus. Avian Dis, 2000, 44 (3): 638-641.

[5] LOCKABY S B, HOERR F J, LAUERMAN L H, et al. Pathogenicity of *Mycoplasma synoviae* in broiler chickens. Vet Pathol, 1998, 35 (3): 178-190.

[6] HINZ K H, BLOME C, RYLL M. Virulence of *Mycoplasma synovia*e strains in experimentally infected broiler chickens. Berl Munch Tierarztl Wochenschr, 2003, 116 (1-2): 59-66.

[7] BRADBURY J M, GARUTI A. Dual infection with *Mycoplasma synoviae* and a tenosynovitis-inducing reovirus in chickens. Avian Pathol, 1978, 7 (3): 407-419.

[8] RECK C, MENIN Á, PILATI C, et al. Características clínicas e anatomo-histopatologicas da infecção experimental mista por *Orthoreovirus aviario* e *Mycoplasma synovia*e em frangos de corte. Pesquisa Veterinária Brasileira, 2012, 32 (8): 687-691.

[9] LANDMAN W J M, FEBERWEE A. Aerosol-induced *Mycoplasma synoviae* arthritis: the synergistic effect of infectious bronchitis virus infection. Avian Pathol, 2004, 33 (6): 591-598.

[10] ORTIZ A, KLEVEN S H. Serological detection of *Mycoplasma synoviae* infection in turkeys. Avian Dis, 1992, 36 (3): 749-752.

[11] RECK C, MENIN Á, CANEVER M F, et al. Rapid detection of *Mycoplasma synoviae* and avian reovirus in clinical samples of poultry using multiplex PCR. Avian Dis, 2013, 57 (2): 220-224.

[12] NOORMOHAMMADI A H, MARKHAM P F, MARKHAM J F, et al. *Mycoplasma synoviae* surface protein MSPB as a recombinant antigen in an indirect ELISA. Microbiology (Society for General Microbiology), 1999, 145 (Pt 8): 2087.

[13] BEN ABDELMOUMEN MARDASSI B, BEN MOHAMED R, GUERIRI I, et al. Duplex PCR to differentiate between *Mycoplasma synoviae* and *Mycoplasma gallisepticum* on the basis of conserved species-specific sequences of their hemagglutinin genes. Journal of Clinical Microbiology, 2005, 43 (2): 948-958.

[14] TROXLER S, MAREK A, PROKOFIEVA I, et al. TaqMan real-time reverse transcription-PCR assay for universal detection and quantification of avian hepatitis e virus from clinical samples in the presence of a heterologous internal control RNA. Journal of Clinical Microbiology, 2011, 49 (4): 1339-1346.

[15] WITTWER C T, HERRMANN M G, GUNDRY C N, et al. Real-Time multiplex PCR assays. Methods (San Diego, Calif.), 2001, 25 (4): 430-442.

[16] MACKAY I M, ARDEN K E, NITSCHE A. Real-time PCR in virology. Nucleic acids research, 2002, 30 (6): 1292-1305.

[17] XIE Z, LUO S, XIE L, et al. Simultaneous typing of nine avian respiratory pathogens using a novel GeXP analyzer-based multiplex PCR assay. Journal of Virological Methods, 2014, 207: 188-195.

[18] XIE Z, XIE L, PANG Y, et al. Development of a real-time multiplex PCR assay for detection of viral pathogens of penaeid shrimp. Archives of virology, 2008, 153 (12): 2245-2251.

[19] VAITOMAA J, RANTALA A, HALINEN K, et al. Quantitative real-time PCR for determination of microcystin synthetase e copy numbers for microcystis and anabaena in Lakes. Applied and Environmental Microbiology, 2003, 69 (12): 7289-7297.

[20] MONTEBUGNOLI L M D, VENTURI M D, GISSI D B D, et al. Clinical and histologic healing of lichenoid oral lesions following amalgam removal: a prospective study. Oral Surgery, Oral Medicine, Oral Pathology and Oral Radiology, 2012, 113 (6): 766-772.

[21] HEIDENREICH A, BELLMUNT J, ZATTONI F, et al. EAU guidelines on prostate cancer. Part 1: screening, diagnosis, and treatment of clinically localised disease. European Urology, 2011, 59 (1): 61-71.

[22] HEIDENREICH A, BELLMUNT J, BOLLA M, et al. EAU guidelines on prostate cancer. Part I: screening, diagnosis, and treatment of clinically localised disease. Actas Urológicas Españolas (English ed.), 2011, 35 (9): 501-514.

[23] WU H, RAO P, JIANG Y, et al. A sensitive multiplex real-time PCR panel for rapid diagnosis of viruses associated with porcine respiratory and reproductive disorders. Molecular and cellular probes, 2014, 28 (5-6): 264-270.

[24] SPRYGIN A V, ANDREYCHUK D B, KOLOTILOV A N, et al. Development of a duplex real-time TaqMan PCR assay with an internal control for the detection of *Mycoplasma gallisepticum* and *Mycoplasma synoviae* in clinical samples from commercial and backyard poultry. Avian Pathol, 2010, 39 (2): 99-109.

[25] JINDAL N, CHANDER Y, PATNAYAK D P, et al. A Multiplex RT-PCR for the detection of astrovirus, rotavirus, and reovirus in Turkeys. Avian Diseases Digest, 2012, 7 (3): e34-e35.

[26] XIE Z X, FADL A A, GIRSCHICK T, et al. Amplification of avian reovirus RNA using the reverse transcriptase-polymerase chain reaction. Avian Dis, 1997, 41 (3): 654-660.

[27] KURSA O, WOŹNIAKOWSKI G, TOMCZYK G, et al. Rapid detection of *Mycoplasma synoviae* by loop-mediated isothermal amplification. Archives of Microbiology, 2015, 197 (2): 319-325.

[28] XIE Z, PENG Y, LUO S, et al. Development of a reverse transcription loop-mediated isothermal amplification assay for visual detection of avian reovirus. Avian Pathol, 2012, 41 (3): 311-316.

[29] WOŹNIAKOWSKI G, NICZYPORUK J S, SAMOREK SALAMONOWICZ E, et al. development and evaluation of cross-priming amplification for the detection of avian reovirus. Journal of Applied Microbiology, 2015, 118 (2): 528-536.

Development of duplex fluorescence-based loop-mediated isothermal amplification assay for detection of *Mycoplasma bovis* and bovine herpes virus 1

Fan Qing, Xie Zhixun, Xie Zhiqin, Xie Liji, Huang Jiaoling, Zhang Yanfang, Zeng Tingting, Zhang Minxiu, Wang Sheng, Luo Sisi, Liu Jiabo, and Deng Xianwen

Abstract

Mycoplasma bovis (MB) and bovine herpes virus 1 (BHV-1) are two important pathogens that cause bovine respiratory disease in the beef feedlot and dairy industries. The aim of this study was to develop and validate a duplex fluorescence-based loop-mediated isothermal amplification (DLAMP) assay for simultaneous detection of MB and BHV-1.Two sets of specific primers for each pathogen were designed to target the unique sequences of the MB *uvrC* gene and the BHV-1 gB gene. The inner primer for BHV-1 was synthesized with the fluorophore FAM at the 5' end to detect the BHV-1 gB gene, and the inner primer for MB was synthesized with the fluorophore CY5 at the 5' end to detect the MB *uvrC* gene. The DLAMP reaction conditions were optimized for rapid and specific detection of MB and BHV-1. The DLAMP assay developed here could specifically detect MB and BHV-1 without cross-reaction with other known non-target bovine pathogens. The sensitivity of this DLAMP assay was as low as 2×10^2 copies for recombinant plasmids containing the MB and BHV-1 target genes. In a detection test of 125 clinical samples, the positive rates for MB, BHV-1 and co-infection were 44.8%, 13.6% and 1.6%, respectively. Furthermore, the sensitivity and specificity of DLAMP were determined as 95%~96.6% and 100%, respectively, of those of field sample detection by the real-time polymerase chain reaction (PCR) assay recommended by the World Organisation for Animal Health. Overall, DLAMP provides a rapid, sensitive and specific assay for the identification of MB and BHV-1 in clinical specimens and for epidemiological surveillance.

Keywords

MB, BHV-1, DLAMP, fluorescent detection

Introduction

Bovine respiratory disease (BRD) is one of the major diseases affecting cattle health and production, and it causes immense economic losses. Its aetiology is multifactorial; many infectious agents are important in the development of BRD including viruses, bacteria and mycoplasmas. *Mycoplasma bovis* (MB) and bovine herpes virus 1 (BHV-1) are two important pathogens of BRD in the cattle industry[1]. Infection with these two pathogens can lead to reduced production, higher levels of morbidity and mortality, and increased veterinary labour costs, as well as economic losses. Infections with MB and BHV-1 are related to the occurence of high fever, pneumonia and polyarthritis, rhinitis, conjunctivitis, genital tract infection, and reproductive problems such as abortion and reduced fertility in cattle[2, 3]. MB and BHV-1 are distributed worldwide. Stress and environmental factors such as weaning, temperature, dust, stocking density, humidity, inadequate nutrition and transportation are important factors in the development of MB outbreaks. Furthermore, MB is difficult to remove thoroughly from an infected farm after an outbreak, and infected cattle can carry the pathogens for months and even years, remaining a source of infection[4, 5]. Similar to most other herpesviruses, inapparent infection is common in BHV-1 carriers. After BHV-1 infection, the virus can enter neural cells and establish a latent infection in sensory ganglia. Latent infections also occur in nonneural sites, such as the tonsils and lymph nodes[1]. These latent pathogens can be reactivated both by stressful conditions and by administration of glucocorticoids. BHV-1-infected bulls may shed virus intermittently in the semen long after the primary infection[6, 7]. Irrespective of the infectious agent involved, clinical manifestations of BRD may present similarly. Moreover, detection of a bacterial pathogen can mask an underlying viral cause, as both bacteria and viruses can co-infect their natural hosts[8]. A survey of exposure to BHV-1 in Canadian feedlots suggested that respiratory disease caused by MB was associated with previous exposure to BHV-1[9]. Cattle co-infected with BHV-1 and MB are regarded as lifelong carriers and potential shedders of these pathogens, and such infected cattle pose a potential risk to herds in the cattle industry. Therefore, differential diagnosis of MB and BHV-1 infections may lead to a better understanding of the epidemiology and natural distribution of MB and BHV-1 infections in the field, which may provide useful information for the control of BRD.

PCR, a highly sensitive and specific test, is considered a routine diagnostic test for MB and BHV-1 infections[10, 11]. LAMP is an alternative molecular genetic method derived from PCR for carrying out reactions under isothermal conditions using a single enzyme. LAMP-amplified products can be easily visualized by the naked eye. Compared with the two primers of conventional PCR, LAMP has four primers that can recognize 6 locations in the target gene; therefore, the LAMP method has higher reaction specificity than that of a conventional PCR assay. LAMP is $10^2 \sim 10^5$ times more sensitive than conventional PCR[12-14].

The aim of the present study was to develop and validate a DLAMP assay for simultaneous detection of MB and BHV-1 in a single reaction. The DLAMP assay was compared to the World Organisation for Animal Health (OIE) recommended real-time PCR assay to assess its application in routine diagnosis of the aetiological agents involved in BRD.

Materials and methods

Pathogens and DNA/RNA extraction

The pathogens used in this study are listed in Table 5-16-1.Ten nasal swabs, 11 whole blood samples, and 21 semen specimens were collected from healthy cattle, and the extracted DNA samples were used as negative controls. The DNA of mycoplasmas, bacteria and DNA viruses was extracted by using the MiniBEST Universal Genomic DNA Extraction Kit version 5.0 (TaKaRa, Dalian, China) according to the manufacturer's instructions. The RNA of RNA viruses was extracted by using the Universal RNA Extraction Kit (TaKaRa) and subsequently reverse-transcribed to cDNA by using PrimeScript Ⅱ 1 st Strand cDNA Synthesis Kit (TaKaRa). DNA/cDNA was stored at −20 ℃ until use. Finally, 2 μL of each DNA/cDNA solution was used as a template for the DLAMP reaction.

Table 5-16-1 Pathogens used and DLAMP assay results

Pathogen	Source	DLAMP result
Mycoplasma bovis (MB)	7 isolate strains from Guangxi, GVRI	+
Bovine herpes virus 1 (BHV-1)	2 reference strains (Barta Nu, BK125), CVCC; 3 isolate strains from Guangxi, GVRI	+
Bovine respiratory syncytial virus (SRSV)	1 isolate strain from Yunnan, YNCIQ	–
Bovine virus diarrhoea virus (BVDV)	14 isolate strains from Guangxi, GVRI; 3 reference strains (NADL, Oregon CV24, BA), CVCC	–
Bovine parainfluenza type 3 (BPI3)	1 isolate strain from Guangxi, GVRI	–
Mycoplasma agalactiae	2 reference strains (PG2, SI), CVCC; 2 isolate strains from Guangxi, GVRI	–
Mycoplasma conjunctivae	1 isolate strain from Guangxi, GVRI	–
Mycobacterium bovis	1 isolate strain from Guangxi, GVRI	
Porcine *Mycoplasma hyopneumoniae*	1 reference strain (354, PV11), CVCC	–
Mycoplasma mycoides subsp. *mycoides*	3 reference strains (Y-goat, C88021, PGI), CVCC	–
Mycoplasma mycoides subsp. *capri*	4 reference strains (C87011, PG3, C87-13, C87001), CVCC	–
Pasteurella haemolytica	1 reference strain (C52-1), CVCC	–
Pasteurella multocida	3 reference strains (P19, P-2225, C467), CVCC	–
Foot-and-mouth disease virus (FMDV)	3 serotypes (A, O, Asia I) of inactivated viruses, YNCIQ	–
Vesicular stomatitis virus (VSV)	2 serotypes (New Jersey, Indiana) of inactivated viruses, YNCIQ	–
Rinderpest virus (RPV)	1 reference strain (AV1711), CVCC	–
Bovine rotavirus (BRV)	8 isolate strains from Guangxi, GVRI; 2 reference strains (NCDV, C486), CVCC	–
Bluetongue virus (BTV)	5 serotypes (4, 8, 9, 17, 18) of inactivated viruses, YNCIQ	–
Peste des petits ruminants virus (PPRV)	1 inactivated virus, YNCIQ	–
Escherichia coli (ETEC)	3 reference strains (C83919, C83920, C83924), CVCC	–

Note: GVRI, Guangxi Veterinary Research Institute; YNCIQ, Yunnan Entry-Exit Inspection and Quarantine Bureau; CVCC, Chinese Veterinary Culture Collection Center.

Primer design

The selected genes, *uvrC* of MB and gB of BHV-1, are highly conserved and were used to design primers to identify these specific pathogens[15-17]. Published sequences of the MB *uvrC* gene and BHV-1 gB gene were downloaded from GenBank and aligned using MegAlign 7.0 software (DNAStar, USA). Primer selection was supported by the software Primer Premier Version 5.0 (Premier Biosoft International, Canada) according to the restricted design rules for LAMP primers. The DLAMP assay included two sets of each of the four specific primers: two outer primers (F3, B3) and two inner primers (FIP = F1c+F2, BIP = B1c+B2). The inner primer FIP was synthesized with a fluorophore at the 5’end. The BHV-1-FIP primer, labelled with 6-carboxyfluorescein (FAM), showed a green colour at an emission wavelength of 520 nm, whereas the MB-FIP primer, labelled with Cyanine 5 (CY5), showed a red colour at an emission wavelength of 694 nm. A BLAST search program on the GenBank website was used to verify oligonucleotide specificity. All primers were synthesized and HPLC-purified by TaKaRa Inc. (TaKaRa). The sequences of the primers are shown in Table 5-16-2 and Figure 5-16-1.

Table 5-16-2 Sequences of primers

Target pathogen	Target gene	Primers		Sequence (5'-3')	Tm /℃
MB	*uvrC*	Outer	MB-F3	CCTGTCGGAGTTGCAATTGTT	61
			MB-B3	CGGTCAACTTCAACTTGAATTTG	60
		Inner	MB-FIP	CY5–TACCGCCATCAGCTATAACTAAGTCATGAGCGCAGTGCTGATGTTG	60/63 [a]
			MB-BIP	TCCCTGTTATTGGATTAGTAAAAAACATATCTAGGTCAATTAAGGCTTTGG	61/59 [a]
BHV-1	gB	Outer	BHV-1-F3	GGACGATGTGTACACGGC	56
			BHV-1-B3	CTCGATCTGCTGGAAGCG	59
		Inner	BHV-1-FIP	FAM–TCGTACGGGTACACCGAGCGTACCGCACGGGCACCT	60/65 [a]
			BHV-1-BIP	TACATGTCGCCCTTTTACGGGCCCGGCGAGTAGCTGGT	56/68 [a]

Note: [a] The inner primers are FIP: F2+F1c, BIP: B2+B1c.

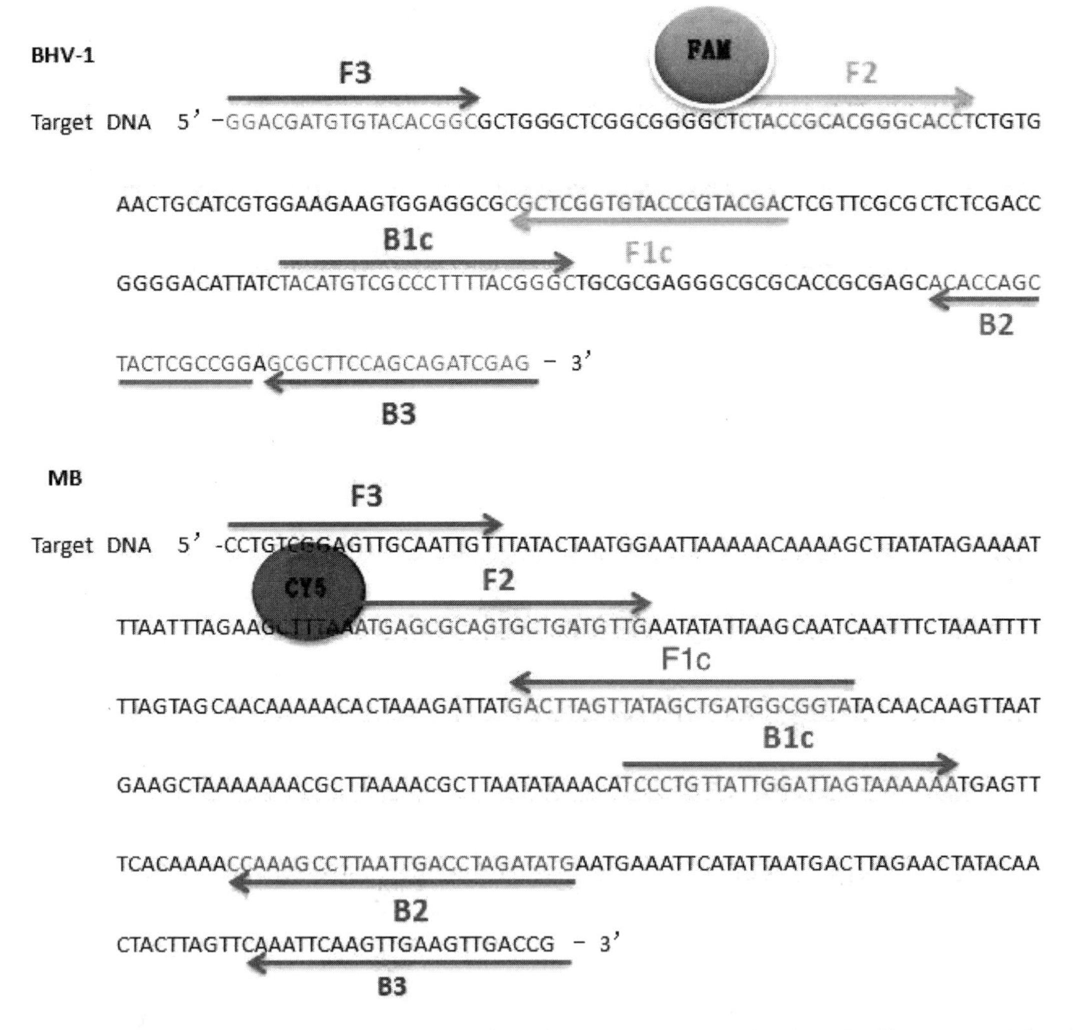

The GenBank accession numbers for MB and BHV-1 are CP023663.1 and KU198480.1, respectively. The nucleotide sequences of the primers are underlined.

Figure 5-16-1 Locations of the primers used in DLAMP

DLAMP assay for simultaneous MB and BHV-1 detection

The DLAMP reaction was performed by using the DNA Amplification Kit (Loopamp, Tokyo, Japan) according to the manufacturer' s instructions. Briefly, 2.5 μL 10 × reaction buffer (200 mM Tris-HCl (pH 8.8), 100 mM KCl, 80 mM $MgSO_4$, 100 mM $(NH_4)_2SO_4$, 1% Tween 20, 8 M betaine, and 14 mM dNTPs), 2 μL of extracted DNA/ cDNA template, 5 pmol F3 and B3 primers, 40 pmol FIP and BIP primers, and 15 units of Bst DNA polymerase (3.0 version) were mixed, and ddH_2O was added to reach a final 25 μL reaction volume. The reaction mixture was incubated at 62 ℃ for 60 min using a Loopamp realtime turbidimeter (LA-320; Eiken Chemical Co., Ltd., Tokyo, Japan), and the reaction was terminated by incubation at 80 ℃ for 5 min.

Analysis of DLAMP products

The DLAMP products were analysed by two methods. The first method was to visually inspect the turbidity of the samples. DNA amplification generated a large amount of white magnesium pyrophosphate precipitate, a byproduct of the DLAMP reaction, which caused turbidity. The turbidity was monitored by the naked eye or by a Loopamp real-time turbidimeter. This method could only identify infected samples; it was unable to differentiate MB from BHV-1. It can be applied in rural areas. The second method was to run agarose gel electrophoresis. A volume of 5 μL of each DLAMP product was subjected to electrophoresis in a 2% agarose gel at 100 V for 60 min, and the fluorescent-dye-conjugated DLAMP products were examined with an image analyser (Universal Hood Ⅲ, 731BR01622, Bio-Rad, USA) to display different colours of electrophoretic bands. Green bands were considered BHV-1, while red bands were considered MB. Since DLAMP products are mixtures of DNA fragments with various lengths, several ladder-like stripes are usually shown on an agarose gel after electrophoresis. No DNA marker is necessary for a DLAMP assay.

Preparation of standards

PCR amplicons containing the full-length sequence of each DLAMP target gene were cloned into the pMD18-T vector (TaKaRa) separately by using the standard procedure, generating the recombinant plasmids pMD18-T-MB and pMD18-T-BHV1. The concentrations of the recombinant plasmids were measured at 260 nm by using a NanoDrop 2000 (ThermoFisher Scientific, Waltham, USA). The copy numbers of the target DNA were calculated based on the following formula: copies/μL=(concentration of plasmid (ng/μL) $\times 6.02 \times 10^{14}$/length of plasmid × 660)[14]. Serial 10-fold dilutions of each target DNA, ranging from 1×10^8 copies/μL to 1 copy/μL, were stored at −20 ℃ until use.

Detection of clinical samples

The present study was conducted on a total of 125 swab samples consisting of 70 nasal swabs and 55 vaginal swabs, which were collected from the same dairy farm in Guangxi, southern China, during 2015 to 2017, where recurrent problems with MB infection were reported[18]. The swab samples were collected from cows displaying typical respiratory-system-related lesions. Following clinical examination, which included assessment of body temperature and respiratory function, the swab samples were eluted in 3.0 mL of PBS. The eluent was centrifuged at 5 000 r/min for 5 min at 4 ℃, and 250 μL supernatant was subjected to DNA extraction using the MiniBEST Universal Genomic DNA Extraction Kit version 5.0 (TaKaRa) according to the manufacturer' s instructions. Finally, 2 μL of extracted DNA was used as a template for the DLAMP test described above in Section *DLAMP assay for simultaneous MB and BHV*-1 *detection*. The samples were also tested for MB and BHV-1 infection by using the single real-time PCR assays recommended by OIE and conventional PCR as references[10, 11, 19].

Results

Specificity of DLAMP

All five BHV-1 strains and seven MB strains tested positive by DLAMP and showed green bands and red bands, respectively, on agarose gels. DLAMP amplification of other common bovine pathogens did not show non-specific products or any inter-assay cross-amplification. Only pathogen-specific targets were amplified by DLAMP (Table 5-16-1, Figure 5-16-2). To validate the specificity of DLAMP, a total of 42 negative samples including 10 nasal swabs, 11 whole blood samples, and 21 semen samples from cattle free from both BHV-1 and MB were tested. All these control samples tested negative for MB and BHV-1 by the DLAMP assay.

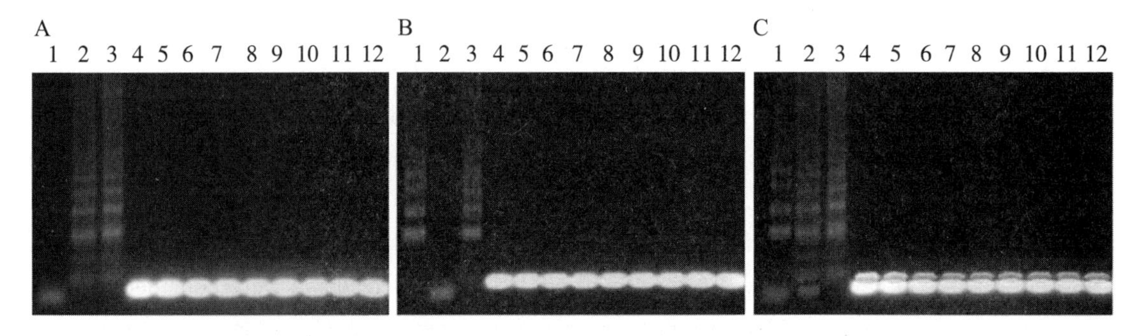

A: 520 nm channel (green); B: 670 nm channel (red); C: Duplex channel (overlapping fluorescence). FAM fluorescence (green) indicates amplified DNA products from BHV-1, whereas CY5 fluorescence (red) indicates amplified DNA products from MB. 1: MB; 2: BHV-1;3: MB+BHV-1; 4: SRSV; 5: BPI3; 6: BVDV; 7: *Mycoplasma agalactiae*; 8: *Mycoplasma conjunctivae*; 9: *Mycobacterium bovis*; 10: *Mycoplasma mycoides* subsp. *mycoides*; 11: *Mycoplasma mycoides* subsp. *capri*; 12: *Pasteurella multocida*. FAM fluorescence (green) indicates amplified DNA products from BHV-1, whereas CY5 fluorescence (red) indicates amplified DNA products from MB.

Figure 5-16-2 Specificity results of DLAMP for BHV-1 and MB

Sensitivity of DLAMP

The sensitivity of the DLAMP assay containing two primer sets for each pathogen as well as fluorescent-dye-labelled primers was evaluated by using serial 10-fold dilutions of standards prepared as described in the previous section. The standards contain the same copy numbers of each recombinant plasmid, which possesses the specific gene sequence of either BHV-1 or MB. Standards containing equivalent amounts of each recombinant plasmid, ranging from 1×10^8 to 1 copy/μL, were prepared from stock by using serial 10-fold dilutions with RNase-free H_2O. Two microliters was used as template in the DLAMP assay for the detection of BHV-1 and MB. The detection limit of the DLAMP assay was 200 copies/μL recombinant plasmid DNA (Figure 5-16-3). All these results demonstrated that the DLAMP assay is capable of simultaneously and sensitively identifying BHV-1 and MB.

Interference of DLAMP

Six artificial samples with various concentrations of plasmid containing the BHV-1 or MB gene were prepared and subjected to the DLAMP assay to assess the interference between higher concentrations and lower concentrations of nucleic acid templates. The six artificial samples were as follows: sample 1, MB (10^8 copies/μL)

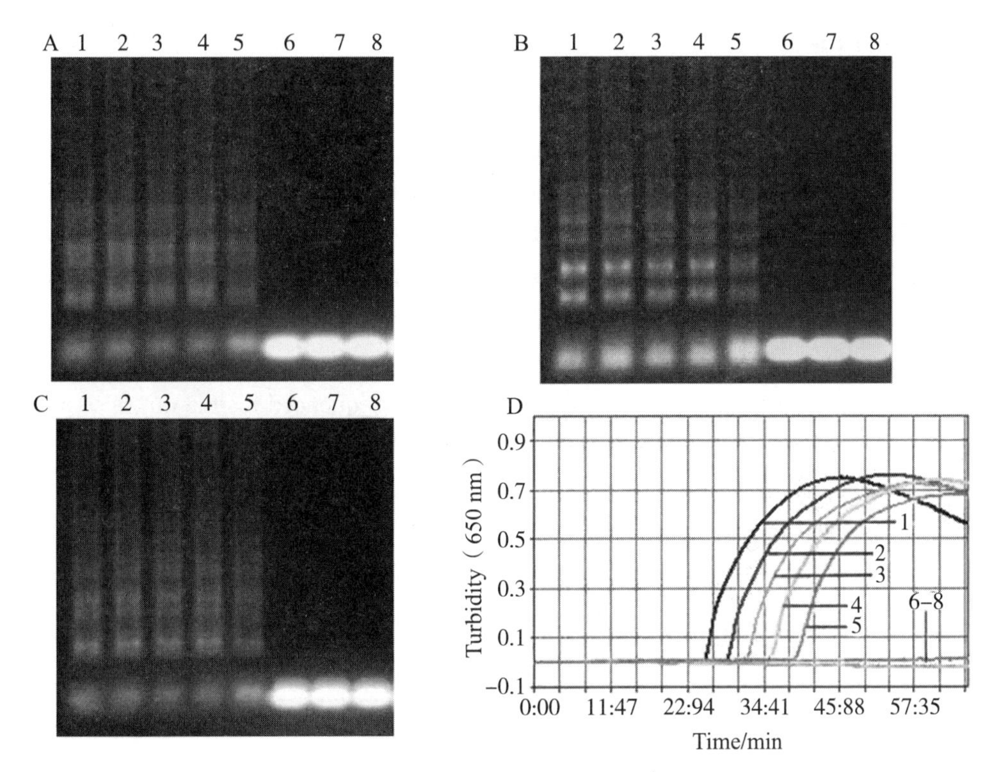

A: 520 nm channel; B: 670 nm channel; C: Duplex channel; D: The turbidity curve. 1: 10^6 copies/μL (standards of equivalent pMD18-T-MB and pMD18-T-BHV-1); 2: 10^5 copies/μL; 3: 10^4 copies/μL; 4: 10^3 copies/μL; 5: 10^2 copies/μL; 6: 10^1 copies/μL; 7: 1 copy/μL; 8: Negative control (color figure in appendix).

Figure 5-16-3 Sensitivity results of DLAMP for BHV-1 and MB

+BHV-1 (10^4 copies/μL); sample 2, MB (10^4 copies/μL)+BHV-1 (10^8 copies/μL); sample 3, MB (10^7 copies/μL) +BHV-1 (10^3 copies/μL); sample 4, MB (10^4 copies/μL)+BHV-1 (10^7 copies/μL); sample 5, MB (10^6 copies/μL) +BHV-1 (10^2 copies/μL); sample 6, MB (10^2 copies/μL)+BHV-1 (10^6 copies/μL). Figure 5-16-4 shows that the amplified bands with corresponding colours were easily observed in the electrophoretogram. These results demonstrated that, in the presence of higher concentrations of the MB DNA template, the amplification of lower concentrations of BHV-1 DNA template was not inhibited in the DLAMP assay, and vice versa.

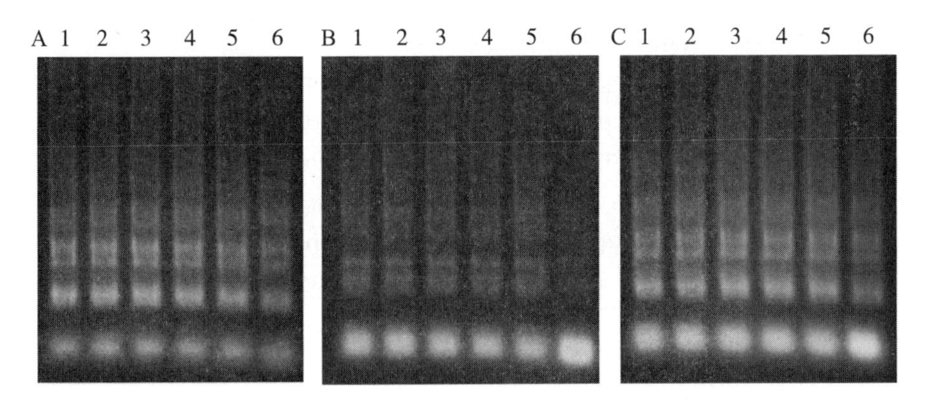

A: 520 nm channel; B: 670 nm channel; C: Duplex channel. 1: MB (10^8 copies/μL)+BHV-1 (10^4 copies/μL); 2: MB (10^4 copies/μL)+BHV-1 (10^8 copies/μL); 3: MB (10^7 copies/μL)+BHV-1 (10^3 copies/μL); 4: MB (10^4 copies/μL)+BHV-1 (10^7 copies/μL); 5: MB (10^6 copies/μL)+BHV-1 (10^2 copies/μL); 6: MB (10^2 copies/μL)+BHV-1 (10^6 copies/μL).

Figure 5-16-4 Interference results of DLAMP for MB and BHV-1

Clinical detection

The performance of the DLAMP assay in clinical specimens was evaluated on 125 swab samples. The OIE-recommended real-time PCR and conventional PCR assays were conducted in parallel to assess the accuracy of the DLAMP assay. Individual test results were summarized in Table 5-16-3; the detection results by DLAMP were in 100% agreement with those of conventional PCR. Identical results were obtained for 122 samples compared by using these three methods. Three discrepant samples were found, including two positive results for MB in real-time PCR that were negative in DLAMP and one sample that was positive for BHV-1 in real-time PCR but negative in DLAMP. To determine whether the three discrepant samples were true positives, the real-time PCR products of the three discrepant samples were sequenced to rule out false positive results. The sensitivity and specificity of DLAMP for detection of MB were 96.6% (56/58) and 100% (65/65), and those for BHV-1 detection by DLAMP were 95% (19/20) and 100% (106/106), respectively, compared to the real-time PCR assay for detection of field samples. This DLAMP assay could detect and differentiate MB from BHV-1 in infected samples.

Table 5-16-3 Evaluation of 125 swab samples for MB and BHV-1 by using DLAMP, real-time PCR and PCR

Pathogen	DLAMP result	Detection rate	Real-time PCR		PCR	
			MB	BHV-1	MB	BHV-1
MB	56	44.80%	58	0	56	0
BHV-1	17	13.60%	0	18	0	17
MB+BHV-1	2	1.60%	2	2	2	2
Total	75	60%	60	20	58	19

Discussion

In recent years, LAMP has drawn a great deal of attention from researchers due to its ease of manipulation and rapid detection speed. However, the use of LAMP for detection of multiplex targets has met with some difficulty due to the methodological limitations of LAMP results for specific target analysis. Whether turbidity is observed, dye is added, or electrophoresis is performed, the results of single LAMP and multiplex LAMP look the same and cannot be distinguished. This feature is in contrast to PCR, in which products of a specific length for each target can be simply distinguished by gel electrophoresis. LAMP products consist of a number of amplified DNA bands of different sizes. The results of LAMP assays commonly display as ladder-like patterns after electrophoresis, making it difficult to distinguish two distinct targets in one multiplex reaction[20-27].

In this study, a DLAMP assay with fluorescently labelled primers was able to identify BHV-1 and MB and to distinguish them from each other in clinical samples with BRD symptoms. Theoretically, LAMP can generate upwards of 10^9 copies from less than one copy of DNA template within an hour. In an experimental study by Socha et al. [28], the sensitivity of LAMP for the BHV-1 gD and gE genes was 2×10^4 copies and 2×10^5 copies, respectively, which was lower than the sensitivity of the DLAMP assay[28]. One reason for this result might be the use of different target genes. In this study, the DLAMP assay utilized the gB gene, whereas Socha et al. chose the gD and gE genes instead. Another reason might be that the DLAMP assay was conducted with optimized Bst DNA polymerase 3.0, which has better performance than that of Bst DNA polymerase[29]. Compared to the OIE-recommended real-time PCR assays for clinical detection, the DLAMP assay missed two MB-positive samples and one BHV-1-positive sample. These three discrepant samples were confirmed by a single-target LAMP assay, which was performed using a set of primers, and the results were consistent with those of the real-time PCR. However, no false positive results were detected in the DLAMP assay. These results demonstrated that this DLAMP assay is specific, although with a slightly decreased sensitivity. It is likely that the two sets of primers used in the DLAMP assay could compete for the reagents in the reaction and interfere with other amplification, resulting in reduced amplification effciency. In contrast, in a typical real-time PCR, only two primers and one probe are used. This difference might explain why the sensitivity of this DLAMP assay is slightly lower than that of real-time PCR. DLAMP is cost-effective compared to real-time PCR, in that it can simultaneously detect two pathogens without expensive instruments. The DLAMP assay needs only a water bath, and the reaction can be finished within 65 min. As co-infection is a common feature of BRD in field outbreaks, the DLAMP assay may be useful for surveillance of the different pathogenic organisms causing BRD without false positive detection[30].

Aside from the detection of multiple pathogens, the DLAMP assay has several advantages. First, DLAMP has all the merits of conventional single-target LAMP assays, including high sensitivity and specificity, lower cost, smaller sample demand, short detection time and simple processing to suit veterinary field diagnostics in rural areas. Second, this DLAMP assay may decrease the risk of generating false positive results. There is a report that strand displacement in the LAMP reaction, starting from randomly existing nicks in the DNA samples, often results in non-specific amplification[31]. In this study, non-specific amplification was observed after a prolonged reaction time of 90 min. Although the real-time turbidimeter could monitor the turbidity of the non-specific amplification curve, the electrophoresis bands of non-specific amplification products do not contain fluorescent colour, showing neither green nor red. In the DLAMP reaction system, the primers with fluorophores require more energy to incorporate than ordinary primers do in the process of amplification. This process can effectively suppress the generation of false positive reactions. Finally, the detection results of DLAMP are accurate and clear, and the final results are determined by the colours of the DLAMP products. This study employed two fluorophores, FAM and CY5, which have different excitation and absorption wavelengths. Two different colours are displayed: the FAM emission wavelength is 520 nm, shown as a green colour, whereas the CY5 emission wavelength is 694 nm, shown as a red colour. Moreover, different fluorophores can be observed in only the selected channels. For example, FAM can be observed in only the 520 nm channel, where CY5 cannot be seen. Fluorescent-dye-conjugated fragments in the DLAMP assay are more specifically and accurately determined compared to judgement of sediment appearance or observation by the naked eye.

In conclusion, the DLAMP assay is a rapid, specific and sensitive duplex method for simultaneous detection of MB and BHV-1. This assay has the potential to be applied to clinical diagnosis and epidemiological screening of MB and BHV-1 co-infection in clinical samples and to monitoring MB and BHV-1 infection status in cattle herds.

References

[1] JONES C, CHOWDHURY S. A review of the biology of bovine herpesvirus type 1 (BHV-1), its role as a cofactor in the bovine respiratory disease complex and development of improved vaccines. Anim Health Res Rev, 2007, 8(2): 187-205.

[2] CERNICCHIARO N, WHITE B J, RENTER D G, et al. Evaluation of economic and performance outcomes associated with the number of treatments after an initial diagnosis of bovine respiratory disease in commercial feeder cattle. Am J Vet Res, 2013, 74(2): 300-309.

[3] TOMAS J. DIVERS S F P. Rebhun's diseases of dairy cattle-2nd edition. Singapore: Elsevier Pte Ltd Press, 2007.

[4] BURKI S, FREY J, PILO P. Virulence, persistence and dissemination of *Mycoplasma bovis*. Vet Microbiol, 2015, 179(1-2): 15-22.

[5] MAUNSELL F P, WOOLUMS A R, FRANCOZ D, et al. *Mycoplasma bovis* infections in cattle. J Vet Intern Med, 2011, 25(4): 772-783.

[6] GRAHAM D A. Bovine herpes virus-1 (BoHV-1) in cattle a review with emphasis on reproductive impacts and the emergence of infection in Ireland and the United Kingdom. Ir Vet J, 2013, 66(1): 15.

[7] NANDI S, KUMAR M, MANOHAR M, et al. Bovine herpes virus infections in cattle. Anim Health Res Rev, 2009, 10(1): 85-98.

[8] CASWELL J L, BATEMAN K G, CAI H Y, et al. *Mycoplasma bovis* in respiratory disease of feedlot cattle. Vet Clin North Am Food Anim Pract, 2010, 26(2): 365-379.

[9] PRYSLIAK T, VAN DER MERWE J, LAWMAN Z, et al. Respiratory disease caused by *Mycoplasma bovis* is enhanced by exposure to bovine herpes virus 1 (BHV-1) but not to bovine viral diarrhea virus (BVDV) type 2. Can Vet J, 2011, 52(11): 1195-1202.

[10] BASHIRUDDIN J B, FREY J, KONIGSSON M H, et al. Evaluation of PCR systems for the identification and differentiation of *Mycoplasma agalactiae* and *Mycoplasma bovis*: a collaborative trial. Vet J, 2005, 169(2): 268-275.

[11] MOORE S, GUNN M, WALLS D. A rapid and sensitive PCR-based diagnostic assay to detect bovine herpesvirus 1 in routine diagnostic submissions. Vet Microbiol, 2000, 75(2): 145-153.

[12] ASHRAF A, IMRAN M, YAQUB T, et al. Development and validation of a loop-mediated isothermal amplification assay for the detection of *Mycoplasma bovis* in mastitic milk. Folia Microbiol (Praha), 2017, 63(3): 373-380.

[13] NOTOMI T, OKAYAMA H, MASUBUCHI H, et al. Loop-mediated isothermal amplification of DNA. Nucleic Acids Res, 2000, 28(12): E63.

[14] XIE Z, LUO S, XIE L, et al. Simultaneous typing of nine avian respiratory pathogens using a novel GeXP analyzer-based multiplex PCR assay. J Virol Methods, 2014, 207: 188-195.

[15] CLOTHIER K A, JORDAN D M, THOMPSON C J, et al. *Mycoplasma bovis* real-time polymerase chain reaction assay validation and diagnostic performance. J Vet Diagn Invest. 2010, 22(6): 956-960.

[16] THOMAS A, DIZIER I, LINDEN A, et al. Conservation of the *uvrC* gene sequence in *Mycoplasma bovis* and its use in routine PCR diagnosis. Vet J, 2004, 168(1): 100-102.

[17] WANG J, O'KEEFE J, ORR D, et al. Validation of a real-time PCR assay for the detection of bovine herpesvirus 1 in bovine semen. J Virol Methods, 2007, 144(1-2): 103-108.

[18] MA C, LI J, QIN Y. Pathogen analysis of bovine respiratory disease complex. China Animal Husbandry & Veterinary Medicine, 2015, 42(9): 2481-2486.

[19] WANG J, O'KEEFE J, ORR D, et al. An international inter-laboratory ring trial to evaluate a real-time PCR assay for the detection of bovine herpesvirus 1 in extended bovine semen. Vet Microbiol, 2008, 126(1-3): 11-19.

[20] AONUMA H, YOSHIMURA A, KOBAYASHI T, et al. A single fluorescence-based LAMP reaction for identifying multiple parasites in mosquitoes. Exp Parasitol, 2010, 125(2): 179-183.

[21] ISEKI H, ALHASSAN A, OHTA N, et al. Development of a multiplex loop-mediated isothermal amplification (mLAMP) method for the simultaneous detection of bovine Babesia parasites. J Microbiol Methods, 2007, 71(3): 281-287.

[22] KOUGUCHI Y, FUJIWARA T, TERAMOTO M, et al. Homogenous, real-time duplex loop-mediated isothermal amplification using a single fluorophore-labeled primer and an intercalator dye: its application to the simultaneous detection of Shiga toxin genes 1 and 2 in Shiga toxigenic Escherichia coli isolates. Mol Cell Probes, 2010, 24(4): 190-195.

[23] KUBOTA R, JENKINS D M. Real-time duplex applications of loop-mediated amplification (LAMP) by assimilating probes. Int J Mol Sci, 2015, 16(3): 4786-4799.

[24] MAHONY J, CHONG S, BULIR D, et al. Multiplex loop-mediated isothermal amplification (M-LAMP) assay for the detection of influenza A/H1, A/H3 and influenza B can provide a specimen-to-result diagnosis in 40 min with single genome

copy sensitivity. J Clin Virol, 2013, 58(1): 127-131.

[25] NURUL NAJIAN A B, ENGKU NUR SYAFIRAH E A, ISMAIL N, et al. Development of multiplex loop mediated isothermal amplification (m-LAMP) label-based gold nanoparticles lateral flow dipstick biosensor for detection of pathogenic Leptospira. Anal Chim Acta, 2016, 903: 142-148.

[26] SONG H, BAE Y, PARK S, et al. Loop-mediated isothermal amplification assay for detection of four immunosuppressive viruses in chicken. J Virol Methods, 2010, 256: 6-11.

[27] YAMAZAKI W, MIOULET V, MURRAY L, et al. Development and evaluation of multiplex RT-LAMP assays for rapid and sensitive detection of foot-and-mouth disease virus. J Virol Methods, 2013, 192(1-2): 18-24.

[28] SOCHA W, ROLA J, URBAN-CHMIEL R, et al. Application of loop-mediated isothermal amplification (LAMP) assays for the detection of bovine herpesvirus 1. Pol J Vet Sci, 2017, 20(3): 619-622.

[29] ZHAO Y, CHEN F, LI Q, et al. Isothermal amplification of nucleic acids. Chem Rev, 2015, 115(22): 12491-12545.

[30] THONUR L, MALEY M, GILRAY J, et al. One-step multiplex real time RT-PCR for the detection of bovine respiratory syncytial virus, bovine herpesvirus 1 and bovine parainfluenza virus 3. BMC Vet Res, 2012, 8: 37.

[31] MITSUNAGA S, SHIMIZU S, OKUDAIRA Y, et al. Improved loop-mediated isothermal amplification for HLA-DRB1 genotyping using RecA and a restriction enzyme for enhanced amplification specificity. Immunogenetics, 2013, 65(6): 405-415.

Development of a visual multiplex fluorescent LAMP assay for the detection of foot-and-mouth disease, vesicular stomatitis and bluetongue viruses

Fan Qing, Xie Zhixun, Wei You, Zhang Yanfang, Xie Zhiqin, Xie Liji, Huang Jiaoling, Zeng Tingting, Wang Sheng, Luo Sisi, and Li Meng

Abstract

Loop-mediated isothermal amplification (LAMP) is a nucleic acid amplification technique that can be used to amplify target genes at a constant temperature, and it has several advantages, including convenience, specificity and sensitivity. However, due to the special interpretation methods of this technology for reaction results, all the previously reported LAMP detection methods have been restricted to identifying a single target, which limits the application of this technology. In this study, we modified conventional LAMP to include a quencher-fluorophore composite probe complementary to the F1c segment of the inner primer FIP; upon strand separation, a gain in the visible fluorescent signal was observed. The probes could be labeled with different fluorophores, showing different colors at the corresponding wavelengths. Therefore, this multiplex LAMP (mLAMP) assay can simultaneously detect 1～3 target sequences in a single LAMP reaction tube, and the results are more accurate and intuitive. In this study, we comprehensively demonstrated a single-reac-tion mLAMP assay for the robust detection of three cattle viruses without nonspecific amplification of other related pathogenic cattle viruses. The detection limit of this mLAMP assay was as low as 526～2 477 copies/reaction for the recombinant plasmids. It is expected that this mLAMP assay can be widely used in clinical diagnosis.

Keywords

multiplex, fluorescent, LAMP assay, FMDV, VSV, BTV

Introduction

Nucleic acid amplification technologies are among the most valuable tools in the field of molecular biology and can achieve the detection and quantitative analysis of trace nucleic acids. Thus, these technologies are widely used in application-oriented fields, such as clinical medical diagnosis, animal and plant quarantine, food safety and transgenic detection[1, 2]. Polymerase chain reaction (PCR) is the most widely used technique for nucleic acid amplification, but thermocycling is necessary to separate the DNA double strand, the cooling cycle must be heated repeatedly through amplification, and the amplification results must be analyzed by agarose gel electrophoresis or an expensive fluorescence quantitative PCR instrument, which requires 2～3 h[3, 4]. This characteristic limits the application of PCR, as the process relies upon specialized laboratories[5].

Currently, preventing and controlling infectious diseases requires efficient point-of-care pathogen detection platforms. The diagnostic platform for field detection should be inexpensive, sensitive, specific, simple, easy to operate, stable and fast; loop-mediated amplification (LAMP) and recombinase polymerase amplification (RPA) technologies should meet those requirements[6-12]. LAMP was first reported in 2000 by Notomi et al.[13]. The LAMP method employs a Bst DNA polymerase with strand displacement activity, and in this method, a set of six specific primers (including loop primers) recognize eight regions on the target sequences to complete amplification at constant temperatures between 60~67 ℃ . The LAMP products were analyzed by three methods. The first method was visual inspection of the turbidity of the LAMP products, which was formed by the accumulation of magnesium pyrophosphate byproducts[14]. The second method was the addition of a fluorescent dye (i. e., SYBR and calcein dyes) to observe the color change[15, 16]. Finally, the LAMP products were analyzed by agarose gel electrophoresis, which easily causes aerosol pollution in laboratories and is currently not recommended. The results of positive samples presented by these three analysis methods are the same regardless of single or multiple reactions. It is impossible to identify the targets responsible for positive results, and differentiation of multiple products cannot be achieved, which limits the application of this technology[17-23]. In the clinic, there are fewer single-pathogen infections, whereas most cases are multipathogen coinfections.

In 2012, Tanner et al.[24] demonstrated for the first time that LAMP technology combined with DARQ (detection of amplification by release of quenching) probes could be used to identify and detect multiple target DNAs in one LAMP reaction. This study improved the previous DARQ LAMP assay by adding reverse transcriptase to the reaction system, optimizing the reaction conditions, and using composite probes similar to DARQ probes to develop a visualized multiplex fluorescent LAMP (mLAMP) assay for the simultaneous detection of foot-and-mouth disease virus (FMDV), vesicular stomatitis virus (VSV) and bluetongue virus (BTV) and to evaluate the specificity, sensitivity and clinical applications of this assay.

Materials and methods

Principle

The standard LAMP primers included the outer primers F3 and B3, inner primers FIP (FIP = F1c+F2) and BIP (BIP = B1c+B2), and loop primers Floop and Bloop. Based on the standard LAMP primer, a probe (termed FD) complementary to F1c was synthesized. The FD probe was modified at the 3' end with a fluorophore, and the FIP was modified at the 5' end with a dark quencher. Before the reaction, FIP and FD were annealed to form a quencher-fluorophore composite probe (FD-FIP). Since FD was complementary to the F1c segment of FIP, the fluorophore was close to the quencher, resulting in fluorescence extinction. During the reaction process, FIP retained the function of the inner primer to guide the amplification, but upon synthesis from the reverse direction guided by BIP, the FD probe was separated, causing the fluorophore to release[24, 25]. LAMP is highly efficient and can amplify a few copies of DNA to 10^9 in less than an hour with greater specificity; thus, a large amount of free FD-fluorophore is released after the reaction, which produces different fluorescent colors under appropriate spectral channels (Figure 5-17-1). Multiplex identification tests can be performed according to the fluorescence color with which the FD probe is labeled. Here, we demonstrated a multiplex fluorescent LAMP assay for the simultaneous identification of three different targets in the same reaction. The fluorophores and detection channels are shown in Table 5-17-1.

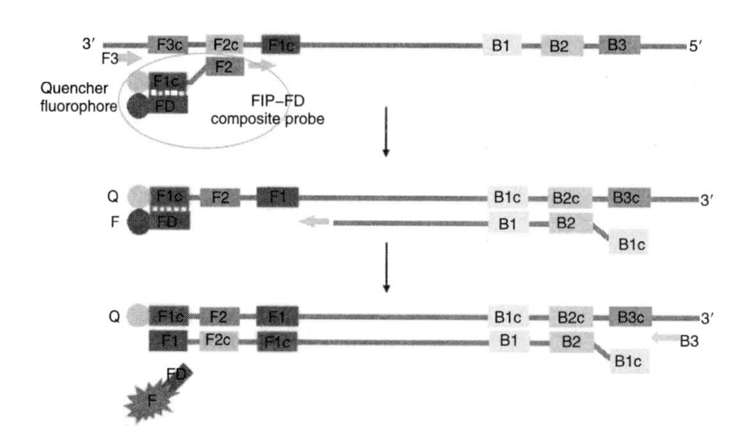

Figure 5-17-1 Schematic presentation of the mechanism of mLAMP (color figure in appendix)

Table 5-17-1 The fluorophores and detection channels

Target	FIP-5’	FD-3’	Wave length (nm)	Color
1	FAM/FITC	BHQ1	520	green
2	Cy5	BHQ3	670	red
3	Cy3	BHQ2	570	blue[a]

Note: [a] The color of Cy3 in the spectrum is orange, and the overlapping color of FAM (green) and Cy5 (red) is also orange. To distinguish them, Cy3 is shown in blue in the output image.

Pathogens and nucleic acid extraction

FMDV serotype A, O and Asia I strains, VSV serotype New Jersey (NJ) and Indiana (IND) strains, BTV serotype 1 and 2 strains, and the viral genomic RNA of peste des petits ruminants virus (PPRV) and epizootic hemorrhagic disease virus (EHDV) were obtained from Yunnan Entry-Exit Inspection and Quarantine Bureau (YNCIQ). The genomic DNA/RNA of bovine viral diarrhea virus (BVDV), swine vesicular disease virus (SVDV), *Mycoplasma bovis* (MB) and infectious bovine rhinotracheitis virus (IBRV) were prepared in our laboratory. Ten nasal swabs and 10 whole blood samples were collected from healthy cattle (FMDV-, VSV- and BTV-free) and were used as negative controls. RNA from the target viruses and negative control were extracted by using the Universal EasyPure RNA Kit (TransGen, China) according to the manufacturer’s instructions and were stored at −80 ℃ until use. Two microliters of DNA/RNA was used as a template for the mLAMP reaction to determine its specificity.

Primer and probe designs

Multiplex LAMP detection of viral cDNA was investigated by using three primer and probe sets targeting FMDV, VSV and BTV. According to the conserved sequence of the FMDV 3D gene, VSV nucleoprotein N gene and BTV VP7 gene, three sets of specificity LAMP primers and FD probes were designed by using Primer Premier 5 and PrimerExplorer V5[26-28]. All oligonucleotide primers and FD probes were synthesized by TaKaRa (TaKaRa, Dalian, China), and their sequences are listed in Table 5-17-2. The fluorophores FAM, Cy5 and Cy3 were used to label the 3’ end of the FD probe, and the corresponding BHQ series quenching group was used to label the 5’ end of the inner primer FIP. FIP-quencher and FD-fluorophore annealing was performed before the reaction to keep the fluorescence in the quenching state. A mixture of 50 μM FIP and 50 μM FD was heated to 90 ℃ for 5 min and slowly cooled to room temperature to form the FD-FIP composite probe, which was stored at −20 ℃.

Table 5-17-2 Primer and FD probe sequences for the mLAMP reaction

Primer	Target gene	Sequence (5'-3')	Tm/℃	Final conc. /μM
FMDV-F3	3D	GTTTGAGGAGGTGTTCCGC	60.2	0.1
FMDV-B3		CATAGTGTCTACGCAGGGC	59.2	0.1
FMDV-FIP		BHQ1-GTAGGCGTGCTCCGTATTCACAGTTTGGCTTCCACCCGAA	64.2/59.5	0.4
FMDV-BIP		AGGGTGGAATGCCATCTGGTTGGTAGAGCACGTGGATGTTGT	59.5/64.9	0.8
FMDV-Floop		GTCTTCAGAATCCACTCGGCA	61.4	0.2
FMDV-Bloop		TCCGCAACAAGCATTATCAAC	60.3	0.2
FMDV-FD1[a]		TGTGAATACGGAGCACGCCTAC-FAM	64.2	0.4
FMDV-FD2[a]		GAATACGGAGCACGCCTAC-FAM	57.1	0.4
FMDV-FD3[a]		TACGGAGCACGCCTAC-FAM	50.9	0.4
FMDV-FD4[a]		GGAGCACGCCTAC-FAM	39.4	0.4
VSV-F3	N	GAACTGAAGACAGCACTTC	55.0	0.1
VSV-B3		CCATCCTCGACTAGACTCTC	57.5	0.1
VSV-FIP		BHQ3-GGATGTAGATGGGAAGCCATTTTGATGGGAAATCAGACCCT	60.0/57.2	0.6
VSV-BIP		ACGGATTACAGAAAGAAACTACTGGAAATCTGGTTGACGCCAC	60.5/57.0	0.8
VSV-Floop		ATCCTCCTCAGCAGAACGGTC	60.1	0.2
VSV-Bloop		ACGGGCTTGAAAATCAGTGC	60.1	0.2
VSV-FD		AAATGGCTTCCCATCTACATCC-Cy5	60.0	0.2
BTV-F3	VP7	GATGGTTCATGCGTGCCG	60.8	0.1
BTV-B3		TCACGCCTGCTTGAGTTTG	60.2	0.1
BTV-FIP		BHQ2-CACATCTCCTCTTGCTCCAGCAAGTAACCGCGGTAGTGTGT	64.1	0.4
BTV-BIP		TTCAGGGTCGTAACGACCCCATGAGTTACCCTGCGCCAT	60.2	0.8
BTV-Floop		TCAGTGACACTTGAATCATATCCG	65.1	0.2
BTV-Bloop		TGGAGAAGAATTGAAAACTTCGC	60.4	0.2
BTV-FD		TGCTGGAGCAAGAGGAGATGTG-Cy3	64.1	0.4

Note: [a] The length of FMDV-FD1 to -FD4 decreases by 3 bases, and the sequence decreases by 1 guanine.

mLAMP reaction

The 25 μL standard mLAMP reaction mixture contained 2 μL nucleic acid template, 12.5 μL WarmStart Multipurpose LAMP/RT-LAMP 2 × Master Mix with UDG (NEB, MA, USA), 1.2 μM FIP, 1.2 μM FD-FIP, 2.4 μM BIP, 0.3 μM F3 and B3, 0.6 μM Floop and Bloop, and 16 U Bst 2.0 WarmStart DNA polymerase, 15 U WarmStart reverse transcriptase. For the multiplex assay, the total primer concentrations corresponded to those described for the standard mLAMP reaction, but each set represented 1/n of the total, in which n is the number of targets; in this study, n was 3. The reactions were performed at 63 ℃ for 75 min for amplification and 80 ℃ for 5 min for termination by using a real-time turbidimeter (LA-320; Eiken Chemical Co., Ltd., Tokyo, Japan).

The mLAMP products were analyzed by using an image analyzer (Universal Hood Ⅲ, 731BR01622, Bio-Rad, USA). The reaction results were interpreted according to the fluorescent color of the reaction tube under the corresponding 1～3 channels as follows: green tubes were considered FMDV-positive and were detected in the 520 nm channel (FAM labeled), red tubes detected in the 670 nm channel were considered VSV-positive (Cy5 labeled), and blue tubes detected in the 570 nm channel were considered BTV-positive (Cy3 labeled). Single posi-tive samples exhibited a single color in one corresponding channel, and multiple positive sam-ples exhibited overlapping colors in two or three channels simultaneously. The turbidity curve was generated by a real-time turbidimeter to interpret the process of amplification. The abscissa represents the reaction time, and the ordinate represents the turbidity intensity, which is the amount of white precipitate of the mLAMP byproduct magnesium pyrophosphate. The turbidity curve can detect only positive results but cannot distinguish which target caused positive results.

Preparation of the RNA standards

The 3D gene of FMDV (GenBank accession number: DQ533483.2), the N gene of VSV (GenBank accession number: M31846.1) and VP7 of BTV (GenBank accession number: AY776331.1) were previously cloned into pEASY-T1, an in vitro transcription vector (TransGen Biotech, China) containing the T7 promoter priming site, generating the recombinant plasmids pEASY-FDMV-3D, pEASY-VSV-N, and pEASY-BTV-VP7. These plasmids were linearized by endonuclease enzymatic digestion using EcoRV. RNA transcription was performed using the In vitro Tran-scription T7 Kit (TaKaRa, Dalian, China). Then, the RNA was purified using the EasyPure RNA Purification Kit (TransGen Biotech, China). The purified RNA transcripts obtained were then quantified by absorbance at 260 nm using a NanoDrop 2000 (ThermoFisher Scientific, Waltham, USA). Each concentration of the purified RNA transcripts was adjusted to 3.0×10^9 copies/μL and mixed equally to prepare the RNA standards by serial 10-fold dilution with RNase-free H_2O. The standards contained the three in vitro-transcribed RNAs with specific gene sequences of the three target viruses, and the concentrations of each RNA ranged from 1×10^8 to 1 copies/μL. Two microliters of standard RNA transcript was used as a template in the mLAMP assay for analysis of the sensitivity of the mLAMP assay.

Real-time RT-PCR

Real-time RT-PCR, which is considered the gold standard, was used in parallel for comparison with the performance of mLAMP in clinical detection. The FMDV-specific primers and probe targeting the 3D gene of FMDV, the VSV-specific primers and probes targeting the L gene of VSV, and the BTV-specific primers and probe targeting the NS3 gene of BTV. The real-time RT-PCR protocols were carried out by using the Transcript Probe One-Step qRT-PCR SuperMix Kit (TransGen Biotech, China) according to the relevant references[29-31].

Clinical sample detection

To evaluate the ability of mLAMP to detect viruses in clinical samples, a total of 111 clinical samples, including 12 vesicular fluid samples, 30 esophageal-pharyngeal (OP) samples, 42 whole-blood samples, 6 vesicular skin samples, and 21 oral swabs, were collected from cattle in Guangxi, China, between January 2020 and July 2022. RNA was extracted as described above and tested using the mLAMP assay. For comparison, RNAs were also tested using conventional FMDV-, VSV- and BTV-specific single real-time RT-PCR methods[29-31]. All positive clinical samples detected by real-time RT-PCR were confirmed by DNA sequencing to rule out false-positive results. The results of mLAMP and real-time RT-PCR assays were compared by measuring the degree of agreement and kappa coefficient (k).

Results

Optimization of the ratio of FD-FIP composite probes to unlabeled FIP primers

To enable multiplexing, we utilized the composite probe mLAMP paradigm, which we believe is the most robust and universal approach for multiplexing with LAMP. However, the composite probe is highly inhibitory toward amplification reactions. A previous publication by Tanner reported that when the FD-FIP composite probe completely replaced the FIP primer, a relatively severe inhibitory effect on amplification was observed. This inhibition was significantly reduced using proportionally unlabeled FIP primers and FD-FIP composite probes[24]. The ratio of FD-FIP composite probes to unlabeled FIP primers is the key to the mLAMP assay and is the foremost condition that should be optimized. This study also confirmed this effect and optimized the working ratio of FD-FIP composite probes to unlabeled FIP primers in the mLAMP system for the detection of three cattle viruses. Figure 5-17-2 shows the impact of the working ratio of the FD-FIP composite probe to unlabeled FIP primer at 0, 25%, 50%, 75%, 100% for FMDV amplification in the mLAMP reaction. The results show that a working ratio of 50% FAM is the optimal working ratio that could effectively reduce the inhibition effect and produce a clear fluorescence signal to discriminate between negative and positive samples. Similarly, the working ratios of Cy5 and Cy3 were also optimized in the same method as described above, and the results showed that 50% Cy3 and 25% Cy5 were the optimal working ratios in the mLAMP reaction.

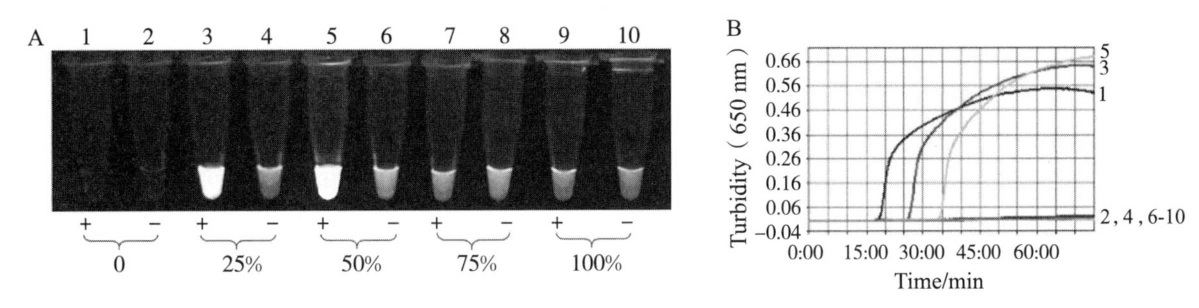

A: The products were imaged by using the 520 nm channel (green); B: The turbidity curve was generated by a real-time turbidimeter. The mLAMP reaction was optimized by using FMDV-specific primers and an FD-FIP composite probe (labeled with FAM and BHQ1). The total amount of FIP in each reaction was maintained at 0.8 μM. Composed of various ratios of FD-FIP to unlabeled FIP as follows: 1: FD-FIP, unlabeled FIP = 0; 3: 25; 5: 50%; 7: 75%; 9: 100%; 2, 4, 6, 8, 10: The negative controls for the corresponding proportions. With the increase in the FD-FIP ratio, the inhibition became more severe with a slower amplification time. At FD-FIP concentrations equal to or higher than 75%, the turbidity curve and fluorescence increase were not observed, suggesting that the reaction was completely inhibited at these concentrations. An equimolar ratio (50%) was used to balance the fluorescence signal and amplification rate for FMDV amplification in the mLAMP reaction.

Figure 5-17-2 The effect of the ratio of FD-FIP composite probe to unlabeled primer on FMDV amplification

Labeling position of the fluorophore/quencher

In mLAMP, the composite probe is composed of an FIP primer with a 5'-end labeled quencher and an FD probe with 3'-end labeled fluorescence. To evaluate the effect of the labeling position of the fluorophore/quencher on the mLAMP reaction, we tested the FMDV-specific composite probe with fluorophores/quenchers labeled at different positions (Figure 5-17-3). The results showed that the fluorophores/quenchers labeling the FIP terminus (positions 1, 2, and 3) could produce a robust green fluorescent signal, indicating that the fluorophore could be used to labeled either FIP or FD. However, the amplification time of the labeling fluorophore at the FD end (positions l and 3) was slightly shorter than that at the FIP end (location 2). The mLAMP reaction was severely inhibited by using fluorophore labeling at the BIP terminal (positions 4 and 5), especially fluorescence labeling at both ends of FIP and BIP (location 5) ; however, the whole reaction was completely inhibited with very few fluorescence increments that could not be achieved with discrimination between positive and negative samples. In conclusion, the composite probes were successfully synthesized with either a 5' quencher or the fluorophore FIP if necessary, which allowed for limited modified oligonucleotide synthesis chemistry.

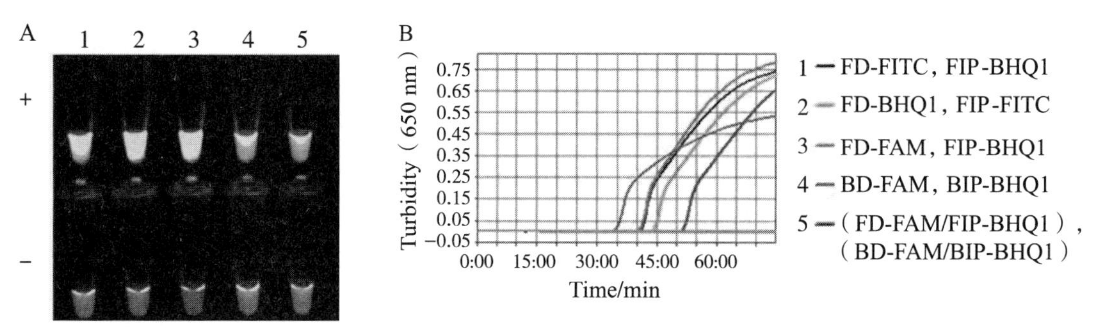

A: 520 nm channel (green); B: The turbidity curve. 1: The FD-FITC/FIP-BHQ1 composite probe is composed of FD labeled with FITC at the 3' end and FIP labeled with BHQ1 at the 5' end, and the amplification product showed a robust fluorescence signal with a shorter initial reaction time; 2: For the FD-BHQ/FIP-FITC composite probe with the quencher and fluorophore position reversed, the initial reaction time was slightly longer than that of composite probe 1; 3: The FD-FAM/FIP-BHQ1 composite probe had a similar fluorescence signal and initial reaction time as composite probe 1; 4: BD-FAM/BIP-BHQ1 composite probe labeled with fluorophore at the BIP terminus; 5: Both the FIP and BIP termini were labeled with fluorophores. (FD-FAM/FIP-BHQ1, BD-FAM/BIP-BHQ1). Composite probes 4 and 5 inhibited mLAMP with a weak fluorescence signal compared to that of 1, 2 and 3.

Figure 5-17-3 The effect of the labeling position of the fluorophore on the mLAMP reaction

Shorter lengths of FD provide faster assay times

During the LAMP reaction process, the reverse direction synthesis chain guided by BIP separates the FD from the composite probe, releasing the fluorophore. Therefore, the binding force between FD and FIP is the main force leading to the inhibition of the composite probe. The more thymine and guanine (GC) that are present in FD, the more hydrogen bonding between FD and FIP, and the more energy needed for FD to detach from FIP. In the case that GC is greater, the inhibition could be reduced by shortening the FD length. To investigate this, four FMDV-specific FD probes with different lengths were designed to compare the effects of different FD lengths on the mLAMP reaction. The length of FMDV-FD1 to -FD4 successively decreases by 3 bases, and the sequence successively decreases by 1 guanine (or thymine) . The results show that the initial reaction time decreases with the decreasing length of FD1 to FD4, but this decrease in time is unequal. The FD4 probe with the shortest length of 13 bp and a Tm value of 39.4 ℃ performed the best and had the shortest initial reaction time (28.3 min). FD1 is completely complementary to 22 bases of FMDV-F1c and has the longest reaction time (the reaction did not start for 57.3 min), and the fluorescence of the FD1 reaction product was slightly weaker than that of the other three probes. Briefly, there were 9 fewer bases in FD4 than in FD1, but reactions with FD4 were 29 min faster than those with FD1 (Figure 5-16-4). Overall, shortening the FD probe length is a good strategy for reducing inhibition and shortening the reaction time.

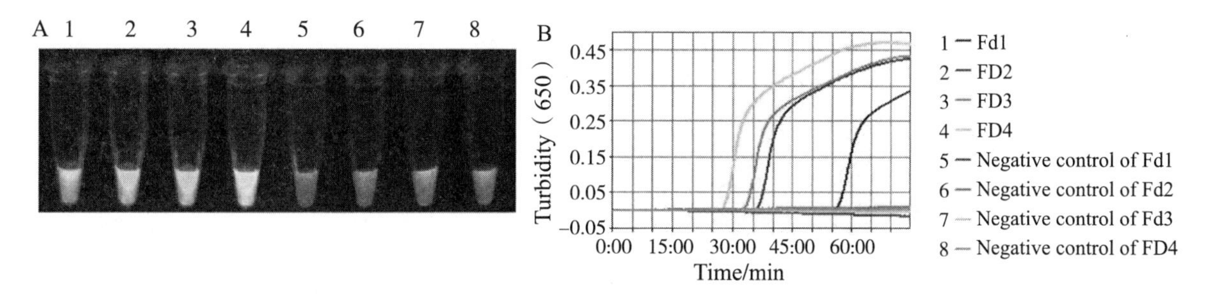

A: 520 nm channel (green); B: The turbidity curve.1: FD1 consists of 22 bases and is completely complementary to FMDV-F1c with a Tm value of 64.2 ℃, and the initial reaction time is 57.3 min; 2: FD2 consists of removing 3 bases from the 5' end of FMDV-FD1 with a Tm value of 57.1 ℃, and the initial reaction time is 37.2 min; 3: FD3 consists of removing 6 bases from the 5' end of FD1 with a Tm value of 50.9 ℃, and the initial reaction time is 34.2 min; 4: FMDV-FD4 consists of removing 9 bases from the 5' end of FMDV-FD1 with a Tm value of 39.4 ℃, and the initial reaction time is 28.3 min; 5-8: Negative controls of FD1-4.

Figure 5-17-4　The effect of FD length on the mLAMP reaction

Bst 2.0 WarmStart DNA polymerase showed more robustness to improve reaction efficiency

NEB commercial premix contains 8 mM magnesium (Mg^{2+}), Bst 2.0 WarmStart DNA polymerase (Bst 2.0 WarmStart) and WarmStart RT × reverse transcriptase. For a single reaction, the premix performed well. However, for multitarget amplification, there were three sets of primers and three composite probes in one reaction system, leading to increased inhibition. The previous section suggested using a 25%～50% ratio of composite probes to unlabeled FIP primers, as the unlabeled FIP primer prevents the reaction from being completely inhibited. However, as one competing reaction component contains three composite probes, these amounts completely inhibited the reaction. The premix performed poorly for multitarget amplification, and an additional enzyme needed to be added to promote multitarget amplification. To determine the optimal enzyme and working amounts, mLAMP reactions containing either Bst 2.0 WarmStart or Bst 3.0 DNA polymerase (Bst 3.0) was used to compare the anti-inhibition effects in mLAMP reactions. The results showed that without the additional enzymes, the mLAMP reaction was completely inhibited with no amplification. With the increase in the working amount of Bst DNA polymerase added, the amplification efficiency increased and the inhibition decreased, which was reflected in the shortening of the initial reaction time and the corresponding increase in fluorescence[25]. In the multiplex assay, better performance was observed for Bst 2.0 WarmStart than Bst 3.0 in NEB commercial premix, in which 16 U of Bst 2.0 WarmStart was added for each reaction with a faster reaction time and higher fluorescence increment (Figure 5-17-5). Overall, it is necessary to supplement enzymes in mLAMP for multiple reactions.

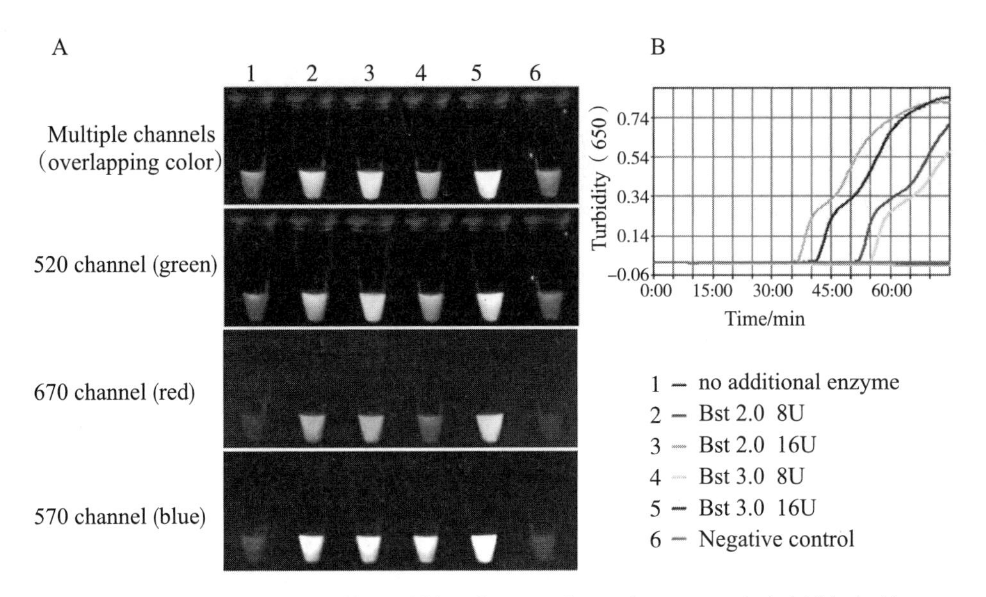

A: 520 nm channel (green); B: The turbidity curve. 1: Without addition of enzyme, the reaction was completely inhibited without any amplification; 2: Each reaction with 8 U Bst 2.0 WarmStart had an initial amplification time of 52 minutes; 3: Each reaction with 16 U Bst 2.0 WarmStart had an initial amplification time of approximately 36 minutes; 4: Each reaction with 3 U Bst 3.0 had an initial amplification time of 55 minutes; 5: Each reaction with 16 U Bst 3.0 had an initial amplification time of 42 minutes; 6: Negative control. Amplification time dependence upon different strand-displacing enzymes with different working amounts. Bst 2.0 WarmStart showed shorter amplification times than Bst 3.0 in the mLAMP reaction.

Figure 5-17-5 The effect of different strand-displacing enzymes on the mLAMP reaction

Analytical specificity of mLAMP

For a successful multiplex assay, accurately distinguishing the target sequences is crucial. To validate the specificity of the mLAMP assay, we next sought to extend the method to detecting triple targets in a single reaction and to determine whether nonspecific amplification of nontarget template nucleic acids occurred. The mLAMP assay was performed using a single template including 3 serotypes of FMDV (A, O, and Asia I), 2 serotypes of VSV (IND and NJ), 2 serotypes of BTV (1 and 2), 6 other related cattle pathogens (PPRV, EHDV, SVDV, BVDV, MB, IBRV) and a mixed sample containing three target viruses to evaluate its cross-amplification. The results showed that the target viruses and the mixed sample generated a turbidity curve of amplification, but this was not observed for the other related cattle pathogens (Figure 5-17-6). Interestingly, the mixed sample containing the three target viruses reacted faster and had a higher turbidity value than that of the single template reaction. This effect was likely due to the turbidity signal in the triple reaction being superposed by multiplex amplification. Each target virus was amplified and showed the corresponding color as expected: the FMDV-positive samples labeled with FAM showed a green color in the 520 channel, the VSV-positive samples labeled with Cy5 showed a red color in the 670 channel, and the BTV-positive samples labeled with Cy3 showed a blue color in the 570 channel. The fluorescence of a single template is visible only in its particular spectral channel and not in other channels. The mixed samples of the three viruses exhibited an overlapping color, which can be observed under the three channels simultaneously. However, tests for the other reference cattle pathogens that are commonly found in cattle were negative, and no cross-amplification was observed[32]. Both the turbidity and fluorescence results indicated that mLAMP has good specificity for detecting FMDV, VSV and BTV without any nonspecific amplification.

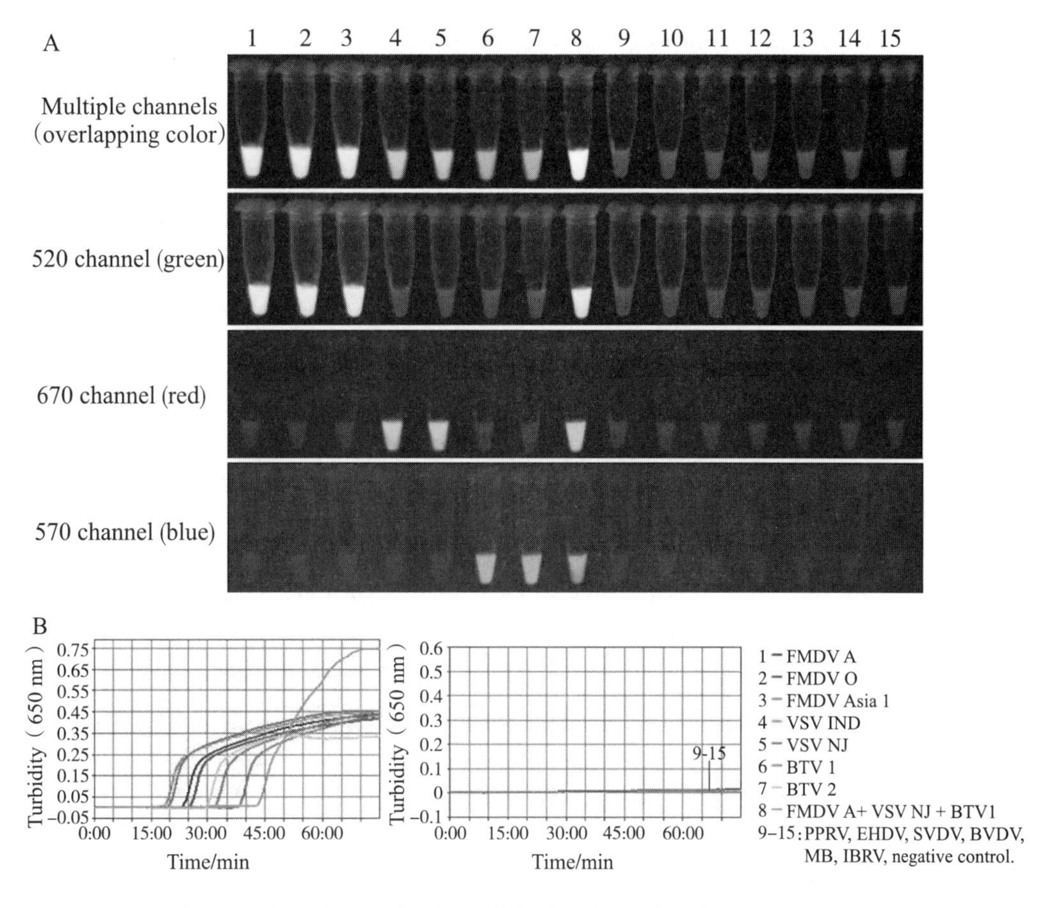

A: The fluorescent mLAMP products were imaged separately using multiple channels; B: The turbidity curve was generated by a real-time turbidimeter to interpret the process of amplification. Green fluorescence (FAM) indicates FMDV-positive amplification, red fluorescence (Cy5) indicates VSV-positive amplification, and blue fluorescence (Cy3) indicates BTV-positive amplification. Overlapping fluorescence indicates multiple positive amplifications.1: FMDV A; 2: FMDV O; 3: FMDV Asia l; 4: VSV IND; 5: VSV ND; 6: BTV 1; 7: BTV 2; 8: FMDV A+VSV IND+BTV; 9-15: PPRV, EHDV, SVDV, BVDV, MB, IBRV, negative control (color figure in appendix).

Figure 5-17-6 Specificity of mLAMP

Analytical sensitivity of mLAMP

To determine the detection limits of mLAMP, each 10-fold serial dilution of the RNA standards was subjected to mLAMP with 10 replicates. The detection limit of mLAMP for the three varieties was determined by using probit regression analysis (SPSS, Inc., Chicago)[33]. The SPSS statistical program generated the probit (predicted proportion positive) versus the template concentration with 95% confidence intervals (CI), as shown in Table 5-17-3. The detection limit of the mLAMP assay was 2 477 copies/reaction for FMDV, 526 copies/reaction for VSV and 913 copies/reaction for BTV. The previous publication showed that the detection limits of conventional real-time RT-PCR were as follows: 11 copies/reaction of the cloned plasmid containing the 3D gene of FMDV, 10 copies/reaction of the cloned plasmid containing the L gene of VSV, and 200 copies/reaction of the cloned plasmid containing the NS3 gene of BTV[29-31]. The sensitivity of this mLAMP for the three cattle viruses was slightly lower than that of conventional single real-time RT-PCR. The reason for this might be that the three sets of primers and probes, with a total of 18 primers and 3 probes, used in the mLAMP assay could compete for the reagents in one reaction and interfere with other amplifications, resulting in reduced amplification efficiency.

Table 5-17-3 Probit analysis to determine the detection limit of mLAMP

Probit[a]	Detection limit, copies/reaction (95% CI)		
	FMDV	VSV	BTV
0.95	2 477 (1 121～29 144)	526 (227～14 160)	913 (368～49 718)

Note: [a] Probit, the predicted proportion of replication from SPSS software.

Comparison of mLAMP and real-time RT-PCR assays in clinical detection

To assess the applicability of mLAMP for the detection of viral RNA in the field, we compared this method with real-time RT-PCR for 111 clinical specimens, and the detection results are shown in Table 5-17-4.The positive real-time RT-PCR products were sent to the BGI (Beijing Genomics Institute, China) for DNA sequence confirmation with real-time RT-PCR primers. The sequencing results showed that all the positive results were true positives. The sensitivity and specificity of mLAMP for the detection of BTV were 100% (15/15) and 100% (96/96), respectively, compared to those of the real-time RT-PCR assay. However, in the detection of FMDV, 21 samples were positive by mLAMP, and 22 samples were positive by real-time RT-PCR. One discrepant sample was found, which was positive for FMDV by real-time RT-PCR but negative by mLAMP. The sequencing results also indicated that the discrepant sample was a true FMDV-positive sample. The sensitivity and specificity of mLAMP for the detection of FMDV were thus 95.5% (21/22) and 100% (89/89), respectively. Since VSV has not been found in China thus far, no VSV-positive samples were detected in this test. No coinfected positive samples were detected in this test. Although no VSV-positive samples were detected due to the geographical limitations of sampling, the results for VSV were consistent with those obtained by real-time RT-PCR. The mLAMP results exhibited 99.1% agreement (kappa coefficient, k=0.98) with the real-time RT-PCR data. The results showed that mLAMP had a positivity rate that was similar to that of real-time RT-PCR.

Table 5-17-4 Performance of the mLAMP assay for detection in clinical samples

Clinical sample	Number	MLAMP/real-time RT-PCR/sequencing		
		FMDV	VSV	BTV
Vesicular fluid	12	6/6/6	0	0
OP samples	30	9/10/10	0	0
Whole blood samples	42	0	0	15/15/15
Vesicular skin samples	6	6/6/6	0	0
Oral swabs	21	0	0	0
Total	111	21/22/22	0	15/15/15

Discussion

Based on the previously developed DARQ LAMP assay for the detection of multiple target DNAs, in this study, we successfully developed a visual mLAMP assay for the identification and detection of three bovine viral RNAs by using composite probes. The analytical sensitivity and specificity and the clinical sample detection ability showed that this technique with composite probes was feasible and could be used for the differential detection of multiple RNA targets. Compared with the DARQ assay, which relies on a real-time instrument to read the reaction results, our mLAMP assay simplifies the method for interpreting the reaction results via observation of the color of the reaction tube after the reaction. This method is more intui-tive and convenient than the DARQ assay. However, the composite probe is highly inhibitory to amplification reactions; thus, additional studies are necessary to broadly apply the composite probe. Here, we also demonstrated four strategies to reduce the inhibitory effect of the composite probes, providing new ideas and solutions for the establishment of mLAMP diagnostic methods in the future.

First, using a ratio of FD-FIP composite probes to unlabeled FIP primers significantly reduced the inhibition effect. Although the fluorescence of the FD-FIP composite probe was extinguished before the reaction, the composite probe still had a certain fluorescence background value. The FD amount was positively correlated with the fluorescence background value, and a higher FD amount was associated with a higher fluorescence background value. As the amount of FD increased, we observed a gradual increase in inhibition and amplification time. Moreover, a high fluorescence background value is unfavorable for discriminating between negative and positive samples after the reaction has occurred, resulting in inaccurate interpretation of the results. In the multiplex assay, each ratio of FD-FIP composite probe to unlabeled FIP primers is the foremost condition and should be first optimized to screen the optimal working ratio. In this study, a 25%～50% FD ratio was optimized to effectively reduce the inhibition effect and generate a clear fluorescence signal after the reaction to discriminate between negative and positive samples.

This effect was likely due to faster target generation with FIP and easier incorporation with the composite probe during exponential amplification, and using a certain amount of composite probe maintains rapid threshold detection with a high fluorescence signal amplitude. This strategy is important for detecting amplification in mLAMP.

Second, increasing the enzyme amount in multiple reactions improves the reaction efficiency and alleviates inhibition. NEB' s Bst 2.0 WarmStart and Bst 3.0 were developed not only to provide faster amplification times but also to provide robustness against inhibitors. The processivity and anti-inhibition effects of these two enzymes were related to the Mg^{2+}content, reaction buffer components and ratio of the FD-FIP composite probe. Previous publications on mLAMP suggested that using Bst 3.0 resulted in a greater robustness against probe inhibition than that of Bst 2.0 WarmStart at a Mg^{2+}concentration of 3 mM[25]. However, in this study, Bst 2.0 WarmStart exhibited a better anti-inhibitory effect than Bst 3.0 in NEB' s commercial premix with 8 mM Mg^{2+}. The results of this study are not consistent with previous publications, but this does not indicate that Bst 2.0 WarmStart is better than Bst 3.0.The optimal enzyme for mLAMP needs to be determined by experimental tests. The primers and probes used in this study combined with the concentration of each component in NEB' s commercial premix (Mg^{2+}8 mM) were more suitable for Bst 2.0 WarmStart. Most likely, the differences between Bst 2.0 WarmStart and Bst 3.0 resulted from the design changes created during the directed evolution of the enzyme and the relationship of varying buffer components (such as monovalent cations and the amount of Mg^{2+}), the melt temperatures for the primers and composite probes, and the resulting processivity of the enzyme. How this impacts the thermodynamic properties of the primer sets in combination with these engineered polymerases is uncertain, i. e., the influence of higher and lower Mg^{2+}amounts on the multiplex assay. The relationship of these properties could be further examined in future studies to better inform optimal primer sets and probe conditions.

Third, compared to the probe designed to label the BIP terminus, the composite probe designed to label the FIP terminus promoted the reaction better. Utilizing composite probes that were synthesized with either a 5' quencher or fluorophore in the FIP orientation resulted in no difference in amplification detection efficiency or inhibition levels. However, labeling the composite probe in both the FIP and BIP orientations completely inhibited the whole reaction. This is due to the superposition of FIP and BIP bidirectional probe inhibition; as a result, FD was not detached from the composite probe. Although the turbidity curve could be monitored, the fluorescence increment was too small to distinguish the positive and negative samples. The mLAMP reactions perform robust amplification of the three targets, with increment of fluorescence signal accompanying decreased concentration of the fluorophore composite probes. This visible increment of the fluorescence signal emphasizes the need for bright fluorophores with high quantum yield and appropriate spectral matching with the fluorescence detection channels. The LAMP-assisted composite probe accommodates any fluorophore that can be quenched and detected, but brighter fluorophores are preferred when detecting more than three targets simultaneously. In these mLAMP reactions, FAM/FITC, Cy5 and Cy3 are bright fluorophores and provide a sharp contrast of fluorescence signals, which are easy to distinguish from background fluorescence.

Finally, an additional strategy for reduced inhibition is shortening the FD length. The main reason for the inhibition of the composite probe was the hydrogen bonding between FD and FIP, which hindered the separation of FD from FIP by the BIP-guided synthesis chain. Therefore, shortening the FD length could reduce the number of GCs in FD, thus reducing the inhibition effect and promoting the reaction. While the modified FD probe is universal for any LAMP, it is important to note that the design of the FD probe is not straightforward. FD should not be too short; otherwise, it cannot form a stable composite probe with FIP, and fluorescence shedding before the reaction will lead to inaccurate detection results. The Tm value is related to the length of FD, and a short FD has a lower Tm value. The LAMP reaction temperature is usually 60～67 ℃, and the design of an FD probe should consider that FD and FIP completely anneal at this temperature without separation. Therefore, each FD probe with different lengths must be validated to screen out the optimal FD, and often, several FD probes must be designed given the complexity.

The composite probe detection assay provides a universal method for multiplex target detection of LAMP amplification without the need for additional primer or probe design; only the addition of a complementary oligonucleotide FD probe to the conventional LAMP primer is needed. After being synthesized from the reverse direction with guidance by BIP, the FD-FIP composite probe is detached, and the fluorophore is released. This detection methodology maybe extendable to other established single-target LAMP methods by the addition of an FD "tail" to an FIP primer in the reaction. The established single-target LAMP method can be modified and combined into a multiplex target LAMP method by using a composite probe. Each mLAMP assay must be optimized for each designed primer set, and each ratio of FD-FIP composite probe, enzyme amounts, and amplification time can vary widely in different assays.

Conclusions

This study combined LAMP technology with a composite probe to develop a visual multiplex fluorescence LAMP assay for the simultaneous identification of multiple target sequences to address the need for point-of-care detection of multiple pathogens.

References

[1] ROTH W K. History and future of nucleic acid amplification technology blood donor testing. Transfus Med Hemother, 2019, 46(2): 67-75.

[2] BAYLIS S A, HEATH A B. World Health Organization collaborative study to calibrate the 3rd International Standard for Hepatitis C virus RNA nucleic acid amplification technology (NAT)-based assays. Vox Sang, 2011, 100(4): 409-417.

[3] FARAJI R, BEHJATI-ARDAKANI M, MOSHTAGHIOUN S M, et al. The diagnosis of microorganism involved in infective endocarditis (IE) by polymerase chain reaction (PCR) and real-time PCR: a systematic review. Kaohsiung J Med Sci, 2017, 34(2): 71-78.

[4] MULLIS K B, FALOONA F A. Specific synthesis of DNA in vitro via a polymerase-catalyzed chain reaction. Methods Enzymol, 1987, 155: 335-350.

[5] LI H M, QIN Z Q, BERGQUIST R, et al. Nucleic acid amplification techniques for the detection of Schistosoma mansoni infection in humans and the intermediate snail host: a structured review and meta-analysis of diagnostic accuracy. Int J Infect Dis, 2021, 112: 152-164.

[6] YAN L, ZHOU J, ZHENG Y, et al. Isothermal amplified detection of DNA and RNA. Mol Biosyst, 2014, 10(5): 970-1003.

[7] KIM J, EASLEY C J. Isothermal DNA amplification in bioanalysis: strategies and applications. Bioanalysis, 2011, 3(2): 227-239.

[8] AHMAD F, HASHSHAM S A. Miniaturized nucleic acid amplification systems for rapid and point-of-care diagnostics: a review. Anal Chim Acta, 2012, 733: 1-15.

[9] CONRAD C C, DAHER R K, STANFORD K, et al. A sensitive and accurate recombinase polymerase amplification assay for detection of the primary bacterial pathogens causing bovine respiratory disease. Front Vet Sci, 2020, 7: 208.

[10] ISLAM M M, KOIRALA D. Toward a next-generation diagnostic tool: A review on emerging isothermal nucleic acid amplification techniques for the detection of SARS-CoV-2 and other infectious viruses. Anal Chim Acta, 2022, 1209: 339338.

[11] BLASER S, DIEM H, VON FELTEN A, et al. From laboratory to point of entry: development and implementation of a loop-mediated isothermal amplification (LAMP)-based genetic identification system to prevent introduction of quarantine insect species. Pest Manag Sci, 2018, 74(6): 1504-1512.

[12] SILVA ZATTI M, DOMINGOS ARANTES T, CORDEIRO THEODORO R. Isothermal nucleic acid amplification techniques for detection and identification of pathogenic fungi: a review. Mycoses, 2020, 63(10): 1006-1020.

[13] NOTOMI T, OKAYAMA H, MASUBUCHI H, et al. Loop-mediated isothermal amplification of DNA. Nucleic Acids Res, 2000, 28(12): E63.

[14] MORI Y, KITAO M, TOMITA N, et al. Real-time turbidimetry of LAMP reaction for quantifying template DNA. J Biochem Biophys Methods, 2004, 59(2): 145-157.

[15] TOMITA N, MORI Y, KANDA H, et al. Loop-mediated isothermal amplification (LAMP) of gene sequences and simple visual detection of products. Nat Protoc, 2008, 3(5): 877-882.

[16] GOTO M, HONDA E, OGURA A, et al. Colorimetric detection of loop-mediated isothermal amplification reaction by using hydroxy naphthol blue. Biotechniques, 2009, 46(3): 167-172.

[17] FAN Q, XIE Z, XIE Z, et al. Development of duplex fluorescence-based loop-mediated isothermal amplification assay for detection of *Mycoplasma bovis* and bovine herpes virus 1. J Virol Methods, 2018, 261: 132-138.

[18] LIU N, ZOU D, DONG D, et al. Development of a multiplex loop-mediated isothermal amplification method for the simultaneous detection of *Salmonella* spp. and Vibrio parahaemolyticus. Sci Rep, 2017, 7: 45601.

[19] CHEN C, ZHAO Q, GUO J, et al. Identification of methicillin-resistant staphylococcus aureus (MRSA) using simultaneous detection of mecA, nuc, and femB by Loop-mediated isothermal amplification (LAMP). Curr Microbiol, 2017, 74(8): 965-971.

[20] DUAN Y B, YANG Y, WANG J X, et al. Simultaneous detection of multiple benzimidazole-resistant beta-tubulin variants of botrytis cinerea using Loop-mediated isothermal amplification. Plant Dis, 2018, 102(10): 2016-2024.

[21] LIU X L, ZHAO X T, MUHAMMAD I, et al. Multiplex reverse transcription loop-mediated isothermal amplification for the simultaneous detection of CVB and CSVd in chrysanthemum. J Virol Methods, 2014, 210: 26-31.

[22] YAMAZAKI W, MIOULET V, MURRAY L, et al. Development and evaluation of multiplex RT-LAMP assays for rapid and sensitive detection of foot-and-mouth disease virus. J Virol Methods, 2013, 192(1-2): 18-24.

[23] KIM M J, KIM H Y. Direct duplex real-time loop mediated isothermal amplification assay for the simultaneous detection of cow and goat species origin of milk and yogurt products for field use. Food Chem, 2018, 246: 26-31.

[24] TANNER N A, ZHANG Y, EVANS T C, JR. Simultaneous multiple target detection in real-time loop-mediated isothermal amplification. Biotechniques, 2012, 53(2): 81-89.

[25] NANAYAKKARA I A, WHITE I M. Demonstration of a quantitative triplex LAMP assay with an improved probe-based readout for the detection of MRSA. Analyst, 2019, 144(12): 3878-3885.

[26] GHAITH D M, ABU GHAZALEH R. Carboxamide and N-alkylcarboxamide additives can greatly reduce non specific amplification in Loop-Mediated Isothermal Amplification for Foot-and-Mouth disease Virus (FMDV) using Bst 3.0 polymerase. J Virol Methods, 2021, 298: 114284.

[27] FOWLER V L, HOWSON E L, MADI M, et al. Development of a reverse transcription loop-mediated isothermal amplification assay for the detection of vesicular stomatitis New Jersey virus: Use of rapid molecular assays to differentiate between vesicular disease viruses. J Virol Methods, 2016, 234: 123-131.

[28] ANTHONY S, JONES H, DARPEL K E, et al. A duplex RT-PCR assay for detection of genome segment 7 (VP7 gene) from 24 BTV serotypes. J Virol Methods, 2007, 141(2): 188-197.

[29] WANG Y, DAS A, ZHENG W, et al. Development and evaluation of multiplex real-time RT-PCR assays for the detection and differentiation of foot-and-mouth disease virus and Seneca Valley virus 1. Transbound Emerg Dis, 2020, 67(2): 604-616.

[30] MULHOLLAND C, MCMENAMY M J, HOFFMANN B, et al. The development of a real-time reverse transcription-polymerase chain reaction (rRT-PCR) assay using TaqMan technology for the pan detection of bluetongue virus (BTV). J Virol Methods, 2017, 245: 35-39.

[31] ZANG J, GAO Z, WANG L, et al. Establishment and application of a TaqM an real-time RT-PCR assay for the detection of vesicular stomatitis virus. Chinese Journal of Veterinary Medicine, 2015, 51(5): 73-76.

[32] ZYRINA N V, ANTIPOVA V N. Nonspecific Synthesis in the reactions of isothermal nucleic acid amplification. Biochemistry (Mosc), 2021, 86(7): 887-897.

[33] SMIEJA M, MAHONY J B, GOLDSMITH C H, et al. Replicate PCR testing and probit analysis for detection and quantitation of Chlamydia pneumoniae in clinical specimens. J Clin Microbiol, 2001, 39(5): 1796-1801.

A multiplex fluorescence-based loop-mediated isothermal amplification assay for identifying chicken parvovirus, chicken infectious anaemia virus, and fowl adenovirus serotype 4

Fan Qing, Xie Zhixun, Zhang Yanfang, Xie Zhiqin, Xie Liji, Huang Jiaoling, Zeng Tingting, Wang Sheng, Luo Sisi, and Li Meng

Abstract

Chicken parvovirus (ChPV), chicken infectious anaemia virus (CIAV) and fowl adenovirus serotype 4 (FAdV-4) are avian viruses that have emerged in recent years and have endangered the global poultry industry, causing great economic loss. In this study, a multiplex fluorescence-based loop-mediated isothermal amplification (mLAMP) assay for detecting ChPV, CIAV and FAdV-4 was developed to simultaneously diagnose single and mixed infections in chickens. Three primer sets and composite probes were designed according to the conserved regions of the NS gene of ChPV, VP1 gene of CIAV and hexon gene of FAdV-4. Each composite probe was labelled with a different fluorophore, which was detached to release the fluorescence signal after amplification. The target viruses were distinguished based on the colour of the mLAMP products. The mLAMP assay was shown to be sensitive, with detection limits of 307 copies of recombinant plasmids containing the ChPV target genes, 749 copies of CIAV and 648 copies of FAdV-4. The assay exhibited good specificity and no cross-reactivity with other symptomatically related avian viruses. When used on field materials, the results of the mLAMP assay were in 100% agreement with those of the previously published PCR assay. The mLAMP assay is rapid, economical, sensitive and specific, and the results of amplification are directly observable by eye. Therefore, the mLAMP assay is a useful tool for the clinical detection of ChPV, CIAV and FAdV-4 and can be applied in rural areas.

Keywords

mLAMP, ChPV, CIAV, FAdV-4, multiplex, differential diagnosis

Introduction

Intensive poultry rearing and the movement of wild birds bring large numbers of birds into close contact with one another, providing favourable conditions for increasing infectious disease transmission and the emergence of new pathogen variants. Chicken parvovirus (ChPV), chicken infectious anaemia virus (CIAV) and fowl adenovirus serotype 4 (FAdV-4) are avian viruses that have emerged in recent years[1-3]. All three of them are DNA viruses that mainly infect chickens under 35 days of age, causing various complications and posing a potential hazard to the poultry industry.

ChPV is one of the causative agents of viral enteritis and is frequently associated with runting stunting syndrome (RSS), which is characterized by significant growth retardation with poor feather development and bone disease, and is considered responsible for the incidence of intestinal diseases and the increase in the mortality rate of sick birds[4-6]. Previously, large-scale surveys of individual cloacal swabs from turkeys and chickens in China revealed the widespread occurrence of ChPV in poultry, with infection rates as high as 38.9%～88.1%[2, 7]. CIAV is an economically important viral pathogen that causes significant immunosuppression and severe anaemia, affecting the poultry industry worldwide. CIAV causes atrophy of chicken systemic lymphoid tissue and bone marrow haematopoietic tissue, leading to immunosuppression and secondary infection with other pathogens. It can be transmitted vertically and horizontally and infects chickens of different ages, but there is a severe clinical form of disease that occurs in young chicks (10～14 days old) with severe morbidity and mortality[8, 9]. Fowl adenoviruses (FAdVs) are ubiquitous in avian populations worldwide with varying degrees of associated clinical disease and can be classified into five species (A, B, C, D, and E) with 12 serotypes (FAdV-1 to 12)[10]. To date, studies have indicated that the FAdV-4 strain is an aetiological agent of hydropericardium hepatitis syndrome (HHS), which is a lethal disease that mainly occurs in 3- to 5-week-old broiler chicks[11].

Outbreaks of FAdV-4 infection have continuously occurred in China in the past decade[3, 12]. Vaccines are currently available for CIAV and FAdV-4, but not for ChPV. However, many studies have reported that the incidence of these three avian diseases has continued to increase annually. In addition, viral coinfection and cocontamination with FAdV-4 and CIAV in attenuated vaccines was observed recently. The three avian diseases have endangered the global poultry industry, causing great economic loss[13-15].

Loop-mediated isothermal amplification (LAMP) is a simple autocycling strand-displacement-based DNA synthesis method that does not require sophisticated equipment. It has emerged as a powerful gene amplification tool for the rapid identification of microbial infections and genetic screening[16, 17]. However, there are some challenges associated with the use of LAMP for the detection of multiplex targets due to methodological limitations in analyzing LAMP results. LAMP products visualized using agarose gel electrophoresis (by colour change after adding dye, or as a white magnesium pyrophosphate precipitate) appear the same in both single LAMP and multiplex LAMP. Therefore, it is difficult to distinguish multiple targets in one reaction[18, 22].

In this study, we report the development of an improved single-tube multiplex fluorescence-based LAMP (mLAMP) assay for identifying ChPV, CIAV and FAdV-4. The mLAMP assay uses three composite probes labelled with different fluorophores, and the multiple results appear in three different colours in the corresponding channel, which can be directly observed by the naked eye without electrophoresis or staining.

Materials and methods

Viruses and nucleic acid extraction

The viruses used in this study are listed in Table 5-18-1; the DNA extracted from five specific pathogen-free (SPF) chicken swab samples was used as a negative control. The viral nucleic acid was extracted using the EasyPure Viral DNA/RNA Kit (TransGen, Beijing, China), and the extracted RNA from RNA viruses (AIV, NDV, IBV, aMPV, and ARV) was reverse transcribed to cDNA by using the PrimeScript™ 1st Strand cDNA Synthesis Kit (TaKaRa, Dalian, China) according to the manufacturer's instructions. The extracted DNA/cDNA was eluted in 30 μL of distilled water and stored at −20 ℃ until use. Finally, 2 μL of each DNA/cDNA solution was used as a template for the mLAMP reaction to evaluate the specificity.

Table 5-18-1 Viruses used and mLAMP assay results

Virus	Source	mLAMP
Chicken parvovirus (ChPV)	18 isolated strains from Guangxi (GX-ChPV-1-18), GVRI	+
Chicken infectious anaemia virus (CIAV)	8 isolated strains from Guangxi (GX1801, GX1804, GX1810, GX1904A, GX1904P, GX1905, GX1805, GX1904B), GVRI	+
Fowl aviadenovirus serotype 4 (FAdV-4)	1 reference strain (AV211), CIVDC; 17 isolated strains from Guangxi (FAdV-4GX2017-001-017), GVRI	+
Fowl aviadenovirus (FAdV)	11 serotypes of FAdV, FAdV (1, 2, 3 and 5-12) reference strains, CIVDC	–
Avian influenza virus (AIV)	3 isolated strains from Guangxi (H3N2, H6N6 and H9N2), GVRI	–
Newcastle disease virus (NDV)	2 reference strains (F48E9, LaSota), GVRI	–
Infectious bronchitis virus (IBV)	1 reference strain (Mass 41), GVRI	–
Avian reovirus (ARV)	1 reference strain (S1133), CIVDC	–
Avian infectious laryngotracheitis virus (ILTV)	1 reference strain (Beijing), GVRI	–
Avian metapneumo virus (aMPV)	1 reference strain (MN-10), GVRI	–
Marek's disease virus (MDV)	1 inactivated vaccine (CVI988), MERIAL, 1 reference strain (GX-MDV1), GVRI	–
Negative control	Five SPF chickens	–

Note: GVRI, Guangxi Veterinary Research Institute; CIVDC, China Institute of Veterinary Drug Control; MERIAL, Merial Animal Health Products Co.

Primer and probe design

The selected genes (namely, the NS gene of ChPV, the VP1 gene of CIAV and the hexon gene of FAdV-4) are highly conserved and were chosen as preferred target regions for the identification of these three DNA viruses[2, 23, 24]. Based on multiple sequence alignment, the primers and probes were designed using Primer Premier 5 and PrimerExplorer V5. In addition to the conventional LAMP primers, probes (termed FD) complementary to the F1C segment of the forward inner primer (FIP) were designed and modified with a fluorophore at the 3' end, and the 5' end of the FIP was modified with a dark quencher; FD and the FIP were annealed before the reaction to form a fluorescence-quenching composite probe (FIP-FD). During the amplification process, the composite probe became detached, and the fluorescence signal was released.

The ChPV-FD probe was labelled with 6-carboxyfluorescein (FAM) at the 3' end, the CIAV-FD probe was labelled with cyanine 5 (Cy5), the FAdV-4-FD probe was labelled with sulfo-cyanine 3 (Cy3), and the 5' end of the FIP was labelled with the corresponding BHQ series quenching group. All primers were synthesized and HPLC-purified by TaKaRa Inc. The sequences of the primers and probes are listed in Table 5-18-1. Before the reaction, the FIP quenching (FIP-Q) and FD fluorescence (FD-F) probes were annealed to quench the fluorescence: a mixture of 50 μM FIP-Q and 50 μM FD-F was heated to 90 ℃ for 5 min and then slowly cooled to room temperature to form the FIP-FD composite probe, which was stored at –20 ℃.

Table 5-18-2 Primer and probe sequences for the mLAMP reaction

Primer	Target gene	Sequence (5'-3')	Tm /℃
ChPV-F3	NS	AGAGGAGGAACCCCCCTAT	60.6
ChPV-B3		CGCTTGCGGTGAAGTCTG	60.3
ChPV-FIP		BHQ1-ACGGGATAAATCCCTGGGACCTTGGTTCTGAATCCGGGCT	60.1/64.9
ChPV-BIP		ACAGAGACCATCGAGCTCCTGGGCTCGTCTGGAAATCCACTC	59.8/65.3
ChPV-Floop		TCTTACCTTCGTTGGCTTTTTCAA	63.0
ChPV-Bloop		AGACGGAGATCCTCAGCGAATC	62.3
ChPV-FD		AGGTCCCAGGGATTTATCCCGT-FAM	60.1
CIAV-F3	VP1	AGGCCACCAACAAGTTCAC	59.9
CIAV-B3		GGTTGATCGGTCCTCAAGTC	59.5
CIAV-FIP		BHQ3-GCAGCCACACAGCGATAGAGTGCCGTTGGAAACCCCTCAC	59.5/65.2
CIAV-BIP		CGCGCTCCCACGCTAAGATCCGGCACATTCTTGAAACCAG	59.0/65.0
CIAV-Floop		ATTGTAATTGCAGCGATACCAATCC	63.8
CIAV-Bloop		ACTGCGGACAATTCAGAAAGCA	62.4
CIAV-FD		CACTCTATCGCTGTGTGGCTGC-Cy5	59.5
FAdV-4-F3	Hexon	GAGGTGAACCTCATGGCC	59.0
FAdV-4-B3		TTGATGCGAGTGAAGGACC	59.0
FAdV-4-FIP		BHQ2-TGGTGGCGTTTCTCAGCATCAGACTTCATGCCCATGGATCAC	65.1/59.7
FAdV-4-BIP		ACGCTCTATACTCGGTGCCCGGTGCGAGCGGGAATGTTG	65.1
FAdV-4-Floop		CTCGAGCTGGTTACTGGTGTTG	60.8/65.6
FAdV-4-Bloop		CTCCACCGCCCTCACCAT	60.3
FAdV-4-FD		CTGATGCTGAGAAACGCCACCA-CY3	65.1

Note: FIP, forward inner primer; BIP, backward inner primer; F3, forward outer primer; B3, backward outer primer; Floop, forward loop primer; Bloop, backward loop primer.

mLAMP reaction

mLAMP amplification and monitoring were performed by a Loopamp real-time turbidimeter (LA-320; Eiken Chemical Co., Ltd., Tokyo, Japan) using the WarmStart Multipurpose LAMP/RT-LAMP 2 × Master Mix Kit (New England Biolabs, Ipswich, MA, USA). Each tube contained a 25 μL reaction mixture composed of 2 μL of DNA template, 12.5 μL of 2 × Master Mix, 16 U of Bst 2.0 WarmStart DNA polymerase (New England Biolabs), 0.4 μM ChPV-FIP, 0.6 μM CIAV-FIP, 0.4 μM FAdV-4-FIP, 0.4 μM ChPV-FIP-FD, 0.2 μM CIAV-FIP-FD, 0.4 μM FAdV-4-FIP-FD, 0.8 μM ChPV-BIP, 0.8 μM CIAVBIP, 0.8 μM FAdV-4-BIP, 0.1 μM ChPV-F3, 0.1 μM CIAV-F3, 0.1 μM FAdV-4-F3, 0.1 μM ChPV-B3, 0.1 μM CIAV-B3, 0.1 μM FAdV-4-B3, 0.2 μM ChPV-Floop, 0.2 μM CIAV-Floop, 0.2 μM FAdV-4-Floop, 0.2 μM ChPV-Bloop, 0.2 μM CIAV-Bloop, and 0.2 μM FAdV-4-Bloop, ddH$_2$O was added to reach the final reaction volume of 25 μL. The reaction conditions were as follows: 63 ℃ for 75 min for amplification and 80 ℃ for 5 min for termination.

The fluorescent mLAMP products were analyzed by an image analyzer (Universal Hood III, 731BR01622, Bio-Rad, Hercules, CA, USA) to distinguish the different colours of the reaction tubes. Green samples detected in the 520 nm channel (FAM-labelled) were considered ChPV-positive, red samples detected in the 670 nm channel were considered CIAV-positive (Cy5-labelled), and blue samples detected in the 570 nm channel were considered FAdV-4-positive (Cy3-labelled). The colour of Cy3 in the spectrum is orange, and the overlap colour of FAM (green) and Cy5 (red) is also orange. To distinguish them, Cy3 was visualized as blue in the output image. Multiple positive samples were observed as mixed colours in two or three channels simultaneously.

DNA standard preparation

The NS gene of ChPV (GX-ChPV-7 strain), the VP1 gene of CIAV (GX1805 strain) and the hexon gene of FAdV-4 (GX-1 strain) were cloned into the pMD18-T vector (TaKaRa) separately following the standard procedure, generating the recombinant plasmids pMD18-T-NS, pMD18-T-VP1, and pMD18-T-hexon, respectively. These plasmids were extracted by using the E. Z. N. A.® Plasmid DNA Mini Kit (Omega Bio-Tek, Norcross, GA, USA), and the concentrations of the recombinant plasmids were determined by measuring the absorbance at 260 nm with a NanoDrop 2 000 (Thermo Fisher Scientific, Waltham, MA, USA). The copy numbers of the target DNA were calculated as previously described[25]. The three recombinant plasmids were mixed in equal volumes and diluted 10-fold to prepare the DNA standard, and 2 μL of the DNA standard was used as a template. For determination of the detection limits, 10 replicates for each serial 10-fold dilution of the standards were used for the mLAMP assay. The detection limit of mLAMP for ChPV, CIAV and FAdV-4 was determined by using probit regression analysis (SPSS Statistics for Windows, Version 17.0, Released 2008, Chicago: SPSS Inc.)[26].

Clinical detection

Three hundred and forty-two samples (cloacal swab, heart, liver and kidney) were collected from chicken farms in Guangxi, China, and were tested to evaluate the ability of mLAMP to detect ChPV, CIAV and FAdV-4 in clinical samples. Cloacal swabs were collected from both healthy and diseased chickens, and tissue samples were collected from diseased and dead chickens with typical symptoms: pericardial effusion, congestive kidney, enlarged liver and bleeding. The swab samples were eluted into 1 mL of sterilized water, and the supernatant was used to extract DNA after centrifugation. The tissue samples were homogenized in PBS, and DNA was extracted from 200 μL of the suspension. DNA extraction was performed as described in the *Viruses and nucleic acid extraction* section. All the extracted DNA was tested using the mLAMP assay and, in parallel, using the ChPV-, CIAV- or FAdV-4-specific conventional PCR assay as previously described[2, 27, 28]. The results of the mLAMP and PCR assays were then compared. All the positive field samples detected based on the PCR products were verified by DNA sequencing to rule out false-positive results.

Results

Specificity of mLAMP

The specificity of the mLAMP assay was verified on a series of viruses, including single or mixed templates of ChPV, CIAV and FAdV-4 and other related avian viruses. The real-time turbidimeter generated a turbidity curve based on the monitored precipitation of the byproduct magnesium pyrophosphate in the mLAMP reaction. The real-time turbidity curve of the mLAMP reaction is shown in Figure 5-18-1 A, and the mLAMP reaction containing three sets of primers could robustly detect targets: single template, duplex template and triple template. Based on turbidity, the triple reaction was the fastest, followed by the duplex reaction, and the single reaction was the slowest. This was because the turbidity signal in the multiple reactions was obtained by superposition of multiple amplifications, so turbidity was detected in the triple reaction the earliest. However, turbidity curves could be generated by either single or multiple reactions. Therefore, for multiple reactions, turbidity curves alone could not distinguish the viruses that caused positive results.

The mLAMP method used end point analysis, regardless of the speed of independent amplification, and the detection provided a robust fluorescent signal for each target, which was interpreted according to the colour of the amplification products. The green fluorescence (FAM) in the 520 nm channel indicated ChPV-positive amplification, the red fluorescence (Cy5) in the 670 nm channel indicated CIAV-positive amplification, and the blue fluorescence (Cy3) in the 570 nm channel indicated FAdV-4-positive amplification. Overlapping fluorescence in multiple channels indicated amplification of multiple positive samples. The viruses could be accurately distinguished in mixed samples containing two or three viruses and displayed corresponding colours in the single and multiple channels simultaneously. Taking Figure 5-18-1 B, lane 4, as an example, the ChPV and CIAV mixed samples simultaneously showed orange colour (overlap of red and green signals) in the multiple channel, green colour in the 520 nm channel, and red colour in the 670 nm channel but no fluorescence in the 570 nm channel. Compared with methods involving the addition of an intercalator and pyrophosphate precipitation, mLAMP could distinguish the targets accurately and intuitively. Based on fluorescence, all 18 ChPV isolate strains tested positive and showed green fluorescence, eight CIAV Guangxi

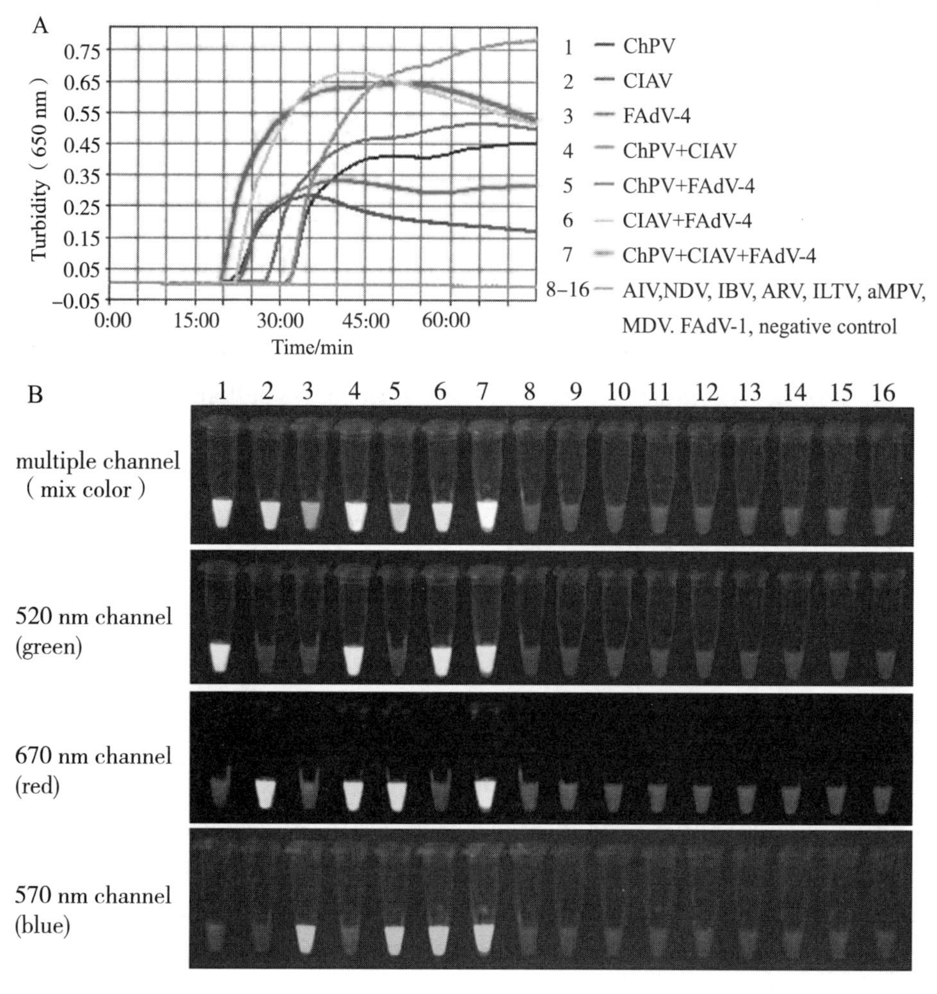

A: The turbidity curve; B: The fluorescent mLAMP products were imaged separately using multiple channels, the single reaction exhibits a single fluorescence signal in the corresponding channel (color figure in appendix).

Figure 5-18-1 Specificity of mLAMP

isolate strains showed red fluorescence, and 18 FAdV-4 strains showed blue fluorescence. There were no cross-reactions with any of the related avian viruses, including 11 other FAdV serotypes (1～3 and 5～12), AIV, NDV, IBV, ARV, ILTV, aMPV and MDV. Only ChPV-, CIAV- and FAdV-4-specific targets were amplified by mLAMP. Similarly, no fluorescence was detected in the negative controls (Table 5-18-1 and Figure 5-18-1). The specificity results showed that the mLAMP assay had good specificity and could accurately distinguish the three target viruses ChPV, CIAV and FAdV-1.

Sensitivity of mLAMP

For determination of the detection limits of mLAMP, each 10-fold serial dilution of the standards, as described in the previous section, was subjected to mLAMP with 10 replicates. The results are shown in Table 5-18-3.The SPSS statistical programme generated the probit (predicted proportion positive) versus the template concentration with a 95% confidence interval (CI), as shown in Table 5-18-4. The detection limit of the mLAMP assay was 307 copies/μL for recombinant plasmids containing the ChPV target genes, 749 copies/μL for CIAV and 648 copies/μL for FAdV-4. All these results demonstrated that the mLAMP assay could simultaneously and sensitively identify ChPV, CIAV and FAdV-4.

Table 5-18-3 Number of positive samples with 10 replicates for each dilution

ChPV standard[a] /(copies/μL)	No. of ChPV-positive samples	CIAV standard /(copies/μL)	No. of CIAV-positive samples	FAdV-4 standard /(copies/μL)	No. of FAdV-4 positive samples	No. of replicates
0.66×10^9	10	1.06×10^9	10	0.95×10^9	10	10
0.66×10^8	10	1.06×10^8	10	0.95×10^8	10	10
0.66×10^7	10	1.06×10^7	10	0.95×10^7	10	10
0.66×10^6	10	1.06×10^6	10	0.95×10^6	10	10
0.66×10^5	10	1.06×10^5	10	0.95×10^5	10	10
0.66×10^4	10	1.06×10^4	10	0.95×10^4	10	10
0.66×10^3	10	1.06×10^3	10	0.95×10^3	10	10
0.66×10^2	8	1.06×10^2	7	0.95×10^2	7	10
0.66×10^1	0	1.06×10^1	0	0.95×10^1	0	10
0.66×10^0	0	1.06×10^0	0	0.95×10^0	0	10

Note: [a] Standards contained DNA plasmids with the specific gene sequences of ChPV, CIAV and FAdV-4. Two microliters of standard were used as a template in the mLAMP assay.

Table 5-18-4 Probit analysis to determine the detection limit of the mLAMP assay

Probit[a]	Template concentration (copies/μL, 95% CI)		
	ChPV	CIAV	FAdV-4
0.95	307 (140～7 557)	749 (310～31 512)	648 (27～29 270)

Note: [a] Probit, predicted proportion of replication from SPSS software, based on 10 replicates of 10 dilutions of three standards listed in Table 5-18-3.

Interference in mLAMP

Seven artificial samples with various concentrations of recombinant plasmids containing the ChPV, CIAV or FAdV-4 conserved gene were prepared and subjected to the mLAMP assay to assess the interference between higher concentrations and lower concentrations of nucleic acid templates. Figure 5-18-2 shows that the fluorescence in the post-amplification reaction tubes was easily observed in the colour-specific channels. These results demonstrated that in the presence of higher concentrations of the DNA template, the amplification of lower concentrations of DNA template was not inhibited in the mLAMP assay, and vice versa.

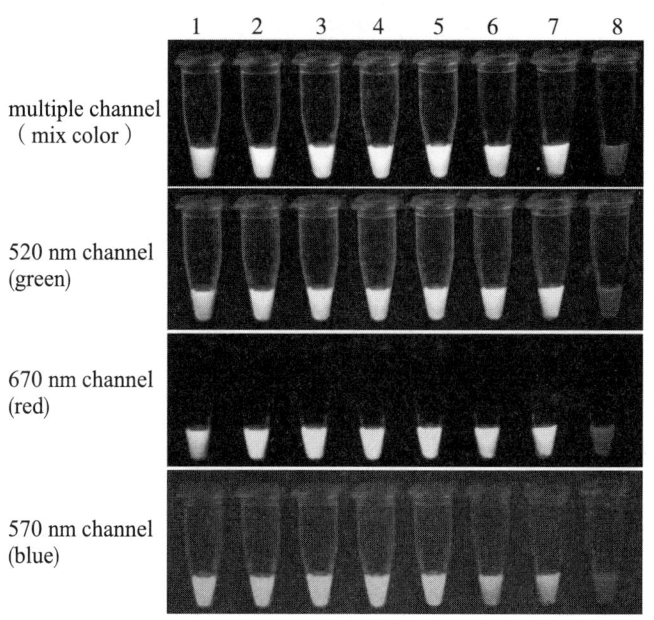

1: ChPV (6.6×10^2 copies/μL)+CIAV (1.06×10^3 copies/μL)+FAdV-4 (0.95×10^8 copies/μL); 2: ChPV (6.6×10^3 copies/μL)+CIAV (1.06×10^9 copies/μL)+FAdV-4 (0.95×10^4 copies/μL); 3: ChPV (6.6×10^5 copies/μL)+CIAV (1.06×10^3 copies/μL)+FAdV-4 (0.95×10^8 copies/μL); 4: ChPV (6.6×10^3 copies/μL)+CIAV (1.06×10^5 copies/μL)+FAdV-4 (0.95×10^7 copies/μL); 5: ChPV (6.6×10^5 copies/μL)+CIAV (1.06×10^8 copies/μL)+FAdV-4 n(0.95×10^4 copies/μL); 6: ChPV (6.6×10^8 copies/μL)+CIAV (1.06×10^3 copies/μL)+FAdV-4 (0.95×10^3 copies/μL); 7: ChPV (6.6×10^8 copies/μL)+CIAV (1.06×10^8 copies/μL)+FAdV-4 (0.95×10^3 copies/μL); 8: Negative control. Each template can be amplified independently without interference from other templates in the same reaction tube.

Figure 5-18-2 Interference results for mLAMP

Analysis of clinical samples

To assess mLAMP for the detection of viral DNA in the field, a comparison was made with conventional PCR for 342 clinical specimens. As seen in Table 5-18-5, the results of the mLAMP assay were in 100% agreement with those of PCR, and all the samples found positive by PCR were verified by sequencing. Both methods identified 54 samples as ChPV single positive, yielding an infection rate of 15.8%, 73 samples as CIAV single positive, yielding an infection rate of 21.3%, and 36 samples as FAdV-4 single positive, yielding an infection rate of 10.5%; 19 coinfected samples were positive for both ChPV and CIAV with an infection rate of 5.6%, six coinfected samples were positive for both ChPV and FAdV-4 with an infection rate of 1.8%, eight coinfected samples were positive for both CIAV and FAdV-4 with an infection rate of 2.3%, and four coinfected samples were positive for all three avian viruses with an infection rate of 1.2%. The positive PCR products were sent to BGI Genomics Institute (Shenzhen, China) for DNA sequence confirmation using PCR primers. The sequencing results showed that all the positive results were true positives. Compared to the PCR assay for detection in field samples, the mLAMP assay showed 100% sensitivity and specificity for the detection of ChPV (83/83 and 259/259, respectively), CIAV (104/104 and 238/238, respectively) and FAdV-4 (54/54 and 288/288, respectively). This mLAMP assay was as easy, rapid, sensitive and specific as the PCR assay.

Table 5-18-5 Performance of the mLAMP assay for detection in clinical samples

Virus	mLAMP result	PCR result	Sequencing result
ChPV	54	54	54
CIAV	73	73	73
FAdV-4	36	36	36
CIAV+ChPV	19	19	19
ChPV+FAdV-4	6	6	6
CIAV+FAdV-4	8	8	8
ChPV+CIAV+FAdV-4	4	4	4
Negative	142	142	142
Total	342	342	342

Discussion

With the intensive development of the poultry industry, the incidence of avian infectious diseases is increasing each year, which has become an important factor restricting the growth and further development of the poultry industry. Epidemics of old infectious diseases, the emergence of new infectious diseases, the increasing number of cases of coinfection and secondary infection in chickens, and the emergence of new serotypes or mutant strains of pathogens have all increased the difficulty of prevention and control of avian diseases. Therefore, it is necessary to develop a high-throughput, rapid and simple method for the detection and differentiation of coinfection to reduce the occurrence of infectious diseases in poultry and ensure the healthy development of the poultry industry.

LAMP technology has drawn increased attention from researchers because it can be easily manipulated and allows rapid detection. Although there have been many studies on multiplex LAMP, these methods have their own disadvantages. The existing multiplex LAMP methods mainly include the following four technical routes. (1) Multiple sets of LAMP primers are placed in one reaction tube for amplification. This method can only determine whether a sample is positive but cannot identify the pathogen that causes the positive reaction, and the test results do not accurately identify the pathogen[29, 30]. (2) Multiplex LAMP amplification is carried out in a real-time PCR instrument. The reaction results are read by a real-time PCR instrument, and this method requires bulky, expensive equipment[31, 32]. (3) A restriction enzyme cleavage site is inserted into the inner primers to construct a multiplex LAMP assay. The multiplex LAMP products are digested by restriction enzymes after amplification and then analyzed by agarose gel electrophoresis. This method is complicated and time-consuming and uses hazardous products[33, 34]. (4) The inner primers for multiplex LAMP are synthesized with a fluorophore. The multiplex LAMP products are analyzed by agarose gel electrophoresis, and the fluorescent dye-conjugated multiplex LAMP products are visualized as electrophoresis bands of different colours. The laboratory environment can be easily contaminated by cap opening[35].

This study improved upon the previous methods and adopted a new technical route: an FD probe complementary to the F1C was synthesized to form a fluorescence-quenching composite probe with the FIP (FIP-FD). The 3' end of the FD probe was modified with a fluorophore (FD-F), and the 5' end of the FIP was modified with a dark quencher. Before the reaction, FIP-Q and FD-F were annealed to form the FIP-FD composite probe with quenched fluorescence. During the mLAMP reaction process, the FIP retained the function of the inner primer to guide amplification, but upon synthesis from the reverse direction guided by the BIP, the FD probe was separated, resulting in the release of the fluorophore. The mLAMP result displays different colours according to the fluorescence of the markers after the reaction and can be read directly by observing the colour of the reaction tube with the naked eye[36].

In this study, an mLAMP assay was successfully established to detect ChPV, CIAV and FAdV-4 simultaneously in one tube in 80 min under isothermal conditions. This assay has several advantages over previous methods. First, this mLAMP assay is easy to perform and has a lower risk of contamination. The LAMP products consist of DNA amplicons of many different sizes, including more than 10^3 times the number of PCR products. The steps of opening the cap and adding the dye can easily lead to aerosol contamination. This assay omits the cap-opening step, thus greatly reducing the risk of laboratory contamination. Second, the detection results of mLAMP are easy to read. The mLAMP assay employs three fluorophores with three different emission wavelengths. It is easy to distinguish the different fluorescence signals with the naked eye: the FAM emission wavelength is 520 nm (green), the Cy5 emission wavelength is 670 nm (red), and the Cy3 emission wavelength is 570 nm (blue). These three colours are in sharp contrast and can be observed using specific channels, which is more reliable and accurate than the observation of white precipitate formation. Finally, in clinical settings, there are few cases of single pathogen infection, and most cases involve multipathogen coinfection. The mLAMP method can identify three viruses at a time to effectively meet the needs of on-site pathogen detection. This study combined LAMP technology with a composite probe to develop a visual multiplex fluorescence-based LAMP assay for the simultaneous identification of multiplex target sequences, providing a new idea for future multiplex LAMP research.

Conclusions

In summary, the mLAMP assay developed in this study is rapid, inexpensive, sensitive, and specific, suggesting that it is a potentially valuable tool for the clinical diagnosis and surveillance of ChPV, CIAV and FAdV-4, and can be applied in rural areas.

References

[1] HUYNH L T M, NGUYEN G V, DO L D, et al. Chicken infectious anaemia virus infections in chickens in northern Vietnam: epidemiological features and genetic characterization of the causative agent. Avian Pathol, 2020, 49(1): 5-14.

[2] ZHANG Y, FENG B, XIE Z, et al. Epidemiological surveillance of parvoviruses in commercial chicken and turkey farms in Guangxi, southern China, During 2014-2019. Front Vet Sci, 2020, 7: 561371.

[3] NIU D, FENG J, DUAN B, et al. Epidemiological survey of avian adenovirus in China from 2015 to 2021 and the genetic variability of highly pathogenic FAdV-4 isolates. Infect Genet Evol, 2022: 105277.

[4] KISARY J, NAGY B, BITAY Z. Presence of parvoviruses in the intestine of chickens showing stunting syndrome. Avian Pathol, 1984, 13(2): 339-343.

[5] KISARY J, AVALOSSE B, MILLER-FAURES A, et al. The genome structure of a new chicken virus identifies it as a parvovirus. J Gen Virol, 1985, 66 (Pt 10): 2259-2263.

[6] NUNEZ L F N, SANTANDER-PARRA S H, DE LA TORRE D I, et al. Molecular characterization and pathogenicity of chicken parvovirus (ChPV) in specific pathogen-free chicks infected experimentally. Pathogens, 2020, 9(8): 606.

[7] ZHANG Y, XIE Z, DENG X, et al. Molecular characterization of parvovirus strain GX-Tu-PV-1, isolated from a Guangxi turkey. Microbiol Resour Announc, 2019, 8(46): e00152-19.

[8] VAGNOZZI A E, ESPINOSA R, CHENG S, et al. Study of dynamic of chicken infectious anaemia virus infection: which

sample is more reliable for viral detection? Avian Pathol, 2018, 47(5): 489-496.

[9] TECHERA C, MARANDINO A, TOMAS G, et al. Origin, spreading and genetic variability of chicken anaemia virus. Avian Pathol, 2021, 50: 311-320.

[10] SUN J, ZHANG Y, GAO S, et al. Pathogenicity of fowl adenovirus serotype 4 (FAdV-4) in chickens. Infect Genet Evol, 2019, 75: 104017.

[11] LI P H, ZHENG P P, ZHANG T F, et al. Fowl adenovirus serotype 4: Epidemiology, pathogenesis, diagnostic detection, and vaccine strategies. Poult Sci, 2017, 96(8): 2630-2640.

[12] VERA-HERNANDEZ P F, MORALES-GARZON A, CORTES-ESPINOSA D V, et al. Clinicopathological characterization and genomic sequence differences observed in a highly virulent fowl Aviadenovirus serotype 4. Avian Pathol, 2016, 45(1): 73-81.

[13] BROWN JORDAN A, BLAKE L, BISNATH J, et al. Identification of four serotypes of fowl adenovirus in clinically affected commercial poultry co-infected with chicken infectious anaemia virus in Trinidad and Tobago. Transbound Emerg Dis, 2019, 66(3): 1341-1348.

[14] LI Y, HU Y, CUI S, et al. Molecular characterization of chicken infectious anemia virus from contaminated live-virus vaccines. Poult Sci, 2017, 96(5): 1045-1051.

[15] SU Q, MENG F, LI Y, et al. Chicken infectious anemia virus helps fowl adenovirus break the protection of maternal antibody and cause inclusion body hepatitis-hydropericardium syndrome in layers after using co-contaminated Newcastle disease virus-attenuated vaccine. Poult Sci, 2019, 98(2): 621-628.

[16] NOTOMI T, OKAYAMA H, MASUBUCHI H, et al. Loop-mediated isothermal amplification of DNA. Nucleic Acids Res, 2000, 28(12): E63.

[17] FAN Q, XIE Z, XIE L, et al. A reverse transcription loop-mediated isothermal amplification method for rapid detection of bovine viral diarrhea virus. J Virol Methods, 2012, 186(1-2): 43-48.

[18] WANG L C, HUANG D, CHEN H W. Simultaneous subtyping and pathotyping of avian influenza viruses in chickens in Taiwan using reverse transcription loop-mediated isothermal amplification and microarray. J Vet Med Sci, 2016, 78(8): 1223-1238.

[19] CHEN C, ZHAO Q, GUO J, et al. Identification of methicillin-resistant staphylococcus aureus (MRSA) using simultaneous detection of mecA, nuc, and femB by Loop-mediated isothermal amplification (LAMP). Curr Microbiol, 2017, 74(8): 965-971.

[20] DUAN Y B, YANG Y, WANG J X, et al. Simultaneous detection of multiple benzimidazole-resistant beta-tubulin variants of botrytis cinerea using Loop-mediated isothermal amplification. Plant Dis, 2018, 102(10): 2016-2024.

[21] FAN Q, XIE Z, XIE Z, et al. Development of duplex fluorescence-based loop-mediated isothermal amplification assay for detection of *Mycoplasma bovis* and bovine herpes virus 1. J Virol Methods, 2018, 261: 132-138.

[22] LIN F, LIU L, HAO G J, et al. The development and application of a duplex reverse transcription loop-mediated isothermal amplification assay combined with a lateral flow dipstick method for Macrobrachium rosenbergii nodavirus and extra small virus isolated in China. Mol Cell Probes, 2018, 40: 1-7.

[23] KOO B S, LEE H R, JEON E O, et al. Genetic characterization of three novel chicken parvovirus strains based on analysis of their coding sequences. Avian Pathol, 2015, 44(1): 28-34.

[24] YUMING F, SHENG Y, WENYU D, et al. Molecular characterization and phylogenetic analysis of fowl adenovirus serotype-4 from Guangdong Province, China. Vet World, 2020, 13(5): 981-986.

[25] XIE Z, LUO S, XIE L, et al. Simultaneous typing of nine avian respiratory pathogens using a novel GeXP analyzer-based multiplex PCR assay. J Virol Methods, 2014, 207: 188-195.

[26] SMIEJA M, MAHONY J B, GOLDSMITH C H, et al. Replicate PCR testing and probit analysis for detection and quantitation of Chlamydia pneumoniae in clinical specimens. J Clin Microbiol, 2001, 39(5): 1796-1801.

[27] WANG L C, ZHANG L, ZHANG Y, et al. Dignosis and preventive treatment of flow adenovirus-4 infection. Anhui Agri. Sci Bull, 2016, 22(10): 116-122.

[28] LI S, FANG M, ZHOU B, et al. Simultaneous detection and differentiation of dengue virus serotypes 1-4, Japanese encephalitis virus, and West Nile virus by a combined reverse-transcription loop-mediated isothermal amplification assay. Virol J, 2011, 8: 360.

[29] YAMAZAKI W, MIOULET V, MURRAY L, et al. Development and evaluation of multiplex RT-LAMP assays for rapid and sensitive detection of foot-and-mouth disease virus. J Virol Methods, 2013, 192(1-2): 18-24.

[30] LIU N, ZOU D, DONG D, et al. Development of a multiplex loop-mediated isothermal amplification method for the simultaneous detection of *Salmonella* spp. and Vibrio parahaemolyticus. Sci Rep, 2017, 7: 45601.

[31] KIM M J, KIM H Y. Direct duplex real-time loop mediated isothermal amplification assay for the simultaneous detection of cow and goat species origin of milk and yogurt products for field use. Food Chem, 2018, 246: 26-31.

[32] ISEKI H, ALHASSAN A, OHTA N, et al. Development of a multiplex loop-mediated isothermal amplification (mLAMP) method for the simultaneous detection of bovine Babesia parasites. J Microbiol Methods, 2007, 71(3): 281-287.

[33] LIU X L, ZHAO X T, MUHAMMAD I, et al. Multiplex reverse transcription loop-mediated isothermal amplification for the simultaneous detection of CVB and CSVd in chrysanthemum. J Virol Methods, 2014, 210: 26-31.

[34] AONUMA H, YOSHIMURA A, KOBAYASHI T, et al. A single fluorescence-based LAMP reaction for identifying multiple parasites in mosquitoes. Exp Parasitol, 2010, 125(2): 179-183.

[35] TANNER N A, ZHANG Y, EVANS T C, JR. Simultaneous multiple target detection in real-time loop-mediated isothermal amplification. Biotechniques, 2012, 53(2): 81-89.

[36] NANAYAKKARA I A, WHITE I M. Demonstration of a quantitative triplex LAMP assay with an improved probe-based readout for the detection of MRSA. Analyst, 2019, 144(12): 3878-3885.

A duplex quantitative real-time PCR assay for the detection of *Haplosporidium* and *Perkinsus* species in shellfish

Xie Zhixun, Xie Liji, Fan Qing, Pang Yaoshan, Deng Xianwen, Xie Zhiqin, Liu Jiabo, and Khan Mazhar I

Abstract

A duplex quantitative real-time polymerase chain reaction (dq-PCR) assay was optimized to simultaneously detect *Haplosporidium* spp. and *Perkinsus* spp. of shellfish in one reaction. Two sets of specific oligonucleotide primers for *Haplosporidium* spp. and *Perkinsus* spp., along with two hydrolysis probes specific for each parasite group, were used in the assay. The dq-PCR results were detected and analyzed using the Light Cycler 2.0 software system. The dq-PCR identified and differentiated the two protozoan parasite groups. The sensitivity of the dq-PCR assay was 200 template copies for both *Haplosporidium* spp. and *Perkinsus* spp. No DNA product was amplified when known DNA from *Marteilia refringens*, *Toxoplasma gondii*, *Bonamia ostreae*, *Escherichia coli*, *Cymndinium* spp., *Mykrocytos mackini*, *Vibrio parahaemolyticus*, and shellfish tissue were used as templates. A total of 840 oyster samples from commercial cultivated shellfish farms from two coastal areas in China were randomly collected and tested by dq-PCR. The detection rate of *Haplosporidium* spp. was 8.6% in the Qingdao, Shandong coastal area, whereas *Perkinsus* spp. was 8.3% coastal oysters cultivated from shellfish farms of Beihai, Guangxi. The dqPCR results suggested that *Haplosporidium* spp. was prevalent in oysters from Qingdao, Shandong, while *Perkinsus* spp. was prevalent in oysters from the coastal areas of Beihai, Guangxi. This dq-PCR could be used as a diagnostic tool to detect *Haplosporidium* spp. and *Perkinsus* spp. in cultivated shellfish.

Keywords

duplex quantitative real-time PCR, *Haplosporidium*, *Perkinsus*, detection

---

Introduction

Haplosporidium spp. and *Perkinsus* spp. are two major protozoan parasite pathogen groups that infect shellfish and have caused mortalities among cultured shellfish since the late 1950s[1-11]. These parasites can cause high mortality rates, leading to economic losses that are detrimental to the shellfish farming industry[9, 12].

Multiple diagnostic methods such as histological examination, electron microscopy, Ray/Mackin fluid thioglycollate medium assay, and histological studies using in situ hybridization are required to detect and differentiate between these protozoan parasites[13-17]. However, these methods are time-consuming, have limited sensitivity, and require a high level of expertise. Molecular assays, such as conventional and real-time PCR, have been used for rapid and sensitive detection of these protozoan parasites[9, 12 18-20]. However, real-time PCR is preferred to conventional PCR in clinical laboratories because there is no need for post amplification handling, thereby leading to faster analysis and quantification of the target genes and limiting cross contamination by amplified products[20-23].

Russell et al.[9] demonstrated that *Haplosporidium* spp. and *Perkinsus* spp. could be detected with greater sensitivity in oysters from the east coast of the United States using multiplex PCR, and that the simultaneous detection of *Haplosporidium* spp. and *Perkinsus* spp. would have useful applications. This work describes the development of a dq-PCR assay for simultaneous detection of *Haplosporidium* spp. and *Perkinsus* spp.

Materials and methods

Pathogens and plasmids

Information related to the plasmids and genomic DNA from different parasites and bacteria used in the present study are listed in Table 5-19-1. DNA isolations from *Escherichia coli* and *Vibrio parahaemolyticus* (GX/O/1/09), as well as gill tissue from oysters, were performed using the DNA extraction kit (Tiangen, Beijing, China) as described by Xie et al.[24]. The 18S ribosomal RNA region of *Haplosporidium* spp. and internal transcribed spacer of *Perkinsus* spp. were amplified using primers listed in Table 5-19-2, and amplicons were cloned into the pMD18-T vector (TaKaRa Dalian, China). Constructed plasmids were transcribed in *E. coli* (Bioteke, Beijing, China) following the manufacturer's protocol. The isolation of plasmids was carried out using a plasmid extraction kit (BioDev, Beijing, China). The recombinant plasmids obtained were sequenced. Sequence data were analyzed using DNASTAR software and were compared with corresponding sequence data in GenBank. The copy number was calculated according to the formula ((copies/μL=$6 \times 10^{23} \times$ DNA concentration (g/μL) /molecular weight (g/mol)) as described by Vaitomaa et al.[25]

Table 5-19-1 Sources of pathogens used and dq-PCR test results

The species of shellfish	Pathogens/field samples	Numbers of sample	Source	Result	
				dr-PCR	
				Haplosporidium spp.	*Perkinsus* spp.
	Haplosporidium costale (7251-07-59)	1	VIMS	+	–
	Haplosporidium nelsoni (7247-07-4)	1	VIMS	+	–
	Perkinsus marinus (5059-8)	1	VIMS	–	+
	Haplosporidium nelsoni (7247-07-4)+ *Perkinsus olseni* (GX/O/1/08)	1	VIMS and BEIHAI	+	+
Oyster	*Perkinsus olseni*	4	BEIHAI	–	+
Mussel	*Perkinsus* spp.	1	BEIHAI	–	+
Oyster	*Haplosporidium* spp.	4	QINGDAO	+	–
Mussel	*Haplosporidium* spp.	1	BEIHAI	+	–
Clam	*Haplosporidium* spp.	1	DALIAN	+	–
	Haplosporidium-pMD18-T	1	GVRI	+	–
	Perkinsus-pMD18-T	1	GVRI	–	+
Human	*Toxoplasma gondii* RH (DNA)	1	LVRI	–	–
Oyster	*Bonamia ostreae* (O-09-01) (DNA)	1	GFRI	–	–
Oyster	*Mykrocytos mackini* (O-10-01) (DNA)	1	GFRI	–	–
Oyster	*Escherichia coli* (GX/O/01/09)	1	GVRI	–	–
Oyster	*Cymndinium* spp. (GX/O/03/08)	1	GFRI	–	–
Oyster	*Vibrio parahaemolyticas* (GX/O/01/08)	1	GVRI		
Oyster	*Marteilia refringens* (GX/O/1/09) (DNA)	1	GVRI	–	–

continued

The species of shellfish	Pathogens/field samples	Numbers of sample	Source	Result	
				dr-PCR	
				Haplosporidium spp.	*Perkinsus* spp.
Oyster	Shellfish tissue (DNA)	1	GVRI	−	−
Oyster	*Perkinsus* spp.+*Haplosporidium* spp.	7	BEIHAI and BEIHAI	+	+
Oyster	*Perkinsus* spp.+*Haplosporidium* spp.	28	BEIHAI and QINGDAO	+	+
Oyster	*Perkinsus* spp.+*Haplosporidium* spp.	3	QINGDAO and BEIHAI	+	+
Oyster	*Perkinsus* spp.+*Haplosporidium* spp.	10	QINGDAO and QINGDAO	+	+

Note: VIMS, Virginia Institute of Marine Science, Gloucester Point, VA, USA; BEIHAI; Beihai, Guangxi coastal area of South China Sea; OINGDAO, Qingdao, Shandong coastal area of Yellow Sea of China; DALIAN, Dalian, Shenyang coastal area of Bohai Sea of China; GVRI, Guangxi Veterinary Research Institute, China; LVRI, Lanzhou Veterinary Research Institute, China; GFRI, Guangxi Fishery Research Institute, China.

Table 5-19-2 Amplified specific sequences for *Haplosporidium* and *Perkinsus* used for cloning

Primer name	Sequence of PCR primers(5'-3')	Size	Identity of amplified region
MSX-A (572)	CGACTTTGGCATTAGGTTTCAGACC	572 bp	18S ribosomal RNA
MSX-B (572)	ATGTGTTGGTGACGCTAACCG		
PerkITS-85 (703)	CCGCTTTGTTTGGATCCC	703 bp	Internal transcribed spacer
PerkITS-750 (703)	ACATCAGGCCTTCTAATGATG		

Oligonucleotide primers and DNA probes

To eliminate primer and probe mismatches, nucleotide sequences available on GenBank were aligned using Megalign software (Figure 5-19-1 A, B). Both sets of primers and hydrolysis probes are listed in Table

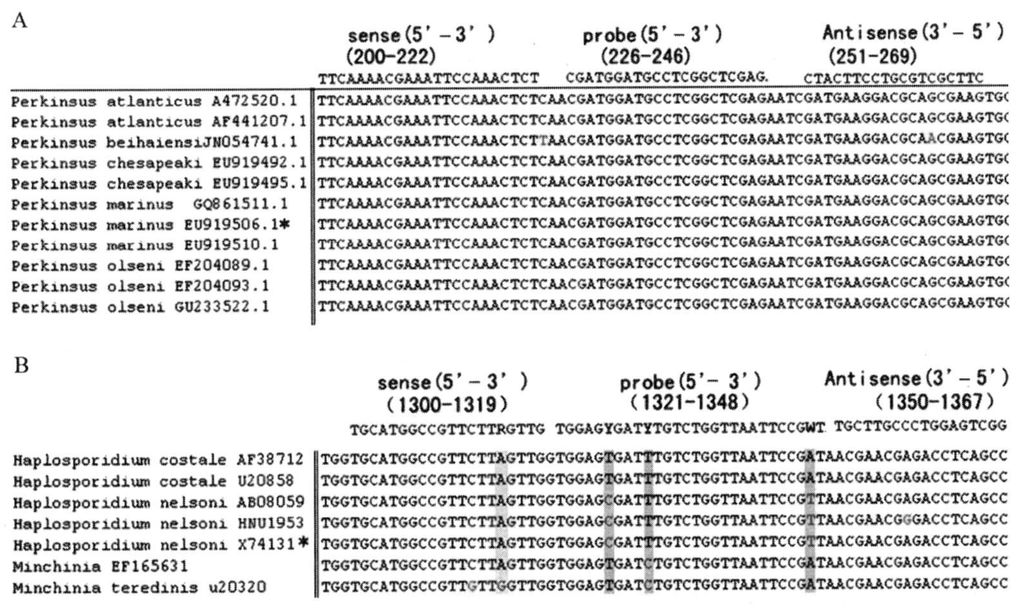

A: *Perkinsus* spp., B: *Haplosporidium* spp.. Primer and probe sequences are showed on the top. Asterisks represent reference sequence.

Figure 5-19-1 Multiple sequence alignment

5-19-3; these were designed to amplify highly conserved sequences of *Haplosporidium* spp. (*Haplosporidium nelsoni* AB080597)and *Perkinsus* spp. (*Perkinsus olseni* EF204089.1) by using the Primer Express Software 3.0 according to the Applied Biosystems guidelines. Both sets of primers and probes were synthesized by TaKaRa, Dalian, China.

Table 5-19-3 Sequence of primers and probes for dq-PCR

Primers and probes	Sequence (5'-3')	Size	Identity of amplified region
Haplosporidium-1241	TGCATGGCCGTTCTTAGTTG	68 bp	18S ribosomal RNA
Haplosporidium-1308	GGCTGAGGTCCCGTTRGT		
Haplosporidium-1262T	ROX-TGGAGYGATYTGTCTGGTTAATTCCGWT-ECLIPSE		
Perkinsus 133	TTCAAAACGAAATTCCAAACTCT	69 bp	Internal transcribed spacer
Perkinsus 201	CTTCGCTGCGTCCTTCATC		
Perkinsus 158T	FAM-CGATGGATGCCTCGGCTCGAG-ECLIPSE		

Note: T, hydrolysis probe.

Duplex real-time PCR assay

The duplex reaction mix contained 1 × real time PCR Premix (Perfect Real Time PCR kit, TaKaRa, Dalian, China); 0.1 μM primer of *Haplosporidium*-1241, *Haplosporidium*-1308, and probe *Haplosporidium*-1262 T; 0.2 μM primer of *Perkinsus* 133, *Perkinsus* 201, and probe *Perkinsus* 158 T; and 2 μL of template. Distilled H_2O was added to bring the final volume to 20 μL. PCR amplification consisted as initial step of 95 ℃ for 30 s (activation of Taq DNA polymerase), followed by 50 cycles of 95 ℃ for 10 s and 60 ℃ for 30 s. Fluorescence was measured at the end of each 60 ℃ incubation. Results of the amplification, detection, and data analysis were performed with the Light Cycler 2.0 System software (Roche, Molecular Biochemical, Mannheim, Germany).

Sensitivity and specificity of the dq-PCR

The sensitivity of the developed dq-PCR was determined by using ten-fold dilutions (1×10^7 to 1×10^0 copies/μL) in triplicate of a genus-specific plasmid containing target genes of both protozoan parasites (*Haplosporidium*-pMD18-T and *Perkinsus*-pMD18-T); DNA extracted from uninfected oyster tissue was used to dilute the plasmid. The dqPCR was compared to the Office International des Epizooties (OIE) recommended conventional PCR for *Haplosporidium* spp.[26] and *Perkinsus* spp.[20] using tenfold dilutions of plasmids as described above. To determine the specificity of this dq-PCR, DNA from *Haplosporidium* spp., *Perkinsus* spp., *Marteilia refringens* (GX/O/1/09), *Toxoplasma gondii*, *Bonamia ostreae*, *Escherichia coli*, *Salmonella* spp., *Mykrocytos mackini*, Vbrio parahaemolyticus, as well as DNA isolated from uninfected oysters was also tested.

Interference assay and repeatability of the dq-PCR

Samples containing different concentrations of plasmids (*Haplosporidium*-pMD18-T and *Perkinsus*-pMD18-T), as noted in the sample list in Table 5-19-4, were tested by the dq-PCR to check for interference with reaction efficiency. To assess the intra- and inter-assay reproducibility, three samples with the same concentration of 1×10^6 copies/μL plasmid (which contained the *Haplosporidium* spp. and *Perkinsus* spp. specific genes) were analyzed simultaneously by the newly developed dq-PCR. The same assay was performed across three different days. The reproducibility was estimated by standard deviation (SD) and coefficient of variability (CV) of the quantification cycle (Cq) average.

Table 5-19-4 Samples used in interference assay

Sample	Concentrations of plasmid
1	*Haplosporidium* spp. (2×10^8 copies)+*Perkinsus* spp. (2×10^2 copies)
2	*Haplosporidium* spp. (2×10^2 copies)+*Perkinsus* spp. (2×10^8 copies)
3	*Haplosporidium* spp. (2×10^8 copies)
4	*Perkinsus* spp. (2×10^2 copies)
5	*Haplosporidium* spp. (2×10^2 copies)
6	*Perkinsus* spp. (2×10^8 copies)

Dq-PCR detection in field samples

A total of 420 oysters were gathered from each sampling point (Beihai, Guangxi in the coastal area in South China Sea, *Crassostrea rivularis*; and Qingdao, Shandong in the coastal area in the Yellow Sea of China, Pacific oyster) at commercially cultivated shellfish farms. The samples were tested by the dq-PCR and the routine PCR (as per OIE recommendations).

Artificial co-infection

In order to emulate mixed infection, we mixed the tissues of 48 *Perkinsus* spp.-positive oysters with *Haplosporidum* spp.-positive tissues from another set of 48 oysters (5 g from each group). We extracted DNA from the mixed tissues using the method describe above.

Results

The duplex real-time PCR assay for the detection of two different protozoan parasites of shellfish was optimized for simultaneous detection of *Haplosporidium* spp. and *Perkinsus* spp.

Sensitivity of dq-PCR

The amplification curve of *Haplosporidium* spp. was shown in the 610 nm channel, and the amplification curve of *Perkinsus* spp. was shown in the 530 nm channel. The sensitivity of the dq-PCR assay was 200 *Haplosporidium* spp. and 200 *Perkinsus* spp. template copies, respectively (Figure 5-19-2 A, B). The efficiency of the assay was 2.22 and 2.17, respectively, while the error rate was 0.034 1 and 0.031 5, respectively. The quantification cycle (Cq) obtained for the different 10-fold dilutions are showed in Table 5-19-5.When compared with conventional PCR results, dq-PCR sensitivity was 10 times higher (Figure 5-19-3). The dq-PCR standard curves are shown in Figure 5-19-4 A, B.

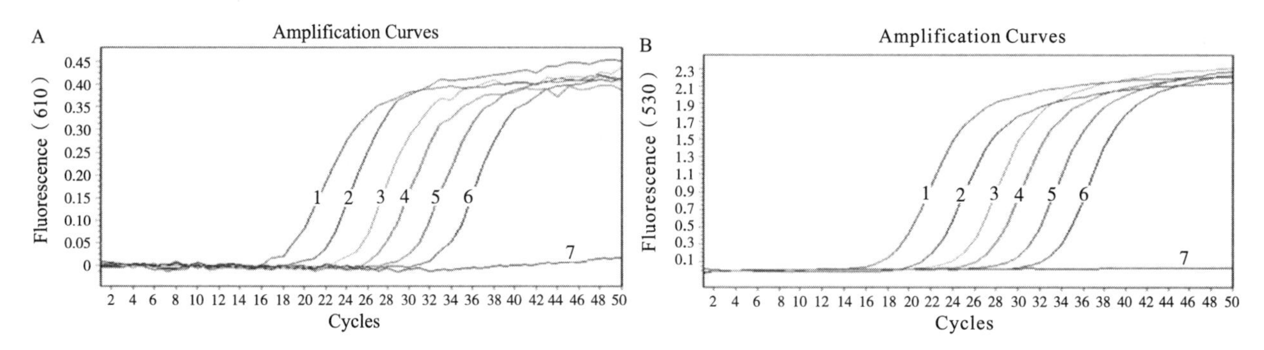

A: *Haplosporidium* spp.; B *Perkinsus* spp.. 1: 2×10^7 copies; 2: 2×10^6 copies; 3: 2×10^5 copies; 4: 2×10^4 copies; 5: 2×10^3 copies; 6: 2×10^2 copies; 7: Negative control.

Figure 5-19-2 Sensitivity of the duplex real-time PCR

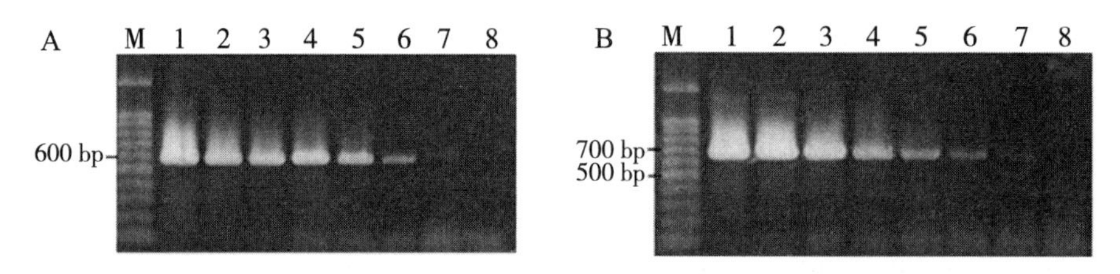

A: *Haplosporidium* spp.; B; *Perkinsus* spp.. 1: 2×10^9 copies; 2: 2×10^8 copies; 3: 2×10^7 copies; 4: 2×10^6 copies; 5: 2×10^5 copies; 6: 2×10^4 copies; 7: 2×10^3 copies; 8: 2×10^2 copies.

Figure 5-19-3 Sensitivity of conventional PCR for *Haplosporidium* spp. and *Perkinsus* spp.

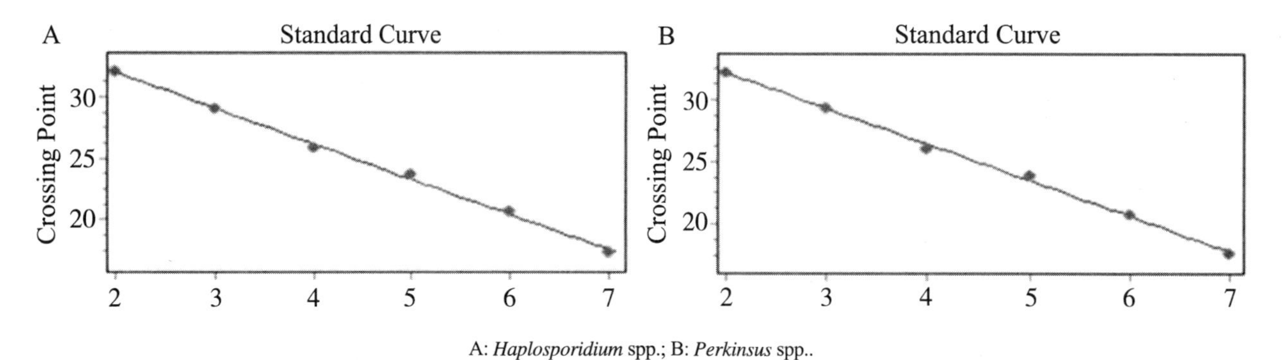

A: *Haplosporidium* spp.; B: *Perkinsus* spp..

Figure 5-19-4 Standard curve

Table 5-19-5 The Cq values of the plasmid tenfold dilution

Tenfold dilutions	2×10^7 copies	2×10^6 copies	2×10^5 copies	2×10^4 copies	2×10^3 copies	2×10^2 copies
Haplosporidium (Cq)	17.28	20.61	23.63	25.81	29.1	32.1
Perkinsus (Cq)	17.67	20.95	23.92	26.24	29.63	32.79

Specificity of dq-PCR

The dq-PCR was able to detect and differentiate between both types of protozoan parasites (Table 5-19-1; Figure 5-19-5 A and B). An amplification product curve was not displayed when samples other than *Haplosporidium* spp. and *Perkinsus* spp. were tested by dq-PCR. The sequences obtained for the amplified products of the dq-PCR aligned with *Haplosporidium* spp. and *Perkinsus* spp. when BLAST analysis was carried out to compare the findings with GenBank data. The amplification curve of *Haplosporidium* spp. was displayed in the 610 nm, different samples displayed only one amplification curve in the 610 nm channel when shellfish were infected only by *Haplosporidium* spp., and no amplification curve was displayed in the 530 nm channel. Only one amplification curve was displayed in the 530 nm channel when shellfish were infected only by *Perkinsus* spp., and no amplification curve was displayed in the 610 nm channel. However, two separate curves in the 610 and 530 nm channels were displayed when shellfish were infected by both parasite groups (amplification curve not shown). The results indicated that the dq-PCR was able to detect and differentiate between each parasite group in field samples infected with both parasites.

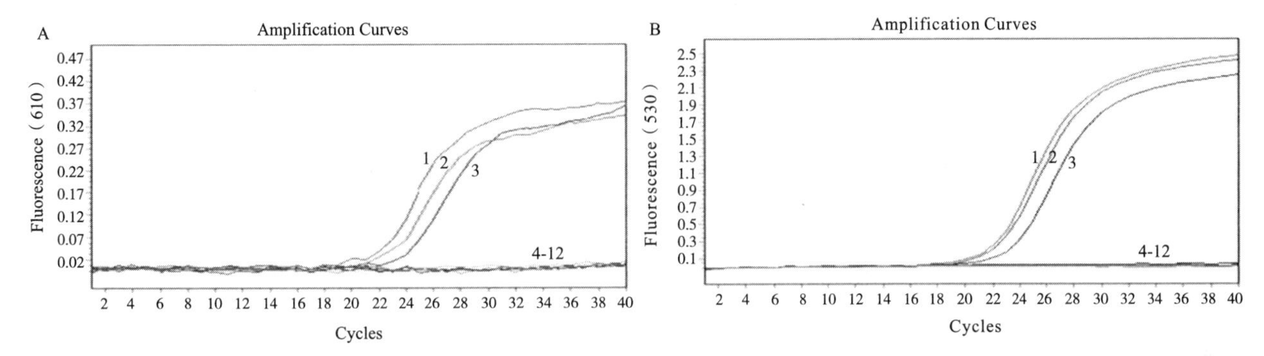

A: Specificity of *Haplosporidium spp.*. 1: *Haplosporidium nelson*; 2: *Haplosporidium nelsoni* plus *Perkinsus olseni*; 3: *Haplosporidium costale*; 4: *Perkinsus* spp.; 5: *Marteilia refringens*; 6: *Vibrio parahaemolyticus*; 7: *Bonamia* spp.; 8: *Mykrocytos mackini*; 9: *Escherichia coli*; 10: *Toxoplasma gondii*; 11: *Cymndinium* spp.; 12: Oyster tissue.

B: Specificity of *Perkinsus* spp.. 1: *Perkinsus olseni*; 2: *Perkinsus olseni* plus *Haplosporidium nelson*; 3: *Perkinsus beihaiensis*; 4: *Haplosporidium* spp.; 5: *Marreiliarefringens*; 6: *Vibrio parahaemolyticus*; 7: *Bonamia* spp.; 8: *Mykrocytos mackini*; 9: *Escherichia coli*; 10: *Toxoplasma gondii*; 11: *Cymndinium* spp.; 12: Oyster tissue.

Figure 5-19-5 Specificity results

Interference assay and repeatability

Since the presence of other templates could alter the efficiency of PCR amplification, different template mixes were tested using the dq-PCR. The results of these experiments showed that there were no systematic deviations in the amplification curves when comparing the signal template with the template mix; moreover, the coefficient of variation was less than 3% (data not shown). No difference in amplification efficiency was observed between the different samples, considering that they had the same concentration and Cq value (Table 5-19-6), as measured by the slopes of the amplification curves during the exponential phase. The results indicated that the sensitivity was not change, and there was a minimal effect of the mixed infection. The detection of three samples (same concentration) tested at the same time is shown in Figure 5-19-6; the coefficient of variation is 0.57% for *Haplosporidium* spp. and 0.78% for *Perkinsus* spp. In addition, the detection of samples with the same concentration noted at different times is shown in Table 5-19-7. The results indicate that the findings produced by the dq-PCR were reproducible.

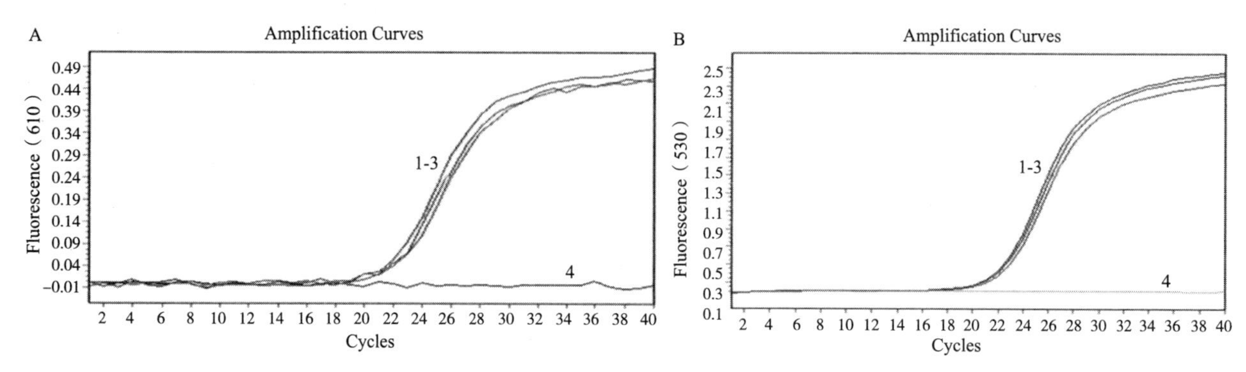

A: Repeatability of the dq-PCR for *Haplosporidium* spp. ; B: Repeatability of the dq-PCR for *Perkinsus* spp. ; 1-3: 2 x 10^6 copies; 4: Negative control.

Figure 5-19-6　Repeatability results

Table 5-19-6　The Cq values of different samples which have the same concentration

Sample	Cycle threshold				
	2×10^8 copies (*Haplosporidium*-pMD18-T+ *Perkinsus*-pMD18-T)	2×10^8copies (*Haplosporidium*-pMD18-T)	2×10^8 copies (*Perkinsus*-pMD18-T)	SD	CV
Haplosporidium	14.27	14.57		0.212	1.47%
Perkinsus	14.63		14.46	0.120	0.83%

Note: SD, Standard deviation; Cq, Quantification Cycle; CV, Coefficient of variation.

Table 5-19-7　The dq-PCR repeatability assays results

Sample	The Cq value/DNA copy number of the same sample in different time				
	Day 1st	Day 4th	Day 7th	SD	CV
Haplosporidium spp.	$20.61/2\times10^6$	$20.54/2\times10^6$	$20.78/2\times10^6$	0.123	0.60%
Perkinsus spp.	$20.75/2\times10^6$	$20.92/2\times10^6$	$21.10/2\times10^6$	0.175	0.84%

Note: SD, Standard deviation; Cq, Quantification Cycle; CV, Coefficient of variation.

Field samples

Detection rates of the dq-PCR assay for 820 cultivated oyster samples from two coastal regions of China are shown in Table 5-19-8. The detection rate of *Haplosporidium* spp. was 8.6% in the Qingdao, Shandong coastal area, whereas the *Perkinsus* spp. detection rate was 8.3% in the coastal region of Beihai, Guangxi. This detection rate was similar to that noted in the routine PCR performed according to the guidelines of the OIE. The range of Cq values for *Haplosporidium* spp. was between 17.8 and 28.53, and the concentration was between 1.71 x 10^7 copies/μL and 3.48 x 10^3 copies/μL. The range of Cq values of the *Perkinsus* spp. was between 20.25 and 30.96, and the concentration was between 3.76 x 10^6 copies/μL and 6.83 x 10^2 copies/μL. Only one sample indicated co-infection of *Haplosporidium* spp. and *Perkinsus* spp. The dq-PCR results suggested that *Haplosporidium* spp. was prevalent in oysters from Qingdao, Shandong and that *Perkinsus* spp. was prevalent in oysters from the Beihai, Guangxi coastal areas; moreover, co-infection by *Haplosporidium* spp. and *Perkinsus* spp. was not common in either Qingdao, Shandong or the Beihai, Guangxi coastal areas. The sequence of 10 positive dq-PCR products matched those of *Haplosporidium* spp. and *Perkinsus* spp. based on the sequence data from GenBank (data not shown).

Table 5-19-8　The detection rate of dq-PCR Oysters infected with *Perkinsus* spp. and *Haplosporidium* spp.

	Perkinsus spp.		*Haplosporidium* spp.	
	Positive samples/total samples	Positive rate	Positive samples/ Total samples	Positive rate
Beihai, Guangxi (*Crassostrea rivularis*)	35/420	8.30%	8/420	1.90%
Qingdao, Shandong (*Pacific oyster*)	13/420	3.10%	36/420	8.60%

Artificial co-infection

Two separate curves in the 610 nm channel and 530 nm channel were displayed separately by the dq-PCR (amplification curve not shown; detection results are shown in Table 5-19-1). Only one amplification curve in the 610 nm channel was displayed when the tested DNA was isolated from the tissue of 48 oysters that tested positive for *Haplosporidium* spp. as per the OIE recommended PCR method. In addition, an amplification curve was not displayed in the 530 nm channel. Only one amplification curve in the 530 nm channel was displayed when testing the DNA isolated from the tissue of 48 oysters that tested positive for *Perkinsus* spp. as per the OIE recommended PCR method. Moreover, an amplification curve was not displayed in the 610 nm channel. The results indicate that dq-PCR is able to detect the presence and differentiate between each parasite group across mixed samples infected with both parasites.

Discussion

The dq-PCR assay described here used PCR primers and hydrolysis probes targeting conserved regions of the genomes of *Haplosporidium* spp. and *Perkinsus* spp. One main advantage of this dq-PCR assay compared to other available tests is the fact that it is able to simultaneously detect and differentiate between two genera of protozoan parasites. By using this approach, it was possible to identify both pathogens in the same reaction vessel, with better detection sensitivity than that noted in conventional detection methods. When different concentrations of *Haplosporidium* spp. and *Perkinsus* spp. were mixed together, the newly developed dq-PCR was able to differentiate between the two protozoan parasite groups. This dq-PCR is especially useful because these protozoan parasite pathogens commonly cause infections in shellfish[9], with the copy numbers of *Haplosporidium* spp. and *Perkinsus* spp. in clinical samples being quite variable.

The duplex assay may be of value for simultaneous detection of shellfish in environmental samples. However, the duplex feature of this assay is optional; the two components can be utilized as single-target assays, or they can be combined into duplex assays without impacting the quality of the results. This makes this assay adaptable to circumstances that may not require the simultaneous detection of both protozoan parasite groups for diagnostic purposes.

Another important feature of this dq-PCR approach is the short turnaround time for results. The confirmatory result of the suspected presence of *Haplosporidium* spp. and *Perkinsus* spp. was obtained within 2 h of receiving the sample in the laboratory. The current methods for laboratory diagnosis of *Haplosporidium* spp. and *Perkinsus* spp. are labor-intensive and may lack the required sensitivity and speed to prevent the infection from worsening. In addition, this dq-PCR does not require the unique expertise involved in morphology-based tests, but it can be performed in any laboratory setting with adequate infrastructure for real-time PCR testing. With this in mind, the dq-PCR assay described here should be a valuable addition to existing laboratory methods for diagnosing *Haplosporidium* spp. and *Perkinsus* spp. infections.

Since the presence of other oligonucleotides, fluorescent probes and DNA template may alter the efficiency of PCR amplification[27]; different samples (2×10^8 copies of *Haplosporidium* spp. plus 2×10^2 copies of *Perkinsus* spp., and 2×10^2 copies of *Haplosporidium* spp. and 2×10^2 copies of *Perkinsus* spp.) were detected by the dq-PCR. The results showed that there was no systematic deviation in the amplification curves and that the presence of other oligonucleotides, through the use of fluorescent probes and a DNA template, did not interfere with PCR amplification in this dq-PCR assay.

Haplosporidium spp. have been reported worldwide from various shellfish samples[3, 5, 8, 18]. *Perkinsus* spp. has previously been reported in oysters of southern China and in the Manila clam, *Ruditapes philippinarum*, of the East China Sea coast[11, 28] . There have been no previous reports of *Haplosporidium* spp. in cultivated *Crassostrea rivularis* in the Beihai, Guangxi, coastal area of south China, or of *Perkinsus* spp. in cultivated Pacific oyster in Qingdao, Shandong, Yellow Sea of China coastal areas. The purpose of this study was to develop a dq-PCR to determine the prevalence of *Haplosporidium* spp. and *Perkinsus* spp. in the oyster population in China. The results suggested that *Haplosporidium* spp. was prevalent in cultivated Pacific oyster of the Qingdao, Shandong and that *Perkinsus* spp. was prevalent in the cultivated *Crassostrea rivularis* in Beihai, Guangxi; in addition, co-infection of *Haplosporidium* spp. and *Perkinsus* spp. was not common in either the Qingdao, Shandong or Beihai, Guangxi coastal areas. We also tested thousands of oysters from the Qingdao, Shandong region, and thousands of oysters from Beihai, Guangxi, according to the convectional PCR method (recommended by the OIE). The results of this investigation (data not shown) were similar regarding the detection rate to the newly developed dq-PCR method.

Only one sample was co-infected with *Haplosporidium* spp. and *Perkinsus* spp. according to the results using the dq-PCR method. Not more than one sample was co-infected in the thousands of oysters examined using the convectional PCR method (as per the OIE recommendations). This may have been the case because the species of oyster (Pacific oyster in the Qingdao region and the *Crassostrea rivularis* in the Beihai region) tested were not suitable for coinfection by *Haplosporidium* spp. and *Perkinsus* spp. In addition, *Haplosporidium* spp. and *Perkinsus* spp. may inhibit the propagation of other pathogens, although this idea requires further investigation.

In conclusion, we have described a fast and sensitive dq-PCR assay for the simultaneous detection of *Haplosporidium* spp. and *Perkinsus* spp. We anticipate that this assay, due to its speed and sensitivity, can be useful for the diagnosis of *Haplosporidium* spp. and *Perkinsus* spp, and for the ultimate control of protozoan parasite group pathogens in cultivated oyster populations.

References

[1] MACKIN J G, OWEN H M, COLLIER A. Preliminary note on the occurrence of a new protistan parasite, *Dermocystidium marinum n.* sp. in Crassostrea virginica (Gmelin). Science, 1950, 111(2883): 328-329.

[2] ANDREWS J D, HEWATT W G. Oyster mortality studies in Virginia. II. The fungus disease caused by *Dermocystidium marina* in oysters in Chesapeake Bay. Ecological Monographs, 1957, 27: 1-26.

[3] HASKIN HH FAU-STAUBER L A, STAUBER LA FAU-MACKIN J A, MACKIN J A. *Minchinia nelsoni n.* sp. *(Haplosporida,* Haplosporidiidae): causative agent of the Delaware Bay oyster epizoötic. Science, 1966, 153: 1414-1416.

[4] KLEINSCHUSTER S, J. P. Sub-clinical infection of oysters (*Crassostrea virginica*) (Gmelin 1791) from Maine by species of the genus Perkinsus (Apicomplexa). Journal of Shellfish Research, 1995, 14: 489-491.

[5] BARBER B J, LANGAN R, HOWELL T L. *Haplosporidium nelsoni* (MSX) epizootic in the Piscataqua River Estuary (Maine/

New Hampshire, USA.). J Parasitol, 1997, 83(1): 148-150.

[6] KWANG S, CHOI., KYUNG I, et al. Report on the Occurrence of Perkinsus sp. in the Manila Clams, Ruditapes philippinarum in Korea. Aquaculture, 1997(10): 227-237.

[7] HAMAGUCHI M, N. S, H. U, et al. *Perkinsus protozoan* infection in short-necked clam Tapes (Ruditapes) philippinarum in Japan. Fish Pathoogy, 1998(33): 473-480.

[8] SUNILA I, KAROLUS J, VOLKJ. A new epizootic of *Haplosporidia* (MSX), a haplosporidian oyster parasite, in Long Island Sound. J Shell Fish Res, 1999, 18: 169-174.

[9] RUSSELL S, FRASCA S, JR., SUNILA I, et al. Application of a multiplex PCR for the detection of protozoan pathogens of the eastern oyster *Crassostrea virginica* in field samples. Dis Aquat Organ, 2004, 59(1): 85-91.

[10] MOSS J A, BURRESON E M, CORDES J F, et al. Pathogens in *Crassostrea ariakensis* and other Asian oyster species: implications for non-native oyster introduction to Chesapeake Bay. Dis Aquat Organ, 2007, 77(3): 207-223.

[11] MOSS J A, XIAO J, DUNGAN C F, et al. Description of P*erkinsus beihaiensis n.* sp., a new *Perkinsus* sp. parasite in oysters of southern China. J Eukaryot Microbiol, 2008, 55(2): 117-130.

[12] PENNA M S, KHAN M, FRENCH R A. Development of a multiplex PCR for the detection of *Haplosporidium nelsoni, Haplosporidium costale* and *Perkinsus marinus* in the eastern oyster (*Crassostrea virginica*, Gmelin, 1971). Mol Cell Probes, 2001, 15(6): 385-390.

[13] RAY S M. A culture technique for the diagnosis of infection with *Dermocystidium marinum* Mackin, Owen and Collier in oysters. Science 1952, 116: 360-361.

[14] WOOD J L, ANDREWS J D. *Haplosporidium costale* (sporozoa) associated with a disease of Virginia oysters. Science 1962, 136: 710-711.

[15] ANDREWS J D, CASTAGNA M. Epizootiology of *Minchinia costalis* in susceptible oysters in seaside bays of Virginia 's eastern shore. Invert Pathol, 1978, 32: 1956-1976.

[16] BUSHEK D, S.E F, S.K. A. Evaluation of methods using Ray's fluid thiglycollate medium for diagnosis of *Perkinsus marinus* infection in the eastern oyster, *Crassostrea virginica*. Annual Review Fish Diseases, 1994, 4: 201-217.

[17] CASAS S M, LA PEYRE J F, REECE K S, et al. Continuous in vitro culture of the carpet shell clam Tapes decussatus protozoan parasite *Perkinsus atlanticus*. Disease of Aquatic Organism, 2002, 52: 217-231.

[18] RENAULT T, STOKES N A, BERTHE F, et al. *Haplosporidiosis* in the Pacific oyster *Crassostrea gigas* from the French Atlantic coast. Diseases of Aquatic Organisms, 2000, 42: 207-214.

[19] KLEEMAN S N, LE ROUX F, BERTHE F, et al. Specificity of PCR and in situ hybridization assays designed for detection of *Marteilia sydneyi* and *M. refringens*. Parasitology, (2002) 125: 131-141.

[20] AUDEMARD C, REECE S, BURRESEN E M. Development of real-time PCR for the detection and quantification of the protistan parasite *Perkinsus marinus*. Applied Environmental Microbiology 2004, 70: 6611-6618.

[21] BELL A S, RANFORD-CARTWRIGHT L C. Real-time quantitative PCR in parasitology. Trends Parasitol, 2002, 18(8): 337-342.

[22] DA SILVA A J, PIENIAZEK N J. Latest advances and trends in PCR-based diagnostic methods//DIONISIO D, Textbook-atlas of intestinal infections in AIDS. Milan: Springer-Verlag, 2003.

[23] MACKAY I M. Real-time PCR in the microbiology laboratory. Clin Microbiol Infect, 2004, 10(3): 190-212.

[24] XIE L, XIE Z, PANG Y, et al. Development of a multiplex PCR for detection of three kinds of protozoa in shellfish. Chin Vet Sci 2009, 39: 941-944.

[25] VAITOMAA J, RANTALA A, HALINEN K, et al. Quantitative real-time PCR for determination of microcystin synthetase e copy numbers for microcystis and anabaena in lakes. Appl Environ Microb, 2003, 69: 7289-7297.

[26] DAY J M, FRANKLIN D E, BROWN B L. Use of competitive PCR to detect and quantify *Haplosporidium nelsoni* infection (MSX disease) in the Eastern Oyster (*Crassostrea virginica*). Mar Biotechnol (NY), 2000, 2(5): 456-465.

[27] XIE Z, XIE L, PANG Y, et al. Development of a real-time multiplex PCR assay for detection of viral pathogens of penaeid

shrimp. Arch Virol, 2008, 153(12): 2245-2251.
[28] ZHANG X, LIANG Y, FAN J, et al. Identification of Perkinsus-like parasite in Manila clam, ruditapes philippinarum using DNA molecular marker at ITS region. Acta Oceanologica Sinica, 2005, 24: 139-144.

Part Ⅲ LAMP Detection Technology

Visual detection of H3 subtype avian influenza viruses by reverse transcription loop-mediated isothermal amplification assay

Peng Yi, Xie Zhixun, Liu Jiabo, Pang Yaoshan, Deng Xianwen, Xie Zhiqin, Xie Liji, Fan Qing, Feng Jiaxun, and Khan Mazhar I

Abstract

Background: Recent epidemiological investigation of different HA subtypes of avian influenza viruses (AIVs) shows that the H3 subtype is the most predominant among low pathogenic AIVs (LPAIVs), and the seasonal variations in isolation of H3 subtype AIVs are consistent with that of human H3 subtype influenza viruses. Consequently, the development of a rapid, simple, sensitive detection method for H3 subtype AIVs is required. The loop-mediated isothermal amplification (LAMP) assay is a simple, rapid, sensitive and cost-effective nucleic acid amplification method that does not require any specialized equipment. Results: A reverse transcription loop-mediated isothermal amplification (RT-LAMP) assay was developed to detect the H3 subtype AIVs visually. Specific primer sets target the sequences of the hemagglutinin (HA) gene of H3 subtype AIVs were designed, and assay reaction conditions were optimized. The established assay was performed in a water bath for 50 min, and the amplification result was visualized directly as well as under ultraviolet (UV) light reflections. The detection limit of the RT-LAMP assay was 0.1 pg total RNA of virus, which was one hundred-fold higher than that of RT-PCR. The results on specificity indicated that the assay had no cross-reactions with other subtype AIVs or avian respiratory pathogens. Furthermore, a total of 176 clinical samples collected from birds at the various live-bird markets (LBMs) were subjected to the H3-subtype-specific RT-LAMP (H3-RT-LAMP). Thirty-eight H3 subtype AIVs were identified from the 176 clinical samples that were consistent with that of virus isolation. Conclusions: The newly developed H3-RT-LAMP assay is simple, sensitive, rapid and can identify H3 subtype AIVs visually. Consequently, it will be a very useful screening assay for the surveillance of H3 subtype AIVs in underequipped laboratories as well as in field conditions.

Keywords

loop-mediated isothermal amplification, H3 subtype, avian influenza

---

Introduction

Influenza A viruses are classified into subtypes consisting of 16 hemagglutinin (HA) and 9 neuraminidase (NA) based on the antigenic differences of the HA and NA proteins, which are surface glycoproteins found on the viral envelope[1, 3]. While the viruses with all subtypes can be detected in wild aquatic waterfowl, only few of the subtypes' influenza A viruses can infect mammalian species[1]. Avian influenza viruses (AIVs) can be divided into two distinct groups based on their virulence: the highly virulent viruses, including H5-and H7-subtype AIVs, are called highly pathogenic avian influenza (HPAI) and cause high mortality and morbidity; all other subtypes are categorized as low pathogenic avian influenza (LPAI), which cause mild or no symptoms in birds[2, 4, 5].

It has been shown that LPAIVs from birds can be a potential source of reassortant human influenza A viruses. In 1957, the virus that caused the influenza pandemic was found to posses three genes from subtype H2N2 from an avian virus and all remaining genes from a circulating human H1N1 virus[1]. Moreover, the Hong Kong influenza virus (H3N2) in 1968 was a reassortant with avian (H3) PB1 and HA genes and six other genes from human (H2N2) virus[6].

Currently, there is no evidence that H5N1 HPAIVs can transmit widely between humans; the human-pandemic influenza strains are all generated by gene reassortment between human influenza viruses, avian influenza viruses and swine influenza viruses[7] . The influenza A viruses spreading in humans recently include seasonal H3N2, H1N1 and pandemic influenza A H1N1 viruses. It's notable that the results of a recent epidemiological investigation into different HA subtype AIVs shows that the H3 subtype is the predominant subtype among LPAIVs, and the seasonal variations in isolation of H3 subtype AIVs are consistent with that of human H3 subtype influenza viruses[8]. Some research predicts that H3 subtype AIVs will get the ability of infecting human directly through gene reassortment[9, 10]. Consequently, it is important to enhance surveillance for H3 subtype AIVs infections and to develop a simple, rapid and sensitive detection method. Loop-mediated isothermal amplification (LAMP) is a new nucleic acid amplification method that uses a set of six primers that target the eight site of a conserved gene, yielding a large white pyrophosphate ion by-product. The result can be viewed by adding a fluorescent metal indicator to the reaction mix before amplification[11-13]. Another significant advantage is that the assay can be performed in a water bath within 30～60 min. Therefore, LAMP is simple, rapid and specific enough to detect pathogenic microorganisms in field conditions, and currently the assay is being used widely for the detection of various viral pathogens[14-22].

The purpose of the study reported here was specifically to enhance the surveillance of the H3 subtype of AIVs by developing and optimizing the LAMP technique for the rapid detection of H3 subtype AIVs in live birds.

Methods

Viral strains and DNA/RNA extraction

The reference strains of AIVs, clinical samples and other avian respiratory pathogens used in this study are listed in Table 5-20-1 and 5-20-2. The Genomic DNA/RNA was extracted from 200 μL samples using a DNA/RNA Miniprep Kit (Axygen Biosciences, Hangzhou, China), according to the protocol suggested by the manufacturer. The DNA and RNA were eluted with 40 μL elution buffer and were stored at −70℃ immediately, until use.

Table 5-20-1 Clinic cloacal swab samples used to evaluate the feasibility of RT-LAMP assay

Number	Sources and identification	Virus isolation	RT-PCR	RT-LAMP
1	A/Duck/Guangxi/N42/2009 (H3N2)	+	+	+
2	A/Duck/Guangxi/M20/2009 (H3N2)	+	+	+
3	A/Duck/Guangxi/04D3/2009 (H3)	+	+	+

continued

Number	Sources and identification	Virus isolation	RT-PCR	RT-LAMP
4	A/Duck/Guangxi/04D8/2009 (H3)	+	+	+
5	A/Duck/Guangxi/017D5/2009 (H3)	+	+	+
6	A/Duck/Guangxi/LZD11/2009 (H3)	+	+	+
7	A/Duck/Guangxi/LZD15/2009 (H3)	+	+	+
8	A/Duck/Guangxi/LZD23/2009 (H3)	+	+	+
9	A/Duck/Guangxi/LZD27/2009 (H3)	+	+	+
10	A/Duck/Guangxi/LZD28/2009 (H3)	+	+	+
11	A/Duck/Guangxi/042D16/2009 (H3)	+	+	+
12	A/Duck/Guangxi/046D6/2009 (H3)	+	+	+
13	A/Duck/Guangxi/046D15/2009 (H3)	+	+	+
14	A/Duck/Guangxi/048D617/2010 (H3)	+	–	+
15	A/Duck/Guangxi/047D20/2010 (H3)	+	+	+
16	A/Duck/Guangxi/047D15/2010 (H3)	+	+	+
17	A/Duck/Guangxi/047D16/2010 (H3)	+	+	+
18	A/Duck/Guangxi/049D7/2010 (H3)	+	+	+
19	A/Duck/Guangxi/054D2/2010 (H3)	+	+	+
20	A/Duck/Guangxi/054D3/2010 (H3)	+	+	+
21	A/Duck/Guangxi/054D4/2010 (H3)	+	+	+
22	A/Duck/Guangxi/054D7/2010 (H3)	+	+	+
23	A/Duck/Guangxi/054D16/2010 (H3)	+	+	+
24	A/Duck/Guangxi/054D19/2010 (H3)	+	+	+
25	A/Duck/Guangxi/054D20/2010 (H3)	+	+	+
26	A/Chicken/Guangxi/055C2/2010 (H3)	+	+	+
27	A/Duck/Guangxi/057D19/2010 (H3)	+	+	+
28	A/Duck/Guangxi/057D6/2010 (H3)	+	+	+
29	A/Duck/Guangxi/WYD1/2010 (H3)	+	+	+
30	A/Duck/Guangxi/066D16/2010 (H3)	+	+	+
31	A/Duck/Guangxi/068D8/2010 (H3)	+	+	+
32	A/Chicken/Guangxi/068C3/2010 (H3)	+	+	+
33	A/Duck/Guangxi/070D8/2010 (H3)	+	–	+
34	A/Duck/Guangxi/070D14/2010 (H3)	+	–	+
35	A/Duck/Guangxi/072D5/2010 (H3)	+	+	+
36	A/Duck/Guangxi/072D4/2010 (H3)	+	+	+
37	A/Duck/Guangxi/072D2/2010 (H3)	+	+	+
38	A/Duck/Guangxi/072D7/2010 (H3)	+	+	+
39	A/Chicken/Guangxi/LF/2007	–	–	–
40	A/Duck/Guangxi/RX/2009	–	–	–
41	A/Chicken/Guangxi/015C7/2009	–	–	–
42	A/Duck/Guangxi/015D7/2009	–	–	–
43	A/Duck/Guangxi/011D3/2009	–	–	–
44	A/Duck/Guangxi/011D10/2009	–	–	–

continued

Number	Sources and identification	Virus isolation	RT-PCR	RT-LAMP
45	A/Duck/Guangxi/016D10/2009	–	–	–
46	A/Duck/Guangxi/016D8/2009	–	–	–
47	A/Duck/Guangxi/022D1/2009	–	–	–
48	A/Chicken/Guangxi/037C10/2009	–	–	–
49	A/Duck/Guangxi/LAD9/2009	–	–	–
50	A/Francolin/Nanning/018B–3/2010	–	–	–
51	A/Francolin/Nanning/020B–7/2010	–	–	–
52	A/Francolin/Nanning/022B12/2010	–	–	–
53	A/Francolin/Nanning/022B13/2010	–	–	–
54	A/Francolin/Nanning/022B7/2010	–	–	–
55	A/Chicken/Guangxi/DXC4/2010	–	–	–
56	A/Chicken/Guangxi/066C10/2010	–	–	–
57	A/Chicken/Guangxi/067C4/2010	–	–	–
58	A/Chicken/Guangxi/067C1/2010	–	–	–
59	A/Duck/Guangxi/068D18/2010	–	–	–
60	A/Duck/Guangxi/01D5/2009	–	–	–
61	A/Duck/Guangxi/02D6/2009	–	–	–
62	A/Duck/Guangxi/02D7/2009	–	–	–
63	A/Duck/Guangxi/04D5/2009	–	–	–
64	A/Duck/Guangxi/010D9/2009	–	–	–
65	A/Duck/Guangxi/011C5/2009	–	–	–
66	A/Duck/Guangxi/011C17/2009	–	–	–
67	A/Duck/Guangxi/011D5/2009	–	–	–
68	A/Duck/Guangxi/014D10/2009	–	–	–
69	A/Duck/Guangxi/014D15/2009	–	–	–
70	A/Duck/Guangxi/014D18/2009	–	–	–
71	A/Duck/Guangxi/016D7/2009	–	–	–
72	A/Duck/Guangxi/017C2/2009	–	–	–
73	A/Duck/Guangxi/017C12/2009	–	–	–
74	A/Duck/Guangxi/019C13/2009	–	–	–
75	A/Duck/Guangxi/026D10/2009	–	–	–
76	A/Duck/Guangxi/024C2/2009	–	–	–
77	A/Duck/Guangxi/024D2/2009	–	–	–
78	A/Duck/Guangxi/028D6/2009	–	–	–
79	A/Goose/Guangxi/027G10/2009	–	–	–
80	A/Duck/Guangxi/025D6/2009	–	–	–
81	A/Duck/Guangxi/026D21/2009	–	–	–
82	A/Duck/Guangxi/027D2/2009	–	–	–
83	A/Duck/Guangxi/033C1/2009	–	–	–
84	A/Duck/Guangxi/033C2/2009	–	–	–
85	A/Duck/Guangxi/033C3/2009	–	–	–

continued

Number	Sources and identification	Virus isolation	RT-PCR	RT-LAMP
86	A/Duck/Guangxi/033C4/2009	–	–	–
87	A/Duck/Guangxi/033C5/2009	–	–	–
88	A/Duck/Guangxi/033C7/2009	–	–	–
89	A/Duck/Guangxi/033C9/2009	–	–	–
90	A/Duck/Guangxi/033C10/2009	–	–	–
91	A/Duck/Guangxi/031G2/2009	–	–	–
92	A/Duck/Guangxi/031G11/2009	–	–	–
93	A/Duck/Guangxi/031G12/2009	–	–	–
94	A/Duck/Guangxi/042D1/2009	–	–	–
95	A/Duck/Guangxi/042D3/2009	–	–	–
96	A/Duck/Guangxi/014D4/2009	–	–	–
97	A/Duck/Guangxi/042D9/2009	–	–	–
98	A/Duck/Guangxi/042D10/2009	–	–	–
99	A/Duck/Guangxi/042D19/2009	–	–	–
100	A/Duck/Guangxi/044D6/2009	–	–	–
101	A/Duck/Guangxi/044D7/2009	–	–	–
102	A/Duck/Guangxi/044D8/2009	–	–	–
103	A/Duck/Guangxi/044D10/2009	–	–	–
104	A/Duck/Guangxi/044D12/2009	–	–	–
105	A/Duck/Guangxi/044D17/2009	–	–	–
106	A/Duck/Guangxi/044C4/2009	–	–	–
107	A/Duck/Guangxi/044C7/2009	–	–	–
108	A/Duck/Guangxi/041D10/2009	–	–	–
109	A/Duck/Guangxi/041D11/2009	–	–	–
110	A/Duck/Guangxi/041D13/2009	–	–	–
111	A/Duck/Guangxi/048D15/2010	–	–	–
112	A/Duck/Guangxi/048D18/2010	–	–	–
113	A/Duck/Guangxi/048D2/2010	–	–	–
114	A/Duck/Guangxi/048D16/2010	–	–	–
115	A/Duck/Guangxi/PXD7/2010	–	–	–
116	A/Duck/Guangxi/PXD5/2010	–	–	–
117	A/Duck/Guangxi/PXD15/2010	–	–	–
118	A/Duck/Guangxi/052D16/2010	–	–	–
119	A/Duck/Guangxi/052D14/2010	–	–	–
120	A/Duck/Guangxi/052D8/2010	–	–	–
121	A/Duck/Guangxi/052D2/2010	–	–	–
122	A/Duck/Guangxi/052D1/2010	–	–	–
123	A/Duck/Guangxi/049D1/2010	–	–	–
124	A/Duck/Guangxi/049D2/2010	–	–	–
125	A/Duck/Guangxi/049D32010	–	–	–
126	A/Duck/Guangxi/049D4/2010	–	–	–

continued

Number	Sources and identification	Virus isolation	RT-PCR	RT-LAMP
127	A/Duck/Guangxi/049D5/2010	–	–	–
128	A/Duck/Guangxi/049D8/2010	–	–	–
129	A/Duck/Guangxi/049D12/2010	–	–	–
130	A/Duck/Guangxi/049D15/2010	–	–	–
131	A/Duck/Guangxi/049D17/2010	–	–	–
132	A/Duck/Guangxi/049D18/2010	–	–	–
133	A/Duck/Guangxi/055D4/2010	–	–	–
134	A/Duck/Guangxi/055D5/2010	–	–	–
135	A/Duck/Guangxi/055D6/2010	–	–	–
136	A/Duck/Guangxi/055D7/2010	–	–	–
137	A/Duck/Guangxi/055D13/2010	–	–	–
138	A/Duck/Guangxi/055D20/2010	–	–	–
139	A/Duck/Guangxi/058D1/2010	–	–	–
140	A/Duck/Guangxi/058D6/2010	–	–	–
141	A/Duck/Guangxi/058D8/2010	–	–	–
142	A/Duck/Guangxi/058D12/2010	–	–	–
143	A/Duck/Guangxi/058D13/2010	–	–	–
144	A/Duck/Guangxi/058D15/2010	–	–	–
145	A/Duck/Guangxi/062D2/2010	–	–	–
146	A/Duck/Guangxi/062D3/2010	–	–	–
147	A/Duck/Guangxi/062D4/2010	–	–	–
148	A/Duck/Guangxi/062D5/2010	–	–	–
149	A/Duck/Guangxi/062D7/2010	–	–	–
150	A/Duck/Guangxi/062D8/2010	–	–	–
151	A/Duck/Guangxi/062D9/2010	–	–	–
152	A/Duck/Guangxi/062D10/2010	–	–	–
153	A/Duck/Guangxi/062D11/2010	–	–	–
154	A/Duck/Guangxi/062D12/2010	–	–	–
155	A/Duck/Guangxi/062D13/2010	–	–	–
156	A/Duck/Guangxi/062D14/2010	–	–	–
157	A/Duck/Guangxi/062D15/2010	–	–	–
158	A/Duck/Guangxi/062D17/2010	–	–	–
159	A/Duck/Guangxi/062D18/2010	–	–	–
160	A/Duck/Guangxi/062D19/2010	–	–	–
161	A/Duck/Guangxi/066D2/2010	–	–	–
162	A/Duck/Guangxi/066D5/2010	–	–	–
163	A/Duck/Guangxi/066D7/2010	–	–	–
164	A/Duck/Guangxi/066D10/2010	–	–	–
165	A/Duck/Guangxi/066D20/2010	–	–	–
166	A/Duck/Guangxi/BSD20/2010	–	–	–
167	A/Duck/Guangxi/BSD12/2010	–	–	–

continued

Number	Sources and identification	Virus isolation	RT-PCR	RT-LAMP
168	A/Duck/Guangxi/BSD18/2010	–	–	–
169	A/Duck/Guangxi/BSD23/2010	–	–	–
170	A/Duck/Guangxi/BSD38/2010	–	–	–
171	A/Duck/Guangxi/070D5/2010	–	–	–
172	A/Duck/Guangxi/072D14/2010	–	–	–
173	A/Duck/Guangxi/072D16/2010	–	–	–
174	A/Duck/Guangxi/072D13/2010	–	–	–
175	A/Duck/Guangxi/072D19/2010	–	–	–
176	A/Duck/Guangxi/072D1/2010	–	–	–

Table 5-20-2　Virus strains used for specificity of H3-RT-LAMP assay

Number	Virus strain	H3-RT-LAMP
1	A/Duck/Guangxi/030D/2009 (H1N1)	–
2	A/Mallard/Alberta/77 (H2N3)	–
3	A/Mallard/Alberta/85 (H3N6)	+
4	A/Duck/Guangxi/N42/2009 (H3N2)	+
5	A/Duck/Guangxi/027D/2009 (H4)	–
6	A/Turkey/GA/209092/02 (H5N2)	–
7	A/Turkey/CA/35621/84 (H5N3)	–
8	A/waterfowl/GA/269452-56/03 (H5N7)	–
9	A/Turkey/MA/40550/87-Bel42 (H5N1)	–
10	A/Turkey/WI/68 (H5N9)	–
11	A/Turkey/Ontario/63 (H6N8)	–
12	A/Chicken/NY/273874/03 (H7N2)	–
13	A/Turkey/Ontario/6118/67 (H8N4)	–
14	A/Duck/Guangxi/RX/09 (H9N2)	–
15	A/Turkey/MN/24838-590/79 (H10N7)	–
16	A/Chicken/Guangxi/43C/09 (H11)	–
17	A/Duck/Alberta/60/76 (H12N5)	–
18	A/Gull/MD/704/77 (H13N6)	–
19	A/Duck/Australia/341/83 (H15N8)	–
20	Newcastle disease virus (LaSota)	–
21	Infectious bronchitis virus (M41)	–
22	Infectious Laryngotracheitis virus	–
23	*Mycoplasma gallisepticum* (S6)	–

Design of primers for the RT-LAMP assay

According to the sequences of the H3 subtype AIVs' hemagglutinin (HA) gene available in GenBank (accession number: JN003630) and the sequences of viruses isolated in China, several primer sets of RT-LAMP assays were designed using the LAMP primer design software Primer Explorer V4. Finally, an optimal set was chosen after many comparable experiments to ensure the highest sensitivity and specificity of the RT-LAMP assay. The RT-LAMP primer set comprising two outer primers (forward primer F3 and backward primer B3), two inner primers (forward inner primer FIP and backward inner primer BIP), and two loop primers (forward loop primer LF and backward loop primer LB) recognized eight sites on the target sequence specific to the HA gene. The details for the primers are shown in Table 5-20-3 and Figure 5-20-1.

Table 5-20-3 Sequences of primers designed for RT-LAMP assay

Primer name	Sequence (5'-3')[a]	Genome position
FIP	GGATTATAGTCTGTTGGCTTCTCC-GAATTC-TGTATGTTCAAGCCTCA	658～720
BIP	GGCCAATCTGGCAGAATAAGC-GATATC-CCATTACTATTGATTACCAGTACGT	750～823
F3	CACAAATCAAGAACAAACCA	635～654
B3	CCGAGGAGCGATTAGGTT	825～842
LF	GGTAGAGACTGTGACTCT	678～695
LB	ATCTATTGGACAGTAGTCAAACCTG	771～795

Note: [a] The restriction enzyme site sequences of EcoR I and EcoR V were added in FIP and BIP primer respectively. Primers of F3, B3 were applied to the RT-PCR assay of H3 subtype AIVs.

632 F3 F2 LFc
AAGCACAAATCAAGAACAAACCAATCTGTATGTTCAAGCCTCAGGAAGAGTC
F1c
ACAGTCTCTACCAGGAGAAGCCAACAGACTATAATCCCAAACATTGGATCTAG
B1c LB
ACCCTGGGTAAGGGGCCAATCTGGCAGAATAAGCATCTATTGGACAGTAGTCA
B2c B3c
AACCTGGGGACGTACTGGTAATCAATAGTAATGGAAACCTAATCGCTCCTCGG

GGCTATTTCAAAACACGTATTGGGAAAAGCTCAATAATGAGATCAGATGCACCT
900
ATTG

Locations of primers binding sequences are underlined and boxed. The referenced sequence A/Duck/Guangxi/N42/2009 (H3N2) can be obtained in GenBank (accession number: JN003630).

Figure 5-20-1 Positions of RT-LAMP primers on HA gene of H3 subtype AIVs

Optimization of the RT-LAMP conditions

The RT-LAMP assay was carried out in a conventional water bath with 25 μL of reaction mixture (Table 5-20-4). The concentration of reactions component comprised primers (synthesized by Invitrogen), $MgSO_4$ (Sigma), Betaine (Sigma), deoxynucleoside triphosphate (dNTP), RNA template, RevertAid ™ M-MuLV Reverse Transcriptase (MBI Fermentas) and Bst DNA polymerase (New England Biolabs) were optimized. To visualize the reaction, 25 μmol/L Calcein (International Laboratory, USA) and 0.5 mmol/L $MnCl_2$ (International Laboratory, USA) had been added previously to the reaction mixture, as suggested by the related reports[13]. The amplification reaction was performed in a water bath at 59 ℃, 60 ℃, 61 ℃, 62 ℃, 63 ℃, 64 ℃, and 65℃ for 30, 45, and 60 min respectively to find the optimal temperature and time. Then the reaction was terminated by heating at 80 ℃ for 5 min. All of the experiments were repeated three times.

RT-PCR

Prior to this study, there was no official primer or procedure for RT-PCR to detect H3 subtyping of AIVs.

The outer primers of RT-LAMP (forward primer F3 and backward primer B3) that were specific to target sequence and had high priming efficiency were applied in RT-PCR assay of H3 subtype AIVs. The reaction condition of RT-PCR was optimized to get the highest sensitivity of amplification. Finally, the RT-PCR assay was carried out in a 50 μL reaction volume containing 10 mM of each dNTP, 5 μL of 5 × RT Reaction Buffer, 200 U of RevertAid™ M-MuLV Reverse Transcriptase, 5 μL of 10 × Taq Buffer, 5 U of Taq DNA polymerase (MBI Fermentas), 10 μM each of B3 and F3 primers (Table 5-20-3), and 1 μL of extracted RNA, DEPC-treated water up to 50 μL.

The reaction was performed in a Thermal Cycler (BIO-RAD) at 42 ℃ for 60 min and at 94 ℃ for 5 min; followed by 30 cycles of 94 ℃ for 40 s, 55 ℃ for 30 s, and 72 ℃ for 30 s; and a final extension for 10 min at 72 ℃. Then the amplified products were analyzed using 1% agarose gel electrophoresis.

Analysis of RT-LAMP assay products

For the visual inspection of the products of the RT-LAMP assay, white magnesium pyrophosphate precipitations in the tube were centrifuged at 8 000 × g for 5 min in a Biofuge centrifuge (Primo R, HERAEUS). Meanwhile, color changes of the reaction mixture were observed directly under daylight and UV light.

Agarose gel electrophoresis analysis and DNA sequencing

Five μL of the RT-LAMP assay product was analyzed using 1.5% agarose gel electrophoresis. To test the specificity of the RT-LAMP assay product, the RT-LAMP amplicons were first digested with EcoR I and EcoR V restriction enzymes using the suggested protocol and then analyzed with 1.5% agarose gel electrophoresis. The products digested by restriction enzymes were purified using an AxyPrep™ DNA Gel Extraction Kit and were sequenced by Invitrogen Company.

Sensitivity and specificity of RT-LAMP assay

The detection limit of RT-LAMP was determined by testing serial 10-fold dilutions of total RNA of A/Duck/Guangxi/N42/2009 (H3N2) respectively, and the same concentration samples were comparatively detected by RT-PCR. All of the experiments were repeated three times. The concentration range of total RNA in the diluted sample was 1 ng/tube to 1 fg/tube, measured by ultraviolet spectrophotometer (Beckman UV-800).

To evaluate the specificity of the primer set used for the RT-LAMP assay, DNA/RNA extracted from different subtype AIVs and from Newcastle disease virus (NDV), infectious bronchitis virus (IBV), infectious laryngotracheitis virus (ILTV), *Mycoplasma gallisepticum* (MG) were detected by RT-LAMP assay. The amplification results were analyzed using 1.5% agarose gel electrophoresis and the color changes of reaction mixture were inspected under daylight and UV light.

Detection of clinical specimens by RT-LAMP assay

A total of 176 cloacal swabs were collected from poultry at LBMs. The clinical specimens were prepared in a viral transport medium, which was made up of 0.05 M phosphate buffered saline (PBS) containing antibiotics of penicillin (10 000 units/mL), streptomycin (10 mg/mL), gentamycin (10 mg/mL), kanamycin (10 mg/mL) and 5% (v/v) fetal bovine serum. Cloacal swabs samples were injected into 9-to 11-day-old embryonated specific pathogen-free (SPF) chicken eggs as previously described[5]. Allantoid fluids were recovered 48～72 h after incubation for virus detection and titration as described[23]. All clinical specimens were tested by RT-LAMP, RT-PCR, and virus isolation, respectively.

Results

The optimal protocol of RT-LAMP assay and inspection of products

The RT-LAMP reaction was optimized in a 25 μL total reaction mixture and performed in a conventional waterbath for 45 min at 63 ℃ and then for 5 min at 80 ℃ to terminate the reaction. The details of optimize reaction solution of RT-LAMP assay are shown in Table 5-20-4.

The positive results of RT-LAMP assay showed typical ladder pattern by 1.5% agarose gel electrophoresis (Figure 5-20-2 A). Furthermore, a large amount of DNA was synthesized, yielding a large white magnesium pyrophosphate by-product, so the white magnesium pyrophosphate on the bottom of the tube could be inspected after centrifugation. Alternatively, a color change in the reaction solution could be seen with the naked eye: the solution changed from orange to green for a positive reaction and remained orange for the negative reaction (Figure 5-20-2 C). In addition, the positive sample showed a strong green fluorescence under UV light (Figure 5-20-2 D). The products of the RT-LAMP assay digested by EcoR I and EcoR V restriction enzymes displayed predicted fragments (Figure 5-20-2 B). These fragments were purified and then confirmed by sequencing analysis (date not shown).

Table 5-20-4 The reaction system of RT-LAMP assay

Reaction component	Reaction volume of H3-RT-LAMP assay
10×Thermopol Buffer	2.5 μL
$MgSO_4$	5 mmol/L
Betaine	1 mmol/L
F3	0.2 μmol/L
B3	0.2 μmol/L
primer FIP	1.6 μmol/L
BIP	1.6 μmol/L
LF	0.8 μmol/L
LB	0.8 μmol/L
dNTP	1.4 mmol/L
Calcein	25 μmol/L
$MnCl_2$	0.5 mmol/L
M-MuLV RTase	200 U
Bst DNA polymerase	8 U
Template RNA	2 μL
Nuclease-free water	Up to 25 μL

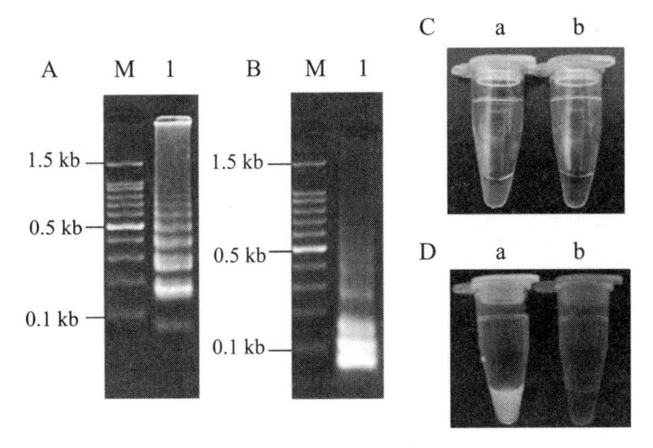

A: The products of RT-LAMP assay confirmed by 1.5% agarose gel electrophoresis; B: The products of RT-LAMP assay digested by EcoR Ⅰ and EcoR Ⅴ restriction enzyme; C: The result of RT-LAMP assay under daylight; D: The result of RT-LAMP assay under UV light. M: DNA marker; 1: Positive sample; a: Positive sample (green); b: Negative sample (color figure in appendix).

Figure 5-20-2 Detection of H3 subtype AIVs by RT-LAMP assay

Specificity and sensitivity of RT-LAMP assay for detection of H3 subtype AIVs

The viral DNA/RNA extracted from the different subtype AIVs and NDV, IBV, ILTV, MG were detected using the RT-LAMP assay to evaluate the specificity. The results of the negative sample showed no typical ladder pattern by 1.5% agarose gel electrophoresis and also had no color change detectable under daylight or UV light, which indicated that RT-LAMP assay had no cross reaction with other viruses (Table 5-20-2). The detection limit of RT-LAMP assay was 0.1 pg total RNA of virus, which was 100-fold higher than that of RT-PCR. The results of the visual detection of sensitivity tests are shown in Figure 5-20-3 A, B and C .

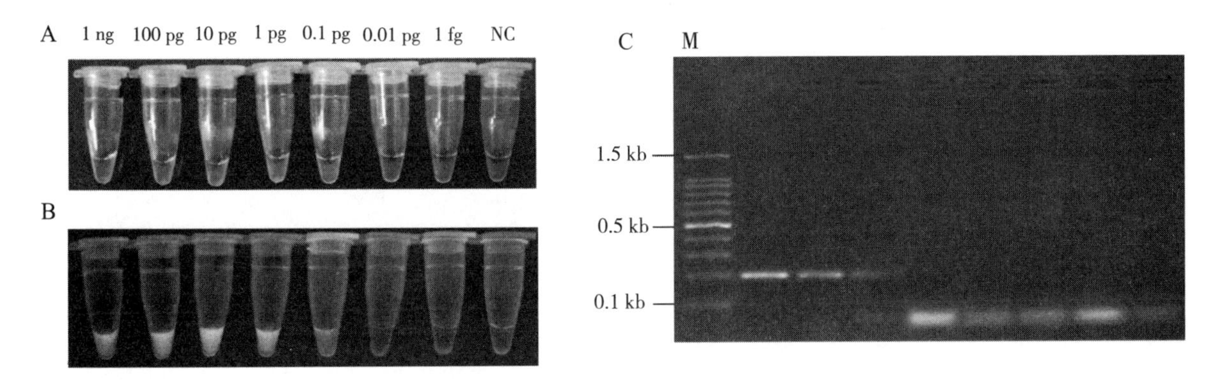

A: The result of sensitivity test for RT-LAMP assay under daylight; B: The result of sensitivity test for RT-LAMP assay under UV light. C: The result of sensitivity test for RT-PCR. The detection range of virus total RNA was 1 ng/tube to 1 fg/tube. M: DNA marker; NC: Negative control (color figure in appendix).

Figure 5-20-3 Comparative sensitivity tests between RT-LAMP and RT-PCR assays for detection of H3 subtype AIVs

Evaluation of RT-LAMP assay with clinical Samples

A total of 176 random cloacal swab samples were collected from poultry at various LBMs and were tested by RT-LAMP, RT-PCR, and virus isolation respectively (Table 5-20-1). The results of virus isolation showed that there were 38 positive samples of H3 subtype AIVs among the 176 cloacal swab samples, which was consistent with that of RT-LAMP assay; however, three positive samples were missed by RT-PCR. The results of statistical analysis showed that comparative detection of H3 subtype AIVs from cloacal swab samples by RT-LAMP was not statistically significant from RT-PCR and virus isolation (Table 5-20-5).

Table 5-20-5 Comparative detection of cloacal swab samples by virus isolation, RT-PCR and RT-LAMP

Results	Virus isolation [a]	RT-PCR [b]	RT-LAMP [c]
Positive	38	35	38
Negative	138	141	138
Total	176	176	176
Sig. (*P* value)	a-b: 0.696	b-a: 0.696	c-a: 1.00C
	a-c: 1.000	b-c: 0.696	c-b: 0.696

Note: [a] Virus isolation; [b] RT-PCR; [c] RT-LAMP. The significant differences among the three detection methods were analyzed by SPSS software. The default significance level is 0.05.

Discussion

The H3 subtype AIVs can provide genes for human influenza virus through gene reassortment, which raises great concerns in terms of its potential threat to human health[6]. Consequently, the development of a rapid, simple, sensitive detection method for H3 subtype AIVs is required. So far, there are several PCR-based methods being used to detect AIVs, but they all need precision instruments to amplify the nucleic acid of target sample; therefore, they can' t be applied in field conditions[24, 25]. The loop-mediated isothermal amplification (LAMP) assay is a nucleic acid amplification method developed by Notomi et al. That does not require any specialized equipment[11] and can be performed in a water bath or a heating block at an isothermal temperature between 60 ℃ to 65 ℃ within 30 to 60 min. In contrast, the conventional PCR method takes at least 1 h.

The LAMP assay relies on autocycling strand displacement DNA synthesis performed by Bst DNA polymerase with a high degree of strand displacement activity, which leads to an excellent sensitivity. The primer set of LAMP assay comprise two outer primers (forward primer F3 and backward primer B3), two inner primers (forward inner primer FIP and backward inner primer BIP), and two loop primers (forward loop primer LF and backward loop primer LB) and it recognizes eight sites on the target sequence specific to HA gene, so it' s specificity is very high. The viral RNA can be detected by the one-step RT-LAMP assay. When this is performed with reverse transcriptase and LAMP reaction mixture, the reverse transcription and DNA amplification can be accomplished at a constant temperature within 60 min.

During amplification, a large amount of white magnesium pyrophosphate is produced, which can be inspected directly with the naked eye. Alternatively, the result of the RT-LAMP assay can be visualized by adding fluorescence reagent (Calcein and $MnCl_2$) into the reaction mixture to observe the change of the reaction color under daylight or UV light[11-13]. Before the amplification reaction, calcein is combined with manganese ion to get the quenching effect, so the reaction solution is orange. Along with LAMP reaction processing, pyrophosphate ions remove manganese ions from calcein, resulting in greater fluorescence, which indicates the presence of the target gene[12].

In the present study, we first developed a rapid and sensitive RT-LAMP assay to visually detect H3 subtype AIVs. According to the sequences of the HA gene of H3 subtype AIVs available in GenBank and the sequences of H3 subtype AIVs isolated in China, we designed several sets of primers and finally chose the optimal set after many comparative experiments to ensure the high specificity and sensitivity of the RT-LAMP assay.

The proportions of the reagents in the reaction mixture were optimized to develop a stable RT-LAMP assay with high sensitivity. The detection limit of the RT-LAMP assay was 0.1 pg total RNA of virus, which was 100-fold higher than that of RT-PCR. The results of the specificity test showed that the assay had no cross reaction with other subtype AIVs and avian respiratory pathogens. Furthermore, a total of 176 cloacal swab samples collected from LBMs were tested using RT-LAMP, RT-PCR, and virus isolation respectively. The results showed that the clinical sensitivity of the RT-LAMP assay was consistent with virus isolation.

Although the RT-LAMP assay has many advantages over the similar nucleic acid amplification method, there is still a problem to note. Because of the high sensitivity of the RT-LAMP assay, a micro amount of RNA contamination in reagents, environment and instruments, such as the pipette, can result in a false positive. Moreover, the RT-LAMP assay has great amplification efficiency so the reaction product can form aerosol to contaminate the surroundings when opening the tube. To avoid the contamination in this study, the fluorescence reagent (Calcein and $MnCl_2$) were added to RT-LAMP reaction mixture before amplification in order to inspect the result of RT-LAMP assay directly with the naked eye.

Conclusions

In this study, the established RT-LAMP assay with high sensitivity was performed in a water bath within only 50 min, and the amplification results were visualized by adding fluorescence reagent. In summary, the newly developed assay can be used as one important method for detecting H3 subtype AIVs in field conditions with no need for specialized equipment.

References

[1] WEBSTER R G, BEAN W J, GORMAN O T, et al. Evolution and ecology of influenza A viruses. Microbiol Rev, 1992, 56(1): 152-179.

[2] HORIMOTO T, KAWAOKA Y. Pandemic threat posed by avian influenza A viruses. Clin Microbiol Rev, 2001, 14(1): 129-149.

[3] FOUCHIER R A, MUNSTER V, WALLENSTEN A, et al. Characterization of a novel influenza A virus hemagglutinin subtype (H16) obtained from black-headed gulls. J Virol, 2005, 79(5): 2814-2822.

[4] ALEXANDER D J. A review of avian influenza in different bird species. Vet Microbiol, 2000, 74(1-2): 3-13.

[5] EDWARDS S. OIE laboratory standards for avian influenza. Dev Biol (Basel), 2006, 124: 159-162.

[6] STROPKOVSKA A, MUCHA V, FISLOVA T, et al. Broadly cross-reactive monoclonal antibodies against HA2 glycopeptide of Influenza A virus hemagglutinin of H3 subtype reduce replication of influenza A viruses of human and avian origin. Acta Virol, 2009, 53(1): 15-20.

[7] KARASIN A I, CARMAN S, OLSEN C W. Identification of human H1N2 and human-swine reassortant H1N2 and H1N1 influenza A viruses among pigs in Ontario, Canada (2003 to 2005). J Clin Microbiol, 2006, 44(3): 1123-1126.

[8] PENG Y, ZHANG W, XUE F, et al. Etiological examination on the low pathogenicity avian influenza viruses with different HA subtypes from poultry isolated in eastern China from 2006 to 2008. Chin J Zoonoses, 2009, 25(2): 119-121.

[9] SONG M S, OH T K, MOON H J, et al. Ecology of H3 avian influenza viruses in Korea and assessment of their pathogenic potentials. J Gen Virol, 2008, 89(Pt 4): 949-957.

[10] CAMPITELLI L, FABIANI C, PUZELLI S, et al. H3N2 influenza viruses from domestic chickens in Italy: an increasing role for chickens in the ecology of influenza. J Gen Virol, 2002, 83(Pt 2): 413-420.

[11] NOTOMI T, OKAYAMA H, MASUBUCHI H, et al. Loop-mediated isothermal amplification of DNA. Nucleic Acids Res, 2000, 28(12): E63.

[12] MORI Y, NAGAMINE K, TOMITA N, et al. Detection of loop-mediated isothermal amplification reaction by turbidity derived from magnesium pyrophosphate formation. Biochem Biophys Res Commun, 2001, 289(1): 150-154.

[13] TOMITA N, MORI Y, KANDA H, et al. Loop-mediated isothermal amplification (LAMP) of gene sequences and simple

visual detection of products. Nat Protoc, 2008, 3(5): 877-882.

[14] ROVIRA A, ABRAHANTE J, MURTAUGH M, et al. Reverse transcription loop-mediated isothermal amplification for the detection of Porcine reproductive and respiratory syndrome virus. J Vet Diagn Invest, 2009, 21(3): 350-354.

[15] PHAM H M, NAKAJIMA C, OHASHI K, et al. Loop-mediated isothermal amplification for rapid detection of Newcastle disease virus. J Clin Microbiol, 2005, 43(4): 1646-1650.

[16] XU J, ZHANG Z, YIN Y, et al. Development of reverse-transcription loop-mediated isothermal amplification for the detection of infectious bursal disease virus. J Virol Methods, 2009, 162(1-2): 267-271.

[17] POON L L, LEUNG C S, CHAN K H, et al. Detection of human influenza A viruses by loop-mediated isothermal amplification. J Clin Microbiol, 2005, 43(1): 427-430.

[18] ITO M, WATANABE M, NAKAGAWA N, et al. Rapid detection and typing of influenza A and B by loop-mediated isothermal amplification: comparison with immunochromatography and virus isolation. J Virol Methods, 2006, 135(2): 272-275.

[19] IMAI M, NINOMIYA A, MINEKAWA H, et al. Development of H5-RT-LAMP (Loop-mediated isothermal amplification) system for rapid diagnosis of H5 avian influenza virus infection. Vaccine, 2006, 24(44-46): 6679-82.

[20] IMAI M, NINOMIYA A, MINEKAWA H, et al. Rapid diagnosis of H5N1 avian influenza virus infection by newly developed influenza H5 hemagglutinin gene-specific Loop-mediated isothermal amplification method. J Virol Methods, 2007, 141(2): 173-180.

[21] MA X J, SHU Y L, NIE K, et al. Visual detection of pandemic influenza A H1N1 Virus 2009 by reverse-transcription loop-mediated isothermal amplification with hydroxynaphthol blue dye. J Virol Methods, 2010, 167(2): 214-417.

[22] KUBO T, AGOH M, MAI LE Q, et al. Development of a reverse transcription-loop-mediated isothermal amplification assay for detection of pandemic (H1N1) 2009 virus as a novel molecular method for diagnosis of pandemic influenza in resource-limited settings. J Clin Microbiol, 2010, 48(3): 728-735.

[23] SWAYNE E D, GLISSON R J, JACKWOOD M W, et al. A laboratory manual for the isolation and identification of avian pathogens. Kennett Square, Pennsylvania: American Association of Avian Pathologist University of Pennsylvania, 1998.

[24] DAS A, SPACKMAN E, SENNE D, et al. Development of an internal positive control for rapid diagnosis of avian influenza virus infections by real-time reverse transcription-PCR with lyophilized reagents. J Clin Microbiol, 2006, 44(9): 3065-3073.

[25] TSUKAMOTO K, ASHIZAWA H, NAKANISHI K, et al. Subtyping of avian influenza viruses H1 to H15 on the basis of hemagglutinin genes by PCR assay and molecular determination of pathogenic potential. J Clin Microbiol 2008, 46(9): 3048-3055.

Rapid detection of Group Ⅰ avian adenoviruses by a loop-mediated isothermal amplification

Xie Zhixun, Tang Yi, Fan Qing, Liu Jiabo, Pang Yaoshan, Deng Xianwen, Xie Zhiqin, Peng Yi, Xie Liji, and Khan Mazhar I

Abstract

A loop-mediated isothermal amplification (LAMP) assay was optimized for the rapid detection of Group Ⅰ avian adenoviruses. A set of six primers was designed from the DNA sequences of hexon genes from Group Ⅰ avian adenovirus. The assay was performed in a water bath for 60 min at 63 ℃, and the amplification result was visualized by adding a fluorescence dye reagent or by inspecting the white sediment. The results showed that the LAMP assay could detect all 12 serotypes of Group Ⅰ avian adenovirus and nine Guangxi Group Ⅰ avian adenovirus isolates. This avian adenovirus Group Ⅰ-specific LAMP assay could detect 238 copies of avian adenovirus. No cross-reactions were detected using the LAMP assay with avian adenoviruses type Ⅱ and Ⅲ or with other avian viruses. The ability of LAMP to detect Group Ⅰ avian adenovirus isolates was further evaluated with 184 cloacal swab samples from poultry. In total, 72 out of 184 cloacal swab samples from poultry were identified as positive by LAMP, whereas 45 out of 184 were identified as positive by conventional PCR test. The Group Ⅰ avian adenovirus specific LAMP results were further confirmed by real-time PCR. This specific LAMP method holds promise as a rapid and specific diagnostic assay for detection of samples from birds suspected of adenovirus infection.

Keywords

avian, Group Ⅰ, adenoviruses, LAMP

Introduction

Avian adenovirus Group Ⅰ, which is classified into 12 serotypes, causes various diseases and syndromes in poultry and has a tremendous economic impact on poultry production worldwide[2, 3, 12]. Routine laboratory methods for detecting and characterizing Group Ⅰ avian adenoviral infection based on viral isolation and serodiagnosis are time-consuming and laborious[1, 7, 18]. Several investigators have reported molecular biology technology-based detection systems for rapid detection of avian adenoviral infection in clinical specimens. Nucleic acid-based assays hold greater promise for detecting Group Ⅰ avian adenoviral infection. Polymerase chain reaction (PCR), a highly sensitive and specific test, is considered the best option for detecting Group Ⅰ avian adenoviruses[6, 11, 19, 20]. Unfortunately, PCR assays require sophisticated equipment and cannot be implemented successfully outside a specialized laboratory.

The recently described loop-mediated isothermal amplification (LAMP) assay amplifies target DNA sequences with high sensitivity and can be completed within 1 h under isothermal conditions[13, 14]. Most importantly, the LAMP technique can amplify specific sequences of DNA under isothermal conditions of between 63 ℃ and 65 ℃. The amplified target DNA is visible to the naked eye[15]. The LAMP assay has been used successfully to detect many disease agents[5, 8, 9, 10, 16, 17]. However, no reports are available on application of the

LAMP method for detection of Group Ⅰ avian adenoviruses. The purpose of the study reported here was to develop and optimize the LAMP technique for rapid detection of Group Ⅰ avian adenoviruses.

Materials and Methods

Virus and tissue culture

Table 5-21-1 describes the 25 avian adenoviruses belonging to Group Ⅰ, five other avian pathogens, and five specific-pathogen-free (SPF) chickens swab samples, as a negative control, used in this study. All Groups Ⅰ, Ⅱ, Ⅲ serotypes, and nine known Group Ⅰ avian adenovirus field isolates from Guangxi, China were propagated in chicken embryo liver cells according to a previously described method[20]. The other avian pathogens were propagated in the allantoic cavity of 10-day-old SPF embryonated chicken eggs; the allantoic fluids were harvested and clarified by centrifugation according to a previously described method[20].

Table 5-21-1 Viruses used

Strains	Serotype	Source [a]	PCR [b]	LAMP [b]
			Agarose gel electrophoresis	Color change with dye
Group Ⅰ				
1.Avian adenovirus	Type 1	CVCC, China	+	+
2.Avian adenovirus	Type 2	CVCC, China	+	+
3.Avian adenovirus	Type 3	CVCC, China	+	+
4.Avian adenovirus	Type 4	CVCC, China	+	+
5.Avian adenovirus	Type 5	CVCC, China	+	+
6.Avian adenovirus	Type 6	CVCC, China	+	+
7.Avian adenovirus	Type 7	CVCC, China	+	+
8.Avian adenovirus	Type 8	CVCC, China	+	+
9.Avian adenovirus	Type 9	CVCC, China	+	+
10.Avian adenovirus	Type 10	CVCC, China	+	+
11.Avian adenovirus	Type 11	CVCC, China	+	+
12.Avian adenovirus	Type 12	CVCC, China	+	+
13.GX-AAV-1	Type 1	GVRI	+	+
14.GX-AAV-2	Type 1	GVRI	+	+
15.GX-AAV-3	Type 1	GVRI	+	+
16.GX-AAV-4	Type 1	GVRI	+	+
17.GX-AAV-5	Type 1	GVRI	+	+
18.GX-AAV-6	Type 1	GVRI	+	+
19.GX-AAV-7	Type 1	GVRI	+	+
20.GX-AAV-8	Type 1	GVRI	+	+
21.GX-AAV-9	Type 1	GVRI	+	+
Group Ⅱ				
22.Turkey hemorrhagic enteritis		GVRI	–	–
23.Marble spleen disease		GVRI	–	–
24.Avian adenovirus splenomegaly		GVRI	–	–

continued

Strains	Serotype	Source [a]	PCR [b]	LAMP [b]
			Agarose gel electrophoresis	Color change with dye
Group Ⅲ				
25.Egg drop syndrome		GVRI	–	–
Other avian viruses				
26.Infectious bursal disease virus		GVRI	–	–
27.Chicken anemia virus		GVRI	–	–
28.Avian reovirus		GVRI	–	–
29.Infectious bronchitis virus		GVRI	–	–
30.Infectious laryngotracheitis virus		GVRI	–	–
Negative control				
31.SPF bird 1		GVRI	–	–
32.SPF bird 2		GVRI	–	–
33.SPF bird 3		GVRI	–	–
34.SPF bird 4		GVRI	–	–
35.SPF bird 5		GVRI	–	–

Note: [a] CVCC, China Veterinary Culture Collection Center; GVRI, Guangxi Veterinary Research Institute. [b] +, positive; –, negative.

Extraction of DNA and RNA

DNA from chicken anemia virus, infectious laryngotracheitis virus, and avian adenoviruses were extracted using phenol : chloroform : isoamyl alcohol (1 : 1 : 24 v/v) according to the Gibco BRL manufacturer's protocol (Gibco BRL, Grand Island, NY) and RNA extractions from avian reovirus, infectious bursal disease viruses, and avian infectious bronchitis virus were performed according to the Trizol LS manufacturer's protocol (Invitrogen, Carlsbad, CA) as previously described[20]. Nucleic acid samples were stored at –20 ℃.

Design of Group Ⅰ avian adenovirus-specific LAMP primers

The oligonucleotide primers for LAMP of avian adenovirus Group Ⅰ isolates were designed based on the sequence of the hexon gene. The sequence of an avian adenovirus type Ⅰ strain was retrieved from GenBank (accession number: Z67970.1) and aligned with the available sequences of other isolates to identify the conserved regions by using LAMP support software (Primer Explored V4). A set of six primers comprising two outer, two inner, and two loop primers was designed. The two outer primers are known as the forward outer primer (F3) and the backward outer primer (B3) and are important for strand displacement. The inner primers are known as the forward inner primer (FIP) and the backward inner primer (BIP), and each has two distinct sequences corresponding to the sense and antisense sequences of the target, one for priming the first stage and the other for self-priming in later stages. FIP contained the F1c (complementary to F1) and F2 sequences. BIP contained the B1c (complementary to B1) and B2 sequences. FIP and BIP were high-performance liquid chromatography-purified primers. Two additional loop primers were designed to accelerate the amplification reaction as described[15]. The loop F and loop B primers were composed of sequences complementary to those between the F1 and F2, and the B1 and B2 regions, respectively. The sequences of the selected primers were compared to an alignment of hexon gene sequences. These primers were synthesized and purified by

Invitrogen Inc. (Dalin, China). Details of the oligonucleotide primers used for amplification of the hexon genes of Group Ⅰ avian adenoviruses are described in Figure 5-21-1.

For evaluation of sensitivity of the LAMP, a comparison was made between three methods: LAMP, conventional PCR, and real-time PCR (RT-PCR) using SYBR Green I (Solarbio Company, Beijing, China).

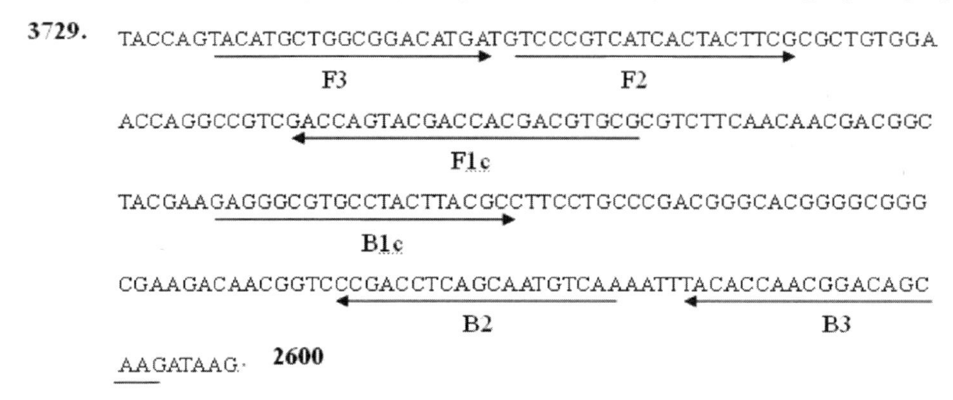

Primer name	sequence(5′-3′)	Genome position
FIP= F2+F1c	CACGTCGTGGTCGTACTGGTC -GTCCCGTCATCACTACTTCG	F1c,2445-2465; F2,2405-2424
BIP=B1c+B2	GAGGGCGTGCCTACTTACGC -TTGACATTGCTGAGGTCGG	B1c,2493-2512; B2,2554-2572
F3	TACATGCTGGCGGACATGA	2385-2403
B3	CTTGCTGTCCGTTGGTGTA	2577-2595
F Loop	GCCTGGTTCCACAGCGC	2424-2440
BLoop	TTCCTGCCCGACGGG	2515-2528

Figure 5-21-1 Details of oligonucleotide primers used for LAMP amplification of the hexon genes of Group Ⅰ avian adenoviruses

Conventional PCR

PCR was performed using two primers specific for the hexon gene, described previously[20], as follows: F1, 5'-CCCTCCCACCGCTAACCA-3' and F2, 5'-CACGTTGCCCTTATCTTGC-3' in accordance with the standard protocol[20]. PCR amplification was performed with the PCR master mix kit (Qiagen Inc., Beijing, China) by using 1 μL (20 ng) of DNA template and 50 pmol of each primer in a 25-μL reaction volume by following the manufacturer's protocol with the following cycling times and temperatures: 94℃ for 5 min and 30 cycles of 94 ℃ for 30 s, 55 ℃ for 30 s, and 72 ℃ for 30 s. After PCR was performed, a 10-μL portion was analyzed by agarose gel electrophoresis on a 1% agarose gel and the DNA was visualized by ethidium bromide staining.

RT-PCR

Real-time quantitative PCR with SYBR Green I was performed[19] using the specific primers (forward primer: 5'-GTGCCTACTTACGCCTTCCIG-3', reverse primer: 5'-CTTATCTTGCTGTCCGTTGGTG-3') to amplify the hexon gene. Individual RT-PCR reactions were performed using 10 μL of SYBR Premix Ex Taq (2×) buffer (TaKaRa Biotech, Dalian, China), 0.5 μL of 10 μM of each primer, 1 μL of DNA template, and water to make up a 20-μL volume reaction mixture. RT-PCR was performed in a Light Cyder 2.0 (Roche Diagnostics, Indianapolis, IN) using the following temperatures: initial reaction at 94 ℃ for 10 s and 40 cycles of 94 ℃ for 5 s, 59 ℃ for 10 s, and 72℃ for 10 s. Following amplification, a melting curve analysis of the amplified DNA was performed at temperatures between 65 ℃ and 95 ℃ with the temperature increasing at a rate of 0.1 ℃/s. At the end of each reaction, cycle threshold was manually set up at the level that reflected the best kinetic PCR parameters, acquiring melting curves through which the possibility of any nonspecific amplification in reactions was analyzed.

LAMP

The LAMP technique was standardized using various concentrations of primers, DNA/RNA templates of positive and negative controls, buffers, and salt concentrations in the LAMP reaction. The optimized 25-μL reaction mixture containing a 1 μL sample template, 1 μL (40 pmol) each of primers BIP and FIP, 1 μL (5 pmol) each of primers F3 and B3, 1 μL (20 pmol) each of primers Floop and Bloop, 2.5 μL of a 10 × reaction mixture (20 mM Tris-HCl, 10 mM KCl, 8 mM $MgSO_4$, 10 mM $(NH_4)_2SO_4$, 0.1% Tween® 20 (Amresco, Solon, OH), 1 M betaine, 8 U of Bst DNA polymerase (New England Biolabs, Beijing, China), and 1.4 mM of each dNTP), and 17.5 μL of distilled water. To determine the optimal temperature for the LAMP assay, the primers and sample mixtures were incubated at 63 ℃ for 20, 40, 60, or 80 min and, at the end of each incubation period, the reaction was terminated by heating at 80 ℃ for 2 min. The reaction temperature was optimized (60, 61.5, 63, 64.5, and 66 ℃) and LAMP was performed for a predetermined period.

Analysis of LAMP product

The LAMP end product (5 μL) was analyzed by gel electrophoresis with a 1% agarose gel and, simultaneously, 2 μL of 100 000 × SYBR Green (Solarbio, Beijing, China) nucleic acid was added to the tube after the reaction. Samples that turned yellow-green were considered positive while samples that turned orange were considered negative.

Evaluation of LAMP

To evaluate its analytic specificity, LAMP was performed on a panel of viral isolates from chickens. The panel included 12 serotypes of Group Ⅰ avian adenoviruses, nine Guangxi isolates of Group Ⅰ avian adenoviruses, three samples from Group Ⅱ , and one from Group Ⅲ of avian adenoviruses and five different avian DNA or RNA viruses. Five cloacal swab samples were collected from SPF chickens and used as a negative control (Table 5-21-1).

To evaluate the sensitivity of LAMP, the virus was serially diluted (10-fold), tested, and compared by conventional and RT-PCR assays. After the reaction, LAMP products were visualized by electrophoresis with a 1% agarose gel. The sensitivity and specificity were compared between assays. For intra- and interassay reproducibility evaluation of the LAMP method, known positive and negative swab samples from SPF chickens were tested in triplicate on three different days (Table 5-21-2).

Table 5-21-2 Result of LAMP reproducibility (agarose gelelectrophoresis)

Reproducibility type	Sample number	Intertest		
		1st day	4th day	7th day
Intratest	1-1	+	+	+
	1-2	+	+	+
	1-3	+	+	+

continued

Reproducibility type	Sample number	Intertest		
		1st day	4th day	7th day
Intratest	2-1	–	–	–
	2-2	–	–	–
	2-3	–	–	–
Intratest	3-1	+	+	+
	3-2	+	+	+
	3-3	+	+	+
Intratest	4-1	+	+	+
	4-2	+	+	+
	4-3	+	+	+
Intratest	5-1	–	–	–
	5-2	–	–	–
	5-3	–	–	–

Note: +, positive; –, negative.

Clinical specimens

After initial validation studies, we determined the reliability of this Group I adenovirus-specific LAMP as a method of viral DNA detection from clinical specimens. Cloacal swabs were randomly collected from 184 chickens at various live bird markets and from small poultry flocks known to be infected with adenoviruses in Guangxi, China and were processed and tested[19, 20].

Results and discussion

Flowing standardization and optimization of the LAMP assay, the optimal concentrations of primers (inner-outer-loop) were revealed to be with 40, 5, and 20 pmol, respectively, resulting in an 8 ∶ 1 ∶ 4 ratio. Detection of gene amplification was accomplished by agarose gel analysis and color change after adding dye at the end of the LAMP reaction (Figure 5-21-2, Figure 5-21-4). The LAMP assay could amplify the target sequences of the hexon genes of Group Ⅰ avian adenoviruses at 63 ℃ in 60 min, as observed by agarose gel electrophoresis. Amplification was observed as a ladder-like pattern on the gel, due to formation of a mixture of stem-loop DNAs with various stem lengths and cauliflower-like structures with multiple loops, formed by annealing between alternately inverted repeats of the target sequence in the same strand (Figure 5-21-2).

The Group Ⅰ avian adenovirus-specific LAMP specifically amplified all 12 serotypes, exhibiting no cross-reactivity with avian adenoviruses Groups Ⅱ and Ⅲ or other avian DNA or RNA viruses (Table 5-21-1, Figure 5-21-2). This specificity was confirmed by agarose gel electrophoresis and a color-change assay (Figure 5-21-3, Figure 5-21-4). We also determined the sensitivity of this method by using a 10-fold serial dilution of a known amount of virus. The detection limit of LAMP was 238 copies, whereas the detection limit of conventional and RT-PCR were 2 380 copies and 238 copies, respectively (Figure 5-21-3, Figure 5-21-4).

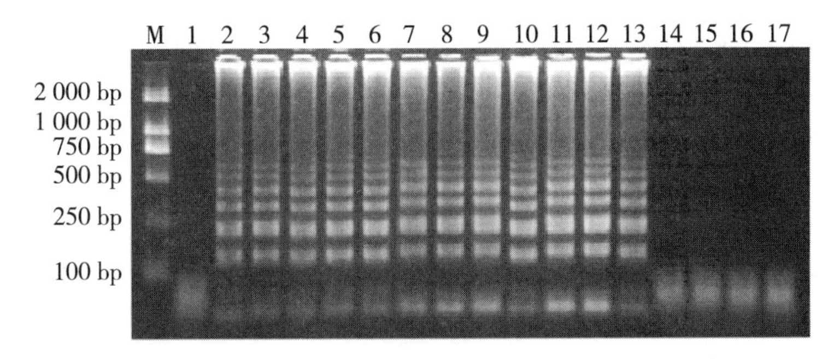

M: DNA marker; 1: Negative control from SPF chickens; 2: AAV1; 3: AAV2; 4: AAV3; 5: AAV4; 6: AAV5; 7: AAV6; 8: AAV7; 9: AAV8; 10: AAV9; 11: AAV10; 12: AAV11; 13: AAV12; 14-17: EDS, ARV, IBDV, and CAV, respectively.

Figure 5-21-2 Specificity of the LAMP assay for Group Ⅰ avian adenoviruses

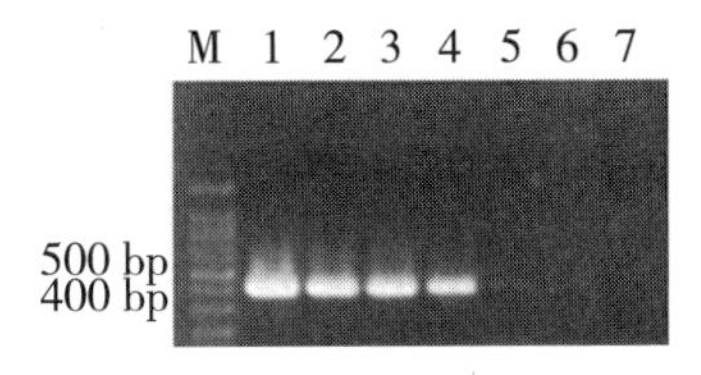

M: DNA marker; 1: 2.38×10^8 copies/tube; 2: 2.38×10^7 copies/tube; 3: 2.38×10^6 copies/tube; 4: 2.38×10^5 copies/tube; 5: 2.38×10^4 copies/tube; 6: 2.38×10^3 copies/tube; 7: 2.38×10^2 copies/tube.

Figure 5-21-3 Sensitivity of the LAMP assay for Group Ⅰ avian adenoviruses

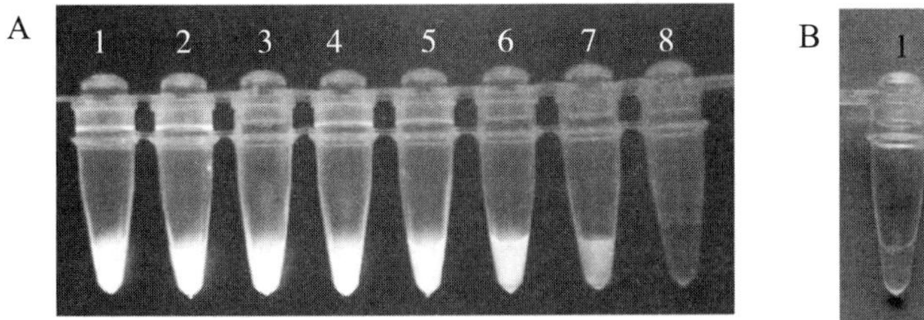

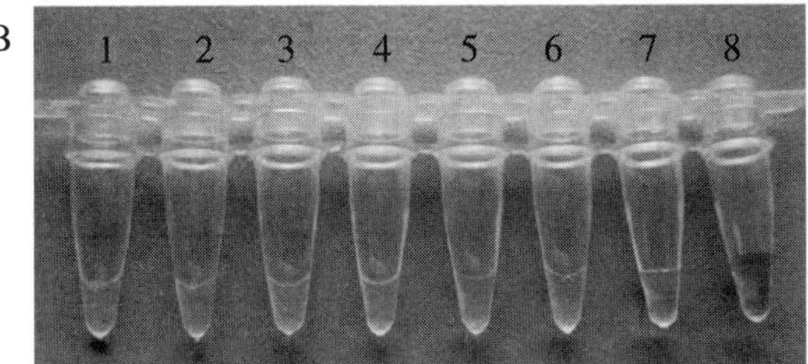

A: The result with ultraviolet light; B: The result without ultraviolet light; 1: 2.38×10^8 copies/tube; 2: 2.38×10^7 copies/tube; 3: 2.38×10^6 copies/tube; 4: 2.38×10^5 copies/tube; 5: 2.38×10^4 copies/tube; 6: 2.38×10^3 copies/tube; 7: 2.38×10^2 copies/tube; 8: 2.38×10^1 copies/tube.

Figure 5-21-4 LAMP analysis using SYBR Green Ⅰ

In comparison with conventional PCR and RT-PCR, using previously described procedures using specific primers[19, 20], 45 out of 184 cloacal swab samples were identified as positive by conventional PCR, whereas 74 out of 184 cloacal swab samples were identified as positive by RT-PCR and 72 out of 184 cloacal swab samples were identified as positive by LAMP. The only discordant samples were the 27 samples that were identified as positive by both LAMP and RT-PCR but negative by conventional PCR (Table 5-21-3).

Table 5-21-3 Comparison of conventional PCR, real-time PCR, and LAMP assays for detection of Group Ⅰ avian adenoviruses from chickens in Guangxi, China

Sample	PCR (+)	Real-time PCR		LAMP (+)
		Amplification curve (+)	Melting curve (+)	
184	45	74	74	72

Note: +, positive.

The results indicate that LAMP is more sensitive than conventional PCR. However, RT-PCR detection was similar to LAMP, indicating the reliability of LAMP. Similarly, the higher detection of Group Ⅰ avian adenovirus from the cloacal swabs from live bird markets and small poultry flocks, as compared to conventional PCR, further validates the usefulness of this technique. The LAMP technique offers several advantages over PCR analysis for detecting Group Ⅰ avian adenovirus-specific DNA[6, 11]. Firstly, the LAMP method amplified DNA with high efficiency under isothermal conditions without a significant influence of the copresence of nontarget DNA. We also report that LAMP was not inhibited by blood serum and plasma heparin, which are known to inhibit PCR amplification[4]. Its detection limit was 238 copies, being 10-fold higher in sensitivity than conventional PCR. Secondly, LAMP is highly specific for the target sequence. This is attributed to recognition of the target sequence by six independent sequences in the initial stage and by four independent sequences during the later stages of the LAMP reaction. This partly alleviates the general problem of backgrounds associated with all nucleic acid amplification method. Thirdly, the LAMP is simple and easy to perform once it is optimized, requiring only six primers, a Bst DNA polymerase, and a regular laboratory water bath. Therefore, we have described an avian adenovirus Group Ⅰ-specific LAMP assay that does not cross-react with avian adenovirus Group Ⅱ and Ⅲ viruses or other avian viruses. Due to its high amplification, the issue of cross-contamination could be a limiting factor as previously described[15].

In conclusion, the Group Ⅰ avian adenovirus-specific LAMP assay developed in this study is simple, rapid, and cost effective as well as sensitive and specific. This assay has potential usefulness for dinical diagnosis and surveillance of the samples from birds suspected of avian adenovirus infections in developing countries, as it does not require the use of sophisticated equipment.

References

[1] ADAIR B M, MCFERRAN J B, CALVERT V M. Development of a microtitre fluorescent antibody test for serological detection of adenovirus infection in birds. Avian Pathol, 1980, 9(3): 291-300.

[2] COWEN B S. Inclusion body hepatitis-anaemia and hydropericardium syndromes: aetiology and control. World Poult Sci J, 1992(48): 247-254.

[3] COWEN B S, ROTHENBACHER H, SCHWARTZ L D, et al. A case of acute pulmonary edema, splenomegaly, and ascites in guinea fowl. Avian Dis, 1988, 32(1): 151-156.

[4] DAHIYA S, SRIVASTAVA R N, HESS M, et al. Fowl adenovirus serotype 4 associated with outbreaks of infectious hydropericardium in Haryana, India. Avian Dis, 2002, 46(1): 230-233.

[5] ENOMOTO Y, YOSHIKAWA T, IHIRA M, et al. Rapid diagnosis of herpes simplex virus infection by a loop-mediated isothermal amplification method. J Clin Microbiol, 2005, 43(2): 951-955.

[6] GANESH K, SURYANARAYANA V V, RAGHAVAN R. Detection of fowl adenovirus associated with hydropericardium hepatitis syndrome by a polymerase chain reaction. Vet Res Commun, 2002, 26(1): 73-80.

[7] HESS M. Detection and differentiation of avian adenoviruses: a review. Avian Pathol, 2000, 29(3): 195-206.

[8] HONG T C, MAI Q L, CUONG D V, et al. Development and evaluation of a novel loop-mediated isothermal amplification method for rapid detection of severe acute respiratory syndrome coronavirus. J Clin Microbiol, 2004, 42(5): 1956-1961.

[9] IWAMOTO T, SONOBE T, HAYASHI K. Loop-mediated isothermal amplification for direct detection of Mycobacterium tuberculosis complex, M. avium, and M. intracellulare in sputum samples. J Clin Microbiol, 2003, 41(6): 2616-2622.

[10] KATO H, YOKOYAMA T, KATO H, et al. Rapid and simple method for detecting the toxin B gene of *Clostridium difficile* in stool specimens by loop-mediated isothermal amplification. J Clin Microbiol, 2005, 43(12): 6108-6112.

[11] MASE M, MITAKE H, INOUE T, et al. Identification of group I-III avian adenovirus by PCR coupled with direct sequencing of the hexon gene. J Vet Med Sci, 2009, 71(9): 1239-1242.

[12] MCFERRAN J B. Adenovirus (group I) infections of chickens//CALNEK B W. Diseases of poultry, 9th ed. Ames: Iowa State University Press, 1991: 553-563.

[13] NOTOMI T, NAGAMINE K, MORI Y, et al. DNA amplification: loop-mediated isothermal amplification (LAMP) of DNA analysis//Demidov V V, Broude N E. DNA amplification. Norfolk, VA: Horizon Bioscience, 2004: 119-121.

[14] NOTOMI T, OKAYAMA H, MASUBUCHI H, et al. Loop-mediated isothermal amplification of DNA. Nucleic Acids Res, 2000, 28(12): e63.

[15] NAGAMINE K, HASE T, NOTOMI T. Accelerated reaction by loop-mediated isothermal amplification using loop primers. Mol Cell Probes, 2002, 16(3): 223-229.

[16] PARIDA M, POSADAS G, INOUE S, et al. Real-time reverse transcription loop-mediated isothermal amplification for rapid detection of West Nile virus. J Clin Microbiol, 2004, 42(1): 257-263.

[17] PHAM H M, NAKAJIMA C, OHASHI K, et al. Loop-mediated isothermal amplification for rapid detection of Newcastle disease virus. J Clin Microbiol, 2005, 43(4): 1646-1650.

[18] SAIFUDDIN M, WILKS C R. Development of an enzyme-linked immunosorbent assay to detect and quantify adenovirus in chicken tissues. Avian Dis, 1990, 34(2): 239-245.

[19] WEN Y L, XIE. Z., Y. W. Development of a SYBR Green I real-time PCR assay for the detection of avian adenovirus group I. Chinee Vet Sci, 2008, 38: 753-756.

[20] XIE Z, FADL A A, GIRSHICK T, et al. Detection of avian adenovirus by polymerase chain reaction. Avian Dis, 1999, 43(1): 98-105.

Development of a reverse transcription loop-mediated isothermal amplification assay for visual detection of avian reovirus

Xie Zhixun, Peng Yi, Luo Sisi, Wang Ying, Liu Jiabo, Pang Yaoshan, Deng Xianwen, Xie Zhiqin, Xie Liji, Fan Qing, Teng Liqiong, and Wang Xiuqing

Abstract

Avian reovirus (ARV) is an important pathogen of poultry and causes significant economic losses to the poultry industry. To develop a rapid and sensitive method for the surveillance of ARV, a reverse transcription loop-mediated isothermal amplification (RT-LAMP) assay was established using a set of six primers specific to the Sl gene segment of ARV. The established assay was performed at 62 ℃ for 60 min in a thermal block, and the result was visualized directly under daylight or ultraviolet light. The detection limit of the RT-LAMP assay was 10 fg total RNA, which was 100-fold higher than that of reverse transcriptase polymerase chain reactions. The specificity of the assay was supported by the lack of cross-reaction with other avian pathogens. Furthermore, viral RNAs of field isolates were successfully detected by the assay. Overall, the newly established RT-LAMP assay is simple, rapid, sensitive, specific, and can visually detect ARV without the use of any specialized equipment.

Keywords

avian reovirus, S1 gene, reverse transcription loop-mediated isothermal amplification (RT-LAMP)

Introduction

Avian reovirus (ARV) belongs to the genus of *Orthoreovirus* of the Reoviridae. It is a non-enveloped, double-stranded RNA virus with 10 linear gene segments [1-3]. The 10 gene segments are named as large (L1, L2, L3), medium (M1, M2, M3), and small (S1, S2, S3, S4) based on their size[4]. ARV infects chicken, turkeys, ducks, and other avian species, and causes several disease syndromes including viral arthritis and tendosynovitis, stunting syndrome, respiratory disease, malabsorption syndrome, and diseases of the central nervous system[5-12]. Moreover, co-infections of ARV with other immune suppressive pathogens including avian reticuloendotheliosis virus, chicken anaemia virus, and avian leucosis virus lead to diminished weight gains, poor feed conversion and reduced marketability of affected birds, which results in considerable economic losses. To enhance surveillance for ARV infection, a simple, rapid, and sensitive detection method is urgently needed.

Several methods such as virus isolation, dot-blot hybridization, reverse transcriptase polymerase chain reaction (RT-PCR), RT-PCR restriction fragment length polymorphism, real-time RT-PCR, and enzyme-linked immunosorbent assay have been used in the diagnosis of ARV infection[13-19]. However, virus isolation is time-consuming and the other methods require specialized equipment, which makes them difficult to apply under field conditions for rapid detection of ARV. Loop-mediated isothermal amplification (LAMP) is a relatively new nucleic acid amplification method, which does not require any specialized equipment[20]. LAMP is performed in a water bath or thermal block within 30 to 60 min using a set of six primers targeting the eight sites of a highly conserved gene. The fluorescence reagents (calcein and $MnCl_2$) are added into the reaction mixture to observe the change of reaction colour under daylight or ultraviolet (UV) light, so the result can be inspected directly and immediately by naked eyes when the amplification is completed[21]. The viral RNA can be detected by one-step reverse transcription loop-mediated isothermal amplification (RT-LAMP) assay performed with reverse transcriptase and LAMP reaction mixture. Currently, the assay is being widely used for the detection of various pathogens including viruses, bacteria and parasites[22-28].

In the present study, a RT-LAMP assay was developed for the rapid detection of ARV using primers specific for the S1 gene segment of ARV. The established RT-LAMP assay was simple, sensitive and can be applied in diagnosis of ARV infection under field conditions.

Materials and methods

Virus strains and DNA/RNA extraction

Reference strains of ARV, ARV field isolates and other pathogens used in this study are listed in Table 5-22-1. ARV reference strains and field isolates were propagated in chicken embryo fibroblast cells. Virus titres were determined as described previously[29]. Avian influenza viruses were grown in 10-day-old specific-pathogen-free chicken embryos. Allantoic fluids were harvested and stored at –70 ℃ . The median egg infectious dose was determined by the Reed &Muench method[30]. Newcastle disease virus, infectious bronchitis virus, infectious laryngotracheitis virus, adenovirus, *Mycoplasma gallisepticum* and *Salmonella enteritidis* were obtained from the China Institute of Veterinary Drug Control (IVDC, Beijing, China) and used directly for DNA/RNA extraction.

The genomic DNA/RNA was extracted from 200 μL virus or bacteria stocks listed in Table 5-22-1 using a DNA/RNA Miniprep Kit (Axygen Biosciences, Hangzhou, China), according to the protocol suggested by the manufacturer. The DNA and RNA were eluted with 40 μL elution buffer. The concentrations of total DNA and RNA were measured by a UV spectrophotometer (Beckman UV-800, Beckman Coulter, Guangzhou, China). DNA and RNA samples were stored at –70 ℃ immediately until use.

Table 5-22-1 ARV strains and other avian pathogens used in the specificity test of RT-LAMP assay

Virus strain/other avian pathogen	Code	Source	Titre	RT-LAMP
ARV reference strain	S1133-1	UConn [a]	$10^{7.5}$ $TCID_{50}$/mL	+
ARV reference strain	S1133-2	IVDC [b]	$10^{8.0}$ $TCID_{50}$/mL	+
ARV reference strain	WVU2937	UConn	$10^{7.0}$ $TCID_{50}$/mL	+
ARV reference strain	1733	UConn	$10^{9.25}$ $TCID_{50}$/mL	+
ARV reference strain	203-M3	UConn	$10^{7.0}$ $TCID_{50}$/mL	+
ARV reference strain	C78	UConn	$10^{7.5}$ $TCID_{50}$/mL	+
ARV field isolate	R1	GVRI [c]	$10^{8.5}$ $TCID_{50}$/mL	+
ARV field isolate	R2	GVRI	$10^{7.3}$ $TCID_{50}$/mL	+
ARV field isolate	R3	GVRI	$10^{7.5}$ $TCID_{50}$/mL	+
ARV field isolate	R4	GVRI	$10^{7.9}$ $TCID_{50}$/mL	+
ARV field isolate	R5	GVRI	$10^{8.55}$ $TCID_{50}$/mL	+

continued

Virus strain/other avian pathogen	Code	Source	Titre	RT-LAMP
ARV field isolate	R6	GVRI	$10^{8.75}$ $TCID_{50}$/mL	+
ARV field isolate	R7	GVRI	$10^{8.4}$ $TCID_{50}$/mL	+
ARV field isolate	R8	GVRI	$10^{8.6}$ $TCID_{50}$/mL	+
ARV field isolate	R9	GVRI	$10^{8.26}$ $TCID_{50}$/mL	+
ARV field isolate	R10	GVRI	$10^{8.64}$ $TCID_{50}$/mL	+
ARV field isolate	Iso1	GVRI	$10^{7.85}$ $TCID_{50}$/mL	+
ARV field isolate	Iso2	GVRI	$10^{7.89}$ $TCID_{50}$/mL	+
ARV field isolate	Iso3	GVRI	$10^{7.5}$ $TCID_{50}$/mL	+
ARV field isolate	Iso4	GVRI	$10^{6.75}$ $TCID_{50}$/mL	+
ARV field isolate	Iso5	GVRI	$10^{8.0}$ $TCID_{50}$/mL	+
ARV field isolate	Iso6	GVRI	$10^{8.25}$ $TCID_{50}$/mL	+
ARV field isolate	Iso7	GVRI	$10^{8.5}$ $TCID_{50}$/mL	+
ARV field isolate	Iso8	GVRI	$10^{7.25}$ $TCID_{50}$/mL	+
ARV field isolate	Iso9	GVRI	$10^{8.3}$ $TCID_{50}$/mL	+
ARV field isolate	Iso10	GVRI	$10^{7.92}$ $TCID_{50}$/mL	+
ARV field isolate	Iso11	GVRI	$10^{7.5}$ $TCID_{50}$/mL	+
H5 subtype AIV	A/Turkey/GA/209092/02 (H5N2)	UConn	$10^{6.0}$ EID_{50}/mL	–
H9 subtype AIV	A/Duck/Guangxi/RX/09 (H9N2)	GVRI	$10^{7.0}$ EID_{50}/mL	–
Newcastle disease virus	LaSota	IVDC	$10^{9.0}$ EID_{50}/mL	–
Infectious bronchitis virus	M41	IVDC	$10^{8.3}$ EID_{50}/mL	–
Infectious laryngotracheitis virus	Beijing	IVDC	$10^{5.0}$ EID_{50}/mL	–
Adenovirus	CELO	IVDC	$10^{6.0}$ $TCID_{50}$/mL	–
M. gallisepticum	S6	IVDC	$10^{8.0}$ CFU/mL	–
S. enteritidis	203	IVDC	$10^{8.0}$ CFU/mL	–

Note: EID_{50}, 50% egg infectious dose; $TCID_{50}$, 50% tissue culture infectious dose; AIV, avian influenza virus. [a] University of Connecticut; [b] China Institute of Veterinary Drug Control; [c] Guangxi Veterinary Research Institute.

Primer design for the RT-LAMP assay

Based on the sequence of the S1 gene segment of ARV available in GenBank (accession number: AF330703.1), a primer set of RT-LAMP assay was designed using the LAMP primer design software, Primer Explorer V4. The RT-LAMP primer set comprised two outer primers (forward primer F3 and backward primer B3), two inner primers (forward inner primer FIP and backward inner primer BIP) and two loop primers (forward loop primer LF and backward loop primer LB). Since each inner primer FIP (F1c+F2) / BIP (B1c+B2) consists of two sequences, which target two specific sites, the six primers recognized eight sites on the target sequence specific to the S1 gene segment. The details of the primers are shown in Table 5-22-2 and Figure 5-22-1. All primers were synthesized by Invitrogen (Guangzhou, China).

Table 5-22-2 Sequences of primers used in the RT-LAMP assay

Primer name	Sequence (5'-3')	Genome position
FIP (F1c + F2)	ACCATGACTCGCGGTAGACTCC-ATCTCGTCGAACGGTTTAGC	1 032 to 1 051
BIP (B1c + B2)	TTCGCCTCCGCTTAGTGTCG-ACAGAAGTAGGGGTCCATGT	1 115 to 1 134
F3	CTCCACTGCCATCTCCAATT	1 001 to 1 020
B3	CCTCCGCCGAGTATGATGT	1 194 to 1 212
LF	TAACACGATCCTGCAGATCTGTAAT	1 053 to 1 077
LB	CTGACGGCGTGGTTTCATTAG	1 135 to 1 155

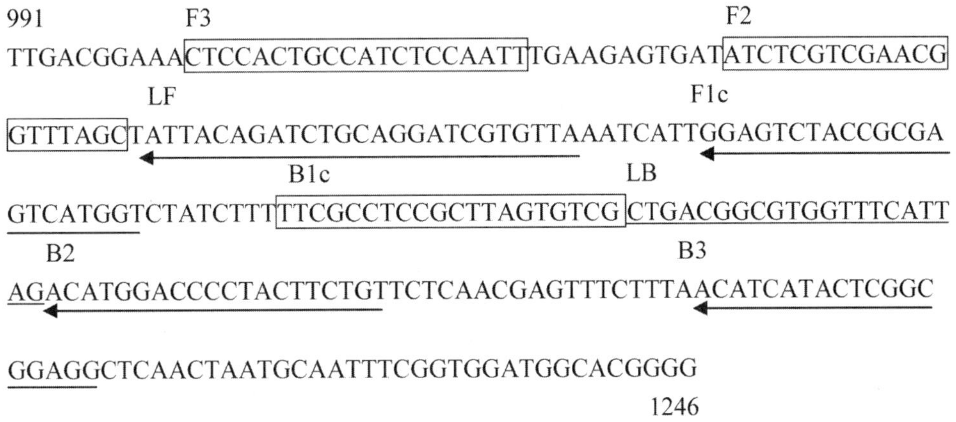

991 F3 F2
TTGACGGAAACTCCACTGCCATCTCCAATTTGAAGAGTGATATCTCGTCGAACG
LF F1c
GTTTAGCTATTACAGATCTGCAGGATCGTGTTAAATCATTGGAGTCTACCGCGA
B1c LB
GTCATGGTCTATCTTTTTCGCCTCCGCTTAGTGTCGCTGACGGCGTGGTTTCATT
B2 B3
AGACATGGACCCCTACTTCTGTTCTCAACGAGTTTCTTTAACATCATACTCGGC

GGAGGCTCAACTAATGCAATTTCGGTGGATGGCACGGGG
1246

Locations of primer binding sequences are underlined and boxed. The sequences of primers LF, F1c, B2, and B3 are reverse and complementary to the Genbank sequence. (GenBank accession number: AF330703.1)

Figure 5-22-1 Positions of RT-LAMP primers on the S1 gene segment of V S1133

Optimization of the RT-LAMP conditions

The RT-LAMP assay was performed in tubes containing 10 × Thermopol Buffer (New England Biolabs, Beijing, China), Bst DNA polymerase (New England Biolabs), $MgSO_4$ (Sigma, Shanghai, China), betaine (Sigma), primers, dNTP (MBI Fermentas, Shenzhen, China), calcein (International Laboratory, South San Francisco, CA, USA), $MnCl_2$ (International Laboratory), RevertAid^{TM} M-MuLV Reverse Transcriptase (MBI Fermentas), template RNA, and nuclease-free water (MBI Fermentas). Based on the previous studies, the different combinations of various concentrations of each component were tested for amplification efficiency. The amplification reaction was performed in a thermal block at 59 ℃, 60 ℃, 61 ℃, 62 ℃, 63 ℃, 64 ℃, and 65 ℃ for 30 min, 45 min, and 60 min, respectively, to find the optimal temperature and time. The reaction was then terminated by heating at 80 ℃ for 5 min. All of the experiments were repeated three times.

Analysis of RT-LAMP products

For the visual inspection of the products of the RT-LAMP assay, the fluorescence reagents (calcein and $MnCl_2$; International Laboratory) were added into the reaction mixture before amplification and colour changes of the reaction mixture were observed directly under daylight and UV light (365 nm). Alternatively, 5 μL of the RT-LAMP products were analysed using 1.5% agarose gel electrophoresis.

Specificity of the RT-LAMP assay for ARV

To evaluate the specificity of the RT-LAMP assay, DNA/RNA extracted from different reference strains and field isolates of ARV and from H5 subtype avian influenza virus, H9 subtype avian influenza virus, Newcastle disease virus, infectious bronchitis virus, infectious laryngotracheitis virus, adenovirus, *M. gallisepticum* and *S. enteritidis* were detected by LAMP or RT-LAMP assay. The amplification results were analysed using 1.5% agarose gel electrophoresis and the colour changes of the reaction mixture were inspected under daylight and UV light.

Sensitivity of the RT-LAMP assay for ARV RNA

The detection limit of the RT-LAMP assay was determined by testing a series of 10-fold dilutions of ARV S1133 RNA, and the same concentration of ARV S1133 RNA was also subjected to RT-PCR. The concentration of total RNA in the diluted samples ranged from 10 ng/tube to 1 fg/tube, as measured by a UV spectrophotometer (Beckman UV-800). The amplification results were analysed using 1.5% agarose gel electrophoresis and the colour changes of the reaction mixture were inspected under daylight and UV light. The experiments were repeated three times.

Reverse transcriptase polymerase chain reaction

The primers of RT-PCR specific to the Sl gene segment of ARV have been described previously[14]. The RT-PCR assay was carried out in a 50 μL reaction volume containing 10 mM of each dNTP, 5 μL of 5 × RT reaction buffer, 200 U RevertAid™ M-MuLV Reverse Transcriptase (MBI Fermentas), 5 μL of 10 × Taq buffer, 5 U of Taq DNA polymerase (MBI Fermentas), 10 μM of each RT-PCR primer, and 1 μL of extracted RNA, with Diethylpyrocarbonate-treated water up to 50 μL. The reaction was performed in a Thermal Cycler (BIO-RAD, Hercules, CA, USA) at 42 ℃ for 60 min and at 94 ℃ for 5 min; followed by 35 cycles of 94 ℃ for 1 min, 55 ℃ for 1 min, and 72 ℃ for 1 min; and a final extension for 10 min at 72 ℃. The amplified products were then analysed using 1% agarose gel electrophoresis.

Results

Optimization of the RT-LAMP assay

After optimization of the reaction conditions, the RT-LAMP assay was carried out in a 25 μL total reaction mixture containing 2.5 μL of 10 × Thermopol Buffer (20 mM Tris-HCl, 10 mM $(NH_4)_2SO_4$, 10 mM KCl, 2 mM $MgSO_4$, 0.1% Triton X-100), 5 mmol/L $MgSO_4$, 1 mmol/L betaine, 0.2 μmol/L F3 primer, 0.2 μmol/L B3 primer, 1.6 μmol/L FIP, 1.6 μmol/L BIP, 0.8 μmol/L LF primer, 0.8 μmol/L LB primer, 1.4 mmol/L dNTP, 25 μmol/L calcein, 0.5 mmol/L $MnCl_2$, 200 U RevertAid™ M-MuLV Reverse Transcriptase, 8 U Bst DNA polymerase, 2 μL template RNA, and nuclease-free water up to 25 μL. The reaction was amplified in a conventional thermal block for 60 min at 60 ℃ and then for 5 min at 80 ℃ to terminate the reaction.

Specificity and sensitivity of the RT-LAMP assay

Six ARV reference strains and 21 ARV field isolates were used in the RT-LAMP assay. Three detection methods were used to visualize the specific amplified products. All of these known ARV RNAs were amplified successfully by the RT-LAMP assay and showed positive results (Table 5-22-1). As shown in lanes 9 to 17 of

Figure 5-22-2 A, a typical DNA ladder pattern was observed in 1.5% agarose gel electrophoresis of the RT-LAMP products. Under daylight, these positive reactions showed green colour (Figure 5-22-2 B, lanes 9 to 17), and a strong green fluorescent colour was observed under UV light (Figure 5-22-2 C, lanes 9 to 17). In contrast, all the negative control samples listed in Table 5-22-1 had no colour change and remained orange under daylight (Figure 5-22-2 B, lanes 1 to 8) or UV light (Figure 5-22-2 B, lanes 1 to 8), suggesting that the RT-LAMP assay is highly specific to ARV. This was further supported by the lack of the typical ladder pattern of amplification products of ARV in all negative control samples (Figure 5-22-2 A, lanes 1 to 8).

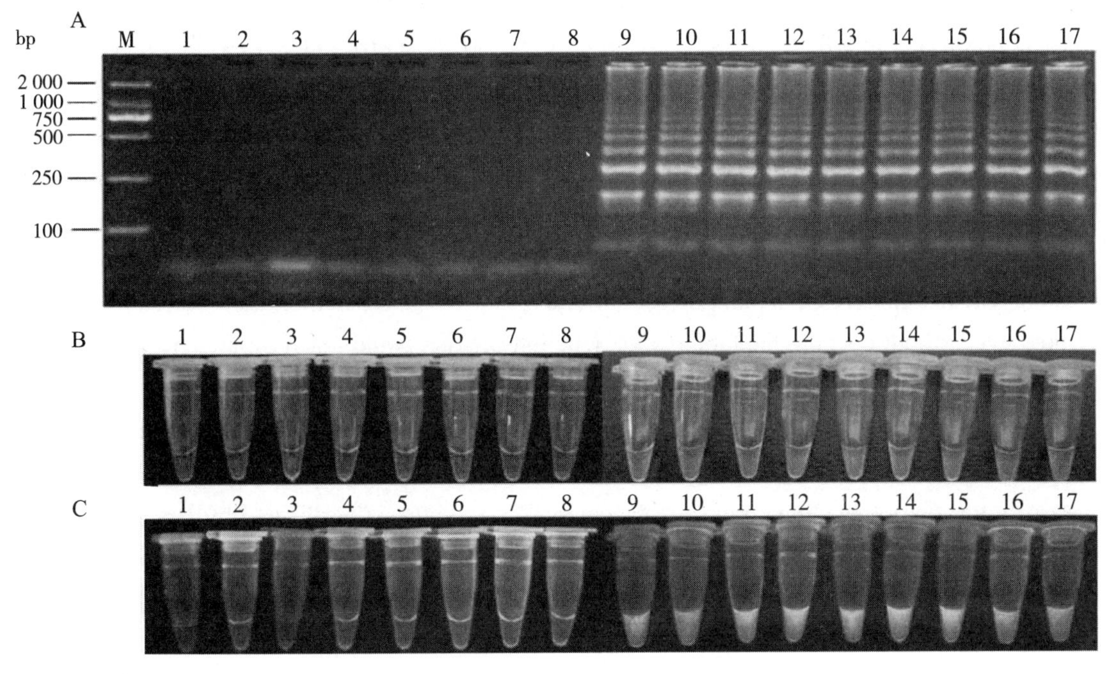

A: Agarose gel electrophoresis of the RT-LAMP products; B: Visualization of the RT-LAMP products under daylight; C: Visualization of the RT-LAMP products under UV light. M: DNA marker; 1: H5 subtype avian influenza virus; 2: H9 subtype avian influenza virus; 3: Newcastle disease virus; 4: Infectious bronchitis virus; 5: Infectious laryngotracheitis virus; 6: Adenovirus; 7: *M. gallisepticum*; 8: *S. enteritidis*; 9: S1133 ARV; 10: 1733 ARV; 11: 203-M3 ARV; 12: C78 ARV; 13: R1 ARV; 14: R2 ARV; 15: ARV Iso1; 16: ARV Iso2; 17: ARV Iso3 (color figure in appendix).

Figure 5-22-2 Specificity of the RT-LAMP assay

The sensitivity of the RT-LAMP assay was then determined and compared with that of RT-PCR in parallel. As shown in Figure 5-22-3 A, the detection limit of the RT-PCR was 1 pg viral RNA. However, the detection limit of the RT-LAMP assay was 10 fg total viral RNA (Figure 5-22-3 B), which was 100-fold higher than that of the RT-PCR (Figure 5-22-3). The RT-LAMP assay was therefore highly specific, sensitive and superior to the RT-PCR for the detection of ARV RNA.

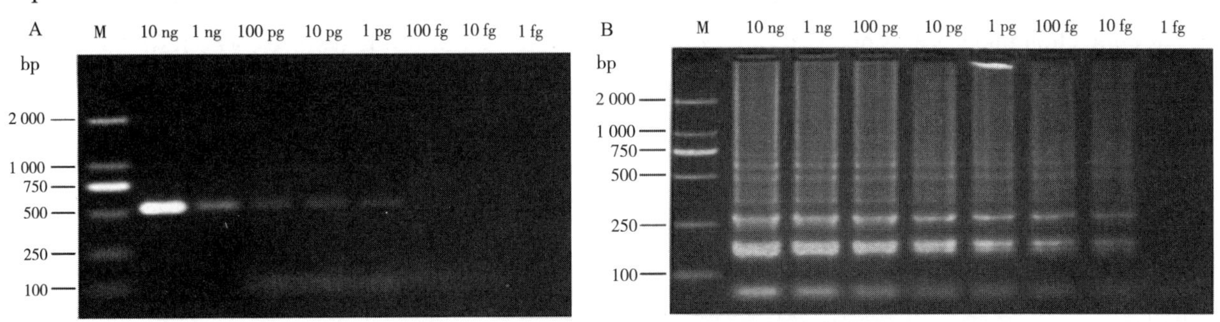

A: Sensitivity of RT-PCR; B: Sensitivity of the RT-LAMP assay as detected by 1.5% agarose gel electrophoresis; M: DNA marker. The amount of total RNA used in the RT-LAMP and RT-PCR is indicated.

Figure 5-22-3 Sensitivity of the RT-LAMP assay in comparison with RT-PCR for detection of ARV

Discussion

In this study, a set of six primers targeting the S1 gene segment of ARV were designed and the reaction conditions were optimized through repeated experiments. The result of the specificity tests indicated that the developed assay can successfully detect the ARV of the reference strain and field isolates and showed no cross-reaction with other pathogenic avian viruses and bacteria. Compared with the RT-PCR, the established RT-LAMP assay offers several advantages. First, the detection limit of the RT-LAMP assay was 10 fg total viral RNA, which was 100-fold higher than that of RT-PCR. Secondly, the amplification reaction of RT-LAMP assay was accomplished in a thermal block for 60 min and no specialized PCR machine is needed. Thirdly, the results can be inspected immediately by the naked eye and it is rapid and easy without the use of gel electrophoresis. However, due to the high sensitivity of the assay, contamination with even the smallest amount of template RNA can lead to a false positive result. Caution needs to be exercised to prevent contamination. One of the common sources of contamination comes from adding the fluorescence reagents to the tubes after amplification to visualize products. To overcome this problem, we added the fluorescence reagent (calcein and $MnCl_2$) to the reaction mixture before amplification. The results of the RT-LAMP assay were inspected directly by the naked eye without opening the tube, so the reaction product could not form an aerosol to contaminate the surroundings.

LAMP is a relatively new nucleic acid amplification method that needs a set of six primers targeting the eight sites of a highly conserved gene sequence. The primer set determines the specificity of the assay. The LAMP assay relies on autocycling strand displacement DNA synthesis performed by Bst DNA polymerase with a high degree of strand displacement activity, yielding a large white pyrophosphate ion by-product. The result of the RT-LAMP assay can be visualized by adding fluorescence reagent (calcein and $MnCl_2$) into the reaction mixture before the amplification reaction. Prior to amplification, calcein is initially combined with manganese ion and the reaction quenched, so the reaction solution is orange. When the amplification reaction proceeds, manganese ion is deprived of calcein by the generated pyrophosphate ion, which results in the emission of fluorescence. The free calcein is also apt to combine with the magnesiumion (Mg^{2+}) in the reaction mixture, leading to stronger fluorescence emission[21]. The appropriate concentration of each reaction reagent such as $MgSO_4$, betaine, dNTP, Bst DNA polymerase, calcein and $MnCl_2$ can influence the amplification efficiency of the LAMP assay, which in turn affects the fluorescence emission. Therefore, it is necessary to optimize the reaction system.

ARV is an important avian pathogen and is distributed worldwide[31]. ARV primarily causes viral arthritis/tendosynovitis in broilers, which is a major cause of leg weakness. Furthermore, ARV is immunosuppressive and enhances the susceptibility of chickens to other avian pathogens[32, 33]. Early detection and surveillance of ARV infection under field conditions are thus critical to the control of the diseases. The RT-LAMP assay reported here is highly sensitive and specific and serves as a rapid screen assay for the early detection of ARV.

References

[1] SPANDIDOS D A, GRAHAM A F. Physical and chemical characterization of an avian reovirus. J Virol, 1976, 19(3): 968-976.

[2] SCHNITZER T J, RAMOS T, GOUVEA V. Avian reovirus polypeptides: analysis of intracellular virus-specified products, virions, top component, and cores. J Virol, 1982, 43(3): 1006-1014.

[3] JOKLIK W K. The Reoviridae. New York: Plenum Publishing Co., 1983.

[4] BENAVENTE J, MARTINEZ-COSTAS J. Avian reovirus: structure and biology. Virus Res, 2007, 123(2): 105-119.
[5] PETEK M, FELLUGA B, BORGHI G, et al. The Crawley agent: an avian reovirus. Arch Gesamte Virusforsch, 1967, 21(3): 413-424.
[6] GERSHOWITZ A, WOOLEY R E. Characterization of two reoviruses isolated from turkeys with infectious enteritis. Avian Dis, 1973, 17(2): 406-414.
[7] GLASS S E, NAQI S A, HALL C F, et al. Isolation and characterization of a virus associated with arthritis of chickens. Avian Dis, 1973, 17(2): 415-424.
[8] MCFERRAN J B, CONNOR T J, MCCRACKEN R M. Isolation of adenoviruses and reoviruses from avian species other than domestic fowl. Avian Dis, 1976, 20(3): 519-524.
[9] PAGE R K, FLETCHER O J, ROWLAND G N, et al. Malabsorption syndrome in broiler chickens. Avian Dis, 1982, 26(3): 618-624.
[10] HIERONYMUS D R, VILLEGAS P, KLEVEN S H. Identification and serological differentiation of several reovirus strains isolated from chickens with suspected malabsorption syndrome. Avian Dis, 1983, 27(1): 246-254.
[11] ROBERTSON M D, WILCOX G E. Avian reovirus. Veterinary Bulletin, 1986, 56: 154-174.
[12] VAN DE ZANDE S, KUHN E M. Central nervous system signs in chickens caused by a new avian reovirus strain: a pathogenesis study. Vet Microbiol, 2007, 120(1-2): 42-49.
[13] LIU H J, GIAMBRONE J J. Characterization of a nonradioactive cloned cDNA probe for detecting avian reoviruses. Avian Dis, 1997, 41(2): 374-378.
[14] XIE Z, FADL A A, GIRSHICK T, et al. Amplification of avian reovirus RNA using the reverse transcriptase-polymerase chain reaction. Avian Dis, 1997, 41(3): 654-660.
[15] XIE Z, QIN C, XIE L, et al. Recombinant protein-based ELISA for detection and differentiation of antibodies against avian reovirus in vaccinated and non-vaccinated chickens. J Virol Methods, 2010, 165(1): 108-111.
[16] LEE L H, SHIEN J H, SHIEH H K. Detection of avian reovirus RNA and comparison of a portion of genome segment S3 by polymerase chain reaction and restriction enzyme fragment length polymorphism. Res Vet Sci, 1998, 65(1): 11-15.
[17] YIN H S, LEE L H. Development and characterization of a nucleic acid probe for avian reoviruses. Avian Pathol, 1998, 27(4): 423-426.
[18] LIU H J, CHEN J H, LIAO M H, et al. Identification of the sigma C-encoded gene of avian reovirus by nested PCR and restriction endonuclease analysis. J Virol Methods, 1999, 81(1-2): 83-90.
[19] GUO K, DORMITORIO T V, OU S C, et al. Development of TaqMan real-time RT-PCR for detection of avian reoviruses. J Virol Methods, 2011, 177(1): 75-79.
[20] MORI Y, NAGAMINE K, TOMITA N, et al. Detection of loop-mediated isothermal amplification reaction by turbidity derived from magnesium pyrophosphate formation. Biochem Biophys Res Commun, 2001, 289(1): 150-154.
[21] TOMITA N, MORI Y, KANDA H, et al. Loop-mediated isothermal amplification (LAMP) of gene sequences and simple visual detection of products. Nat Protoc, 2008, 3(5): 877-882.
[22] HONG T C, MAI Q L, CUONG D V, et al. Development and evaluation of a novel loop-mediated isothermal amplification method for rapid detection of severe acute respiratory syndrome coronavirus. J Clin Microbiol, 2004, 42(5): 1956-1961.
[23] PHAM H M, NAKAJIMA C, OHASHI K, et al. Loop-mediated isothermal amplification for rapid detection of Newcastle disease virus. J Clin Microbiol, 2005, 43(4): 1646-1650.
[24] POON L L, LEUNG C S, CHAN K H, et al. Detection of human influenza A viruses by loop-mediated isothermal amplification. J Clin Microbiol, 2005, 43(1): 427-430.
[25] CURTIS K A, RUDOLPH D L, OWEN S M. Rapid detection of HIV-1 by reverse-transcription, loop-mediated isothermal amplification (RT-LAMP). J Virol Methods, 2008, 151(2): 264-270.
[26] ROVIRA A, ABRAHANTE J, MURTAUGH M, et al. Reverse transcription loop-mediated isothermal amplification for the detection of Porcine reproductive and respiratory syndrome virus. J Vet Diagn Invest, 2009, 21(3): 350-354.

[27] XU J, ZHANG Z, YIN Y, et al. Development of reverse-transcription loop-mediated isothermal amplification for the detection of infectious bursal disease virus. J Virol Methods, 2009, 162(1-2): 267-271.

[28] YAMAZAKI W, TAGUCHI M, ISHIBASHI M, et al. Development of a loop-mediated isothermal amplification assay for sensitive and rapid detection of Campylobacter fetus. Vet Microbiol, 2009, 136(3-4): 393-396.

[29] TRAN A, BERARD A, COOMBS K M. Avian reoviruses: propagation, quantification, and storage. Curr Protoc Microbiol, 2009, Chapter 15: Unit15C 2.

[30] REED L, MUENCH H. A simple method for estimating fifty percent endpoints. . American Journal of Hygiene, 1938, 27: 493-497.

[31] VAN DER HEIDE L. The history of avian reovirus. Avian Dis, 2000, 44(3): 638-641.

[32] JONES R C. Avian reovirus infections. Rev Sci Tech, 2000, 19(2): 614-625.

[33] CUI Z, MENG S, JIANG S, et al. Serological surveys of chicken anemia virus, avian reticuloendotheliosis virus and avian reovirus infections in white meattype chickens in China. Acta Veterinaria et Zootechnica Sinica, 2006, 37: 152-155.

Reverse-transcription, loop-mediated isothermal amplification assay for the sensitive and rapid detection of H10 subtype avian influenza viruses

Luo Sisi, Xie Zhixun, Xie Liji, Liu Jiabo, Xie Zhiqin, Deng Xianwen, Huang Li, Huang Jiaoling, Zeng Tingting, and Khan Mazhar I

Abstract

Background: The H10 subtype avian influenza viruses (H10N4, H10N5 and H10N7) have been reported to cause disease in mammals, and the first human case of H10N8 subtype avian influenza virus was reported in 2013. Recently, H10 subtype avian influenza viruses (AIVs) have been followed more closely, but routine diagnostic tests are tedious, less sensitive and time consuming, rapid molecular detection assays for H10 AIVs are not available. Methods: Based on conserved sequences within the HA gene of the H10 subtype AIVs, specific primer sets of H10 subtype of AIVs were designed and assay reaction conditions were optimized. A reverse-transcription loop-mediated isothermal amplification (RT-LAMP) assay was established for the rapid detection of H10 subtype AIVs. The specificity was validated using multiple subtypes of AIVs and other avian respiratory pathogens, and the limit of detection (LOD) was tested using concentration gradient of *in vitro*-transcribed RNA. Results: The established assay was performed in a water bath at 63 ℃ for 40 min, and the amplification result was visualized directly as well as under daylight reflections. The H10-RT-LAMP assay can specifically amplify H10 subtype AIVs and has no cross-reactivity with other subtypes AIVs or avian pathogens. The LOD of the H10-RT-LAMP assay was 10 copies per μL of *in vitro*-transcribed RNA. Conclusions: The RT-LAMP method reported here is demonstrated to be a potentially valuable means for the detection of H10 subtype AIV and rapid clinical diagnosis, being fast, simple, and low in cost. Consequently, it will be a very useful screening assay for the surveillance of H10 subtype AIVs in underequipped laboratories as well as in field conditions.

Keywords

avian influenza, H10 subtype, RT-LAMP, amplification

Introduction

The influenza A viruses belong to the family Orthomyxoviridae and are classified into 18 hemagglutinin (HA) and 11 neuraminidase (NA) subtypes based on their antigenic properties[1, 2]. Avian influenza viruses (AIVs) can be divided into two distinct groups based on their virulence: highly pathogenic avian influenza viruses (HPAIVs) and low pathogenic avian influenza viruses (LPAIVs). HPAIVs generally cause high morbidity and mortality in poultry flock, even directly infect human or cause death such as H5N1 in Hong-Kong in 1997[3], but the H7N9 of HPAIVs broke out and infected human was low pathogenic in poultry in 2013[4]; LPAIVs such as H9N2 cause mild or no symptoms in poultry, wild birds and human, and most AIV subtypes are LPAIVs[5, 6].

Most H10 subtype AIVs belong to LPAIVs and have been brought into sharp focus since the winter of 2013. H10N8 was reported to cause human disease for the first time in December 2013 following the emergence of the H7N9 and H5N1 subtype AIVs that cause serious disease in humans and have become a threat to public health[7, 8]. As of 15 February 2014, three cases of human infection with A (H10N8) virus have been confirmed in Jiangxi Province, China, of whom two died[9]. H10N8 had been reported only two cases that water samples from the Dongting Lake in 2007 and a duck from Guangdong in 2012 in China before human cases[10, 11]. Currently, H10 subtype AIVs are generally well-known as human infection and followed more focus. Indeed, H10 subtype AIVs have been reported previously, and some of these viruses can cause disease in mammals. H10N4 was confirmed to cause interstitial pneumonia in minks[12], and H10N5 has been found in pigs[13]. H10N7 led to increased mortality and decreased egg production in chickens on a commercial poultry farm in Australia[14, 15]. Rapid detection is very important for the prevention and control of AIVs. Methods for the detection of the H5, H7 and H9 subtypes of AIVs have been developed and applied to surveillance and clinical care[16]. Currently, rapid assays for the detection and surveillance of H10 subtype AIVs are not available, and the development of an effective and applicable assay is urgently needed.

The loop-mediated isothermal amplification (LAMP) technique was invented in 2000[17] and provides a powerful gene amplification tool with simple conditions requiring only a water bath at 58～65 ℃ and a 30～60 min incubation. LAMP amplification relies on the Bst DNA polymerase with highly auto-cycling strand displacement DNA synthesis activity. RT-LAMP assay is being used widely for the detection of AIVs such as H1, H3 and H9 subtypes and various viral pathogens[18-23]. In recent years, improvements in the LAMP assay, such as visual detection of amplified product by adding calcein or hydroxynaphthol blue (HNB)[24], which only require a temperature-controlled water bath and effectively avoid nucleic acid production pollution, have made it easier to apply in primary clinical setting or for field use. In this study, we developed an effective RT-LAMP assay with calcein and $MnCl_2$ [25] , to visual detect the H10 subtype AIVs, which might be suitable in the surveillance of the H10 subtype of AIVs for the rapid detection of H10 subtype AIVs in poultry and wild birds.

Materials and methods

Design of primers

The HA gene sequences of the H10 subtype AIVs were available downloaded from the NCBI Influenza Virus Resource Database and were compared and analyzed using the DNAstar software to identify conserved regions. Several primer sets were designed within the conserved regions using the LAMP primer design software (Primer Explorer V4) for used in the H10-RT-LAMP assays. The specificity of the primer sets was evaluated using NCBI BLAST to confirm that no cross-reactivity with sequences from other avian pathogens would occur. Finally, an optimal set of primers was chosen after many rounds of analysis to ensure the highest specificity and sensitivity of the H10-RT-LAMP assay. The RT-LAMP primer set consists of two outer primers (F3 and B3), two inner primers (FIP and BIP, FIP = F1c+F2, BIP = B1c+B2), and two loop primers (LF and LB). The high-performance liquid chromatography (HPLC)-purified primer information is shown in Table 5-23-1.

Table 5-23-1 Primer information

Name	Primer sequences (5'-3')	Genome position
FIP = F1c+F2	CATTGTGTGAGAAGGTGATATTA-ACAATTTTGTTCCGGTTGTTG	F1c, 794-772 F2, 681-701
BIP = B1c+B2	GATAGCACCGAGYCGAGTTAG-ACAATTATTGTCTATTGGTGCAT	B1c, 802-822 B2, 880-858
F3	TGGGACACAATCAYTGTCCA	634-653
B3	TATAGAACCCCCTCTCCAAA	913-894
LF	GAAAATCAATCCGCCCACTT	746-727
LB	ATTGGAAGAGGATTGGGGAT	830-849

Note: Genome position according to the HA gene of A/duck/Guangdong/E1/2012 (H10N8) (GenBank accession number: JQ924786.1).

Viral strains and DNA/RNA extraction

The reference and isolate strains or inactive strains (H5, H7, H14, H15 and H16) of AIV, human influenza virus (H1N1, H3N2, and Influenza B viruses), other avian pathogens and clinical samples used in this study are listed in Table 5-23-2 and Table 5-23-3.The genomic DNA/RNA was extracted from 200 μL samples using a DNA/RNA Miniprep Kit (TaKaRa, Dalian, China) according to the protocol suggested by the manufacturer. The DNA and RNA were eluted with 40 μL of elution buffer and immediately stored at −80 ℃ until use.

Table 5-23-2 Sources of pathogens analyzed and H10-RT-LAMP assay results

Number	Virus strain	Source	Titre	H10-RT-LAMP	GenBank accession or laboratory references
1	A/Duck/Guangxi/030D/2009 (H1N1)	GVRI	$10^{5.5}$ EID_{50}/mL	–	KC608160
2	A/Duck/Guangxi/N42/2009 (H3N2)	GVRI	$10^{5.4}$ EID_{50}/mL	–	JN003630
3	A/Duck/Guangxi/GXd–4/2009 (H6N6)	GVRI	$10^{6.8}$ EID_{50}/mL	–	JX304746.1
4	A/Turkey/Ontario/6118/1967 (H8N4)	UC	$10^{6.3}$ EID_{50}/mL	–	GU053171.1
5	A/Duck/Guangxi/RX/2009 (H9N2)	GVRI	$10^{6.9}$ EID_{50}/mL	–	KF768236.1
6	A/Duck/Guangxi/LAD9/2009 (H9N8)	GVRI	$10^{7.1}$ EID_{50}/mL	–	KF768214.1
7	A/Turkey/MN/3/1979 (H10N7)	UC	$10^{6.7}$ EID_{50}/mL	+	GU186627.2
8	A/Duck/Alberta/60/1976 (H12N5)	UC	$10^{6.3}$ EID_{50}/mL	–	AB288334
9	A/Gull/Maryland/704/1977 (H13N6)	UC	$10^{6.5}$ EID_{50}/mL	–	D90308.1
10	A/Mallard duck/Astrakhan/263/1982 (H14N5)	UC	$10^{6.2}$ EID_{50}/mL	–	CY014604.1
11	A/wedge–tailed shearwater/Western Australia/ 2576/1979 (H15N9)	UC	$10^{5.4}$ EID_{50}/mL	–	CY006010.1
12	A/shorebird/Delaware/168/2006 (H16N3)	UC	$10^{6.0}$ EID_{50}/mL	–	EU030976.1
13	Newcastle disease virus (NDV): LaSota	GVRI	$10^{8.0}$ EID_{50}/mL	–	JF950510.1
14	Infectious bronchitis virus (IBV): M41	GVRI	$10^{8.3}$ EID_{50}/mL	–	DQ834384
15	Infectious laryngotracheitis virus (ILTV)	GVRI	$10^{5.8}$ EID_{50}/mL	–	NC_006623.1
16	A/Mallard/Alberta/77 (H2N3)	UC	$10^{6.0}$ EID_{50}/mL	–	Peng Y et al. (2011) [19]
17	A/Turkey/GA/209092/02 (H5N2)	UC	$10^{7.8}$ EID_{50}/mL	–	Peng Y et al. (2011) [19]
18	A/Turkey/CA/35621/84 (H5N3)	UC	$10^{6.5}$ EID_{50}/mL	–	Peng Y et al. (2011) [19]
19	A/Waterfowl/GA/269452–56/03 (H5N7)	UC	$10^{7.0}$ EID_{50}/mL	–	Peng Y et al. (2011) [19]
20	A/Turkey/WI/68 (H5N9)	UC	$10^{7.2}$ EID_{50}/mL	–	Peng Y et al. (2011) [19]

continued

Number	Virus strain	Source	Titre	H10-RT-LAMP	GenBank accession or laboratory references
21	A/Chicken/NY/273874/03 (H7N2)	UC	$10^{7.0}$ EID_{50}/mL	–	Peng Y et al. (2011) [19]
22	A/Duck/HK/876/80 (H10N3)	HKU	$10^{6.0}$ EID_{50}/mL	+	Xie Z et al. (2006) [16]
23	A/Chicken/Guangxi/90C/2011 (H1N2)	GVRI	$10^{6.3}$ EID_{50}/mL	–	HA and HI test
24	A/Duck/Guangxi/070D/2010 (H4N6)	GVRI	$10^{6.2}$ EID_{50}/mL	–	HA and HI test
25	A/Chicken PA/3979/97 (H7N2)	PU	$10^{6.6}$ EID_{50}/mL	–	HA and HI test
26	A/Duck/PA/2099/12 (H11N9)	PU	$10^{6.0}$ EID_{50}/mL	–	HA and HI test
27	A/Guangxi/1415/15 (H1N1)	GXCDC	$10^{5.8}$ EID_{50}/mL	–	HA and HI test
28	A/Guangxi/1241/14 (H1N1)	GXCDC	$10^{6.0}$ EID_{50}/mL	–	HA and HI test
29	A/Guangxi/1420 /15 (H3N2)	GXCDC	$10^{5.6}$ EID_{50}/mL	–	HA and HI test
30	A/Guangxi/1632/15 (H3N2)	GXCDC	$10^{6.2}$ EID_{50}/mL	–	HA and HI test
31	B/Guangxi/1418/15	GXCDC	$10^{5.8}$ EID_{50}/mL	–	HA and HI test
32	B/Guangxi/1470/15	GXCDC	$10^{5.6}$ EID_{50}/mL	–	HA and HI test

Note: GGVRI, Guangxi Veterinary Research Institute, Nanning, Guangxi, China; UC, University of Connecticut, USA; PU University of Pennsylvania, USA; HKU, University of Hong Kong, China; GXCDC, Guangxi Provincial Center for Disease Control and Prevention, China.

Table 5-23-3 Results of two assays for the detection of H10 subtype AIVs in clinical samples

Sample type	Total samples	Number of positive samples	
		LAMP	Viral isolation
Chicken	198	0	0
Duck	611	4	4
Geese	234	2	2
Francolin	134	1	1
Pigeon	119	1	1
Total	1 296	8	8

RT-LAMP reaction

The H10-RT-LAMP assay was performed in a 25 μL reaction containing 2.5 μL 10 × Thermopol Reaction Buffer (New England Biolabs, Beijing, China), 8 U Bst DNA polymerase (large fragment; New England Biolabs, Beijing, China), 200 U reverse transcriptase M-MLV (TaKaRa, Dalian, China), primers (for FIP and BIP, 1.6 μM each, for F3 and B3, 0.2 μM each, and for LF and LB, 0.8 μM of each), and 2 μL of the template RNA. Based on previous studies[18-23], the final concentration ranges of the three primary reagents, dNTPs (TaKaRa, Dalian, China) (0.2 to 1.6 mM), betaine (Sigma-Aldrich) (0.2 to 1.4 M), and $MgSO_4$ (Sigma-Aldrich) (2 to 9 mM), were optimized for amplification efficiency. The amplification reaction was tested at 58 to 65℃ for 30 to 60 min to find the optimal temperature and incubation time. Finally, the reaction was terminated by heating at 80 ℃ for 5 min. All the experiments were repeated three times.

Real-time monitoring of turbidity at the desired temperature was achieved by using an LA-320CE turbidimeter (Eiken Chemical Co. Ltd. Tochigi, Japan) that recorded the optical density (OD) at 650 nm every 6 s. Turbidity values greater than 0.1 were considered positive. In addition to the turbidity meter, reaction results could be observed visually based on a color changes of green for positive and orange for negative results by using 25 μM calcein (Eiken Chemical, Tochigi, Japan) combined with 0.5 mM $MnCl_2$ (Eiken Chemical, Tochigi, Japan).

In vitro transcription

Following the extraction of RNA from the A/duck/HK/876/80 (H10N3) strain, the HA gene was amplified by RT-PCR using previously reported primers[26]. The purified PCR product was cloned into the pGEM T-easy vector and transformed into *Escherichia coli* DH5 α cells. The plasmid, named H10-pGEM, was sequenced and linearized using the restriction endonuclease *Spe* I. The plasmid was then used as a template for the RiboMax T7 (Promega, Madison, WI) *in vitro* transcription system according to the manufacturer's instructions. After DNase treatment to eliminate residual DNA, the RNA product was purified using a RNA purification kit (Tiangen, Beijing), and its concentration was determined spectrophotometrically using a NanoDrop 2000 (Thermo Scientific, USA). The *in vitro* transcribed RNA was sequenced to further evaluate the direction of insert and the length. The right RNA transcribed from the HA gene from the H10 subtype AIVs was 10-fold serially diluted from 10^4 copies/μL to 1 copy/μL as standard quantitatively.

Specificity and sensitivity of the assay

To evaluate the specificity of the primer set used for the H10-RT-LAMP assay, DNA/RNA extracted from AIV subtypes H1, H2, H3, H4, H5, H6, H7, H8, H9, H10, H11, H12, H13, H14, H15, H16, human influenza viruses (H1N1, H3N2 and influenza B viruses), Newcastle disease virus (NDV), infectious bronchitis virus (IBV), and infectious laryngotracheitis virus (ILTV) were analyzed using the H10-RT-LAMP assay.

A 10-fold dilution series vitro-transcribed RNA ranging from 10^4 copies/μL to 1 copy/μL served as the standard to test the sensitivity of this assay. All reactions were performed in triplicate.

Clinical sample detection

This study was approved, and the protocol sample collection was conducted by the Animal Ethics Committee of the Guangxi Veterinary Research Institute, which supervises all live bird markets (LBMs) in Guangxi. Oral and cloacal swab samples were gently collected with permission from the owners of LBMs, and the fowls were not anesthetized before sampling and were observed for 30 min after sampling before being returned to their cages. The clinical samples were prepared in a viral transport medium, which was made up of 0.05 M phosphate buffered saline (PBS) containing antibiotics of penicillin (10 000 units/mL), streptomycin (10 mg/mL), gentamycin (10 mg/mL), kanamycin (10 mg/mL) and 5 % (v/v) fetal bovine serum, and were placed in the ice box.

A total of 1 296 clinical swab samples (Table 5-23-3) including healthy chickens, duck, geese, francolins and pigeons were collected from five LBMs named Xiuxiang (No.1～246), Beihu (No.247～525), Sulu (No.526～813), Danchun (No.814～1102) and Jiangqiao (No.1103～1296) in Nanning in Guangxi and assayed by RT-LAMP and virus isolation. Virus isolations were prepared by inoculating specific-pathogen-free (SPF) embryonated chicken eggs and were tested using a hemagglutination assay (HA) and a hemagglutination inhibition (HI) assay described previously[27].

Results

Optimization of H10-RT-LAMP

Following standardization and optimization, the optimal of final concentrations were 1.4 mM each dNTP, 0.8 M betaine, 6 mM $MgSO_4$.The H10-RT-LAMP reactions were incubated at 63 ℃ for 40 min and then heat-inactivated at 80 ℃ for 5 min to terminate the reaction.

Specificity and sensitivity of the H10-RT-LAMP assay

As expected, the turbidity value of the H10 subtype AIVs gradually increased at 20 min. We observed that the increased turbidity curve appeared only when H10 subtype AIVs was used as the template that indicated a positive result, while straight lines were observed for the other avian pathogens (H1-H9 AIV, H11-H16 AIV, human influenza viruses, NDV, IBV and ILTV) indicating a lack of amplification (Figure 5-23-1 A). A green color was observed for the H10 subtype AIVs, and the other reactions remained the same orange color as before the incubation (Figure 5-23-1 B). The results of the detection assay for each strain are shown in Table 5-23-2. All the results suggest that this newly developed RT-LAMP assay only amplifies H10 subtype AIVs and shows no cross-reactivity with other avian pathogens and human influenza viruses (H1N1, H3N2 and influenza B virus), thus validating the high specificity of this H10-RT-LAMP assay. The sensitivity of the test was established by dilution series vitro-transcribed RNA standard samples. The LOD for the RT-LAMP was 10 copies/μL (Figure 5-23-2 A and B). Direct visualization of the Calcein dye-aided color change not only determined amplification but also did not require expensive instruments, needing only a temperature-controlled water bath, and the calcein dye results were consistent with the turbidity data.

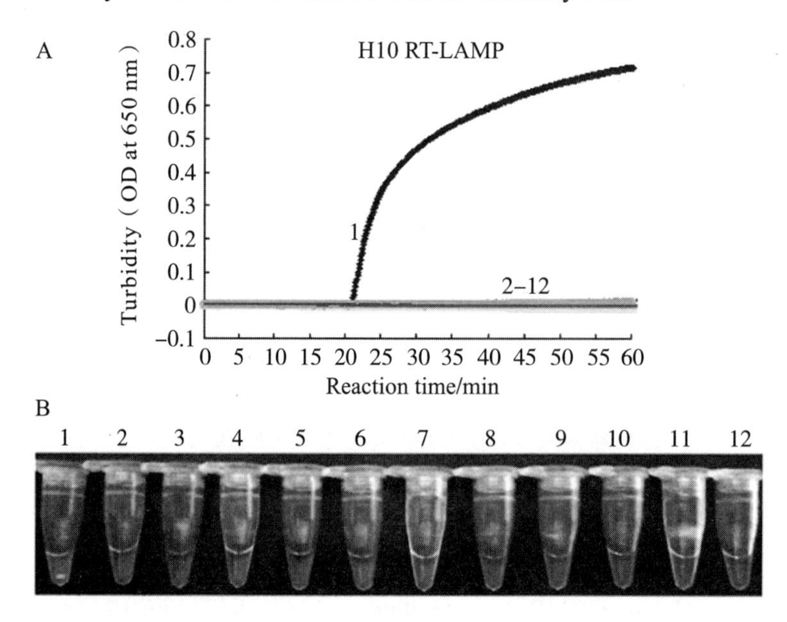

A: The turbidity curve of the RT-LAMP products; B: Visualization of the RT-LAMP products. 1: H10N3; 2: H1N1; 3: H3N2; 4: H5N3; 5: H6N8; 6: H7N2; 7: H9N2; 8: H11N9; 9: NDV; 10: IBV; 11: ILTV; 12: blank control. (Refer to the colors of C and D in the appendix Figure 5-20-2 for the colors of positive and negative samples in the LAMP assay.)

Figure 5-23-1 Specificity results of the H10-RT-LAMP assay

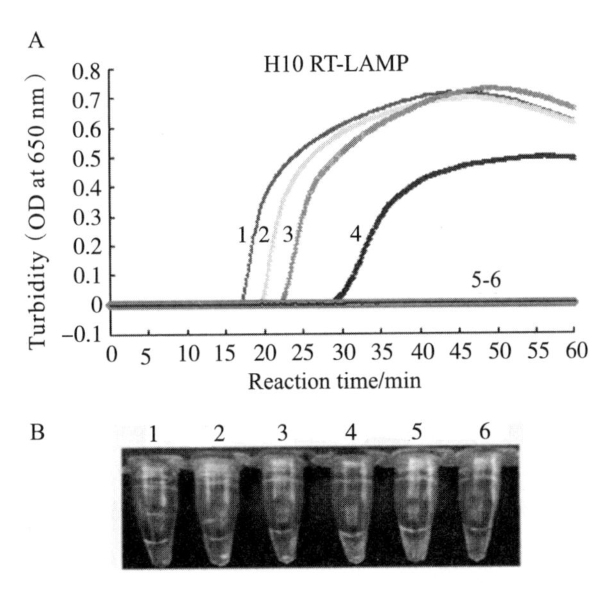

A: The turbidity curves of the RT-LAMP products from a 10-fold dilution series of RNA; B: Visualization of the RT-LAMP products from a 10-fold dilution series of RNA. 1: 10^4 copies/μL; 2: 10^3 copies/μL; 3: 10^2 copies/μL; 4: 10 copies/μL; 5: 1 copy/μL; 6: blank control. (Refer to the colors of C and D in the appendix Figure 5-20-2 for the colors of positive and negative samples in the LAMP assay.)

Figure 5-23-2 Sensitivity results of the H10-RT-LAMP assay

Clinical sample detection

To evaluate the sensitivity of the H10-RT-LAMP assay in clinical practice, a total of 1 296 oral and cloacal swab samples were collected and subjected to methods of detection in parallel: H10-RT-LAMP, and viral isolation. Of the 1 296 oral and cloacal swab samples tested, 8 were confirmed to be positive for H10 subtype AIVs for both RT-LAMP and viral isolation (Table 5-23-3).

Discussion

According to the World Organisation for Animal Health (OIE), HPAIVs includes two characteristic i) high pathogenic in vivo and ii) presence of multiple basic amino acids at the HA cleavage site. Significantly, there were two H10 subtype AIVs (A/turkey/England/384/79 (H10N4) and A/mandarin duck/Singapore/805/F-72/7/93 (H10N5)) that lacked multi-basic amino acids at the HA cleavage site, a typical marker for HPAIVs, but are highly virulent in chickens according to the test in vivo, and been reported as HPAIVs by both the OIE and the EU definitions[28, 27]. Some findings have suggested that the HA molecule of H10 subtypes may have an affinity for human cell receptors and may also help the virus to adapt to multiply in humans[29]. Therefore, similar to H7N9, the first sign of a poultry infection could be the emergence of clinical cases of H10N8 in humans, which would then lead to the identification of silent outbreaks in poultry[8]. Therefore, the development of a simple, rapid, visual and effective assay for H10 subtype AIVs diagnosis would significantly aid efforts to control this disease. We need to enhance surveys and collect samples monthly, quarterly and annually to monitor the H10 subtype AIVs.

Virus isolation is a gold standard and fundamental diagnostic method for the detection of AIVs, the results are accurate and reliable, but the procedure is time consuming labor intensive to inoculate embryonated chicken egg for 2～5 days and harvest allantoic fluid for HI identification requiring the positive serum of multiple subtypes AIVs. Other molecular biological diagnostic techniques, such as conventional reverse transcription polymerase chain reaction (RT-PCR) and quantitative real-time RT-PCR (qRT-PCR) have been not reported for H10 subtype AIVs.

We see several obvious advantages of the RT-LAMP assay: (1) The specificity and sensitivity are greatly enhanced by six primers recognizing eight independent target sequence regions, whereas conventional RT-PCR primers only recognize two independent regions. (2) The RT-LAMP assay only requires small amounts of cDNA to accomplish efficient amplification. Furthermore, reverse transcription can proceed at the same temperature, 58～67 ℃ , as the LAMP reaction. Therefore, the two reactions can run at the same time, making LAMP a one-step assay for RNA-based pathogens that require reverse transcription for molecular detection. (3) RT-LAMP is simple and easy to perform and only requires a temperature-controlled water bath, making it suitable for primary clinical settings or field use. Therefore, we believe that this simple, fast, and effective assay might be applied to the field detection of H10 subtype AIVs infections. The RT-LAMP assay has many advantages as mentioned above, there is still a problem to note. The RT-LAMP assay has great amplification efficiency so the reaction product can form aerosol to contaminate the surroundings when opening the tube. To avoid the contamination in this study, the fluorescence reagent (Calcein and $MnCl_2$) were added to RT-LAMP reaction mixture before amplification in order to inspect the result of RT-LAMP assay directly with the naked eye.

In this study, we used calcein's color change to visually determine results; we also monitored turbidity using the Loopamp Realtime Turbidimeter. To our knowledge, this is the first study to explore the use of H10 AIVs LAMP technology in a diagnostic test for H10 AIVs. This assay and its analysis are simple, specific, sensitive and rapid, do not require special equipment, and can be applied to detect H10 AIVs in clinical samples.

Conclusions

In summary, this newly developed H10-RT-LAMP assay is a specific, sensitive, rapid, and cost-effective assay for the rapid detection and epidemiological surveillance of the H10 subtype AIVs. It can be applied to the rapid visual diagnosis of clinical samples and can also provide an effective tool to prevent and control H10 subtype AIVs disease.

References

[1] CHAN J F, TO K K, TSE H, et al. Interspecies transmission and emergence of novel viruses: lessons from bats and birds. Trends in Microbiology, 2013, 21(10):544-555.

[2] TONG S, ZHU X, LI Y, et al. New world bats harbor diverse influenza A viruses. PLOS Pathogens, 2013, 9(10):e1003657.

[3] BEIGEL J H, FARRAR J, HAN A M, et al. Avian influenza A (H5N1) infection in humans. The New England Journal of Medicine, 2005, 353(13):1374-1385.

[4] GAO R, CAO B, HU Y, et al. Human infection with a novel avian-origin influenza A (H7N9) virus. The New England Journal of Medicine, 2013, 368(20):1888-1897.

[5] LI S, ZHOU Y, ZHAO Y, et al. Avian influenza H9N2 seroprevalence among pig population and pig farm staff in Shandong, China. Virology Journal, 2015, 12:34.

[6] ALEXANDER D J. A review of avian influenza in different bird species. Veterinary Microbiology, 2000, 74(1-2):3-13.

[7] CHEN H, YUAN H, GAO R, et al. Clinical and epidemiological characteristics of a fatal case of avian influenza A H10N8 virus infection: a descriptive study. Lancet, 2014, 383(9918):714-721.

[8] To K K, TSANG A K, CHAN J F, et al. Emergence in China of human disease due to avian influenza A (H10N8)-cause for concern?. The Journal of Infection, 2014, 68(3):205-215.

[9] QI W, ZHOU X, SHI W, et al. Genesis of the novel human-infecting influenza A (H10N8) virus and potential genetic diversity of the virus in poultry, China. Euro Surveillance, 2014, 19(25).

[10] ZHANG H, XU B, CHEN Q, et al. Characterization of an H10N8 influenza virus isolated from Dongting lake wetland. Virology Journal, 2011, 8:42.

[11] JIAO P, CAO L, YUAN R, et al. Complete genome sequence of an H10N8 avian influenza virus isolated from a live bird market in southern China. Journal of Virology, 2012, 86(14):7716.

[12] ENGLUND L, HARD A S C. Two avian H10 influenza A virus strains with different pathogenicity for mink (Mustela vison). Archives of Virology, 1998, 143(4):653-666.

[13] WANG N, ZOU W, YANG Y, et al. Complete genome sequence of an H10N5 avian influenza virus isolated from pigs in central China. Journal of Virology, 2012, 86(24):13865-13866.

[14] VIJAYKRISHNA D, DENG Y M, SU Y C, et al. The recent establishment of North American H10 lineage influenza viruses in Australian wild waterfowl and the evolution of Australian avian influenza viruses. Journal of Virology, 2013, 87(18):10182-10189.

[15] ARZEY G G, KIRKLAND P D, ARZEY K E, et al. Influenza virus A (H10N7) in chickens and poultry abattoir workers, Australia. Emerging Infectious Diseases, 2012, 18(5):814-816.

[16] XIE Z, PANG Y S, LIU J, et al. A multiplex RT-PCR for detection of type A influenza virus and differentiation of avian H5, H7, and H9 hemagglutinin subtypes. Molecular and Cellular Probes, 2006, 20(3-4):245-249.

[17] NOTOMI T, OKAYAMA H, MASUBUCHI H, et al. Loop-mediated isothermal amplification of DNA. Nucleic Acids Research, 2000, 28(12):E63.

[18] PENG Y, XIE Z X, GUO J, et al. Visual detection of H1 subtype and identification of N1, N2 subtype of avian influenza virus by reverse transcription loop-mediated isothermal amplification assay. Chinese Journal of Virology, 2013, 29(2):154-161.

[19] PENG Y, XIE Z, LIU J, et al. Visual detection of H3 subtype avian influenza viruses by reverse transcription loop-mediated isothermal amplification assay. Virology Journal, 2011, 8:337.

[20] CHEN H T, ZHANG J, SUN D H, et al. Development of reverse transcription loop-mediated isothermal amplification for rapid detection of H9 avian influenza virus. Journal of Virological Methods, 2008, 151(2):200-203.

[21] XIE L, XIE Z, ZHAO G, et al. A loop-mediated isothermal amplification assay for the visual detection of duck circovirus. Virology Journal, 2014, 11:76.

[22] FAN Q, XIE Z, XIE L, et al. A reverse transcription loop-mediated isothermal amplification method for rapid detection of bovine viral diarrhea virus. Journal of Virological Methods, 2012, 186(1-2):43-48.

[23] XIE Z, FAN Q, LIU J, et al. Reverse transcription loop-mediated isothermal amplification assay for rapid detection of Bovine Rotavirus. BMC Veterinary Research, 2012, 8:133.

[24] PARIDA M, POSADAS G, INOUE S, et al. Real-time reverse transcription loop-mediated isothermal amplification for rapid detection of West Nile virus. Journal of Clinical Microbiology, 2004, 42(1):257-263.

[25] MORI Y, NAGAMINE K, TOMITA N, et al. Detection of loop-mediated isothermal amplification reaction by turbidity derived from magnesium pyrophosphate formation. Biochemical and Biophysical Research Communications, 2001, 289(1):150-154.

[26] HOFFMANN E, STECH J, GUAN Y, et al. Universal primer set for the full-length amplification of all influenza A viruses. Archives of Virology, 2001, 146(12):2275-2289.

[27] EDWARDS S. OIE laboratory standards for avian influenza. Developments in Biologicals, 2006, 124:159-162.

[28] WOOD G W, BANKS J, STRONG I, et al. An avian influenza virus of H10 subtype that is highly pathogenic for chickens, but lacks multiple basic amino acids at the haemagglutinin cleavage site. Avian Pathology, 1996, 25(4):799-806.

[29] VACHIERI S G, XIONG X, COLLINS P J, et al. Receptor binding by H10 influenza viruses. Nature, 2014, 511(7510):475-477.

A loop-mediated isothermal amplification assay for the visual detection of duck circovirus

Xie Liji, Xie Zhixun, Zhao Guangyuan, Liu Jiabo, Pang Yaoshan, Deng Xianwen, Xie Zhiqin, Fan Qing, and Luo Sisi

Abstract

Duck circovirus (DuCV) infection in farmed ducks is associated with growth problems or retardation syndromes. Rapid identification of DuCV infected ducks is essential to control DuCV effectively. Therefore, this study aims to develop of an assay for DuCV to be highly specific, sensitive, and simple without any specialized equipment. A set of six specific primers was designed to target the sequences of the Rep gene of DuCV, and A loop-mediated isothermal amplification (LAMP) assay were developed and the reaction conditions were optimized for rapid detection of DuCV. The LAMP assay reaction was conducted in a 62 ℃ water bath condition for 50 min. Then the amplification products were visualized directly for color changes. This LAMP assay is highly sensitive and able to detect twenty copies of DuCV DNA. The specificity of this LAMP assay was supported by no cross-reaction with other duck pathogens. This LAMP method for DuCV is highly specific and sensitive and can be used as a rapid and direct diagnostic assay for testing clinical samples.

Keywords

duck circovirus, loop-mediated isothermal amplification, visual detection

---

Background

The duck circovirus (DuCV) was reported initially in Germany in 2003, which was detected from two female 6-week-old female Mulard ducks with feathering disorders, poor body condition, immunosuppression and low weight[1]. Subsequently, DuCV infection was confirmed in Hungary[2], United Kingdom[3], United States[4] and the mainland of China[5, 6]. Histopathologic examination of the bursa of Fabricius (BF) demonstrated lymphocyte depletion, necrosis and histiocytosis. DuCV was detected in Mulard ducks, as well as other duck species of Muscovy, Mule, Cherry valley, Pekin and Pockmark ducks[4-8]. DuCV is a small (15～16 nm in diameter), round, nonenveloped, single-stranded DNA virus with a circular genome of approximately 1.9 kb. The genome contains two major open reading frames (ORFs), designated ORF V1 (Rep gene) and ORF C1 (Cap gene)[1].

Virus isolation is a fundamental diagnostic method, but no in vivo culture system is yet available for the propagation of DuCV[9, 10]. Other diagnostic techniques, such as conventional polymerase chain reaction (PCR)[2], nested PCR[11], real-time PCR[2, 10], ELISA[9] and in situ hybridization (ISH) have been developed. However, these techniques usually need more time and require specialized equipment. For example, PCR requires agarose gel analysis for the detection of amplification products and must be performed in specialized laboratories and ISH requires several days to complete the assay.

Recently, a new technique known as loop-mediated isothermal amplification (LAMP) has been described[12].

It can be used to amplify specific target DNA sequences with high sensitivity and, advantageously, it can be completed within 30 to 60 min under isothermal conditions, without the need of a thermal cycler and/or specialized laboratory[12]. This technique eliminates the heat denaturation step for the DNA synthesis used in conventional PCR and relies instead on auto-cycling strand displacement DNA synthesis, which is achieved by a DNA polymerase with high strand displacement activity and a set of specially designed primers: two inner primers and two outer primers. Another important feature of LAMP is the color change, which is visible to the naked eye and results can be obtained in 60 min at a constant set temperature. The LAMP assay has been previously used to successfully detect other viral pathogens[13-19]. The objective of the present study is to develop and optimize a LAMP assay for the detection of DuCV.

Methods

Virus strains and DNA/RNA extraction

The DuCV strains and the other duck pathogen strains used in this study are listed in Table 5-24-1. The genomic DNA/RNA was extracted from 200 μL of the viruses listed in Table 5-24-1 using an EasyPure Viral DNA/RNA kit (TransGen, Beijing, China), according to the manufacturer's protocol. The DNA and RNA were eluted with 50 μL elution buffer. The concentrations of total DNA and RNA were measured by UV spectrophotometry (Beckman UV800, Beckman Coulter, USA). DNA and RNA samples were stored immediately at −70 ℃ until required.

Table 5-24-1 Pathogens strains and LAMP assay results

Pathogen strain/other duck pathogens	Source	Virus information
DuCV (GX1006)	GVRI	Original positive tissue specimen
DuCV (GX1008)	GVRI	Original positive tissue specimen
DuCV (GX1104)	GVRI	Original positive tissue specimen
DuCV (GX1105)	GVRI	Original positive tissue specimen
DuCV (GX1209)	GVRI	Original positive tissue specimen
DuCV (GX1208)	GVRI	Original positive tissue specimen
Muscovy duck parvovirus (AV238)	CIVDC	Duck embryo cultured virus
Avian influenza virus subtype H5 (Inactivated)	HVRI	Chick embryo cultured virus
Avian influenza virus subtype H9	GVRI	Chick embryo cultured virus
Duck plague virus	GVRI	Duck embryo cultured virus
Duck paramyxovirus	GVRI	Duck embryo cultured virus
Gosling parvovirus	GVRI	Duck embryo cultured virus
Duck hepatitis virus (AV2111)	CIVDC	Duck embryo cultured virus
Negative tissue (Duck spleen and liver)	GVRI	

Note: GVRI, Guangxi Veterinary Research Institute, China; CIVDC, China Institute of Veterinary Drugs Control, China; HVRI, Harbin Veterinary Research Institute, China; +, Positive; −, Negative.

Primer design for the LAMP assay

Primer design for DuCV LAMP was based on the published Rep gene sequence of the DuCV strain 33753-52 (GenBank accession number: DQ100076). The DuCV strain 33753-52 sequence was aligned with the available sequences of twenty viral isolates to identify the conserved regions.

A set of six specific primers for the DuCV LAMP assay was designed using LAMP primer design software, Primer Explorer V4. The LAMP primer set comprised two outer primers (forward primer F3 and backward primer B3), two inner primers (forward inner primer FIP and backward inner primer BIP) and two loop primers (forward loop F and backward loop B). The outer primers (F3 and B3) were used in the initial steps of the LAMP reactions but later, during the isothermal cycling, only the inner primers were used for strand displacement DNA synthesis. The loop primers were designed to accelerate the amplification reaction as previously described[17]. Since each inner primer (FIP (F1c+F2) and BIP (B1c+B2)) consists of two sequences, targeting two specific sites, the six primers consequently recognized eight sites on the target sequence specific to the Rep gene segment. The details of the primers are shown in Table 5-24-2. All primers were purchased from Invitrogen (Guangzhou, China).

Table 5-24-2 Oligonucleotide primers used for RT-LAMP assay

Primer name	Sequence (5'-3')	Genome position
FIP = F1c + F2	TTCAGGAATCCCTGAAGGTGATCGTCGGMGAGGAVAAGG	F1c, 203～222, F2, 167～185
BIP = B1c + B2	GCGMGAGCTGCCGCCCTTCTTCVTCAGATCCCCGG	B1c, 236～252, B2, 292～309
F3	AGTTBTGCACGCTCGACAAT	135～154
B3	GTCGACTCTTTGGMGCAATA	320～339
Loop F	AGGYGTVCCVTTCGCGC	186～202
Loop B	AGGAAGAGCCTGGCTCTC	268～285

Note: Genome position according to the DuCV complete genome sequence (GenBank accession number: DQ100076). Abbreviations are as follows: M, A or C; V, A or C or G; B, T or G or C.

Optimization of the DuCV LAMP conditions

The DuCV LAMP assay was performed in tubes containing 10 × Thermopol® Reaction Buffer (New England Biolabs, Beijing, China), large fragment Bst DNA polymerase (New England, Beijing, China), dNTPs (TaKaRa, Dalian, China), primers, betaine (Sigma-Aldrich, Shanghai, China), $MgSO_4$ (Sigma-Aldrich, Shanghai, China), calcein (International Laboratory USA San Francisco, USA), $MnCl_2$ (International Laboratory USA San Francisco, USA), template DNA/RNA and nuclease-free water. Based on the previous studies, different combinations of various concentrations of each component (dNTPs (0.4～1.6 mmol/L), betaine (0.8～1.4 mmol/L), $MgSO_4$ (2～9 mmol/L)) were tested for amplification efficiency. The amplification reaction was performed in a thermal block between 59 ℃ to 65 ℃ within 40 to 80 min, to ascertain the optimal incubation temperature and time. At the end of each incubation period, the reaction was terminated by heating at 80 ℃ for 5 min. All of the experiments were repeated three times.

Analysis of DuCV LAMP products

For a visual inspection of the LAMP assay products, fluorescence reagents (calcein and $MnCl_2$; International Laboratory, USA) were added to the reaction mixture before amplification and a color change of the reaction mixture was noted upon successful amplification. Samples that turned green were considered positive, while samples that remained orange were considered negative. Alternatively, 3 μL of the DuCV LAMP products were analyzed by 1.5% agarose gel electrophoresis. The presence of a smear or a pattern of multiple bands with different molecular weights indicated a positive result[13].

Evaluation of DuCV LAMP assays

To evaluate the specificity of the assay, DNA/RNA samples extracted from different virus strains including DuCV, Muscovy duck parvovirus, avian influenza virus subtypes H5 and H9, duck plague virus, duck paramyxovirus, Gosling parvovirus and duck hepatitis virus were tested by the DuCV LAMP assay (Table 5-24-1). The detection limit of the LAMP to DuCV was assessed and compared with conventional PCR [5] using a series of ten-fold dilutions of the DuCV template.

Briefly, DNA extracted from the DuCV strain GX1006 was used in the PCR reaction to amplify the Rep gene. The amplified product of the Rep gene was cloned into the pMD18-T cloning vector (TaKaRa, Dalian, China) according to the manufacturer's directions. The recombinant plasmids were sequenced. The sequence data were analyzed using DNASTAR software and were compared with the corresponding sequence data in GenBank. The copy number was calculated according to the following formula: (copies/μL=$6 \times 10^{23} \times$ DNA concentration, g/μL) /molecular weight, g/mol, as described by Xie et al.[20]. A series of ten-fold dilutions (1×10^7 to 1×10^0 copies/μL) were used to assess the sensitivity of the DuCV LAMP assay.

Detection of clinical samples by DuCV LAMP assay

A total of 181 clinical samples were collected from each sampling point (Nanning, Yulin, Hengxian, Ningming, Shanglin, Dongxing and Liuzhou) on commercial duck farms in Guangxi, China (Table 5-24-3). Spleen and liver (0.1 g) tissues were homogenized in 500 μL sterile saline, centrifugation at 12 000 g for 15 min, and then 200 μL of the supernatant were used to extract DNA by using an EasyPure Viral DNA/RNA kit (Trans Gen, Beijing, China), according to the manufacturer's protocol. The extracted DNAs were tested by both the DuCV LAMP assay and real-time PCR as previously described by Fringuelli et al.[2].

Table 5-24-3 Detection results of clinical samples by DuCV LAMP assay

Location of samples	LAMP		Real-time PCR	
	Positive samples/total samples	Positive rate	Positive samples/total samples	Positive rate
Nanning	5/29	17.24%	5/29	17.24%
Yulin	4/77	5.19%	4/77	5.19%
Hengxian	0/7	0%	0/7	0%
Ningming	0/13	0%	0/13	0%
Shanglin	1/38	2.63%	1/38	2.63%
Dongxing	1/11	9.09%	1/11	9.09%
Liuzhou	0/6	0%	0/6	0%
Total	11/181	6.08%	11/181	6.08%

Results

DuCV LAMP assay

After optimization of the reaction conditions, the DuCV LAMP assay was carried out in a 25 μL reaction mixture containing 2.5 μL 10 × Thermopol® Reaction Buffer (20 mM Tris-HCl, 10 mM $(NH_4)_2SO_4$, 10 mM KCl, 2 mM $MgSO_4$, 0.1% Triton X-100), 8 U Bst DNA polymerase, 1 mmol/L betaine, 7 mmol/L $MgSO_4$, 1.4 mmol/L dNTPs, 0.2 mmol/L F3 primer, 0.2 mmol/L B3 primer, 0.8 mmol/L forward loop (LF) primer,

0.8 mmol/L backward loop (LB) primer, 1.6 mmol/L forward inner primer (FIP), 1.6 mmol/L backward inner primer (BIP), 25 mmol/L calcein, 0.5 mmol/L $MnCl_2$, 2 μL template DNA and dH_2O to make the final volume up to 25 μL. The initial color of the reaction solution, prior to amplification, was orange. The optimal reaction time and incubation temperature were found to be 50 min at 62 ℃.

Specificity and sensitivity of DuCV LAMP assay

The optimized DuCV LAMP assay was used to specifically amplify six Guangxi field DuCV strains. Test results (Figure 5-24-1 A and B) showed that this technique exhibited no cross-reactivity with other avian viral pathogens tested including Muscovy duck parvovirus, Avian influenza virus subtype H5, Avian influenza virus subtype H9, Duck plague virus, Duck paramyxovirus, Gosling parvovirus and Duck hepatitis virus. Six DuCV field strains were tested by the DuCV LAMP assay and all final products of the LAMP assay yielded a positive green color (Figure 5-24-1 B, lanes 1 to 6) and showed a typical DNA ladder pattern after 1.5% agarose gel electrophoresis (Figure 5-24-1 A, lanes 1 to 6). The products of the DuCV LAMP assay for the other avian viral pathogens remained negative orange color (no color change) (Figure 5-24-1 B, lanes 7 to 13); furthermore, the other viral pathogens lacked the typical DNA ladder pattern, showing amplification did not occur (Figure 5-24-1 A, lanes 7 to 13).

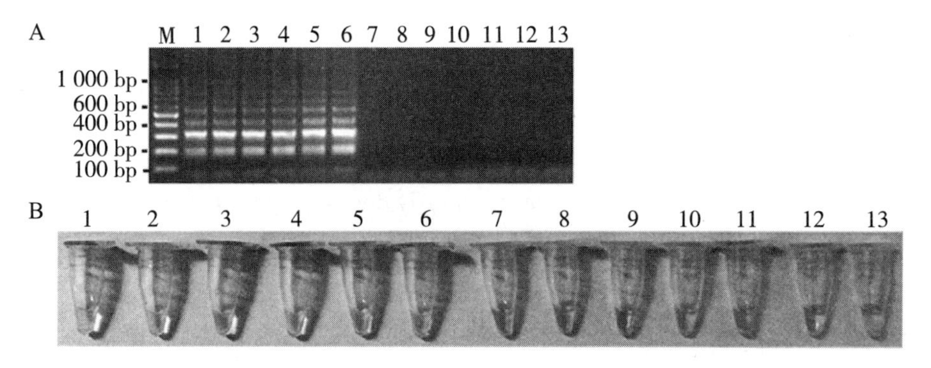

A: Agarose gel electrophoresis of the LAMP products; M: DNA marker; B: Visualization of the LAMP products; 1: DuCV GX1006; 2: DuCV GX1008; 3: DuCV GX1104; 4: DuCV GX1105; 5: DuCV GX1209; 6: DuCV GX1208; 7: Muscovy duck parvovirus AV238; 8: Avian influenza virus subtype H5; 9: Avian influenza virus subtype H9; 10: Duck plague virus; 11: Duck paramyxovirus; 12: Gosling parvovirus; 13: Duck hepatitis virus AV2111 (refer to the colors of C and D in the appendix Figure 5-20-2 for the colors of positive and negative samples in the LAMP assay).

Figure 5-24-1 Specificity of DuCV LAMP assay

The sensitivity of the DuCV LAMP assay was then determined and compared with that of conventional PCR in parallel. As shown in Figure 5-24-2 A, the detection limit of the DuCV LAMP assay is twenty copies (Figure 5-24-2 A and B). However, the detection limit of conventional PCR was 2×10^3 copies (Figure 5-24-3). This indicates that the sensitivity of the DuCV LAMP assay was 100-fold higher than that of conventional PCR. Our results showed that the DuCV LAMP assay is highly specific, sensitive and superior to the conventional PCR for the detection of DuCV.

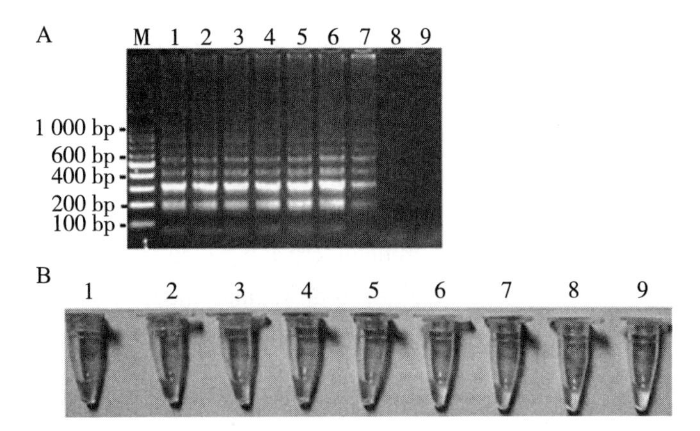

A: Agarose gel electrophoresis of the LAMP products; M: DNA marker; B: Visualization of the LAMP products. 1: 2×10^7 copies/tube; 2: 2×10^6 copies/tube; 3: 2×10^5 copies/tube; 4: 2×10^4 copies/tube; 5: 2×10^3 copies/tube; 6: 2×10^2 copies/tube; 7: 20 copies/tube; 8: 2 copies/tube; 9: Negative control (refer to the colors of C and D in the appendix Figure 5-20-2 for the colors of positive and negative samples in the LAMP assay).

Figure 5-24-2 Sensitivity of the DuCV LAMP assay

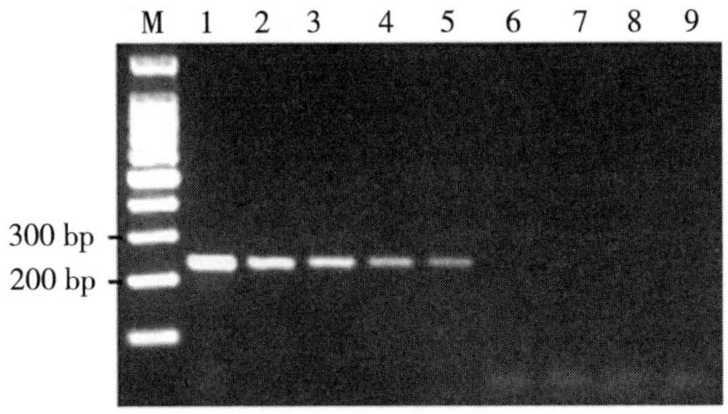

M 1 2 3 4 5 6 7 8 9

300 bp

200 bp

M: DNA marker; 1: 2×10^7 copies/tube; 2: 2×10^6 copies/tube; 3: 2×10^5 copies/tube; 4: 2×10^4 copies/tube; 5: 2×10^3 copies/tube; 6: 2×10^2 copies/tube; 7: 20 copies/tube; 8: 2 copies/tube; 9: Negative control.

Figure 5-24-3 Sensitivity of conventional PCR

Detection of clinical samples by DuCV LAMP assay

The DuCV LAMP assay was used to test 181 clinical samples, which were also tested by real-time PCR for a comparison of the two assays. Eleven duck samples (6.08%) were tested positive and 170 (93.92%) were tested negative by both methods (Table 5-24-3). The DuCV LAMP and real-time PCR yielded 100% of agreement in testing these samples. However, DuCV LAMP was noted to be a quicker, easier and more cost-efficient method compared to real-time PCR.

Discussion

DuCV can cause feathering disorders, poor body condition, immunosuppression and low weight in ducks[1]. Consequently, the development of a rapid, simple and sensitive detection method for DuCV is essential.

In this study, a set of six primers targeting the Rep gene segment of DuCV were designed and the reaction conditions for LAMP were optimized through repeated experiments. The concentrations of each reaction reagent, i.e., $MgSO_4$, betaine, dNTP, Bst DNA polymerase, calcein and $MnCl_2$, can influence the amplification efficiency of the LAMP assay, which subsequently affects the fluorescence emission of the resulting solution. Therefore, it is necessary to optimize the reaction system to obtain the appropriate reagent concentrations, reaction temperature and amplification time. The results of the specificity tests indicated that the DuCV

LAMP assay can successfully detected the DuCV field strains and showed no cross reaction with other viral pathogens. The established DuCV LAMP assay has several advantages in compared with conventional PCR. First, the detection limit of the DuCV LAMP assay is twenty copies, which is 100-fold higher than that of conventional PCR. Secondly, the amplification reaction of DuCV LAMP assay can be accomplished in a conventional laboratory water bath for 50 min and does not require a PCR machine or any other specialized equipment. Thirdly, the results can be inspected immediately and the color change is visible to the naked eye, which makes a rapid and easy determination of the test result without the need of electrophoresis analysis.

Nevertheless, the main challenge in the development of a DuCV LAMP assay is the huge genetic heterogeneity of this virus[4, 18, 21], with FIP and BIP being the most important primers in the LAMP assay that need to be conserved across the many strains. Another significant challenge in the development of a DuCV LAMP assay is its high amplification rate, which can result in potential cross-contamination issues. To investigate this issue, we added a dye (calcein with $MnCl_4$) into the reaction system before the amplification in order to eliminate the chance of contamination. Prior to amplification, calcein combines with the manganese ions (Mn^{2+}) and the reaction solution turns orange. When the positive LAMP amplification reaction proceeds, the generated pyrophosphate ions remove the manganese ions (Mn^{2+}) from calcein, resulting in the emission of fluorescence from calcein. The free calcein may then combine with magnesium ions (Mg^{2+}) in the reaction mixture, leading to stronger fluorescence emission[22]. By the way, after the positive LAMP amplification reaction proceeds, the generated pyrophosphate ions combine with manganese ions (Mn^{2+}), and generate manganous pyrophosphate. By centrifuge, the manganous pyrophosphate precipitated at the bottom of the tube (white sediment at the bottom of the tube). And the negative LAMP did not have the white sediment. The generated manganous pyrophosphate is another way to determine the LAMP result by the white sediment besides examination of the color by naked eye.

A LAMP assay for Goose circovirus detection has been previously reported[23], but the sequence of the Goose circovirus and DuCV was different. To our knowledge, this is the first study to explore the use of DuCV LAMP technology in a diagnostic test for DuCV. The method and analysis is simple, specific, sensitive and rapid, furthermore, it can be used to detect DuCV in clinical samples.

Conclusions

A simple, rapid, highly sensitive and specific DuCV LAMP assay for the detection of DuCV has been developed and established in our present study. This technique has the potential to be applied in clinical or field conditions as it does not require the use of sophisticated equipment.

References

[1] HATTERMANN K, SCHMITT C, SOIKE D, et al. Cloning and sequencing of Duck circovirus (DuCV). Archives of virology, 2003, 148(12): 2471-2480.

[2] FRINGUELLI E, SCOTT A N, BECKETT A, et al. Diagnosis of duck circovirus infections by conventional and real-time polymerase chain reaction tests. Avian Pathol, 2005, 34(6): 495-500.

[3] BALL N W, SMYTH J A, WESTON J H, et al. Diagnosis of goose circovirus infection in Hungarian geese samples using polymerase chain reaction and dot blot hybridization tests. Avian Pathol. 2004, 33(1):51-58.

[4] BANDA A, GALLOWAY-HASKINS R I, SANDHU T S, et al. Genetic analysis of a duck circovirus detected in commercial Pekin ducks in New York. Avian Dise, 2007, 51(1): 90-95.

[5] ZHANG X, JIANG S, WU J, et al. An investigation of duck circovirus and co-infection in Cherry valley ducks in Shandong

Province, China. Veterinary Microbiology, 2009, 133(3): 252-256.

[6] XIE L J, XIE Z X, ZHAO G Y, et al. Complete genome sequence analysis of a duck circovirus from Guangxi pockmark ducks. Journal of Virology, 2012, 86(23): 13136.

[7] SOIKE D, ALBRECHT K, HATTERMANN K, et al. Novel circovirus in Mulard ducks with developmental and feathering disorders. The Veterinary Record, 2004, 154(25): 792-793.

[8] WAN C H, FU G H, SHI S H, et al. Epidemiological investigation and genome analysis of duck circovirus in southern China. Virologica Sinica, 2011, 26(5):(5): 289-296.

[9] LIU S N, ZHANG X X, ZOU J F, et al. Development of an indirect ELISA for the detection of duck circovirus infection in duck flocks. Veterinary Microbiology, 2010, 145(1-2): 41-46.

[10] WAN C, HUANG Y, CHENG L, et al. The development of a rapid SYBR Green I-based quantitative PCR for detection of Duck circovirus. Virology Journal, 2011, 8: 465.

[11] HALAMI M Y, NIEPER H, M LLER H, et al. Detection of a novel circovirus in mute swans (*Cygnus olor*) by using nested broad-spectrum PCR. Virus Research, 2008, 132(1-2): 208-212.

[12] NOTOMI T, OKAYAMA H, MASUBUCHI H, et al. Loop-mediated isothermal amplification of DNA. Nucleic acids research, 2000, 28(12): E63.

[13] XIE Z X, TANG Y, FAN Q, et al. Rapid detection of group I avian adenoviruses by a loop-mediated isothermal amplification. Avian Dis, 2011, 55(4): 575-579.

[14] ENOMOTO Y, YOSHIKAWA T, IHIRA M, et al. Rapid diagnosis of herpes simplex virus infection by a loop-mediated isothermal amplification method. Journal of Clinical Microbiology, 2005, 43(2): 951-955.

[15] PENG Y, XIE Z X, LIU J B, et al. Visual detection of H3 subtype avian influenza viruses by reverse transcription loop-mediated isothermal amplification assay. Virology Journal, 2011, 8: 337.

[16] NEMOTO M, IMAGAWA H, TSUJIMURA K, et al. Detection of equine rotavirus by reverse transcription loop-mediated isothermal amplification (RT-LAMP). The Journal of Veterinary Medical Science, 2010, 72(6): 823-826.

[17] ROVIRA A, ABRAHANTE J, MURTAUGH M, et al. Reverse transcription loop-mediated isothermal amplification for the detection of Porcine reproductive and respiratory syndrome virus. Journal of Veterinary Diagnostic Investigation, 2009, 21(3): 350-354.

[18] XIE Z X, FAN Q, LIU J B, et al. Reverse transcription loop-mediated isothermal amplification assay for rapid detection of Bovine Rotavirus. BMC Veterinary Research, 2012, 8: 133.

[19] FAN Q, XIE Z X, XIE L J, et al. A reverse transcription loop-mediated isothermal amplification method for rapid detection of bovine viral diarrhea virus. Journal of Virological Methods, 2012, 186(1-2): 43-48.

[20] XIE Z X, XIE L J, PANG Y S, et al. Development of a real-time multiplex PCR assay for detection of viral pathogens of penaeid shrimp. Archives of Virology, 2008, 153(12): 2245-2251.

[21] WANG D, XIE X, ZHANG D, et al. Detection of duck circovirus in China: A proposal on genotype classification. Veterinary Microbiology, 2011, 147(3-4): 410-415.

[22] TOMITA N, MORI Y, KANDA H, et al. Loop-mediated isothermal amplification (LAMP) of gene sequences and simple visual detection of products. Nature Protocols, 2008, 3(5): 877-882.

[23] WOŹNIAKOWSKI G, KOZDRUŃ W, SAMOREK-SALAMONOWICZ E. Loop-mediated isothermal amplification for the detection of goose circovirus. Virology Journal, 2012, 9: 110.

A reverse transcription loop-mediated isothermal amplification method for rapid detection of bovine viral diarrhea virus

Fan Qing, Xie Zhixun, Xie Liji, Liu Jiabo, Pang Yaoshan, Deng Xianwen, Xie Zhiqin, Peng Yi, and Wang Xiuqing

Abstract

A reverse transcription loop-mediated isothermal amplification (RT-LAMP) assay was developed and optimized to detect bovine viral diarrhea viral (BVDV) RNA. The RT-LAMP assay is highly sensitive and able to detect 4.67×10^0 copies of BVDV RNA. Additionally, the RT-LAMP method is capable of detecting both genotypes of BVDV. No cross-reaction with other bovine viruses was observed. The ability of RT-LAMP to detect BVDV RNA from bovine fecal swabs was also evaluated. Of the 88 fecal swabs, 38 were found to be positive by RT-LAMP assay, whereas 39 were positive by real-time RT-PCR. Taken together, the BVDV specific RT-LAMP method is highly specific and sensitive and can be used as a rapid and direct diagnostic assay for testing clinical samples.

Keywords

bovine viral diarrhea virus (BVDV), 5' untranslated region (5' UTR), reverse transcription loop-mediated isothermal amplification (RT-LAMP)

Introduction

Bovine viral diarrhea virus (BVDV) is a positive-sense, single stranded RNA virus with a genome size of approximately 12.5 kb. BVDV is a member of the genus *Pestivirus* in the family Flaviviridae. BVDV is classified into two biotypes, cytopathogenic and noncytopathogenic, based on the presence or absence of the cytopathogenic effects in cell cultures. In addition, there are two major genotypes of BVDV (BVDV1 and BVDV2) based on the genetic relatedness[1].

BVDV has a high prevalence rate and low mortality, leading to significant economic losses[2]. BVDV infected animals may develop fever, mild diarrhea, and leukopenia. Infection of pregnant animals with noncytopathogenic BVDV during the first trimester may cause abortion, stillborn, or persistently infected claves[3]. Noncytopathogenic BVDV may spontaneously mutate to the cytopathogenic biotype, resulting in the onset of fatal mucosal disease[4]. Calves infected persistently are a major source of virus shedding since they usually do not exhibit any apparent clinical signs.

Identification of persistently infected calves, in combination with a proper vaccination program, is essential to the successful control of BVDV. The currently available diagnostic methods for BVDV include virus isolation, immunoassay, electron microscopy (EM), nuclei acid hybridization, and reverse transcription-polymerase chain reaction (RT-PCR)[5-8]. Although RT-PCR is a highly sensitive and specific test for detecting BVDV RNA, it has the intrinsic disadvantage of requiring a high-precision instrument such as a thermocycler for amplification and a time-consuming and complicated detection method (gelelec-trophoresis unit). In recent years, many studies have demonstrated the potential application of loop-mediated isothermal amplification (LAMP) or reverse transcription (RT)-LAMP assay for rapid detection of viral DNA or RNA[9-16]. Like RT-PCR, the RT-LAMP technique amplifies the target viral RNA sequence[17]. In this study, a RT-LAMP assay for detection of BVDV RNA is developed. Since the 5'untranslated region (5'UTR) of BVDV is among the most conserved regions and has been chosen as a preferred target region for detection of BVDV RNA by RT-PCR[15, 18], a set of six primers was designed to amplify 6 target sequences at the 5'UTR of the BVDV genome for the RT-LAMP assay.

Materials and methods

BVDV reference strains, field isolates, and other bovine pathogens

BVDV reference strains, field isolates, and other bovine pathogens used in the study are listed in Table 5-25-1.Three BVDV genotype 1 (BVDV1) reference strains, 13 BVDV1 field isolates, and 1 BVDV genotype 2 (BVDV2) reference strain were used to develop and optimize the RT-LAMP conditions. The BVDV reference virus strains were propagated in Madin-Darby bovine kidney (MDBK) cells in Dulbecco's modification of Eagle's medium (DMEM) supplemented with 10% fetal calf serum (Shijiqing, China, free of BVDV and antibody to BVDV). Thirteen field isolates were isolated from calves with typical BVDV clinical signs such as diarrhea, miscarriages and stillborn. For virus isolation, a total of 1 g of fresh aseptically collected liver tissues was homogenized in 4 mL of phosphate-buffered saline (PBS, pH 7.2). A 1 : 10 dilution of the supernatants was then inoculated onto MDBK monolayer cultures and incubated for 6 days. The presence or absence of cytopathic effect was recorded. After 3 times of freezing and thawing, culture supernatants were collected after centrifugation and stored at −70 ℃ for further characterization with indirect immunoperoxidase (IPX) test and electron microscopy (EM). All 13 field isolates were BVDV positive as revealed by both positive IPX staining and the typical BVDV morphology (40～60 nm in diameter) under an electron microscope.

Five other bovine pathogens, 2 negative tissue samples (1 healthy bovine nasal swab and 1 healthy bovine blood sample), 28 fecal negative samples, and 5 individual MDBK cell culture samples were included as negative controls to test the specificity of the assay (Table 5-25-1). All the BVDV positive and negative samples were validated by both the BVDV antibody test kit (IDEXX, USA) and an RT-PCR assay.

Table 5-25-1 Virus and samples used in the RT-LAMP

Name	Source	RT-LAMP result	
		Agarose gel electrophoresis	Color change after adding dye
BVDV1			
OregonCV24	CVCC, China	+	+
NADL	CVCC, China	+	+
AV68	CVCC, China	+	+
GX-BVDV1	GVRI	+	+
GX-BVDV2	GVRI	+	+
GX-BVDV3	GVRI	+	+
GX-BVDV4	GVRI	+	+
GX-BVDV5	GVRI	+	+
GX-BVDV6	GVRI	+	+

continued

Name	Source	RT-LAMP result	
		Agarose gel electrophoresis	Color change after adding dye
GX-BVDV7	GVRI	+	+
GX-BVDV8	GVRI	+	+
GX-BVDV9	GVRI	+	+
GX-BVDV10	GVRI	+	+
GX-BVDV11	GVRI	+	+
GX-BVDV12	GVRI	+	+
GX-BVDV13	GVRI	+	+
BVDV2			
GX-041	GVRI	+	+
Other bovine pathogens			
Bovine rotavirus (NCDV strain, G6P10 genotype)	CVCC, China	–	–
Mycobacterium bovis (MB332, Guangxi field isolate)	CVCC, China	–	–
Classical swine fever virus (Guangming strain, AV64, North American genotype)	CVCC, China	–	–
Classical swine fever virus (Shimen strain, AV1411, North American genotype)	CVCC, China	–	–
Classical swine fever virus (79105strain, AV63, North American genotype)	CVCC, China	–	–
Infective bovine rhinotracheitis virus (Bartha Nu/67strain, AV20)	CVCC, China	–	–
Bovine Corona virus (GX-BC-125, Guangxi field isolate)	GVRI	–	–
Negative tissue samples			
Healthy bovine nasal swab	GVRI	–	–
Healthy bovine blood sample	GVRI	–	–
Positive tissue sample			
BVDV bovine nasal swab	GVRI	+	–
BVDV bovine blood sample	GVRI	+	–
Fecal negative samples			
A/Holstein cow/Guangxi/NN1732/2009	NN	–	–
A/Holstein cow/Guangxi/NN3363/2009	NN	–	–
A/Holstein cow/Guangxi/NN4523/2009	NN	–	–
A/Holstein cow/Guangxi/NN4462/2009	NN	–	–
A/Holstein cow/Guangxi/NN21/2009	NN	–	–
A/Holstein cow/Guangxi/NN12/2009	NN	–	–
A/Water buffalo/Guangxi/NN789/2009	NN	–	–
A/Water buffalo/Guangxi/NN35/2009	NN	–	–
A/Water buffalo/Guangxi/NN4520/2009	NN	–	–
A/Water buffalo/Guangxi/NN7/2009	NN	–	–
A/Water buffalo/Guangxi/NN3620/2009	NN	–	–
A/Water buffalo/Guangxi/NN49/2009	NN	–	–

continued

Name	Source	RT-LAMP result	
		Agarose gel electrophoresis	Color change after adding dye
A/Water buffalo/Guangxi/NN332/2009	NN	–	–
A/Water buffalo/Guangxi/NN28/2009	NN	–	–
A/Water buffalo/Guangxi/NN0137/2009	NN	–	–
A/Water buffalo/Guangxi/NN703/2009	NN	–	–
A/Water buffalo/Guangxi/NN46/2009	NN	–	–
A/Yellow cow/Guangxi/LZ789/2010	LZ	–	–
A/Yellow cow/Guangxi/LZ45/2010	LZ	–	–
A/Yellow cow/Guangxi/LZ733/2010	LZ	–	–
A/Yellow cow/Guangxi/LZ719/2010	LZ	–	–
A/Yellow cow/Guangxi/LZ776/2010	LZ	–	–
A/Yellow cow/Guangxi/LZ782/2010	LZ	–	–
A/Yellow cow/Guangxi/LZ713/2010	LZ	–	–
A/Yellow cow/Guangxi/LZ708/2010	LZ	–	–
A/Yellow cow/Guangxi/LZ730/2010	LZ	–	–
A/Yellow cow/Guangxi/LZ744/2010	LZ	–	–
A/Yellow cow/Guangxi/LZ760/2010	LZ	–	–
Cell cultures			
MDBK1	GVRI	–	–
MDBK2	GVRI	–	–
MDBK3	GVRI	–	–
MDBK4	GVRI	–	–
MDBK5	GVRI	–	–

Note: CVCC, Chinese Veterinary Culture Collection Center; GVRI, Guangxi Veterinary Research Institute; NN, Jinguang Diary Farm, Nanning, Guangxi; LZ, Huangshi Cattle Farm, Liuzhou, Guangxi; +, positive; –, negative.

RNA/DNA extraction

The genomic viral RNA was extracted from 250 μL of BVDV-infected culture supenatant by using the TRIZOL RNA extract reagent (Invitrogen, America) in accordance with the manufacture' s protocol. DNA was extracted using phenol ∶ chloroform ∶ isoamyl alcohol (1 ∶ 1 ∶ 24, v/v/v) according to the Gibco BRL manufacturer' s protocol (Gibco BRL, Grand Island, NY, USA). The extracted RNA and DNA were eluted in distilled water and stored at –70 ℃ until use.

Primer design

A multiple sequence alignment was performed for 40 randomly selected BVDV1 isolates and 15 BVDV2 isolates from GenBank database. The 5'UTR of BVDV, which is sufficiently conserved among the randomly selected BVDV isolates, was chosen for primer design. Primer Explored V4 software was used to design the RT-LAMP primers. The two outer primers are known as the forward outerprimer (F3) and the backward outer primer (B3), which helps in strand displacement. The inner primers are known as the forward inner primer (FIP) and the backward inner primer (BIP), respectively. Each inner primer has two distinct sequences corresponding to the sense and antisense sequences of the target, one for priming during the early stage and the other for self-priming during late stage of LAMP. FIP contains F1C (complementary to F1) and F2 sequence. BIP contains the B1C sequence (complementary to B1) and B2 sequence. Both FIP and BIP were high-performance liquid chromatography purified. FIP and BIP contained 3 R (C+G) and 1 Y (A+T) to facilitate the amplification of both BVDV1 and BVDV2 (Table 5-25-2 and Figure 5-25-1). All primers were synthesized by Invitrogen (Guangzhou, China). The nucleotide sequences and locations of the primers for both Oregon CV24 (GenBank accession number: 0911605.1) and New York 93 (GenBank accession number: AF502399.1) are shown in Table 5-25-2 and Figure 5-25-1.

Table 5-25-2 Primers used in RT-LAMP

Primer name	Type	Genome position	Sequence (5'-3')
F3	Forward outer	97～119	AAGAGGCTAGCCATGCCCTTAGT
B3	Reverse outer	448～470	CTGCCTGGTCGTAAACAGGTTCC
FIP [a]	Forward inner	F1C, 185～208	GTCGAAC**R**ACTGACGACTA**R**CCTG
	(FIP=F1C+F2)	F2, 163～182	TGGATGGCTTAAG**R**CCTGAG
BIP [a]	Reverse inner	B1C, 326～346	TGATAGGGTGCTGCAGAGGCC
	(BIP=B1C+B2)	B2, 363～388	CATGTGCCATGTACAGCAGAG**Y**TTTT

Note: The positions of the primers are based on the nucleotide sequence of Oregon CV24 (GenBank accession number: AF0911605.1). a Each inner primer of RT-LAMP contains two connected primers. The R and Y are bolded.

```
                 F3
Oregon CV24: 97AAGAGGCTAGCCATGCCCTTAGTAGGACTAGCATAGCGAGGGGG
NewYork 93:  96 AAGAGGCTAGCCATGCCCTTAGTAGGACTAGCAAAAGTAGGGGA
                                                     F2
Oregon CV24:  GTAGCAACAGTGGTGAGTTCGTTGGATGGCTTAAGCCCTGAGTAC
NewYork 93:   CTAGCGGTAGCAGTGAGTTCGTTGGATGGCTTAAGTCCTGAGTAC
                  F1c
Oregon CV24:  AGGGTAGTCGTCAGTGGTCCGACGCCTTAACA—TGAAG—GTCTC
NewYork 93:   AGGGGAGTCGTCAGTGGTTCGACACTCCATTAGTCGAGGAGTCTC

Oregon CV24:  GAGATGCCACGTGGACGAGGGCACGCCCAAAGCACATCTTAGCC
NewYork 93:   GAGATGCCATGTGGACGAGGGCATGCCC ACGGCACA TCTTAACC

Oregon CV24:  CGAGCGGGGGTCGCTCGGACGAAAACAGTTTGATCAACTGCTAC
NewYork 93:   CA TGCGGGGGTTGCA TGGGTGAAAGCGCTATTCGTGGC- GTTAT
                                   B1c
Oregon CV24:  GAAT ACAGCCTGATAGGGTGCTGCAGAGGCCCACTGTATTGCTAC
NewYork 93:   GACACAGCCCTGATAGGGTGT TGCAGAGGCCTGCTATTCCGCTAG
                         B2
Oregon CV24   TAAAAATCTCTGCTGTACATGGCACATGGAA TTGATTACAAATGA
NewYork 93:   TAAAAAACTCTGCTGTACATGGCACATGGAGTTGTTTTCAAATGA

Oregon CV24:  ACTCTTATACAAAACATACAAACAAAAACCCGTCGGGGTGGAGG
NewYork 93:   ACTTTTATACAAAACATATAAACAAAAACCAGCAGGCGTCGTGG
                  B3
Oregon CV24:  AACCTGTTTACGAC—CAGGCAG   470
NewYork 93:   AACCTGTTTACGACG TCAACGG  474
```

The GenBank accession numbers for Oregon CV24 and New York 93 are 0911605.1 and 502399.1, respectively. The nucleotide sequences of primers are underlined. The mismatched nucleotides between the two strains are marked in red.

Figure 5-25-1 Locations of the primers used in RT-LAMP

RT-PCR

RT-PCR was performed as described previously[19]. Briefly, RT-PCR was performed with the Quant One Step RT-PCR Kit (Qiagen Inc, Beijing, China) by using 1 μL (20 ng) of RNA template and 50 pmol of each primer in a 25 μL reaction volume by following the manufacturer's protocol with the following cycling times and temperatures: 94 ℃ for 5 min and 30 cycles of 94 ℃ for 30 s, 55 ℃ for 30 s, and 72 ℃ for 30 s. RT-PCR product was analyzed by agarose gel electrophoresis.

Real-time RT-PCR

Real-time RT-PCR was performed using the following primers and probe, which detects a 102 bp product of the 5'UTR of BVDV. The primers are 5'-TAGCCATGCCCTTAGTAGGACT-3' and 5'-GAACCACTGACGACTACCCTGT-3'. The probe is FAM-CAGTGGTGAGTTCGTTGGATGGCT-BHQ1.Real-time RT-PCR amplification was carried out with the Real-time One Step RT-PCR Kit (TaKaRa, Dalian, China) as described previously[15]. Briefly, 10 μL of 2 × One Step RT buffer Ⅲ, 0.5 μL of TaKaRa Ex Taq HS (5 U/μL), 0.5 μL of Prime Script RT Enzyme Mix Ⅱ (5 U/μL), 0.5 μL of each primer (4 μM), 0.5 μL of probe (4 μM), 1 μL of RNA template, and 7 μL of DNase/RNase-free water were added into a 0.2 mL tube. Real-time PCR was performed in a Light Cycler 2.0 (Roche Diagnostic, Indianapolis, IN) using the following conditions: reverse transcription reaction at 42 ℃ for 5 min, 95 ℃ for 10 s, and 40 cycles of 95 ℃ for 5 s, 60 ℃ for 30 s. Cycle threshold (CT) was manually set up to reflect the best kinetic parameter.

RT-LAMP

The RT-LAMP reaction was performed in a 25 μL reaction mixture containing 1.4 μmol of each deoxyribonucleotide triphosphate, 0.8 mmol of betaine (Sigma Chemical Co., Beijing, China), 2.5 μL of 10 × Thermo buffer, 8 mmol $MgSO_4$, 8 U of Bst DNA polymerase (large fragment; New England Biolabs), 0.125 U of enhanced Avian myeloblastosis virus reverse transcriptase (TaKaRa, Dalian, China), and 1 μL of the extracted target RNA. Various concentrations of the FIP, BIP, F3, B3, loop F, and loop B primers were used to optimize the ratio of primers. To determine the optimal temperature and incubation time for the RT-LAMP assay, the reaction mixtures were incubated in a water bath at 60 ℃, 61.5 ℃, 63 ℃, 64.5 ℃, and 66 ℃ for 20, 40, 60 and 80 min, respectively. The reaction was terminated by heating at 80 ℃ for 2 min.

Analysis of RT-LAMP product

The RT-LAMP products were analyzed by three methods. The first and the most direct method was to visually inspect the turbidity of the samples formed due to the accumulation of magnesium pyrophosphate, a byproduct of the DNA amplification reaction[11, 20]. The second method was to add a fluorescent dye such as SYBR Green Ⅰ (Solarbio, Beijing, China) or Genefinder™ (Boiv, Xiamen, China) to the samples and to observe the color change under an ultraviolet (UV) hand lamp at a 365 nm wavelength. Samples which were yellow-green were considered positive, while samples which were orange were considered negative. Samples were compared with a negative control to allow for background fluorescence[14, 15]. Finally, the RT-LAMP products were detected by agarose gel electrophoresis. The RT-LAMP reaction generates a combination of DNA fragments of different sizes. The presence of a smear or a pattern of multiple bands of different molecular weights was considered a positive result. A molecular marker was used to estimate the sizes of amplified products.

Specificity and sensitivity of RT-LAMP

To evaluate the specificity of the RT-LAMP assay, experiments were performed initially on a panel of reference BVDV virus strains including both genotype 1 and 2 (Table 5-25-1). In addition, bovine rotavirus (BRV), *Mycobacterium bovis*[21], classical swine fever virus (CSFV), infectious bovine rhinotracheitis virus (BoHV-1), bovine coronavirus (BC), bovine negative fecal samples, healthy bovine nasal swab and blood sample, and MDBK cells were included as negative controls.

The detection limit of RT-LAMP was tested and compared with conventional RT-PCR and real-time RT-PCR by using the same templates at identical RNA concentrations (4.67×10^8 to 4.67×10^{-1} copies of RNA/μL). Additionally, the healthy cattle fecal swab samples spiked with known amount of viral RNA of BVDV strain Oregon CV24 were quantitated by real-time RT-PCR and RT-LAMP. Briefly, BVDV RNA derived from Oregon CV24 was serially diluted from 4.67×10^8 to 4.67×10^{-1} copies of RNA/μL and added to the healthy fecal samples. RNAs extracted from these artificial positive samples were subjected to RT-LAMP and real-time PCR.

Fecal specimens

A total of 88 fecal swab samples were collected from calves with diarrhea, which came from different dairy farms in the Guangxi (Table 5-25-3). A written informed consent was obtained from each participating farm owner. The veterinarians of the participating farms collected fecal swab samples from calves of 6 to 48 months old. No official review and approval of the animal protocol by Guangxi Veterinary Research Institute was needed. The samples were diluted in 1 mL of sterilized water, followed by RNA extraction as described in Section *RNA/DNA extraction*. BVDV-specific RT-LAMP assay was performed as describe in Section *RT–LAMP*. The results of BVDV specific RT-LAMP were compared with the results of conventional RT-PCR and real-time RT-PCR. The real-time RT-PCR products were cloned into a PMD18-T (TaKaRa, Dalian, China) vector and sequenced. The phylogenetic analysis of the 102 bp real-time RT-PCR products of these BVDV positive samples was performed using MegAlign (DNAStar 5.0).

Table 5-25-3 Comparison of RT-PCR, real-time RT-PCR, and RT-LAMP methods for detection of BVDV from clinical fecal samples

Location of samples	Number of samples	Number of positive samples		
		RT-PCR	Real-time RT-PCR	RT-LAMP
Nanning	23	9	13	12
Liuzhou	15	5	7	7
Fangcheng	10	1	2	2
Shangsi	9	2	3	3
Guilin	20	6	9	9
Hengxian	11	4	5	5
Total	88	30	39	38

Results

Optimization of RT-LAMP

Following standardization and optimization, the optimal ratio of primer concentrations for RT-LAMP reaction was found to be 8 : 1 : 4 (1.6, 0.2 and 0.8 mmol) for inner : outer : loop primers. The RT-LAMP assay amplified a 228 bp target sequence of the 5' UTR of BVDV after incubation at 63 ℃ in 60 min. The RT-

LAMP products were observed as a ladder-like pattern on the agarose gel. This is due to the formation of a mixture of stem-loop DNAs with various stem lengths and cauliflower-like structures by annealing between alternately inverted repeats of the target sequence in the same strand.

Sensitivity and specificity of RT-LAMP

The RT-LAMP specifically detected BVDV Oregon CV24, NADL, AV68, 13 Guangxi field isolates and GX-041.No cross-reactivity with other bovine viruses was observed (Table 5-25-1). This specificity was further confirmed by agarose gel electrophoresis and the color change after adding a fluorescent dye (data not shown). The sensitivity of this method was determined by using a 10-fold serial dilution of BVDV RNA. The detection limit of RT-LAMP and real-time RT-PCR were 4.67×10^0 copies, whereas the detection limit of RT-PCR was 4.67×10^3 copies (Figure 5-25-2). The detection limit of RT-LAMP for spiking negative samples was 4.67×10^1 copies, whereas the detection limit of real-time RT-PCR for spiking negative samples was 4.67×10^3 copies (data not shown).

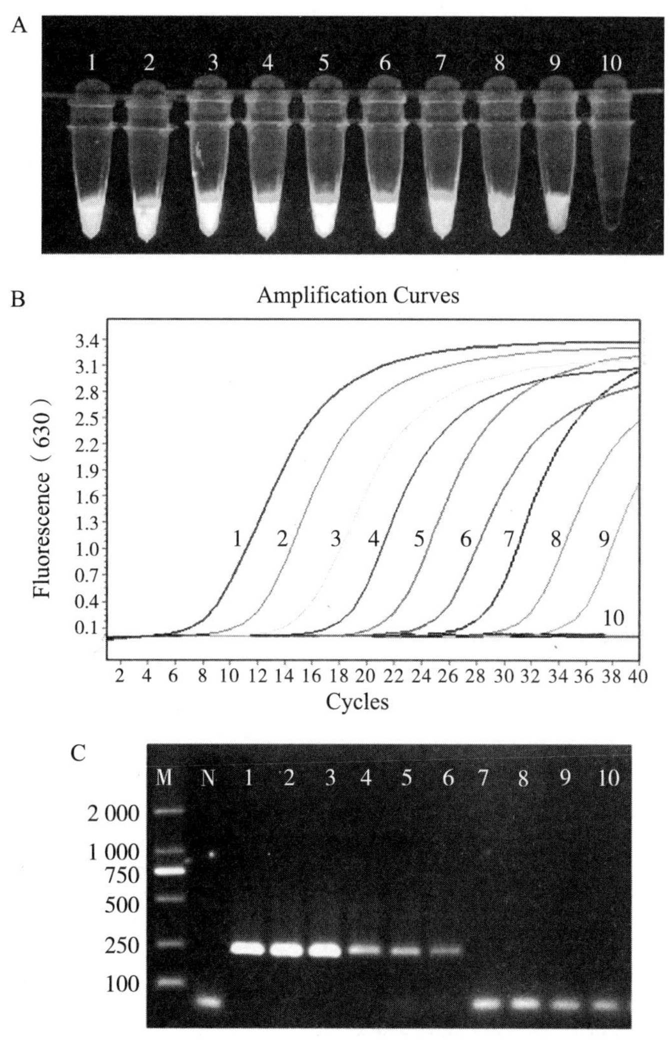

A: Results of the RT-LAMP products were viewed under ultraviolet (UV) light; B: Results of the real-time RT-PCR analysis; C: Results of the RT-PCR analysis. M: DNA marker; N: Negative control; 1: 4.67×10^8 copies/tube; 2: 4.67×10^7 copies/tube; 3: 4.67×10^6 copies/tube; 4: 4.67×10^5 copies/tube; 5: 4.67×10^4 copies/tube; 6: 4.67×10^3 copies/tube; 7: 4.67×10^2 copies/tube; 8: 4.67×10^1 copies/tube; 9: 4.67×10^0 copies/tube; 10: 4.67×10^{-1} copies/tube. All experiments were repeated three times and similar results were obtained.

Figure 5-25-2　The sensitivities of RT-LAMP, RT-PCR and real-time RT-PCR assays

Detection of BVDV from clinical samples

The sensitivity of the RT-LAMP method in detecting BVDV RNA from 88 clinical fecal samples was compared with those of RT-PCR and real-time RT-PCR methods. Results are shown in Table 5-25-3.Thirty of 88 fecal swab samples (34.1%) were positive by RT-PCR analysis, whereas 39 of 88 fecal swab samples (44.3%) were positive by real-time RT-PCR, and 38 of 88 (43.2%) were positive by RT-LAMP (Table 5-25-3). Thirty samples (34.1%) were positive by three methods. Eight samples (9.1%) were positive by both real-time RT-PCR and RT-LAMP, but negative by RT-PCR analysis. No sample (0%) was positive by RT-PCR and negative by RT-LAMP. All RT-PCR positive samples are also RT-LAMP positive. However, a sample was positive by real-time PCR, but negative by RT-LAMP. All the real-time RT-PCR positive samples were true positive as evidenced by DNA sequence analysis, suggesting that no nonspecific amplification in RT-LAMP reaction occurred. A phylogenetic analysis of the 102 bp real-time RT-PCR products of 39 samples in comparison with those of Oregon CV24 and New York 93 is shown in Figure 5-25-3.

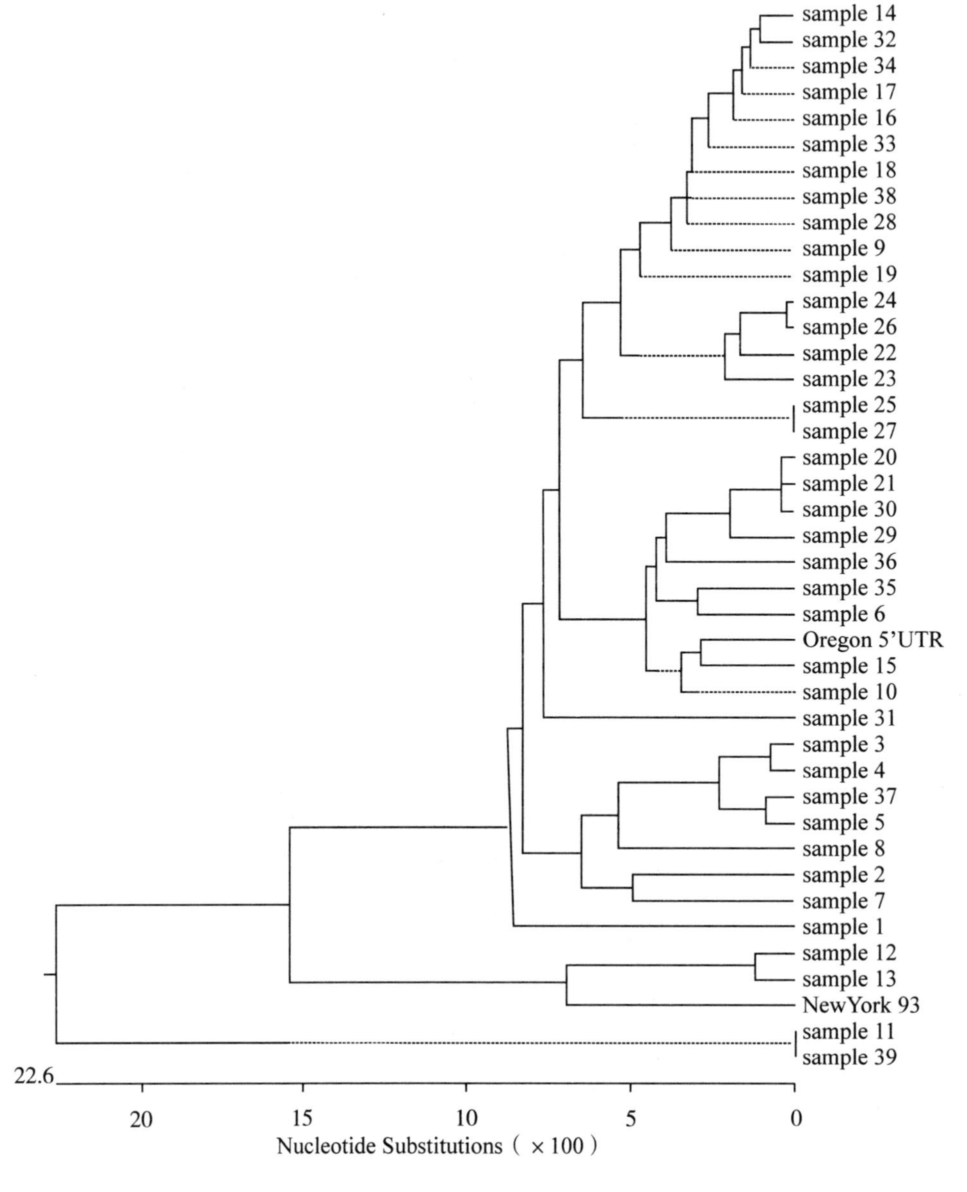

Figure 5-25-3 The phylogenetic analysis of the 102 bp real-time RT-PCR products of the 39 BVDV positive field samples in comparison with Oregon CV24 and New York 93 Reference strains by using MegAlign (DNAStar 5.0)

Discussion

RT-LAMP has been used in detecting viral RNA molecules due to its simplicity and high sensitivity for a number of viruses including avian influenza virus, classical swine fever virus, West Nile virus, and porcine reproductive and respiratory syndrome virus[9, 12, 14, 22]. In this study, an RT-LAMP method was developed for detection of both BVDV1 and BVDV2.The RT-LAMP method exhibited the same sensitivity as real-time RT-PCR and did not detect classical swine fever virus and other bovine viruses. Similarly, an RT-RAMP method for classical swine fever virus did not detect BVDV[22], suggesting the specificity of each RT-LAMP assay. Twenty-one BVDV1 isolates and 17 BVDV2 isolates were selected at random from the GenBank database and the sequence homology in the 5' UTR region upon which the RT-LAMP primers were designed was analyzed. It is estimated that the RT-LAMP would detect 95% of BVDV1 and 70.6% of BVDV2. The lower detection rate for BVDV2 is primarily due to the higher genetic variations observed in the 5' UTR of BVDV2 compared to BVDV1 isolates. All 39 BVDV positive samples confirmed by real-time RT-PCR were clustered within the BVDV1 genotype as revealed by DNA sequence analysis of the 102 bp PCR product (Figure 5-25-3). The RT-LAMP method will be particularly useful in detecting BVDV from cattle herds in China since BVDV1 is the major type of BVDV circulating in the cattle herds[23, 24]. BVDV2 was first isolated from cattle in China in 2010[25]. Due to the inherent genetic heterogeneity of BVDV, it remains to be a challenge to develop a universal molecular test that is capable of detecting all BVDV isolates.

The BVDV specific RT-LAMP method is 1 000 times more sensitive than RT-PCR in detecting BVDV RNA. In contrast, real-time RT-PCR has the same sensitivity as RT-LAMP (4.67 copies of viral RNA). The RT-LAMP failed to detect a true BVDV positive clinical fecal sample, which was confirmed by real-time RT-PCR and DNA sequencing. There were 8 fecal samples positive by RT-LAMP and real-time RT-PCR but negative by RT-PCR. It is possible that the 8 fecal samples may have had few viral RNA molecules, which were below the detection limit of RT-PCR analysis. It is interesting to note that the primers for RT-PCR (100～342 nt), real-time RT-PCR (104～205 nt), and RT-LAMP (97～470 nt) described in the study detected the same 5' UTR region of BVDV Oregon CV24 (GenBank accession number: AF0911605.1).

SYBR Green Ⅰ and GeneFinder™ were used for the detection of the amplified products and their detection efficiency was compared. Both fluorescent dyes give consistent results. Some previous studies indicated that false positive results could occur due to the high amplification efficiency, short target sequences, or contaminations associated with adding the fluorescent dye[20]. The false positive results associated with contamination can be avoided by adding SYBR green Ⅰ directly to the reaction mixture.

In conclusion, the BVDV-specific RT-LAMP is a simple, rapid, highly sensitive and specific assay. This technique has potential application in both clinical diagnosis and field surveillance of BVDV since it does not require the use of sophisticated equipment.

References

[1] RIDPATH J F, BOLIN S R, DUBOVI E J. Segregation of bovine viral diarrhea virus into genotypes. Virology, 1994, 205(1): 66-74.

[2] HOUE H. Epidemiology of bovine viral diarrhea virus. Vet Clin North Am Food Anim Pract, 1995, 11(3): 521-547.

[3] MAHONY T J, MCCARTHY F M, GRAVEL J L, et al. Genetic analysis of bovine viral diarrhoea viruses from Australia. Vet Microbiol, 2005, 106(1-2): 1-6.

[4] MEYERS G, RUMENAPF T, TAUTZ N, et al. Insertion of cellular sequences in the genome of bovine viral diarrhea virus.

Arch Virol Suppl, 1991, 3: 133-142.

[5] DEREGT D, CARMAN P S, CLARK R M, et al. A comparison of polymerase chain reaction with and without RNA extraction and virus isolation for detection of bovine viral diarrhea virus in young calves. J Vet Diagn Invest, 2002, 14(5): 433-437.

[6] RIDPATH J F, HIETALA S K, SORDEN S, et al. Evaluation of the reverse transcription-polymerase chain reaction/probe test of serum samples and immunohistochemistry of skin sections for detection of acute bovine viral diarrhea infections. J Vet Diagn Invest, 2002, 14(4): 303-307.

[7] FULTON R W, HESSMAN B, JOHNSON B J, et al. Evaluation of diagnostic tests used for detection of bovine viral diarrhea virus and prevalence of subtypes 1a, 1b, and 2a in persistently infected cattle entering a feedlot. J Am Vet Med Assoc, 2006, 228(4): 578-584.

[8] YOUSSEF B Z. Comparative study between ELISA, immuno-diffusion and cell bound immuno assay techniques for detection of anti-bovine viral diarrhea antibodies in calves of some farms in alexandria and behira governorates. J Egypt Public Health Assoc, 2006, 81(1-2): 29-41.

[9] PARIDA M, POSADAS G, INOUE S, et al. Real-time reverse transcription loop-mediated isothermal amplification for rapid detection of West Nile virus. J Clin Microbiol, 2004, 42(1): 257-263.

[10] ENOMOTO Y, YOSHIKAWA T, IHIRA M, et al. Rapid diagnosis of herpes simplex virus infection by a loop-mediated isothermal amplification method. J Clin Microbiol, 2005, 43(2): 951-955.

[11] CHO H S, KANG J I, PARK N Y. Detection of canine parvovirus in fecal samples using loop-mediated isothermal amplification. J Vet Diagn Invest, 2006, 18(1): 81-84.

[12] CHEN L, FAN X Z, WANG Q, et al. A novel RT-LAMP assay for rapid and simple detection of classical swine fever virus. Virol Sin, 2010, 25(1): 59-64.

[13] KOMIYAMA C, SUZUKI K, MIURA Y, et al. Development of loop-mediated isothermal amplification method for diagnosis of bovine leukemia virus infection. J Virol Methods, 2009, 157(2): 175-179.

[14] ROVIRA A, ABRAHANTE J, MURTAUGH M, et al. Reverse transcription loop-mediated isothermal amplification for the detection of Porcine reproductive and respiratory syndrome virus. J Vet Diagn Invest, 2009, 21(3): 350-354.

[15] FAN Q, XIE Z, LIU J, et al. Establishment of real-time flourescent quantitative PCR for detection of bovine viral diarrhea virus. Progress in Veterinary Medicine, 2010, 31(10): 10-14.

[16] YIN S, SHANG Y, ZHOU G, et al. Development and evaluation of rapid detection of classical swine fever virus by reverse transcription loop-mediated isothermal amplification (RT-LAMP). J Biotechnol, 2010, 146(4): 147-150.

[17] NOTOMI T, OKAYAMA H, MASUBUCHI H, et al. Loop-mediated isothermal amplification of DNA. Nucleic Acids Res, 2000, 28(12): E63.

[18] LETELLIER C, KERKHOFS P, WELLEMANS G, et al. Detection and genotyping of bovine diarrhea virus by reverse transcription-polymerase chain amplification of the 5' untranslated region. Vet Microbiol, 1999, 64(2-3): 155-167.

[19] HAMEL A L, WASYLYSHEN M D, NAYAR G P. Rapid detection of bovine viral diarrhea virus by using RNA extracted directly from assorted specimens and a one-tube reverse transcription PCR assay. J Clin Microbiol, 1995, 33(2): 287-291.

[20] THEKISOE O M, BAZIE R S, CORONEL-SERVIAN A M, et al. Stability of loop-mediated isothermal amplification (LAMP) reagents and its amplification efficiency on crude trypanosome DNA templates. J Vet Med Sci, 2009, 71(4): 471-475.

[21] GOLEMBA M D, PARRENO V, JONES L R. Simple procedures to obtain exogenous internal controls for use in RT-PCR detection of bovine pestiviruses. Mol Cell Probes, 2008, 22(3): 212-214.

[22] CHEN H T, ZHANG J, SUN D H, et al. Development of reverse transcription loop-mediated isothermal amplification for rapid detection of H9 avian influenza virus. J Virol Methods, 2008, 151(2): 200-203.

[23] LI Y, LIU Z, WU Y. Isolation and identification of bovine viral diarrhea virus-mucosal disease virus strain Changchun 184. Chin J Vet Sci, 1983, 3(2): 546-553.

[24] XUE F, ZHU Y M, LI J, et al. Genotyping of bovine viral diarrhea viruses from cattle in China between 2005 and 2008. Vet Microbiol, 2010, 143(2-4): 379-383.
[25] XIE Z, TANG Y, FAN Q, et al. Rapid detection of Group I avian adenoviruses by a Loop-mediated isothermal amplification. Avian Dis, 2011, 11(6): 575-579.

Reverse transcription loop-mediated isothermal amplification assay for rapid detection of bovine rotavirus

Xie Zhixun, Fan Qing, Liu Jiabo, Pang Yaoshan, Deng Xianwen, Xie Zhiqin, Xie Liji, and Khan Mazhar I

Abstract

Background: Bovine rotavirus (BRV) infection is common in young calves. This viral infection causes acute diarrhea leading to death. Rapid identification of infected calves is essential to control BRV successfully. Therefore development of simple, highly specific, and sensitive detection method for BRV is needed. Results: A reverse transcription loop-mediated isothermal amplification (RT-LAMP) assay was developed and optimized for rapid detection of BRV. Specific primer sets were designed to target the sequences of the VP6 gene of the neonatal calf diarrhea virus (NCDV) strain of BRV. The RT-LAMP assay was performed in a water bath for 60 min at 63 ℃, and the amplification products were visualized either directly or under ultraviolet light. This BRV specific RT-LAMP assay could detect 3.32 copies of subtype A BRV. No cross-reactions were detected with other bovine pathogens. The ability of RT-LAMP to detect bovine rotavirus was further evaluated with 88 bovine rectal swab samples. Twenty-nine of these samples were found to be positive for BRV using RT-LAMP. The BRV-specific RT-LAMP results were also confirmed by real-time RT-PCR assay. Conclusions: The bovine rotavirus-specific RT-LAMP assay was highly sensitive and holds promise as a prompt and simple diagnostic method for the detection of group A bovine rotavirus infection in young calves.

Keywords

BRV, VP6 gene, RT-LAMP

Background

Rotavirus, a double-stranded RNA virus, is a member of the family Reoviridae. Rotaviruses are classified into six, or possibly seven serogroups[1, 2]. Rotavirus infections with group A are the major cause of acute diarrhea among newborn animals and humans leading to death[3, 4]. Identification of infected calves, in combination with a proper vaccination program, is essential to control BRV successfully. The current methods for the detection and characterization of BRV, which include virus isolation, immunoassay, electron microscopy, and nucleic acid hybridization, are time consuming and laborious[5-7]. On the other hand, the polymerase chain reaction (PCR) has been used successfully for the detection and characterization of BRV[8-10]. Unfortunately, PCR assays require sophisticated equipment, which is costly to maintain, and must be performed in specialized laboratories.

The recently described loop-mediated isothermal amplification (LAMP) can amplify specific target DNA sequences with high sensitivity and can be completed within 60 min under isothermal conditions without the need of a thermal cycler and specialized laboratory[11]. This technique eliminates the heat denaturation step for DNA synthesis used in conventional PCR, and relies instead on auto-cycling strand displacement DNA synthesis achieved by a DNA polymerase with high strand displacement activity and a set of two specially designed inner and two outer primers. Another important feature of LAMP is a resulting color change following the addition of a fluorescent dye, making it visible to the naked eye. The LAMP assay has been used successfully to detect many pathogens[12-19] and in using reverse transcriptase, it has been further adapted for the detection of RNA viruses[15, 16]. The objective of the study reported here was to develop and optimize the reverse transcription LAMP (RT-LAMP) assay for the detection of group A BRV in calves.

Methods

Cells and virus strains

Table 5-26-1 lists the pathogen strains used in this study and describes the following: ten group A BRV strains, five other bovine pathogens other than BRV, four negative controls, three normal bovines rectal swab samples (one each from normal bovine nasal mucus and blood sample). Nasal mucus swabs and blood samples were taken from all bovine following routine pre slaughter examination for the normal animals to be slaughter. The collection of these samples could be considered as part of regular and routine examination, therefore no official review and approval of the Guangxi Veterinary Research Institute was needed. The viruses were propagated using rhesus monkey epithelial cell line (MA-104) and grown in Eagle's minimal essential medium (MEM) according to previously described methods[6]. MEM was supplemented with 10% fetal calf serum free from BRV and BRV antibodies (Shijiqing, China). The BRV strains were titrated by plaque assay in MA-104 cells in accordance to the published protocols[6, 16].

Table 5-26-1 Pathogens used and RT-LAMP assay results

	Source	RT-LAMP result	
		Agarose gel electrophoresis	Color change after adding dye
Bovine rotavirus (BRV)			
NCDV G6P[1]	CVCC	+	+
BRV014 G6P[5]	CVCC	+	+
GX-BRV-1 G6P[11]	GVRI	+	+
GX-BRV-2 G6P[5]	GVRI	+	+
GX-BRV-3 G10P[11]	GVRI	+	+
GX-BRV-4 G6P[11]	GVRI	+	+
GX-BRV-5 G6P[11]	GVRI	+	+
GX-BRV-6 G6P[5]	GVRI	+	+
GX-BRV-7 G6P[5]	GVRI	+	+
GX-BRV-8 G10P[11]	GVRI	+	+
Other bovine pathogens			
Bovine virus diarrhea (BVDV)	CVCC	–	–
Mycobacterium bovis (MB)	GVRI	–	–
Classical swine fever virus (CSFV)	CVCC	–	–
Infective bovine rhinotracheitis virus (IBRV)	CVCC	–	–
Bovine Coronavirus (BCV)	GVRI	–	–

continued

	Source	RT-LAMP result	
		Agarose gel electrophoresis	Color change after adding dye
Negative control			
Rectal swab of normal bovine 1	GVRI	–	–
Rectal swab of normal bovine 2	GVRI	–	–
Rectal swab of normal bovine 3	GVRI	–	–
Nasal swab of normal bovine	GVRI	–	–
Blood of normal bovine	GVRI	–	–
cells			
cell-1	GVRI	–	–
cell-2	GVRI	–	–
cell-3	GVRI	–	–
cell-4	GVRI	–	–
cell-5	GVRI	–	–

Note: CVCC, China Veterinary Culture Collection Center; GVRI, Guangxi Veterinary Research Institute; +, positive; –, negative.

DNA/RNA extraction

Genomic viral RNA, which includes BRV and other bovine virus strains from sample cultures, was extracted from infected MA-104 cell culture supernatant, as well as from swab sample cells from normal bovine using the TRIZOL RNA extraction kit in accordance with the manufacturer’ s protocol (Invitrogen, Carlsbad, CA, USA). Mycobacterium DNA was extracted using phenol：chloroform isoamyl alcohol (1：1：24 v/v) according to the Gibco BRL manufacturer’s protocol (Gibco BRL, Grand Island, New York, USA). All nucleic acid samples were stored at –70 ℃ until use.

Design of group A specific BRV RT-LAMP primers

Primer design for group A specific bovine rotavirus-RT-LAMP was based on the published sequence of strain neonatal calf diarrhea virus (NCDV) (GenBank, accession number: K02254.1). The NCDV sequence was aligned with the available sequences of 21 isolates (Figure 5-26-1 A and B) to identify the conserved regions using Primer Explored V4 soft ware. A set of six primers comprised of two outer, two inner, and two loop primers was designed and is shown in Table 5-26-2.The inner primers, which are known as the forward inner primer (FIP) and the backward inner primer (BIP), each have two distinct sequences corresponding to the sense and antisense sequences of the target, one for priming the first stage, and the other for self-priming in later stages in the reaction. FIP contains F1c (complementary to F1), and the F2 sequence. BIP contains the B1c sequence (complementary to B1), and the B2 sequence. The outer primers (F3 and B3) were used in the initial steps of LAMP reactions, but later during the isothermal cycling only the inner primers were used for strand displacement DNA synthesis. Two additional loop primers (loopF and loopB) were designed to accelerate the amplification reaction as described[18]. The sequences of the selected primers were compared to VP6 gene sequences[20-22]. All the primers were synthesized and purified by Invitrogen Inc (Guangzhou, China).

Table 5-26-2 Details of oligonucleotide primers used for RT-LAMP assay

Primer name	sequence (5'-3')	Genome position
FIP [a] =F2+F1c	GTTGTGATCTGTTCAACGTGAATGA-gaattc-GGGTTTTACATTTCATAAACCAA	F1c, 510～534 F2, 467～489
BIP [b] =B2+B1c	GCTCATGATAACTTGATGGGgatatc-GCACATGAGTAGTCGAATCC	B1c, 537～556 B2, 597～616
F3	TTGCAAAATAGAAGACAAAGAAC	444～466
B3	GTTGCGTATTAGCTGGCG	625～642
Loop F	CTGAATAAGGGAAAATGTTTGGTTT	483～507
loop B	GATGTGGCTCAATGCGGG	560～577

Note: Genome position according to the bovine rotavirus complete genome sequence (GenBank accession number: K02254.1). [a] FIP is composed of F2 and F1c, and they are linked by a sequences of EcoR I of -gaattc-. [b] BIP is composed of B2 and B1c, and they are linked by a sequences of EcoR V of -gatatc-.

```
A                  F2 →                                              ← F1c
Consensus       GGGTTTTACATTTCATAAACCAAACATTTTCCCTTATTCAGCGTCATTCACGTTGAACAGATCACAACC
21 Sequences     470       480       490       500       510       520       530
K02254.1        GGGTTTTACATTTCATAAACCAAACATTTTCCCTTATTCAGCTTCATTCACGTTGAACAGATCACAACC
AF317128.1      GGGTTTTACATTTCATAAACCAAACATATTCCCATATTCAGCTTCATTCACGTTGAACAGATCACAACC
AF317127.1      GGGTTTTACATTTCATAAACCAAACATTTTCCCTTATTCAGCATCATTCACGTTGAACAGATCACAACC
AF317126.1      TGGTTTCACATTTCATAAACCAAACATTTTCCCTTATTCAGCTTCATTCACGTTGAACAGATCACAACC
AF411322.2      GGGTTTTACATTTCATAAACCAAACATTTTCCCTTATTCAGCGTCATTCACGTTGAACAGATCACAACC
EU873011.1      GGGTTTTACATTTCATAAACCAAACATTTTCCCTTATTCAGCGTCATTCACGTTGAACAGATCACAACC
EU873012.1      GGGTTTTACATTTCATAAACCAAACATTTTCCCTTATTCAGCGTCATTCACGTTGAACAGATCACAACC
HM988973.1      GGGTTTTACATTTCATAAACCAAACATTTTCCCTTACTCAGCGTCATTCACGTTGAACAGATCACAACC
HM988974.1      GGGTTTTACATTTCATAAACCAAACATTTTCCCTTACTCAGCGTCATTCACGTTGAACAGATCACAACC
EF200565.1      GGGTTTTACATTTCATAAACCAAATATTTTCCCTTATTCAGCATCATTCACGTTGAACAGATCACAACC
EF200566.1      AGGTTTTACATTTCATAAACCAAACATTTTCCCTTATTCAGCGTCATTTACGTTGAACAGATCACAACC
EF200568.1      GGGTTTTACATTTCATAAACCAAATATTTTCCCTTATTCAGCATCATTCACGTTGAACAGATCACAACC
EF200571.1      GGGTTTTACATTTCATAAACCAAATATTTTCCCTTATTCAGCATCATTCACGTTGAACAGATCACAACC
EF592587.1      AGGTTTTACATTTCATAAACCAAATATTTTCCCTTATTCAGCATCATTCACGTTGAACAGATCACAACC
EF592588.1      AGGTTTTACATTTCATAAACCAAATATTTTCCCTTATTCAGCATCATTCACGTTGAACAGATCACAACC
EF592589.1      AGGCTTTACATTTCATAAACCAAATATTTTCCCTTATTCAGCATCATTCACGTTGAACAGATCACAACC
GU384194.1      GGGTTTTACATTTCATAAACCAAACATTTTCCCTTATTCAGCTTCATTCACGTTGAACAGATCACAACC
GU984757.1      AGGTTTTACATTTCATAAACCAAACATTTTCCCTTATTCAGCGTCATTTACGTTGAACAGATCACAACC
GU984758.1      AGGTTTTACATTTCATAAACCAAACATTTTCCCTTATTCAGCGTCATTTACGTTGAACAGATCACAACC
GU984759.1      AGGTTTTACATTTCATAAACCAAACATTTTCCCTTATTCAGCGTCATTTACATTGAACAGATCACAACC
AB374146.1      GGGTTTCACATTTCATAAACCAAACATTTTCCCTTATTCAGCATCATTCACGTTGAACAGATCACAACC

B                  B1c →                                                 ← B2
Consensus       GCTCATGATAACTTGATGGGTACGATGTGGCTCAATGCGGGATCAGAAATTCAGGTCGCTGGATTCGACTACTCATGTGC
21 Sequences      540       550       560       570       580       590       600       610
K02254.1        GCTCATGATAACTTGATGGGTACGATGTGGCTCAATGCGGGATCAGAAATTCAGGTCGCTGGATTCGACTACTCATGTGC
AF317128.1      GCTCATGATAACTTGATGGGAACGATGTGGTTAAACGCAGGATCAGAAATCCAAGTAGCTGGATTCGACTACTCATGTGC
AF317127.1      GCTCATGACAACTTGATGGGTACGATGTGGCTCAATGCAGGATCAGAAATTCAGGTCGCTGGATTCGACTACTCATGTGC
AF317126.1      GCTCATGATAACTTGATGGGTACGATGTGGCTCAATGCTGGGTCAGAAATACAGGTGGCTGGATTCGACTACTCATGTGC
AF411322.2      GCTCATGATAACTTGATGGGTACGATGTGGCTCAATGCGGGATCAGAAATTCAGGTCGCTGGATTCGACTACTCATGTGC
EU873011.1      GCTCATGATAACTTGATGGGTACGATGTGGCTCAATGCGGGATCAGAAATTCAGGTCGCTGGATTCGACTACTCATGTGC
EU873012.1      GCTCATGATAACTTGATGGGTACGATGTGGCTCAATGCGGGATCAGAAATTCAGGTCGCTGGATTCGACTACTCATGTGC
HM988973.1      GCTCATGATAACTTGATGGGTACGATGTGGCTCAATGCGGGATCAGAAATTCAGGTCGCTGGATTCGACTACTCATGTGC
HM988974.1      GCTCATGATAACTTGATGGGTACGATGTGGCTCAATGCGGGATCAGAAATTCAGGTCGCTGGATTCGACTACTCATGTGC
EF200565.1      GCTCATGATAACTTGATGGGTACGATGTGGCTAAATGCGGGATCAGAAATTCAGGTCGCTGGATTCGACTACTCGTGTGC
EF200566.1      GCTCATGATAACTTGATGGGTACGATGTGGCTTAATGCGGGATCAGAAATTCAGGTCGCTGGATTCGACTACTCGTGTGC
EF200568.1      GCTCATGATAACTTGATGGGTACGATGTGGCTAAACGCGGGATCAGAAATTCAGGTCGCTGGATTCGACTACTCGTGTGC
EF200571.1      GCTCATGATAACTTGATGGGTACGATGTGGCTAAACGCGGGATCAGAAATTCAGGTCGCTGGATTCGACTACTCGTGTGC
EF592587.1      GCTCATGATAACTTGATGGGTACGATGTGGCTAAATGCGGGATCAGAAATTCAGGTCGCTGGATTCGACTATTCGTGTGC
EF592588.1      GCTCATGATAACTTGATGGGTACGATGTGGCTAAATGCGGGATCAGAAATTCAGGTCGCTGGATTCGACTACTCGTGTGC
EF592589.1      GCTCATGATAACTTGATGGGTACGATGTGGCTAAATGCGGGATCAGAAATTCAGGTCGCTGGATTCGACTACTCGTGTGC
GU384194.1      GCTCATGATAACTTGATGGGTACGATGTGGCTCAATGCGGGATCAGAAATCCAGGTCGCTGGATTCGACTACTCATGTGC
GU984757.1      GCTCATGATAACTTGATGGGTACGATGTGGCTCAATGCGGGATCAGAAATTCAGGTCGCTGGATTCGACTACTCATGTGC
GU984758.1      GCTCATGATAACTTGATGGGTACGATGTGGCTCAATGCGGGATCAGAAATTCAGGTCGCTGGATTCGACTACTCATGTGC
GU984759.1      GCTCATGATAACTTGATGGGTACGATGTGGCTCAATGCGGGATCAGAAATTCAGGTCGCTGGATTCGACTACTCATGTGC
AB374146.1      GCTCATGATAACTTGATGGGTACGATGTGGCTCAATGCGGGATCAGAAATTCAGGTCGCTGGATTCGACTATTCATGTGC
```

Figure 5-26-1 Multiple sequence alignment of BRV VP6 genes

Real-time RT-PCR

The sensitivity of the RT-LAMP method was compared to real-time RT-PCR using the two primers and one of the following probes: F5 (5' -TCATTTCAGTTGATGAGACCACC-3'), F6 (5' -ATTCAATTCTAAGC GTGAGTCTAC-3'), or HEX-AATATGACACCAGCGGTAGCGGCBHQ1.This real time RT-PCR amplifies a 112-bp target sequence of the VP6 gene of group A BRV[10]. Real-time RT-PCR amplification was carried out using the Real-time one step RT-PCR Kit (TaKaRa, Dalian, China) as described in previously published protocol[10]. After the real-time RT-PCR was performed, cycle threshold (CT) was manually setup to reflect the best kinetic PCR parameters, such that any nonspecific amplification in reaction could be analyzed.

Optimization of the RT-LAMP condition

The RT-LAMP assay was optimized using various concentrations of primers, buffers, salt and RNA/ DNA templates for positive and negative controls. The reaction mixture was optimized to 25 μL, containing primers in various concentrations, 1.4 mM of each deoxyribonucleotide triphosphate, 0.8 M of betaine (Sigma Chemical Co., Beijing, China), 2.5 μL of 10 × Thermo buffer, 8 mM $MgSO_4$, 8 U of *Bst* DNA polymerase (large fragment; New England Biolabs), 0.125 U of enhanced avian myeloblastosis virus reverse transcriptase, and 2 μL of the extracted target RNA. To determine the optimal duration for the RT-LAMP assay, the primer and reverse transcribed sample mixtures were incubated in a 63 ℃ water bath for 20, 40, 60, and 80 min. At the end of each incubation period, the reaction was terminated by heating at 80 ℃ for 5 min. The reaction temperature was optimized using 61 ℃ , 62 ℃ , 63 ℃ , and 64 ℃ .

Analysis of RT-LAMP product

In order to analyze the amplified products, three detection methods were evaluated. First, turbidity: the accumulation of magnesium pyrophosphate, a byproduct of the DNA amplification reaction, increases the turbidity of the sample. Turbidity was evaluated by visual inspection of the samples, comparing them to a negative control sample. Second, color change: 1 mL of 10 000 × SYBR Green I nucleic acid stain was added to the tube after the reaction. Samples turning yellow-green were considered positive, while samples turning orange were negative. Samples showing fluorescence under an ultraviolet hand lamp at a 365-nm wavelength were considered positive as described previously[12]. Samples were compared with a negative control to account for background fluorescence. Third, gel electrophoresis: run on a 1% agarose gel, RT-LAMP reaction end product yields a combination of DNA fragments of varying sizes[11]. Therefore, the presence of a smear or a pattern of multiple bands of different molecular weights indicates a positive result. A molecular marker was used to estimate amplified product size.

Lastly, a restriction enzyme analysis: EcoR I and EcoR V restriction of the RT-LAMP product confirms reaction specificity[14]. Briefly, digestion reactions were performed using 3 μL of RT-LAMP product with 12 U of restriction enzyme in 25 μL total volume. The size of the digested fragments was estimated by agarose gel electrophoresis as described[19].

Evaluation of RT-LAMP

To evaluate specificity, the RT-LAMP test was performed on a panel of viral isolates from bovine's reference viruses, (Table 5-26-1). The panel included three rectal and one nasal swab samples from a normal bovine, one blood sample from a normal bovine, one cell sample (repeated five times) as a negative control, ten strains of BRV, and five different bovine DNA and RNA pathogens (Table 5-26-1).

The detection limit of RT-LAMP was tested and compared with real-time RT-PCR using triplicate templates at identical concentration. RNA transcripts corresponding to the VP6 of NCDV strain were generated for use as standards in the analysis of sensitivity of the assay. Briefly, RNA was extracted from NCDV strain using the TRIZOL RNA extract reagent. The purified RNA was resuspended in distilled water and used in the RT-PCR reaction. The amplified product of VP6 was cloned into the pGM-T vector (TaKaRa, Dalian, China) according to the manufacturer' s directions and sequenced to verify its identity. The recombinant plasmid pGM-T-VP6 was linearized by digestion with restriction enzyme Not Ⅰ, gel purified, and used as a template with a Ribo Max T7 In Vitro Transcription System (Promega, Madison, Wisconsin, USA) according to the manufacturer' s protocol. The length of RNA transcripts was verified by agarose gel electrophoresis. The RNA of VP6 was quantitated using UV spectrophotometry at 260 nm, and calculated copy numbers were calculated from the concentration as described previously[20]. A series of 10-fold dilutions were used to test the assay' s sensitivity of BRV RT-LAMP (Figure 5-26-2).

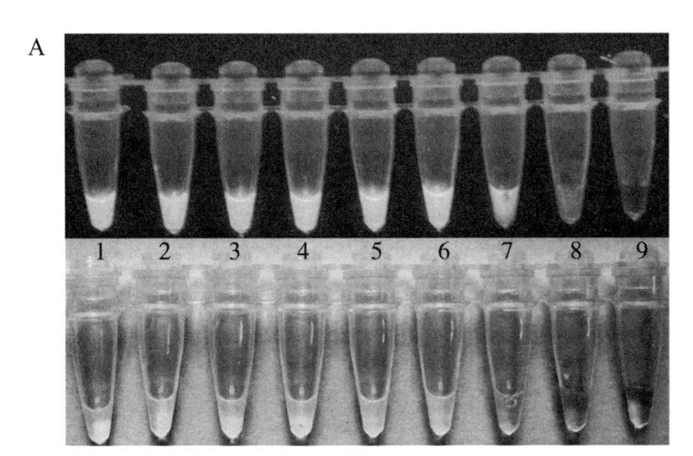

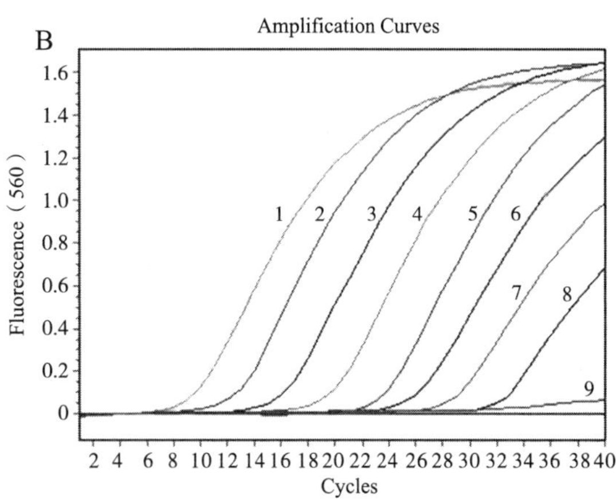

A: RT-LAMP product; B: Real-time RT-PCR. 1: 3.32×10^7 copies/tube; 2: 3.32×10^6 copies/tube; 3: 3.32×10^5 copies/tube; 4: 3.32×10^4 copies/tube; 5: 3.32×10^3 copies/tube; 6: 3.32×10^2 copies/tube; 7: 3.32×10^1 copies/tube; 8: 3.32×10^0 copies/tube; 9: 3.32×10^{-1} copies/tube (color figure in appendix).

Figure 5-26-2 Sensitivity of RT-LAMP-and real-time RT-PCR

Detection of clinical specimen by RT-LAMP assay

After validation studies, we determined the reliability of the group A specific BRV-RT-LAMP as a method of viral RNA detection for clinical specimens. Written informed consent was obtained from each participating farm owner. On participating farms, the veterinarian collected rectal swab samples from the calves. The rectal swab samples were taken from calves between 3 to 180 days age, and were considered as part of regular and routine clinical-diagnostic care. No official review and approval of Guangxi Veterinary Research Institute was needed. A total of 88 rectal swab samples were collected from calves with acute diarrhea from different dairy farms in the Guangxi (Table 5-26-3). The samples were placed into 1 mL of sterilized water, and processed as described previously[12]. BRV-specific RT-LAMP assay was performed as described above. The results of group A specific BRV RT-LAMP were compared with the results of BRV real-time RT-PCR. Restriction enzyme analysis of RT-LAMP products and its sequencing were used to assess the reliability of the methods for the rapid detection of BRV.

Table 5-26-3 Comparison of real-time RT-PCR and RT-LAMP for the detection of BRV in clinical samples

Location of samples	Number of samples	Number Real-time RT-PCR	Number of positive samples for assay RT-LAMP	Sequencing
Nanning	24	7	7	correct
Liuzhou	15	3	3	correct
Fangcheng	10	3	3	correct
Shangsi	13	5	5	correct
Guilin	20	10	10	correct
Hengxian	16	1	1	correct
Total	88	29	29	correct

Note: RT-LAMP-positive samples were all confirmed to be BRV by sequencing the RT-LAMP-products restriction analysis of VP6 genes.

Results

The optimal protocol of RT-LAMP assay and inspection of products

Following standardization and optimization, the optimal ratio of primer (inner-outer-loop) concentrations for the RT-LAMP reaction was found to be 8 ∶ 1 ∶ 4 equivalents to 1.6, 0.2 and 0.8 mM. Gene amplification was detected by an increase in turbidity, as well as adding dye for color change indication (Figure 5-26-3). Restriction enzyme analysis performed with EcoR I and EcoR V on the RT-LAMP product validated no nonspecific reaction in RT-LAMP assay. The RT-LAMP assay amplified the 199-bp target sequence of the VP6 gene of BRV (Figure 5-26-4). For the reproducible sensitive and specific results of RT-LAMP assay, the optimal reaction time and incubation temperature was found to be 60 min at 63 ℃.

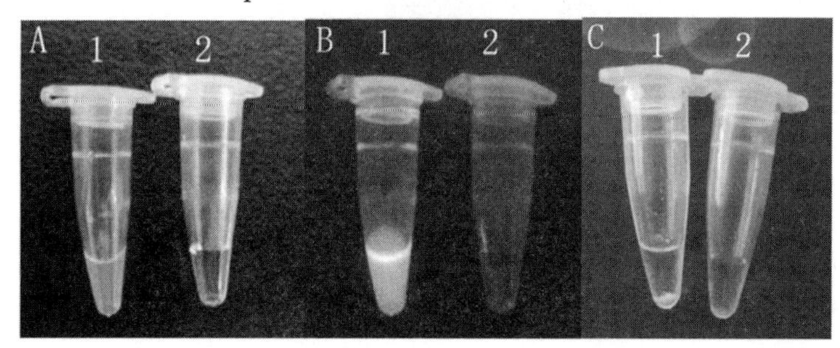

A: Fluorescent dye added seen without ultraviolet light; B: Fluorescent dye added seen with ultraviolet light; C: By turbidity with white sediment. 1: Positive control sample; 2: Negative control sample (color figure in appendix).

Figure 5-26-3 Detection of BRV RT-LAMP product

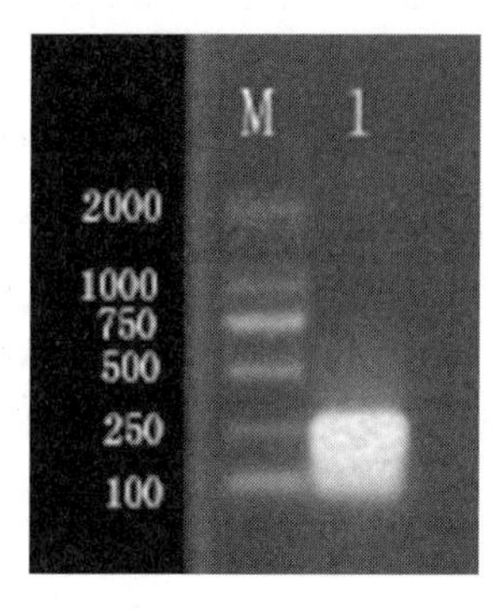

M: DNA marker; 1: RT-LAMP products digested with EcoR I and EcoR V.

Figure 5-26-4 Restriction analysis of RT-LAMP product

Specificity and sensitivity of RT-LAMP assay

The bovine rotavirus-specific RT-LAMP assay specifically amplified strains NCDV-014, and 8 Guangxi field bovine rotavirus strains, which have been isolated from the Guangxi dairy farms, and exhibits no cross-reactivity with other pathogens (Table 5-26-1). This specificity was confirmed by agarose gel electrophoresis (Figure 5-26-5) and a color change assay (Figure 5-26-3). We also determined the assay sensitivity using a 10-fold dilution series. The detection limit of RT-LAMP was 3.32 copies (Figure 5-26-2 A). Similarly, the detection limit of real-time RT-PCR analysis was 3.32 copies (Figure 5-26-2 B). The results indicate that RT-LAMP is as sensitive as real-time RT-PCR, both of which can detect 3.32 copies of bovine rotavirus VP6 gene.

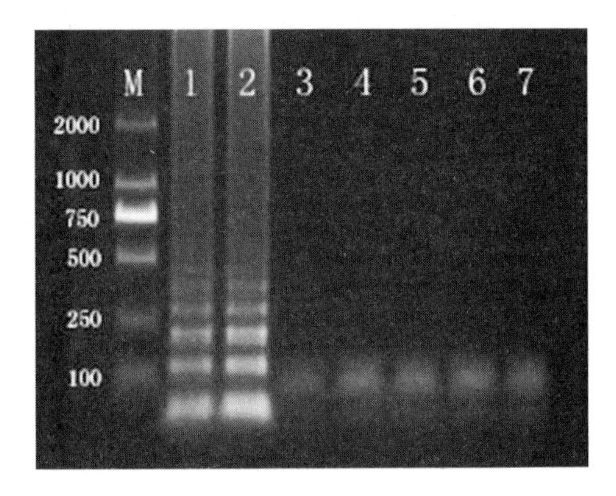

M: DNA marker; 1: 014 strain; 2: NCDV strain; 3: BVDV; 4: IBRV; 5: CSFV; 6: MB; 7: BCV.

Figure 5-26-5 The detection of RT-LAMP product assessed by 1% agarose gel electrophoresis

Evaluation of RT-LAMP assay with clinical samples

In comparison with real-time RT-PCR, using the above described procedure with specific primers, 29 rectal swab samples (33.0%) were found positive by both real-time RT-PCR analysis and RT-LAMP, and 59 (70%) were found negative by both tests. As such, the coincidence of real-time RT-PCR and RT-LAMP was 100%. However, RT-LAMP is a quicker, easier, and more cost efficient method than real-time RT-PCR. Restriction enzyme revealed 199-bp target sequence of the VP6 gene of BRV, and sequencing analysis results showed that the clonal sequences of 29 samples were VP6 of BRV. The results indicated no nonspecific amplification in RT-LAMP reaction occurred (data not shown).

Discussion

The RT-LAMP was shown here to be specific and sensitive to detect rotavirus in rectal swab samples of calves with acute diarrhea. The BRV-specific RT-LAMP primers and real time probe were designed using VP6 genes sequencing of 21 BRV isolates (Figure 5-26-1 A and B) from the Genbank data. VP6 a group-specific gene of BRV is the most immunogenic viral protein and considered to be very conserved among group A BRV and suitable to use as conserved gene specific for BRV[21-23]. The RT-LAMP was determined to be specific, as no cross-reactions were observed when other bovine pathogens were tested.

The RT-LAMP technique offers several advantages over PCR for the detection of bovine rotavirus RNA. Firstly, the RT-LAMP demonstrates high efficiency under isothermal conditions without a significant influence of non-target DNA and demonstrates sensitivity identical to that of real-time RT-PCR assays. Secondly, the RT-LAMP is easy to perform once the appropriate primers are selected and optimized. The reaction requires only four primers, a Bst DNA polymerase, and a temperature adjustable conventional laboratory water bath. Thirdly, the results of RT-LAMP are visible to the naked eye without the need for electrophoresis. This is attributable to recognition of the target sequence by six primers in the initial stage and by four primers during the later stages of the RT-LAMP reaction. The RT-LAMP system possesses specificity in addition to high sensitivity and may be used in large-scale molecular surveys of BRV infections in the field.

Nevertheless, RT-LAMP is not without caveats: First, the main challenge in development of molecular tests for BRV is the large genetic heterogeneity of this virus[3, 4]. This emphasizes the need for constant updating of primers of molecular diagnostic tests for BRV, with FIP and BIP being the most important primers in the entire assay, and they need to be conserved across many strains. Second, due to its high amplification production, the issue of potential cross-contamination problems needs to be addressed. We have added dye into the reaction system before the actual amplification in order to eliminate contamination.

It should be pointed out that in utilizing the VP6 gene target for the optimization of specific primers, we would expect this BRV RT-LAMP assay to meet the practical needs of BRV detection in the field around the world. While the RT-LAMP developed here needs to be validated against rotaviruses of groups B and C and other groups, these bovine rotavirus groups have not yet been identified in cattle in southern China. To the authors' knowledge, this is the first study that has explored the use of RT-LAMP technology in a diagnostic test for BRV. The test was simple, specific, and rapid; it was able to detect BRV from rectal swabs from BRV-infected calves.

Conclusions

In this study, the established RT-LAMP assay with high sensitivity and specificity was performed in a water bath within only 60 min, and the amplification results were visualized by the naked eye after adding a fluorescent reagent. The newly developed assay can be used as one important tool for detecting BRV in calves under field conditions with no need for specialized equipment.

References

[1] PRASAD B V, CHIU W. Structure of rotavirus. Curr Top Microbiol Immunol. 1994, 185: 9-29.

[2] PESAVENTO J B, CRAWFORD S E, ESTES M K, et al. Rotavirus proteins: structure and assembly. Curr Top Microbiol Immunol, 2006, 309: 189-219.

[3] WOODE G N, BRIDGER J C, JONES J M, et al. Morphological and antigenic relationships between viruses (rotaviruses) from acute gastroenteritis of children, calves, piglets, mice, and foals. Infect Immun, 1976, 14(3): 804-810.

[4] KALICA A R, SERENO M M, WYATT R G, et al. Comparison of human and animal rotavirus strains by gel electrophoresis of viral RNA. Virology, 1978, 87(2): 247-255.

[5] DE BEER M, PEENZE I, DA COSTA MENDES V M, et al. Comparison of electron microscopy, enzyme-linked immunosorbent assay and latex agglutination for the detection of bovine rotavirus in faeces. J S Afr Vet Assoc, 1997, 68(3): 93-96.

[6] OJEH C K. Isolation and propagation of bovine rotavirus in cell culture. Rev Elev Med Vet Pays Trop, 1984, 37(4): 400-405.

[7] AL-YOUSIF Y, ANDERSON J, CHARD-BERGSTROM C, et al. Development, evaluation, and application of lateral-flow immunoassay (immunochromatography) for detection of rotavirus in bovine fecal samples. Clin Diagn Lab Immunol, 2002, 9(3): 723-725.

[8] GUTIERREZ-AGUIRRE I, STEYER A, BOBEN J, et al. Sensitive detection of multiple rotavirus genotypes with a single reverse transcription-real-time quantitative PCR assay. J Clin Microbiol, 2008, 46(8): 2547-2554.

[9] ZHU W, DONG J, HAGA T, et al. Rapid and sensitive detection of bovine coronavirus and group a bovine rotavirus from fecal samples by using one-step duplex RT-PCR assay. J Vet Med Sci, 2011, 73(4): 531-534.

[10] FAN Q, XIE Z, LIE J, et al. Detection of bovine rotavirus by TaqMan based real-time reverse transcription polymerase chain reaction assay. China Animal Husbandry & Veterinary Medicine, 2011, 38: 105- 108.

[11] NOTOMI T, OKAYAMA H, MASUBUCHI H, et al. Loop-mediated isothermal amplification of DNA. Nucleic Acids Res, 2000, 28(12): E63.

[12] XIE Z, TANG Y, FAN Q, et al. Rapid detection of Group I avian adenoviruses by a Loop-mediated isothermal amplification. Avian Dis, 2011: 575-579.

[13] SAITO R, MISAWA Y, MORIYA K, et al. Development and evaluation of a loop-mediated isothermal amplification assay for rapid detection of *Mycoplasma pneumoniae*. J Med Microbiol, 2005, 54(Pt 11): 1037-1041.

[14] ENOMOTO Y, YOSHIKAWA T, IHIRA M, et al. Rapid diagnosis of herpes simplex virus infection by a loop-mediated isothermal amplification method. J Clin Microbiol, 2005, 43(2): 951-955.

[15] PENG Y, XIE Z, LIE J, et al. Visual detection of H3 subtype avian influenza viruses by reverse transcription loop-mediated isothermal amplification assay. Virology Journal, 2011(8): 337.

[16] NEMOTO M, IMAGAWA H, TSUJIMURA K, et al. Detection of equine rotavirus by reverse transcription loop-mediated isothermal amplification (RT-LAMP). J Vet Med Sci, 2010, 72(6): 823-836.

[17] GRAY J, DESSELBERGER U. Rotaviruses: methods and protocols. Totowa, N J: Humana Press, 2000: 262.

[18] NAGAMINE K, HASE T, NOTOMI T. Accelerated reaction by loop-mediated isothermal amplification using loop primers. Mol Cell Probes, 2002, 16(3): 223-239.

[19] ROVIRA A, ABRAHANTE J, MURTAUGH M, et al. Reverse transcription loop-mediated isothermal amplification for the detection of Porcine reproductive and respiratory syndrome virus. J Vet Diagn Invest, 2009, 21(3): 350-354.

[20] CHEN N H, CHEN X Z, HU D M, et al. Rapid differential detection of classical and highly pathogenic North American Porcine Reproductive and Respiratory Syndrome virus in China by a duplex real-time RT-PCR. J Virol Methods. 2009, 161(2): 192-198.

[21] TSUNEMITSU H, KAMIYAMA M, KAWASHIMA K, et al. Molecular characterization of the major capsid protein VP6 of bovine group B rotavirus and its use in seroepidemiology. J Gen Virol, 2005, 86(Pt 9): 2569-2575.

[22] MEDICI M C, ABELLI L A, MARTINELLI M, et al. Molecular characterization of VP4, VP6 and VP7 genes of a rare G8P[14] rotavirus strain detected in an infant with gastroenteritis in Italy. Virus Res, 2008, 137(1): 163-167.

[23] TOSSER G, DELAUNAY T, KOHLI E, et al. Topology of bovine rotavirus (RF strain) VP6 epitopes by real-time biospecific interaction analysis. Virology, 1994, 204(1): 8-16.

Part Ⅳ Nanopartical Electrochemical Sensor Detection Technology

Ultrasensitive electrochemical immunoassay for avian influenza subtype H5 using nanocomposite

Xie Zhixun, Huang Jiaoling, Luo Sisi, Xie Zhiqin, Xie Liji, Liu Jiabo, Pang Yaoshan, Deng Xianwen, and Fan Qing

Abstract

We report a novel electrochemical immunosensor that can sensitively detect avian influenza virus H5 subtype (AIV H5) captured by graphene oxide-H5-polychonal antibodies-bovine serum albumin (GO-PAb-BSA) nanocomposite. The graphene oxide (GO) carried H5-polychonal antibody (PAb) were used as signal amplification materials. Upon signal amplification, the immunosensor showed a 256-fold increase in detection sensitivity compared to the immunosensor without GO-PAb-BSA. We designed a PAb labeling GO strategy and signal amplification procedure that allow ultrasensitive and selective detection of AIV H5. The established method responded to 2^{-15} HA unit/50 μL H5, with a linear calibration range from 2^{-15} to 2^{-8} HA unit/50 μL. In summary, we demonstrated that the immunosenser has a high specificity and sensitivity for AIV H5, and the established assay could be potentially applied in the rapid detection of other pathogenic microorganisms.

Keywords

electrochemical immunoassay, avian influenza subtype H5, graphene oxide

Introduction

H5N1 influenza virus is highly pathogenic in poultry, wild birds, and has occasionally infected humans with serious and fatal outcomes[1]. Since 2003, the WHO has reported H5N1 in more than 46 countries for animal cases and 15 countries for human cases with 650 people infected and 386 dead[2]. A variety of technologies for diagnosing avian influenza virus (AIV) have been developed, such as virus isolation, serologic assays, enzyme-linked immunosorbent assay, and polymerase chain reaction (PCR) based assays[3-9]. However, there are some disadvantages with these diagnostic methods making them less ideal in practical applications. For example, these methods either poor in specificity, low in sensitivity, time consuming, or requiring a well equipped laboratory and highly trained technicians[10-12].

Electrochemical immunosensors are particularly attractive due to their high sensitivity, capacity for quick analysis, easy for pretreatment, small analyte volume, simple instrumentation, minimal manipulation and wide range of uses[13-15]. Several electrochemical immunosensors have been developed and extensively applied to detect antigens[16-17]. In order to meet the increasing demand for early and ultrasensitive detection of biomarkers, various signal amplification technologies using nanomaterials have been developed[18-20]. Graphene oxide (GO) monolayers made from carbon atoms packed into dense honeycomb crystal structures, have unique nanostructures and properties that render them suitable as electrochemical biosensors. For example, they are in good colloidal condition, have a large surface area and their manufacturing costs are low[21-22]. Our present work is motivated by the promising applications of BSA functionalized GO in signal amplification for ultrasensitive detection of AIV H5.

Materials and methods

Avian pathogens and culture conditions

The avian pathogens used in this study are listed in Table 5-27-1. Inactivated H5N1 was provided by the Harbin Veterinary Research Institute, China. Inactivated H5N2, H5N9, H7N2 were provided by the Pennsylvania State University, US. Aside from the H5 and H7 subtypes, all other AIV subtypes, Newcastle disease virus (NDV), and infectious bronchitis virus (IBV), respectively were propagated in the allantoic cavity of 10-day-old specificpathogen-free (SPF) embryonated chicken eggs, whereas infectious laryngotracheitis virus (ILT) was propagated on the chorioallantoic membrane in 10-day-old SPF embryonated chicken eggs as described elsewhere[4, 23]. The allantoic fluids from embryonated eggs infected with AIV, NDV, ILT and IBV were harvested after incubation at 37 ℃ for 36 h[4, 23]. *Mycoplasma galliepticum* (MG) was propagated in Frey's broth and incubated at 37 ℃ as previously described[24]. H5-polychonal antibodies and H5-monoclonal antibodies were purchased from Abcam (Cambridge, UK). Graphite powder (<45 mm), chloroauric acid ($HAuCl_4$), 1-ethyl-3 (3-dimethylaminopropyl) carbodimide hydrochloride (EDC), sodium chloroacetate ($ClCH_2COONa$), bovine serum albumin (BSA, 96%~99%), N-hydroxysuccinimide-activated hexa (ethylene glycol) undecane thiol (NHS) were all acquired from Sigma-Aldrich. All other reagents were of analytical reagent grade and used without further purification. Phosphate buffered solution (PBS; 10 mmol/L), at various pH values were prepared by mixing stock solutions of NaH_2PO_4 and Na_2HPO_4.

Table 5-27-1 Sources of pathogens used and electrochemical immunoassay assay results

Avian pathogen samples	Source	H5 electrochemical immunoassay
Inactivated H5N1 AIV Re-1	HVRI	+
Inactivated H5N2/chicken/QT35/87	PU	+
Inactivated H5N9/chicken/QT35/98	PU	+
H1N3 AIV Duck/HK/717/79-d1	HKU	-
H2N3 AIV Duck/HK/77/76	HKU	-
H3N6 AIV Duck/HK/526/79/2B	HKU	-
H4N5 AIV Duck/HK/668/79	HKU	-
H6N8 AIV Duck/HK/531/79	HKU	-
Inactivated H7N2/chicken/PA/3979/97	PU	-
H8N4 AIV Turkey/ont/6118/68	HKU	-
H9N6/Duck/HK/147/77	HKU	-
H10N3 AIV Duck/HK/876/80	HKU	-
H11N3 AIV Duck/HK/661/79	HKU	-
H12N5 AIV Duck/HK/862/80	HKU	-
H13N5 AIV Gull/MD/704/77	HKU	-
NDV	GVRI	-

continued

Avian pathogen samples	Source	H5 electrochemical immunoassay
IBV	GVRI	–
ILTV	GVRI	–
MG	GVRI	–
Liver, lung and small intestine of SPF chicken		–

Note: HVRI, Harbin Veterinary Research Institute, China; HKU, The University of Hong Kong, China; GVRI, Guangxi Veterinary Research Institute, China; PU, Pennsylvania State University, USA.

Instruments

For electrochemical studies, we employed a CHI660D electrochemical workstation (Shanghai CH Instruments, Shanghai, China) with a standard three-electrode cell that contained a platinum wire auxiliary electrode, a saturated calomel reference electrode (SCE) and a working electrode (the modified electrode as working electrode). All potential values given refer to SCE. All experiments were performed at room temperature (25 ± 0.5 ℃).

Synthesis of graphene oxide

Graphene oxide (GO) was prepared by modified Hummers method[25]. Typically, 1.0 g of graphite powder and 2.5 g of $NaNO_3$ were added to 100 mL of concentrated H_2SO_4 and stirred for 1 h. The mixture was continuously stirred and ice-cooled as 5 g of $KMnO_4$ was slowly added. The mixed slurry was then stirred at 35 ℃ for 20 h. After that, 100 mL of deionized water was added slowly to the reacted slurry and then stirred at 85 ℃ for another 2 h. Next, 300 mL of deionized water was added to the reacted slurry. Then, 6 mL of 30% H_2O_2 was added; the slurry immediately turned into a bright yellow solution and bubbles appeared. The resultant solution was stirred for 2 h and then allowed to precipitate for 24 h; after that, the supernatant was decanted. The resultant yellow slurry was centrifuged and washed with 500 mL of 0.5 mol/L HCl. After stirring for 2 h, the solution was centrifuged, washed again, before further washing with deionized water until the pH of the solution increased to neutral (pH 7.0). The remaining dark-yellow solid was dried under vacuum at 40 ℃ for 48 h and ground to a fine powder. The drying process for GO was conducted at low temperatures because GO slowly decomposes (deoxygenates) above 60～80 ℃. 1.0 mg of GO fine powder was added into 1 mL of deionized water and stirred for 30 min to obtain 1.0 mg/mL GO aqueous solution. Then 1.0 mg/mL GO aqueous solution was placed into an ice bath and sonicated. The ice bath was changed after each treatment to make sure that sample temperature was below 5 ℃. Finally, the resultant sample was centrifuged at 12 000 r/min for 10 min; the upper solution was used in the experiments.

Preparation of GO-PAb-BSA bioconjugates

To convert hydroxyl and epoxide groups to carboxylic groups, 50 mg of NaOH and 50 mg of $ClCH_2COONa$ was added to 1 mL of a 1 mg/mL GO suspension, which was followed by bath sonication for 1 h. After these treatments, the resulting product, GO-COOH, was neutralized with dilute hydrochloric acid and purified by repeated rinsing and centrifugation until the product was well-dispersed in deionized water. The GO-COOH suspension was then dialyzed against distilled water for over 48 h to remove any ions. For the preparation of GO-PAb-BSA bioconjugates, 400 μL GO (0.1 mg/mL) was activated with 10 μL EDC (5 mg/

mL) and 20 μL NHS (3 mg/mL) in PBS buffer (pH 5.2) and activated for 30 min. The mixture was centrifuged at 13 000 r/min for 10 min, and the supernatant was discarded. The buffer wash was repeated to remove excess EDC and NHS. The resulting functionalized mixture was dispersed in 1.0 mL of PBS buffer (pH 7.4) and sonicated for 5 min to obtain a homogeneous suspension. Then, 1 mL of PAb (1 μg/mL) and 2 mL of BSA (0.25% (w/v)) were added to the suspension, and the mixture was stirred overnight at 4 ℃ . The reaction mixture was washed with PBS and centrifuged at 13 000 r/min for 5 min, three times. The supernatant was discarded. The resulting mixture was redispersed in 1.0 mL of PBS (pH 7.4) and stored at 4 ℃ .

Fabrication of the immunosensor

The gold electrode (GE; Ø=3 mm) was initially polished with 0.05 mm alumina to obtain a mirror-like surface before being rinsed with distilled water and placed into an ultrasonic bath to remove any physically adsorbed substances. Next, the electrode was placed into an electrochemical cell with 0.05 M H_2SO_4 and chemically cleaned until the background signal stabilized. The clean electrode was thoroughly rinsed with ddH_2O, dried with nitrogen gas, quickly immersed in the 5 mM thiourea solution and incubated at room temperature for 24 h. To prepare the gold nanoparticle-modified surface, a potential scan was applied to the thiourea-gold electrode, which started at 0 V (scan rate of 50 mV/s) and held at the peak potential −0.2 V for 60 s in a solution of 1% $HAuCl_4$. After that, the electrode was washed with ddH_2O and immersed in PBS solution (pH 7.4) containing 10 μg/mL H5-monoclonal antibodies (MAb), and immobilized at 4 ℃ overnight. Finally, the modified electrode was incubated in 0.25% BSA solution for I h at 37 ℃ to block any remaining active sites on the gold nanoparticle (AuNP) monolayer, and thus avoiding non-specific adsorption. The finished immunosensor was stored at 4 ℃. The procedures used for construction of the immunosensor are shown in Figure 5-27-1.

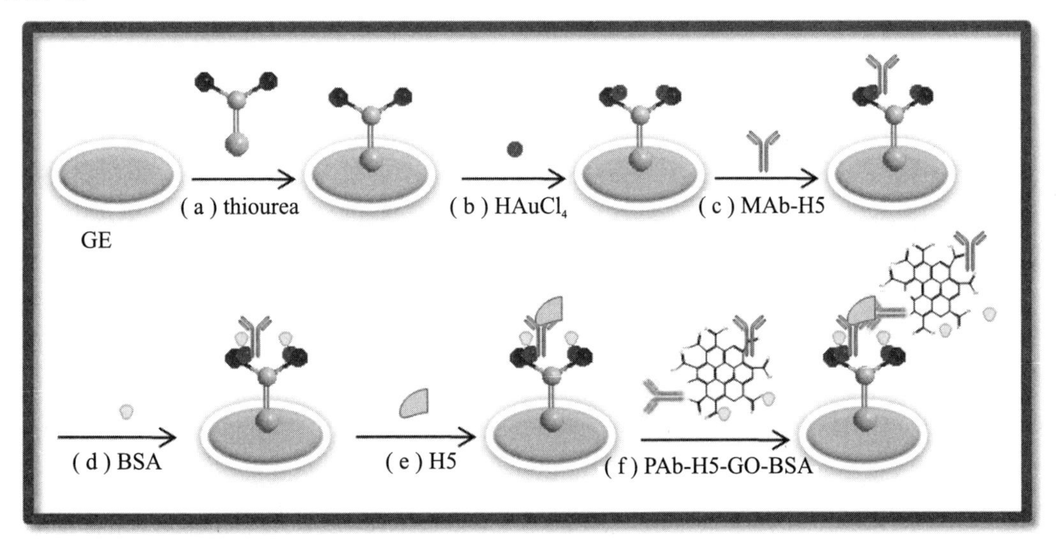

Figure 5-27-1 Immunosensor fabrication process (color figure in appendix)

Immunoassay for detection of H5 antigen

A sandwich immunoassay was used for detection of AIV H5. First, the immunosensor, MAb-AuNPs-thiourea-GE, was incubated with 100 μL of various concentrations of H5 antigen for 30 min, washed with PBS buffer. Next, the electrode was incubated with 200 μL of GO-PAb-BSA bioconjugates for 40 min, washed with

PBS buffer to remove non-specific adsorption conjugates. Finally, electrochemical detection was performed in the presence of 5 mM $[Fe(CN)_6]^{3-/4-}$ and 0.01 M PBS (containing 0.1 M KCl; pH 7.0).

Criteria for judgement of positive or negative samples

Eighteen negative samples from SPF chickens were measured by the immunosensor, and the average current was calculated. The principle of statistics was used to distinguish positive from negative.

Performance of immunosensor

Avian influenza virus H5 was diluted to 2^{-15}～2^{2} HA unit/50 μL with PBS, then use immunosensor to test different concentrations of the virus from high to low. All of the experiments were repeated three times, and the average currents of each concentrations were calculated.

Specificity of immunosensor assay

To evaluate the specificity of the immunosensors, some nontarget samples such as H1N3, H2N3, H3N6, H4N5, H6N8, H7N2, H8N4, H9N6, H10N3, H11N3, H12N5, H13N5, NDV, ILTV, IBV and MG were tested by using the developed immunosensors. The immunosensor was incubated with 100 μL samples for 30 min, washed with PBS buffer, then incubated with 200 μL of GO-PAb-BSA bioconjugates for 40 min, and washed with PBS buffer. Lastly, the electrochemical detection was performed in the presence of 5 mM $[Fe(CN)_6]^{3-/4-}$and 0.01 M PBS (containing 0.1 M KCl; pH 7.0). All of the experiments were repeated three times. The H5 antigen was used as a positive control, and tissue from the SPF chickens were used as negative controls.

Results

Electrochemical characteristics of different electrodes

Electrochemical impedance spectroscopy (EIS) is regarded as an effective technique for probing the features of surface modified electrodes. The impedance spectrum includes a semicircle portion and a linear portion. The semicircle diameter corresponds to the electron-transfer resistance (R_{et}), and the linear part corresponds to the diffusion process. As shown in Figure 5-27-2, for the bare GE, we observed a linear part at low frequencies (curve a), suggesting a very low R_{et} to redox probe $[Fe(CN)_6]^{3-/4-}$. After the bare GE was modified with thiourea, the resistance for the redox probe increased (curve b). R_{et} then decreased when AuNPs were adhered (curve c), proving that AuNPs promote electron transfer and enhance the conductivity of the electrode. Subsequently, when the MAb was loaded on the surface of the AuNPs, the EIS showed a large increase in diameter (curve d), indicating that the antibody forms an additional barrier and further prevents transfer between the redox probe and the electrode surface. The result is consistent with the notion that the hydrophobic layer of protein insulates the conductive support and hinders the interfacial electron transfer. After BSA was used to block non-specific sites, R_{et} increased in a similar manner (curve e), possibly attributed to the same reason as when H5 antigen was loaded. R_{et} increased (curve f) after the resulting immunosensor was incubated in H5 antigen at 2^{-4} HA unit/50 μL, which indicates the formation of hydrophobic immunocomplex layer embarrassing the electron transfer. R_{et} further increased (curve g) after the resulting immunosensor was incubated in GO-PAb-BSA bioconjugates. The impedance change obtained after the modifying process implies that thiourea, MAb, BSA, H5 and GO-PAb-BSA have been assembled successively onto the GE electrode.

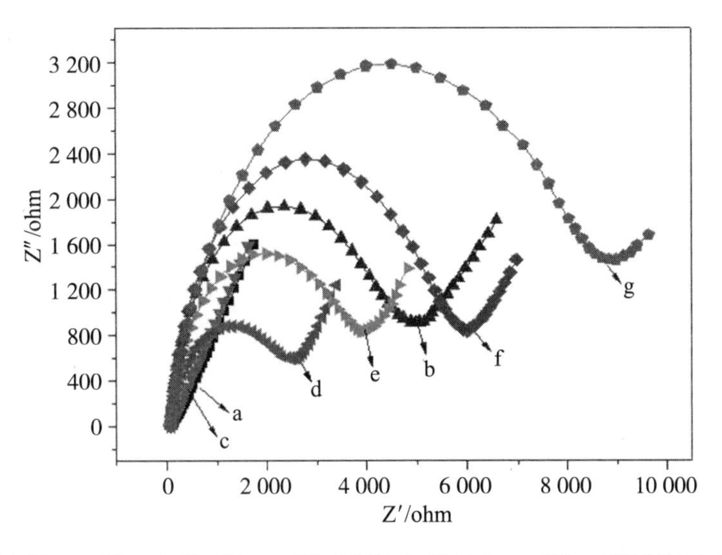

a: Bare gold electrode (GE); b: Thiourea-GE; c: AuNPs-thiourea-GE; d: MAb-AuNPs-thiourea-GE; e: BSA-MAb-AuNPs-thiourea-GE; f: H5-BSA-MAb-AuNPs-thiourea-GE; g: GO-PAb-BSA-H5-BSA-MAb-AuNPs-thiourea-GE. Supporting electrolyte was 5 mM $[Fe (CN)_6]^{3-/4-}$, 0.1 M KCl and 0.01 M PBS (pH 7.0). The frequency range was between 0.1 and 100 000 Hz (AC 5 mV, DC 0.24 V vs SCE).

Figure 5-27-2 EIS of the different electrodes

Cyclic voltammetry (CV) technique was used to study the assembly process of the modified electrode. The CV scans of the different modified electrodes are shown in Figure 5-27-3. A well-defined redox wave is shown in curve a, corresponding to the reversible redox reaction of ferricyanide ions on the bare GE electrode. The redox peaks then apparently disappeared after thiourea was coated onto the electrode surface, owing to a thiourea film that greatly obstructed electron and mass transfer (curve b). When the bare GE was modified with AuNPs, the peak current of the system gradually increased (curve c), potentially due to AuNPs effectively increasing the surface area and active sites of the electrode. The immobilization of MAb on the electrode surface resulted in a decreased peak current (Figure 5-27-3 d), which suggests that MAb severely reduces the surface area and active sites needed for electron transfer. The peak current slightly decreased (Figure 5-27-3 e) after BSA was used to block non-specific sites.

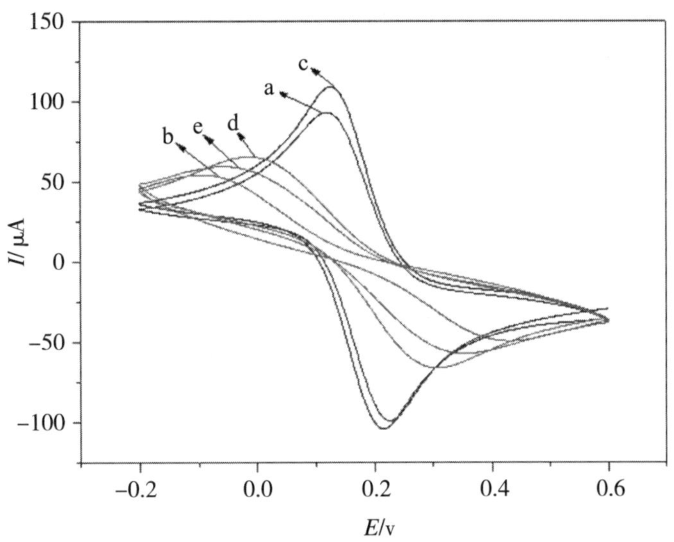

a: Bare gold electrode (GE); b: Thiourea-GE; c: AuNPs-thiourea-GE; d: MAb-AuNPs-thiourea-GE; e: BSA-MAb-AuNPs-thiourea-GE; f: H5-BSA-MAb-AuNPs-thiourea-GE. Supporting electrolyte was 5 mM $[Fe (CN)6]^{3-/4-}$, 0.1 M KCl and 0.01 M PBS (pH 7.0). Scan rate was 50 mV/s.

Figure 5-27-3 Cyclic voltammograms of the electrode at different stages

The signal amplification was also confirmed by differential pulse voltammetry (DPV) measurements. As shown in Figure 5-27-4, a 7-fold increase in the change current was observed with GO-PAb-BSA-H5-BSA-MAb-AuNPs-thiourea-GE (I_2) compared with H5-BSA-MAb-AuNPs-thiourea-GE (I_1). This can be explained as GO-PAb-BSA introduced more proteins onto the electrode surface, preventing further transfer from the redox probe to the electrode surface.

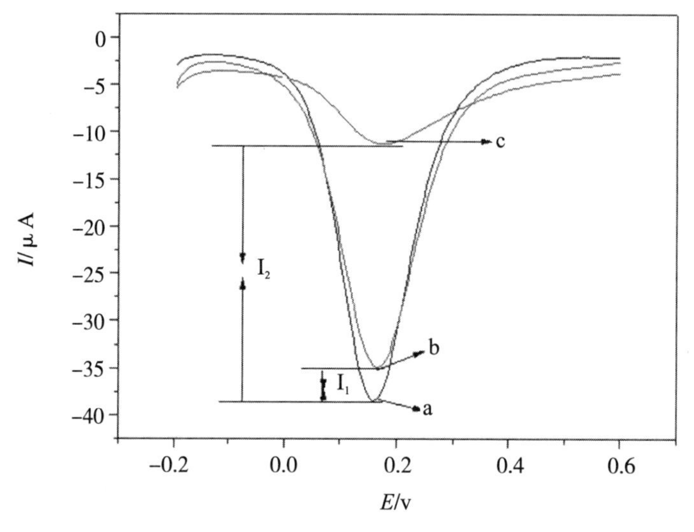

a: BSA-MAb-AuNPs-thiourea-GE; b: H5-BSA-MAb-AuNPs-thiourea-GE; c: GO-PAb-BSA-H5-BSA-MAb-AuNPs-thiourea-GE. Supporting electrolyte was 5 mM $[Fe(CN)_6]^{3-/4-}$, 0.1 M KCl and 0.01 M PBS (pH 7.0).

Figure 5-27-4 Differential pulse voltammograms of the immunosensor measurement process

Optimization of analytical conditions

The effect of pH on the immunosensor was investigated between pH 6.0 and pH 8.0. As shown in Figure 5-27-5, increasing the pH from 6.0 to 7.0 resulted in an increased peak current; further increases in pH resulted in the peak current decreasing. These results showed that the maximum current response occurred at pH 7.0. Therefore, PBS at pH 7.0 was used throughout this study.

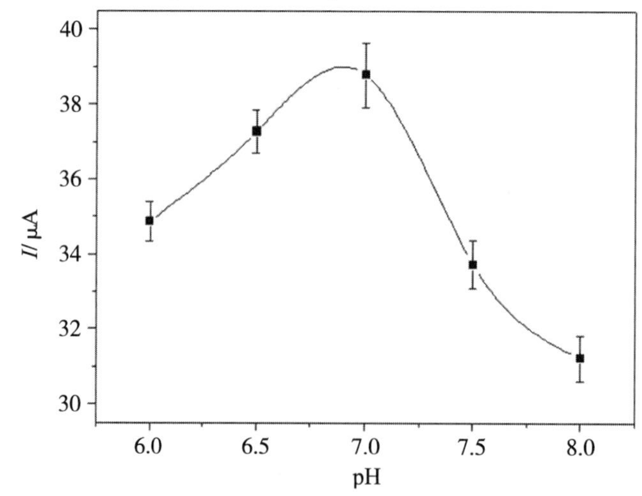

The scan rate was 50 mV/s. Initial potential and the end potential were 0.6 V and –0.2 V, respectively.

Figure 5-27-5 The effect of pH on the current response of the immunosensor

The incubation time is an important parameter for both capturing H5 antigen and specifically recognizing GO-PAb-BSA. We showed that the electrochemical response decreased with increasing H5 antigen incubation time and tended to reach a steady value after 30 min (curve a, Figure 5-27-6), indicating thorough capture of the antigens on the electrode surface. In the second immunoassay incubation step, the current also decreased upon increasing incubation time and it reached a plateau at 40 min, which indicates that binding sites between the antigen and detecting antibody were saturated (curve b, Figure 5-27-6).

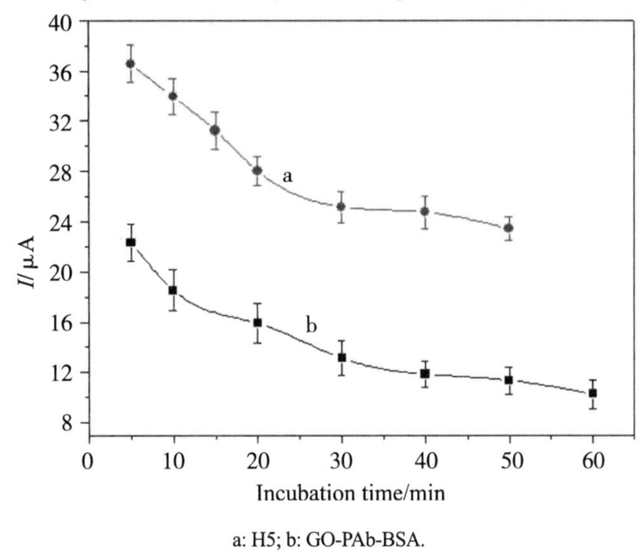

a: H5; b: GO-PAb-BSA.

Figure 5-27-6 Influence of the incubation time on the current response

Criteria for judgement of positive or negative samples

According to the principle of statistics, if the current of a sample is more than the critical value (critical value = average current of the negative sample+3 × standard deviation), the sample is considered as positive. Eighteen samples were tested negative by immunosensor (the average current: −38.46, standard deviation: 0.38), the critical value is −37.32. When a current of a sample is more than −37.32, the sample will be considered positive for the H5 antigen.

Performance of immunosensor

Immunosensor performance was evaluated according to its ability to detect H5 using the DPV technique (in 5 mM $[Fe(CN)_6]^{3-/4-}$, 0.1 M KCl and 0.01 M PBS) , under optimized sandwich-type immunoreaction conditions. As expected for a sandwich mechanism, the DPV peak current density of the immunosensor decreased with increasing H5 concentrations. Figure 5-27-7 illustrates the calibration plots of the cathodic peak current in response to varying H5 concentrations. According to the amplification effect of the GO-PAb-BSA, the linear range spans H5 concentrations of 2^{-15} to 2^{-8} HA unit/50 μL with a detection limit of 2^{-15} HA unit/50 μL (Figure 5-27-7). For comparison, the current response of the immunosensor was also recorded without GO-PAb-BSA amplification. In the absence of GO-PAb-BSA amplification, the linear range spans H5 concentrations from 2^{-7} to 2^{0} HA unit/50 μL.

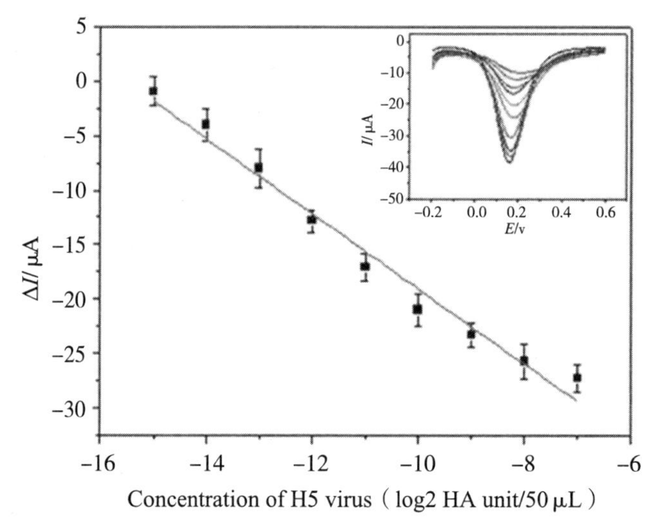

Inset: DPV of the GO-PAb-BSA modified immunosensor with various concentrations of H5 (from top to bottom: 0, 2^{-15}, 2^{-14}, 2^{-13}, 2^{-12}, 2^{-11}, 2^{-10}, 2^{-9}, 2^{-8}, and 2^{-7} HA unit/ 50 μL, in pH 7.0 PBS containing 5 mM $K_4Fe(CN)_6$, 5 mM $K_3Fe(CN)_6$ and 0.1 M KCl).

Figure 5-27-7 The relationship between antigen concentrations and the sensor response to current

Specificity study

We tested our developed immunosensor against some nontarget samples such as H1N3, H2N3, H3N6, H4N5, H6N8, H7N2, H8N4, H9N6, H10N3, H11N3, H12N5, H13N5, NDV, ILTV, IBV and MG. The experimental procedure was the same as that used for the H5 target. Figure 5-27-8 demonstrates that all the non-target samples produced baseline signals similar to the negative control, tissues fiom the SPF chickens. The results indicate that our developed immunosensor has good specificity for the target H5.

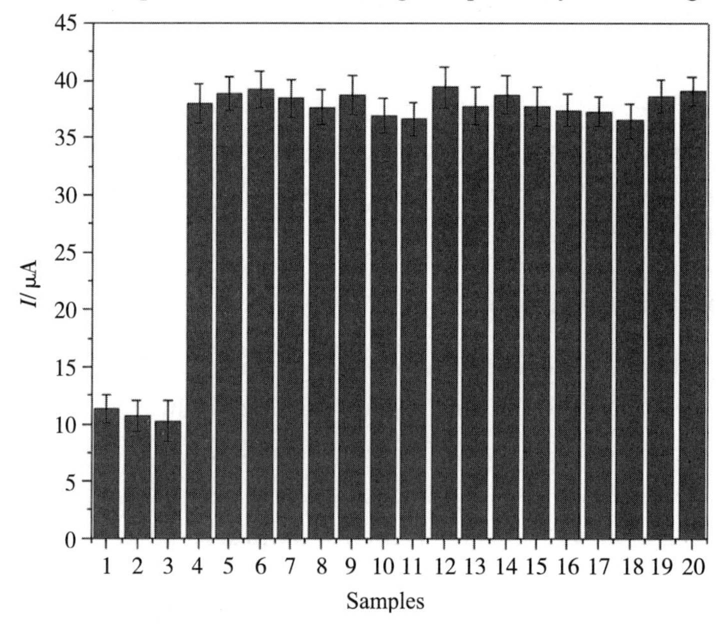

1: H5N1; 2: H5N2; 3: H5N9; 4: H1N3; 5: H2N3; 6: H3N6; 7: H4N5; 8: H6N8; 9: H7N2; 10: H8N4; 11: H9N6; 12: H10N3; 13: H11N3; 14: H12N5; 15: H13N5; 16: NDV; 17: IBV; 18: ILTV; 19: MG; 20: SPF chicken.

Figure 5-27-8 Selectivity of the electrochemical immunosensor

Discussion

Electrochemical immunosensors have been proven as an inexpensive and simple analytical method with remarkable detection sensitivity, and ease of miniaturization[26]. Various types of electrochemical immunosensors have been reported, including, amperometric, potentiometric, capacitive and impedance immunosensors. Amperometric electrochemical immunosensors have been considered as one of the most potential approaches for a higher sensitivity, less complicated instrumentation and broad linear range[27]. Therefore, in this study we investigated an amperometric electrochemical immunosensor for detecting AIV H5.

Because of the large specific surface area, high surface free energy, good biocompatibility and suitability, many kinds of nanomaterials have been widely used in electrochemical immunosensors, including metal nanoparticles (gold, silver), semiconductor nanoparticles and electroactive component-loaded nanovehides (silica nanoparticle, polymer beads, and liposome beads). Compared to the traditional metal ion labels, enzyme labels and redox probe labels, the common characteristic of nanomaterial labels in electrochemical immunosensors lies in their ability to provide signal amplification[28].

For our immunosensor fabrication, thiourea was self-assembled on a gold electrode (GE) via Au-S covalent bonds, which then yielded an interface containing amine groups ready for the electrochemical reduction of $HAuCl_4$, preparation of AuNP modified GE surfaces and H5 monoclonal antibody attachment. AuNPs were used to promote electron transfer between proteins and electrodes. In particular, AuNPs have been extensively used as an immobilizing matrix for retaining the bioactivity of antibodies.

Our present work highlights the promising applications of BSA functionalized GO in signal amplification for ultrasensitive detection of AIV H5. Herein, GO was employed as a nanocarrier for BSA and antibody co-immobilization. Enhanced sensitivity for the AIV H5 was based on the following signal amplification strategy: first, the high specific surface area of GO allowed multiplel binding events of BSA and second, GO and BSA hindered the diffusion of ferricyanide toward the electrode surface.

Conclusions

In summary, we have successfully designed a PAb labeled GO immunosensor and demonstrated its use in the ultrasensitive and selective detection of H5. Enhanced sensitivity was achieved by using GO as a nanocarrier to link BSA and PAb at a high ratio. Our immunosensor detected H5 antigen efficiently over a broad linear range and with a high sensitivity. We anticipate that this method may be extended for determination of other proteins and may have a promising potential in dinical applications.

References

[1] WANG R H, WANG Y, LASSITER K, et al. Interdigitated array microelectrode based impedance immunosensor for detection of avian influenza virus H5N1.Talanta, 2009, 79: 159-164.

[2] WORLD HEALTH ORGANIZATION (WHO). Cumulative number of confirmed human cases of avian influenza A (H5N1) reported to WHO. (2014-01-24) [2014-04-14]. http://www.who.int/influenza/human_animal_interface/EN_GIP_20140124CumulativeNumberH5N1cases. pdf?ua = 1.

[3] BAI H, WANG R H, HARGIS B, et al. A SPR aptasensor for detection of avian influenza virus H5N1.Sensors, 2012, 12: 12506-12518.

[4] XIE Z X, PANG Y S, LIU J B, et al. A multiplex RTPCR for detection of type A influenza virus and differentiation of avian H5, H7, and H9 hemagglutinin subtypes. Mol Cell Probes, 2006, 20: 245-249.

[5] YUEN K Y, CHAN P K S, PEIRIS M, et al. Clinical features and rapid viral diagnosis of human disease associated with avian influenza A H5N1 virus. Lancet, 1998, 351 (9101): 467-471.

[6] VELUMANI S, DU Q, FENNER B J, et al. Development of an antigen-capture ELISA for detection of H7 subtype avian influenza from experimentally infected chickens. J Virol Methods, 2008, 147 (2): 219- 225.

[7] QI X, LI X H, RIDER P, et al. Molecular Characterization of Highly Pathogenic H5N1 Avian Influenza A Viruses Isolated from Raccoon Dogs in China. PLOS ONE, 2009, 4 (3): e4682.

[8] LEE C W, SUAREZ D L. Application of real-time RT-PCR for the quantitation and competitive replication study of H5 and H7 subtype avian influenza virus. J Virol Methods, 2004, 119 (2): 151-158.

[9] CHANTRATITA W, SUKASEM C, KAEWPONGSRI S, et al. Qualitative detection of avian influenza A (H5N1) viruses: A comparative evaluation of four real-time nucleic acid amplification methods. Mo Cell Probes, 2008, 22 (5-6): 287-293.

[10] ROWE T, ABERNATHY R A, PRIMMER J H, et al. Detection of antibody to avian influenza A (H5N1) virus in human serum by using a combination of serologic assays. J Clin Microbiol, 1999, 37: 937-943.

[11] FOUCHIER R A, BESTEBROER T M, HERFST S, et al. Detection of influenza A viruses from different species by PCR amplification of conserved sequences in the matrix gene. J Clin Microbiol, 2000, 38: 4096-4101.

[12] CHEN Y, XU F, FAN X, et al. Evaluation of a rapid test for detection of H5N1 avian influenza virus. J Virol Methods, 2008, 154: 213-215.

[13] TIAN J N, HUANG J L, ZHAO Y C, et al. Electrochemical immunosensor for prostate-specific antigen using a glassy carbon electrode modified with a nanocomposite containing gold nanoparticles supported with starch-functionalized multi-walled carbon nanotubes. Microchim Acta, 2012, 178 (1/2): 81-88.

[14] HAQUE A M J, PARK H, SUNG D, et al. An electrochemically reduced graphene oxide-based electrochemical immunosensing platform for ultrasensitive antigen detection. Anal Chem, 2012, 84: 1871-1878.

[15] LIU X Q, ZHAO R X, MAO W L, et al. Detection of cortisol at a gold nanoparticle/protein G-DTBP-scaffold modified electrochemical immunosensor. Analyst, 2011, 136: 5204-5210.

[16] WANG G F, ZHANG G, HUANG H, et al. Graphene-Prussian blue/gold nanoparticles based electrochemical immunoassay of carcinoembryonic anti gen. Anal Methods, 2011, 3: 2082-2087.

[17] SU B L, TANG J, CHEN H F, et al. Thionine/nanogold multilayer film for electrochemical immunoassay of alpha-fetoprotein in human serum using biofunctional double-codified gold nanoparticles. Anal Methods, 2010, 2: 1702-1709.

[18] AMANO Y, CHENG Q. Detection of influenza virus: Traditional approaches and development of biosensors. Anal Bioanal Chem, 2004, 381: 156-164.

[19] LI Q F, ZENG L X, WANG J C, et al. Magnetic mesoporous organic-inorganic $NiCo_2O_4$ hybrid nanomaterials for electrochemical immunosensors. ACS Appl Mater Inter, 2011, 3: 1366-1373.

[20] DU D, WANG L M, SHAO Y Y, et al. Functionalized graphene oxide as a nanocarrier in a multienzyme labeling amplification strategy for ultrasensitive electrochemical immunoassay of phosphorylated p53 (S392). Anal Chem, 2011, 83: 746-752.

[21] LI T, YANG M H, LI H. Label-free electrochemical detection of cancer marker based on graphene-cobalt hexacyanoferrate nanocomposite. Journal of Electroanalytical Chemistry, 2011, 655: 50-55.

[22] CHEN D, FENG H B, LI J H. Graphene Oxide: preparation, functionalization, and electrochemical applications. Chem Rev, 2012, 112: 6027-6053.

[23] SWAYNE D, SENNE D A, BEARD. Avian influenza//SWAYNE D E, GLISSON J R, JACKSON M W, et al. A laboratory manual for the isolation and identification of avian pathogens, American association of avian pathologists. 4th ed. Pennsylvania: Kennett Square. 1998.

[24] FREY M I, HANSON R P, ANDERSON D P. A medium for the isolation of avian mycoplasma. Am J Vet Res, 1968, 29: 2163-2171.

[25] MUSZYNSKI R, SEGER B, KAMAT P V. Decorating graphene sheets with gold nanoparticles. J Phys Chem C, 2008, 112

(14): 5263-5266.
[26] VASHIST S K, ZHENG D, AL-RUBEAAN K, et al. Advances in carbon nanotube based electrochemical sensors for bioanalytical applications. Biotechnol Adv, 2011, 29: 169-188.
[27] ZHUO Y, YU R J, YUAN R, et al. Enhancement of carcinoembryonic antibody immobilization on gold electrode modified by gold nanoparticles and SiO_2/Thionine nanocomposite. J of Electroanal Chem, 2009, 628: 90-96.
[28] LIU G D, LIN Y H. Nanomaterial labels in electrochemical immunosensors and immunoassays. Talanta, 2007, 74: 308-317.

Silver nanoparticles coated graphene electrochemical sensor for the ultrasensitive analysis of avian influenza virus H7

Huang Jiaoling, Xie Zhixun, Xie Zhiqin, Luo Sisi, Xie Liji, Huang Li, Fan Qing, Zhang Yanfang, Wang Sheng, and Zeng Tingting

Abstract

A new, highly sensitive electrochemical immunosensor with a sandwich-type immunoassay format was designed to quantify avian influenza virus H7 (AIV H7) by using silver nanoparticle-graphene (AgNPs-G) as trace labels in clinical immunoassays. The device consists of a gold electrode coated with gold nanoparticle-graphene nanocomposites (AuNPs-G), the gold nanoparticle surface of which can be further modified with H7-monoclonal antibodies (MAbs). The immunoassay was performed with H7-polyclonal antibodies (PAbs) that were attached to the AgNPs-G surface (PAb-AgNPs-G). This method of using PAb-AgNPs-G as detection antibodies shows high signal amplification and exhibits a dynamic working range of $1.6 \times 10^{-3} \sim 16$ ng/mL, with a low detection limit of 1.6 pg/mL at a signal-to-noise ratio of 3σ. In summary, we showed that this novel immunosensor is highly specific and sensitive to AIV H7, and the established assay could potentially be applied to rapidly detect other pathogenic microorganisms.

Keywords

graphene, electrochemical immmunosensor, nanolabel, sensitivity enhancement, avian influenza virus H7

Introduction

Since 1959, avian influenza virus (AIV) H7 infections have been observed frequently; one example is AIV H7N1 infection in poultry during the 1999-2000 epidemic in Italy[1]. In the Netherlands, an AIV H7N7 outbreak not only impacted the poultry industry but also infected 89 people in 2003[2]. In March 2013, AIV H7N9 was first isolated from three patients in China, and since then, more than 450 human cases of H7N9 infection have been reported, with 165 deaths[3, 4].

Rapid diagnosis and time-monitoring of potential AIV H7 outbreaks are among the first important steps in disease prevention and control. Currently, several methods are available for AIV H7 detection, such as polymerase chain reaction (PCR)-based assays[5, 6], enzyme-linked immunosorbent assays (ELISAs)[7], reverse transcription loop-mediated isothermal amplification (RT-LAMP)[8], and nucleic acid sequence-based amplification (NASBA)[9]. However, the disadvantages with these diagnostic methods make them less than ideal for practical applications.

In recent years, electrochemical immunosensors have attracted considerable interest due to their intrinsic advantages, such as high sensitivity, low cost, low power requirements and high compatibility with advanced micromachining technologies[10, 11]. Antigens/antibodies are typically used as capture probes in electrochemical immunosensors, but antigens/antibodies usually exhibit weak electrochemical behavior, so a bioactive enzymelabeled method occupies an important position in the sandwichtype electrochemical immunosensor[12, 13]. However, trouble arises because enzyme activity, such as that of horseradish peroxidase, is not stable. Silver nanoparticles are a type of metal nanoparticle that exhibits good electrochemical behavior, such as exhibiting long-term stability at room temperature. Therefore, several research groups have developed silver nanoparticles as labels to stabilize immunosensor signals[14, 15]. However, low sensitivity results when only using silver nanoparticles to label the detection antibody. Thus, exploring a new trace label that is based on the electrochemical principle would be valuable[15, 16].

Recently, graphene, a single layer of carbon atoms in a closely packed honeycomb two-dimensional lattice, has been used as trace for signal amplification of electrochemical immunosensors due to its high electrical conductivity, high surface-to volume ratio, high electron transfer rate and exceptional thermal stability[17-20]. In this work, graphene was prepared and used both for the immobilization of monoclonal antibodies (MAbs) and as a tracer to label polyclonal antibodies (PAbs) to fabricate electrochemical immunosensors. For MAb immobilization, a graphene-chitosan (G-Chi) homogeneous composite was dispersed in an acetic acid solution, and then the gold nanoparticles (AuNPs) were synthesized in situ at the composite MAb immobilized on the AuNP surface. For the preparation of the tracer to label PAbs, AgNPs were synthesized in situ at the G-Chi composite and conjugate PAb. The aim of this study was to develop a new conductive nanolabel with highly amplified properties for sandwich-type electrochemical immunoassays.

Materials and methods

Reagents and materials

Graphite powder (<45 mm), hydrochloroauric acid ($HAuCl_4$), silver nitrate ($AgNO_3$), $NaNO_3$, H_2SO_4 and $KMnO_4$ were purchased from Guoyao Group Chemical Reagents Co., Ltd., Shanghai. H7-polyclonal antibodies and H7-monoclonal antibodies were purchased from Abcam (Cambridge, UK). Bovine serum albumin (BSA) was supplied by Beijing Dingguo Biotechnology Co., Ltd (China). All chemicals were used without further purification. Deionized water (>18.2 MΩ/cm) was used in all experiments. Phosphate-buffered saline (PBS; 10 mmol/L), at various pH values was prepared by mixing stock solutions of NaH_2PO_4 and Na_2HPO_4.

Instruments

For electrochemical studies, we employed a CHI660D electrochemical workstation (Shanghai CH Instruments, Shanghai, China) with a standard three-electrode cell that contained a platinum wire auxiliary electrode, a saturated calomel reference electrode (SCE) and a working electrode (the modified electrode as working electrode). All potential values given refer to SCE. All experiments were performed at room temperature (25 ± 0.5 ℃).

Graphene synthesis

Graphene oxide (GO) was prepared using a modified Hummers method[21]. Briefly, 1.0 g of graphite powder and 2.5 g of $NaNO_3$ were added to 100 mL of concentrated H_2SO_4 and stirred for 1 h. The mixture was continuously stirred and ice-cooled as 5 g of $KMnO_4$ was slowly added. The mixed slurry was then stirred at 35 ℃ for 20 h. After that, 100 mL of deionized water was added slowly to the reacted slurry and then stirred at 85 ℃ for another 2 h. Next, 300 mL of deionized water was added to the reacted slurry. Then, 6 mL of 30% H_2O_2

was added; the slurry immediately turned into a bright yellow solution, and bubbles appeared. The resultant solution was stirred for 2 h and then allowed to precipitate for 24 h; after that, the supernatant was decanted. The resultant yellow slurry was centrifuged and washed with 500 mL of 0.5 mol/L HCl. After stirring for 2 h, the solution was centrifuged and washed again before further washing with deionized water until the pH of the solution increased to neutral (pH 7.0). Graphene was obtained by reduction of GO using $NaBH_4$ as a reducing agent at 85 ℃ for 3 h[22].

Preparation of AuNP-G nanocomposites

Gold nanoparticle-graphene nanocomposites (AuNPs-G) were prepared according to a previously reported method [23] , with certain modifications. In short, 0.5 wt.% of a chitosan solution was first prepared by dissolving chitosan power in a 1.0% (v/v) acetic acid solution with stirring for 1 h at room temperature until completely dispersed. Then, 1 mg of graphene was added to 1 mL of the above chitosan solution, which was then ultrasonicated for 2 h and stirred for 24 h at room temperature. The resultant black suspension appeared to be homogeneous and stable. Herein, $Au^{3+}$was used as an oxidant and could be reduced to AuNPs by chitosan at 80 ℃ , In this work, 0.5 mL of 1 mM $HAuCl_4$ was added to the resultant graphene-chitosan (G-Chi) supernatant under vigorous stirring at room temperature for 4 h. The homogeneous mixture was then incubated at 80 ℃ for 1 h with stirring. The AuNP-G nanocomposites were obtained when a pink solution appeared and did not change.

Preparation of AgNP-G nanocomposites

Silver nanoparticle-graphene nanocomposites (AgNPs-G) were prepared according to the method that was mentioned above, with certain modifications. One mL of 1 mM $AgNO_3$ was initially added to a 1-mL graphene aqueous dispersion (1 mg/mL) with stirring for 2 h at room temperature. Then, 1 mL of the above chitosan solution was added, and the solution was then stirred for 5 h at room temperature. Afterward, the suspension was put in a water bath and reacted for 1 h at 80 ℃. The resulting dispersion was continuously stirred at room temperature for 12 h. The chitosan-enwrapped AgNPs-G were collected by centrifugation and washed with water.

Preparation of PAb-AgNP-G bioconjugates

To convert hydroxyl and epoxide groups (chitosan contains hydroxyl and epoxide groups, and graphene contains epoxide groups) to carboxylic groups, 50 mg of NaOH and 50 mg of $ClCH_2COONa$ were added to 1 mL of a 1 mg/mL AgNPs-G suspension, followed by bath sonication for 1 h. After these treatments, the resulting product, AgNPs-G-COOH, was neutralized with dilute hydrochloric acid and purified by repeated rinsing and centrifugation until the product was well dispersed in deionized water. The AgNPs-G-COOH suspension was then dialyzed against distilled water for over 48 h to remove any ions. To prepare PAb-AgNPs-G bioconjugates, 400 μL AgNPs-G (0.1 mg/mL) was activated with 10 μL 1-ethyl-3 (3-dimethylaminopropyl) carbodiimide hydrochloride (EDC) (5 mg/mL) and 20 μL N-hydroxysuccinimide-activated hexa (ethylene glycol) undecane thiol (NHS) (3 mg/mL) in PBS buffer (pH 5.2) for 30 min[24]. The mixture was centrifuged at 10 000 r/min for 10 min, and the supernatant was discarded. The buffer wash was repeated to remove excess EDC and NHS. The resulting functionalized mixture was dispersed in 1.0 mL of PBS buffer (pH 7.4) and sonicated for 5 min to obtain a homogeneous suspension. Then, 1 mL of PAb (1μg/

mL) and 2 mL of BSA (0.25% (w/v)) were added to the suspension, and the mixture was stirred overnight at 4 ℃. The reaction mixture was washed with PBS and centrifuged at 10 000 r/min for 5 min three times. The supernatant was discarded. The resulting mixture was redispersed in 1.0 mL of PBS (pH 7.4) and stored at 4 ℃.

Fabrication of the immunosensor

The gold electrode (GE; Ø=3 mm) was initially polished with 0.05 mm alumina to obtain a mirror-like surface before being rinsed with distilled water and placed into an ultrasonic bath to remove any physically adsorbed substances. Next, the electrode was placed into an electrochemical cell with 0.05 M H_2SO_4 and chemically cleaned until the background signal stabilized. The clean electrode was thoroughly rinsed with ddH_2O and dried with nitrogen gas, and 6 μL of the above solution of AuNP-G nano-composites was pipetted onto the surface of the clean GE. The casting solution was allowed to dry at 4 ℃ overnight. Then, the modified electrode (AuNPs-G-GE) was washed with water and immersed in the PBS solution (pH 7.4) containing 10 μg/mL (200 μL) H7-monoclonal antibodies (MAb) and immobilized at 4 ℃ overnight. Finally, the resulting electrode was incubated in BSA solution (0.25%, w/w) for approximately 1 h at 37 ℃ to block possible remaining active sites and to avoid non-specific adsorption. The completed immunosensor (MAb-AuNPs-G-GE) was stored at 4 ℃ when not in use. The procedures used for construction of the immunosensor are shown in Figure 5-28-1.

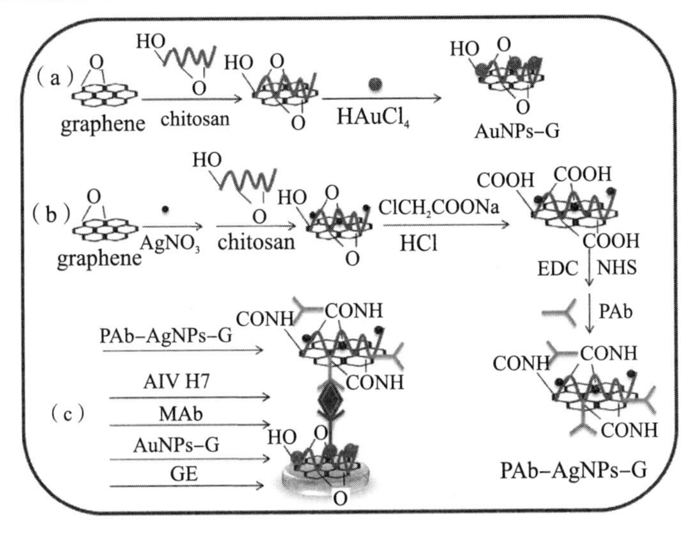

Figure 5-28-1 Immunosensor fabrication process

Immunoassay for the detection of AIV H7

A sandwich immunoassay was used to detect AIV H7. First, the immunosensor, MAb-AuNPs-G-GE, was incubated with 100 μL of various concentrations of AIV H7 (A/chicken/BD135/2013, H7N9, supported by China Agricultural University) for 30 min and then washed with PBS buffer. Next, the electrode was incubated with 200 μL of PAb-AgNPs-G bioconjugates for 40 min and washed with PBS buffer to remove non-specific adsorption conjugates. Finally, the AgNPs deposition on GE was taken out and placed in a 1 mol/L KCl solution with a platinum wire auxiliary electrode and SCE as counter and reference electrodes, respectively. Linear sweep voltammetry (LSV) was then performed from −0.15 to 0.25 V at a 50 mV/s scanning rate to record the stripping currents for AIV H7 detection.

Results and discussion

Transmission electron microscopy (TEM) characterization of the nanocomposites

In this study, graphene-based immunoassays contained two types of composites: the AuNPs-G platform and the AgNPs-G probe. Figure 5-28-2 A and B shows the AuNPs-G and AgNPs-G. Figure 5-28-2 A reveals that a relatively well-dispersed AuNP was successfully and stably assembled onto the surface of graphene, providing an efficient surface for MAb immobilization through the formation of a covalent bond between Au atoms of the AuNPs-G and amine groups of the antibody. As shown in Figure 5-28-2 B, AgNPs-G was covered with well-distributed AgNPs, which may produce electrochemical signals to detect AIV H7.

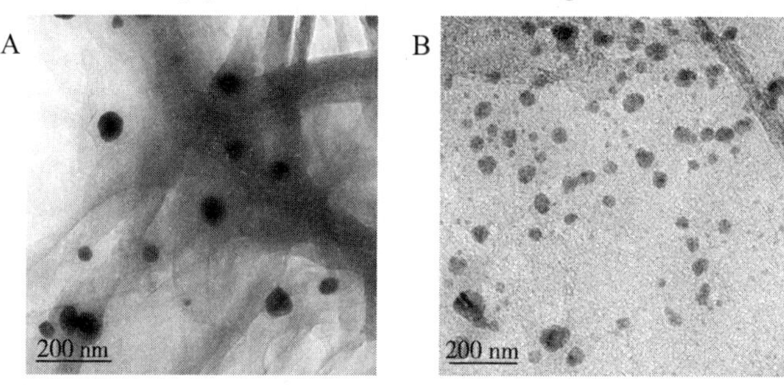

A: AuNPs-G; B: AgNPs-G.

Figure 5-28-2 TEM images

Electrochemical characterization of the immunosensor

Cyclic voltammetry (CV) is an effective and convenient technique for probing the features of the modified electrode surface. Here, CV was used to investigate electrochemical behaviors after each assembly step. The CVs of the modified electrodes in 5 mM $[Fe(CN)_6]^{3-/4-}$ solution are presented in Figure 5-28-3. The redox label $[Fe(CN)_6]^{3-/4-}$ revealed a reversible CV at the bare GE (Figure 5-28-3 a). After the GE was modified with G-Chi and AuNP-G composites, the peak current of the system gradually increased (Figure 5-28-3 b, c). When the AuNP-G composites were immobilized on GE, the current response of the system reached its maximum. This result may occur because the AuNPs can effectively increase the surface area and active sites of the electrode, and graphene can enhance the electrical conductivity, which made it easy for the $[Fe(CN)_6]^{3-/4-}$ to spread to the surface of GE. To improve the sensitivity of the immunosensor, AuNP-G composites were used to modify GE in this work. After MAbs had been immobilized on the electrode surface, the peak current clearly decreased (Figure 5-28-3 d), suggesting that the protein MAb severely reduced the effective area and active sites for electron transfer between $[Fe(CN)_6]^{3-/4-}$ and GE. The peak current decreased in the same way after BSA was used to block non-specific sites (Figure 5-28-3 e). When the immunosensor was incubated with AIV H7 for 30 min, a dramatic decrease in current was observed (Figure 5-28-3 f). This decrease was attributed to the formation of the MAb-AIV H7 immunocomplex, which acts as the inert electron and mass transfer blocking layer and hinders diffusion of ferricyanide toward the electrode surface.

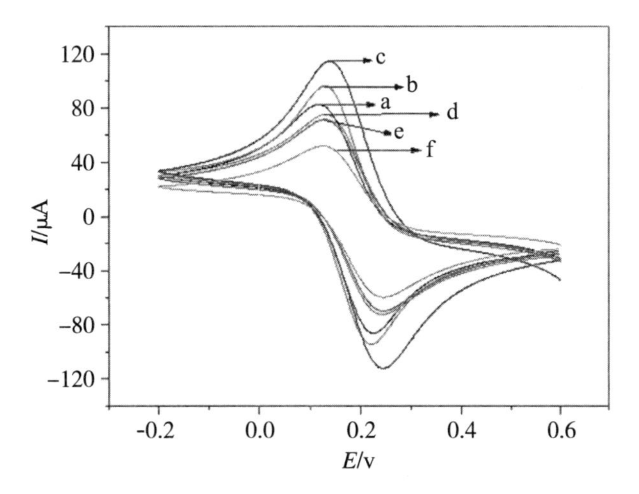

a: Bare gold electrode (GE); b: G-Chi-GE; c: AuNPs-G-GE; d: MAb-AuNPs-G-GE; e: BSA-MAb-AuNPs-G-GE; f: AIV H7-BSA-MAb-AuNPs-G-GE. Supporting electrolyte, 5 mM $[Fe(CN)_6]^{3-/4-}$ +0.1 M KCl+0.01 M PBS (pH 7.0); Scan rate 50 mV.

Figure 5-28-3 Cyclic voltammograms of the electrode at different stages

The electrochemical characteristics and amplification performances of the AIV H7 immunosensor were investigated using linear sweep voltammetry (LSV), and the results are shown in Figure 5-28-4. The curves in Figure 5-28-4 show the LSV plots of the BSA-MAb-AuNPs-GGE-modified gold electrode. No anodic peak can be observed because of the lack of substances with electrochemical activity in the working potential range of the working solution, which provides a low and reproducible background current. After the PAb-AgNPs-G were adsorbed onto the electrode via sandwich immunoreactions, a stable anodic peak at 0.066 V VS. SCE (curve b in Figure 5-28-4) was detected. These results suggest the efficient redox activity of PAb-AgNP-G bioconjugates.

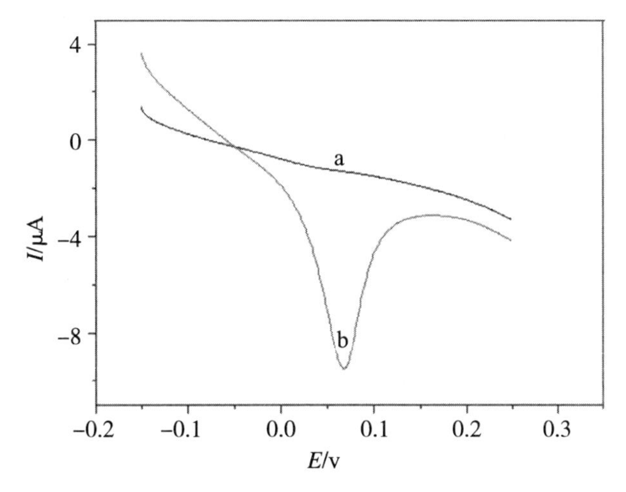

a: BSA-MAb-AuNPs-G-GE; b: PAb-AgNPs-G-AIV H7-BSA-MAb-AuNPs-G-GE. The sample included 100 μL of 16 ng/mL (A/chicken/BD135/2013, H7N9).

Figure 5-28-4 Linear sweep voltammetry of the immunosensor measurement process

Optimization of reaction conditions

The incubation time is important for both capturing AIV H7 and specifically recognizing PAb-AgNPs-G. We showed that the electrochemical response increased with increasing AIV H7 incubation time and tended to reach a steady value after 30 min (Figure 5-28-5 a), indicating thorough capture of the antigens on the electrode surface. In the second immunoassay incubation step, the current also increased upon increasing incubation time and reached a plateau at 30 min, which indicates that binding sites between the antigen and detecting antibody were saturated (Figure 5-28-5 b). Each test was repeated three times.

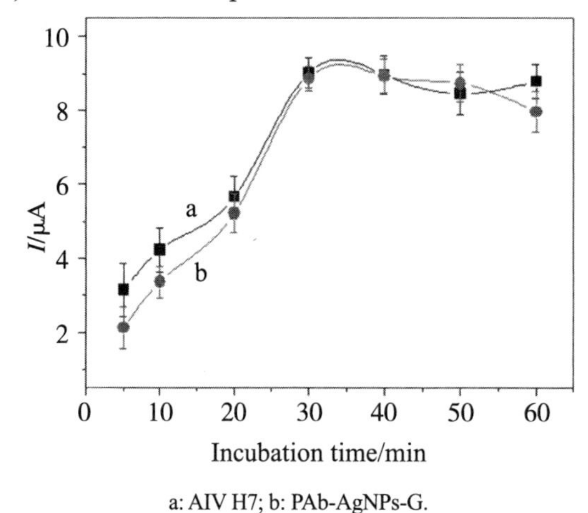

a: AIV H7; b: PAb-AgNPs-G.

Figure 5-28-5 Influence of incubation time on the current response of immunosensor

Analytical properties of the immunosensor

Under these optimized experimental conditions, the immunosensor reacted with AIV H7 at different concentrations for LSV determination, and each concentration was tested five times. As shown in Figure 5-28-6, the LSV peak current of the immunosensor increased with increasing antigen concentrations. In the 1.6×10^{-3}~16 ng/mL range, the equation is I=3.098+1.290 7 lg ($C_{AIV\ H7}$), and the correlation coefficient was 0.995 8, with a detection limit of 1.6 pg/mL at a signal-to-noise ratio of 3 σ (where σ is the standard deviation of the blank, n =12). The sensitivity of the proposed method was not only superior to the traditional ELISA method, but it was also comparable to some reported novel immunosensors (Table 5-28-1)[25-32]. The possible explanations may be as follows: (1) Chitosan has abundant amino groups, and it exhibits good biocompatibility[33] and an excellent film-forming ability that originate from its protonation. It is soluble in slightly acidic solution and stable from the insolubility in solutions with pH over pKa (6.3) [34]. Thus, chitosan is a highly suitable matrix for immobilizing bioactive molecules and constructing biosensors. Graphene-AuNPs-chitosan composites and graphene-AgNPs-chitosan composites were obtained here. (2) In taking advantage of

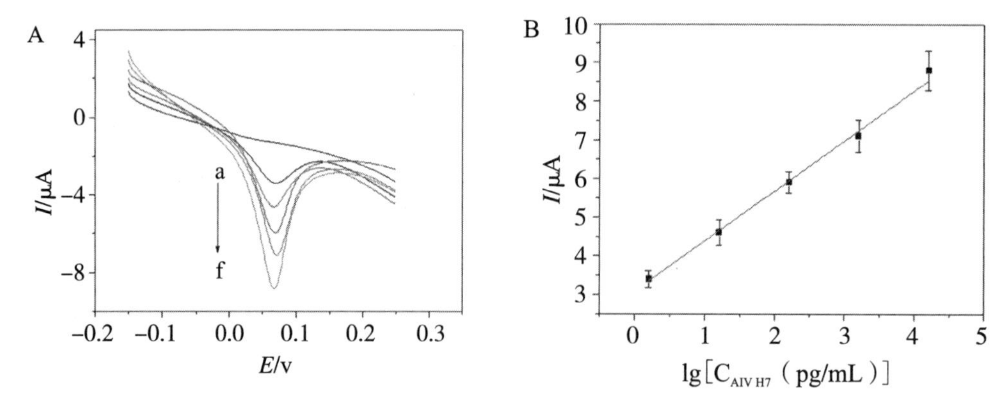

A: Typical LSV signals in the presence of different AIV H7 concentrations; a: 0; b: 1.6; c: 16; d: 160; e: 1 600; f: 16 000 pg/mL. B: The relationship between antigen concentration and sensor response to current.

Figure 5-28-6 Sensitivity results of immunosensor

the high surface-to-volume ratio of graphene, multiple AgNPs are immobilized on the graphene surface, which significantly improves the response signal; and (3) graphene-AuNPs-chitosan composite films can be further modified with MAbs.

Table 5-28-1 Comparison of different methods for determining the AIV antigen

Technique	Linear range	Detection limit	Reference
ELISA	–	10 ng/mL	[25]
DAS-ELISA	–	2.5 ng/mL	[26]
Resonance light scattering	0.5～50 ng/mL	0.15 ng/mL	[27]
Fluoroimmunoassay	8～510 ng/mL	0.15 ng/mL	[28]
Indirect fluorescence immunoassay	0.27～12 ng/mL	0.09 ng/mL	[29]
Bifunctional magnetic nanobeads of an electrochemical immunosensor	0.01～20 ng/mL	6.8 pg/mL	[30]
Real-time reverse transcription-polymerase chain reaction	–	3.2×10-hemagglutination units	[31]
Reverse transcription-loop-mediated isothermal amplification	–	42.47 copies/reaction	[32]
Silver nanoparticle-coated graphene electrochemical sensor	1.6×10^{-3}～16 ng/mL	1.6 pg/mL	This work

Selectivity, precision, and stability

We tested our developed immunosensor against some nontarget samples, such as inactivated H7N9, H7N2, H7N7, H1N1, H2N3, H3N2, H4N6, H6N8, H8N4, H9N2, H10N3, H11N9, H12N5, H13N5, H14N5, H15N9, H16N3, ILTV (infectious laryngotracheitis virus), IBV (infectious bronchitis virus) and NDV (Newcastle disease virus). The experimental procedure was the same as that used for the H7 target. The concentrations were calculated according to the calibration curve. The results are shown in Table 5-28-2: all results of the AIV H7 sample tests were positive, and all results of non-target sample tests were negative, standard deviations (RSDs) were 3.0%～6.7%. These results indicate that our developed immunosensor has good specificity for the target AIV H7.

The precision of the immunosensor was investigated using the LSV of intra-and inter-assays. The intra-assay precision of immunosensors was evaluated by detecting three concentration levels five times per run. The LSVs of the intra-assay were 5.3%, 4.7% and 4.3% at 16, 160 and 1 600 pg/mL AIV H7, respectively. Similarly, the LSVs of the inter-assay were 5.7%, 6.1% and 7.2% at 16, 160 and 1 600 pg/mL AIV H7, respectively. Thus, the precision and reproducibility of the immunosensors were acceptable.

The stability of the synthesized AgNP-G bioconjugates was examined. When not in use, they could be stored in pH 7.0 PBS containing 0.1% NaN_3 at 4 ℃ for at least 2 weeks without obvious signal changes. Moreover, the bioconjugate retained 91.2% of its initial response after a storage period of 30 days. We speculate that the slow decrease of response was mainly attributed to the gradual deactivation of the immobilized biomolecules on the surface of AgNPs-G.

Conclusions

In this work, we designed a new sandwich-type immunosensor based on PAb-AgNP-G bioconjugates for the sensitive detection of AIV H7.The immunosensor displayed a wide range of linear responses (1.6×10^{-3}～16 ng/mL) and a low detection limit (1.6 pg/mL). The immobilized MAb molecules exhibit an excellent electrochemical response that is selective to AIV H7 in PBS, pH 7.0. More importantly, this approach is well suited for biomedical sensing and clinical applications. Hopefully, the immobilized technique and the detection methodology can be further developed for other pathogens. Considering the application for clinical analysis, the time of analysis is a critical factor; therefore, how to develop simpler and more effective detection methods remains a challenge.

Table 5-28-2 The results of the specificity test

No. [a]	Avian pathogen	Source	LSV signal		Results
			Mean/μA, *n*=3	RSD/%	
(1)	A/Chicken/BD135/2013 (H7N9)	CAU	8.12	4.2	+
(2)	A/Chicken PA/3979/97 (H7N2)	PU	8.37	4.7	+
(3)	A/Chicken/NY/273874/03 (H7N2)	UCONN	7.81	4.5	+
(4)	A/Duck/HK/47/76 (H7N2)	UHK	7.71	5.9	+
(5)	A/Duck/42846/07 (H7N7)	PU	7.99	6.3	+
(6)	A/Duck/Guangxi/030D/2009 (H1N1)	GVRI	1.64	3.0	–
(7)	A/Duck/HKJ77/76 d77/3 (H2N3)	UHK	1.54	3.6	–
(8)	A/Duck/Guangxi/M20/2009 (H3N2)	GVRI	1.38	6.3	–
(9)	A/Duck/Guangxi/070D/2010 (H4N6)	GVRI	1.43	3.1	–
(10)	A/Chicken/QT35/98 (H5N9)	PU	1.81	3.9	–
(11)	A/Duck/Guangxi/GXd–6/2010 (H6N8)	GVRI	1.72	4.1	–
(12)	A/Turkey/Ontario/6118/68 (H8N4)	UHK	1.36	5.2	–
(13)	A/Chicken/Guangxi/DX/2008 (H9N2)	GVRI	1.97	3.7	–
(14)	A/Duck/HK/876/80 (H10N3)	UHK	1.51	3.0	–
(15)	A/Duck/PA/2099/12 (H11N9)	PU	1.46	6.5	–
(16)	A/Duck/HK/862/80 (H12N5)	UHK	1.59	6.7	–
(17)	A/Gull/Md/704/77 (H13N5)	UHK	1.61	5.4	–
(18)	A/Mallard/Astyakhan/263/82 (H14N5)	UCONN	1.39	6.1	–
(19)	A/Shearwater/Western Australia/2576/79 (H15N9)	UCONN	1.36	4.5	–
(20)	A/Shorebird/Delaware/168/06 (H16N3)	CIVDC	1.67	5.1	–
(21)	ILTV (Beijing)	CIVDC	1.56	3.3	–
(22)	IBV (Mass41)	CIVDC	1.38	5.4	–
(23)	NDV (F48E9)	CIVDC	1.86	3.9	–
(24)	(1)+(10)		8.23	6.4	+
(25)	(1)+(13)		8.07	5.9	+
(26)	(1)+(18)+(23)		8.12	6.8	+
(27)	(2)+(8)+(9)		7.81	6.3	+
(28)	(2)+(15)+(17)		7.77	4.6	+
(29)	(2)+(18)+(22)+(23)		7.93	3.9	+
(30)	(30) (6)+(7)+(8)		1.96	4.7	–
(31)	(31) (12)+(14)+(16)		1.67	5.8	–
(32)	(32) (18)+(19)+(20)		1.59	6.0	–

Note: +, positive; –, negative; PU, Pennsylvania State University, USA; CAU, China Agricultural University; UHK, University of Hong Kong, China; GVRl, Guangxi Veterinary Research Institute; CIVDC, China Institute of Veterinary Drug Control; and UCONN, University of Connecticut, USA. [a] Containing 10 ng/mL AIV H7 and 20 ng/mL (or U/mL) of interfering virus.

References

[1] BUSANI L, VALSECCHI M G, ROSSI E, et al. Risk factors for highly pathogenic H7N1 avian influenza virus infection in poultry during the 1999–2000 epidemic in Italy. Vet J, 2009, 181: 171-177.

[2] FOUCHIER R A, SCHNEEBERGER P M, ROZENDAAL F W, et al. Avian influenza A virus (H7N7) associated with human conjunctivitis and a fatal case of acute respiratory distress syndrome. Proc Natl Acad Sci USA, 2004, 101: 1356-1361.

[3] WHO. Risk assessment -Human infections with avian influenza A(H7N9) virus. (2014-10-02) [2016-02-01]. http://www.who.int/influenza/human_animal_interface/influenza_h7n9/riskassessment_h7n9_2Oct14.pdf?ua=1.

[4] CHEN Y, LIANG W, YANG S, et al. Human infections with the emerging avian influenza A H7N9 virus from wet market poultry: clinical analysis and characterisation of viral genome. Lancet, 2013, 381: 1916-1925.

[5] XIE Z X, PANG Y S, LIU J B, et al. A multiplex RT-PCR for detection of type A influenza virus and differentiation of avian H5, H7, and H9 hemagglutinin subtypes. Mol Cell Probes, 2006, 20: 245-249.

[6] LEE C W, SUAREZ D L. Application of real-time RT-PCR for the quantitation and competitive replication study of H5 and H7 subtype avian influenza virus. J Virol Methods, 2004, 119: 151-158.

[7] VELUMANI S, DU Q, FENNER BJ, et al. Devel-development of an antigen-capture ELISA for detection of H7 subtype avian influenza from experimentally infected chickens. J Virol Methods, 2008, 147: 219-225.

[8] BAO H M, WANG X R, ZHAO Y H, et al. Development of a reverse transcription loop-mediated isothermal amplification method for the rapid detection of avian influenza virus subtype H7. J Virol Methods, 2012, 179: 33-37.

[9] COLLINS R A, KO L S, FUNG K Y, et al. Rapid and sensitive detection of avian influenza virus subtype H7 using NABSA. Biochem Bioph Res Co, 2003, 300: 507-515.

[10] BAHADIR E B, SEZGINTÜRK M K. Applications of electrochemical immunosensors for early clinical diagnostics. Talanta, 2015, 132: 162-174.

[11] AKTER R, RHEE C K, RAHMAN M A. Sensitivity enhancement of an electrochemical immunosensor through the electrocatalysis of magnetic bead-supported non-enzymatic labels. Biosens Bioelectron, 2014, 54: 351-357.

[12] YANG M H, ZHANG J L, CHEN X. Competitive electrochemical immunosensor for the detection of histamine based on horseradish peroxidase initiated deposition of insulating film. J Electroanal Chem, 2015, 736: 88-92.

[13] ZHONG Z Y, WU W, WANG D, et al. Nanogold-enwrapped graphene nanocomposites as trace labels for sensitivity enhancement of electrochemical immunosensors in clinical immunoassays: Carcinoembryonic antigen as a model. Biosens Bioelectron, 2010, 25: 2379-2383.

[14] LI Y Y, HAN J, CHEN R H, et al. Label electrochemical immunosensor for prostate-specific antigen based on graphene and silver hybridized mesoporous silica. Anal Biochem, 2015: 469: 76-82.

[15] HUANG J L, TIAN J N, ZHAO Y C, et al. Ag/Au nanoparticles coated graphene electrochemical sensor for ultrasensitive analysis of carcinoembryonic antigen in clinical immunoassay. Sensor Actuat B: Chem, 2015, 206: 570-576.

[16] SUN G Q, ZHANG L N, ZHANG Y, et al. Multiplexed enzyme-free electrochemical immunosensor based on ZnO nanorods modified reduced graphene oxide-paper electrode and silver deposition-induced signal amplification strategy. Biosens Bioelectron, 2015, 71: 30-36.

[17] CASTRO R K, ARAUJO J R, VALASKI R, et al. New transfer method of CVD-grown graphene using a flexible, transparent and conductive polyaniline-rubber thin film for organic electronic applications. Chem Eng J, 2015, 273: 509-518.

[18] XIE Z X, HUANG J L, LUO S S, et al. Ultrasensitive electrochemical immunoassay for avian influenza subtype H5 using nanocomposite. PLOS ONE, 2014, 9: e94685.

[19] SAMANMAN S, NUMNUAM A, LIMBUT W, et al. Highly-sensitive label-free electrochemical carcinoembryonic antigen immunosensor based on a novel Au nanoparticles-graphene-chitosan nanocomposite cryogel electrode. Anal Chim Acta, 2015, 853: 521-532.

[20] ZHONG G X, WANG P, FU F H, et al. Electrochemical immunosensor for detection of topoisomerase based on graphene—gold nanocomposites. Talanta, 2014, 125: 439-445.

[21] MUSZYNSKI R, SEGER B, KAMAT P V. Decorating graphene sheets with gold nanoparticles. J Phys Chem C, 2008, 112: 5263-5266.

[22] ZHAO Y C, ZHAN L, TIAN J N, et al. Enhanced electrocatalytic oxidation of methanol on Pd/polypyrrole-grapheme in alkaline medium. Electrochim Acta, 2011, 56: 1967-1972.

[23] HUANG K J, NIU D J, XIE W Z, et al. A disposable electrochemical immunosensor for carcinoembryonic antigen based on nano-Au/multi-walled carbon nanotubes-chitosans nanocomposite film modified glassy carbon electrode. Anal Chim Acta, 2010, 659: 102-108.

[24] RAGHAV R, SRIVASTAVA S. Immobilization strategy for enhancing sensitivity of immunosensors: L-Asparagine-AuNPs as a promising alternative of EDC–NHS activated citrate-AuNPs for antibody immobilization. Biosens Bioelectron, 2016, 78: 396-403.

[25] MU B, HUANG X, BU P, et al. Influenza virus detection with pentabody-activated nanoparticles. J Virol Methods, 2010, 169: 282-289.

[26] ZHANG A, JIN M L, LIU F F, et al. Development and Evaluation of a DAS-ELISA for rapid detection of avian influenza viruses. Avian Dis, 2006, 50: 325-330.

[27] ZOU X, HUANG H, GAO Y, et al. Detection of avian influenza virus based on magnetic silica nanoparticles resonance light scattering system. Analyst, 2012, 137: 648-653.

[28] CHEN L P, SHENG Z H, ZHANG A D, et al. Quantum-dots-based fluoroimmunoassay for the rapid and sensitive detection of avian influenza virus subtype H5N1. Luminescence, 2010, 25: 419-423.

[29] LI X P, LU D L, SHENG Z H, et al. A fast and sensitive immunoassay of avian influenza virus based on label-free quantum dot probe and lateral flow test strip. Talanta, 2012, 100: 1-6.

[30] WU Z, ZHOU C H, CHEN J J, et al. Zhang, Bifunctional magnetic nanobeads for sensitive detection of avian influenza A (H7N9) virus based on immunomagnetic separation and enzyme-induced metallization. Biosens Bioelectron, 2015, 68: 586-592.

[31] KANG X P, WU W L, ZHANG C T, et al. Detection of avian influenza A/H7N9/2013 virus by real-time reverse transcription-polymerase chain reaction. J Virol Methods, 2014, 206: 140-143.

[32] NAKAUCHI M, TAKAYAMA I, TAKAHASHI H, et al. Development of a reverse transcription loop-mediated isothermal amplification assay for the rapid diagnosis of avian influenza A (H7N9) virus infection. J Virol Methods, 2014, 204: 101-104.

[33] LIU Y, WANG M K, ZHAO F, et al. The direct electron transfer of glucose oxidase and glucose biosensor based on carbon nanotubes/chitosan matrix. Biosens Bioelectron, 2005, 21: 984-988.

[34] SORLIER P, DENUZIERE A, VITON C, et al. Relation between the degree of acetylation and the electrostatic properties of chitin and chitosan. Biomacromolecules, 2001, 2: 765-772.

Electrochemical immunosensor with Cu(I)/Cu(II)-chitosan-graphene nanocomposite-based signal amplification for the detection of Newcastle disease virus

Huang Jiaoling, Xie Zhixun, Huang Yihong, Xie Liji, Luo Sisi, Fan Qing, Zeng Tingting, Zhang Yanfang, Wang Sheng, Zhang Minxiu, Xie Zhiqin, and Deng Xianwen

Abstract

An electrochemical immunoassay for the ultrasensitive detection of Newcastle disease virus (NDV) was developed using graphene and chitosan-conjugated Cu(I)/Cu(II) (Cu(I)/Cu(II)-Chi-Gra) for signal amplification. Graphene (Gra) was used for both the conjugation of an anti-Newcastle disease virus monoclonal antibody (MAb/NDV) and the immobilization of anti-Newcastle disease virus polyclonal antibodies (PAb/NDV). Cu(I)/Cu(II) was selected as an electroactive probe, immobilized on a chitosan-graphene (Chi-Gra) hybrid material, and detected by differential pulse voltammetry (DPV) after a sandwich-type immune response. Because Gra had a large surface area, many antibodies were loaded onto the electrochemical immunosensor to effectively increase the electrical signal. Additionally, the introduction of Gra significantly increased the loading amount of electroactive probes (Cu(I)/Cu(II)), and the electrical signal was further amplified. Cu(I)/Cu(II) and Cu(I)/Cu(II)-Chi-Gra were compared in detail to characterize the signal amplification ability of this platform. The results showed that this immunosensor exhibited excellent analytical performance in the detection of NDV in the concentration range of $10^{0.13}$ to $10^{5.13}$ EID_{50}/0.1 mL, and it had a detection limit of $10^{0.68}$ EID_{50}/0.1 mL, which was calculated based on a signal-to-noise (S/N) ratio of 3. The resulting immunosensor also exhibited high sensitivity, good reproducibility and acceptable stability.

Keywords

electrochemical immunosensor, Newcastle disease virus, signal amplification

Introduction

Newcastle disease virus (NDV) is a viral disease of poultry that belongs to avian paramyxovirus 1. It is a single-strand, non-segmented, and negative-sense RNA virus[1], and it is a great threat to the poultry industry[2]. The first important step in NDV prevention and control is to develop a rapid and sensitive method for diagnosis. Currently, several methods for detecting NDV, included virus isolation[3], reverse transcription polymerase chain reaction (RT-PCR)[4], real-time RT-PCR[5], immunochromatographic strip (ICS) tests[6], and reverse transcription loop-mediated isothermal amplification (RT-LAMP) assays[7], have been reported. However, these diagnostic methods had some disadvantages; for example, virus isolation is the gold standard for the detection of NDV, but the procedure is time-consuming. For RT-PCR, appropriate laboratory facilities and a trained technician are needed. Real-time RT-PCR requires complicated operations as well as expensive reagents and equipment. Therefore, these diagnostic methods are limited in practical applications.

Electrochemical immunosensors are powerful tools that have good specificity, high sensitivity, good precision, and simple instrumentation; give rapid and reliable responses; and are relatively low cost. Their use in clinical diagnosis, food analysis, environmental monitoring and archaeological studies should be highly valuable[8]. Furthermore, electrochemical immunosensors are based on antibody-antigen reactions. Therefore, immobilizing antibodies or antigens on a transducer as a biorecognition element plays a very important role in the construction of electrochemical immunosensors. Different methods for immobilizing antibodies/antigens on a transducer, including chemical and physical adsorption, have been discussed[9]. It has been reported that chitosan (Chi) is a suitable matrix for immobilizing biorecognition elements due to its biocompatibility, hydrophilicity, mouldability, chemical reactivity, and biodegradability[10]. However, Chi is non-conductive and has low solubility in different solutions; thus, many kinds of nanomaterials have been combined with Chi to increase its conductivity for the fabrication of electrochemical immunosensors[11]. Modifying transducers with conductive materials enhances the electron transfer between the electrode surface and electrolyte[10, 12, 13]. Furthermore, modifying them with nano-materials provides a rougher surface that enables the biorecognition element to attach closely to the electrode surface. Many kinds of nanomaterials, including Gra[14], multi-walled carbon nanotubes[15], gold nanoparticles[12], magnetic nanoparticles[16], quantum dots[17] and hybrid nanostructures [18], have been used in immunosensors.

Gra has a one-atom-thick planar structure composed of sp^2-hybridized carbon atoms packed in a honey-comb-like lattice[19]. Due to this unique structure, Gra has an exceptionally high surface-to-volume ratio, electrical conductivity, and thermal conductivity and good mechanical properties[20]. Gra has been used to improve the sensitivity and stability of immunosensors many times[21, 22]. However, the direct immobilization of protein molecules on Gra is difficult. As previously mentioned, Chi can easily immobilize protein molecules and form a film on transducers. Due to these properties, nanocomposites consisting of Chi and Gra are an ideal immunosensor material, and our group successfully synthesized a silver nanoparticle-chitosan-graphene composite to construct an electrochemical immunosensor[23].

However, copper is much less expensive than silver nanoparticles, and Cu(Ⅱ) ions can be adsorbed by Chi from aqueous solutions via chelation because of its unique three-dimensional structure[24]. Additionally, the synthesis of CuO (Cu(Ⅱ)) and Cu_2O (Cu(Ⅰ)) using Chi as a stabilizing and reducing agent has been reported[25-27]. Furthermore, Cu(Ⅱ) ions provide a good stripping voltammetric signal[28]. In addition, Cu(Ⅰ) has a direct band gap of 2.0 eV and is a p-type semiconductor that is very important in superconductors and electrode materials[26, 27]. As previously mentioned, Cu(Ⅰ) and Cu(Ⅱ) can be used as electroactive materials. The more electroactive a material carried by an immunosensor is, the more sensitive the immunoassay is. Therefore, in this study, Gra, which has a high loading capacity, was used to load a large amount of electroactive probes on an immunosensor. Hybrid Cu(Ⅰ)/Cu(Ⅱ)-modified Gra effectively amplifies signals. In this work, a sandwich-type electrochemical immunosensor was designed using a gold nanoparticle-chitosan-graphene (AuNP-Chi-Gra) nanocomposite as the platform and a Cu(Ⅰ)/Cu(Ⅱ)-chitosan-graphene (Cu(Ⅰ)/Cu(Ⅱ)-Chi-Gra) nanocomposite as the label for detecting NDV with a low detection limit ($10^{0.68}$ EID_{50}/0.1 mL) and high sensitivity in a relatively wide linear range (from $10^{0.13}$ to $10^{5.13}$ EID_{50}/0.1 mL). The developed immunosensor shows potential for applications in the clinical screening of other pathogenic microorganisms and point-of-care diagnostics.

Materials and methods

Reagents and materials

MAb/NDV and PAb/NDV were purchased from Abcam (Cambridge, UK). Copper sulfate ($CuSO_4$), hydrochloroauric acid ($HAuCl_4$), graphite powder (< 45 mm), $KMnO_4$, $NaNO_3$ and H_2SO_4 were supplied by the Guoyao Group Chemical Reagents Co., Ltd., Shanghai. Bovine serum albumin (BSA) was purchased from Sigma (USA). All chemicals used were of analytical reagent grade. Double-distilled deionized water was used in all experiments. In addition, 10 mmol/L PBS (pH=7.4) was prepared by mixing stock solutions of 10 mmol/L NaH_2PO_4 and 10 mmol/L Na_2HPO_4.

Instruments

SEM was performed on a HITACHI UHR FE-SEM SU8000 Series (SU8020) instrument. FT-IR spectra were collected on a Nicolet IS10 instrument. XPS analysis was performed on an X-ray photoelectron spectrometer (ESCALAB 250Xi, Thermo Scientific). A CHI660D electrochemical workstation (Beijing CH Instruments, Beijing, China) with a standard three-electrode cell (a working electrode, an SCE as the reference electrode and a platinum wire as the auxiliary electrode) was employed to study the electrochemical characteristics. Electrochemical detection was performed at room temperature (25 ± 0.5 ℃).

Gra synthesis

A modified Hummers method was used to prepare Gra oxide[29]. In short, $NaNO_3$ (2.5 g) and graphite powder (1.0 g) were added to concentrated H_2SO_4 (100 mL) and stirred for 2 h. $KMnO_4$ (5 g) was slowly added to the mixture under continuous stirring, and the mixture was then cooled with ice. Next, the mixture was stirred at 35 ℃ for 24 h. Double-distilled deionized water (100 mL) was slowly added to the reacted slurry, which was then stirred at 80 ℃ for another 3 h. Next, more double-distilled deionized water (300 mL) was added to the reacted slurry. Then, 6 mL of H_2O_2 (30%) was added (bubbles appeared, and the slurry immediately turned bright yellow). The resulting solution was continuously stirred for 3 h and then precipitated for 24 h at room temperature. The supernatant was subsequently decanted. The resulting yellow slurry was washed with 0.5 mol/L HCl (500 mL) and centrifuged. The solution was washed with double-distilled deionized water and centrifuged until the pH of the solution was neutral (pH=7.0). Gra oxide was obtained after the solution was ultrasonicated for 2 h. To obtain Gra, Gra oxide was reduced at 95 ℃ for 3 h using $NaBH_4$, as a reducing agent.

Preparation of the Chi-Gra nanocomposite

Chi-Gra was prepared according to a previously reported method[23]. Briefly, Chi powder was dissolved in a 1.0% (v/v) acetic acid solution under stirring for 0.5 h at room temperature until it was completely dispersed. The Chi solution (0.5 wt. %) was thus prepared. Then, Gra (10 mg) was added to the Chi solution (10 mL), ultrasonicated for 1 h, and stirred for 24 h at 25 ℃ . Finally, the Chi-Gra nanocomposite was obtained.

Preparation of the AuNP-Chi-Gra nanocomposite

The AuNP-Chi-Gra nanocomposite was prepared as previously described[23, 30]. Furthermore, 0.5 mL of $HAuCl_4$, (1 mM) was added to Chi-Gra (5 mL) under stirring at 25 ℃ for 4 h. Then, the solution was incubated at 80 ℃ for 1 h with vigorous stirring. $Au^{3+}$was subsequently reduced to AuNPs by Chi at 80 ℃ . Finally, the

AuNP-Chi-Gra nanocomposite was obtained.

Preparation of the Cu(Ⅰ)/Cu(Ⅱ)-Chi-Gra nanocomposite

The Cu(Ⅰ)/Cu(Ⅱ)-Chi-Gra nanocomposite was prepared according to the method used to prepare the AuNP-Chi-Gra nanocomposite with certain modifications. $CuSO_4 \cdot 5H_2O$ was used as the source of copper. First, 10 mg of $CuSO_4 \cdot 5H_2O$ was added to 5 mL of the Chi-Gra nanocomposite under continuous stirring at 25 ℃ for 8 h. Then, the mixture was incubated at 95 ℃ for 4 h under continuous stirring. Finally, the Cu(Ⅰ)/Cu(Ⅱ)-Chi-Gra nanocomposite was obtained.

Preparation of PAb/NDV-Cu(Ⅰ)/Cu(Ⅱ)-Chi-Gra nanocomposite bioconjugates

First, 5 mL of the Cu(Ⅰ)/Cu(Ⅱ)-Chi-Gra nanocomposite obtained from the above preparation method was centrifuged (12 000 r/min, 10 min), the supernatant was discarded, and the residue was washed with double-distilled deionized water three times to remove the excess Chi, $Cu^{2+}$and SO_4^{2-} that did not combine with Gra. Then, 5.0 mL of a PBS buffer (pH=7.4) was added to the residue to disperse the Cu(Ⅰ)/Cu(Ⅱ)-Chi-Gra nanocomposite, and the mixture was sonicated for 10 min to obtain a homogeneous suspension. Next, 1 mL of PAb/NDV (10 μg/mL) was added to the homogeneous suspension, and the mixture was vigorously stirred for 5 min at 4 ℃. Then, 1 mL of 1% glutaraldehyde was slowly added to the solution under continuous stirring. The solution was subsequently incubated at 4 ℃ for 8 h. The reaction mixture was washed with PBS (pH=7.4) and centrifuged (12 000 r/min, 10 min) three times. The supernatant was discarded, the resulting mixture was dispersed in PBS (5.0 mL, pH=7.4), and 1 mL of a 2.0% (w/v) BSA solution was added to the suspension, which was then incubated at 4 ℃ for 8 h. The obtained PAb/NDV-Cu(Ⅰ)/Cu(Ⅱ)-Chi-Gra nanocomposite was stored at 4 ℃ for further use.

Fabrication of the electrochemical immunosensor

First, 0.05 mm alumina was used to polish a GCE (Ø=3 mm) until it had a mirror-like surface. Then, the GCE was rinsed with double-distilled deionized water and ultrasonicated in baths of double-distilled deionized water, ethyl alcohol, and double-distilled deionized water to remove any physically adsorbed substances. Next, the GCE was placed in H_2SO_4 (0.05 M) and chemically cleaned until the background signal stabilized. Finally, the GCE was thoroughly rinsed with double-distilled deionized water and dried with nitrogen gas to obtain a clean GCE.

Figure 5-29-1 shows the procedures used to construct the immunosensor. The process was as follows: the AuNP-Chi-Gra (8 μL) nanocomposite was dropped onto the clean GCE surface, dried at 4 ℃ overnight to obtain the modified electrode (AuNP-Chi-Gra-GCE), washed with double-distilled deionized water, immersed in a 1 μg/mL (200 μL) MAb/NDV PBS solution (pH=7.4) and incubated at 4 ℃ for 8 h. The resulting electrode (MAb/NDV-AuNP-Chi-Gra-GCE) was immersed in a 1.0% (w/w) BSA solution for 1 h at 37 ℃ to block the remaining active sites. The final modified electrode was stored at 4 ℃ when not in use.

Electrochemical immunosensor detection

A well-known sandwich immunoassay was used to detect NDV. First, the MAb/NDV-AuNP-Chi-Gra-GCE immunosensor was incubated with 15 μL of the sample for 30 min and then washed with a PBS buffer (pH=7.4) to remove non-specifically adsorbed conjugates. Next, the modified electrode was incubated with

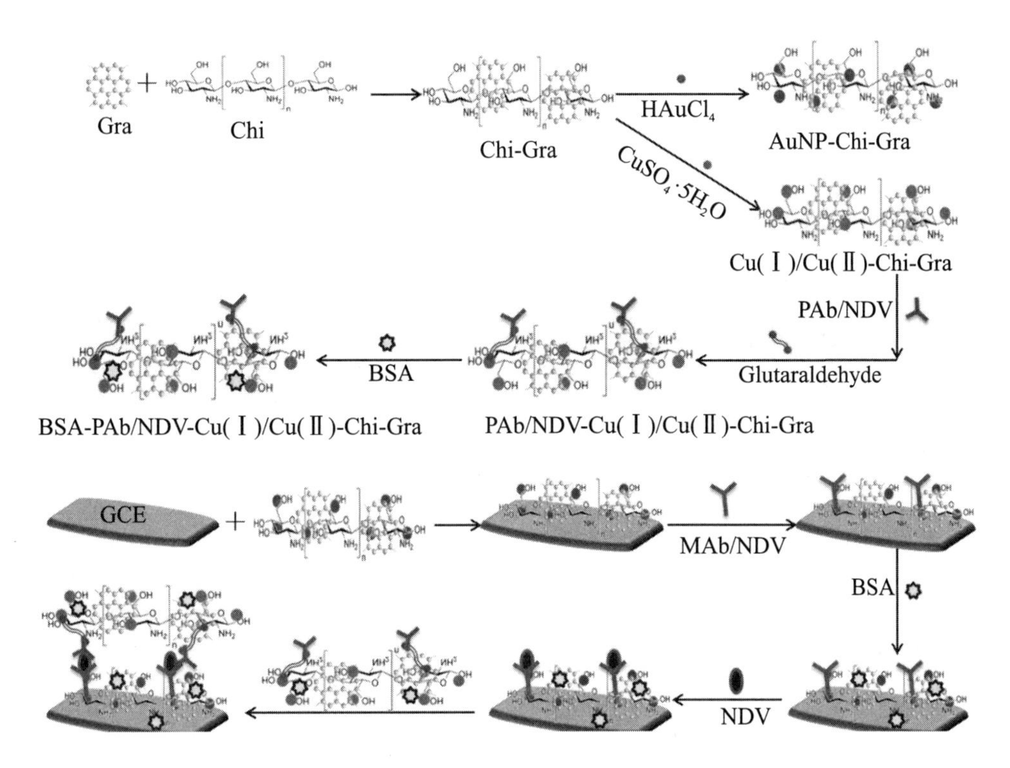

Figure 5-29-1　Preparation procedures of AuNP-Chi-Gra, Cu(I)/Cu(II)-Chi-Gra and the immunosensor (color figure in appendix)

200 μL of the PAb/NDV-Cu(I)/Cu(II)-Chi-Gra nanocomposite for 40 min and washed with a PBS buffer (pH=7.4). Finally, the resulting electrode was placed in a 0.01 mol/L PBS (pH=7.4) KCl solution, and DPV experiments were performed (−0.3 to 0.4 V, 50 mV/s) to detect NDV.

Results and discussion

Morphological characterization of the nanocomposites

Figure 5-29-2 shows scanning electron microscopy (SEM) images and energy dispersive spectrometry (EDS) analyses of Gra, Chi-Gra and Cu(I)/Cu(II)-Chi-Gra. The image of Gra confirms that its structure had many folds (A). After Gra was modified with Chi, the folded structure was filled with Chi, and the surface of the Chi-Gra composite became smooth (B). The presence of Chi on Gra was confirmed by EDS analysis (E). N was observed in the sample because Chi is a natural, biocompatible polymer with many amino groups. Interestingly, the Cu(I)/Cu(II)-Chi-Gra nanocomposite exhibited many upturned folded edges and had a porous matrix (C). Due to this characteristic structure, the exposed surface of the Cu(I)/Cu(II)-Chi-Gra nanocomposite was larger than those of the Chi-Gra composite and Gra. The active surface area increased, resulting in a high surface/volume ratio for antibody immobilization. Furthermore, this porous structure facilitated electrochemical signal amplification. The successful incorporation of Cu(I)/Cu(II) into the Chi-Gra surface was also confirmed by EDS analysis (F).

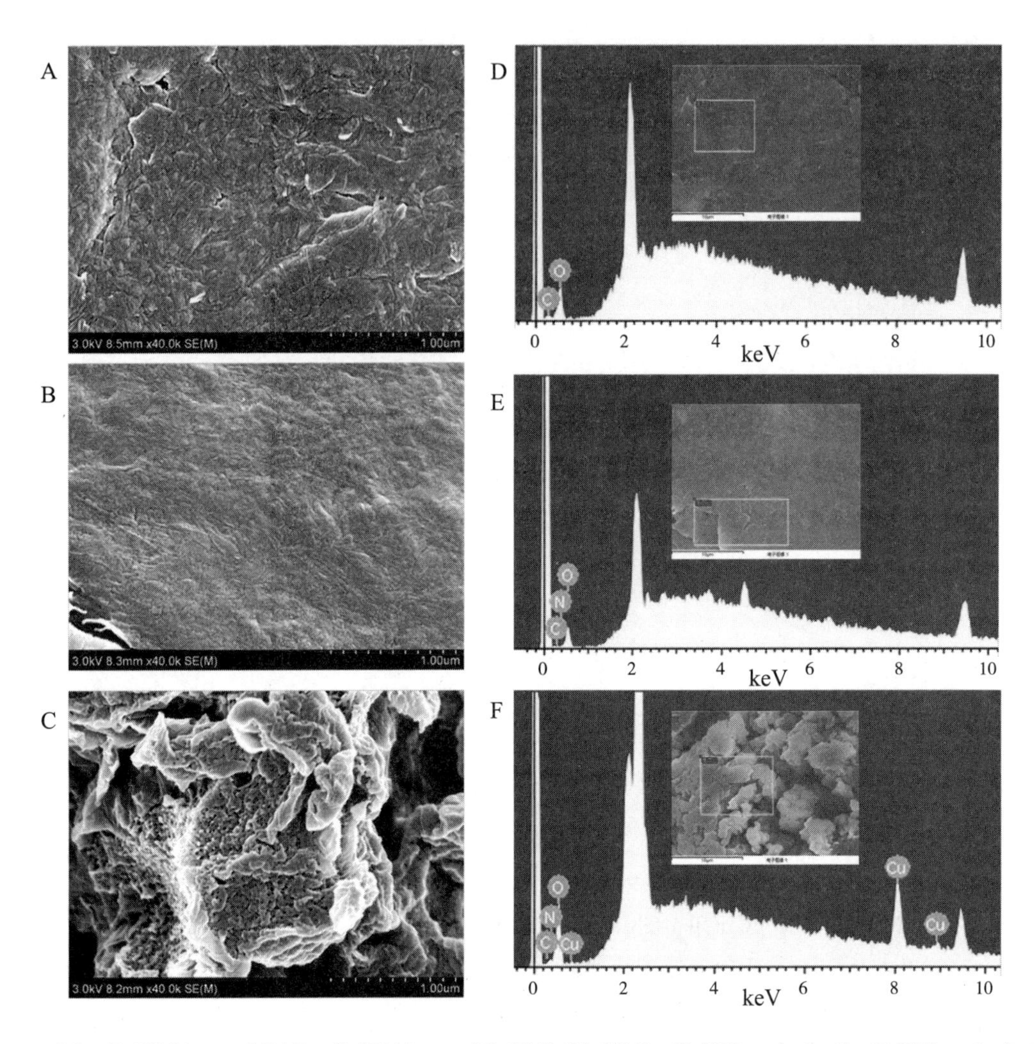

A: SEM image of Gra; B: SEM image of Chi-Gra; C: SEM image of Cu(Ⅰ)/Cu(Ⅱ)-Chi-Gra; D: EDS results for Gra; E: EDS results for Chi-Gra; F: EDS results for Cu(I)/Cu(Ⅱ)-Chi-Gra.

Figure 5-29-2 SEM and EDS characterization of the nanocomposites

Chemical characterization of the nanocomposites

Fourier transform infrared (FT-IR) spectra of Chi, Gra, Chi-Gra, $CuSO_4$ and Cu(Ⅰ)/Cu(Ⅱ)-Chi-Gra are presented in Figure 5-29-3. As shown in Figure 5-29-3A, the stretching vibrations of the -OH bonds in Chi were observed at 3 425 cm^{-1}, and this band overlapped with the $-NH_2$ stretching peaks[31]. The signals originating from the C-H stretching vibrations were observed at approximately 2 920 cm^{-1} and 2 878 cm^{-1}[32]. The NH_2 group and γ-NH_2 bending vibrations appeared at 1 653 cm^{-1} and 1 596 cm^{-1}, respectively[33]. Furthermore, the peak at 1 424 cm^{-1} was attributed to the OH bending vibration. The stretching vibrations of the C-C-O bonds in the Chi backbone were observed at approximately 1 154 cm^{-1}, 1 081 cm^{-1} and 1 034 cm^{-1}. As shown in Figure 5-29-3A, the characteristic absorption bands of pure Gra appeared at 1 555 cm^{-1}, 1 459 cm^{-1}, and 1 420 cm^{-1} (benzene ring backbone stretching vibrations); 1 659 cm^{-1} (C=0 stretching vibration); 2 916 cm^{-1} (C-H stretching vibration); and 3 406 cm^{-1} (O-H stretching vibration). Chi adsorption on Gra resulted in the appearance of the characteristic absorption bands of pure Gra in the FT-IR spectrum of Chi-Gra (Figure 5-29-3 A), but compared with pure Gra, the characteristic absorption bands of Chi-Gra had lower intensities, which helped confirm that Chi was successfully adsorbed on Gra. Comparing the spectra of Chi-Gra and Cu(Ⅰ)/Cu(Ⅱ)-Chi-Gra (Figure

5-29-3 B) revealed some changes in the intensities and shifts in the peaks. Furthermore, the main absorption peaks of pure $CuSO_4$ (Figure 5-29-3 B) were also observed in the FT-IR spectrum of Cu(Ⅰ)/Cu(Ⅱ)-Chi-Gra (Figure 5-29-3 B), providing evidence of the interaction between $CuSO_4$, and Chi-Gra. Chi-Gra binds Cu^{2+} well because Chi-Gra contains many negatively charged groups (carboxylic (O=C–OH), hydroxyl (-C–OH) and carbonyl (–C=O)) that can strongly interact with the positively charged Cu^{2+} ion in $CuSO_4$.

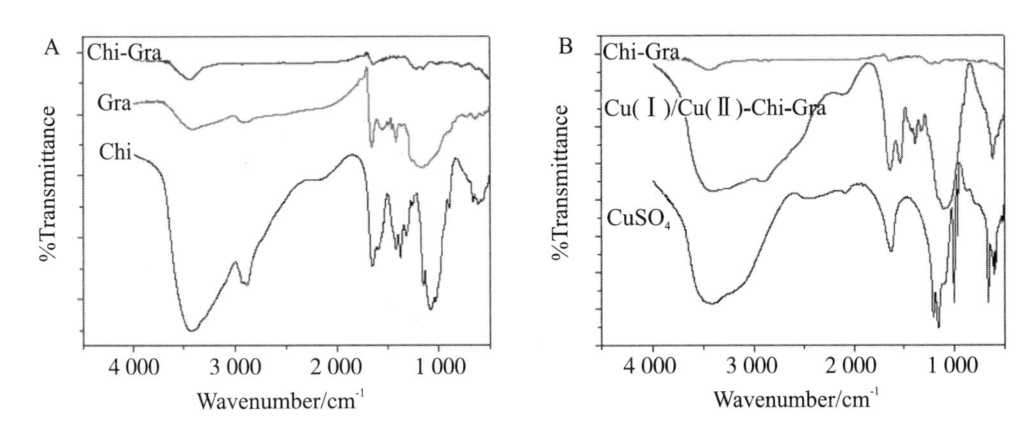

Figure 5-29-3 FT-IR spectra of the nanocomposites

In addition, X-ray photoelectron spectroscopy (XPS) was used to identify the valence state of Cu. The XPS spectrum of Cu(Ⅰ)/Cu(Ⅱ)-Chi-Gra is shown in Figure 5-29-4A. The formation of Cu_2O was confirmed by the presence of the Cu $2p_{3/2}$ peak at 931.73 eV and the Cu $2p_{1/2}$ peak at 951.39 eV[34]. Furthermore, the presence of Cu $2p_{3/2}$ and Cu $2p_{1/2}$ peaks with binding energies of 933.26 eV and 953.14 eV, respectively, proved the formation of CuO[34]. The presence of $CuSO_4$ was confirmed by the Cu $2p_{3/2}$ peak at 934.91 eV and Cu $2p_{1/2}$ peak at 954.62 eV[35]. In addition, to obtain a clearer XPS survey, 10 times the amount of $CuSO_4$ was added to Chi-Gra to prepare rich[Cu(Ⅰ)/Cu(Ⅱ)]-Chi-Gra, and the XPS spectrum of rich[Cu(Ⅰ)/Cu(Ⅱ)]-Chi-Gra shown in Figure 5-29-4B confirmed that the valence states of the Cu element were Cu^{+}[Cu(Ⅰ)] and Cu^{2+}[Cu(Ⅱ)] . The concentration of Cu(Ⅰ) in rich[Cu(Ⅰ)/Cu(Ⅱ)]-Chi-Gra was higher than that in Cu(Ⅰ)/Cu(Ⅱ)-Chi-Gra because the ability of Chi to chelate Cu^{2+} is stronger than the ability of Chi to reduce Cu^{2+} to Cu^{+}. Additionally, the presence of Cu4, Cu4′, Cu5 and Cu5′ in rich[Cu(Ⅰ)/Cu(Ⅱ)-Chi-Gra might be due to the different Cu^{2+}-chelating abilities ofthe various functional groups in Chi-Gra. Under competitive conditions, functional groups with a stronger Cu^{2+}-chelating ability chelate Cu^{2+} first, and functional groups with a weaker Cu^{2+}-chelating ability chelate Cu^{2+} last. When the amount of Cu^{2+} is too low, the functional groups with a weaker Cu^{2+}-chelating ability lose Cu^{2+}, but these functional groups can chelate Cu^{2+} when a sufficient amount of Cu^{2+} is present. Therefore, Cu4, Cu4′, Cu5 and Cu5′ were present in rich [Cu(Ⅰ)/Cu(Ⅱ)]-Chi-Gra, but absent in Cu(Ⅰ)/Cu(Ⅱ)-Chi-Gra.

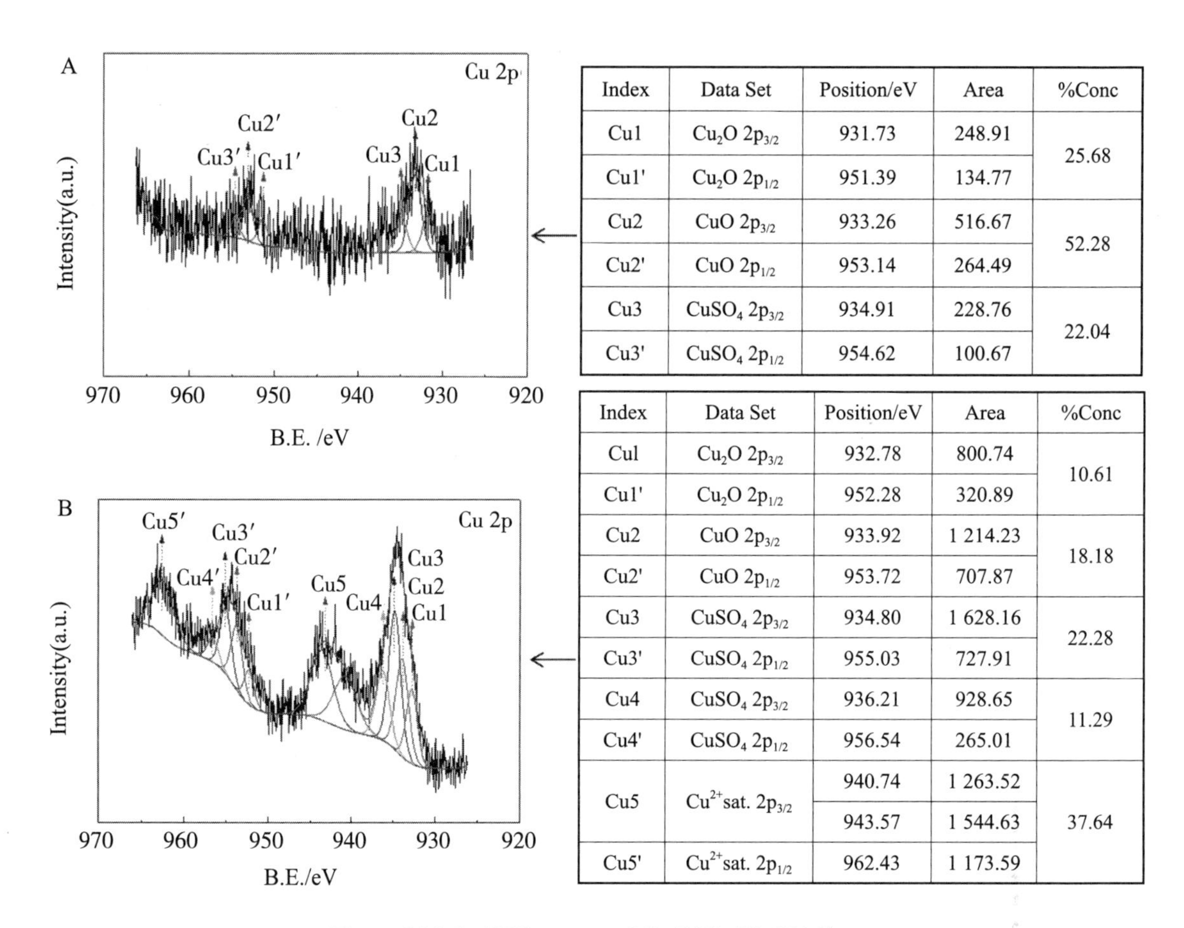

Index	Data Set	Position/eV	Area	%Conc
Cul	Cu_2O $2p_{3/2}$	931.73	248.91	25.68
Cul'	Cu_2O $2p_{1/2}$	951.39	134.77	
Cu2	CuO $2p_{3/2}$	933.26	516.67	52.28
Cu2'	CuO $2p_{1/2}$	953.14	264.49	
Cu3	$CuSO_4$ $2p_{3/2}$	934.91	228.76	22.04
Cu3'	$CuSO_4$ $2p_{1/2}$	954.62	100.67	

Index	Data Set	Position/eV	Area	%Conc
Cul	Cu_2O $2p_{3/2}$	932.78	800.74	10.61
Cul'	Cu_2O $2p_{1/2}$	952.28	320.89	
Cu2	CuO $2p_{3/2}$	933.92	1 214.23	18.18
Cu2'	CuO $2p_{1/2}$	953.72	707.87	
Cu3	$CuSO_4$ $2p_{3/2}$	934.80	1 628.16	22.28
Cu3'	$CuSO_4$ $2p_{1/2}$	955.03	727.91	
Cu4	$CuSO_4$ $2p_{3/2}$	936.21	928.65	11.29
Cu4'	$CuSO_4$ $2p_{1/2}$	956.54	265.01	
Cu5	Cu^{2+}sat. $2p_{3/2}$	940.74	1 263.52	37.64
		943.57	1 544.63	
Cu5'	Cu^{2+}sat. $2p_{1/2}$	962.43	1 173.59	

Figure 5-29-4　XPS spectra of Cu(I)/Cu(II)-Chi-Gra

Electrochemical characterization of the immunosensor

Cyclic voltammetry (CV) was used to investigate the surface of the glassy carbon electrode (GCE) during the process. The electrochemical behaviour was monitored in 5 mM $[Fe(CN)_6]^{3-/4-}$ (1 : 1) and 0.01 M phosphate-buffered saline (PBS) (pH=7.4, containing 0.1 M KCl) in the potential range of –0.2 to 0.6 V at a scan rate of 50 mV/s, and the results are shown in Figure 5-29-5A. A pair of well-defined voltammetric peaks was obtained for the bare GCE (curve a-1). Coating the bare GCE with AuNP-Chi (curve a-2) and AuNP-Chi-Gra (curve a-3) caused an increase in the redox peak current. A comparison of the curves indicated that the AuNPs and Gra had good conductivity and electrocatalytic effects. After attaching MAb/NDV to the modified GCE (curve a-4), the current decreased. This decrease can be explained by the following two factors: (1)AuNP-Chi-Gra could conjugate MAb/NDV via Au-S covalent bonds, and (2) electron transfer was hindered by MAb/NDV. Subsequently, BSA was used to block the immunosensorand the redox peaks decreased even further (curve a-5), because BSA is hydrophobic and electron transfer was further inhibited.

To investigate the immunosensor detection programme, CV was performed in 0.01 mmol/L PBS (pH=7.4) containing 0.1 mmol/L KCl, and the results are shown in Figure 5-29-5 B. For CV curve b-l in Figure 5-29-5B, which was obtained with the BSA-MAb/NDV-AuNP-Chi-Gra film-modified GCE, the background current was low, and no CV redox waves were observed because of the absence of electrochemically active substances in the working solution. After the immunosensor was incubated with $10^{5.13}$ EID_{50}/0.1 mL NDV and sandwiched for the immunoreaction with PAb/NDV-Cu(Ⅰ)/Cu(Ⅱ)-Chi-Gra, stable redox peaks were observed at 0.13 and –0.08 V vs. saturated calomel electrode (SCE) (curve b-2 in Figure 5-29-5B), due to the redox reaction of

Cu(Ⅰ)/Cu(Ⅱ). The peak at 0.13 V was caused by the oxidation of Cu(Ⅰ) to Cu(Ⅱ), and the reduction of Cu(Ⅱ) to Cu(Ⅰ)produced the peak at −0.08 V. These results indicated the efficient redox activity of Cu(Ⅰ)/Cu(Ⅱ)-functionalized Gra.

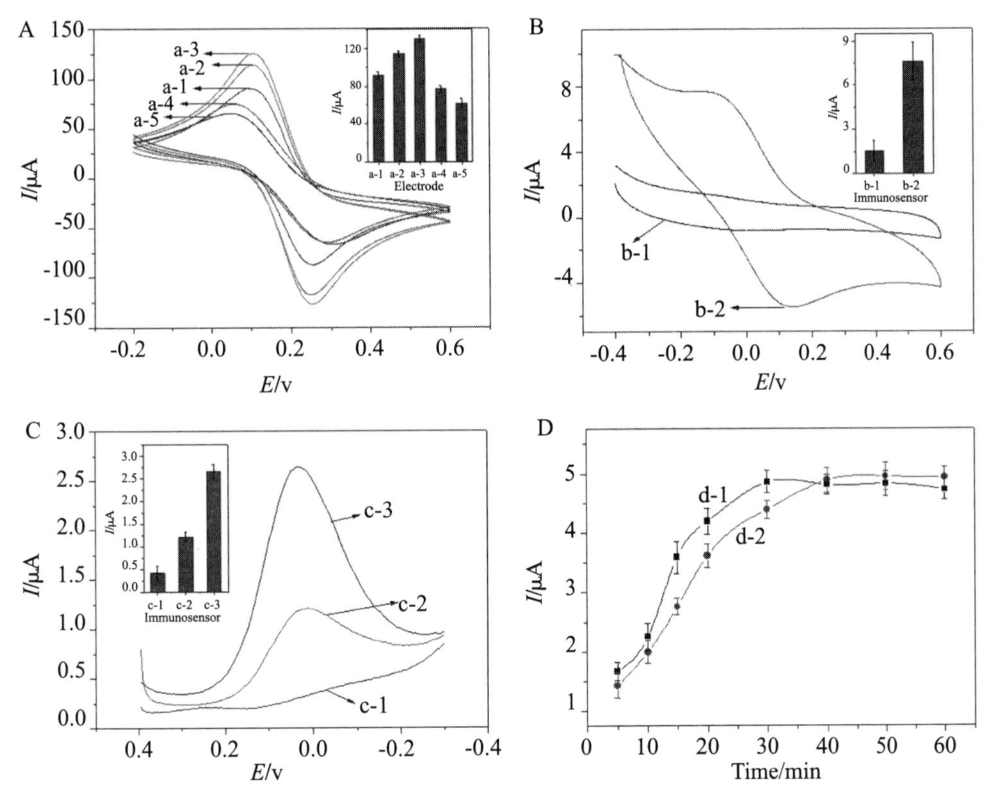

A: CV curves of the electrode at different stages obtained at a scan rate of 50 mV/s. a-1: GCE; a-2: AuNP-Chi-GCE; a-3: AuNP-Chi-Gra-GCE; a-4: MAb/NDV-AuNP-Chi-Gra-GCE, and a-5: BSA-MAb/NDV-AuNP-Chi-Gra-GCE. The supporting electrolyte was 5 mM $[Fe(CN)_6]^{3-/4-}$+0.1 M KCl+0.01 M PBS (pH=7.4).
B: CV curves of the immunosensor measurement process. b-1: BSA-MAb/NDV-AuNP-Chi-Gra-GCE; b-2: PAb/NDV-Cu(Ⅰ)/Cu(Ⅱ)-Chi-Gra-NDV-BSA-MAb/NDV-AuNP-Chi-Gra-GCE. The supporting electrolyte was 0.1 M KCl+0.01 M PBS (pH=7.4). The sample included 15 μL of $10^{5.13}$ EID_{50}/0.1 mL NDV (F48E9).
C: DPV of the immunosensor measurement process. c-1: BSA-MAb/NDV-AuNP-Chi-Gra-GCE; c-2: PAb/NDV-Cu(Ⅰ)/Cu(Ⅱ)-Chi-NDV-BSA-MAb/NDV-AuNP-Chi-Gra-GCE, and c-3: PAb/NDV-Cu(Ⅰ)/Cu(Ⅱ)-Chi-Gra-NDV-BSA-MAb/NDV-AuNP-Chi-Gra-GCE. The supporting electrolyte was 0.1 M KCl+0.01 MPBS (pH=7.4). The sample included 15 μL of $10^{2.13}$ EID_{50}/0.1 mL NDV (F48E9).
D: Influence of the incubation time on the current response of the immunosensor to NDV (d-1) and PAb/NDV-Cu(Ⅰ)/Cu(Ⅱ)-Chi-Gra (d-2).

Figure 5-29-5 Electrochemical characterization of the immunosensor

Comparison of different signal amplification strategies

Signal amplification strategies are very important for immunosensors. Two signal label materials (PAb/NDV-Cu(Ⅰ)/Cu(Ⅱ)-Chi-Gra and PAb/NDV-Cu(Ⅰ)/Cu(Ⅱ)-Chi) were prepared, and differential pulse voltammetry (DPV) was performed from −0.3 to 0.4 V at a 50 mV/s scan rate using a $10^{2.13}$ EID_{50}/0.1 mL sample to evaluate the effects of the signal amplification materials. The results are shown in Figure 5-29-5. As indicated by curve c-1, in the absence of a signal labelling material, a low background current was obtained, and no anodic peak was observed for the immunosensor. In contrast, the immunosensor conjugated with PAb/NDV-Cu(Ⅰ)/Cu(Ⅱ)-Chi-Gra (curve c-3) exhibited a greater current shift than the immunosensor conjugated with PAb/NDV-Cu(Ⅰ)/Cu(Ⅱ)-Chi (curve c-2). The increase in the current shift was due to the use of Gra, which has with a high surface/volume ratio, as the carrier, leading to the immobilization of Cu(Ⅰ)/Cu(Ⅱ) on the GCE and facilitating electrochemical signal amplification. These results confirmed that the immunosensor with Gra could load more of the electroactive signal labelling material and PAb/NDV than the immunosensor without Gra. Accordingly, the signal of the immunosensor was greatly amplified by using Gra.

Optimization of the experimental conditions

During NDV capture and the specific reaction with the signal labelling material (PAb/NDV-Cu(Ⅰ)/Cu(Ⅱ)-Chi-Gra), the incubation time is an important factor. Thus, the incubation times of NDV and PAb/NDV-Cu(Ⅰ)/Cu(Ⅱ)-Chi-Gra were optimized separately. To optimize the NDV incubation time, different incubation times (5, 10, 15, 20, 30, 40, 50, and 60 min) were used, and after incubation with NDV, the immunosensors were incubated with PAb/NDV-Cu(Ⅰ)/Cu(Ⅱ)-Chi-Gra for 60 min. Finally, the immunosensors were used for DPV detection. Each test was repeated five times. The results are shown in Figure 5-29-5 D, curve d-1. As the NDV incubation time was increased up to 30 min, the electrochemical response increased; after 30 min, a constant value was reached, indicating that the immunoreaction was complete, and all the NDV in the sample was captured by the immunosensor. Thus, the optimal incubation time for NDV was 30 min.

To optimize the PAb/NDV-Cu(Ⅰ)/Cu(Ⅱ)-Chi-Gra incubation time, the immunosensors were first incubated with NDV ($10^{5.13}$ EID_{50}/0.1 mL) for 30 min and then incubated with PAb/NDV-Cu(Ⅰ)/Cu(Ⅱ)-Chi-Gra for 5, 10, 15, 20, 30, 40, 50, and 60 min, respectively. Finally, the immunosensors were used for DPV detection. Each test was repeated five times. The results are shown in Figure 5-29-5 D, curve d-2. In the second immunoreaction step, as the PAb/NDV-Cu(Ⅰ)/Cu(Ⅱ)-Chi-Gra incubation time was increased, the electrochemical response current increased, reaching a steady-state value at 40 min, which indicates that the reaction between NDV and PAb/NDV-Cu(Ⅰ)/Cu(Ⅱ)-Chi-Gra was complete. Thus, the optimal incubation time for PAb/NDV-Cu(Ⅰ)/Cu(Ⅱ)-Chi-Gra was 40 min. Compared with NDV, PAb/NDV-Cu(Ⅰ)/Cu(Ⅱ)-Chi-Gra required more time to complete the reaction, which might be due to the greater steric hindrance of PAb/NDV-Cu(Ⅰ)/Cu(Ⅱ)-Chi-Gra.

Analytical performance of the immunosensor

The response of the prepared immunosensor was measured at different concentrations of NDV (F48E9) under the optimal experimental conditions. The results are shown in Figure 5-29-6 A. The electrochemical response current increased as the concentration of NDV increased, and the peak of the electrochemical response current was proportional to the concentration in the range of $10^{0.13}$ to $10^{5.13}$ EID_{50}/0.1 mL. The linear regression equation, which is shown in Figure 5-29-6 B, was I (μA) =0.75 log EID_{50}/0.1 mL+1.05, with a correlation coefficient of 0.970 75, and the limit of determination for NDV was $10^{0.68}$ EID_{50}/0.1 mL, which was calculated based on a signal-to-noise ratio of 3 (S/N=3). These results demonstrated that the immunosensor was sensitive enough to quantitatively monitor NDV.

The results for the immunosensor with PAb/NDV-Cu(Ⅰ)/Cu(Ⅱ)-Chi-Gra as the signal label were compared with those for the immunosensor with PAb/NDV-Cu(Ⅰ)/Cu(Ⅱ)-Chi as the signal label, and the results obtained with PAb/NDV-Cu(Ⅰ)/Cu(Ⅱ)-Chi are shown in Figure 5-29-6 C. The electrochemical response current increased linearly with increasing NDV concentration, and the calibration curve in the range of $10^{0.13}$ to $10^{5.13}$ EID_{50}/0.1 mL (Figure 5-29-6 D) was: I (μA) =0.15 log EID_{50}/0.1 mL+1.10. The limit of determination for NDV was $10^{2.09}$ EID_{50}/0.1 mL (S/N=3). This result indicated that Gra can improve the immunosensor sensitivity. In addition, as shown in Figure 5-29-6 C (curve c-2), the background signal was high when PAb/NDV-Cu(Ⅰ)/Cu(Ⅱ)-Chi was used as the signal label because without Gra, the excess Chi could not be removed from PAb/NDV-Cu(Ⅰ)/Cu(Ⅱ)-Chi by centrifugation, and the excess Chi chelated with Cu(Ⅰ)/Cu(Ⅱ) was attached to the GCE by non-specific binding.

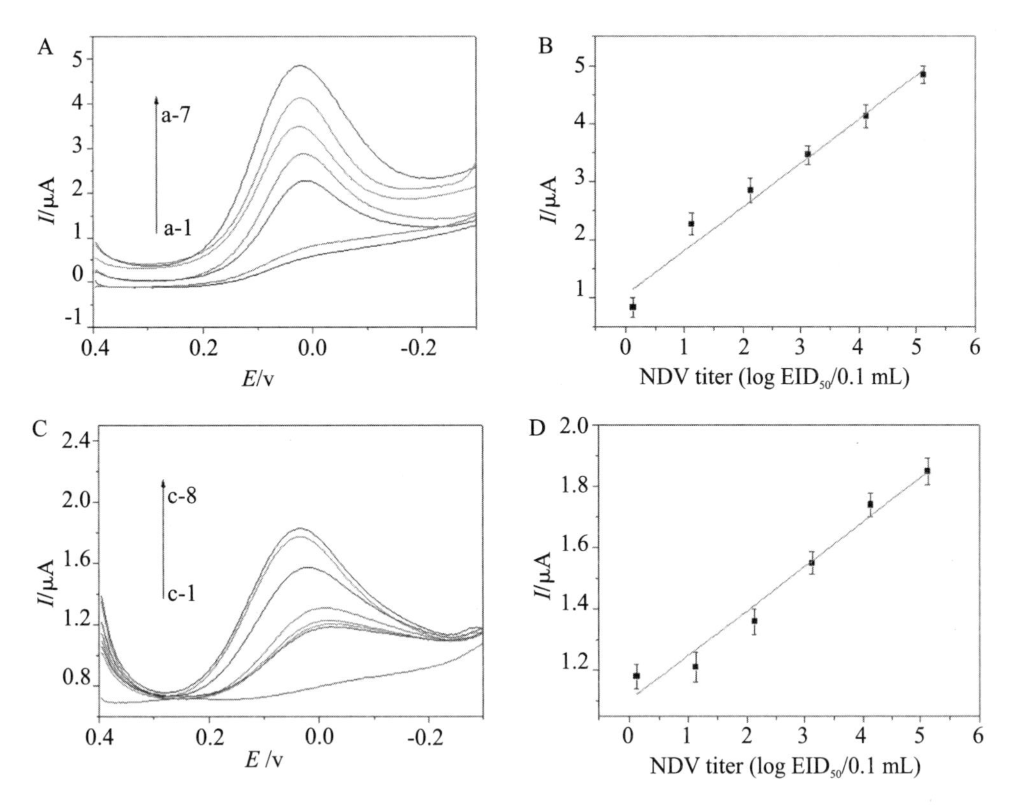

A: Typical DPV signals acquired in the presence of different concentrations of NDV with PAb/NDV-Cu(I)/Cu(II)-Chi-Gra as the label. a-1: 0; a-2: $10^{0.13}$ EID_{50}/0.1 mL; a-3: $10^{1.13}$ EID_{50}/0.1 mL; a-4: $10^{2.13}$ EID_{50}/0.1 mL; a-5: $10^{3.13}$ EID_{50}/0.1 mL; a-6: $10^{4.13}$ EID_{50}/0.1 mL; and a-7: $10^{5.13}$ EID_{50}/0.1 mL.
B: Relationship between the antigen concentration and sensor current response corresponding to A.
C: Typical DPV signals before incubation with NDV (c-l) and in the presence of different concentrations of NDV (c-2: 0; c-3: $10^{0.13}$ EID_{50}/0.1 mL; c-4: $10^{1.13}$ EID_{50}/0.1 mL; c-5: $10^{2.13}$ EID_{50}/0.1 mL; c-6: $10^{3.13}$ EID_{50}/0.1 mL; c-7: $10^{4.13}$ EID_{50}/0.1 mL; and c-8: $10^{5.13}$ EID_{50}/0.1 mL) with PAb/NDV-Cu(I)/Cu(II)-Chi as the label.
D: Relationship between the antigen concentration and sensor current response corresponding to C. Error bar = ± standard deviation.

Figure 5-29-6 Sensitivity results of immunosensor

Comparison of methods

The results of a comparative study between the designed method and other methods for NDV detection are summarized in Table 5-29-1(a). The table shows that the developed electrochemical immunosensor has acceptable sensitivity and advantages over the other methods in terms of rapid detection, intuitiveness, user-friendliness and cost.

Selectivity, repeatability, reproducibility and stability of the immunosensor

Selectivity is a significant parameter for an immunosensor. Therefore, to determine the selectivity of the fabricated immunosensor, some possible interferents, including aviadenovirus group I (AAV, $10^{6.37}$ EID_{50}/0.1 mL), infectious bronchitis virus (IBV, $10^{7.02}$ EID_{50}/0.1 mL), infectious laryngotracheitis virus (ILTV, $10^{5.84}$ EID_{50}/0.1 mL), avian influenza virus subtype H7 (AIV H7, $10^{6.45}$ EID_{50}/0.1 mL), avian reovirus (ARV, $10^{6.51}$ EID_{50}/0.1 mL), infectious bursal disease (IBD, $10^{7.34}$ EID_{50}/0.1 mL), glucose (1.0 μg/mL), vitamin C (1.0 μg/mL) and BSA (1.0 μg/mL), were investigated. The results are depicted in Figure 5-29-7 A. When the fabricated immunosensor was exposed to possible interferents (Figure 5-29-7 A , samples a-2 to a-10): AAV, IBV, ILTV, AIV H7, ARV, IBD, glucose, vitamin C, and BSA), the detection currents were as low as that for the negative control (Figure 5-29-7 A, sample a-1: ddH_2O). The immunosensor exhibited a higher signal when incubated with a sample including NDV (Figure 5-29-7 A, samples a-11, a-16) than when incubated with samples containing the possible interferents (Figure 5-29-7 A, samples a-2 to a-10). Additionally, the responses of the fabricated immunosensor to $10^{5.13}$ and $10^{3.13}$ EID_{50}/0.1 mL NDV solutions containing other interfering substances were measured (Figure 5-29-7 A , samples a-12 to a-15, a-17 to a-20), and the current variation due to the interfering substances was less than 5% of that obtained without interferences. The results show that the developed immunosensor had good selectivity for NDV.

Under the optimal experimental conditions, equivalently prepared immunosensors were used to detect $10^{3.13}$ EID_{50} NDV 20 times to evaluate the repeatability of the developed immunosensor, and the results are shown in Figure 5-29-7 B. The relative standard deviation was 2.58%, demonstrating the good repeatability of the immunosensor. The reproducibility of the immunosensor was evaluated by preparing six different batches of the immunosensor independently. A series of six different batches of the immunosensor were prepared for the detection of $10^{2.13}$ EID_{50} NDV, and the results are shown in Figure 5-29-7 C. The relative standard deviation was found to be 2.84%, showing the excellent reproducibility.

Long-term storage stability tests show the robustness of an immunosensor. The current responses of the developed immunosensor were periodically checked to evaluate its stability. The immunosensor was stored in PBS (pH=7.4) at 4 ℃ when it was not in use. Every week, electrochemical measurements were performed with the developed immunosensor, and the average value was calculated based on five assays. The results shown in Figure 5-29-7 D indicated that the immunosensor response current decreased by only 4.1% after 2 weeks.

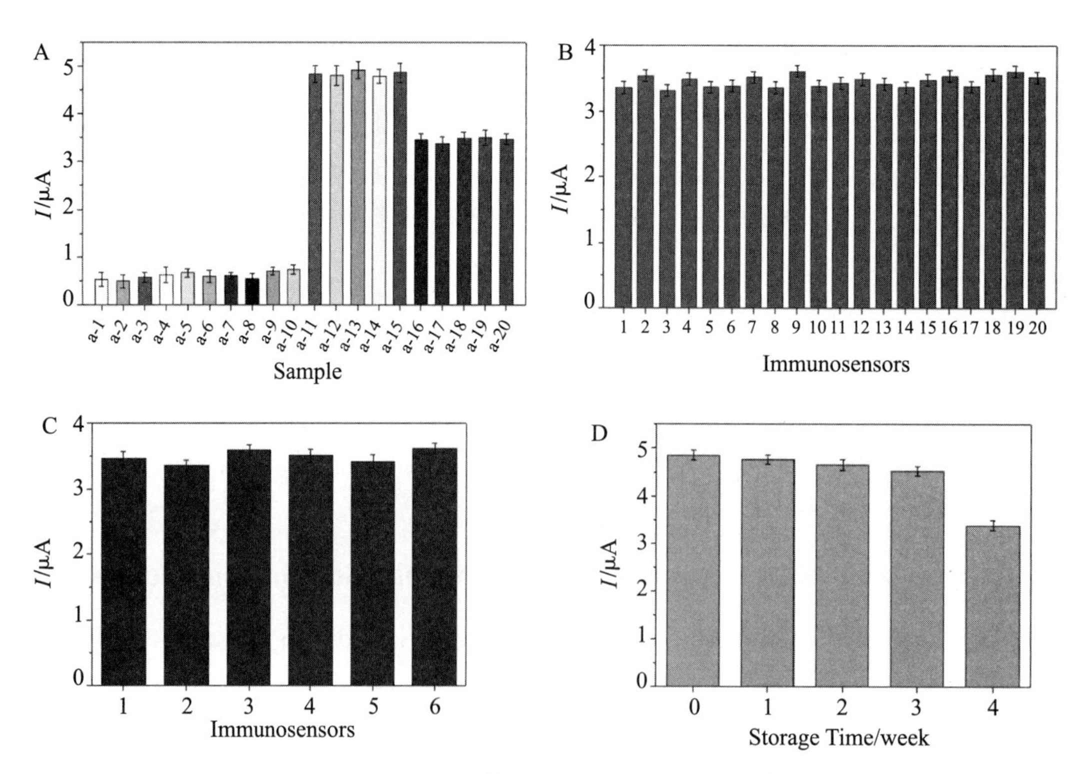

A: Selectivity of the immunosensor. a-1: ddH_2O; a-2: AAV ($10^{6.37}$ EID_{50}/0.1 mL); a-3: IBV ($10^{7.02}$ EID_{50}/0.1 mL); a-4: ILTV ($10^{5.84}$ EID_{50}/0.1 mL); a-5: AIV H7 ($10^{6.45}$ EID_{50}/0.1 mL); a-6: ARV ($10^{6.51}$ EID_{50}/0.1 mL); a-7: IBD ($10^{7.34}$ EID_{50}/0.1 mL); a-8: Glucose (1.0 μg/mL); a-9: Vitamin C (1.0 μg/ mL); a-10: BSA (1.0 μg/mL); a-11: NDV ($10^{5.13}$ EID_{50}/0.1 mL); a-12: NDV ($10^{5.13}$ EID_{50}/0.1 mL)+AAV ($10^{6.37}$ EID_{50}/0.1 mL); a-13: NDV ($10^{5.13}$ EID_{50}/0.1 mL)+IBV ($10^{7.02}$ EID_{50}/0.1 mL); a-14: NDV ($10^{5.13}$ EID_{50}/0.1 mL)+AIV H7 ($10^{6.54}$ EID_{50}/0.1 mL); a-15: NDV ($10^{5.13}$ EID_{50}/0.1 mL)+ARV ($10^{6.51}$ EID_{50}/0.1 mL); a-16: NDV ($10^{3.13}$ EID_{50}/0.1 mL); a-17: NDV ($10^{3.13}$ EID_{50}/0.1mL)+ILTV ($10^{5.84}$ EID_{50}/0.1 mL); a-18: NDV ($10^{3.13}$ EID_{50}/0.1 mL)+IBD ($10^{7.34}$ EID_{50}/0.1 mL); a-19: NDV ($10^{3.13}$ EID_{50}/0.1 mL)+vitamin C (1.0 μg/mL); a-20: NDV ($10^{3.13}$ EID_{50}/0.1 mL)+BSA (1.0 μg/mL). B: Repeatability. C: reproducibility. D: Storage stability of the immunosensor.

Figure 5-29-7 Performance of the immunosensor

After four weeks, the immunosensor current response decreased by 9.5% relative to its initial current, which indicated that the immunosensor had acceptable storage stability.

Application of the proposed immunosensor for the detection of NDV

Oral and cloacal swab samples, which were gently collected from fowls at different live bird markets in Guangxi, were used as clinical samples. A viral transport medium composed of 0.05 mmol/L PBS containing 10 mg/mL gentamycin, 10 mg/mL kanamycin, 10 mg/mL streptomycin, 5% (v/v) foetal bovine serum and 10 000 units/mL penicillin was used to prepare the clinical samples, and the clinical samples were placed in an ice box.

With the permission of the owners of the live bird markets, a total of 120 clinical samples were collected from chickens, the samples were assayed using the proposed immunosensor, and seven NDV-positive samples were detected. Virus isolation [3] was employed to confirm the test results. The positive results detected by the developed immunosensor were in agreement with the results of virus isolation, and the results are summarized in Table 5-29-1 b, c. To test the recovery by the proposed immunosensor, NDV standards were added to the clinical samples that had been confirmed as positive. The results (Table 5-29-1 d) showed that the fabricated immunosensor had acceptable recovery (96.28%～104.49%). Considering the acceptable recovery in real samples, the immunosensor was found to be practical for sample detection.

Table 5-29-1　Comparison of the proposed immunosensor with other sensors for NDV detection (a); results of clinical samples (b); analysis data sheet of positive samples (c); recovery results of clinical samples with different concentrations of NDV (d)

(a) Method	Detection time	Detection limit	References		
Virus isolation	4～7 days	1 EID_{50}/mL	[3]		
RT-PCR	5 h	$10^{4.0}$ EID_{50}/0.1 mL	[4]		
Real-time RT-PCR	3 h	10^{1} EID_{50}/mL	[5]		
ICS	15 min	$10^{4.9}$ EID_{50}/0.1 mL	[6]		
RT-LAMP	3 h	1.3 Haemagglutination units	[7]		
Proposed immunosensor	70 min	$10^{0.68}$ EID_{50}/0.1 mL	This study		
(b) Method	Total number of samples	Number of positive samples	Positive rate/%		
Proposed immunosensor	120	7	5.8		
Virus isolation	120	7	5.8		
(c) Number	Results of the proposed immunosensor			Results of virus isolation	
	Measured concentration (EID_{50}/0.1 mL)	Average (EID_{50}/0.1 mL)	RSD (% , n=5)		
1	40.74, 39.90, 41.27, 39.15, 42.65	40.74	3.28	Positive	
2	92.47, 90.73, 93.04, 91.38, 94.31	92.39	1.52	Positive	
3	107.46, 105.92, 108.17, 110.29, 109.67	108.30	1.61	Positive	

continued

(c) Number	Results of the proposed immunosensor			Results of virus isolation	
	Measured concentration (EID_{50}/0.1 mL)	Average (EID_{50}/0.1 mL)	RSD (% , n=5)		
4	367.41, 370.35, 361.91, 374.34, 354.73	365.75	2.09	Positive	
5	409.32, 417.93, 406.78, 423.32, 428.46	417.16	2.19	Positive	
6	742.16, 737.59, 731.81, 749.19, 728.94	737.94	1.10	Positive	
7	1 490.28, 1 481.38, 1 463.57, 1 447.34, 1 452.85	1 467.08	1.25	Positive	
(d) Number	Initial NDV concentration in sample (EID_{50}/0.1 mL)	Added NDV amount (EID_{50}/0.1 mL)	Total found		Recovery rate (%, n=5)
			Average (EID_{50}/0.1 mL)	RSD (% , n=5)	
1	40.74	50	87.36	2.74	96.28
2	92.51	100	190.83	2.38	99.13
3	108.30	500	610.17	1.76	100.31
4	365.75	1 000	1 363.72	1.47	99.85
5	417.16	5 000	5 421.03	2.39	100.07
6	737.94	10 000	11 219.82	3.56	104.49
7	1 467.08	50 000	50 734.94	2.71	98.58

Conclusions

In summary, AuNP-Chi-Gra was used as a platform, and PAb/NDV-Cu(Ⅰ)/Cu(Ⅱ)-Chi-Gra was used as a label for signal amplification in this work. Based on the well-known sandwich immunoreaction, a novel electrochemical immunosensor was developed for the quantitative detection of NDV. It exhibited a linear response over a wide range ($10^{0.13}$ to $10^{5.13}$ EID_{50}/mL), had a low detection limit ($10^{0.68}$ EID_{50}/0.1 mL), and was more sensitive than an immunosensor with PAb/NDV-Cu(Ⅰ)/Cu(Ⅱ)-Chi as the signal label (the limit of detection for NDV was $10^{2.09}$ EID_{50}/0.1 mL). This newly designed immunosensor might have widespread application potential because it had acceptable reproducibility, selectivity and stability; could be obtained by a facile fabrication procedure; and was ultrasensitive for the detection of NDV.

References

[1] BROWN V R, BEVINS S N. A review of virulent Newcastle disease viruses in the United States and the role of wild birds in viral persistence and spread. Vet Res, 2017, 48(1): 68.

[2] GANAR K, DAS M, SINHA S, et al. Newcastle disease virus: current status and our understanding. Virus Res, 2014, 184: 71-81.

[3] LIU X F, WAN H Q, NI X X, et al. Pathotypical and genotypical characterization of strains of Newcastle disease virus isolated from outbreaks in chicken and goose flocks in some regions of China during 1985-2001. Arch Virol, 2003, 148(7): 1387-1403.

[4] ALI A, REYNOLDS D L. A multiplex reverse transcription-polymerase chain reaction assay for Newcastle disease virus and avian pneumovirus (Colorado strain). Avian Dis, 2000, 44(4): 938-943.

[5] FRATNIK STEYER A, ROJS O Z, KRAPEZ U, et al. A diagnostic method based on MGB probes for rapid detection and simultaneous differentiation between virulent and vaccine strains of avian paramyxovirus type 1. J Virol Methods, 2010, 166(1-2): 28-36.

[6] LI Q, WANG L, SUN Y, et al. Evaluation of an immunochromatographic strip for detection of avian avulavirus 1 (Newcastle disease virus). J Vet Diagn Invest, 2019, 31(3): 475-480.

[7] LI Q, XUE C, QIN J, et al. An improved reverse transcription loop-mediated isothermal amplification assay for sensitive and specific detection of Newcastle disease virus. Arch Virol, 2009, 154(9): 1433-1440.

[8] HOSU O, SELVOLINI G, CRISTEA C, et al. Electrochemical immunosensors for disease detection and diagnosis. Curr Med Chem, 2018, 25(33): 4119-4137.

[9] FELIX F S, ANGNES L. Electrochemical immunosensors—A powerful tool for analytical applications. Biosens Bioelectron. 2018, 102: 470-478.

[10] AYDIN E B, AYDIN M, SEZGINTÜRK M K. Electrochemical immunosensor based on chitosan/conductive carbon black composite modified disposable ITO electrode: An analytical platform for p53 detection. Biosens Bioelectron, 2018, 121: 80-89.

[11] AYDIN M, AYDIN E B, SEZGINTÜRK M K. A highly selective electrochemical immunosensor based on conductive carbon black and star PGMA polymer composite material for IL-8 biomarker detection in human serum and saliva. Biosens Bioelectron, 2018, 117: 720-728.

[12] WANG B, JI X, ZHAO H, et al. An amperometric β-glucan biosensor based on the immobilization of bi-enzyme on Prussian blue-chitosan and gold nanoparticles-chitosan nanocomposite films. Biosens Bioelectron, 2014, 55: 113-119.

[13] BHATTARAI J K, NEUPANE D, NEPAL B, et al. Preparation, modification, characterization, and biosensing application of nanoporous gold using electrochemical techniques. Nanomaterials, 2018, 8(3): 171.

[14] SUN D, LI H, LI M, et al. Electrochemical immunosensors with AuPt-vertical graphene/glassy carbon electrode for alpha-fetoprotein detection based on label-free and sandwich-type strategies. Biosens Bioelectron, 2019, 132: 68-75.

[15] REZAEI B, SHOUSHTARI A M, RABIEE M, et al. An electrochemical immunosensor for cardiac Troponin I using electrospun carboxylated multi-walled carbon nanotube-whiskered nanofibres. Talanta, 2018, 182: 178-186.

[16] SUN B, GOU Y, MA Y, et al. Investigate electrochemical immunosensor of cortisol based on gold nanoparticles/magnetic functionalized reduced graphene oxide. Biosens Bioelectron, 2017, 88: 55-62.

[17] TUTEJA S K, CHEN R, KUKKAR M, et al. A label-free electrochemical immunosensor for the detection of cardiac marker using graphene quantum dots (GQDs). Biosens Bioelectron, 2016, 86: 548-556.

[18] BHARDWAJ H, PANDEY M K, RAJESH, et al. Electrochemical Aflatoxin B1 immunosensor based on the use of graphene quantum dots and gold nanoparticles. Mikrochim Acta, 2019, 186(8): 592.

[19] STANKOVICH S, DIKIN D, DOMMETT G, et al. Graphene-based composite materials. Nature, 2006, 442: 282-286.

[20] GEIM A K, NOVOSELOV K S. The rise of graphene. Nat. Mater, 2007, 6: 183-191.

[21] LIU J, WANG J, WANG T, et al. Three-dimensional electrochemical immunosensor for sensitive detection of carcinoembryonic antigen based on monolithic and macroporous graphene foam. Biosens Bioelectron, 2015, 65: 281-286.

[22] LI L, ZHANG L, YU J, et al. All-graphene composite materials for signal amplification toward ultrasensitive electrochemical immunosensing of tumor marker. Biosens Bioelectron, 2015, 71: 108-114.

[23] HUANG J, XIE Z, XIE Z, et al. Silver nanoparticles coated graphene electrochemical sensor for the ultrasensitive analysis of avian influenza virus H7. Anal Chim Acta, 2016, 913: 121-127.

[24] GUZMAN J, SAUCEDO I, REVILLA J, et al. Copper sorption by chitosan in the presence of citrate ions: influence of metal speciation on sorption mechanism and uptake capacities. Int J Biol Macromol, 2003, 33(1-3): 57-65.

[25] DI TOCCO A, ROBLEDO S N, OSUNA Y, et al. Development of an electrochemical biosensor for the determination of triglycerides in serum samples based on a lipase/magnetite-chitosan/copper oxide nanoparticles/multiwalled carbon nanotubes/pectin composite. Talanta, 2018, 190: 30-37.

[26] SINGH J, SRIVASTAVA M, ROYCHOUDHURY A, et al. Bienzyme-functionalized monodispersed biocompatible cuprous oxide/chitosan nanocomposite platform for biomedical application. J Phys Chem B, 2013, 117(1): 141-152.

[27] WANG H, ZHANG Y, WANG Y, et al. Facile synthesis of cuprous oxide nanowires decorated graphene oxide nanosheets nanocomposites and its application in label-free electrochemical immunosensor. Biosens Bioelectron, 2017, 87: 745-751.

[28] OLIVEIRA P R, LAMY-MENDES A C, REZENDE E I, et al. Electrochemical determination of copper ions in spirit drinks using carbon paste electrode modified with biochar. Food Chem, 2015, 171: 426-431.

[29] YU H, ZHANG B, BULIN C, et al. High-efficient synthesis of graphene oxide based on improved Hummers method. Sci. Rep, 2016, 6: 36143.

[30] HUANG K J, NIU D J, XIE W Z, et al. A disposable electrochemical immunosensor for carcinoembryonic antigen based on nano-Au/multi-walled carbon nanotubes-chitosans nanocomposite film modified glassy carbon electrode. Anal Chim Acta, 2010, 659(1-2): 102-108.

[31] KUMAR S, KIM H, GUPTA M K, et al. A new chitosan–thymine conjugate: Synthesis, characterization and biological activity. Int J Biol Macromol, 2012, 50(3): 493-502.

[32] SHANMUGASUNDARAM N, RAVICHANDRAN P, REDDY P N, et al. Collagen-chitosan polymeric scaffolds for the in vitro culture of human epidermoid carcinoma cells. Biomaterials, 2001, 22(14): 1943-1951.

[33] QI L, XU Z, JIANG X, et al. Preparation and antibacterial activity of chitosan nanoparticles. Carbohyd Res, 2004, 339 (16): 2693-2700.

[34] MORALES J, ESPINOS J P, CABALLERO A, et al. XPS study of interface and ligand effects in supported Cu_2O and CuO nanometric particles. J Phys Chem B, 2005, 109(16):7758-7765.

[35] QIU H, ZHANG S, PAN B, et al. Effect of sulfate on Cu(II) sorption to polymer-supported nano-iron oxides: behavior and XPS study. J Colloid Interface Sci, 2012, 366(1):37-43.

Au/Fe_3O_4 core-shell nanoparticles are an efficient immunochromatography test strip performance enhancer—a comparative study with Au and Fe_3O_4 nanoparticles

Huang Jiaoling, Xie Zhixun, Xie Liji, Xie Zhiqin, Luo Sisi, Deng Xianwen, Huang Li, Zeng Tingting, Zhang Yanfang, Wang Sheng, and Zhang Minxiu

Abstract

Immunochromatography test strips that use metal particles constructed from Au, Fe_3O_4, and Au/Fe_3O_4 nanoparticles were developed for the rapid detection of avian influenza virus subtype H7 (AIV H7). The principle of this immunochromatography test strip was based on a sandwich immunoreaction in which AIV H7 antigens bind specifically to their corresponding antibodies on a nitrocellulose membrane. An antibody-metal (Au, Fe_3O_4 or Au/ Fe_3O_4) nanoparticle conjugate was used as a label and coated onto a glass fiber membrane, which was used as a conjugate pad. To create a test and a control zone, an anti-H7 polyclonal antibody and an anti-IgG antibody were immobilized onto the nitrocellulose membrane, respectively. Positive samples displayed brown/red lines in the test and control zones of the nitrocellulose membrane, whereas negative samples resulted in a brown/red line only in the control zone. The limit of detection (LOD) of the Au/Fe_3O_4 nanoparticle-based immunochromatography test strips was found to be $10^{3.5}$ EID_{50} (EID_{50}: 50% Egg Infective Dose), which could be visually detected by the naked eye within 15 min. In addition, 200 clinical samples were tested using the Au/Fe_3O_4 nanoparticle-based immunochromatography test strip to estimate its performance, and seven were positive for AIV H7. In summary, the Au/Fe_3O_4 nanoparticle-based immunochromatography test strip offers a simple and cost-effective tool for the rapid detection of AIV H7.

Keywords

immunochromatography test strips, Fe_3O_4, Au, core-shell nanopartical

---

Introduction

Avian influenza virus subtype H7 (AIV H7) has been frequently observed; for example, there was an AIV H7N2 outbreak in the Northeastern United States in 2002[1], and in the Netherlands, an AIV H7N7 outbreak not only impacted the poultry industry but also infected 89 people in 2003[2]. In March 2013, the first case of human infection with avian influenza A H7N9 virus was reported by the Chinese Centers for Disease Control and Prevention, and since then, more than 450 human cases of H7N9 infection have been reported[3, 4]. A variety of technologies for diagnosing AIV H7, such as virus isolation and identification[5], reverse transcription-polymerase chain reaction (RT-PCR)-based assays[6], real-time reverse transcription-polymerase chain reaction (real-time RT-PCR)[7], enzyme-linked immunosorbent assays (ELISAs)[8] and reverse transcription loop-mediated isothermal amplification (RT-LAMP)[9], have been developed. However, the disadvantages of these diagnostic methods, including the fact that they are time-consuming and require several experimental steps, including incubation and washing steps, make them less than ideal for practical applications. Therefore, it is necessary to explore simple, sensitive and rapid methods for the detection of virus AIV H7.

In the early 1980s, researchers developed immunochromatography test strips, which combine the advantages of chromatography and immunoassays into a single method. In this technique, the reaction between antibody and antigen occurs after chromatographic separation through a nitrocellulose membrane using capillary flow. Immunochromatography test strips are rapid, intuitive, user-friendly, inexpensive, and easily used by non-skilled personnel[10-12]. Au nanoparticles have been the most widely used labels in immunochromatography test strips[12-17] due to their long-term stability, easily controllable size distribution, and good compatibility with biological molecules, such as antibodies, antigens, proteins, DNAs, and RNAs. However, this method is generally only used for analyzing high concentrations of analytes. These limitations of Au nanoparticles have resulted in an increased use of various reporters that employ other nanoparticles as labels, such as magnetic nanoparticles(Fe_3O_4)[18], organic fluorophores[19] and quantum dots[20]. Several studies have demonstrated that Fe_3O_4 particlelabeled detection systems specifically improve lateral flow assay sensitivity[18, 21]. Moreover, Fe_3O_4 particles are easily and rapidly separated using a magnet during the labeling process. However, the Fe_3O_4 particle surface must be modified prior to labeling.

Therefore, we hypothesized that Au/Fe_3O_4 core-shell nanoparticles could combine the advantages of Au nanoparticles and Fe_3O_4 nanoparticles and avoid the above-mentioned disadvantages of each of these particles. In the current study, we used Au/Fe_3O_4 core-shell nanoparticles as a label to develop a novel immunochromatography test for the detection of AIV subtype H7 (AIV H7) and compared these results to those obtained with Au and Fe_3O_4 nanoparticles.

Materials and methods

Preparation of Au, Fe_3O_4 and Au/Fe_3O_4 core-shell nanoparticles

Fe_3O_4 and Au/Fe_3O_4 nanoparticles were prepared as previously described[22]. Briefly, 4.64 g of $FeCl_3{\cdot}6H_2O$ and 1.71 g of $FeSO_4{\cdot}7\ H_2O$ (Guoyao Group Chemical Reagents Co., Ltd., Shanghai, China) were dissolved in 250 mL of deionized water at a Fe^{2+}/Fe^{3+} molar ratio of 1∶2, and 2 mL of 0.2 mol/L sulfuric acid was added to prevent the solution from undergoing Fe^{2+} oxidation. Ammonia (Guoyao Group Chemical Reagents Co., Ltd.) (25%) was added until the pH reached 9.0~9.5, and the solution was stirred for 30 min at room temperature. After amination, the mixture was heated to 80 ℃ and incubated for 30 min. A black suspension was produced, and this mixture was sonicated for 10 min. The deposit was separated using a magnet and rinsed with hot water until the fluid became neutral in color, yielding Fe_3O_4 nanoparticles. The Fe_3O_4 nanoparticles were then placed in a flask for preparation of the Au/Fe_3O_4 nanoparticles.

A total of 0.229 g of sodium citrate (Guoyao Group Chemical Reagents Co., Ltd.) was dissolved in 100 mL of deionized water, and the solution was heated to 99 ℃ using a water bath under vigorous stirring. A 1 mL volume of the prepared Fe_3O_4 suspension was then added to this solution. Finally, 5 mL of 10 mmol/L hydrochloroauric acid (Guoyao Group Chemical Reagents Co., Ltd.) was added dropwise to the solution, and the reaction was allowed to occur for 15 min. The water bath was then removed, and the suspension was continually stirred for 15 min. The solids were removed again using a magnet and rinsed with deionized water as the claret-red suspension was cooled to ambient temperature. An Au/Fe_3O_4 suspension was then prepared by dispersing the solids in 20 mL of water, which was maintained at 4 ℃ . Additionally, Au nanoparticles were obtained via the reduction of hydrochloroauric acid with sodium citrate using the same procedure but without addition of the Fe_3O_4 suspension.

Preparation of the antibody-metal nanoparticle conjugates

The antibody-metal nanoparticle conjugates (antibody-Au, antibody-Fe_3O_4 and antibody-Au/Fe_3O_4 nanoparticle conjugates) were prepared according to a previously reported method[23, 24] with slight modifications. The Au/Fe_3O_4 nanoparticle solution was adjusted to pH 8.5 with 0.1 mol/L K_2CO_3, and 50 μL of anti-H7 monoclonal antibodies (Abcam, Cam-bridge, UK) (1 mg/mL) was added dropwise to 10 mL of the Au/Fe_3O_4 nanoparticle solution. The mixture was incubated for 30 min at room temperature, and 1 mL of 10% bovine serum albumin (BSA) (Beijing Dingguo Biotechnology Co., Ltd., Beijing China) solution was then added to block the residual surface of the Au nanoparticles. The obtained solution was separated using a magnet, and after the supernatant was discarded, 1 mL of 1% BSA solution was added to the Au/Fe_3O_4 conjugate for resuspension. The separation and suspension processes were repeated twice, and the precipitate was resuspended in 2 mL of storage buffer (the storage buffer consisted of 50 mmol/L sodium phosphate buffer (pH 7.8) containing 5% polyvinylpyrrolidone (w/v), 1.25% sucrose (w/v), 0.05% PEG8000 (w/v), 0.2% BSA (w/v), and 0.05% Tween-20 (v/v)) and stored at 4 ℃ until use. The antibody-Au and antibody-Fe_3O_4 nanoparticle conjugates were obtained using the same procedure, but the antibody-Au nanoparticle conjugate solution was additionally centrifuged for 30 min at 12 000 × g and 4 ℃ .

Pretreatment of the sample pad, conjugate pad, and nitrocellulose membrane

The sample pad, which was made from glass fiber (Shanghai Kinbio Tech Co., Ltd., Shanghai, China), was saturated with a buffer (pH 7.4) containing 20 mmol/L sodium borate, 1% (w/v) sucrose, 1% (w/v) BSA, 0.5% (v/v) Tween-20, and 0.05% (w/v) NaN_3 in water for 40 min, dried at 37 ℃ and stored in its dried condition until use.

The conjugate pad (Shanghai Kinbio Tech Co., Ltd.), which was composed of polyester fiber, was immersed in a solution containing 2% (w/v) BSA, 3% (w/v) sucrose and 0.05% (w/v) NaN_3 in water for 1 h and then dried at 37 ℃ . After 3 μL per strip of the antibody-Au, antibody-Au/Fe_3O_4 or antibody-Fe_3O_4 nanoparticle conjugate was placed onto the polyester fiber to be used as the conjugate pad, the conjugate pad was dried for 1 h at 37 ℃ and stored in its dried condition until use.

The nitrocellulose membrane (Shanghai Kinbio Tech Co., Ltd.) was treated with a buffer (pH 7.4) containing 10 mmol/L phosphate-buffered saline (PBS) solution, 3% (w/v) BSA, and 0.5% Tween-20 for 1 h and subjected to three 5 min washes with 0.1% Tween-20 in 10 mmol/L PBS. The membrane was then dried at 37 ℃ and stored in its dried condition until use.

Assembly and analysis of the immunochromatography test strip

The immunochromatography test strip contains four main elements: a sample pad, a conjugate pad, a nitrocellulose membrane, and an absorbent pad. The strip was positioned in such a way that the ends of the elements overlapped, ensuring continuous flow of the developing solution from the sample pad to the absorbent pad via capillary action. As shown in Figure 5-30-1 A, the nitrocellulose membrane was pasted onto the center of the backing plate. The conjugate pad was also pasted onto the plate such that it overlapped the nitrocellulose membrane by 2 mm. The sample pad was pasted onto the same end such that its margin justified to the conjugate pad. The absorbent pad was then pasted onto the other end of the nitrocellulose membrane with the same 2 mm overlap. Afterward, 1 mg/mL anti-H7 polyclonal antibodies (Abcam) (ie., the test line) and a goat anti-mouse IgG antibody (Beijing Dingguo Biotechnology Co., Ltd., Beijing China) (i. e., the control line) were immobilized onto the nitrocellulose membrane using an automatic dispenser (BioDot XYZ3000, USA). In particular, the anti-H7 polyclonal antibodies and the goat anti-mouse IgG antibody were dispensed onto the nitrocellulose membrane as the test line and the control line, respectively, with a width of 1 mm and a volume of 1 μL/cm. The entire assembled plate was then cut into 3.5 mm strips and stored in its dried condition until use.

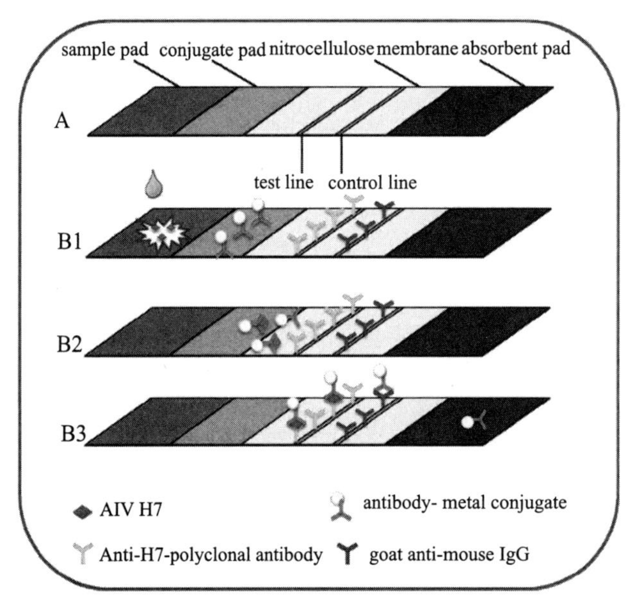

B1: A sample containing AIV H7 is applied to the sample pad. B2: AIV H7 combines with the antibody-metal conjugate and migrates along the nitrocellulose membrane by capillary action. B3: The formed complexes continue to migrate along the membrane and are captured by the other type of anti-H7-polyclonal antibodies to form antibody-metal-AIV H7-anti-H7-polyclonal antibody complexes on the test line. As the liquid sample continues to migrate, the excess antibody-metal-AIV H7 complexes are captured by the secondary antibody (goat anti-mouse IgG), resulting in the accumuation of metal on the control line. The excess antibody-metal conjugates continue to migrate toward the absorption pad.

Figure 5-30-1 Schematic illustration of the test strip and the detection of AIV H7 using metal-based immunochromatography test strips

In total, 100 μL of sample solution containing the desired concentration of AIV H7 was added to the sample pad. Negative samples from specific-pathogen-free (SPF) chickens were used as controls. Both the sample and control solutions migrated toward the absorption pad via capillary action. After 15 min, the results could be assessed by eye: a negative result was indicated by a colorless test line, whereas a positive result presented as a brown/red test line. Both negative and positive results exhibited a red control line because the goat anti-mouse IgG is a nonspecific antibody that binds to all types of mouse antibodies.

Results and discussion

Principle of the method

In this study, the principle of the Au, Fe_3O_4 and Au/Fe_3O_4 core-shell nanoparticles is based on the sandwich immunoassay, as illustrated in Figure 5-30-1. The immunochromatography test strip consists of a sample pad, a conjugate pad, a nitrocellulose membrane, and an absorption pad (Figure 5-30-1 A). All the components were assembled onto a plastic adhesive backing card. Sample solution containing the desired concentration of AIV H7 was added to the sample pad as illustrated in Figure 5-30-1 B1.The sample solution then migrated into the conjugation pad and bound to the antibody-metal nanoparticle conjugates based on the antibody-antigen interaction (Figure 5-30-1 B2). The formed complexes continued to migrate along the membrane and were captured by anti-H7 polyclonal antibodies immobilized beforehand on the nitrocellulose membrane to form antibody-metal-AIV H7-anti-H7 polyclonal antibody complexes, which resulted in the accumulation of metal nanoparticles on the test line. As the liquid sample continued to migrate, the excess antibody-metal-AIV H7-anti-H7 polyclonal antibody complexes were captured by the secondary antibody (goat anti-mouse IgG antibody), which resulted in the accumulation of metal nanoparticles on the control line. After 15 min, there were metal nanoparticles on the test line and the control line (Figure 5-30-1 B3). The results could be assessed by eye: a negative result was indicated by a colorless test line, whereas a positive result presented as a red/brown test line. Both negative and positive results exhibited a red/brown control line because the goat antimouse IgG is a nonspecific antibody that binds to all types of mouse antibodies. The excess antibody-metal conjugates continued to flow into the absorption pad at the end of the strip. If no metal nanoparticles accumulated on the control line, the test strip was invalidated and discarded. The fast immunoreaction and wash-free immunochromatography test strip detection made the assay rapid and easy to use.

Optimization of the antibody-Au/Fe_3O_4 conjugates and the pretreatment solution

The Au/Fe_3O_4 nanoparticle solution was adjusted to pH 8.5 with 0.1 mol/L K_2CO_3, and different volumes (i. e., 6, 5, 4, 3, and 2 μL) of anti-H7 monoclonal antibodies (1 mg/mL) were added to 1 mL of the Au/Fe_3O_4 nanoparticle solution to obtain different antibody-Au/Fe_3O_4 conjugates. These different antibody-Au/Fe_3O_4 conjugates were used for the construction of immunochromatography test strips for testing the same sample. The results are presented in Figure 5-30-2. The color intensity detected using 5 μL of anti-H7 monoclonal antibodies yielded the best result.

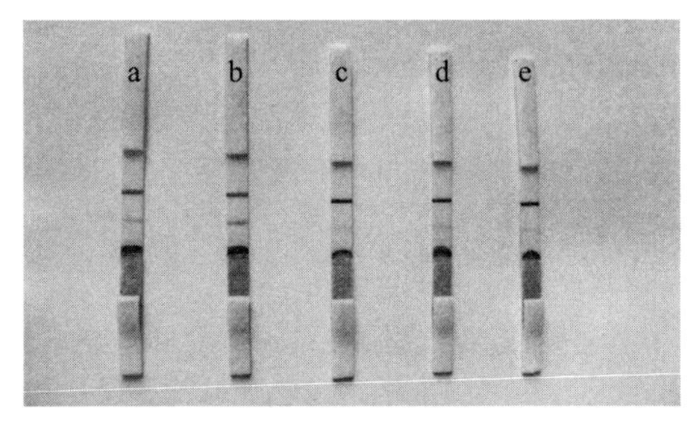

a: 6 μL; b: 5 μL; c: 4 μL; d: 3 μL; e: 2 μL.

Figure 5-30-2 Different volumes of H7 monoclonal antibodies (1 mg/mL) to 1mL of Au/Fe_3O_4 nanoparticle solution

The specificity of the immunochromatography test strip might be influenced by many factors, such as storage of the antibody-Au/Fe_3O_4 nanoparticle conjugates and pretreatment of the sample pad, conjugate pad and nitrocellulose membrane. In our study, the antibody-Au/Fe_3O_4 nanoparticle conjugates were stored in storage buffer. The sample pad, conjugate pad and nitrocellulose membrane were pretreated according to the method described above. The effect of storage buffer was evaluated using sodium phosphate buffer instead of storage buffer to store the antibody-Au/Fe_3O_4 nanoparticle conjugates to detect negative samples from SPF chickens, which revealed nonspecific absorption, with two brown lines appearing on the nitrocellulose membrane (Figure

5-30-3 a). The sample pad, conjugate pad and nitrocellulose membrane without pretreatment were used to construct the immunochromatography test strip for detecting negative samples from SPF chickens, which revealed nonspecific absorption, with two brown lines appearing on the nitrocellulose membrane (Figure 5-30-3 b-e). In contrast, the immunochromatography test strip developed in this study generated a colorless test line due to its good specificity (Figure 5-30-3 f).

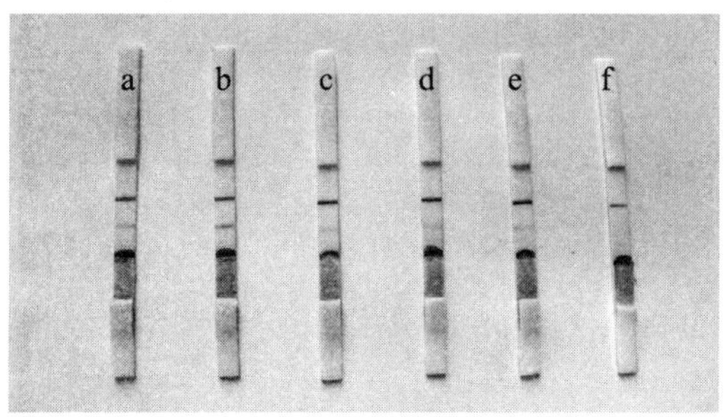

a: Antibody-Au/Fe_3O_4 nanoparticle conjugates were stored in sodium phosphate buffer; b: The sample pad; conjugate pad and nitrocellulose membrane were not pretreated; c: The sample pad was not pretreated; d:The conjugate pad was not pretreated; e: The nitrocellulose membrane was not pretreated; f: The immunochromatography test strip developed in this study.

Figure 5-30-3 Optimization of the specificity of the immunochromatography test strip

Detection of AIV H7 using the immunochromatography test strip

Under optimal conditions, the size of the test strip was 3.5 mm in width and 8 cm in length with a sample volume of 100 μL. The results of the Au/Fe_3O_4 core-shell nanoparticle-based immunochromatography strip test demonstrated that only one brown control line appeared for a negative test (i. e., negative samples from SPF chickens). In contrast, AIV H7 tested positive and formed brown test and control lines on the nitrocellulose membrane (Figure 5-30-4). The results of the Au nanoparticle-based immunochromatography strip test (Figure 5-30-5) were similar to the Au/Fe_3O_4 core-shell nanoparticle-based immunochromatography strip test: only the color differed (red). However, the Fe_3O_4 nanoparticle-based immunochromatography strip test was considered invalid and was discarded because no Fe_3O_4 accumulated on the control line when negative samples were detected and no Fe_3O_4 accumulated on the control and test lines when positive samples were detected.

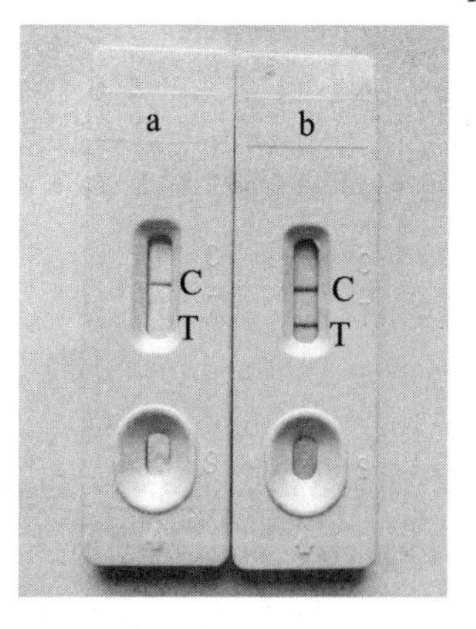

C: Control line; T: Test line. a: A negative result (i. e., samples from SPF chickens); b: A positive result (i. e., $10^{5.5}$ EID_{50} of AIV H7 (A1Duck/HK/47/76 (H7N2)).

Figure 5-30-4 Results of the Au/Fe_3O_4 nanoparticle-based immunochromatography test strip for AIV H7

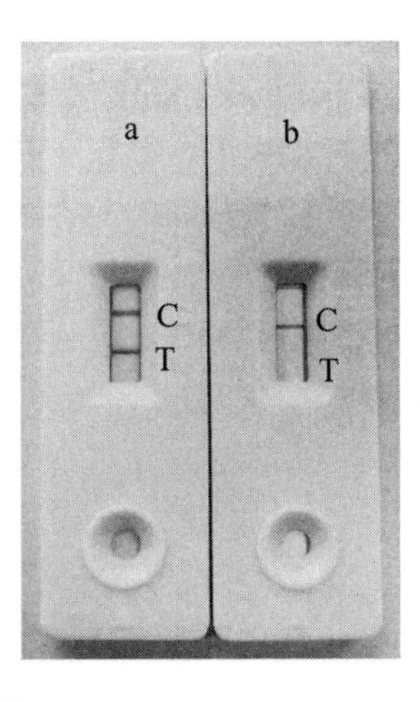

C: Control line; T: Test line. a: A positive result (i. e., $10^{5.5}$ EID_{50} of AIV H7 (A/Duck/HK/47/76, H7N2)); b: A negative result (i. e., samples from SPF chickens).

Figure 5-30-5 Results of the Au nanoparticle-based immunochromatography test strip for AIV H7

The most likely reason for this result is that the Fe_3O_4 nanoparticles failed to label the anti-H7 monoclonal antibodies, whereas the anti-H7 monoclonal antibodies successfully immobilized onto the Au/Fe_3O_4 and Au nanoparticles because the Au/Fe_3O_4 and Au nanoparticles but not the Fe_3O_4 nanoparticles are highly compatible with antibodies. If Fe_3O_4 nanoparticles were selected as the labeling material for targeting antibodies to fabricate the immunochromatography test strip, they would need to be modified with effective functional groups (e. g., a carbonyl group). Therefore, the Fe_3O_4 nanoparticle-based immunochromatography test strip would be more difficult to fabricate than the Au/Fe_3O_4 and Au nanoparticle-based immunochromatography test strips.

The abilities of the Au/Fe_3O_4 and Au nanoparticle-based immunochromatography test strips to detect AIV H7 were evaluated at concentrations between $10^{6.5}$ EID_{50} and $10^{2.5}$ EID_{50}. The experiments were repeated three times, and the results are presented in Figure 5-30-6. Two brown lines appeared at concentrations of AIV H7 as low as $10^{3.5}$ EID_{50}, whereas only the control line appeared brown on the nitrocellulose membrane at concentrations less than $10^{3.5}$ EID_{50}. Therefore, the limit of detection (LOD) of this method was $10^{3.5}$ EID_{50} for AIV H7 after evaluation by the naked eye within 15 min. When the antibody-Au nanoparticles conjugate was used as a label, the detection limit for AIV H7 was $10^{4.5}$ EID_{50} (Figure 5-30-7). Compared with the antibody-Au nanoparticle conjugate, the use of the antibody-Au/Fe_3O_4 conjugate as the label is more sensitive. A comparison study between the Au/Fe_3O_4 immunochromatography test strip and other methods for AIV H7 detection is summarized in Table 5-30-1. It the results indicated that the developed Au/Fe_3O_4 immunochromatography test strip has acceptable sensitivity, with an advantage in rapid detection, intuitive, user-friendly and inexpensive and can truly achieve point-of-care testing. Our ongoing research will focus on the application of the Au/Fe_3O_4 immunochromatography test strip for detection of other disease in the field.

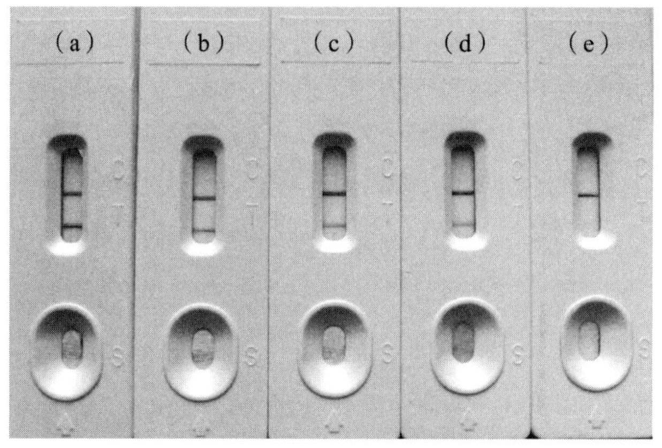

a: $10^{6.5}$ EID_{50}; b: $10^{5.5}$ EID_{50}; c: $10^{4.5}$ EID_{50}; d:$10^{3.5}$ EID_{50}; e: $10^{2.5}$ EID_{50}.

Figure 5-30-6 Detection of AIV H7 using the Au/ Fe_3O_4 nanoparticle-based immunochromatography strip test with AIV H7

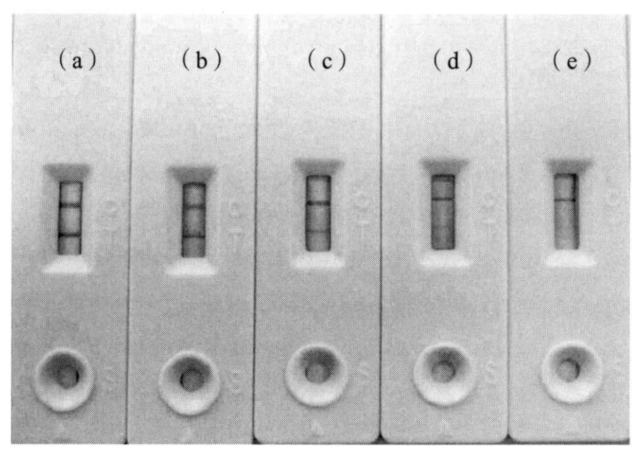

a: $10^{6.5}$ EID_{50}; b: $10^{5.5}$ EID_{50}; c: $10^{4.5}$ EID_{50}; d:$10^{3.5}$ EID_{50}; e: $10^{2.5}$ EID_{50}.

Figure 5-30-7 Detection of AIV H7 using the Au nanoparticle-based immunochromatography strip test with AIV H7

Table 5-30-1 A comparison study between the Au/Fe_3O_4 immunochromatography test strip and other methods for AIV H7 detection

Methods	Detection time	Detection limit	Advantages	Disadvantages	Reference
Virus isolation and identification	5～7 days	1 EID_{50}/mL	"Gold standard", sensitive, accurate	Labor-intensive and time consuming procedure	[5]
RT-PCR	5 h	100 pg	Good in sensitive	Require expensive equipment, appropriate laboratory facilities and a trained technician	[6]
Real-time RT-PCR	3 h	3.2×10^{-4} HAUs	Good in specificity and sensitivity	Expensive and complicated operation	[7]
ELISAs	3 h	10^3 $TCID_{50}$	Good in specificity	Many experimental steps including incubation and washing steps	[8]
RT-LAMP	1.5 h	42.47 copies/reaction	Simple instruments, good in sensitivity	High rate of false positives	[9]
Au/Fe_3O_4 immunochromatography test strip	15 min	$10^{3.5}$ EID_{50}	Rapid, low-cost, intuitive, user-friendly and truly realize point-of-care testing	Other disease detection will be needed for the on-going research	This study

Note: HAUs, Hemagglutination units; $TCID_{50}$, 50% Tissue Culture Infective Dose.

We also evaluated the shelf life of the Au/Fe_3O_4 nanoparticle-based immunochromatography test strip for field use. All Au/ Fe_3O_4 nanoparticle-based immunochromatography test strips were stored at 4 ℃ for 90 days to test the sample concentration of $10^{4.5}$ EID_{50} and $10^{3.5}$ EID_{50} AIV H7. Negative samples from SPF chickens were used as controls. The results demonstrated that the Au/Fe_3O_4 nanoparticle-based immunochromatography test strip can be stored at 4 ℃ for 90 days and still continue to detect AIV H7 effectively.

To evaluate the specificity of the Au/Fe_3O_4 nanoparticle-based immunochromatography test strip, certain non-target samples, such as inactivated H7N9, H7N2, H7N7, H1N1, H2N3, H3N2, H4N6, H6N8, H8N4, H9N2, H10N3, H11N9, H12N5, H13N5, H14N5, H15N9, H16N3, infectious tracheitis virus (ILTV), infectious bronchitis virus (IBV) and Neweastle disease virus (NDV), were tested. The experimental procedure was the same as that used for the H7 target. The experiments were repeated three times. The results are shown in Table 5-30-2. All the results of the AIV H7 sample tests were positive, and all results of the non-target sample tests were negative. These results demonstrate that the test strip is specific for AIV H7.

Table 5-30-2 Specificity of the Au/Fe_3O_4 nanoparticle-based immunochromatography test strip

Avian pathogen	Source	Concentration		Results
		Virus titer (EID_{50})	HA titer	
A/Chicken/BD135/2013 (H7N9)	CAU	$10^{5.5}$		+
A/Chicken PA/3979/97 (H7N2)	PU	10^{4}		+
A/Chicken/NY/273874/03 (H7N2)	UCONN	$10^{4.5}$		+
A/Duck/HK/47/76 (H7N2)	UHK	$10^{5.5}$		+
A/Duck/42846/07 (H7N7)	PU	10^{4}		+
A/Duck/Guangxi/030D/2009 (H1N1)	GVRI		128	–
A/Duck/HK/77/76 d77/3 (H2N3)	UHK		64	–
A/Duck/Guangxi/M20/2009 (H3N2)	GVRI		64	–
A/Duck/Guangxi/070D/2010 (H4N6)	GVRI		32	–
A/Chicken/QT35/98 (H5N9)	PU		128	–
A/Duck/Guangxi/GXd–6/2010 (H6N8)	GVRI		64	–
A/Turkey/Ontario/6118/68 (H8N4)	UHK		128	–
A/Chicken/Guangxi/DX/2008 (H9N2)	GVRI		256	–
A/Duck/HK/876/80 (H10N3)	UHK		64	–
A/Duck/PA/2099/12 (H11N9)	PU		64	–
A/Duck/HK/862/80 (H12N5)	UHK		128	–
A/Gull/Md/704/77 (H13N5)	UHK		64	–
A/Mallard/Astrakhan/263/82 (H14N5)	UCONN		64	–
A/Shearwater/Western Australia/2576/79 (H15N9)	UCONN		64	–
A/Shorebird/Delaware/168/06 (H16N3)	CIVDC		64	–
ILTV (Beijing)	CIVDC	10^{6}		–
IBV (Mass41)	CIVDC	$10^{5.2}$		–
NDV (F48E9)	CIVDC		128	–

Note: PU, Pennsylvania State University, USA; CAU, China Agricultural University; UHK, University of Hong Kong, China; GVRI, Guangxi Veterinary Research Institute; CIVDC, China Institute of Veterinary Drug Control; UCONN, University of Connecticut, USA.

Application of the Au/Fe_3O_4 nanoparticle-based immunochromatography test strip for the detection of AIV H7

The clinical samples were prepared in a viral transport medium composed of 0.05 M PBS containing penicillin (10 000 units per mL), streptomycin (10 mg/mL), gentamycin (10 mg/mL), kanamycin (10 mg/mL) and 5% (v/v) fetal bovine serum and were placed in an icebox.

A total of 200 clinical swab samples were collected from chickens with permission from the owners of the live bird markets, and the samples were assayed using the optimized Au/Fe_3O_4 nanoparticle-based immunochromatography test strip: seven AIV H7-positive samples were detected and confirmed by virus isolation (virus isolations were prepared by inoculating SPF embryonated chicken eggs and were tested using a hemagglutination assay (HA) and a hemagglutination inhibition (HI) assay as described previously[25]), with the positive results being 100% comparable to those for virus isolation. Our immunochromatography test strip uses an Au/Fe_3O_4 core-shell nanoparticle whose Au nanoparticle shell makes it perfectly biocompatible and whose magnetic nanoparticle Fe_3O_4 core ensures it can be rapidly separated by a magnet. In contrast, magnetic-based immunochromatography strips require modification of the magnetic particle surface before labeling.

Conclusions

A novel approach for the rapid detection of AIV H7 using an immunochromatography test strip was successfully developed. We used Au/Fe_3O_4 core-shell nanoparticles as the label; these particles are easily and rapidly separated using a magnet during the labeling process, and the Au/Fe_3O_4 surface requires no modification prior to labeling. This assay, which had an LOD of $10^{3.5}$ EID_{50}, provided a 10-fold lower LOD compared with an assay that used an antibody-Au conjugate label (LOD: $10^{4.5}$ EID_{50}). The assay specifically detected AIV H7N2 and recombinant HA proteins of H7 subtypes, including H7N7 and H7N9, but did not react with non-H7 subtypes, including H1N1, H2N3, H3N2, H4N6, H6N8, H8N4, H9N2, H10N3, H11N9, H12N5, H13N5, H14N5, H15N9, H16N3, ILTV, IBV and NDV. Our assay was also successfully applied for the detection of AIV H7 in clinical samples through a single step and within 15 min. The results also demonstrated that the test strip can be stored at 4 ℃ for 90 days and continue to detect AIV H7 effectively. Therefore, the Au/Fe_3O_4 immunochromatography strip test reported herein, which is a rapid, simple and low-cost method, is a potentially valuable means for the detection and rapid clinical diagnosis of AIV H7. Consequently, it will be a very useful screening assay for the surveillance of AIV H7 in underequipped laboratories.

References

[1] COLLINS R A, KO L S, FUNG K Y, et al. Rapid and sensitive detection of avian influenza virus subtype H7 using NASBA. Biochem Biophys Res Commun, 2003, 300(2): 507-515.

[2] FOUCHIER R A, SCHNEEBERGER P M, ROZENDAAL F W, et al. Avian influenza A virus (H7N7) associated with human conjunctivitis and a fatal case of acute respiratory distress syndrome. Proc Natl Acad Sci U S A, 2004, 101(5): 1356-1361.

[3] WORLD HEALTH ORGANIZATION (WHO). WHO risk assessment human infections with avian influenza A (H7N9) virus. (2014-10-02) [2018-04-16], http://www.who.int/influenza/human_animal_interface/influenza_h7n9/riskassessment_h7n9_20ct14.pdf?ua=1.

[4] CHEN Y, LIANG W, YANG S, et al. Human infections with the emerging avian influenza A H7N9 virus from wet market poultry: clinical analysis and characterisation of viral genome. Lancet, 2013, 381(9881): 1916-1925.

[5] CHARLTON B, CROSSLEY B, HIETALA S. Conventional and future diagnostics for avian influenza. Comp Immunol Microbiol Infect Dis, 2009, 32(4): 341-350.

[6] XIE Z, PANG Y S, LIU J, et al. A multiplex RT-PCR for detection of type A influenza virus and differentiation of avian H5, H7, and H9 hemagglutinin subtypes. Mol Cell Probes, 2006, 20(3-4): 245-249.

[7] KANG X, WU W, ZHANG C, et al. Detection of avian influenza A/H7N9/2013 virus by real-time reverse transcription-polymerase chain reaction. J Virol Methods, 2014, 206: 140-143.

[8] VELUMANI S, DU Q, FENNER B J, et al. Development of an antigen-capture ELISA for detection of H7 subtype avian influenza from experimentally infected chickens. J Virol Methods, 2008, 147(2): 219-225.

[9] NAKAUCHI M, TAKAYAMA I, TAKAHASHI H, et al. Development of a reverse transcription loop-mediated isothermal amplification assay for the rapid diagnosis of avian influenza A (H7N9) virus infection. J Virol Methods, 2014, 204: 101-104.

[10] WANG L, LU D, WANG J, et al. A novel immunochromatographic electrochemical biosensor for highly sensitive and selective detection of trichloropyridinol, a biomarker of exposure to chlorpyrifos. Biosens Bioelectron, 2011,26(6): 2835-2840.

[11] SHIM W B, DZANTIEV B B, EREMIN S A, et al. One-step simultaneous immunochromatographic strip test for multianalysis of ochratoxin a and zearalenone. J Microbiol Biotechnol, 2009, 19(1): 83-92.

[12] ZHOU Y, ZHANG Y, PAN F, et al. A competitive immunochromatographic assay based on a novel probe for the detection of mercury (II) ions in water samples. Biosens Bioelectron, 2010, 25(11): 2534-2538.

[13] KANG K, CHEN L, ZHAO X, et al. Development of rapid immunochromatographic test for hemagglutinin antigen of H7 subtype in patients infected with novel avian influenza A (H7N9) virus. PLOS ONE, 2014, 9(3): e92306.

[14] MENG K, SUN W, ZHAO P, et al. Development of colloidal gold-based immunochromatographic assay for rapid detection of *Mycoplasma suis* in porcine plasma. Biosens. Bioelectron, 2014, 55: 396-399.

[15] LI D, WEI S, YANG H, et al. A sensitive immunochromatographic assay using colloidal gold-antibody probe for rapid detection of pharmaceutical indomethacin in water samples. Biosens Bioelectron, 2009, 24(7): 2277-2280.

[16] ZHOU Y, PAN F G, LI Y S, et al. Colloidal gold probe-based immunochromatographic assay for the rapid detection of brevetoxins in fishery product samples. Biosens Bioelectron, 2009, 24(8): 2744-2747.

[17] PREECHAKASEDKIT P, PINWATTANA K, DUNGCHAI W, et al. Development of a one-step immunochromatographic strip test using gold nanoparticles for the rapid detection of Salmonella typhi in human serum. Biosens Bioelectron, 2012, 31(1): 562-566.

[18] PECK R B, SCHWEIZER J, WEIGL B H, et al. A magnetic immunochromatographic strip test for detection of human papillomavirus 16 E6. Clin Chem, 2006, 52(11): 2170-2172.

[19] KHREICH N, LAMOURETTE P, BOUTAL H, et al. Detection of Staphylococcus enterotoxin B using fluorescent immunoliposomes as label for immunochromatographic testing. Anal Biochem, 2008, 377(2): 182-188.

[20] YANG Q, GONG X, SONG T, et al. Quantum dot-based immunochromatography test strip for rapid, quantitative and sensitive detection of alpha fetoprotein. Biosens Bioelectron, 2011, 30(1): 145-150.

[21] HANDALI S, KLARMAN M, GASPARD AN, et al. Development and evaluation of a magnetic immunochromatographic test to detect Taenia solium, which causes taeniasis and neurocysticercosis in humans. Clin Vaccine Immunol, 2010, 17(4): 631-637.

[22] LI J, GAO H, CHEN Z, et al. An electrochemical immunosensor for carcinoembryonic antigen enhanced by self-assembled nanogold coatings on magnetic particles. Anal Chim Acta, 2010, 665(1): 98-104.

[23] ZHU J, ZOU N, ZHU D, et al. Simultaneous detection of high-sensitivity cardiac troponin I and myoglobin by modified sandwich lateral flow immunoassay: proof of principle. Clin Chem, 2011, 57(12): 1732-1738.

[24] LIU M, JIA C, HUANG Y, et al. Highly sensitive protein detection using enzyme-labeled gold nanoparticle probes. Analyst, 2010, 135(2): 327-331.

[25] EDWARDS S. OIE laboratory standards for avian influenza. Dev Biol, 2006, 124: 159-162.

An enzyme-free sandwich amperometry-type immunosensor based on Au/Pt nanoparticle-functionalized graphene for the rapid detection of avian influenza virus H9 subtype

Huang Jiaoling, Xie Zhixun, Li Meng, Luo Sisi, Deng Xianwen, Xie Liji, Fan Qing, Zeng Tingting, Zhang Yanfang, Zhang Minxiu, Wang Sheng, Xie Zhiqin, and Li Dan

Abstract

Avian influenza virus H9 subtype (AIV H9) has contributed to enormous economic losses. Effective diagnosis is key to controlling the spread of AIV H9. In this study, a nonenzymatic highly electrocatalytic material was prepared using chitosan (Chi)-modified graphene sheet (GS)-functionalized Au/Pt nanopartices (GS-Chi-Au/Pt), followed by the construction of a novel enzyme-free sandwich electrochemical immunosensor for the detection of AIV H9 using GS-Chi-Au/Pt and graphene-chitosan (GS-Chi) nanocomposites as a nonenzymatic highly electrocatalytic material and a substrate material to immobilize capture antibodies (avian influenza virus H9-monoclonal antibody, AIV H9/MAb), respectively. GS, which has a large specific surface area and many accessible active sites, permitted multiple Au/Pt nanoparticles to be attached to its surface, resulting in substantially improved conductivity and catalytic ability. Au/Pt nanoparticles can provide modified active sites for avian influenza virus H9-polyclonal antibody (AIV H9/PAb) immobilization as signal labels. Upon establishing the electrocatalytic activity of Au/Pt nanoparticles on graphene towards hydrogen peroxide (H_2O_2) reduction for signal amplification and optimizing the experimental parameters, we developed an AIV H9 electrochemical immunosensor, which showed a wide linear range from $10^{1.37}$ EID_{50}/mL to $10^{6.37}$ EID_{50}/mL and a detection limit of $10^{0.82}$ EID_{50}/mL. This sandwich electrochemical immunosensor also exhibited high selectivity, reproducibility and stability.

Keywords

electrochemical immunosensor, Au/Pt nanoparticles, electrocatalysis, avian influenza virus H9 subtype

Introduction

Since avian influenza virus H9 subtype (AIV H9) was isolated from affected chickens in 1994, AIV H9 has become widespread in poultry[1, 2]. AIV H9 mainly causes poor weight gain, difficulty breathing and reduced egg production, leading to substantial economic losses in the poultry industry[3]. Therefore, rapid, sensitive and specific assays are urgently needed to screen for and control AIV H9 infection.

In recent years, electrochemical immunosensors have attracted increasing attention in clinical diagnosis due to their simple operation, rapid response, low instrumentation cost, excellent sensitivity of electrochemical techniques and high specificity of immunoreactions. In particular, sandwich-type electrochemical immunosensors have been widely applied in clinical diagnosis because double signal amplification with the substrate material and detection antibody labels enhance the electrochemical immunosensor sensitivity[4, 5]. The signal amplification strategy is the most important factor for developing sandwich-type electrochemical immunosensors, and it largely relies on employing various signal amplification labels in combination with secondary antibodies to form conjugated immunocomplexes[6-8]. Nanomaterials have attracted considerable research interest because they not only show enzyme-mimetic activity but also have some improved properties over native enzymes, such as good stability, easy synthesis and cost effectiveness[9, 10]. Among the variety of nanomaterials, noble metal nanostructures have been widely used as labels to fabricate enzyme-free electrochemical immunosensors due to their excellent stability, high electrocatalytic activity, simple synthesis and easy storage[11-13]. In particular, Au and Pt nanoparticles have aroused wide interest due to their excellent electrochemical catalytic properties towards the reduction of H_2O_2, and thus, Au and Pt nanoparticles have been used to electrochemically catalyse the reduction of H_2O_2 and increase the sensitivity of electrochemical immunosensors[14-16]. Moreover, bimetallic nanomaterials usually show much higher catalytic activity than their monometallic counterparts because of the synergistic effect[17, 18].

A graphene sheet (GS) is a material with a two-dimensional monolayer carbonaceous structure that shows remarkable mechanical stiffness, a high specific surface area, high thermal conductivity and fast electron transport and has been utilized in different applications in various fields, including biomedical applications, energy conversion and storage systems, electrocatalysts and disease diagnosis[19-23]. GSs are usually used as electrode modification materials and signal amplifier materials in electrochemical sensor applications because of their unique physicochemical and biological properties[24-26]. Nonetheless, GSs displayed poor hydrophilicity and difficulty immobilizing biometric molecules, which generally adversely affect the performance of electrochemical sensors. Hence, GSs have been modified with organic molecules or inorganic molecules to avoid this problem. We have found that GS-chitosan (GS-Chi) nanocomposites are easily immobilizing biometric molecules and are better dispersed in water[27]. Interestingly, Chi is rich in amino groups[28], and GS-Chi might improve the loading capacity and dispersion of Au/Pt through covalent binding between noble metal nanoparticles and amino (-NH_2) groups. In addition, Au/Pt might also enable the facile conjugation of capture antibodies due to the formation of stable Au-N and Pt-N bonds between Au/Pt and -NH_2 residues on antibodies[29]. Therefore, the combination of Au/Pt and GS-Chi may ideally enhance the signal response of the electrochemical immunosensor.

Substrate materials with effective immobilization of capture antibodies and high conductivity to facilitate transfer of electrons between the electrode and electrolyte are another important factor to improve the quality of electrochemical immunosensors. The GS-Chi nanocomposite was employed as the electrode substrate material. Due to the large specific surface area of the GS-Chi nanocomposite, more capture antibodies were stably immobilized onto the surface of the GS-Chi nanocomposite via glutaraldehyde. The superior conductivity of the GS-Chi nanocomposite enhanced electron transfer from the electrolyte to the electrode surface, further enhancing the signal response of the electrochemical immunosensor[28].

In the present study, a GS-Chi nanocomposite was used as a substrate material to immobilize capture antibodies (AIV H9/MAbs) and as a carrier to load catalytically active material (Au/Pt) and detection antibodies (AIV H9/PAbs) to fabricate an enzyme-free sandwich-type electrochemical immunosensor for AIV H9 detection. Due to its very large specific surface area and excellent electrical conductivity, GS-Chi not only increased the amounts of AIV H9/MAbs, Au/Pt and AIV H9/PAbs immobilized but also improved the electron transfer efficiency. A substantial increase in sensitivity towards the reduction of H_2O_2 was obtained with the synergetic effect of GS-Chi-Au/Pt. The established electrochemical immunosensor exhibited excellent analytical performance for AIV H9 detection and could potentially be applied as an electrochemical immunosensor for the detection of other pathogens.

Materials and methods

Reagents and Materials

Chloroplatinic acid hexahydrate ($H_2PtCl_2 \cdot 6H_2O$), chloroauric acid tetrahydrate ($HAuCl_4 \cdot 4H_2O$) and bovine serum albumin (BSA) were obtained from Sigma-Aldrich Chemical Co. (St. Louis, MO, USA). Graphite powder ($<$45 mm), $NaNO_3$, H_2SO_4 and $KMnO_4$ were supplied by Guoyao Group Chemical Reagents Co., Ltd. (Shanghai, China). All chemicals were analytical reagent grade. Double-distilled deionized water (ddH_2O) prepared with a Millipore water purification system was used throughout the experiments. Phosphate-buffered saline (PBS, pH 7.0), which was prepared with 10 mmol/L NaH_2PO_4 and 10 mmol/L Na_2HPO_4 and containing 0.9% NaCl, was used as washing buffer. Electrolytes with different pH values were prepared by mixing different volumes of 10 mmol/L NaH_2PO_4 and 10 mmol/L Na_2HPO_4 and contained 0.1 mol/L KCl. Viral transport medium was prepared with PBS containing 5% (v/v) foetal bovine serum, streptomycin (10 mg/mL), kanamycin (10 mg/ mL), gentamycin (10 mg/mL) and penicillin (10 000 units/mL).

Viruses and Antibodies

The AIVs used in this study included AIV subtypes H9, H1, H2, H3, H4, H5 (H5N9 subtype), H6, H7, H8, H10, H11, H12, H13, H14, H15 and H16. All the viruses were collected and stored in a −80 ℃ freezer in our laboratory prior to use (Table 5-31-1). AIV H9/PAbs and AIV H9/MAbs were prepared by our group[30].

Table 5-31-1 Viruses used in this study

Avian pathogen	Source	Viral titre
A/Duck/Guangxi/030D/2009 (AIVH1)	GVRI	$10^{5.61}$ $EID_{50}mL^{-1}$
A/Duck/HK/77/76d77/3 (AIVH2)	UHK	$10^{6.29}$ $EID_{50}mL^{-1}$
A/Chicken/Guangxi/015C10/2009 (AIVH3)	GVRI	$10^{7.43}$ $EID_{50}mL^{-1}$
A/Duck/Guangxi/070D/2010 (AIVH4)	GVRI	$10^{5.45}$ $EID_{50}mL^{-1}$
Inactivated A/Chicken/QT35/98 (AIVH5)	PU	128 HAUs
A/Chicken/Guangxi/121/2013 (AIVH6)	GVRI	$10^{6.19}$ $EID_{50}mL^{-1}$
A/ChickenPA/3979/97 (AIVH7)	PU	256 HAUs
A/Turkey/Ontario/6118/68 (AIVH8)	UHK	$10^{6.23}$ $EID_{50}mL^{-1}$
A/Chicken/Guangxi/116C4/2012 (AIVH9)	GVRI	$10^{6.37}$ $EID_{50}mL^{-1}$
A/Duck/HK/876/80 (AIVH10)	UHK	$10^{5.73}$ $EID_{50}mL^{-1}$
A/Duck/PA/2099/12 (AIVH11)	PU	$10^{6.47}$ $EID_{50}mL^{-1}$
A/Duck/HK/862/80 (AIVH12)	UHK	$10^{7.12}$ $EID_{50}mL^{-1}$
A/Gull/Md/704/77 (AIVH13)	UHK	$10^{5.37}$ $EID_{50}mL^{-1}$
A/Mallard/Astrakhan/263/82 (AIVH14)	UCONN	$10^{6.72}$ $EID_{50}mL^{-1}$
A/Shearwater/Western Australia/2576/79 (AIVH15)	UCONN	$10^{5.81}$ $EID_{50}mL^{-1}$
A/Shorebird/Delaware/168/06 (H16N3)	CIVDC	$10^{5.63}$ $EID_{50}mL^{-1}$

Note: GVRI, Guangxi Veterinary Research Institute; UHK, University of Hong Kong, China; PU, Pennsylvania State University, USA, UCONN, University of Connecticut, USA; CVDC, China Institute of Veterinary Drug Control; HAUs, Haemagglutination units.

Apparatus

Fourier-transform infrared (FT-IR) spectra were measured with a Nicolet SI10 FT-IR Spectrometric Analyser (USA) using KBr pellets. All electrochemical measurements were performed using a PARSTAT 4000A instrument (Princeton Applied Research, USA). A standard three-electrode system composed of a modified glassy carbon electrode (GCE, Ø=3 mm) as the working electrode, a saturated calomel electrode (SCE) as the reference electrode and a platinum wire electrode as the counter electrode was used. The nanomaterials were characterized using transmission electron microscopy (TEM) and energy-dispersive X-ray spectroscopy (EDS) elemental analysis (Tecnai G2 F30 S-TWIN, FIE, USA).

Synthesis of GS-Chi-Au/Pt-AIV H9/PAbs

First, GSs were prepared using the method reported by Hummers, with some modifications[31]. Briefly, 1.0 g of graphite powder and 2.5 g of $NaNO_3$ were added to 100 mL of concentrated H_2SO_4 and stirred for 2 h at room temperature. Then, 5 g of $KMnO_4$ were slowly added to this mixture with continuous stirring and cooled with ice at the same time. The mixture was reacted at 35 ℃ for 24 h with continuous stirring. Then, 100 mL of ddH_2O was slowly added to the mixture and stirred at 80 ℃ for 3 h, and then, 300 mL of ddH_2O was added. Subsequently, 6 mL of 30% H_2O_2 was added to the mixture, many bubbles appeared, and the mixed solution immediately turned bright yellow. The resulting solution was stirred for 3 h, and then, the supernatant was decanted after precipitation for 24 h at room temperature. The obtained yellow slurry was washed with 500 mL of 0.5 mol/L HCl and centrifuged. Next, the slurry was washed with ddH_2O by centrifugation until the pH of the supernatant was approximately 7.0. Then, 100 mL of ddH_2O was added and ultrasonicated for 2 h, and GS oxide was obtained. Ten millilitres of 1.0% $NaBH_4$ was added dropwise at 95 ℃ while stirring and incubated for 3 h, and GS oxide was reduced. After centrifugation, the supernatant was decanted, and the precipitate was washed with ddH_2O three times and then dried in a vacuum drying oven at 90 ℃ for 8 h to obtain GS.

Second, GS-Chi was prepared using our previously described method[32]. Briefly, 0.05 mg of Chi powder was added to 100 mL of a 1.0% (v/v) acetic acid solution while stirring until it was completely dispersed, and a 0.5 wt% Chi solution was obtained. Then, 100 mg of GSs was added to the 100 mL 0.5 wt% Chi solution, ultrasonicated for 1 h, continuously stirred for 24 h at room temperature, and GS-Chi was obtained.

Third, GS-Chi-Au/Pt was prepared using the following procedure: 1 mL of $HAuCl_4$ (10 mmol/L) and 1 mL of K_2PtCl_4 (10 mmol/L) were added to 20 mL of the GS-Chi (1 mg/mL) solution, and the mixture was stirred at 25 ℃ for 3 h. Then, the mixture was heated to 80 ℃ in a water bath and stirred for 1 h. Au^{3+} and Pt^{2+} were reduced to Au/Pt by Chi at 80 ℃. Finally, the GS-Chi-Au/Pt nanocomposite was generated. Different proportions of GS-Chi-Au/Pt were prepared using the method mentioned above by adding different volumes of $HAuCl_4$ (10 mmol/L) and K_2PtCl_4 (10 mmol/L) to 20 mL of the GS-Chi (1 mg/mL) solution.

Finally, the as-prepared GS-Chi-Au/Pt (10 mL) was mixed with 500 μL of 10 μg/mL AIV H9/PAb (0.5 μg/mL), and the mixture was oscillated in a shaking water bath at 4 ℃ overnight. The product was centrifuged and washed three times with PBS (pH 7.0) to remove the unbound AIV H9/PAb. The immunocomplex was redispersed into 10 mL of PBS (pH 7.0) containing 1 wt% BSA. Figure 5-31-1 shows the procedure used to prepare GS-Chi-Au/Pt-AIV H9/PAbs.

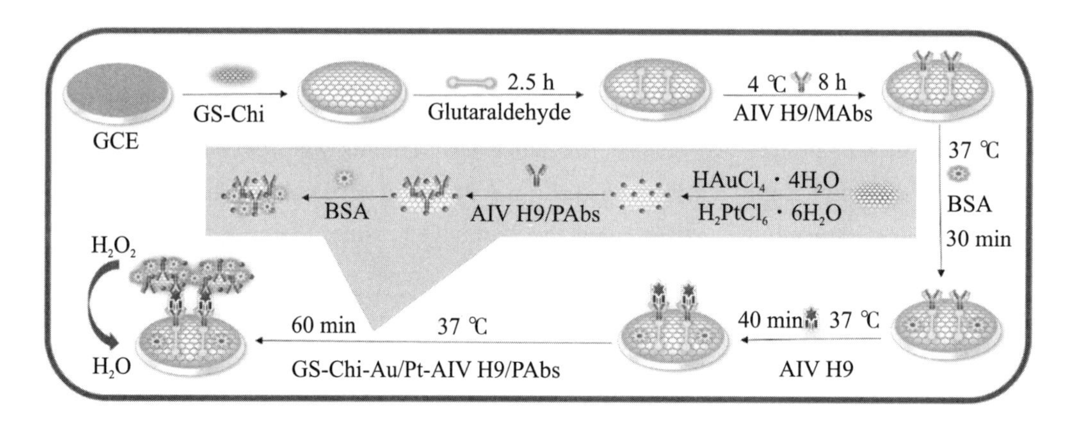

Figure 5-31-1 Schematic illustrating the electrochemical immunosensor (color figure in appendix)

Fabrication of the electrochemical immunosensor

The process used to fabricate the electrochemical immunosensor is shown in Figure 5-31-1. The GCE was successively polished with 1.0 μm, 0.3 μm, and 0.05 μm alumina polishing powders and rinsed with ddH_2O after each polishing step. The GCE was sonicated in ddH_2O, ethanol, and ddH_2O for 5 min each.

Next, 8 μL of GS-Chi (1 mg/mL) was dropped onto the surface of the clean GCE and dried at room temperature. Then, the modified GCE was incubated with 8 μL of 5% glutaraldehyde for 2.5 h and washed with ddH_2O. Eight microlitres of AIV H9/MAbs (10 μg/mL) was deposited onto the modified GCE and incubated at 4 ℃ for 8 h. Excess AIV H9/MAbs were removed by washing three times with PBS (pH 7.0). Subsequently, 8 μL of BSA (1 wt%) was dropped on the integrated AIV H9/MAbs-GS-Chi-GCE and reacted for 30 min at 37 ℃ to block the nonspecific adsorption sites of AIV H9/MAbs-GS-Chi-GCE. Then, the electrode was washed with PBS (pH 7.0) to remove unbound BSA. Finally, the immunosensor was steeped in 1 mL of PBS (pH 7.0) containing 1 wt% BSA and stored at 4 ℃ when not in use.

Electrochemical measurement

First, the immunosensor was incubated with 8 μL of varying concentrations of AIV H9 or clinical samples for 40 min at 37 ℃ . Second, the unbound AIV H9 and interfering substances were removed by washing the sensor three times with PBS, and 8 μL of GS-Chi-Au/Pt-H9/PAbs was incubated on the immunosensor at 37 ℃ for 60 min. Finally, the immunosensor was rinsed to remove unbound GS-Chi-Au/Pt-H9/PAbs, and amperometric i-t measurements were performed. We minimized the response of common interfering components and reduced the background current by selecting –0.4 V as the working potential for the amperometric i-t measurements. After the background current was stabilized (approximately 50 s later), 10 mmol/L H_2O_2 was added to the electrolyte, unless indicated otherwise, and the current change was monitored and recorded. Electrochemical impedance spectroscopy (EIS) was performed at frequencies ranging from 0.01 Hz to 100 kHz using an amplitude of 10 mV at 0.2 V.

Clinical sample preparation

Oral and cloacal swab samples were gently collected from chickens at different live poultry markets in Guangxi and used as clinical samples. The oral swab and the cloacal swab from the same chickens were placed together into a collection tube that contained 1 mL of viral transport medium and counted as a single clinical sample, and the clinical samples were placed in an ice box for transport to our laboratory. Then, cotton swabs were repeatedly shaken and wiped in the transport medium, leading to viral transfer to the transport medium. The cotton swabs were discarded, and the obtained solutions were stored at –70 ℃. Before detection, the samples were repeatedly frozen and thawed 3 times and then centrifuged at 4 000 r/min for 5 min. The obtained supernatant was analysed using the fabricated electrochemical immunosensor according to the steps described in section *Electrochemical Measurement.*

Results and discussion

Characterization of the GS-Chi-Au/Pt nanocomposite

FT-IR spectroscopy was performed to confirm that GS was modified with Chi. The FT-IR spectrum of Chi presented all of the characteristic bands for Chi, including C-C-O bond stretching vibrations at 1 034 cm^{-1}, 1 081 cm^{-1} and 1 154 cm^{-1}; $-NH_2$ bending vibrations at 1 596 cm^{-1} and 1 653 cm^{-1}; and C-H stretching at 2 920 cm^{-1} and 2 878 cm^{-1}. In addition, $-NH_2$ and -OH bond stretching vibrations were observed at 3 425 cm^{-1} (Figure 5-31-2 A)[33]. Additionally, the FT-IR spectrum of GS preserved all the characteristic bands of GS: benzene ring backbone stretching vibrations at 1 420 cm^{-1}, 1 459 cm^{-1} and 1 555 cm^{-1}; C=O stretching vibration at 1 659 cm^{-1}; C-H stretching vibration at 2 916 cm^{-1}; and O-H stretching vibration at 3 406 cm^{-1} (Figure 5-31-2 A)[34]. FT-IR spectra of the GS-Chi combination presented all the characteristic bands of GS and Chi, except that the characteristic absorption bands of GS-Chi had different intensities from those of GS and Chi, which indicates that the surfaces of GS were successfully modified with Chi.

Moreover, Raman spectroscopy was employed to confirm the successful synthesis of GS-Chi. As shown in Figure 5-31-2 B, the Raman spectrum obtained from GS showed two prominent bands: one was the D-band at 1 349 cm^{-1} due to stretching of sp^3 carbons, and the other was the G-band at 1 582 cm^{-1} due to stretching of sp^2 carbons. These two peaks were also observed in GS-Chi[35], revealing that GS-Chi still retains the basic structure of GS. In addition, previous studies indicate that the D/G-band ratio changes in the presence of oxygenated groups at the upper and lower surfaces, as well as the edges of sheets. Here, the D/G-band ratios of GS and GS-Chi were 1.086 and 1.095, respectively. This difference might be due to the increased number of oxygenated groups at the upper and lower surfaces of GS after it was modified with Chi. In addition, the 2D band is usually used to confirm the presence of a few layers of GS. The value of the 2D band decreased as the number of layers GS increased. Here, the value of the 2D band of GS was slightly higher than the value of the 2D band of GS-Chi, which clearly indicates that Chi was successfully adsorbed onto the GS surface.

Figure 5-31-2 C shows the structural features of GS-Chi. From the TEM image, GS-Chi, which was present as thin, wrinkled, rippled and flake-like structures, was readily observed. A TEM image of Chi-Au/Pt is shown in Figure 5-31-2 D. The obtained Chi-Au/Pt had a uniform globular morphology. Figure 5-31-2 E shows the TEM image of GS-Chi-Au/Pt. Clearly, Au/Pt was bound to the GS. Here, Au^{3+} and Pt^{2+} ions were adsorbed by GS-Chi from aqueous solutions via chelation and then reduced to Au/Pt nanoparticles by Chi. GS-Chi-Au/Pt was formed, and Chi was used as a stabilizing agent and reducing agent. However, we found that the Au/Pt nanoparticles immobilized on the surface of GS-Chi became larger than the Au/Pt nanoparticles dispersed in the Chi solution. The potential explanation is that Chi agglomerated when a large amount Chi was adsorbed to the surface of GS, and the agglomeration of Chi would cause the Au/Pt nanoparticles to agglomerate. Figure 5-31-2 F shows the HRTEM image of Chi-Au/Pt, and Figure 5-31-2 G-I shows the TEM elemental mapping

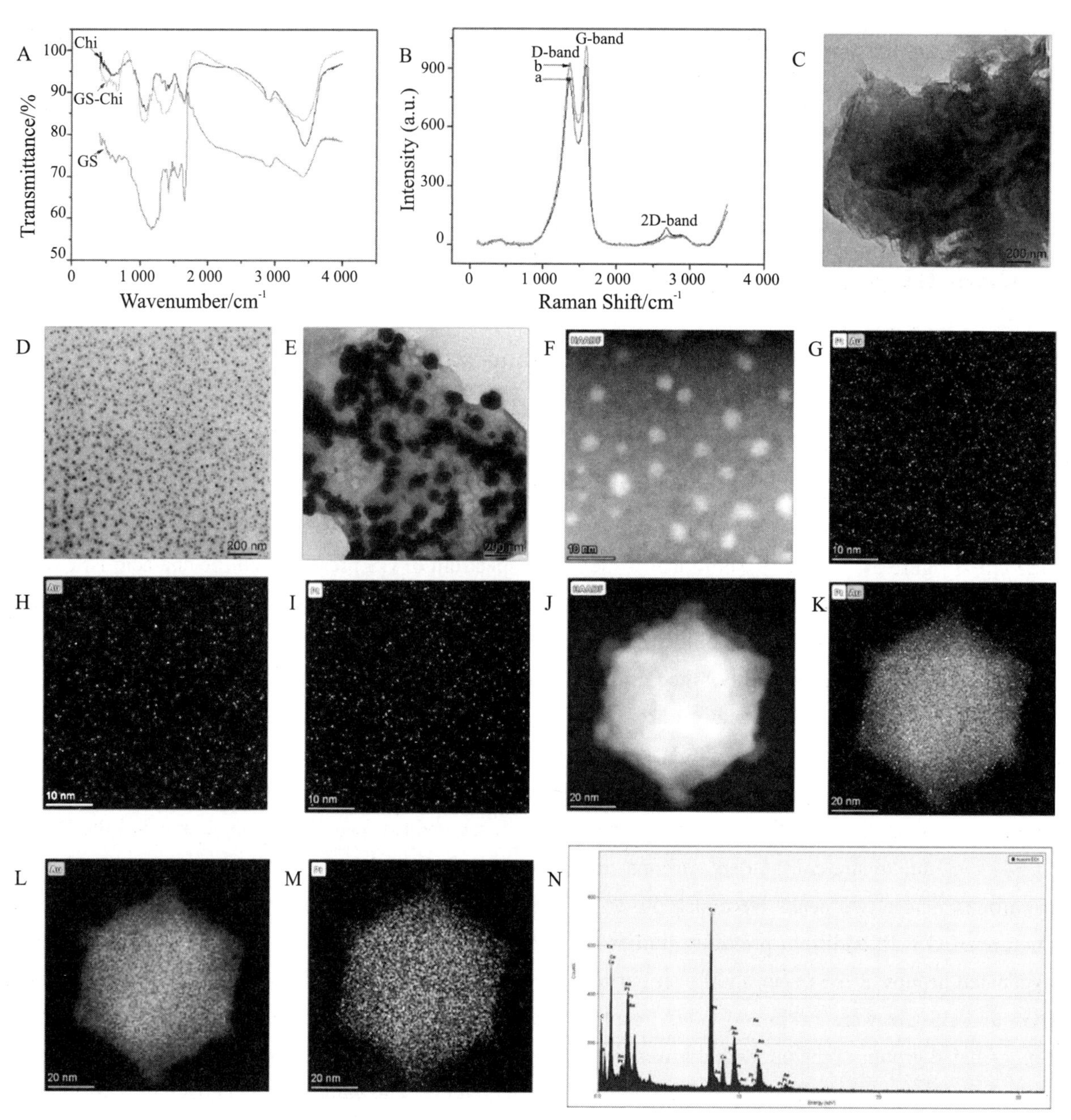

A: FT-R spectra of Chi, GS, and GS-Chi; B: Raman spectra of GS and GS-Chi; C: TEM image of GS-Chi; D: TEM image of Chi-Au/Pt; E: TEM image of GS-Chi-Au/Pt; F: HRTEM image of Chi-Au/Pt; G: Overlay on the image of Chi-Au/Pt; H: Au elemental mapping image of Chi-Au/Pt; I: Pt elemental mapping image of Chi-Au/Pt; J: GS-Chi-Au/Pt; K: Overlay on an image of GS-Chi-Au/Pt; L: Au elemental mapping image of GS-Chi-Au/Pt; M: Pt elemental mapping image of GS-Chi-Au/Pt; N: EDS elemental analysis of GS-Chi-Au/Pt (color figure in appendix).

Figure 5-31-2 Characterization of the related nanocomposites

image of Chi-Au/Pt. The diameter of Au/Pt was approximately 1～5 nm, and the Au and Pt elements were distributed uniformly randomly and relatively independently. Figure 5-31-2 J shows the HRTEM image of GS-Chi-Au/Pt, and Figure 5-31-2 K-M shows the TEM elemental mapping image of GS-Chi-Au/Pt. The diameter of Au/Pt was approximately 60 nm, and Au and Pt elements were distributed uniformly randomly and centralized. In addition, the EDS spectrum of GS-Chi-Au/Pt confirmed the presence of Chi, Au, and Pt elements, indicating that Au/Pt had been loaded on the surface of GS. Cu was observed because the samples were fixed on the Cu network for testing.

Electrochemical characterization of the immunosensor

Chronocoulometry was employed to investigate the effective surface areas of the bare GCE and GS-Chi-GCE (Figure 5-31-3). The effective surface areas of the electrode were calculated using Eq. (1), which was devised by Anson [36] .

$$Q(t)=2nFAcD^{1/2}t^{1/2}/\pi^{1/2}+Q_{dl}+Q_{ads} \tag{1}$$

Here, 0.1 mM $K_3[Fe(CN)_6]$ was used as a model complex, and the diffusion coefficient D was 7.6×10^{-6} cm^2/s. In Eq. (1), A is the effective surface area of the electrode, Q_{dl} is the double layer charge, n, F and c have their usual meanings, and Q_{ads} is the faradic charge. The effective surface areas of the bare GCE and GS-Chi-GCE were calculated to be 0.087 cm^2 and 0.301 cm^2 from the slope of the Q versus $t^{1/2}$ curve (Figure 5-31-3 B). The effective surface area of GS-Chi-GCE was 3.5 times larger than that of the bare GCE. The effective surface area of GCE increased substantially after modification with GS-Chi.

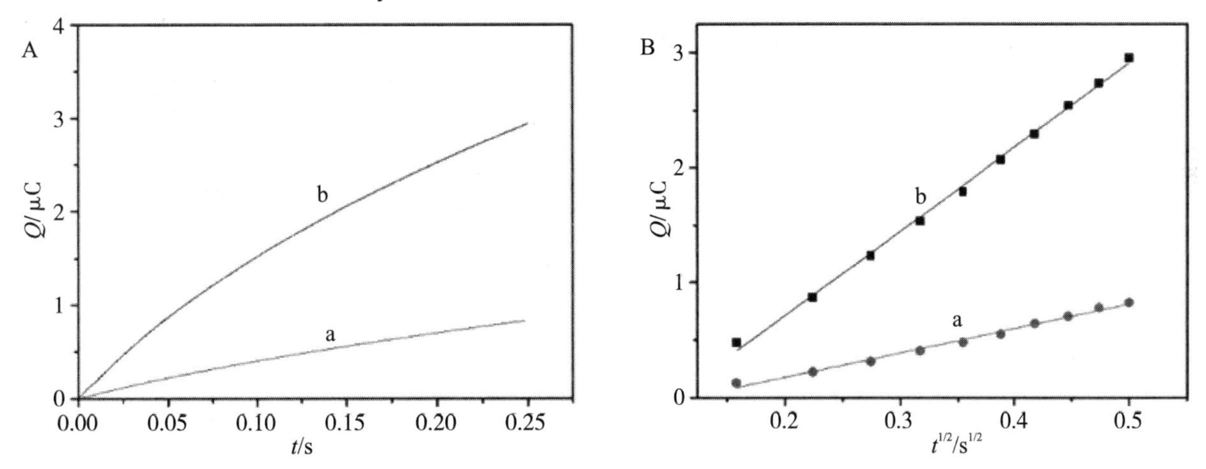

A: Plot of Q-t curve of the GCE (a) and GS-Chi-GCE (b) in 0.1 mM $K_3[Fe(CN)_6]$ containing 1.0 M KCl; B: Corresponding plot of Q-$t^{1/2}$ on GCE (a) and GS-Chi-GCE (b).

Figure 5-31-3 Electrochemical characterization of the immunosensor

In this study, the step-by-step assembly process of the GCE was investigated by performing EIS in an electrolyte (pH 7.0) containing 5 mM $[Fe(CN)_6]^{3-/4-}$ and recorded from 0.01 Hz to 100 kHz using an amplitude of 10 mV at 0.2 V. Figure 5-31-4 A shows Nyquist plots of EIS during electrode modification. In the Nyquist plots, the linear portion occurs at low frequencies and is related to electrochemical behaviour limited by diffusion. The semicircular portion occurs at high frequencies, and the semicircle diameter corresponds to the electron transfer resistance (R_{et}) [37, 38] . The results are shown in Figure 5-31-4 A and B. The bare GCE exhibited a perfect semicircle (Figure 5-31-4 A, curve a) with a R_{et} of 1 141 Ω (%RSD) =2.97 (Figure 5-31-4 B, columnar a), while a very small semicircle (Figure 5-31-4 A, curve b) with a R_{et} of 103 Ω (%RSD) =1.87 (Figure 5-31-4B, columnar b) was observed after modification with GS-Chi because of the large electron transfer-promoting effect and excellent conductivity of GS. When AIV H9/MAbs, BSA and AIV H9 ($10^{6.37}$ EID_{50}/mL) were successively loaded onto the modified electrode, and the relevant R_{et} increased sequentially (1 526 Ω (%RSD) =2.73, 1 934Ω (%RSD) =3.21 and 3 434 Ω (%RSD) =3.21, respectively) (Figure 5-31-4 B, columns c-e) because the resistance increased due to the poor conducting power of AIV H9/MAbs, BSA and AIV H9. These results revealed the successful construction of the modified electrode. The electron transfer rate constant (K_e) was successfully obtained from Eq. (2), which shows the relationship between R_{et} and K_{et} [39] , as follows:

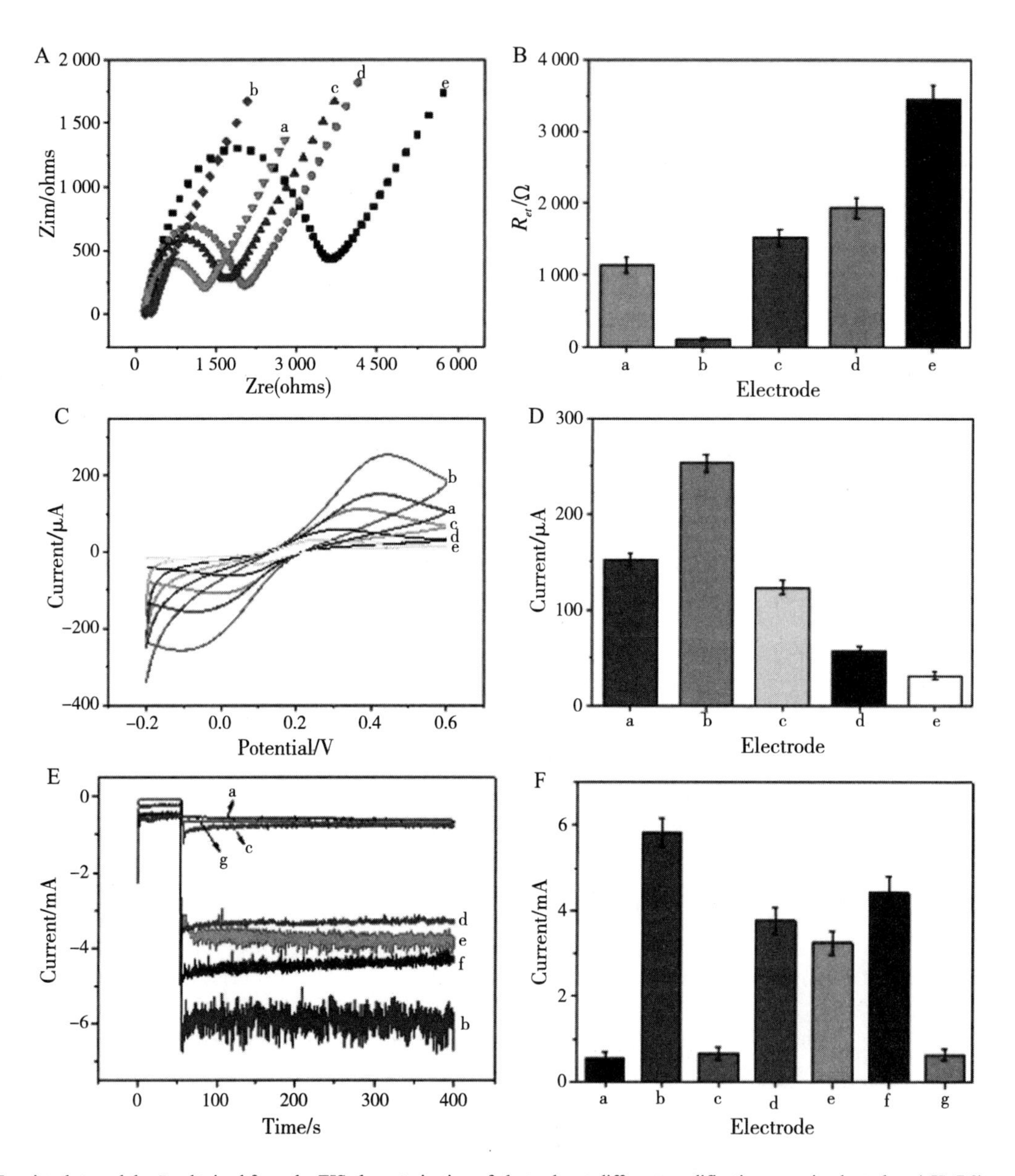

A and B: Nyquist plots and the R_{et} obtained from the EIS characterization of electrodes at different modification steps in electrolyte (pH=7.0) containing 5 mM $[Fe(CN)_6]^{3-/4-}$: (a) GCE, (b) GS-Chi-GCE, (c) AIV H9/MAbs-GS-Chi-GCE, (d) BSA-AIV H9/MAbs-GS-Chi-GCE, (e) AIV H9-BSA-AIV H9/MAbs-GS-Chi-GCE.

C and D: CV of (a) GCE, (b) GS-Chi-GCE, (c) AIV H9/MAbs-GS-Chi-GCE, (d) BSA-AIV H9/MAbs-GS-Chi-GCE, and (e) AIV H9-BSA-AIV H9/MAbs-GS-Chi-GCE.

E and F: Amperometric i-t curves and measured responses of the immunosensor for the detection of $10^{6.37}$ EID_{50}/mL AIV H9 (a) without labels; (b) with GS-Chi-Au/Pt-AIV H9/PAbs, (c) GS-Chi-AIV H9/PAbs, (d) GS-Chi-Au-AIV H9/PAbs, (e) Chi-Au/Pt-AIV H9/PAbs and (f) GS-Chi-Au/Pt-AIV H9/PAbs as labels; and (g) without incubation with AIV H9 but with GS-Chi-Au/Pt-AIV H9/PAbs as labels at −0.4 V towards the reduction of 10 mM H_2O_2 in electrolyte at pH=7.0.

Figure 5-31-4 Electrochemical characterization of the electrodes

$$K_{et}=\frac{RT}{n^2F^2R_{et}AC_{\text{redox}}} \tag{2}$$

A corresponds to the geometrical area of the electrode surface, while redox is the concentration of $[Fe(CN)_6]^{3-/4-}$. The K_{et} values were calculated to be 6.62×10^{-5} cm/s, 73.69×10^{-5} cm/s, 4.92×10^{-5} cm/s, 3.87×10^{-5} cm/s and 2.18×10^{-5} cm/s for GCE, GS-Chi-GCE, AIV H9/MAbs-GS-Chi-GCE, BSA-AIV H9/ MAbs-GS-Chi-GCE and AIV H9-BSA-AIV H9/MAbs-GS-Chi-GCE, respectively.

In addition, cyclic voltammetry (CV) is an effective technique for evaluating the successful modification of electrodes. The results are shown in Figure 5-31-4 C and D. The CV curve of the GCE electrode showed a reversible redox label of $[Fe(CN)_6]^{3-/4-}$, while the peak current of GS-Chi-GCE increased substantially due to the large electron transfer-promoting effect and excellent conductivity of GS. After AIV H9/MAbs, BSA and $10^{6.37}$ EID_{50}/mL AIV H9 were immobilized on the modified electrode, the relevant peak current decreased in turn. The CV results were consistent with the EIS results, indicating the successful construction of the AIV H9 immunosensor.

Amperometric i-t curve measurements were employed to investigate the electrocatalytic performance of the proposed immunosensor. The results are shown in Figure 5-31-4 E and F. The electrocatalytic activity of GS-Chi (curve a) was ignored because it was weaker than that of GS-Chi-Au/Pt (curve b). During the immunosensor detection process, the electrocatalytic performance between GS-Chi-H9/PAbs, GS-Chi-Au-H9/PAbs, Chi-Au/Pt-H9/PAbs, GS-Chi-Pt-H9/PAbs and GS-Chi-Au/Pt-H9/PAbs as labels was compared. The immunosensors from the same batch were incubated with $10^{6.37}$ EID_{50}/mL AIV H9 and then incubated with the two different types of labels. Finally, amperometric i-t measurements were performed in electrolyte (pH 7.0) with 10 mmol/L H_2O_2. As expected, the immunosensor using GS-Chi-Au/Pt-H9/PAbs (curve b) as the label displayed a much higher amperometric change than that using GS-Chi-H9/PAbs (curve c), GS-Chi-Au-H9/PAbs (curve d), Chi-Au/Pt-H9/PAbs (curve e) and GS-Chi-Pt-H9/PAbs (curve f) as the labels. The high sensitivity was mainly attributed to the large specific surface area, enhanced electrical conductivity and fast electron transport of GS and the excellent catalytic activity of Au/Pt. When no AIV H9 was immobilized on the electrode, the peak current was similar to curve a after the immunosensor was reacted with GS-Chi-Au/Pt-H9/PAbs (curve g), indicating that our developed immunosensor has good selectivity.

Optimization of the method

The following parameters were optimized: (a) GS-Chi concentration; (b) AIV H9/MAb concentration; (c) AIV H9/PAb concentration; (d) ratio of Au, Pt and GS; (e) pH value of the electrolyte; (f) H_2O_2 concentration; (g) incubation time of AIV H9; and (h) incubation time of GS-Chi-Au/Pt-AIV H9/PAbs bioconjugates.

The concentration of GS-Chi is a critical factor because it plays an important role in immobilization of AIV H9/MAbs and enhancing electron transfer; in other words, the concentration of GS-Chi will affect the amperometric response. Therefore, optimization of the concentration of GS-Chi is necessary. Figure 5-31-5 A shows currents of the amperometric i-t curve when the GCE was modified with different concentrations of GS-Chi to detect $10^{6.37}$ EID_{50}/mL AIV H9 using GS-Chi-Au/Pt-AIV H9/PAbs as labels in electrolyte (pH=7.0) containing 10 mM H_2O_2. As shown in this figure, a gradual increase in GS-Chi concentrations from 0.25 to 1.0 mg/mL resulted in remarkable increases in the current response. During this period, the GS-Chi increased, resulting in an abundance of AIV H9/MAbs anchoring sites on the surface of the electrode and a gradual increase in current responses. Nevertheless, the current response signal decreased as the GS-Chi concentration was increased from 1 to 2 mg/mL. These results indicated that 1.0 mg/mL GS-Chi was an optimal condition. The current changes indicated that a suitable concentration of GS-Chi can efficiently enhance the conductivity, while excessive GS-Chi was loaded on the GCE as a substrate, which increased the interface electron transfer resistance of the electrode.

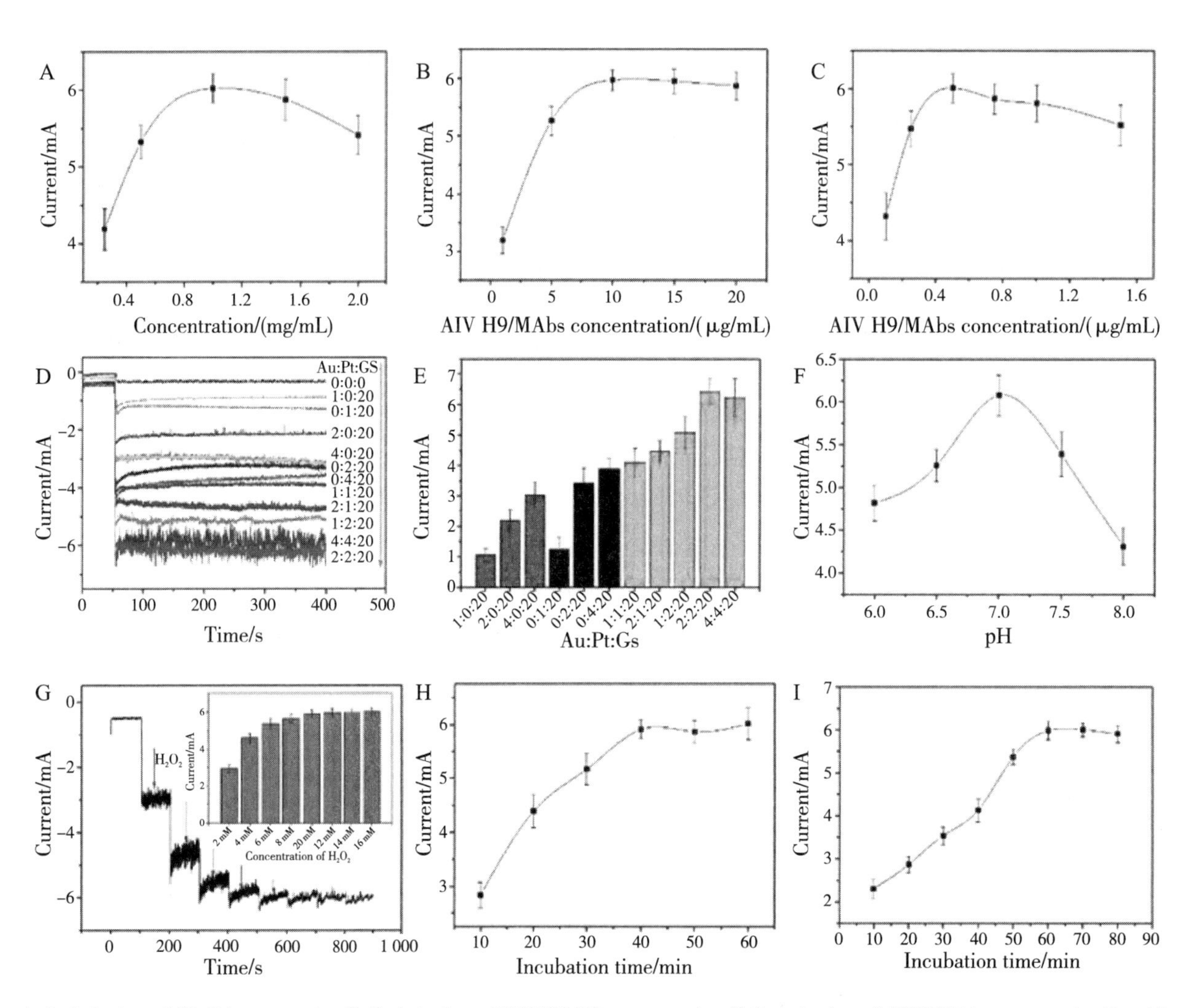

A: Optimizations of GS-Chi concentration; B: Optimizations of AIV H9/MAbs concentration; C: Optimizations of AIV H9/PAbs concentration; D and E: Optimizations of ratio of Au, Pt and GS; F: Optimizations of pH values of electrolyte; G: Optimizations of the concentration of H_2O_2; H: Optimizations of incubation time of AIV H9; I: Optimizations of incubation time of GS-Chi-Au/Pt-AIV H9/PAbs bioconjugates.

Figure 5-31-5 Optimization of parameters

Concentrations ranging from 1 to 20 μg/mL and 0.1 to 1.5 μg/mL were investigated for AIV H9/MAbs and AIV H9/PAbs, respectively. The current response gradually increased with increasing AIV H9/MAbs (Figure 5-31-5 B) and AIV H9/PAbs concentrations (Figure 5-31-5 C) and then reached maximum values at 10 μg/mL and 0.5 μg/mL, respectively. The plateau signal occurred because of over saturation of the electrode surface, resulting in a limited number of conjugated AIV H9/MAbs/AIV H9/AIV H9/PAbs complexes that could be formed on the surface of the electrode. Thus, 10 and 0.5 μg/mL were the optimal concentrations for AIV H9/MAbs and AIV H9/PAbs, respectively.

Figure 5-31-5 D and E display the effect of the ratio of Au, Pt and GS (Au:Pt:GS) in GS-Chi-Au/Pt-AIV H9/PAbs on the current responses of the electrochemical immunosensor. The results showed that the signal responses increased with the ratio of Au, Pt and GS ranging from 1 : 0 : 20 to 4 : 0 : 20 in the absence of Pt and 0 : 1 : 20 to 0 : 4 : 20 in the absence of Au. In the presence of Au and Pt, the signal responses increased with the ratio of Au, Pt and GS in the range from 1 : 1 : 20 to 2 : 2 : 20, and then the signal responses levelled off. In addition, Au and Pt together exhibit much higher catalytic ability than their monometallics because of the synergistic effect. Therefore, the ratio (2 : 2 : 20) of Au (0.1 mg/mL), Pt (0.1 mg/mL) and GS (1 mg/mL) was chosen as the optimized signal label composition in this study.

The pH value of the electrolyte is an important factor for electrochemical immunosensors. Figure 5-31-5 F shows the effect of the pH value in the electrolyte on the current responses. The current responses were enhanced with increasing pH values from 6.0 to 7.0 and then decreased when the pH value was above 7.0. The optimal amperometric response was achieved at pH=7.0. The reason for the fluctuation of the signal response may be that alkaline or acidic environments disrupt the stability of antibody-antigen binding. Thus, pH=7.0 in the electrolyte was selected for use in subsequent experiments.

Figure 5-31-5 G shows the result of optimization of the concentration of H_2O_2. As shown in this figure, the current responses were enhanced when the H_2O_2 concentration was increased from 1 to 10 mM, and then the current responses levelled off after 10 mM. This is because 10 mM H_2O_2 reaches saturation, and excessive H_2O_2 will not join in the catalytic reaction. Therefore, the optimal concentration of H_2O_2 was 10 mM. In addition, the incubation time is a great influencing factor for both the antigen and the antibody. As shown in Figure 5-31-5 H and I, the current responses gradually increased with the incubation time of AIV H9 and GS-Chi-Au/Pt-AIV H9/PAbs until the immunoreaction finished at 40 min and 60 min, respectively. The results indicated that the immunoreaction of AIV H9/MAbs with AIV H9 will be finished after 40 min, and the immunoreaction of AIV H9 with GS-Chi-Au/Pt-AIV H9/PAbs will be finished after 40 min. Therefore, 40 min and 60 min were chosen as the optimal incubation time points for AIV H9 and GS-Chi-Au/Pt-AIV H9/PAbs, respectively.

Detection of AIV H9 with the immunosensor

Under the optimized reaction conditions, the detection capability of the proposed electrochemical immunosensor was determined by constructing an amperometric i-t curve at an initial potential of –0.4 V in 10 mL of electrolyte (pH 7.0) containing 10 mM H_2O_2 to detect AIV H9 at a concentration ranging from $10^{6.37}$ EID_{50}/mL to $10^{1.37}$ EID_{50}/mL. As shown in Figure 5-31-6 A, the current signal responses of the electrochemical immunosensor were enhanced with increasing concentrations of AIV H9. The difference in current signal intensity between the blank (curve a) and the AIV H9 curve indicated that the nonspecific adsorption on the electrochemical immunosensor and the effects of the background current of GS-Chi were negligible. The calibration plot showed a good linear relationship between the change in the current signal and the logarithm values of the AIV H9 concentration (Figure 5-31-6 B). The linear regression equation of the calibration curve was I (mA) =0.951 7 lgEID_{50} mL^{-1}+0.193 6, with a correlation coefficient of R=0.993 2 and a low detection limit of $10^{0.82}$ EID_{50}/mL(S/N=3). In addition, the analytical performance of this immunosensor was superior to that of the previous methods developed for the detection of AIV H9 (Table 5-31-2). The reasons are attributable to the following aspects: (1) the GS-Chi-Au/Pt nanocomposites used as signal labels had superior electrocatalytic capability towards the reduction of H_2O_2 because the planar electric transport properties were enhanced by electron-electron correlation assistance, and (2) GS-Chi used as a matrix not only had remarkable electroconductivity that promoted electron transfer towards the surface of the electrode but also provided a large specific surface area to firmly bind AIV H9/PAbs, which improved the antigen-antibody reaction to detect AIV H9.

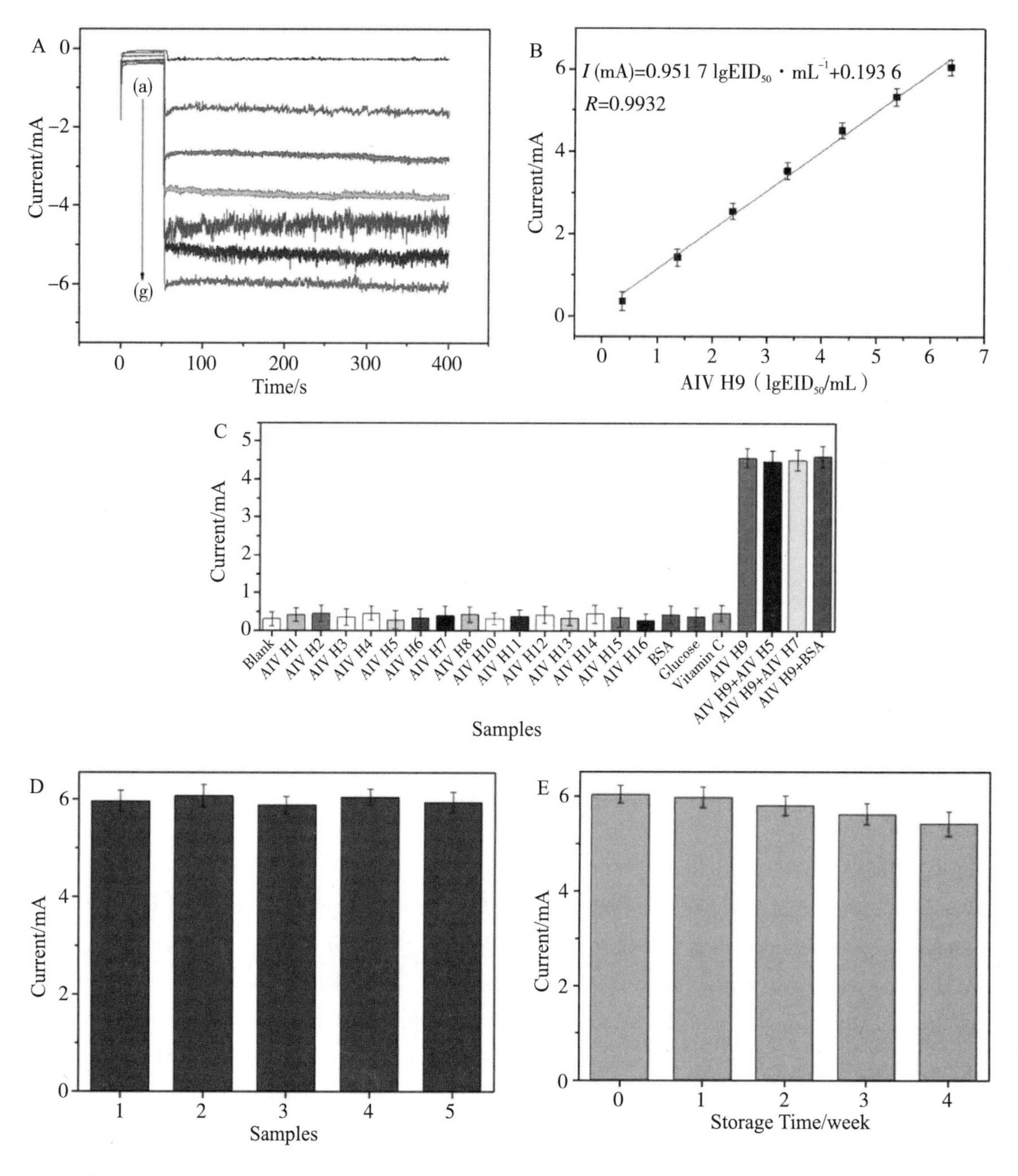

A: Amperometric i-t current response of the immunosensor towards the addition of 10 mM H_2O_2, for detecting the following different concentrations of AV H9. a: $10^{0.37}$ EID$_{50}$/mL; b: $10^{1.37}$ EID$_{50}$/mL; c: $10^{2.37}$ EID$_{50}$/mL; d: $10^{3.37}$ EID$_{50}$/mL; e: $10^{4.37}$ EID$_{50}$/mL; f: $10^{5.37}$ EID$_{50}$/mL; g: $10^{6.37}$ EID$_{50}$/mL. B: Calibration curve of the immunosensor response to different concentrations of AIV H9. C: Specificity of the immunosensor towards the target AIV H9 and other interfering substances. D: Reproducibility of the developed immunosensor in the presence of $10^{6.37}$ EID$_{50}$/mL. E: Long term stability of the proposed immunosensor.

Figure 5-31-6 Sensitivity, selectivity, reproducibility and stability of the immunosensor

Table 5-31-2 Comparison of the proposed immunosensor with other methods for AIV H9 detection

Method	Detection time	Detection limit	References
Virus isolation	4 days	1 EID_{50}/mL	[6]
RT-PCR	6 h	100 EID_{50}/mL	[7]
Real-time RT-PCR	4 h	1 EID_{50} /mL	[8]
RT-LAMP	3 h	10 copies per reaction	[9]
ELISA	2 h	$10^{-2.3}$ $TCID_{50}$	[10]
Proposed immunosensor	1.5 h	$10^{0.82}$ EID_{50}/mL	This study

Note: $TCID_{50}$, 50% Tissue culture infective dose.

Selectivity, reproducibility and stability of the immunosensor

The selectivity of the immunosensor played a crucial role in detecting target samples without separation. Several interfering substances, including AIV subtype H1 ($10^{5.61}$ EID_{50}/mL), H2 ($10^{6.29}$ EID_{50}/mL), H3 ($10^{7.43}$ EID_{50}/mL), H4 ($10^{5.45}$ EID_{50}/mL), H5 (128 HAUs), H6 ($10^{6.19}$ EID_{50}/mL), H7 (256 HAUs), H8 ($10^{6.23}$ EID_{50}/mL), H10 ($10^{5.73}$ EID_{50}/mL), H11 ($10^{6.47}$ EID_{50}/mL), H12 ($10^{7.12}$ EID_{50}/mL), H13 ($10^{5.37}$ EID_{50}/mL), H14 ($10^{6.72}$ EID_{50}/mL), H15 ($10^{5.81}$ EID_{50}/mL), H16 ($10^{5.63}$ EID_{50}/mL), BSA (1.0 μg/mL), glucose (1.0 μg/mL) and vitamin C (1.0 μg/mL), were used in this work to evaluate the selectivity of the developed immunosensor. The results are shown in Figure 5-31-6 C. The current responses of the mixtures of AIV H9 ($10^{4.37}$ EID_{50}/mL) with other possible interfering substances were similar to those of AIV H9 ($10^{4.3}$ EID_{50}/mL) alone, while nonspecific substances and blank solution showed similar current responses, indicating that the proposed electrochemical immunosensor had high selectivity for AIV H9 detection.

The reproducibility of the proposed electrochemical immunosensor was investigated by recording the current responses in the presence of the same concentrations of $10^{6.37}$ EID_{50}/mL AIV H9. The results are shown in Figure 5-31-6 D. The RSD of the intra-assay reproducibility of the immunosensor, which was 1.6%, is shown as an error bar and indicated that the developed immunosensor has good reproducibility.

The long-term stability of the fabricated electrochemical immunosensor was further investigated by storing BSA-AIV H9/MAbs-GS-Chi-GCE at 4 ℃ when not in use and then successively incubating it with AIV H9 ($10^{6.37}$ EID_{50}/mL) and GS-Chi-Au/Pt-AIV H9/PAbs bioconjugates. As shown in Figure 5-31-6 E, the current responses decreased by less than 4% after 2 weeks of storage. The current responses eventually retained 89.72% of the initial value. Based on the results, the long-term stability of the proposed electrochemical immunosensor for AIV H9 detection was acceptable.

Analysis of AIV H9 in clinical samples

Ninety-eight clinical samples were obtained from chickens with the permission of the host of the live poultry markets, and clinical samples were detected using the developed immunosensor. Three AIV H9-positive samples were identified. The results were confirmed by virus isolation. The results are shown in Table 5-31-3 a and b. All the results obtained from the clinical samples detected using the proposed immunosensor were consistent with the results obtained through virus isolation[40].

Standard spiking was also employed to evaluate the accuracy and practical applicability of the developed immunosensor. A series of different concentrations of AIV H9 (100.00, 150.00, 200.00, 250.00, 300.00 and 400.00 EID_{50}/mL) was added to the three AIV H9-positive samples (from the aforementioned clinical samples that were identified as the AIV H9-positive sample), followed by detection using the fabricated electrochemical immunosensor according to the steps listed above in *Electrochemical Measurement* section. The recovery was calculated as the ratio between the Found and Added AIV H9. The results are shown in Table 5-31-3 c. The recoveries were obtained from 96.02% to 104.63% with an RSD of 2.17%~4.18% (n=5), indicating that the developed immunosensor was feasible for AIV H9 detection in clinical samples.

Table 5-31-3 Results from clinical samples (a), results from the analysis of positive samples (b), and results for the recovery of different concentrations of AIV H9 from clinical samples (c)

(a)

Method	Total number of samples	Number of positive samples	Positivity rate/%
Proposed immunosensor	98	3	3.06
Virus isolation	98	3	3.06

(b)

No	Results from the proposed immunosensor			Results of virus isolation
	Measured concentration/(EID_{50}/mL)	Average /(EID_{50}/mL)	RSD/% , n=5	
1	60.24, 57.31, 62.17, 63.54, 64.49	61.55	4.63	Positive
2	141.25, 137.86, 143.74, 136.47, 142.18	140.3	2.17	Positive
3	192.67, 188.34, 193.28, 187.59, 195.73	191.52	1.8	Positive

(c)

No	Initial AIV H9 concentration in the sample /(EID_{50}/mL)	Added amount of AIV H9 / (EID_{50}/mL)	Total found		Recovery rate /%, n=5
			Average/(EID_{50}/mL)	RSD/%, n=5	
1	61.55	100.00	169.03	3.47	104.63
2	61.55	150.00	206.93	2.73	97.82
3	140.3	200.00	326.74	4.07	96.02
4	140.3	250.00	381.37	2.17	97.71
5	191.52	300.00	497.31	4.18	101.18
6	191.52	400.00	584.62	3.75	98.83

Conclusions

In this study, we synthesized Au/Pt nanoparticle-functionalized GS (GS-Chi-Au/Pt) with high electrocatalytic activity towards H_2O_2 reduction. We have developed an enzyme-free sandwich electrochemical immunosensor utilizing GS-Chi as the matrix platform and GS-Chi-Au/Pt as the signal amplification label. This assay system is capable of detecting AIV H9 in clinical samples. Benefiting from the favourable cooperation of GS-Chi-Au/Pt with good H_2O_2 catalysis, ideal conductivity of GS-Chi and efficient antibody immobilization, the established immunosensor exhibited good sensitivity, specificity, reproducibility, accuracy and stability. Importantly, this developed method not only expands the application of GS-Chi-Au/Pt in electrochemical biosensors but also provides a novel nonenzymatic method for the accurate determination of other biomolecules in clinical diagnosis.

References

[1] VENKATESH D, POEN M J, BESTEBROER T M, et al. Avian influenza viruses in wild birds. virus evolution in a multihost

ecosystem. J Virol, 2018, 92e00433.

[2] CHENG K L, WU J, SHEN W L, et al. Avian influenza virus detection rates in poultry and environment at live poultry markets, Guangdong, China. Emerg Infect Dis, 2020, 26: 591-595.

[3] SWIETON E, OLSZEWSKA M, GIZA A, et al. Evolution of H9N2 low pathogenic avian influenza virus during passages in chickens. Infect Genet Evol, 2019, 75: 103979.

[4] LI W, YANG Y, MA C, et al. A sandwich-type electrochemical immunosensor for ultrasensitive detection of multiple tumor markers using an electrical signal difference strategy. Talanta, 2020, 219: 121322.

[5] XU L, LIU Z, LEI S, et al. A sandwich-type electrochemical aptasensor for the carcinoembryonic antigen via biocatalytic precipitation amplification and by using gold nanoparticle composites. Microchim Acta, 2019, 186 (7): 473.

[6] SHEN C, WANG L, ZHANG H, et al. An electrochemical sandwich immunosensor based on signal amplification technique for the determination of alpha-fetoprotein. Front Chem,2020, 8: 589560.

[7] ZHENG S, LI M, LI H, et al. Sandwich-type electrochemical immunosensor for carcinoembryonic antigen detection based on the cooperation of a gold-vertical graphene electrode and gold@ silica-methylene blue. J Mater Chem B, 2020, 8: 298-307.

[8] TANG D, SU B, TANG J, et al. Nanoparticle-based sandwich electrochemical immunoassay for carbohydrate antigen 125 with signal enhancement using enzyme-coated nanometer-sized enzyme-doped silica beads. Anal Chem, 2010, 82: 1527-1534.

[9] YU W, SANG Y, WANG T, et al. Electrochemical immunosensor based on carboxylated single-walled carbon nanotube-chitosan functional layer for the detection of cephalexin. Food Sai Nutr, 2020, 8: 1001-1011.

[10] ZHANG Q, LI L, QIAO Z, et al. Electrochemical conversion of Fe_3O_4 magnetic nanoparticles to electroactive Prussian blue analogues for self-sacrificial label biosensing of avian influenza virus H5N1. Anal Chem, 2017, 89 (22): 12145-12151.

[11] XIAO J, HU X, WANG K, et al. A novel signal amplification strategy based on the competitive reaction between 2D Cu-TCPP (Fe) and polyethyleneimine (PEI) in the application of an enzyme-free and ultrasensitive electrochemical immunosensor for sulfonamide detection. Biosens Bioelectron, 20201, 50: 111883.

[12] SHEN W J, ZHUO Y, CHAI Y Q, et al. Enzyme-free electrochemical immunosensor based on host-guest nanonets catalyzing amplification for procaleitonin detection. ACS Appl Mater Interfaces, 2015, 7: 4127-4134.

[13] YANG Q, WANG P, MA E, et al. A sandwich-type electrochemical immunosensor based on Au@ Pd nanodendrite functionalized MoO_2 nanosheet for highly sensitive detection of HBsAg. Bioelectrochemistry, 2021, 138: 107713.

[14] ZHAO L, LI S, HE J, et al. Enzyme-free electrochemical immunosensor configured with Au-Pd nanocrystals and N-doped graphene sheets for sensitive detection of AFP. Biosens Bioelectron, 2013, 49: 222-225.

[15] JIN L, MENG Z, ZHANG Y, et al. Ultrasmall Pt nanoclusters as robust peroxidase mimics for colorimetric detection of glucose in human serum. ACS Appl Mater Interfaces, 20179, (11): 10027-10033.

[16] SAXENA R, FOUAD H, SRIVASTAVA S. Gold nanoparticle based electrochemical immunosensor for detection of T3 hormone. J Nanosci Nanotechnol, 2020, 20: 6057-6062.

[17] LI Y, ZHANG Y, LI F, et al. Ultrasensitive electrochemical immunosensor for quantitative detection of SCCA using Co_3O_4@ CeO_2-AuaPt nanocomposite as enzyme-mimetic labels. Biosens Bioelectron, 2017, 92: 33-39.

[18] LV H, LI Y, ZHANG X, et al. Enhanced peroxidase-like properties of Au@Pt DNs/NG/$Cu^{2+}$and application of sandwich-type electrochemical immunosensor for highly sensitive detection of CEA. Biosens Bioelectron, 2018, 112: 1-7.

[19] ZHANG F, YANG K, LIU G, et al. Recent advances on graphene: synthesis, properties and applications. Compos Part A Appl S1, 2022, 60: 107051.

[20] ATAR N, YOLA M L. A novel QCM immunosensor development based on gold nanoparticles functionalized sulfur-doped graphene quantum dot and h-ZnS-CdS NC for Interleukin-6 detection. Anal Chim Acta, 2021, 1148: 338202.

[21] KARAMAN C, KARAMAN O, ATAR N, et al. Tailoring of cobalt phosphide anchored nitrogen and sulfur co-doped three dimensional graphene hybrid: boosted electrocatalytic performance towards hydrogen evolution reaction. Electrochim Acta, 2021, 380: 138262.

[22] KARAMAN C, KARAMAN O, YOLA B B, et al. A novel electrochemical aflatoxin B1 immunosensor based on gold nanoparticledecorated porous graphene nanoribbon and Ag nanocube-incorporated MoS2 nanosheets New J Chem, 2021, 45: 11222-11233.

[23] AKGA A, KARAMAN O, KARAMAN C, et al. A comparative study of CO catalytic oxidation on the single vacancy and di-vacancy graphene supported single-atom iridium catalysts: A DFT analysis. Surf Nterfaces, 2021, 25: 101293.

[24] KARAMAN O, OZCAN N, KARAMAN C, et al. Electrochemical cardiac troponin l immunosensor based on nitrogen and boron-doped graphene quantum dots electrode platform and Ce-doped SnO_2/SnS_2 signal amplification. Mater Today Chem, 2022, 23: 100666.

[25] KARAMAN C, BÖLÜKBASI Ö S, YOLA B B, et al. Electrochemical neuron-specific enolase (NSE) immunosensor based on $CoFe_2O_4$@Ag nanocomposite and AuNPs@MoS_2/rGO. Anal Chim Acta, 2022, 1200: 339609.

[26] YEN P I, SAHOO S K, CHIANG Y C, et al. Using different ions to tune graphene stack structures from sheet-to onion-1ike during plasma exfoliation, with supercapacitor applications. Nanoscale Res Lett, 2019, 14: 141.

[27] HUANG J L, XIE Z X, HUANG Y H, et al. Electrochemical immunosensor with Cu(Ⅰ)/Cu(Ⅰ) -chitosan-graphene nanocomposite-based signal amplification for the detection of new castle disease virus. Sci Rep, 2020, 10: 13869.

[28] AYDIN E B, AYDIN M, SEZGINTÜRK M K. Electrochemical immunosensor based on chitosan/conductive carbon black composite modified disposable ITO electrode: an analytical platform forp53 detection. Biosens Bioelectron, 2018, 121: 80-89.

[29] SUN D, LI H, LI M, et al. Electrochemical immunosensors with Au/Pt-vertical graphene/glassy carbon electrode for alpha-fetoprotein detection based on label-free and sandwich-type strategies. Biosens Bioelectron, 2019, 132: 68-75.

[30] DENG X W, XIE Z X, XIE Z Q, et al. Preparation and identification of monoclonal antibodies against hemagglutinin of H9 subtype avian influenza virus. J South Agric, 2016, 47679-683.

[31] YU H, ZHANG B, BULIN C, et al. High-efficient synthesis of graphene oxide based on improved Hummers method. Sci Rep, 2016, 6: 36143.

[32] HUANG J L, XIE Z X, XIE Z Q, et al. Silver nanoparticles coated graphene electrochemical sensor for the ultrasensitive analysis of avian influenza virus H7. Anal Chim Acta, 2016, 913: 121-127.

[33] KUMAR S, KOH J, KIM H, et al. A new chitosanthymine conjugate: synthesis, characterization and bio-logical activity. Int J Biol Macromol, 2012, 50: 493-502.

[34] PANDEY S, KARAKOTI M, DHALI S, et al. Bulk synthesis of graphene nanosheets from plastic waste: an invincible method of solid waste management for better tomorrow. Waste Manag, 2019, 88: 48-55.

[35] KHAN Q A, SHAUR A, KHAN TA, et al. Characterization of reduced graphene oxide produced through a modified Hoffman method. Cogent Chem, 2017, 3: 1298980.

[36] ANSON F C. Application of potentiostatic current integration to the study of the adsorption of cobalt (Ⅱ) ethylenedinitrilo (tetraacetate) on mercury electrodes. Anal Chem, 1964, 36 (4): 932-934.

[37] SOARES C, TENREIRO MACHADO J A, LOPES A M, et al. Electrochemical impedance spectroscopy characterization of beverages. Food Chem, 2020, 302: 125345.

[38] MA F, YAN J, SUN L, et al. Electrochemical impedance spectroscopy for quantization of matrix Metalloproteinase-14 based on peptides inhibiting its homodimerization and heterodimerization. Talanta, 2019, 205: 120142.

[39] JAMPASA S, LAENGEE P PATARAKUL K, et al. Electrochemical immunosensor based on gold-labeled monoclonal anti-LipL32 for leptospirosis diagnosis. Biosens Bioelectron, 2019, 142: 111539.

[40] LUO S, XIE Z, XIE Z, et al. Surveillance of live poultry markets for low pathogenic avian influenza viruses in Guangxi Province, southern China, from 2012-2015. Sci Rep, 2017, 7(1):17577.

Explore how immobilization strategies affected immunosensor performance by comparing four methods for antibody immobilization on electrode surfaces

Huang Jiaoling, Xie Zhixun, Xie Liji, Luo Sisi, Zeng Tingting, Zhang Yanfang, Zhang Minxiu, Wang Sheng, Li Meng, Wei You,Fan Qing, Xie Zhiqin, Deng Xianwen, and Li Dan

Abstract

Among the common methods used for antibody immobilization on electrode surfaces, which is the best available option for immunosensor fabrication? To answer this question, we first used graphene-chitosan-Au/Pt nanoparticle (G-Chi-Au/PtNP) nanocomposites to modify a gold electrode (GE). Second, avian reovirus monoclonal antibody (ARV/MAb) was immobilized on the GE surface by using four common methods, which included glutaraldehyde (Glu), 1-ethyl-3-(3-dimethylaminopropyl)-carbodimide/N-hydroxysuccinimide (EDC/NHS), direct incubation or cysteamine hydrochloride (CH). Third, the electrodes were incubated with bovine serum albumin, four different avian reovirus (ARV) immunosensors were obtained. Last, the four ARV immunosensors were used to detect ARV. The results showed that the ARV immunosensors immobilized via Glu, EDC/NHS, direct incubation or CH showed detection limits of $10^{0.63}$ EID_{50}/mL, $10^{0.48}$ EID_{50}/mL, $10^{0.37}$ EID_{50}/mL and $10^{0.46}$ EID_{50}/mL ARV (S/N=3) and quantification limits of $10^{1.15}$ EID_{50}/mL, and $10^{1.00}$ EID_{50}/mL, $10^{0.89}$ EID_{50}/mL and $10^{0.98}$ EID_{50}/mL ARV (S/N=10), respectively, while the linear range of the immunosensor immobilized via CH ($0\sim10^{5.82}$ EID_{50}/mL ARV) was 10 times broader than that of the immunosensor immobilized via direct incubation ($0\sim10^{4.82}$ EID_{50}/mL ARV) and 100 times broader than those of the immunosensors immobilized via Glu ($0\sim10^{3.82}$ EID_{50}/mL ARV) or EDC/NHS ($0\sim10^{3.82}$ EID_{50}/mL ARV). And the four immunosensors showed excellent selectivity, reproducibility and stability.

Keywords

electrochemical immunosensor, immobilization strategies, graphene

---

Introduction

Electrochemical immunosensors are a novel detection method that are combined with electrochemical analysis and immunoassays and are widely used in the diagnosis of various pathogens due to their advantages of simplicity, rapidity, and high sensitivity in point-of-care testing[1, 2]. The sensitivity and linear range of electrochemical immunosensors are affected by the electrode surface modification and antibody immobilization strategy[3, 5]. Electrochemical immunosensors can be categorized as "label-free" and "sandwich" types[5, 6]. "Sandwich" immunosensors waste time and energy compared to the "label-free" type, in which the electrochemical signal molecule is immobilized on the electrode or dissolved in an electrolyte in only one step to detect the target[7, 8]. Therefore, it is more desirable to construct "label-free" electrochemical immunosensors.

Substrate materials with large specific surface areas and high conductivities are important in improving the capabilities of electrochemical immunosensors. The large specific surface area enhances antibody surface loading[9] , and the high conductivity enhances electron transfer from the target molecules to the surface of electrode[10] . Different nanoparticles with large specific surface areas and high electronic transmission capabilities were used to modify electrode surfaces and to increase the sensitivity and linear range of "label-free" electrochemical immunosensors, and this approach has been reported many times[11, 12]. Graphene (G) has shown potential for application in electrochemical immunosensors because of its low manufacturing cost, superior conductivity and large specific surface area[13]. Among metal nanomaterials, gold nanoparticles (AuNPs) and platinum nanoparticles (PtNPs) are most widely used in electrochemical immunosensors because of their excellent performance and excellent conductivity[14, 15]. Additionally, G has been functionalized and adsorbed on chitosan (Chi) through π-π stacking to form graphene-chitosan (G-Chi) hybrid materials[16]. In the G-Chi hybrid materials, Chi chelates $Au^{3+}$and Pt^{2+}metal ions and acts as a reducing agent to convert the Au^{2+} and Pt^{2+} ions into AuNPs and PtNPs[17, 18]; further, G-Chi hybrid materials can be loaded with substantial amounts of AuNPs and PtNPs because of the large specific surface area of G. Hence, we designed a "label-free" electrochemical immunosensor based on G-Chi-Au/PtNP nanocomposites in this work.

In addition, effective immobilization of antibodies is an essential step in constructing electrochemical immunosensors and constitute another important factor in improving the performance of the electrochemical immunosensors[19] . The sensitivities and linear ranges of electrochemical immunosensors are limited by the antibody immobilization strategy chosen[20] . Various antibody immobilization strategies, including glutaraldehyde (Glu) cross-linking, 1-ethyl-3-(3-dimethylaminopropyl)-carbodiimide/N-hydroxysuccinimide (EDC/NHS) chemistry, direct incubation and cysteamine hydrochloride (CH), have been exploited by different research groups[21-26], but it is not known which approach is best. To answer this question, four immobilization strategies were compared in the present study: (1) Glu immobilization, (2) EDC/NHS immobilization, (3) direct immobilization and (4) CH immobilization. The results showed that the linear range obtained with the CH immobilization strategy was 10 times broader than that realized with the direct immobilization strategy and 100 times broader than those seen with the Glu immobilization strategy and EDC/NHS immobilization strategy, and their detection limits were similar.

Methods

Bovine serum albumin (BSA), chloroauric acid tetrahydrate ($HAuCl_4 \cdot 4H_2O$) and chloroplatinic acid hexahydrate ($H_2PtCl_6 \cdot 6H_2O$) were procured from Sigma-Aldrich Chemical Co. (St. Louis, MO, USA.). Graphite powder ($<$45 mm), potassium ferrocyanide ($K_4[Fe(CN)_6]$), potassium ferricyanide ($K_3[Fe(CN)_6]$), Glu, EDC, NHS, CH, H_2SO_4 and $KMnO_4$ were obtained from Guoyao Group Chemical Reagents Co., Ltd. (Shanghai, China).

Viruses and antibodies

ARV (S1133, China Institute of Veterinary Drug Control), AIV H3 (A/Duck/ Guangxi/M20/2009, Guangxi Veterinary Research Institute), AIV H9 (A/Chicken/Guangxi/DX/2008, Guangxi Veterinary Research Institute), NDV (F48E9, China Institute of Veterinary Drug Control), LTV (ILT/13, China Institute of Veterinary Drug Control), IBV (Mass41, China Institute of Veterinary Drug Control) and IBDV (China Institute of Veterinary Drug Control) were collected and stored in a −80 ℃ freezer in our laboratory prior to use. ARV/MAbs were prepared by our group[27].

Synthesis of G-Chi-Au/PtNP-ARV/MAbs

First, G was obtained by the Hummers method with slight modifications[28]. In short, 2.5 g $NaNO_3$ and 1.0 g graphite powder were added to 100 mL concentrated H_2SO_4 under continuous stirring and reacted at room temperature for 2 h. Then, the obtained mixture was cooled in an ice bath, 5 g $KMnO_4$ was slowly added under continuous stirring, and the mixture was kept for 24 h at 35 ℃. Then, 100 mL ddH_2O was added to the mixture under continuous stirring, the mixture was kept for 3 h at 80 ℃, and 300 mL ddH_2O was added. Subsequently, 6 mL 30% H_2O_2 was added, the mixture solution turned bright yellow, and many bubbles appeared. After continuous stirring for 3 h, the obtained solution was precipitated for 24 h at room temperature, the supernatant was decanted, 500 mL 0.5 mol/L HCl was added to the slurry, and the mixture was washed by centrifugation. Next, the obtained slurry was washed with ddH_2O until the pH of the supernatant was approximately 7.0. Then, 100 mL ddH_2O was added, and G oxide was obtained after ultrasonication for 2 h. The G oxide was heated to 95 ℃ in a water bath, 10 mL 1.0% $NaBH_4$, was added under continuous stirring, and the mixture was kept for 3 h, washed with ddH_2O three times, and dried in a vacuum drying oven at 90 ℃ for 8 h to obtain G.

Second, G-Chi was prepared according to our previous report[29]. Briefly, 0.05 mg Chi powder was added to 100 mL 1.0% (v/v) acetic acid solution under continuous stirring at room temperature and maintained for 0.5 h. Then, 100 mg of G was added, the mixture was continuously stirred for 24 h, and G-Chi was collected by centrifugation and washed with ddH_2O.

Third, 1 mL 10 mmol/L $HAuCl_4$, 1 mL 10 mmol/L K_2PtCl_4, and 20 mL 1mg/mL G-Chi solution were mixed together, and the mixture was stirred at room temperature for 3 h. Then, the mixture was heated to 80 ℃ under continuous stirring for 1 h to obtain the G-Chi-Au/PtNP nanocomposite.

Fabrication of the electrochemical immunosensor

The step-by-step fabrication of the electrochemical immunosensor is illustrated in Figure 5-32-1. A gold electrode (GE) was polished with 1.0 μm, 0.3 μm, and 0.05 μm alumina polishing powders, rinsed with ddH_2O, and cleaned by sonication in ddH_2O, ethanol, and ddH_2O for 5 min each. Subsequently, the GE was dried by flushing with nitrogen gas.

Next, 8 μL of prepared G-Chi-Au/PtNP was dropped onto the surface of the clean GE and dried at room temperature, and G-Chi-Au/PtNP-GE was obtained. ARV/MAbs was immobilized onto G-Chi-Au/PtNP-GE by four different strategies: (1) G-Chi-Au/PtNP-GE was incubated with 10 μL of 5% Glu for 3 h and washed with PBS three times, and then 8 μL of 100 μg/mL ARV/MAbs was deposited onto G-Chi-Au/PtNP-Glu-GE and incubated at 4 ℃ for 8 h; (2) G-Chi-Au/PtNP-GE was anodized in 0.5 mol/L NaOH solution with a potential of +1.3 V for 40 s to increase the number of -COOH groups on its surface. The anodized G-Chi-Au/PtNP-GE was incubated with 10 μL of solution that contained 50 mmol/L EDC and 30 mmol/L NHS in MES (pH 4.7) at room temperature for 1 h. -COOH groups were converted to amine-reactive NHS esters in this step to prepare for ARV/MAb immobilization. Then, after washing with PBS (pH 7.4) three times to remove the unreacted EDC/NHS, 8 μL 100 μg/mL ARV/MAbs was deposited onto the NHS-activated surface of G-Chi-Au/PtNP-EDC/NHS-GE, which was then incubated at 4 ℃ for 8 h; (3) 8 μL 100 μg/mL ARV/MAbs was deposited onto G-Chi-Au/PtNP-GE and incubated at 4 ℃ for 8 h without any further modification; and (4) G-Chi-Au/PtNP-GE was incubated with 10 μL 2 mg/mL CH at room temperature in the dark for 4 h and washed with PBS three times, after which 8 μL 100 μg/mL ARV/MAbs was deposited onto G-Chi-Au/PtNP-CH-GE, and the material was incubated at 4 ℃ for 8 h. All the electrodes with immobilized ARV/MAbs prepared via above four strategies were washed with PBS (pH 7.4) to remove physically adsorbed or excess ARV/MAbs, incubated with 1% BSA in 0.01 mol/L PBS (pH 7.4) at room temperature for 1 h to block the free active sites on the electrodes and washed three time with PBS (pH7.4). The obtained immunosensors were denoted GE-G-Chi-Au/PtNP-Glu-ARV/MAb-BSA, GE-G-Chi-Au/PtNP-EDC/NHS-ARV/MAb-BSA, GE-G-Chi-Au/PtNP-ARV/MAb-BSA and GE-G-Chi-Au/PtNP-CH-ARV/MAb-BSA, respectively, and stored at 4 ℃ until use.

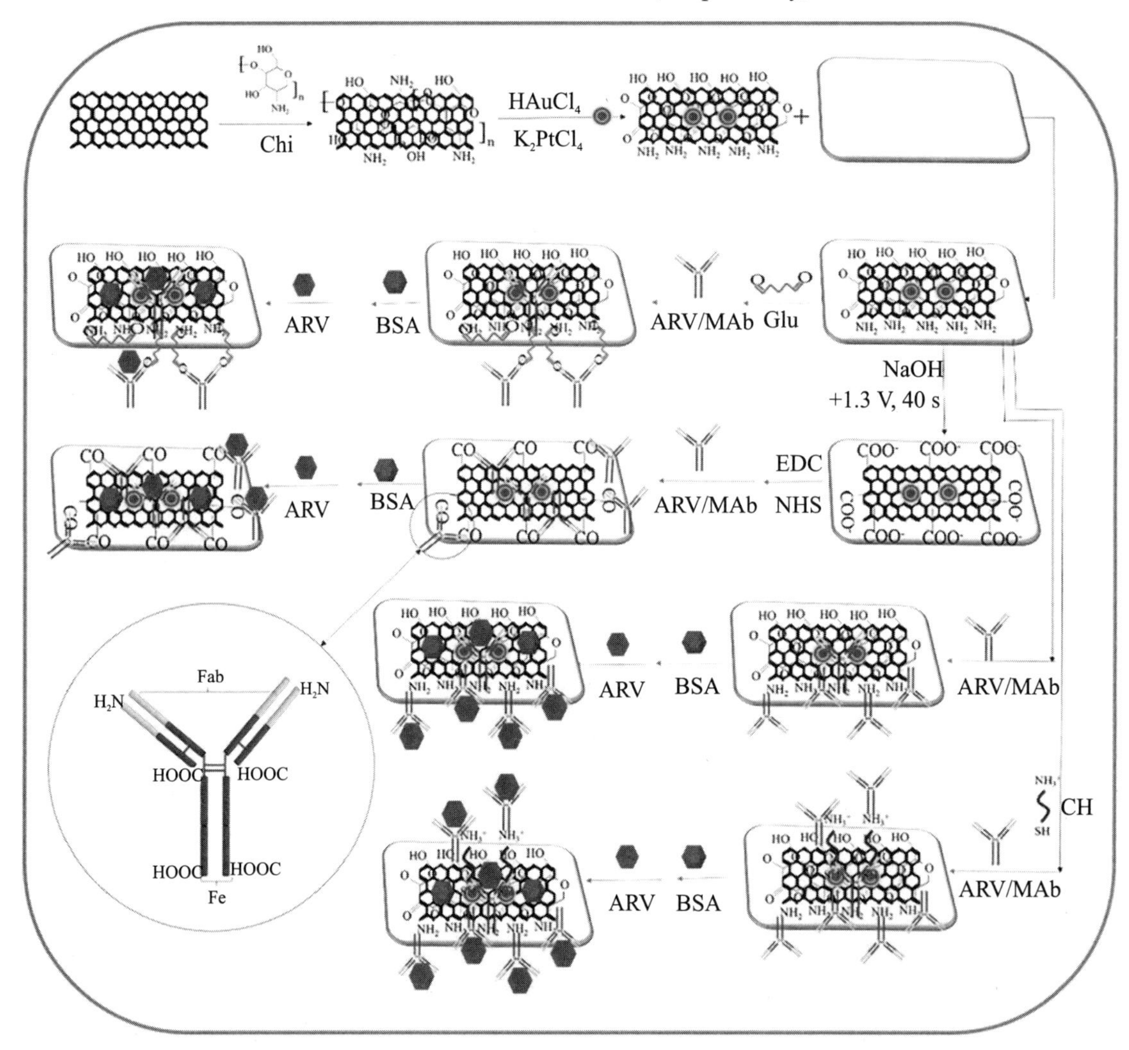

Figure 5-32-1 Schematic illustration of the electrochemical immunosensors (color figure in appendix)

Electrochemical analysis

The immunosensors were incubated with 8 μL of different concentrations of ARV at 37 ℃ for 30 min and washed with PBS (pH 7.4). Then, the sensors were subjected to EIS measurements in buffer containing 5.0 mM $K_3[Fe(CN)]_6$, and $K_4[Fe(CN)]_6$, in PBS (pH 7.4).

Results and discussion

Nanoparticle synthesis and characterization

A transmission electron microscopy (TEM) micrograph of G-Chi, which has a thin, wrinkled, rippled and flake-like structure, is shown in Figure 5-32-2 A. Figure 5-32-2 B shows the TEM micrograph of G-Chi-Au/PtNP which indicates that Au/Pt was loaded on the surface of G-Chi. In addition, energy dispersive spectroscopy (EDS) elemental analysis of G-Chi-Au/PtNP was employed to determine the presence of Au and Pt, which confirmed that Au/PtNP had been loaded on the surface of G-Chi (Figure 5-32-2 C). The mechanism for formation of G-Chi-Au/PtNP involved Au^{3+} and Pt^{2+} adsorption from aqueous solution due to chelation by G-Chi and then reduction to Au/Pt nanoparticles by Chi. Chi was used as both a stabilizing agent and reducing agent.

A

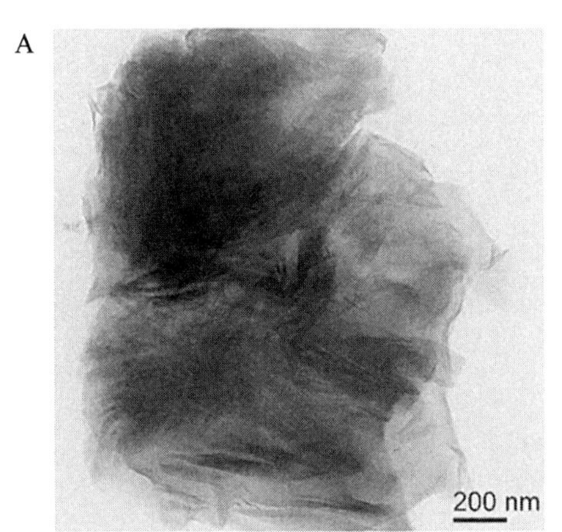

B

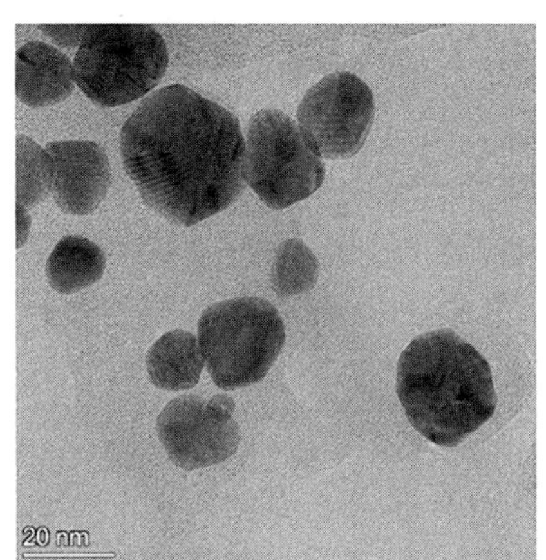

C

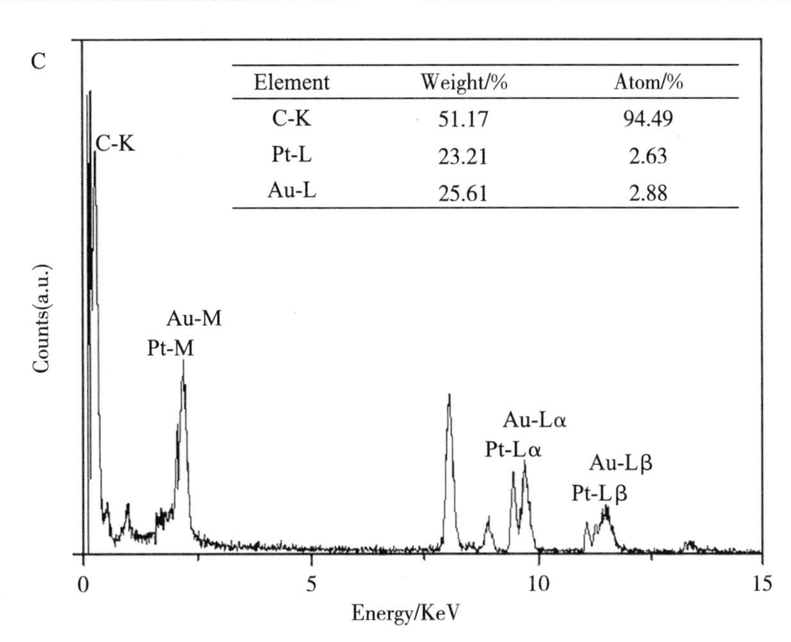

A: TEM images of G-Chi; B: TEM images of G-Chi-Au/PtNP; C: EDS elemental analysis of G-Chi-Au/PtNP.

Figure 5-32-2 Characterization of the investigated nanocomposites

Electrochemical characterization

Electrochemical impedance spectroscopy (EIS) is an effective technique for probing the features of surface modified electrodes, and it sensitively analyzes the interactions of analytes with modified electrodes and produces measurable electric signals. More important, EIS is more sensitive than either amperometric or voltammetric methods[30, 31]. In typical EIS Nyquist plots, a semicircle appears in the high-frequency region, while a line appears in the low-frequency region, and the diameter of the semicircle corresponds to the electron transfer resistance (R_{et}). In brief, the resistance on the surface of the electrode can be estimated by determining the semicircle diameter. Here, EIS was employed to characterize which material is better for electrode modification, and the results are shown in Figure 5-31-3. The diameter of the semicircle in the Nyquist plot of GE corresponds to an impedance of 1 561 Ω, which was decreased to 1 167 Ω, 950 Ω and 439 Ω upon modification of the GE with G-Chi-PtNP, G-Chi-AuNP and G-Chi-Au/PtNP, respectively, due to the high conductivities of G, AuNP and PtNP. These results showed that G-Chi-Au/PtNP exhibited the fastest electron transfer, and it was selected as the material for GE modification.

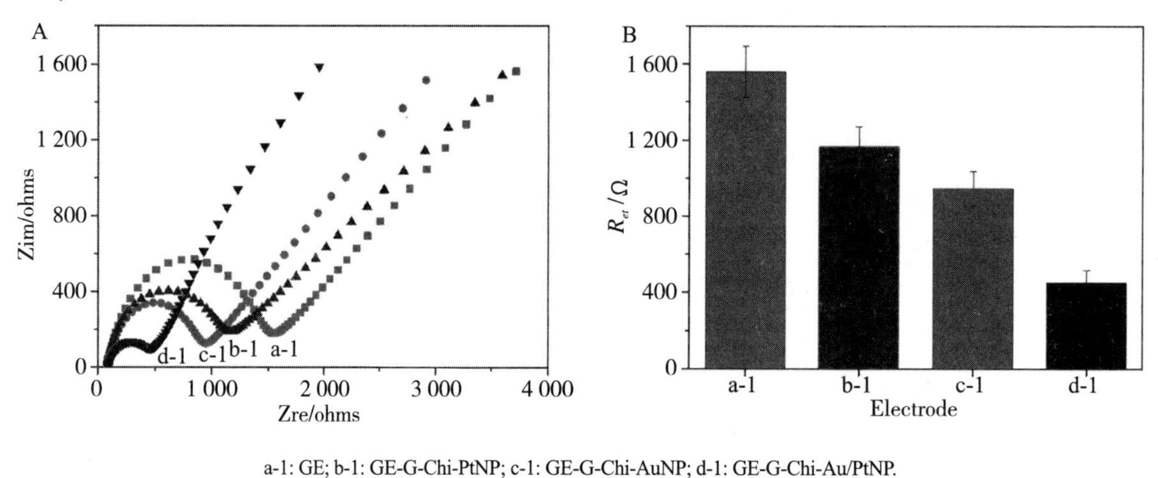

a-1: GE; b-1: GE-G-Chi-PtNP; c-1: GE-G-Chi-AuNP; d-1: GE-G-Chi-Au/PtNP.

Figure 5-32-3 Nyquist plots and semicircle diameters from EIS characterization of electrodes modified with different materials

In addition, EIS was employed to survey the layer-by-layer modification of GE. Figure 5-32-4 A and F shows that after the ARV/MAb was immobilized on GE-G-Chi-Au/PtNP by four different methods, the R_{et}s of the GE-G-Chi-Au/PtNP-Glu-ARV/MAb, GE-G-Chi-Au/PtNP-ARV/MAb, GE-G-Chi-Au/PtNP-EDC/NHS-ARV/MAb and GE-G-Chi-Au/PtNP-CH-ARV/MAb increased to 965 Ω, 1 232 Ω, 1 519 Ωand 1 778 Ω, respectively, because ARV/MAb is protein with poor electrical conductivity and impedes electron transfer to the surface of the electrode. After GE-G-Chi-Au/PtNP-Glu-ARV/MAb, GE-G-Chi-Au/PtNP-ARV/MAb, GE-G-Chi-Au/PtNP-EDC/NHS-ARV/MAb and GE-G-Chi-Au/PtNP-CH-ARV/MAb were blocked with BSA, their R_{et} values were further increased to 1 598 Ω, 2 078 Ω, 2 331 Ω and 2 670 Ω, respectively (Figure 5-32-4 B, F), because BSA is a protein. More important, these results show that the ability of the four different methods for ARV/MAb immobilization decreased in the order CH>EDC/NHS>direct incubation>Glu. After GE-G-Chi-Au/PtNP-Glu-ARV/MAb-BSA, GE-G-Chi-Au/PtNP-ARV/MAb-BSA, GE-G-Chi-Au/PtNP-EDC/NHS-ARV/MAb-BSA and GE-G-Chi-Au/PtNP-CH-ARV/MAb-BSA were incubated with $10^{6.82}$ EID_{50}/mL ARV, the ARV was adsorbed on the electrodes via a specific response with ARV/MAb, and the corresponding R_{et}s were further increased to 4 841 Ω, 7 468 Ω, 5 547 Ω, and 9 417 Ω (Figure 5-32-4 C, F) because electron transfer to the surface of the electrode was impeded by the ARV protein, and the results demonstrated that the ability of the four different immunosensors to combine ARV decreased in the order CH>direct incubation>EDC/NHS>Glu. Most importantly, the different relative orders for ARV/MAb immobilization and the ARV immobilization demonstrated that the number of ARVs adsorbed on the immunosensors was not only related to the number of ARVs/MAbs but was also affected by the method used to immobilize the ARVs/MAbs.

In addition, BSA was immobilized on GE-G-Chi-Au/PtNP by four different methods to evaluate the ability of the four methods to immobilize protein again, the results are shown in Figure 5-32-4 D and F. The R_{et}s of the GE-G-Chi-Au/PtNP-Glu-BSA, GE-G-Chi-Au/PtNP-BSA, GE-G-Chi-Au/PtNP-EDC/NHS-BSA and

GE-G-Chi-Au/PtNP-CH-BSA were increased to 1 196 Ω, 1 613 Ω, 2 159 Ω and 2 537 Ω because the BSA protein adsorbed on the surface of GE-G-Chi-Au/PtNP impeded electron transfer to the surface of the electrode. These results showed that the abilities of the four different methods to immobilize BSA decreased in the order CH>EDC/NHS>direct incubation>Glu, which corresponds to the results for ARV/MAb immobilization on GE-G-Chi-Au/PtNP by the four different methods. To evaluate the specificities of the immunosensors, GE-G-Chi-Au/PtNP-Glu-BSA, GE-G-Chi-Au/PtNP-BSA, GE-G-Chi-Au/PtNP-EDC/NHS-BSA and GE-G-Chi-Au/PtNP-CH-BSA (which were not yet immobilized with ARV/MAb) were incubated with $10^{6.82}$ EID_{50}/mL ARV, the results were showed in Figure 5-32-4 E and F. The results showed that the values of $R_{et}s$ of GE-G-Chi-Au/PtNP-Glu-BSA, GE-G-Chi-Au/PtNP-BSA, GE-G-Chi-Au/PtNP-EDC/NHS-BSA and GE-G-Chi-Au/PtNP-CH-BSA were hardly change after incubation with ARV, which suggested that the GE-G-Chi-Au/PtNP-Glu-ARV/MAb-BSA, GE-G-Chi-Au/PtNP-ARV/MAb-BSA, GE-G-Chi-Au/PtNP-EDC/NHS-ARV/MAb-BSA and GE-G-Chi-Au/PtNP-CH-ARV/MAb-BSA had high specificities.

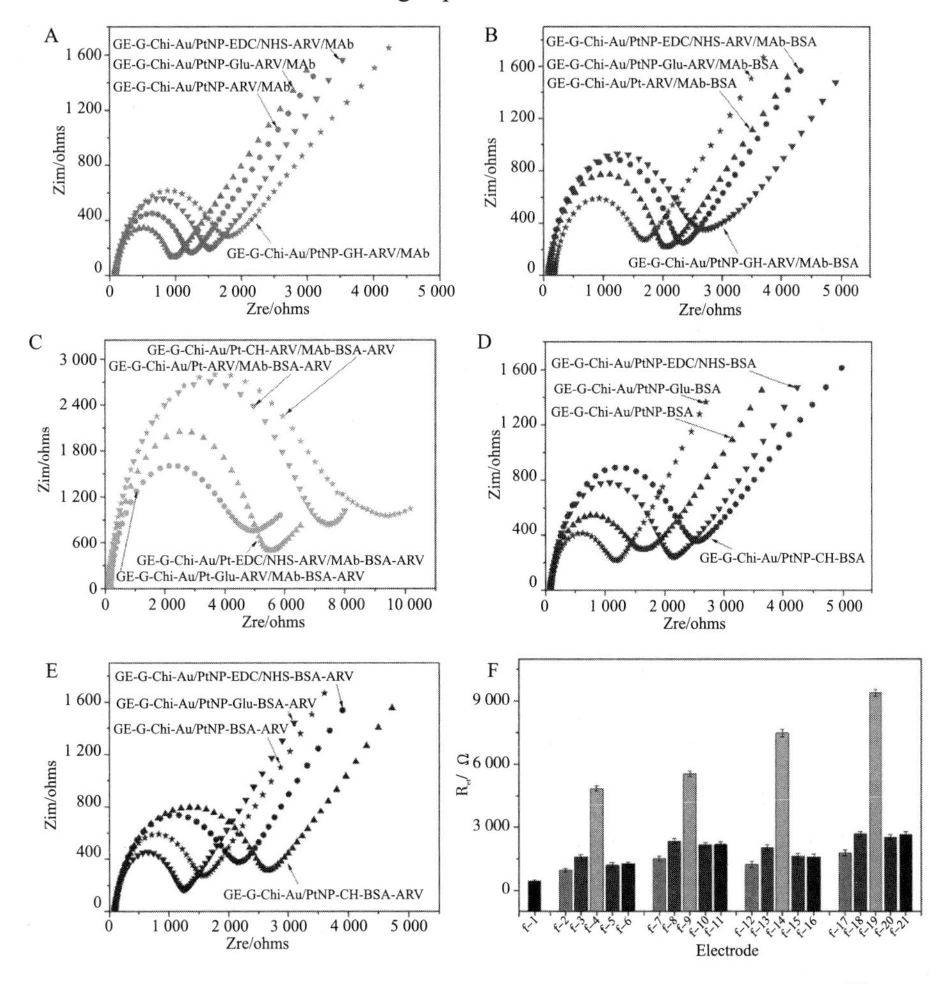

A: After ARV/MAb immobilization on GE-G-Chi-Au/PtNP; B: After blocking with BSA; C: After incubation with $10^{6.82}$ EID_{50}/mL ARV; D: After blocking with BSA without ARV/MAb immobilization; E: After incubation with $10^{6.82}$ EID_{50}/mL ARV without ARV/MAb immobilization; F: R_{et} from EIS characterization of electrodes at different modification steps; f-1: GE-G-Chi-Au/PtNP; f-2: GE-G-Chi-Au/PtNP-Glu-ARV/MAb; f-3: GE-G-Chi-Au/PtNP-Glu-ARV/MAb-BAS; f-4:GE-G-Chi-Au/PtNP-Glu-ARV/MAb-BAS-ARV; f-5: GE-G-Chi-Au/PtNP-Glu-BAS; f-6: GE-G-Chi-Au/PtNP-Glu-BAS-ARV; f-7: GE-G-Chi-Au/PtNP-EDC/NHS-ARV/MAb; f-8: GE-G-Chi-Au/PtNP-EDC/NHS-ARV/MAb-BSA; f-9: GE-G-Chi-Au/PtNP-EDC/NHS-ARV/MAb-BSA-ARV; f-10: GE-G-Chi-Au/PtNP-EDC/NHS-BSA; f-11: GE-G-Chi-Au/PtNP-EDC/NHS-BSA-ARV; f-12: GE-GChi-Au/PtNP-ARV/MAb; f-13: GE-G-Chi-Au/PtNP-ARV/MAb-BSA; f-14: GE-G-Chi-Au/PtNP-ARV/MAb-BSA-ARV; f-15: GE-G-Chi-Au/PtNP-BSA; f-16: GE-G-Chi-Au/PtNP-BSA-ARV; f-17: GE-G-Chi-Au/PtNP-CH-ARV/MAb; f-18: GE-G-Chi-Au/PtNP-CH-ARV/MAb-BSA; f-19: GE-G-Chi-Au/PtNP-CH-ARV/MAb-BSA-ARV; (f-20): GE-G-Chi-Au/PtNP-CH-BSA; (f-21): GE-G-Chi-Au/PtNP-CH-BSA-ARV.

Figure 5-32-4 Nyquist plots for EIS characterization of electrodes after different modification steps in electrolyte (pH=7.0) containing 5 mM $[Fe(CN)_6]^{3-/4-}$

Sensitivity, saturation and extended dynamic range are important factors used to evaluate the performance of an immunosensor. To obtain this information for the immunosensors, the GE-G-Chi-Au/PtNP-Glu-ARV/MAb-BSA, GE-G-Chi-Au/PtNP-EDC/NHS-ARV/MAb-BSA, GE-G-Chi-Au/PtNP-ARV/MAb-BSA and GE-G-Chi-Au/PtNP-CH-ARV/MAb-BSA were used to detect ARV at a concentration of $10^{6.82}$ EID_{50}/mL to $10^{4.82}$ EID_{50}/mL. The results are shown in Figure 5-32-5. R_{et} increased with increasing concentrations of ARV due to the increased formation of the antigen-antibody complex, which is a nonconducting biomolecule. As observed, GE-G-Chi-Au/PtNP-Glu-ARV/MAb-BSA showed saturation beyond $10^{3.82}$ EID_{50}/mL ARV and a linear range of $10^{3.82}$ EID_{50}/mL ARV; the linear regression equation of the calibration curve was expressed as R_{et} (Ω) =765 lgEID_{50}/mL+1 464 with a correlation coefficient of R^2=0.961 75, a low detection limit of $10^{0.63}$ EID_{50}/mL ARV (S/N=3) and a quantification limit of $10^{1.15}$ EID_{50}/mL ARV (S/N=10) (Figure 5-32-5 A, B). GE-G-Chi-Au/PtNP-EDC/NHS-ARV/MAb-BSA showed saturation beyond $10^{3.82}$ EID_{50}/mL ARV and a linear range of 0～$10^{2.82}$ EID_{50}/mL ARV; the linear regression equation of the calibration curve was expressed as R_{et} (Ω) =8 291g EID_{50}/mL+2 406 with a correlation coefficient of R^2=0.987 85, a low detection limit of $10^{0.48}$ EID_{50}/mL ARV (S/N=3) and a quantification limit of $10^{1.00}$ EID_{50}/mL ARV (S/N=10) (Figure 5-32-5 C, D). GE-G-Chi-Au/PtNP-ARV/MAb-BSA showed saturation beyond $10^{4.82}$ EID_{50}/mL ARV and a linear range of 0～$10^{4.82}$ EID_{50}/mL ARV; the linear regression equation of the calibration curve was expressed as R_{et} (Ω) =1 123 lgEID_{50}/mL+1 938 with a correlation coefficient of R^2=0.993 47, a low detection limit of $10^{0.37}$ EID_{50}/mL ARV (S/N=3) (Figure 5-32-5 E, F) and a quantification limit of $10^{0.89}$ EID_{50}/mL ARV (S/N=10). GE-G-Chi-Au/PtNP-CH-ARV/MAb-BSA showed saturation beyond $10^{5.82}$ EID_{50}/mL ARV and a linear range of 0～$10^{5.82}$ EID_{50}/mL ARV; the linear regression equation of the calibration curve was expressed as R_{et} (Ω) =1 084 lg EID_{50}/mL+2 335 with a correlation coefficient of R^2=0.967 3, a low detection limit of $10^{0.46}$ EID_{50}/mL ARV (S/N=3) and a quantification limit of $10^{0.98}$ EID_{50}/mL ARV (S/N=10) (Figure 5-32-5 G, H).

These results show that the sensitivities of the four immunosensors were slightly different, while saturation beyond the extended dynamic range of GE-G-Chi-Au/PtNP-CH-ARV/MAb-BSA was 10 times that of GE-G-Chi-Au/PtNP-ARV/MAb-BSA and 100 times those of GE-G-Chi-Au/PtNP-Glu-ARV/MAb-BSA and GE-G-Chi-Au/PtNP-EDC/NHS-ARV/MAb-BSA. These results were attributed to the different methods of ARV/MAb immobilization. ARV/MAb is a Y-shaped protein that consists of two light and two heavy chains linked by disulfide bonds, and ARV/MAb is monomeric (IgG) and contains the F_{ab} region and F_c region (Figure 5-32-1). The F_{ab} region, which participates in antigen binding, consists of an amino end, while the F_c region, which is the stem of the Y shape, has a carboxylend group[32, 33]. The region (F_{ab} or F_c) of ARV/MAb attached to the electrode depended on immobilization strategy used. For GE-G-Chi-Au/PtNP-Glu-ARV/MAb-BSA, Glu was used as the coupling agent to connect amine groups on the surface of the electrode and amine groups (F_{ab} region) on ARV/MAb. The EDC/NHS-based immobilization strategy for GE-G-Chi-Au/PtNP-EDC/NHS-ARV/MAb-BSA activated carboxyl groups on the surface of the electrode and allowed NHS ester groups to act as intermediates leading to covalent attachment of the activated carboxyl groups with amino groups on the surface of the electrode present in the F_{ab} region on ARV/MAb. The antibodies in GE-G-Chi-Au/PtNP-ARV/MAb-BSA and GE-G-Chi-Au/PtNP-CH-ARV/MAb-BSA were immobilized onto the electrode via the tail end of the F_c region through covalent attachments between electrode surface amino groups and carboxyl groups on ARV/MAb, which resulted in accessibility for antigen binding because it was located in an orthogonal position. However, attachment of ARV/MAb onto the electrode through the F_{ab} region led to a conformation providing difficult antibody-antigen binding because of steric hindrance. Therefore, GE-G-Chi-Au/PtNP-ARV/MAb-BSA and GE-G-Chi-Au/PtNP-CH-ARV/MAb-BSA showed higher saturation levels for ARV than GE-G-Chi-

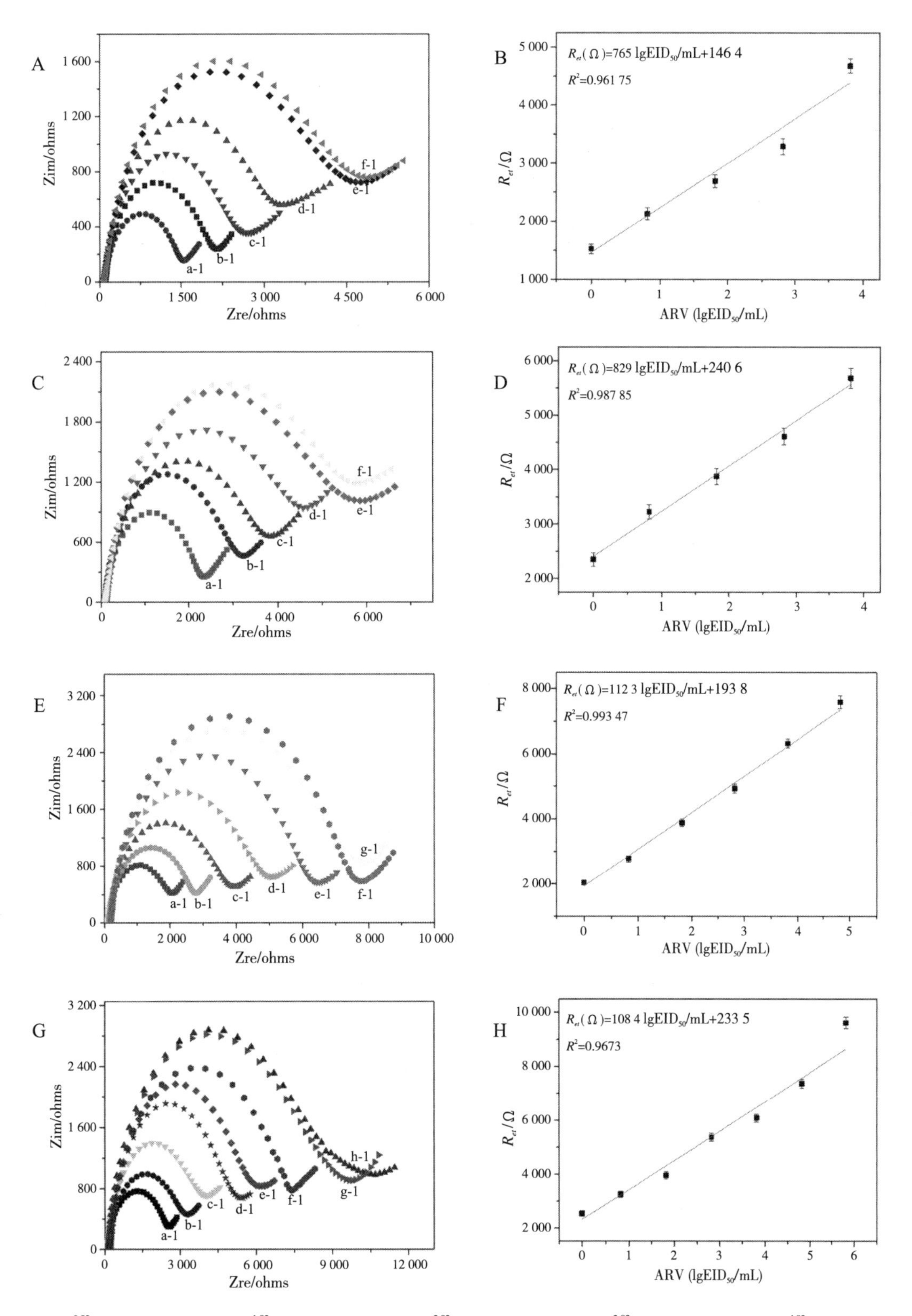

a-1: 0; b-1: $10^{0.82}$ EID_{50}/mL ARV; c-1: $10^{1.82}$ EID_{50}/mL ARV; d-1: $10^{2.82}$ EID_{50}/mL ARV; e-1: $10^{3.82}$ EID_{50}/mL ARV; f-1: $10^{4.82}$ EID_{50}/mL ARV; g-1: $10^{5.82}$ EID_{50}/mL ARV; f-1: $10^{6.82}$ EID_{50}/mL ARV.

Nyquist plots for different concentrations of ARV: A: GE-G-Chi-Au/PtNP-Glu-ARV/MAb-BSA; C: GE-G-Chi-Au/PtNP-EDC/NHS-ARV/MAb-BSA; E: GE-G-Chi-Au/PtNP-ARV/MAb-BSA; G: GE-G-Chi-Au/PtNP-CH-ARV/MAb-BSA.

Calibration curve with different concentrations of ARV, Error bar = RSD (n = 5): B: GE-G-Chi-Au/PtNP-Glu-ARV/MAb-BSA; D: GE-G-Chi-Au/PtNP-EDC/NHS-ARV/MAb-BSA; F: GE-G-Chi-Au/PtNP-ARV/MAb-BSA; F: GE-G-Chi-Au/PtNP-ARV/MAb-BSA; H: GE-G-Chi-Au/PtNP-CH-ARV/MAb-BSA.

Figure 5-32-5 Sensitivity results of immunosensors

Au/PtNP-Glu-ARV/MAb-BSA and GE-G-Chi-Au/PtNP-EDC/NHS-ARV/MAb-BSA, because GE-G-Chi-Au/PtNP-ARV/MAb-BSA and GE-G-Chi-Au/PtNP-CH-ARV/MAb-BSA provided more active sites for antigen binding when ARV was present in high concentration. Although antibodies of GE-G-Chi-Au/PtNP-ARV/MAb-BSA and GE-G-Chi-Au/PtNP-CH-ARV/MAb-BSA were immobilized onto the electrode through the F_c region, GE-G-Chi-Au/PtNP-CH-ARV/MAb-BSA showed a higher saturation level for ARV because it featured more amine groups with which to adsorb more ARV/MAb.

The sensitivity of an immunosensor depends on changes in charge transfer resistance per unit change in ARV concentration per unit area. The four different immunosensors were modified with GE-G-Chi-Au/PtNP. So their electron transfer abilities on the surfaces of the four different immunosensors were similar. In the detection process, antigen-antibody complex formation resulted in similar changes in charge transfer resistance per unit change in ARV concentration per unit area, which provided similar sensitivities for the four different immunosensors. While the different strategies for antibody immobilization led to slight sensitivity differences among the four different immunosensors, GE-G-Chi-Au/PtNP-ARV/MAb-BSA showed the highest sensitivity among the four different immunosensors because ARV/MAb was immobilized on the surface of electrode by direct incubation (we did not use any non-conducting organics as linkers) ; ARV/MAb in the other three immunosensors was immobilized on the electrode surface with Glu, EDC/NHS and CH as linker, respectively. Glu, EDC/NHS and CH with poor electrical conductivities were attached to the electrodes, and the electron transfer abilities decreased, so the sensitivities of the immunosensors were decreased.

Selectivity, reproducibility and stability of the four immunosensors

The selectivity of GE-G-Chi-Au/PtNP-Glu-ARV/MAb-BSA, GE-G-Chi-Au/PtNP-EDC/NHS-ARV/MAb-BSA, GE-G-Chi-Au/PtNP-ARV/MAb-BSA and GE-G-Chi-Au/PtNP-CH-ARV/MAb-BSA played a crucial role in detecting target samples without separation. To evaluate the selectivities of the four different immunosensors, avian influenza virus H3 subtype (AIV H3, $10^{4.71}$ EID_{50}/mL), avian influenza virus H9 subtype (AIV H9, $10^{3.74}$ EID_{50}/mL), Newcastle disease virus (NDV, $10^{4.53}$ EID_{50}/mL), laryngotracheitis virus (LTV, $10^{3.86}$ EID_{50}/mL), infectious bronchitis virus (IBV, $10^{4.36}$ EID_{50}/mL), infectious bursal disease virus (IBDV, $10^{4.67}$ EID_{50}/mL), BSA (1.0 μg/mL) and vitamin C (1.0 μg/mL) were used as interfering substances. As shown in Fig 5-32-6, the $R_{et}s$ of the samples with the interfering substances were almost the same as that of the blank control, and the mixtures of ARV ($10^{2.82}$ EID_{50}/mL) with other possible interfering substances showed similar R_{et} values to those of ARV ($10^{2.82}$ EID_{50}/mL), indicating that GE-G-Chi-Au/PtNP-Glu-ARV/MAb-BSA, GE-G-Chi-Au/PtNP-EDC/NHS-ARV/MAb-BSA, GE-G-Chi-Au/PtNP-ARV/MAb-BSA and GE-G-Chi-Au/PtNP-CH-ARV/MAb-BSA had high selectivity for ARV detection.

The reproducibility of GE-G-Chi-Au/PtNP-Glu-ARV/MAb-BSA, GE-G-Chi-Au/PtNP-EDC/NHS-ARV/MAb-BSA, GE-G-Chi-Au/PtNP-ARV/MAb-BSA and GE-G-Chi-Au/PtNP-CH-ARV/MAb-BSA was investigated by EIS at the same concentration of $10^{3.82}$ EID_{50}/mL ARV. The results are shown in Figure 5-32-7. The RSDs of GE-G-Chi-Au/PtNP-Glu-ARV/MAb-BSA, GE-G-Chi-Au/PtNP-EDC/NHS-ARV/MAb-BSA, GE-G-Chi-Au/PtNP-ARV/MAb-BSA and GE-G-Chi-Au/PtNP-CH-ARV/MAb-BSA were 2.04%, 2.12%, 1.30% and 1.92%, respectively, indicating that the four immunosensors have good reproducibility.

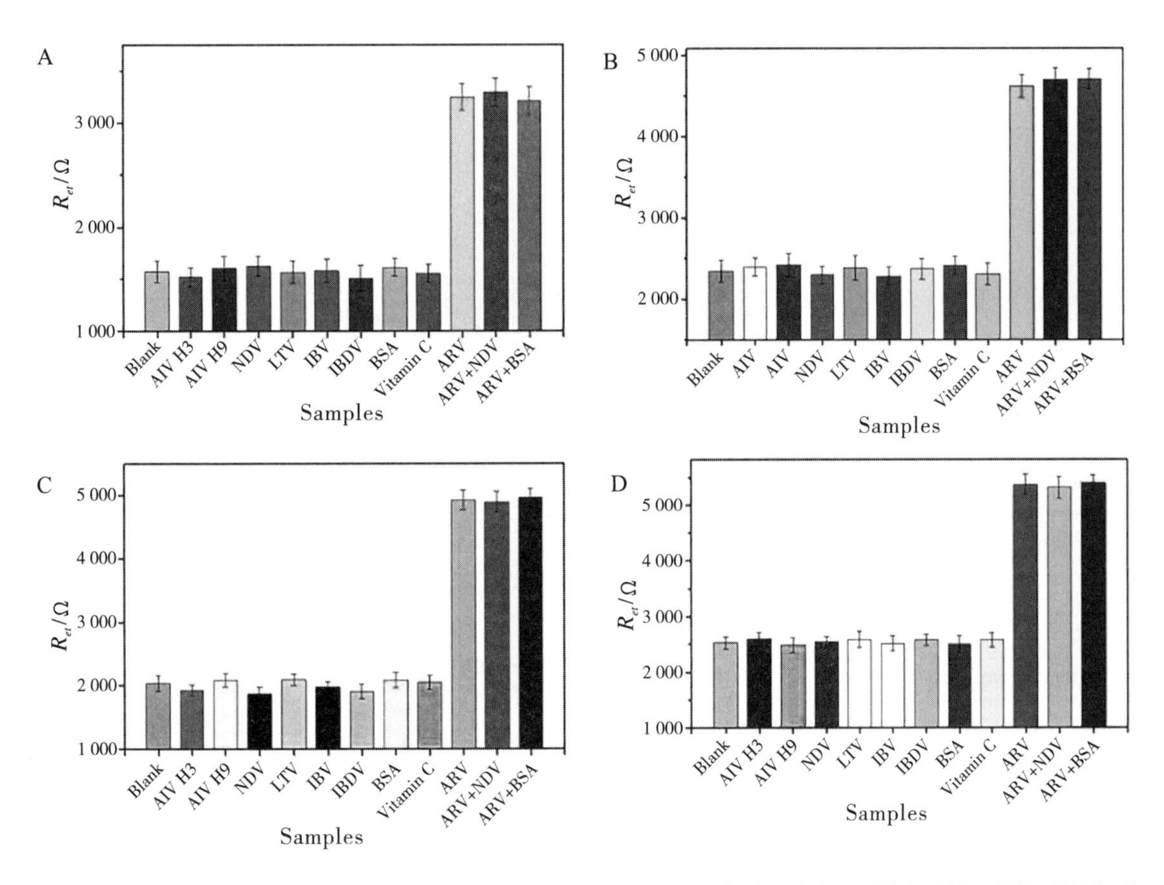

A: GE-G-Chi-Au/PtNP-Glu-ARV/MAb-BSA; B: GE-G-Chi-Au/PtNP-EDC/NHS-ARV/MAb-BSA; C: GE-G-Chi-Au/PtNP-ARV/MAb-BSA; D: GE-G-Chi-Au/PtNP-CH-ARV/MAb-BSA. Error bar=RSD (n=5).

Figure 5-32-6 Specificity of the immunosensors toward the target (ARV) and other interfering substances

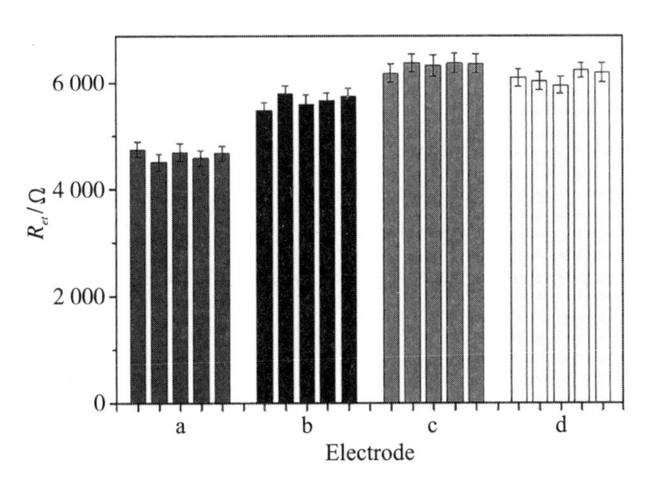

a: GE-G-Chi-Au/PtNP-Glu-ARV/MAb-BSA; b: GE-G-Chi-Au/PtNP-EDC/NHS-ARV/MAb-BSA; c: GE-G-Chi-Au/ PtNP-ARV/MAb-BSA; d: GE-G-Chi-Au/PtNP-CH-ARV/MAb-BSA. Error bar=RSD (n=5).

Figure 5-32-7 Reproducibility of the four immunosensors in the presence of $10^{3.82}$ EID_{50} /mL ARV

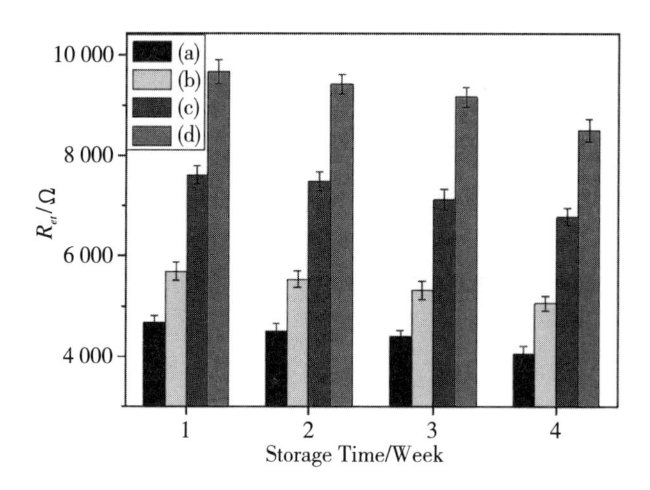

a: GE-G-Chi-Au/PtNP-Glu-ARV/MAb-BSA; b: GE-G-Chi-Au/PtNP-EDC/NHS-ARV/MAb-BSA; c: GE-G-Chi-Au/PtNP-ARV/MAb-BSA; d: GE-G-Chi-Au/PtNP-CH-ARV/MAb-BSA.

Figure 5-32-8 Long-term stability of the proposed immunosensors

Conclusions

In this work, four different antibody immobilization strategies (Glu as a coupling agent to connect amine groups on the surface of the electrode and amine groups present in ARV/MAb; EDC/NHS chemistry to activate carboxyl groups on the surface of the electrode for covalent attachment with amine groups present at the F_{ab} region on ARV/MAb; immobilization of ARV/MAb directly onto the amine group ending the electrode; and immobilization of ARV/MAb directly onto the amine group ending the electrode after it was further modified with CH) were used to construct four different immunosensors (GE-G-Chi-Au/PtNP-Glu-ARV/MAb-BSA, GE-G-Chi-Au/PtNP-EDC/NHS-ARV/MAb-BSA, GE-G-Chi-Au/PtNP-ARV/MAb-BSA and GE-G-Chi-Au/PtNP-CH-ARV/MAb-BSA) to study the effect of immobilization strategy on the immunosensor performance. The EIS results showed that the extended dynamic range of GE-G-Chi-Au/PtNP-CH-ARV/MAb-BSA was 10 times that of GE-G-Chi-Au/PtNP-ARV/MAb-BSA and 100 times that of GE-G-Chi-Au/PtNP-Glu-ARV/MAb-BSA and GE-G-Chi-Au/PtNP-EDC/NHS-ARV/MAb-BSA, indicating that the immunosensors that immobilized the antibody via the F_c region obtained an extended dynamic response. In addition, the sensitivities, selectivities, reproducibilities and stabilities of the immunosensors were almost completely unaffected by the antibody immobilization strategies used in this work.

References

[1] LI J, WANG C, WANG W, et al. Dual-mode immunosensor for electrochemiluminescence resonance energy transfer and electrochemical detection of rabies virus glycoprotein based on Ru(bpy)32+-Loaded dendritic mesoporous silica nanoparticles. Anal Chem, 2022, 94(21): 7655-7664.

[2] LEE T H, CHEN L C, WANG E, et al. Development of an electrochemical immunosensor for detection of cardiac troponin I at the Point-of-Care. Biosensors, 2021, 11(7): 210.

[3] ALVES N J, MUSTAFAOGLU N, BILGICER B. Oriented antibody immobilization by site-specific UV photocrosslinking of biotin at the conserved nucleotide binding site for enhanced antigen detection. Biosens Bioelectron, 2013, 49: 387-393.

[4] BANIUKEVIC J, HAKKI BOYACI I, GOKTUG BOZKURT A, et al. Magnetic gold nanoparticles in SERS-based sandwich immunoassay for antigen detection by well oriented antibodies. Biosens Bioelectron, 2013, 43: 281-288.

[5] ABOLHASAN R, KHALILZADEH B, YOUSEFI H, et al. Ultrasensitive and label free electrochemical immunosensor for

detection of ROR1 as an oncofetal biomarker using gold nanoparticles assisted LDH/rGO nanocomposite. Sci Rep, 2021, 11(1): 14921.

[6] ORTEGA F G, REGIART M D, RODRÍGUEZ-MARTÍNEZ A, et al. Sandwich-Type Electrochemical Paper-Based Immunosensor for Claudin 7 and CD81 Dual Determination on Extracellular Vesicles from Breast Cancer Patients. Anal Chem, 2021, 93(2): 1143-1153.

[7] BISWAS S, LAN Q, XIE Y, et al. Label-Free Electrochemical Immunosensor for Ultrasensitive Detection of Carbohydrate Antigen 125 Based on Antibody-Immobilized Biocompatible MOF-808/CNT. ACS Appl Mater Interfaces, 2021, 13(2):3295-3302.

[8] LIU X, LIN L Y, TSENG F Y, et al. Label-free electrochemical immunosensor based on gold nanoparticle/polyethyleneimine/reduced graphene oxide nanocomposites for the ultrasensitive detection of cancer biomarker matrix metalloproteinase-1. Analyst, 2021, 146(12): 4066-4079.

[9] HUANG J L, XIE Z X, HUANG Y H, et al. Electrochemical immunosensor with Cu(I)/Cu(II)-chitosan-graphene nanocomposite-based signal amplification for the detection of Newcastle disease virus. Sci Rep, 2020, 10(1): 13869.

[10] DU X, ZHENG X, ZHANG Z, et al. A Label-Free Electrochemical Immunosensor for Detection of the Tumor Marker CA242 Based on Reduced Graphene Oxide-Gold-Palladium Nanocomposite. Nanomaterials, 2019, 9(9): 1335.

[11] CHEN Z, LI B, LIU J, et al. A label-free electrochemical immunosensor based on a gold-vertical graphene/TiO2 nanotube electrode for CA125 detection in oxidation/reduction dual channels. Mikrochim Acta, 2022, 189(7): 257.

[12] EHZARI H, AMIRI M, SAFARI M. Enzyme-free sandwich-type electrochemical immunosensor for highly sensitive prostate specific antigen based on conjugation of quantum dots and antibody on surface of modified glassy carbon electrode with core-shell magnetic metal-organic frameworks. Talanta, 2020, 210: 120641.

[13] CHAUDHARY M, VERMA S, KUMAR A, et al. Graphene oxide based electrochemical immunosensor for rapid detection of groundnut bud necrosis orthotospovirus in agricultural crops. Talanta, 2021, 235: 122717.

[14] WANG W J, CHOU M C, LEE Y J, et al. A simple electrochemical immunosensor based on a gold nanoparticle monolayer electrode for neutrophil gelatinase-associated lipocalin detection. Talanta, 2022, 246: 123530.

[15] HUANG Y, WEN Q, JIANG J H, et al. A novel electrochemical immunosensor based on hydrogen evolution inhibition by enzymatic copper deposition on platinum nanoparticle-modified electrode. Biosens Bioelectron, 2008, 24(4): 600-605.

[16] CHOOSANG J, KHUMNGERN S, THAVARUNGKUL P, et al. An ultrasensitive label-free electrochemical immunosensor based on 3D porous chitosan-graphene-ionic liquid-ferrocene nanocomposite cryogel decorated with gold nanoparticles for prostate-specific antigen. Talanta, 2021, 224: 121787.

[17] SEIDI F, REZA SAEB M, HUANG Y, et al. Thiomers of chitosan and cellulose: Effective biosorbents for detection, removal and recovery of metal ions from aqueous medium. Chem Rec, 2021, 21: 1876-1896.

[18] HUANG L, HUANG W, SHEN R, et al. Chitosan/thiol functionalized metal-organic framework composite for the simultaneous determination oflead and cadmium ions in food samples. Food Chem, 2020, 330: 127212.

[19] SHEN Y, SHEN G, ZHANG Y. Label-free electrochemical immunosensor based on ionic liquid containing dialdehyde as a novel linking agent for the antibody immobilization. ACS Omega, 2018, 3: 11227-11232.

[20] SHEN Y, ZHANG Y, LIU M, et al. A simple and sensitive electrochemical immunosensor based on thiol aromatic aldehyde as a substrate for the antibody immobilization. Talanta, 2015, 141: 288-292.

[21] HALÁMEK J, HEPEL M, SKLADAL P. Investigation of highly sensitive piezoelectric immunosensors for 2,4-dichlorophenoxyacetic acid. Biosens Bioelectron, 2001, 16: 253-260.

[22] SITU C, WYLIE A R, DOUGLAS A, et al. Reduction of severe bovine serum associated matrix effects on carboxymethylated dextran coated biosensor surfaces. Talanta, 2008, 76: 832-836.

[23] JOHARI-AHAR M, RASHIDI M R, BARAR J, et al. An ultra-sensitive impedimetric immunosensor for detection of the serum oncomarker CA-125 in ovarian cancer patients. Nanoscale, 2015, 7: 3768-3779.

[24] PUERTAS S, DE GRACIA VILLA M, MENDOZA E, et al. Improving immunosensor performance through oriented

immobilization of antibodies on carbon nanotube composite surfaces. Biosens Bioelectron, 2013, 43: 274-280.

[25] ZHU Q, DU J, FENG S, et al. Highly selective and sensitive detection of glutathione over cysteine and homocysteine with a turn-on fluorescent biosensor based on cysteamine-stabilized CdTe quantum dots. Spectrochim Acta A Mol Biomol Spectrosc, 2022, 267: 120492 .

[26] ZHANG Y, LI Y, WU W, et al. Chitosan coated on the layers'glucose oxidase immobilized on cysteamine/Au electrode for use as glucose biosensor. Biosens Bioelectron, 2014, 60: 271-276.

[27] XIE Z Q, XIE Z X, LIU J B, et al. Preparation and identification of monoclonal antibody against reovirus S1733. China Anim Husb Vet Med, 2012, 39: 47-51 .

[28] YU H, ZHANG B, BULIN C, et al. High-efficient synthesis of graphene oxide based on improved Hummers method. Sei Rep, 2016, 6: 36143.

[29] HUANG J L, XIE Z X, XIE Z Q, et al. Silver nanoparticles coated graphene electrochemical sensor for the ultrasensitive analysis of avian influenza virus H7. Anal Chim Acta, 2016, 913: 121-127.

[30] ONDA K, NAKAYAMA M, FUKUDA K, et al. T Cell impedance measurement by Laplace transformation of charge or discharge current-voltage. J Electrochem Soc, 2006, 153: A1012-A1018.

[31] PARK J Y, LEE Y S, CHANG B Y, et al. (R)-lipo-diaza-18-crown-6 self-assembled monolayer as a selective serotonin receptor. Anal Chem, 2009, 81(10): 3843-3850.

[32] MIMURA Y, ASHTON P R, TAKAHASHI N, et al. Contrasting glycosylation profiles between Fb and Fe of a human IgG protein studied by electrospray ionization mass spectrometry. J Immunol Methods, 2007, 326: 116-126 .

[33] MAHAN A E, TEDESCO J, DIONNE K, et al. A method for high-throughput,sensitive analysis of IgG F.and Fab glycosylation by capillary electrophoresis. J Immunol Methods, 2015, 417: 34-44.

A sandwich amperometric immunosensor for the detection of fowl adenovirus group I based on bimetallic Pt/Ag nanoparticle-functionalized multiwalled carbon nanotubes

Huang Jiaoling, Xie Zhixun, Luo Sisi, Li Meng, Xie Liji, Fan Qing, Zeng Tingting, Zhang Yanfang, Zhang Minxiu, Xie Zhiqin,Wang Sheng, Li Dan, Wei You, Li Xiaofeng, Wan Lijun, and Ren Hongyu

Abstract

An enzyme-free sandwich amperometric immunosensor based on bimetallic Pt/Ag nanoparticle (Pt/AgNPs)-functionalized chitosan (Chi)-modified multiwalled carbon nanotubes (MWCNTs) as dual signal amplifiers and Chi-modified MWCNTs (MWCNTs-Chi) as substrate materials was developed for ultrasensitive detection of fowl adenovirus group I (FAdV-I). MWCNTs have a large specific surface area, and many accessible active sites were formed after modification with Chi. Hence, MWCNTs-Chi, as a substrate material for modifying glassy carbon electrodes (GCEs), could immobilize more antibodies (fowl adenovirus group I-monoclonal antibody, FAdV-I/MAb). Multiple Pt/AgNPs were attached to the surface of MWCNTs-Chi to generate MWCNTs-Chi-Pt/AgNPs with high catalyticability for the reaction of H_2O_2 and modified active sites for fowl adenovirus group I-polyclonalantibody (FAdV-I/PAb) binding. Amperometric i-t measurements were employed to characterize the recognizability of FAdV-I. Under optimal conditions, and the developed immunosensor exhibited a wide linear range ($10^{0.93}$ EID_{50}/mL to $10^{3.43}$ EID_{50}/mL), a low detection limit ($10^{0.67}$ EID_{50}/mL) and good selectivity, reproducibility and stability. This immunosensor can be used in clinical sample detection.

Keywords

amperometric immunosensor, Pt/Ag nanoparticle, multiwalled carbon nanotubes, fowl adenovirus group I

Introduction

Avian adenoviruses were separated into 3 groups (I-III). Group I, the so-called fowl adenovirus group I (FAdV-I), comprises 12 serotypes (FAdV-1, FAdV-2, FAdV-3, FAdV-4, FAdV-5, FAdV-6, FAdV-7, FAdV-8a, FAdV-8b, FAdV-9, FAdV-10, and FAdV-11) and mainly infects turkeys, chickens, ducks, geese and pigeons[1, 2]. FAdV-I can be transmitted horizontally and vertically through feces and progeny, respectively[3]. It has been demonstrated that infections caused by FAdV-I are associated with hydropericardium, hepatitis and runting-stunting, resulting in apathy, weight loss, and mortality[4]. FAdV-I has been reported in many countries, and FAdV-I infections cause enormous economic burdens in poultry farming[4, 5]. Therefore, rapid, specific and sensitive detection methods are needed to screen for and control FAdV-I infection. Methods such as virus isolation and identification, polymerase chain reaction (PCR)-based assays and loop-mediated isothermal amplification (LAMP) have been proposed [6-9]. However, these diagnostic methods require either highly qualified personnel or sophisticated instrumentation.

Amperometric immunosensors are particularly useful for detecting various diseases because of their rapid detection, small analyte volume, high specificity and low detection limits for analyzing complex clinical samples with relatively simple instruments and procedures[10, 11]. Amperometric immunosensors mainly include label-free and sandwich amperometric immunosensors. Label-free amperometric immunosensors, that is, direct immunosensors, work by measuring the signal changes arising directly from immune reactions without requiring labeling; these immunosensors are used for fast, real-time detection[12]. However, when other proteins or antigens are present in a dinical sample, nonspecific binding of the other proteins or antigens on the surface of the substrate can occur, and a small signal is generated, leading to an increase in the background signal, which results in a decrease in sensitivity[10]. In sandwich immunosensors, the antigen is sandwiched between the primary and detection antibodies. The primary antibody, which is immobilized on a solid substrate surface and used to capture the antigen from the sample, is known as the capture antibody. Detection antibodies, known as labeled antibodies, are used to attach to labels such as enzymes and nanomaterials that can be used to amplify signals[13]. Sandwich immunosensors based on signals generated from labels have several advantages, such as decreased nonspecific adsorption and increased sensitivity[14]. In general, compared with label-free amperometric immunosensors, sandwich-type amperometric immunosensors result in more sensitive assays with a wide detection range because of signal amplification with detection antibody labels[13, 14]. Sandwich-type amperometric immunosensors, which convert immune responses triggered by probe target immunocomplexes into readable current signals, are powerful analytical devices.

Multiwalled carbon nanotubes (MWCNTs) have attracted the attention of researchers in recent decades because of their outstanding electrical conductivity and large specific surface areas; hence, they have been widely applied in catalysis, adsorption and energy storage systems[15, 16]. In this work, MWCNTs with sufficient transport channels were employed to enhance the current response of a developed amperometric immunosensor. Metal nanoparticles have attracted interest because of their unique physical, chemical and electronic properties and potential applications in amperometric sensors[17, 18]. The excellent conductivity of metal nanoparticles enhances the peak current by transferring electrons from the redox centers in target molecules to the electrode surfaces[19]. Among metal nanoparticles, noble metal nanoparticles (such as gold nanoparticles, platinum nanoparticles and silver nanoparticles), which have good biocompatibility, high catalytic activity and superior electrical conductivity, have been widely used in amperometric immunosensors[13, 20]. In addition, bimetallic metal nanoparticles have attracted specific interest because they exhibit distinct unique characteristics, such as enhanced electrical conductivity high catalytic ability and long-term stability, unlike single metal nanoparticles because of the synergistic effects between bimetallic metal nanoparticles[21, 22]. Platinum nanoparticle and silver nanoparticle catalysts have been used in immunosensors in recent years because of their unique catalytic abilities. Platinum silver alloy nanoparticles, which are catalysts with synergistic effects between Ag and Pt nanoparticles, exhibit extraordinary electrocatalytic activity for H_2O_2 reduction[23]. In particular, MWCNTs with more active sites and a larger functional surface area were utilized to immobilize Pt/AgNPs to load more active probes bound to biomolecules[24]. Pt/AgNPs can combine with MWCNTs to form composites, which can prevent the agglomeration of Pt/AgNPs and ensure that the nanoparticle size is maintained. Furthermore, MWCNTs have a fast electron transfer rate and a large specific surface area, which can improve immunosensor properties[15]. Inspired by this, bimetallic Pt/AgNPs immobilized on the surface of MWCNTs were developed as a signal amplification label in this work, and the signal was effectively amplified based on the dual catalytic effect of Pt/AgNPs and MWCNTs on H_2O_2.

Another crucial factor in obtaining sensitive electrical signals is the immobilization of biomolecules on highly conductive substrates. MWCNTs, which have remarkable electron transfer capability, are promising substrate materials[25]. However, MWCNTs have poor solubility, which is a major barrier to developing immunosensors based on MWCNTs. To optimize the use of MWCNTs for immunosensor applications, it is necessary to functionalize MWCNTs with biomaterials. Among the various biomaterials, chitosan (Chi) is a popular matrix for immobilizing immunosensor elements due to its nontoxicity, excellent film properties, biocompatibility and availability of amine/hydroxyl groups; additionally, Chi can be used to immobilize biomolecules for biosensor applications[26]. However, due to the nonconductive nature of Chi, its use in the preparation of an electrochemical immunosensor has been limited. To increase the conductivity of Chi, a variety of nanomaterials with high conductivity have been incorporated with Chi[27]. Among them, MWCNTs have a low cost, large surface area and conductive properties. More importantly, Chi can covalently attach to MWCNTs via π-π interactions to form MWCNTs-Chi nanobiocomposites. This approach involves the incorporation of MWCNTs with a polymeric backbone, which is helpful for increasing the solubility of MWCNTs without decreasing their conductivity[28], this approach can not only synergistically promote electron conduction but also effectively immobilize biomolecules such as antibodies and maintain their bioactivity[29]. Based on the above advantages, MWCNTs-Chi nanobiocomposites as substrates provide a suitable microenvironment for immunosensors.

In this study, bimetallic Pt/AgNPs immobilized on the surface of MWCNTs were used as a label to promote dual amplification of the current signal. MWCNTs-Chi were used as a substrate to provide a stable microenvironment for maintaining biomolecular activity and effectively increasing the electron transfer rate on the surface of the electrode. With good coordination of bimetallic Pt/AgNPs immobilized on the surface of MWCNTs-Chi, the established electrochemical immunosensor showed remarkable analytical performance and yielded convincing results in clinical sample assays, demonstrating its good potential for application in disease detection.

Materials and methods

Reagents and instruments

Chloroplatinic acid hexahydrate ($H_2PtCl_6 \cdot 6H_2O$), silver nitrite ($AgNO_3$), bovine serum albumin (BSA, 98% purity) and MWCNTs were purchased from Sigma-Aldrich Chemical Co. (St. Louis, MO, USA). Chi (Mw 1.5×10^{-5}, deacetylation degree$\geqslant$90%), $NaNO_3$, H_2SO_4 and $KMnO_4$ were obtained from Guoyao Group Chemical Reagents Co., Ltd. (Shanghai, China). Deionized water (ddH_2O_2-18 MQ resistance) prepared by a Millipore water purification system was used to prepare the buffers and solutions. All the chemicals were of analytical reagent grade. Phosphate-buffered saline (PBS) was prepared with 10 mmol/L Na_2HPO_4, 10 mmol/L NaH_2PO_4 and 0.9% NaCl.

Apparatus

All the electrochemical measurements were conducted with a PARSTAT 4000A instrument (Princeton Applied Research, USA) in this work. A glassy carbon electrode (GCE, Ø=3 mm), a saturated calomel electrode (SCE) and a platinum wire electrode were used as the working electrode, reference electrode and counter electrode, respectively, composing a standard three-electrode system. Energy-dispersive X-ray spectroscopy (EDS) elemental analysis and transmission electron microscopy (TEM, TecnaiG2 F30 S-TWIN,

FiE, USA) were employed to characterize the nanomaterials.

Viruses and antibodies

Fowl adenovirus group I-monoclonal antibody (FAdV-I/MAb) and fowl adenovirus group I-polylonal antibody (FAdV-I/PAb) were prepared by our research group[30]. In brief, codon optimization was performed on the FAdV-4-100K gene sequence, and a sequence fragment encoding a truncated FAdV-4-100K protein with strong antigenicity was amplified by PCR and cloned into pET-32a (+) to construct the vector for prokaryotic expression. The recombinant protein was purified. Following the immunization of BALB/c mice with the recombinant protein, the serum was collected as a polyclonal antibody. Splenocytes were harvested and fused with SP2/0 cells, and the positive cell lines were screened by indirect enzyme-linked immunosorbent assay (ELISA) and used to prepare monoclonal antibodies. FAdV-1, FAdV-2, FAdV-3, FAdV-4, FAdV-5, FAdV-6, FAdV-7, FAdV-8a, FAdV-8b, FAdV-9, FAdV-10 and FAdV-Il were purchased from the China Institute of Veterinary Drugs Control. Chicken infectious anemia virus (CIAV), infectious bronchitis virus (IBV), Newcastle disease virus (NDV), egg drop syndrome virus (EDSV), avian reovirus (ARV) and the avian influenza virus H9 subtype (AIV H9) were collected and stored in our laboratory prior to use.

Synthesis of MWCNTs-Chi

MWCNTs-Chi was prepared by following the method described above. First, 0.5 g of Chi was added to 100 mL of 1.0% (v/v) acetic acid solution, and the mixture was magnetically stirred at room temperature for 0.5 h. A 0.5 wt% Chi solution was obtained. Second, 100 mg of MWCNTs was added to 100 mL of Chi solution obtained by the above preparation, ultrasonicated for 0.5 h, and continuously magnetically stirred for 48 h at room temperature. A stable MWCNTs-Chi (1mg/mL) suspension was subsequently obtained.

Synthesis of MWCNTs-Chi-Pt/AgNPs-FAdV/PAb

MWCNTs-Chi-Pt/AgNPs-FAdV/PAb were prepared by following the method described above. First, 1 mL of $AgNO_3$ (10 mmol/L) and 1 mL of K_2PtCl_4 (10 mmol/L) were added to the above-prepared MWCNTs-Chi (30 mL) suspension, and the mixture was continuously magnetically stirred for 5 h at room temperature. Second, the mixture was heated in a water bath to 90 ℃ and kept for 0.5 h for further reaction ($Pb^{2+}$and Ag^+ were reduced to Pt/AgNPs by Chi here); then, MWCNTs-Chi-Pt/AgNPs were obtained. Different proportions of MWCNTs-Chi-Pt/AgNPs were prepared by adding different volumes of $AgNO_3$ (10 mmol/L) and K_2PtCl_4 (10 mmol/L) to 30 mL of the MWCNTs-Chi (1 mg/mL) suspension by the method mentioned above.

Third, the as-prepared MWCNTs-Chi-Pt/AgNPs (10 mL) were mixed with 1 mL of 1 μg/mL FAdV-I/PAb, and the mixture was vibrated in a shaker at 4 ℃ for 12 h. The product was subsequently washed with PBS (pH 7.4) containing 1 wt% BSA and centrifuged at 10 000 r/min for 5 min, after which the supernatant was discarded to remove the unbound FAdV-I/PAb. Finally, the obtained MWCNTs-Chi-Pt/AgNPs-FAdV/PAb were redispersed into 10 mL of PBS (pH 7.4) containing 1 wt% BSA.

Fabrication of an electrochemical immunosensor

First, a glass carbon electrode (GCE, Ø=3 mm) was polished to a mirror surface with 0.05 μm Al_2O_3 polishing powder, washed with ddH_2O and cleaned via ultrasonication in ddH_2O, CH_3CH_2OH and ddH_2O for 5 min each.

Second, a 0.5 mol/L H_2SO_4 solution was injected with N_2 for 5 min to remove oxygen, after which the GCE was scanned via cyclic voltammetry (CV) in H_2SO_4 solution at a scanning speed of 50 mV/s and a voltage range of −0.3～+1.5 V until the CV current was stable. Afterward, the GCE was washed with ddH_2O three times and dried with N_2 for further modification.

Third, 10 μL of 1 mg/mL MWCNTs-Chi was dropped on the surface of the GCE and dried at 37 ℃. Next, 10 μL of 10 μg/mL FAdV-I/MAb was deposited onto the electrode, incubated at 4 ℃ for 8 h, and then washed with PBS (pH 7.4) three times to remove the unbound FAdV-I/MAb.

Fourth, 10 μL of 1 wt% BSA, which was used as a nonspecific binding blocker, was deposited onto the electrode and incubated for 1 h at 37 ℃ to eliminate nonspecific binding, after which the modified electrode was washed with PBS (pH7.4) three times to remove the excess BSA. The obtained immunosensor (GCE-MWCNTs-Chi-FAdV-I/MAb-BSA) was stored at 4 ℃ until use.

Detection of the electrochemical immunosensor

The procedure for the electrochemical immunosensor preparation is shown in Figure 5-33-1. First, 10 μL of solution containing different concentrations of FAdV-I was incubated with the developed immunosensor for 30 min to ensure specific binding of FAdV-I to FAdV-I/MAb, after which the mixture was washed with PBS (pH 7.4) three times to remove unbound FAdV-I and other interfering substances. Second, 10 μL of the developed immunosensor was incubated with MWCNTs-Chi-Pt/AgNPs-FAdV/PAb at 37 ℃ for 40 min and then washed with PBS (pH 7.4) three additional times to remove the excess MWCNTs-Chi-Pt/AgNPs-FAdV/PAb. Finally, amperometric i-t measurements were performed, and the voltage was set to −0.1 V based on minimized interfering components and background current. After the signal stabilized (approximately 50 s later), H_2O_2 (5 mmol/L) was added to PBS (10 mL, pH 7.4) under continuous stirring, and the current change was recorded. The mechanism of signal production by catalytic H_2O_2 was as follows[31]: $H_2O_2+e^- \rightarrow OH_{ad}+OH^-$; $OH_{ad}+e^- \rightarrow OH$; $2H^++2OH^- \rightarrow 2H_2O$. In the H_2O_2 reduction process, the key step is OH_{ad} formation, which helps to amplify the signal for sensitive detection of FAdV-I, and MWCNTs-Chi-Pt/AgNPs can provide abundant catalytic active centers for OH_{ad} adsorption[32]. Electrochemical impedance spectroscopy (EIS) was performed in a $[Fe(CN)_6]^{3-/4-}$ (2.5 mM) and 0.1 M KCl mixed solution with an amplitude of 10 mV at 0.2 V (open circuit potential) and frequencies ranging from 0.01 Hz to 100 kHz.

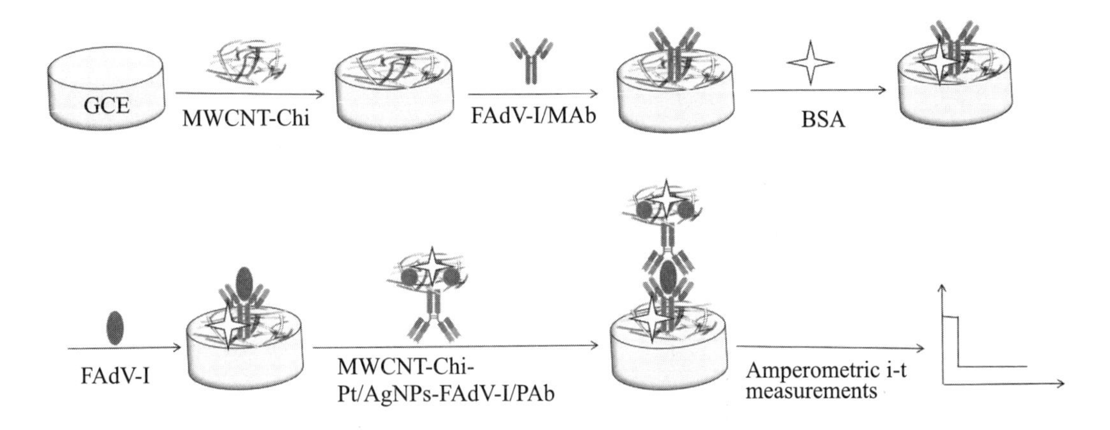

Figure 5-33-1 Schematic illustrating the exploited electrochemical immunosensor

Results and discussion

Characterization of the nanomaterials

Transmission electron microscopy (TEM) was employed to characterize the features of the MWCNTs-Chi and MWCNTs-Chi-Pt/AgNPs. Figure 5-33-2 A shows the typical tubular structure of MWCNTs. Figure 5-33-2 B shows TEM images of the MWCNTs-Chi-Pt/AgNPs. Pt/AgNPs were uniformly distributed on the surface of MWCNTs-Chi. Moreover, energy dispersive X-ray spectroscopy (EDS) was employed to confirm the presence of Pt and Ag in the MWCNTs-Chi-Pt/AgNPs, and the results showed that the Pt/AgNPs were successfully loaded on the surface of the MWCNTs (Figure 5-33-2 C). Here, we suspect that Pt^{2+} and Ag^{+} ions in aqueous solution were adsorbed onto the MWCNTs-Chinanocomposite via chelation and reduced to their corresponding

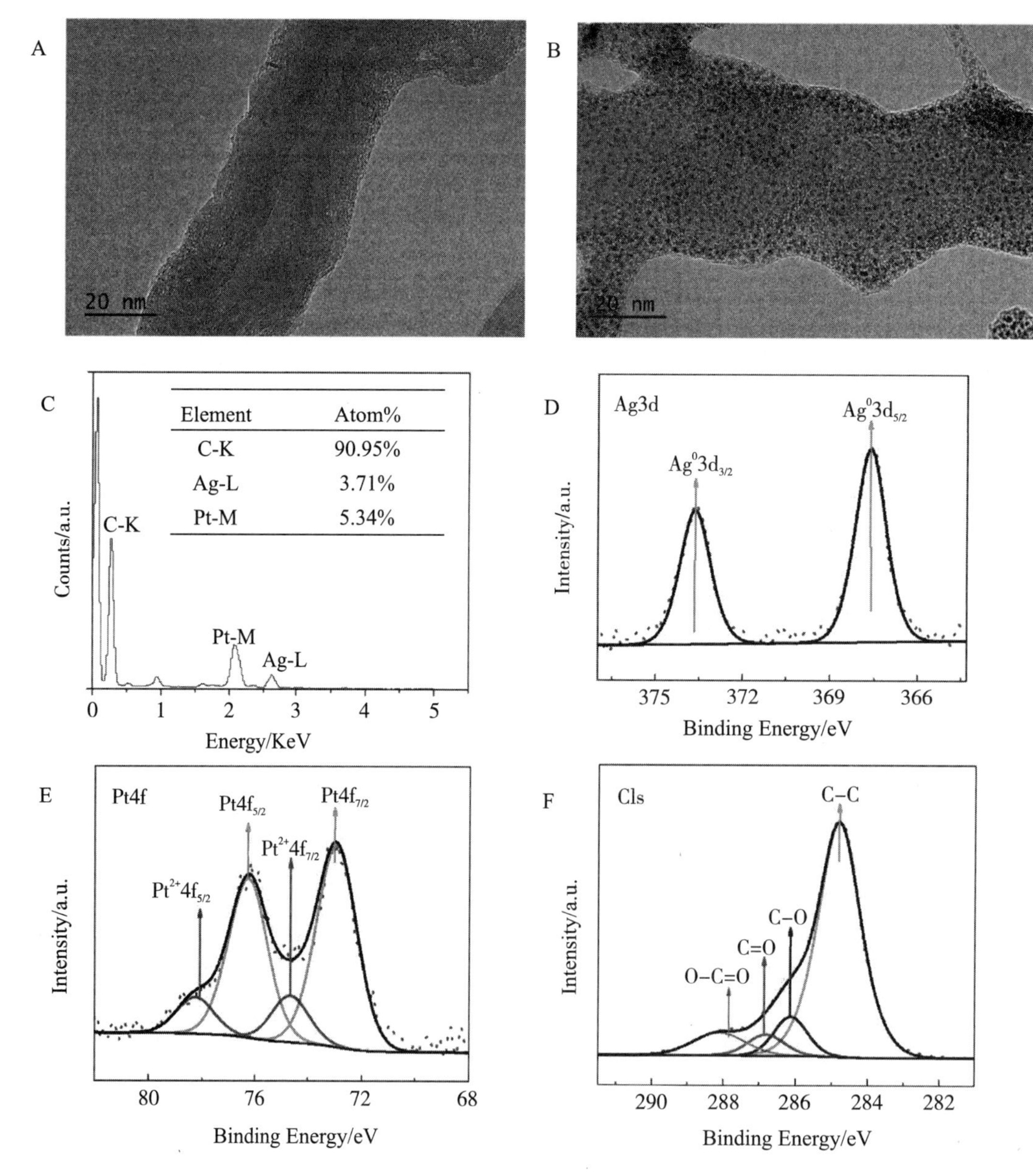

A: TEM image of MWCNTs-Chi; B: TEM image of MWCNTs-Chi-Pt/AgNPs; C: EDS elemental analysis of MWCNTs-Chi-Pt/AgNPs; D: High-resolution Ag 3d; E: High-resolution Pt 4f; F: High-resolution C 1s XPS spectra of MWCNTs-Chi-Pt/AgNPs.

Figure 5-33-2 Characterization of the related nanocomposites

metal nanomaterials by Chi[33, 34]. Chi acted as a reducing agent and stabilizing agent. The MWCNTs acted as carriers to load more Pt/AgNPs and prevent the agglomeration of Pt/AgNPs to ensure that their size was consistent with that of the nanoparticles.

X-ray photoelectron spectroscopy (XPS) was employed to survey the surface features of the MWCNTs-Chi-Pt/AgNPs. The XPS spectrum revealed diffraction peaks at 367.62 ($3d_{5/2}$) and 373.63 ($3d_{3/2}$) eV (Figure 5-33-2 D), which corresponded well to Ag^0, and no obvious peak attributed to Ag^+ was observed, which indicated that metallic Ag^0 was the predominant species[35]. The Ag 3d peak shifted to a lower value of 367.62 eV relative to that of pure Ag (368.3 eV) [36] because of the modification of Ag by alloying with Pt. These results confirmed the formation of the Pt/AgNPs alloy during the synthesis of MWCN Ts-Chi-Pt/AgNPs[37].

As shown in Figure 5-33-2 E, the binding energies at 72.94 and 76.27 eV corresponded to Pt $4f_{7/2}$ and Pt $4f_{5/2}$, respectively, which were attributed to Pt^{2+}, while the binding energies at 74.64 and 78.25 eV were assigned to Pt^0. These results demonstrated that metallic Pt was formed and that the partial reduction of platinum atoms was incomplete[38].

The high-resolution C1s XPS spectrum of the MWCNTs-Chi-Pt/AgNPs is shown in Figure 5-33-2 F. Four diffraction peaks were observed at 284.78, 286.08, 286.7 and 288.13 eV, which were assigned to the covalent C-C, C-O, C=O, and O-C=O bonds, respectively[39]. The diffraction peak intensity ratio of C-C was much higher for MWCNTs-Chi-Pt/AgNPs than for MWCNTs-Chi (Figure 5-33-3), demonstrating the efficient reduction of MWCNTs-Chi[40].

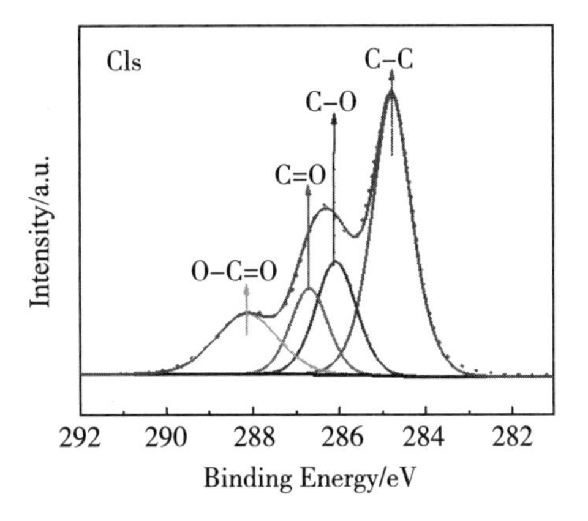

Figure 5-33-3 High-resolution C1s XPS spectrum of MWCNT-Chi

Electrochemical characterization of the immunosensor

In this study, EIS was employed to investigate the step-by-step assembly process of the immunosensor in a mixed solution of $[Fe(CN)_6]^{3-/4-}$ (2.5 mM) and 0.1 M KCl with an amplitude of 10 mV at 0.2 V and frequencies ranging from 0.01 Hz to 100 kHz. The Nyquist plots obtained by the EIS method were fitted to the effective impedance of the electrode, which is usually expressed as the Nyquist diameter of the semicircle of the graph, which is also known as the electron transfer resistance (R_{et}). According to the EIS Nyquist plots, the low-frequency region usually appears as a line that reflects the diffusion process on the electrode surface. The high-frequency region appears as a semicircle that corresponds to R_{et} Figure 5-125 A shows the EIS Nyquist plots obtained during the step-by-step electrode assembly process. In these plots, a semicircle with an R_{et} of 1 039 Ω (%RSD) =3.29 was observed for the bare GCE (Figure 5-33-4 A, curve a-1 and B, column a-1), and a smaller semicircle with an R_{et} of 418 Ω (%RSD) =2.47 was obtained after the GCE was modified with MWCNTs-Chi (Figure 5-33-4 A, curve-1 and B, column b-1) because the excellent conductivity of the MWCNTs promoted electron transfer on the surface of the electrode. Then, the electrode was modified by FAdV-I/MAb, BSA and $10^{3.43}$ EID_{50}/mL FAdV-I. The results (Figure 5-33-4 B, columns c-1 to e-1) showed that the obtained relevant R_{et} values sequentially increased to 1 575 Ω (%RSD=3.14), 2 630 Ω (%RSD=2.91) and 3 584 Ω (%RSD=3.21) because FAdV-I/MAb, BSA and FAdV-I, respectively, have poor conductivity of protein molecules. These results revealed that each step of the electrode modification was successful. The Randles equivalent circuit model is shown in the illustration.

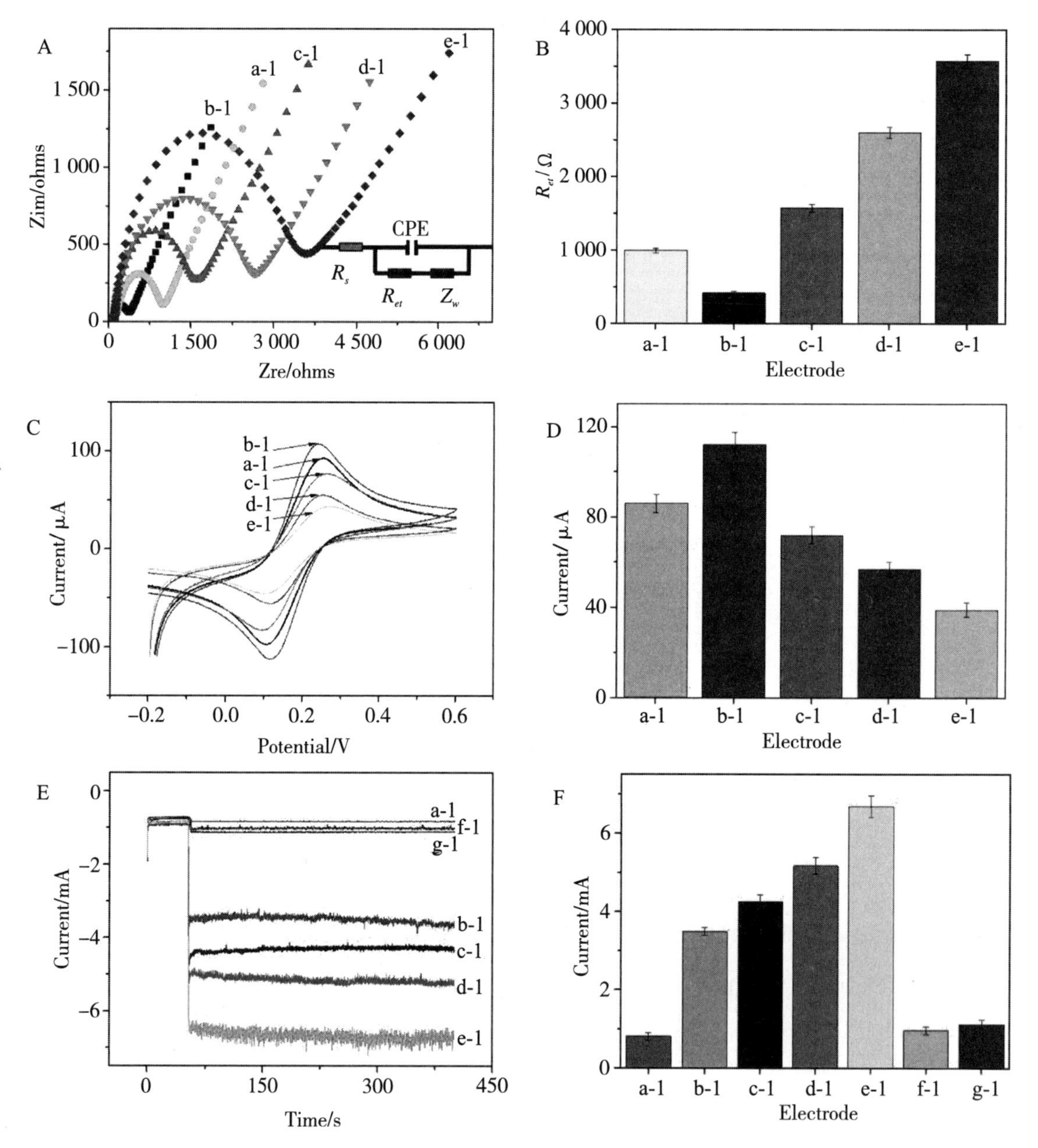

A and B: EIS Nyquist plots and R_{et} values of electrodes at different modification steps. a-1: GCE; b-1: GCE-MWCNTs-Chi; c-1: GCE-MWCNTs-Chi-FAdV-I/MAb; d-1: GCE-MWCNTs-Chi-FAdV-I/MAb-BSA; e-1: GCE-MWCNTs-Chi-FAdV-I/MAb-BSA-FAdV-I.

C and D: CV curves of electrodes at different modification steps. a-1: GCE; b-1: GCE-MWCNTs-Chi; c-1: GCE-MWCNTs-Chi-FAdV-I/MAb; d-1: GCE-MWCNTs-Chi-FAdV-I/MAb-BSA; e-1: GCE-MWCNTs-Chi-FAdV-I/MAb-BSA-FAdV-I.

E and F: Amperometric i-t curves of the immunosensor for the detection of FAdV-I. a-1: Blank; b-1: MWCNTs-Chi-PtNPs-FAdV-I/PAb; c-1: MWCNTs-Chi-AgNPs-FAdV-I/PAb; d-1: Chi-Pt/AgNPs-FAdV-I/PAb; d-1: MWCNTs-Chi-Pt/AgNPs-FAdV-I/PAb as labels; f-1: Immunosensor without incubation of FAdV-I/MAb; g-1: Without incubation of FAdV-I/MAb but with incubation of FAdV-I and MWCNTs-Chi-Pt/AgNPs-FAdV-I/PAb as labels.

Figure 5-33-4 Electrochemical characterization of the immunosensor

Furthermore, cyclic voltammetry (CV) is another valid technique for verifying the modification of electrodes. The CV data of the GCE electrode demonstrated the effectively reversible redox properties of $[Fe(CN)_6]^{3-/4-}$ (Figure 5-33-4 C, D, curve and column a-1), and the peak current of the GCE-MWCNTs-Chi increased due to the good conductivity of the MWCNTs (Figure 5-33-4 C, D, curve and column b-1). After FAdV-I/MAb, BSA and FAdV-I ($10^{3.43}$ EID_{50}/mL) were successively adsorbed on the modified electrode, the corresponding peak currents decreased (Figure 5-33-4 C, D, curve and columns c-1 to e-1). These results are

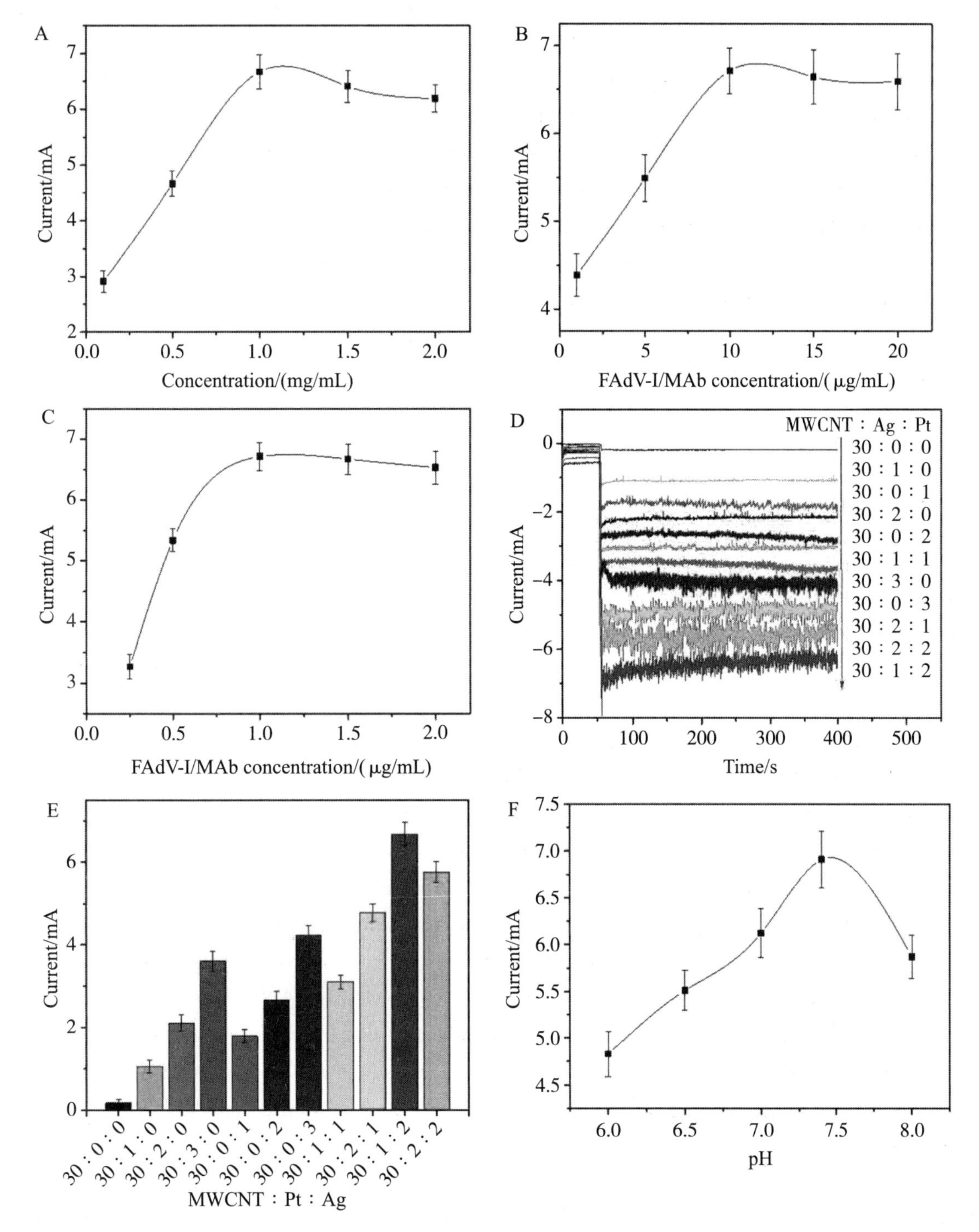

A: MWCNTs-Chi concentration; B: FAdV-I/MAb concentration; C: FAdV-I/PAb concentration; D and E: Ratio of MWCNTs to Ag and Pt; F: PH of the electrolyte.

Figure 5-33-5 Optimizations of the immunosensor

consistent with the EIS results, indicating that the FAdV-I immunosensor was successfully constructed.

Amperometric i-t curve measurements were used to evaluate the electrocatalytic performance of the developed immunosensor. The results showed that the electrocatalytic activity of MWCNTs-Chi-FAdV-I/MAb-BSA (Figure 5-33-4 E, F, curve and column a-1) could be ignored because it was much weaker than that of MWCNTs-Chi-Pt/AgNPs-FAdV-I/PAb (Figure 5-33-4 E, F, curve and column e-1). In addition, the electrocatalytic performance of MWCNTs-Chi-PtNPs-FAdV-I/PAb, MWCNTs-Chi-AgNPs-FAdV-I/PAb, Chi-Pt/AgNPs-FAdV-I/PAb, and MWCNTs-Chi-Pt/AgNPs-FAdV-I/PAb as labels during the detection process were compared. The immunosensor using MWCNTs-Chi-Pt/AgNPs-FAdV-I/PAb (Figure 5-33-4 E, F curve and column e-1) as the label had a much higher current change response than did that using MWCNTs-Chi-PtNPs-FAdV-I/PAb (Figure 5-33-4 E, F, curve and column b-1), MWCNTs-Chi-AgNPs-FAdV-I/PAb (Figure 5-33-4 E, F, curve and column c-1) and Chi-Pt/AgNPs-FAdV-I/PAb (Figure 5-33-4 E, F, curve d-1) as the labels. The higher current change response obtained from MWCNTs-Chi-Pt/AgNPs-FAdV-I/ PAb was attributed mainly to the excellent catalytic activity of Pt/AgNPs and the excellent electrical conductivity and large specific surface area of the MWCNTs. When the electrode was not immobilized with FAdV-I/MAb and was directly blocked with BSA, the obtained peak current (Figure 5-33-4 E, F, curve and column f-1) was similar to the background signal (Figure 5-33-4 E, F, curve and column a-1); after it was reacted with $10^{3.43}$ EID_{50}/mL FAdV-I and MWCNTs-Chi-Pt/AgNPs-FAdV-I/PAb, the obtained peak current (Figure 5-33-4 E, F, curve and column g-1) was also similar to the background signal. These results indicate that the developed immunosensor has excellent selectivity.

Optimization of the method

MWCNTs-Chi, which are conductive substrates, are important for enhancing electron transfer between the surface of the electrode and electrolyte and immobilizing FAdV-I/MAb. Hence, the concentration of MWCNTs-Chi was optimized because it affects the amperometric response. GCEs were modified with different concentrations of MWCNTs-Chi to detect $10^{3.43}$ EID_{50}/mL FAdV-I. The currents obtained from the amperometric i-t curve are shown in Figure 5-33-5 A. The results showed that the current response increased as the MWCNTs-Chi concentration increased from 0.1 to 1.0 mg/mL. Here, the current response increased due to the increase in the MWCNTs-Chi concentration, directly resulting in an increase in the number of FAdV-I/MAb anchoring sites on the electrode. When the concentration of MWCNTs-Chi increased from 1 to 2 mg/mL, the current response signal reached a plateau. Therefore, 1.0 mg/mL MWCNTs-Chi was the optimal concentration.

The concentrations of FAdV-I/MAb and FAdV-I/PAb were investigated, and the results are shown in Figure 5-33-5 B, C. The corresponding response currents gradually increased with increasing FAdV-I/MAb and FAdV-I/PAb concentrations, and the corresponding response currents reached maximum values at 10 μg/mL and 1.0 μg/mL, respectively. As the concentrations of FAdV-I/MAb and FAdV-I/PAb continued to increase, the corresponding response currents reached a plateau, indicating that the concentrations of FAdV-I/MAb and FAdV-I/PAb were oversaturated. Therefore, the optimal concentrations of FAdV-I/MAb and FAdV-I/PAb were 10 μg/mL and 1.0 μg/mL, respectively.

In this work, the MWCNTs-Chi-Pt/AgNPs-FAdV-I/PAb nanocomposite was used as a signal label, and the response current was affected by the ratio of MWCNT, Ag and Pt (MWCNT : Ag : Pt). The results are shown in Figure 5-33-5 D, E. When Ag was present, the corresponding signal increased as the ratio of MWCNT to Ag to Pt increased from 30 : 1 : 0 to 30 : 3 : 0. When Pt was present, the corresponding signal increased as the ratio of MWCNT to Ag and Pt increased from 30 : 0 : 1 to 30 : 0 : 3. When both Ag and Pt were present, the corresponding signal increased as the ratio of MWCNT to Ag to Pt increased from 30 : 1 : 1 to 30 : 1 : 2, and the corresponding signal decreased as the ratio of MWCNT to Ag to Pt increased to 30 : 2 : 2. Furthermore, the presence of Ag and Pt resulted in a greater catalytic ability for H_2O_2 than for single metal nanoparticles because of the synergistic effects between Ag and Pt. Hence, a MWCNT : Ag : Pt ratio of 30 : 1 : 2 was the best choice in this study.

The signal of electrochemical immunosensors is usually affected by the pH of the electrolyte. In this work, the pH in the electrolyte was optimized, and the results are shown in Figure 5-33-5 F. The corresponding currents increased as the pH increased from 6.0 to 7.4, and the corresponding currents decreased when the pH of the electrolyte was above 7.4. Hence, pH=7.4 was the optimal electrolyte. The stability of the antigen-antibody combination may be affected by acidic or alkaline solutions, leading to changes in the signal.

The concentration of H_2O_2 was optimized, and the results are shown in Figure 5-33-6 A, B. The corresponding currents increased when the concentration of H_2O_2 increased from 1 to 5 mM, and the corresponding currents plateaued as the concentration of H_2O_2 continued to increase because H_2O_2 reached saturation. Hence, 5 mM H_2O_2 was chosen as the optimal concentration for addition to the electrolyte.

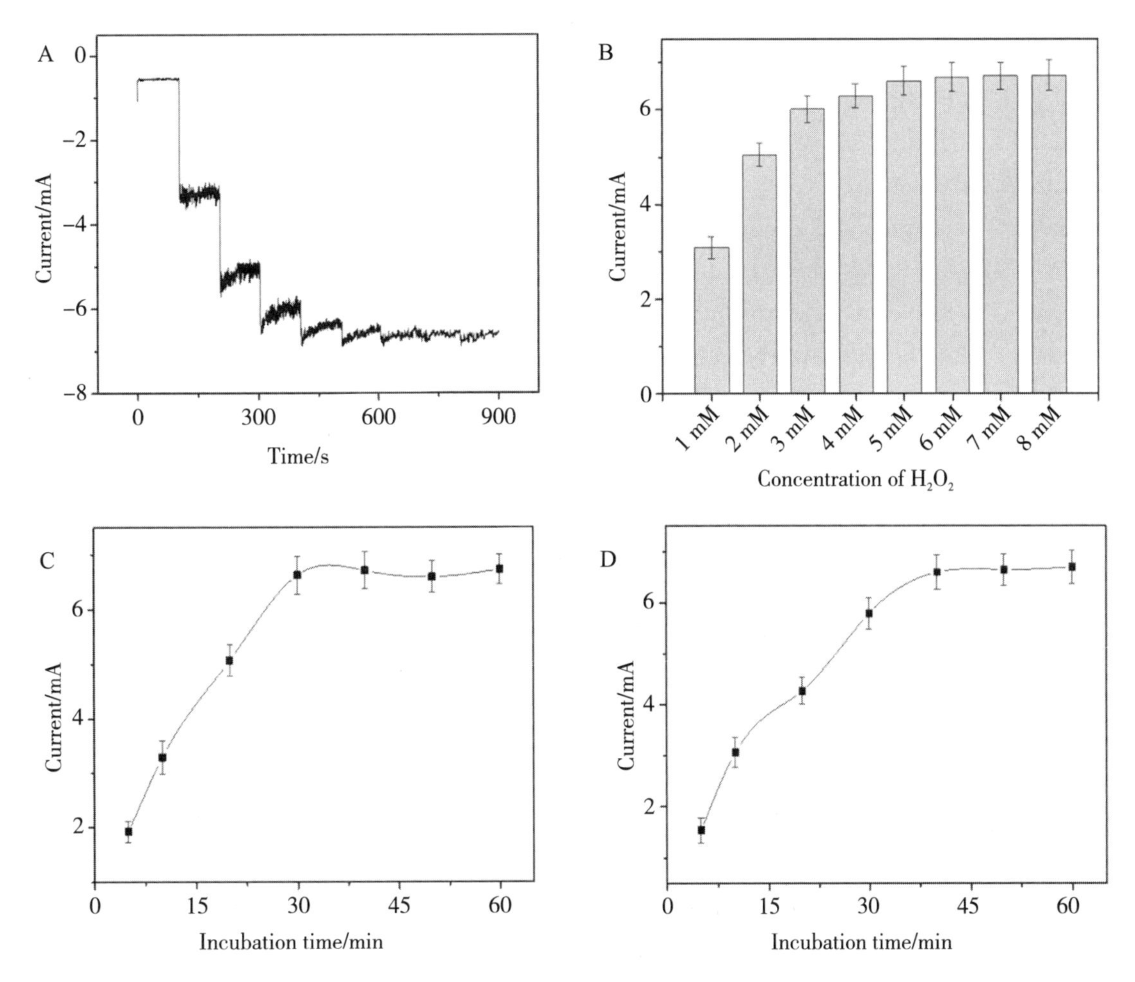

A and B: The concentration of H_2O_2; C: Incubation time of FAdV-I; D: Incubation time of the MWCNTs-Chi-Pt/AgNPs-FAdV-I/PAb nanocomposite.

Figure 5-33-6 Optimization of immunosensor

Furthermore, the incubation time of the sample (FAdV-I) and signal label materials (MWCNTs-Chi-Pt/AgNPs-FAdV-I/PAb nanocomposite) considerably influenced the immunosensor. The corresponding currents increased with increasing incubation time for 30 min and 40 min for FAdV-I and MWCNTs-Chi-Pt/AgNPs-FAdV-I/PAb, respectively, (5-33-6 C, D). The corresponding currents plateaued as the incubation time of FAdV-I and MWCNTs-Chi-Pt/AgNPs-FAdV-I/PAb was extended because the immunoreaction was complete. Hence, the optimal incubation times for FAdV-I and MWCNTs-Chi-Pt/AgNPs-FAdV-I/PAb were 30 min and 40 min, respectively.

Detection of FAdV-I with the immunosensor

The selected immunosensor was used to detect FAdV-I at concentrations ranging from $10^{3.93}$ to $10^{0.93}$ EID_{50}/mL to evaluate its detection capability. Figure 5-33-7 A shows that the corresponding signal intensity increased with increasing concentrations of FAdV-I. Furthermore, the current signal of the blank (curve a-1) was significantly lower than the current signal of FAdV-I (curves b-1-h-1), indicating that the background current of MWCNTs-Chi-FAdV-I/MAb-BSA and nonspecific adsorption on the surface of the immunosensor were negligible. The calibration plot shown in Figure 5-33-7 B shows a good linear relationship between the logarithm values of the FAdV-I concentration and the change in the current signal. The linear regression equation I (mA) =2.152 1 lgEID_{50}/mL -1.185 2 (correlation coefficient of R=0.991 6) and alow detection limit ($10^{0.67}$ EID_{50}/mL (S/N=3)) were obtained. The excellent analytical performance of the exploited immunosensor was attributed to the following factors: (1) MWCNTs-Chi, which has remarkable electroconductivity and a large specific surface area, were used to modify the electrode and not only promoted electron transfer between the surface of the electrode and electrolyte but also provided enough active sites for FAdV-I/MAb binding; (2) the MWCNTs-Chi-Pt/AgNPs-FAdV-I/PAb nanocomposite had excellent electrocatalytic capability for H_2O_2 reduction due to electron-electron correlation assistance; (3) FAdV-I/MAb, which has good specificity, was used as a capture antibody to ensure the specificity of the developed immunosensor; and (4) FAdV-I/PAb, which has more antigen-binding sites as a detection antibody, can enhance the sensitivity of the proposed immunosensor.

Selectivity, reproducibility and stability of the immunosensor

The specificity of the immunosensor plays an important role in the analysis of clinical samples. Interfering substances that may be present in the clinical sample, including IBV ($10^{4.79}$ EID_{50}/mL), CIAV ($10^{5.31}$ EID_{50}/mL), NDV ($10^{6.73}$ EID_{50}/mL), EDSV ($10^{5.17}$ EID_{50}/mL), ARV ($10^{7.18}$ EID_{50}/mL), AIV H9 ($10^{6.47}$ EID_{50}/ mL), BSA (1.0 μg/mL), glucose (1.0 μg/mL) and vitamin C (1.0 μg/mL), were employed to evaluate the specificity of the exploited immunosensor (Figure 5-33-7 C). The current responses of the abovementioned interfering substances were similar to those of the blank solution but markedly lower than those of FAdV-I (including 12 serotypes FAdV-1 ($10^{5.75}$ EID_{50}/mL), FAdV-2 ($10^{4.38}$ EID_{50}/mL), FAdV-3 ($10^{6.21}$ EID_{50}/mL), FAdV-4 ($10^{4.92}$ EID_{50}/mL), FAdV-5 ($10^{5.47}$ EID_{50}/mL), FAdV-6 ($10^{6.35}$ EID_{50}/mL), FAdV-7 ($10^{4.63}$ EID_{50}/mL), FAdV-8a ($10^{6.27}$ EID_{50}/mL), FAdV-8b ($10^{7.35}$ EID_{50}/mL), FAdV-9 ($10^{5.79}$ EID_{50}/mL), FAdV-10 ($10^{6.24}$ EID_{50}/mL), and FAdV-11 ($10^{5.62}$ EID_{50}/mL)). The current responses of the mixtures of FAdV-4 ($10^{3.43}$ EID_{50}/mL) with possible interfering substances (such as NDV, ARV and BSA) were similar to those of FAdV-I. These results indicated that the applied immunosensor had high specificity for FAdV-I detection.

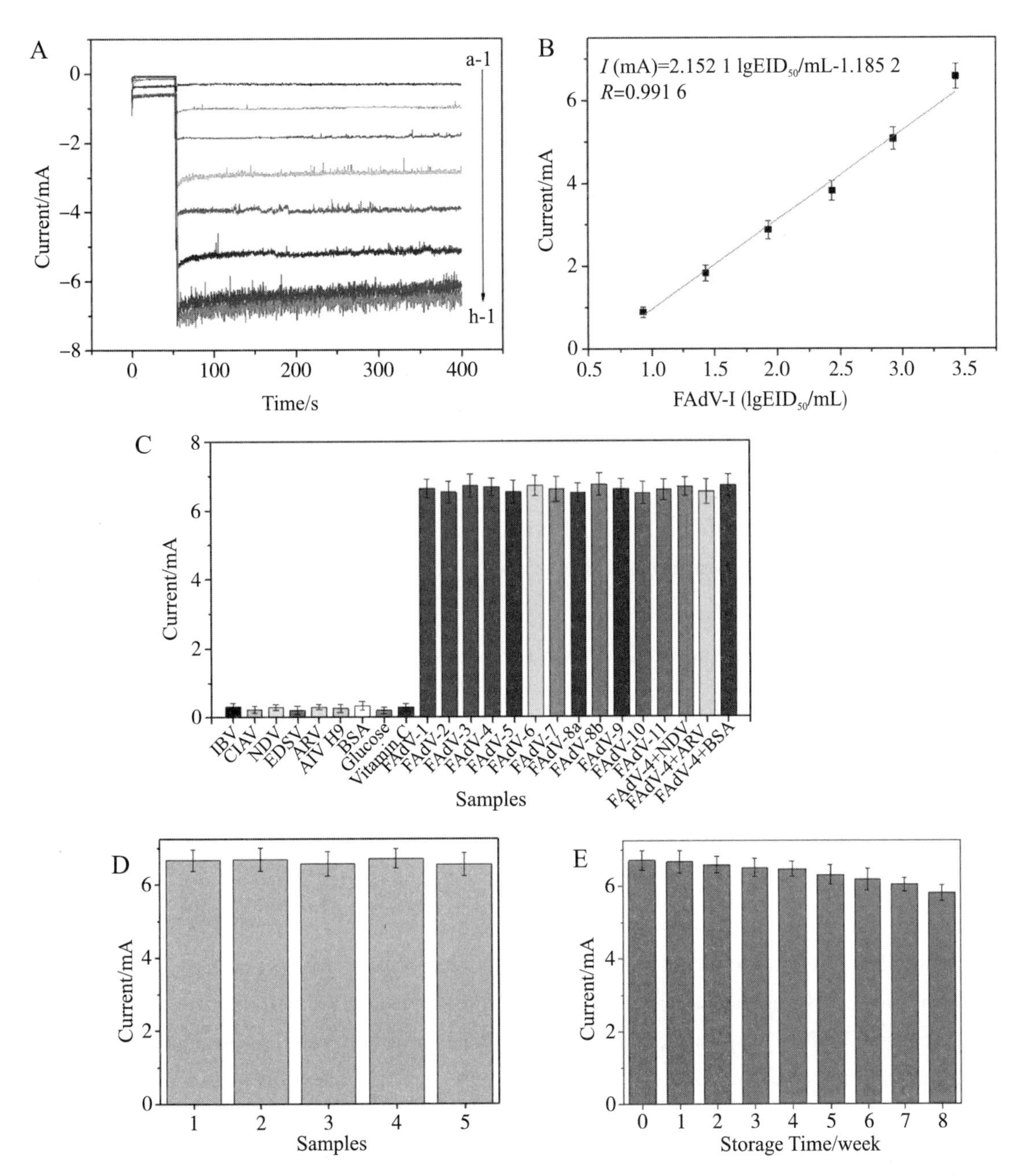

A: Amperometric i-t current response of the developed electrochemical immunosensor for detecting the following concentrations of FAdV-I. a-1: $10^{0.43}$ EID_{50}/mL; b-1: $10^{0.93}$ EID_{50}/mL; c-1: $10^{1.43}$ EID_{50}/mL; d-1: $10^{1.93}$ EID_{50}/mL; e-1: $10^{2.43}$ EID_{50}/mL; f-1: $10^{2.93}$ EID_{50}/mL; g-1: $10^{3.43}$ EID_{50}/mL; h-1: $10^{3.93}$ EID_{50}/mL.

B: Calibration curve of the response of the developed electrochemical immunosensor to different concentrations of FAdV-I.

C: Specificity of the developed electrochemical immunosensor for the target FAdV-I and other interfering substances.

D: Reproducibility of the proposed electrochemical immunosensor in the presence of $10^{3.43}$ EID_{50}/mL FAdV-I.

E: Long-term stability of the developed electrochemical immunosensor.

Figure 5-33-7 Sensitivity, selectivity, reproducibility and stability of the immunosensor

The reproducibility of the developed immunosensor was evaluated by using different batches of the immunosensor to detect the same concentrations of $10^{3.43}$ EID_{50}/mL FAdV-I. The results (Figure 5-33-7 D) showed that the intra- and interassay RSD values were all less than 5.0%, as shown by the error bars, and these results indicated that the applied immunosensor had good reproducibility.

The storage life of the developed immunosensor was surveyed by storing GCE-MWCNTs-Chi-FAdV-I/MAb-BSA and MWCNTs-Chi-Pt/AgNPs-FAdV-I/PAb nanocomposites at 4 ℃ until use. Then, GCE-MWCNTs-Chi-FAdV-I/MAb-BSA were successively incubated with $10^{3.43}$ EID_{50}/mL FAdV-I and MWCNTs-Chi-Pt/AgNPs-FAdV-I/PAb nanocomposites. As shown in Figure 5-33-7 E, the current response remained at 90.01% of that of the new preparation after storage for 7 weeks. Therefore, the storage life of the developed immunosensor was acceptable.

Clinical sample analysis

The developed immunosensor was used to analyze 125 clinical samples collected from chickens in small poultry flocks in Guangxi, China, that were suspected to be infected with adenoviruses, and the results were 100% consistent with the results of real-time PCR and 98.3% consistent with the results of loop-mediated isothermal amplification (LAMP), which were both developed by our group[41, 42] , to prove that the developed immunosensor can be used for real samples. The results are shown in Table 5-33-1. Fifty-nine of the 125 samples (cloacal swab samples) were identified as positive by the proposed immunosensor, which was consistent with the results of real-time PCR, whereas 58 of the 125 samples were identified as positive by LAMP.

Table 5-33-1 Comparison of real-time PCR, LAMP and immunosensor methods for detecting FAdV-I in chickens in Guangxi, China

Method	Total number of sample	Number of positive samples	Positivity rate /%
Proposed immunosensor	125	59	47.2
Real-time PCR	125	59	47.2
LAMP	125	58	46.4

Furthermore, the proposed immunosensor was used to monitor the recovery of cloacal swab samples after treatment with different concentrations of FAdV-I to evaluate its accuracy and practicality. A series of different concentrations of FAdV-I were added by standard addition methods to the six aforementioned clinical samples that were identified as FAdV-I-positive samples; these concentrations were chosen randomly and detected by the fabricated immunosensor. The results are summarized in Table 5-33-2. The recoveries, which were calculated as the ratio between the amount of FAdV-I found and the amount added, ranged from 98.37% to 104.41%, with an RSD ranging from 2.63% to 4.71% (n=5). These results indicated that the fabrication method was appropriate for FAdV-I detection in the samples.

Table 5-33-2 Results of the recovery of different concentrations of FAdV-I from clinical samples

No.	Initial FAdV-I concentration in the sample/(EID_{50}/mL)	Added amount of FAdV-I/(EID_{50}/mL)	Total found		Recovery rate /% (n=5)
			Average/(EID_{50}/mL)	RSD/% (n=5)	
1	107.61	100.00	211.49	3.72	101.86
2	275.42	200.00	487.24	2.85	102.48
3	427.25	400.00	813.75	3.54	98.37
4	531.74	500.00	1 059.31	4.71	102.67
5	693.21	600.00	1 350.27	2.63	104.41
6	721.35	700.00	1 463.79	3.28	102.99

Conclusions

In this work, an enzyme-free sandwich amperometric immunosensor was developed for the detection of FAdV-I by using MWCNTs-Chi, which has ideal conductivity and efficient antibody immobilization, as the matrix platform and a MWCNTs-Chi-Pt/AgNPs-FAdV-I/PAb nanocomposite, which has excellent electrocatalytic activity toward H_2O_2 reduction, as the signal amplification label. The developed amperometric immunosensor showed high specificity, good sensitivity, accuracy, reproducibility and stability and was capable of detecting FAdV-I in clinical samples. More importantly, the developed method not only enables the application of MWCNTs-Chi-Pt/AgNPs nanocomposites in amperometric electrochemical immunosensors but also provides a new enzyme-free method for the determination of other biomolecules in clinical diagnosis.

References

[1] MO J. Historical investigation of fowl adenovirus outbreaks in South Korea from 2007 to 2021: a comprehensive review. Viruses, 2021, 13(11): 2256.

[2] ZHUANG Q, WANG S, ZHANG F, et al. Molecular epidemiology analysis of fowl adenovirus detected from apparently healthy birds in eastern China. BMC Vet Res, 2023, 19(1): 5.

[3] LV L, LU H, WANG K, et al. Emerging of a novel natural recombinant fowl adenovirus in China. Transbound Emerg Dis, 2021, 68(2): 283-288.

[4] SCHACHNER A, MATOS M, GRAFL B, HESS M. Fowl adenovirus-induced diseases and strategies for their control—a review on the current global situation. Avian Pathol, 2018, 47(2): 111-126.

[5] YAMAGUCHI M, MIYAOKA Y, HASAN M A, et al. Isolation and molecular characterization of fowl adenovirus and avian reovirus from breeder chickens in Japan in 2019-2021. J Vet Med Sci, 2022, 84(2): 238-243.

[6] HESS M. Detection and differentiation of avian adenoviruses: a review. Avian Pathol, 2000, 29(3): 195-206.

[7] GANESH K, SURYANARAYANA V V, RAGHAVAN R. Detection of fowl adenovirus associated with hydropericardium hepatitis syndrome by a polymerase chain reaction. Vet Res Commun. 2002;26(1):73-80.

[8] NICZYPORUK J S, KOZDRUŃ W, CZEKAJ H, et al. Detection of fowl adenovirus D strains in wild birds in Poland by Loop-Mediated Isothermal Amplification (LAMP) BMC Vet Res, 2020: 16(1):58.

[9] XIE Z, FADL A A, GIRSHICK T, et al. Detection of Avian adenovirus by polymerase chain reaction. Avian Dis, 1999, 43: 98-105.

[10] MISTRY K K, LAYEK K, MAHAPATRA A, et al. A review on amperometric-type immunosensors based on screen-printed electrodes. Analyst, 2014, 139(10): 2289-2311.

[11] BALAHURA L R, STEFAN-VAN STADEN R I, VAN STADEN J F, et al. Advances in immunosensors for clinical applications. Journal of Immunoassay and Immunochemistry, 2018, 40(1): 40-51.

[12] WANG R, FENG J J, LIU W D, et al. A novel label-free electrochemical immunosensor based on the enhanced catalytic currents of oxygen reduction by AuAg hollow nanocrystals for detecting carbohydrate antigen 199. Biosens Bioelectron, 2017, 96: 152-158.

[13] PEI F, WANG P, MA E, et al. A sandwich-type amperometric immunosensor fabricated by Au@Pd NDs/Fe^{2+}-CS/PPy NTs and Au NPs/NH2-GS to detect CEA sensitively via two detection methods. Biosens Bioelectron, 2018, 122: 231-238.

[14] LI J, YANG H, CAI R, TAN W. Ultrahighly sensitive sandwich-type electrochemical immunosensor for selective detection of tumor biomarkers. ACS Appl Mater Interfaces, 2022, 14(39): 44222-44227.

[15] LI N, WANG Y, CAO W, et al. An ultrasensitive electrochemical immunosensor for CEA using MWCNT-NH2 supported PdPt nanocages as labels for signal amplification. J Mater Chem B, 2015, 3(9): 2006-2011.

[16] YOLA M L, ATAR N. Novel voltammetric tumor necrosis factor-alpha (TNF-α) immunosensor based on gold nanoparticles involved in thiol-functionalized multi-walled carbon nanotubes and bimetallic Ni/Cu-MOFs. Anal Bioanal Chem, 2021,

413(9): 2481-2492.

[17] EHZARI H, AMIRI M, SAFARI M. Enzyme-free sandwich-type electrochemical immunosensor for highly sensitive prostate specific antigen based on conjugation of quantum dots and antibody on surface of modified glassy carbon electrode with core-shell magnetic metal-organic frameworks. Talanta, 2020, 210: 120641.

[18] NANDHAKUMAR P, KIM G, PARK S, et al. Metal Nanozyme with ester hydrolysis activity in the presence of ammonia-borane and its use in a sensitive immunosensor. Angew Chem Int Ed Engl, 2020, 59(50): 22419-22422.

[19] AHMADI A, KHOSHFETRAT S M, MIRZAEIZADEH Z, et al. Electrochemical immunosensor for determination of cardiac troponin I using two-dimensional metal-organic framework/Fe_3O_4-COOH nanosheet composites loaded with thionine and pCTAB/DES modified electrode. Talanta, 2022, 237: 122911.

[20] SUN D, LI H, LI M, et al. Electrochemical immunosensors with AuPt-vertical graphene/glassy carbon electrode for alpha-fetoprotein detection based on label-free and sandwich-type strategies. Biosens Bioelectron, 2019, 132: 68-75.

[21] FENG J, LI Y, LI M, et al. A novel sandwich-type electrochemical immunosensor for PSA detection based on PtCu bimetallic hybrid (2D/2D) rGO/g-C3N4. Biosens Bioelectron, 2017, 91: 441-448.

[22] GU X, WANG K, TIAN S, et al. A SERS/electrochemical dual-signal readout immunosensor using highly-ordered Au/Ag bimetallic cavity array as the substrate for simultaneous detection of three β-adrenergic agonists. Talanta, 2023, 254: 124159.

[23] WANG P, PEI F, MA E, et al. The preparation of hollow AgPt@Pt core-shell nanoparticles loaded on polypyrrole nanosheet modified electrode and its application in immunosensor. Bioelectrochemistry, 2020, 131: 107352.

[24] GÜNER A, CEVIK E, SENEL M, et al. An electrochemical immunosensor for sensitive detection of Escherichia coli O157: H7 by using chitosan, MWCNT, polypyrrole with gold nanoparticles hybrid sensing platform. Food Chem, 2017, 229: 358-365.

[25] SHARMA A, KAUSHAL A, KULSHRESTHA S. A Nano-Au/C-MWCNT based label free amperometric immunosensor for the detection of capsicum chlorosis virus in bell pepper. Arch Virol, 2017, 162: 2047-2052.

[26] NARWAL V, KUMAR P, JOON P, et al. Fabrication of an amperometric sarcosine biosensor based on sarcosine oxidase/chitosan/CuNPs/e-MWCNT/Au electrode for detection of prostate cancer. Enzyme Microb Technol, 2018, 113,44-51.

[27] AYDIN E B, AYDUN M, SEZGINTÜRK M K. Electrochemical immunosensor based on chitosan/conductive carbon black composite modified disposable ITO electrode: An analytical platform for p53 detection. Biosens Bioelectron, 2018, 121: 80-89.

[28] KAUSHIK A, SOLANKI P R, PANDEY M K, et al. Carbon nanotubes — chitosan nanobiocomposite for immunosensor. Thin Solid Films, 2010, 519(3): 1160-1166.

[29] JIAN Y, ZHANG J, YANG C, et al. Biological MWCNT/chitosan composite coating with outstanding anti-corrosion property for implants. Colloids Surf. B Biointerfaces, 2023, 225: 113227.

[30] XIE L J, DENG X W, XIE Z X, et al. Preparation of the monoclonal antibody against 100K protein of serotype 4 fowl adenovirus. Chin Vet Sci, 2023, 53: 1152-1157.

[31] WEI H, WANG E K. Nanomaterials with enzyme-like characteristics (nanozymes): next-generation artificial enzymes. Chemical Society Reviews, 2013, 42: 6060-6093.

[32] LI F, FENG J, GAO Z, et al. Facile synthesis of Cu_2O@TiO_2-PtCu nanocomposites as a signal amplification strategy for the insulin detection. ACS Appl Mater Interfaces. 2019, 11(9): 8945-8953.

[33] TWU Y K, CHEN Y W, SHIH C M. Preparation of silver nanoparticles using chitosan suspensions. Powder Technol, 2008, 185: 251-257.

[34] ADLIM M, BAKAR M A, LIEW K Y, et al. Synthesis of chitosan-stabilized platinum and palladium nanoparticles and their hydrogenation activity. Journal of Molecular Catalysis A Chemical, 2004, 212(1-2): 141-149.

[35] LV J J, ZHENG J N, CHEN L L, et al. Facile synthesis of bimetallic alloyed Pt-Pd nanocubes on reduced graphene oxide with enhanced eletrocatalytic properties. Electrochimica Acta, 2014, 143: 36-43.

[36] SCHAEFER S, MUENCH F, MANKEL E, et al. Double-walled Ag-Pt nanotubes fabricated by galvanic replacement and dealloying: effect of composition on the methanol oxidation activity. Nano, 2015, 10(6): 1550085.
[37] LV J J, FENG J X, LI S S, et al. Ionic liquid crystal-assisted synthesis of PtAg nanoflowers on reduced graphene oxide and their enhanced electrocatalytic activity toward oxygen reduction reaction. Electrochimica Acta, 2014, 133: 407-413.
[38] ZHAO D, YAN B, XU B Q. Proper alloying of Pt with underlying Ag nanoparticles leads to dramatic activity enhancement of Pt electrocatalyst. Electrochemistry Communications, 2008, 10(6): 884-887.
[39] LI F, GUO Y, CHEN M, et al. Comparison study of electrocatalytic activity of reduced graphene oxide supported Pt-Cu bimetallic or Pt nanoparticles for the electrooxidation of methanol and ethanol. International Journal of Hydrogen Energy, 2013, 38(33): 14242-14249.
[40] ZHAO Y, ZHAN L, TIA J, NIE S, et al. Enhanced electrocatalytic oxidation of methanol on Pd/polypyrole-graphene in alkaline medium. Electrochim Acta, 2010, 56: 1967-1972.
[41] WEN YL, X ZHX, WANG Y, et al. Development of a SYBR Green Ⅰ real-time PCR assay for the detection of aviadenovirus group Ⅰ. Chinese Veterinary Science, 2008, 38: 753-756.
[42] XIE Z, TANG Y, FAN Q, et al. Rapid detection of group I avian adenoviruses by a loop-mediated isothermal amplification. Avian Dis, 2011, 55(4): 575-579.

Part V Polymerase Chain Reaction and Reverse Transcriptase-Polymerase Chain Reaction

Amplification of avian reovirus RNA using the reverse transcriptase-polymerase chain reaction

Xie Zhixun, Fadl Amin A, Girshick Theodore, and Khan Mazhar I

Abstract

A reverse transcriptase-polymerase chain reaction method was developed for the detection of avian reovirus. The origin of primers was from the S1 gene of the avian reovirus genome. A reovirus-specific 532-base pair cDNA product was amplified by these primers from six reference strains and 23 field isolates of avian reoviruses, but not from seven different avian pathogenic viruses and bacteria. As little as 1 pg of avian reovirus RNA was detected using gel electrophoresis and Southern blot hybridization.

Keywords

avian, reovirus, reverse transcriptase-polymerase chain reaction, RNA, primers

Introduction

Reovirus infections are prevalent worldwide in chickens, turkeys, and other avian species[6, 17, 23, 31, 32]. Reoviruses have been isolated from a variety of tissues in chickens affected by assorted disease conditions including viral arthritis/tenosynovitis, stunting syndrome, respiratory disease, and malabsorption syndrome[7, 8, 20, 21, 24, 32]. Economic losses caused by reovirus infections are frequently the result of crippling from viral arthritis and tenosynovitis and a general lack of performance that includes diminished weight gains, poor feed conversion, and reduced marketability of affected birds[23].

Various methods have been developed for the diagnosis of reovirus infections, such as virus isolation in cell culture, localization of the virus in infected tissue by electron microscopy, fluo-rescence assay, and the detection of antibody or viral antigen by enzyme-linked immunosorbent assay (ELISA), agar immunodiffusion, viral neutralization, and western immunoblot[2, 3, 7, 8, 10, 12, 13, 15, 17, 24, 28, 32]. Although these methods are useful for detection and characterization of avian reovirus, they are laborious and time-consuming. The polymerase chain reaction (PCR) method for in vitro amplification of target gene sequences[25] has been applied as a rapid diagnostic tool for the detection of a range of avian viral pathogens[9, 11, 14, 16, 22, 30]. This method is not only more rapid but also is more sensitive and specific than other diagnostic procedures. No reports have been made of the application of the PCR technique to reovirus diagnosis in chickens. Here we describe the development of the reovirus-specific reverse transcriptase-polymerase chain reaction (RT-PCR) for the detection of a specific RNA region found only in avian reovirus and not in other pathogenic avian viruses and bacteria.

Materials and methods

Virus and tissue culture

Table 5-34-1 describes the 29 avian reovirus strains/isolates and seven other avian pathogens used in this study. All avian reoviruses were propagated in chicken embryo fibroblasts (CEFs) and plaque purified as described previously[32]. The Newcastle disease virus (NDV) and infectious bronchitis virus (IBV) were propagated in specific-pathogen-free[SPF] chicken embryonating eggs as previously described[1, 5]. Adenovirus was propagated in chick embryo kidney cells as described[17]. The chicken anemia virus was propagated in MDCC-MSB1 cells as described[18]. *Mycoplasma gallisepticum* (MG) was grown in Frey's media[4] and *Salmonella enteritidis* (SE) and *Proteus* spp.were grown in LB broth medium[26].

Table 5-34-1 Viral and bacterial isolates used

Isolates	Code	Source
Reovirus		
Reference strain	S1133	University of Connecticut
Reference strain	S1133	SPAFAS, Inc.
Reference strain	1733	SPAFAS, Inc
Reference strain	WVU 2937	SPAFAS, Inc
Reference strain	203-M3	SPAFAS, Inc
Reference strain	C78	SPAFAS, Inc.
Field isolate	S1	University of Connecticut
Field isolate	S3	University of Connecticut
Field isolate	T1	University of Connecticut
Field isolate	T2	University of Connecticut
Field isolate	T3	University of Connecticut
Field isolate	T4	University of Connecticut
Field isolate	T5	University of Connecticut
Field isolate	T6	University of Connecticut
Field isolate	77: 96-p7	University of Connecticut
Field isolate	250-p7	University of Connecticut
Field isolate	305-p7	University of Connecticut
Field isolate	93: 1046-p8	University of Connecticut
Field isolate	93: 1046-p9	University of Connecticut
Field isolate	94: 96-p12	University of Connecticut
Field isolate	95: 502-p2	University of Connecticut
Field isolate	95: 1031-p2	University of Connecticut
Field isolate	95: 1102-p2	University of Connecticut
Field isolate	95: 2124-p2	University of Connecticut
Field isolate	96: 818	University of Connecticut

continued

Isolates	Code	Source
Field isolate	S5	University of Connecticut
Field isolate	2408-P2	University of Delaware
Field isolate	Miss B-P3	University of Delaware
Field isolate	1733-P2	University of Delaware
Infectious bronchitis virus	Mass-41	University of Connecticut
Newcastle disease virus	B-1	Washington State University
Chicken anemia virus	Cux-1	SPAFAS, Inc
Adenovirus	CELO	SPAFAS, Inc
Mycoplasma gallisepticum	S6	University of California
Salmonella enteritidi	203	University of Maine
Proteus spp.		University of Connecticut

RNA extraction

RNA extraction from avian reovirus isolates and other avian viruses (IBV and NDV) was performed according to the protocol described by the manufacturer's RNA extracting kit reagents (Trizol LS, Life Technologies, Bethesda, MD). Briefly, 250 μL of tissue/cell culture sample of virus was treated with 750 μL of Trizol Ls (Life Technologies) reagent and incubated at room temperature for 5 min. After incubation, 200 μL of chloroform was added by inverting the tube a few times; the mixture was then incubated at room temperature for 15 min. The mixture was microfuged (4 ℃, 10 000 × g, 15 min), the aqueous phase was transferred to a new tube, and 500 μL of isopropyl alcohol was added. The mixture was then incubated at room temperature for 10 min and centrifuged (10 000 × g, 10 min, 4 ℃), and the pellet was washed once with 500 μL of 70% ethanol. After centrifugation (10 000 × g, 30 s, 4 ℃) to recover RNA, the pellet was air dried and resuspended in 50 μL diethylpyrocarbonate (DEPC)-treated water. The concentration of the RNA was determined by spectrophotometry as previously described[26] and the RNA was stored at −70 ℃ until use. Chromosomal DNA from MG, chicken anemia virus (CAV), adenovirus, *Proteus* spp., and SE was extracted and purified by a previously described method[19].

Oligonucleotide primers

The sequences of the 24-mer oligonucleotide MK87 (GGT GCG ACT GCT GTA TTT GGT AAC) and the 21-mer oli-gonucleotide MK88 (AAT GGA ACG ATA GCG TGT GGG), which were used as primers in the PCR, were selected on the basis of the published sequence data of the S1 gene of the avian reovirus S1133 strain[27]. The S1 gene of this avian reovirus is dissimilar to the mammalian reovirus counterpart[27]. These primers flanked a 532-base pair (bp) DNA sequence containing the S1 gene. These primers were synthesized on a model 380B DNA synthesizer (Applied Biosystem, Inc., Foster City, CA) with the assistance of the University of Connecticut Biotechnology Center. The primers were desalted through a Sephadex G-25 column (Pharmacia, Inc., Piscataway, NJ). The concentrations of the primers were determined by spectrophotometry as previously described[26] and then divided into 50 μL volumes and stored at −20 ℃.

Reverse transcription (RT) and PCR amplification

The reverse transcription and PCR amplification were conducted in a thermal cycler (Model 480, Perkin Elmer Cetus, Norwalk, CT). The reverse transcriptase reaction was carried out using a GeneAmp RNA PCR kit (Perkin Elmer Cetus, Norwalk, CT). A reaction volume of 25 μL of reverse transcription mixture contained 4 μL of 80 mM $mgCl_2$, 2 μL of 10 × PCR buffer (500 mM KCl, 200 mM Tris-HCl[pH 8.4], 0.5 mg/mL nuclease-free bovine serum albumin), 2 μL of 10 mM each dinucleoside triphosphate (dNTP), 1 μL (20 units) RNase inhibitor, 1 μL (0.5 μm) random hexamers, 1 μL (50 units) Moloney murine leukemia virus (MuLV) reverse transcriptase, and 200 ng RNA or DNA. A 25 μL total reaction volume was obtained by adding sterile DEPC-treated water. The RT reaction was performed by incubating the mixture at room temperature for 10 min, and then using a thermal cycler setting of 42 ℃ for 60 min, denaturing at 99 ℃ for 5 min, and then cooling at 5 ℃ for 5 min for one cycle. The PCR was carried out as described[25] using reagents from a GeneAmp PCR kit (Perkin Elmer Cetus). A reaction volume of 100 μL of PCR mixture that contained 4 μL of 80 mM $mgCl_2$, 8 μL of 10 × PCR buffer, 600 ng each of MK87 and MK88 primers, and 0.5 μL (2.5 units) of AmpliTaq DNA polymerase was added to the above RT reaction mixture (25 μL). A 100 μL total reaction volume was obtained by adding sterile distilled water. We performed the denaturation step for 1 min at 94 ℃, the annealing for 1 min at 55 ℃, and the extension step for 1 min at 72 ℃, for 35 cycles. The initial denaturation step was conducted for 5 min at 94 ℃. The final extension step was conducted at 72 ℃ for 10 min.

Sensitivity and specificity of the avian reovirus RT-PCR

Two hundred nanograms of genomic RNA from each of the 29 different strains/isolates of reoviruses and RNA or DNA from seven other avian viruses and bacteria (Table 5-34-1) were subjected to the avian reovirus-specific RT-PCR. Tenfold dilutions from 10 ng to 1 fg of genomic RNA from avian reovirus S1133 were also subjected to reovirus RT-PCR to determine the sensitivity of the test.

Detection of RT-PCR products

Gel electrophoresis and Southern blot hybridization[29] were used to detect PCR products. A volume of 20 μL of PCR products was electrophoretically separated on a 1.0% agarose gel, stained with ethidium bromide, and exposed to ultraviolet light to determine the presence and size of the amplified DNA product. For Southern blot hybridization, amplified PCR products on aga rose gels were transferred to Zeta probe membranes (BioRad, Richmond, CA) as described previously[29]. Twenty nanograms of PCR-amplified cDNA specific for reovirus and *Hind*Ⅲ-digested lambda DNA (Life Technologies, Bethesda, MD) were labeled simultaneously with alpha P-dATP (3 000 Ci/ mM) using a DECAprime labeling kit (Ambion, Inc., Austin, TX). The membrane was allowed to hybridize with ^{32}P-labeled PCR-amplified cDNA-specific reovirus and *Hind*Ⅲ lambda DNA as a probe in 1 mM ethylenediaminetetraacetic acid (EDTA), 0.5 M NaH_2PO_4 (pH 7.2), and 7% sodium dodecyl sulfate (SDS) at 65 ℃ for 18 h. The membrane was washed twice in 1 mM EDTA, 40 mM NaH_2PO_4 (pH 7.2), and 1% SDS. The membrane was allowed to dry and was exposed on X-OMAT AR film (Eastman Kodak Co., Rochester, NY).

Results and discussion

All 29 strains/isolates of avian reovirus were positive in the reovirus RT-PCR, with all positive products consisting of the expected 532-bp DNA band. Representative reovirus RT-PCR results are shown in Figure

5-34-1 A and B. In clear contrast, the RNA or DNA from four other avian pathogenic viruses and three bacteria were not amplified under the same RT-PCR condi-tions (Figure 5-34-1 A). Autographs of radiolabeled Southern blot membrane with PCR products demonstrated that the probe binds to amplified products of reovirus but not to those of other viruses and bacteria (Figure 5-34-1 B). The minimum amount detected by the avian reovirus-specific RT-PCR was 1 pg of RNA as determined by gel electrophoresis and Southern blot hybrid-ization (Figure 5-34-2 A, B).

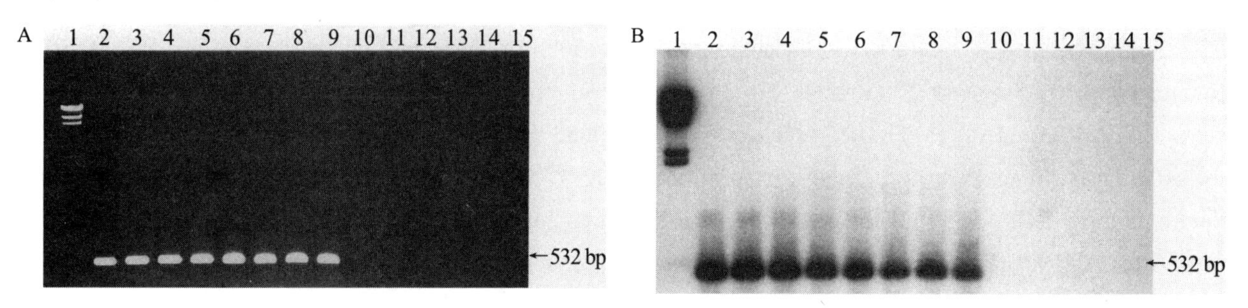

A: Agarose gel electrophoresis of reovirus-specific cDNA products amplified by RT-PCR; B: Southern blot hybridization of reovirus-specific RT-PCR products probed with 32P-labeled S1133 reovirus-amplified cDNA probe and DNA marker labeled with ^{32}P. Lane 1: molecular weight marker lambda DNA digested with *Hind* Ⅲ ; lane 2: S1133; lane 3: 203-M3; lane 4: WVU 2937; lane 5: 1733; lane 6: S1; lane 7: 96 ∶ 818; lane 8: Miss B-P3; lane 9: C78; lane 10: Newcastle disease virus B-1; lane 11: infectious bronchitis virus Mass-41; lane 12: adenovirus CELO; lane 13: chicken anemia virus Cux-l; lane 14: *Mycoplasma gallisepticum* S6; lane 15: *Salmonella enteritidis* 203. The arrow identifies the 532 bp product of target DNA that was amplified by reovirus-specific MK87 and MK88 primers.

Figure 5-34-1 Representative reovirus RT-PCR results

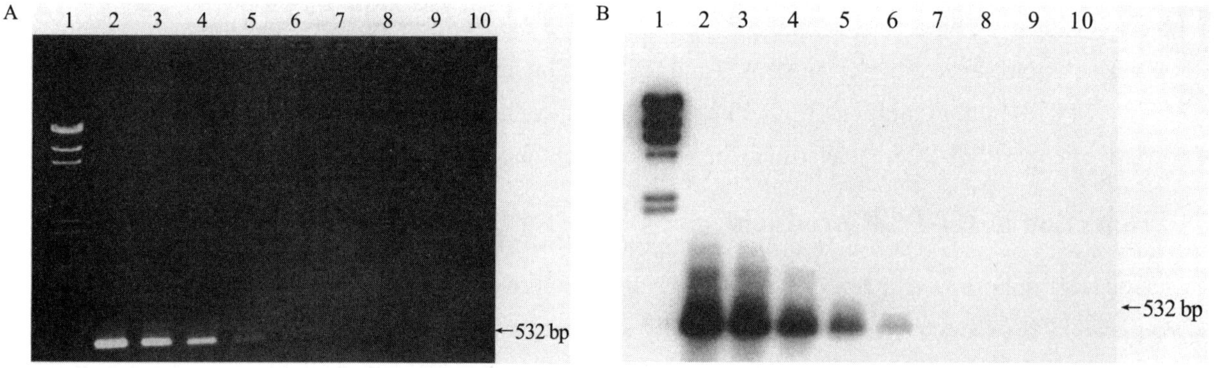

A: Sensitivity of reverse transcriptase-polymerase chain reaction (RT-PCR) products showing on the agarose gel electrophoresis; B: Southern blot hybridization of RT-PCR products probed with ^{32}P-labeled S1133 reovirus-amplified cDNA probe and DNA marker labeled with ^{32}P. Lane 1: *Hind*Ⅲ digested lambda DNA marker; lane 2: 10 ng; lane 3: 1 ng; lane 4: 100 pg; lane 5: 10 pg; lane 6: 1 pg; lane 7: 100 fg; lane 8: 10 fg; lane 9: 1 fg; lane 10: PCR buffer. The arrow identifies the 532 bp product of target DNA that was amplified by reovirus-specific MK87 and MK88 primers.

Figure 5-34-2 Sensitivity of PCR

The need for a rapid, sensitive, and specific test for the laboratory diagnosis of viral diseases of poultry has long been recognized. The diagnosis of avian reovirus infection is currently generally made by virus isolation, a laborious and time-consuming process. The RT-PCR method described here is able to detect as little as 1 pg of RNA (equivalent to $10^2\sim10^3$ virion particles). The detection of such a small amount of RNA may enable viral RNA to be amplified directly from clinical or environmental samples and may avoid the extra step of growing the virus in the cell culture. We have not tested mammalian reovirus as a heterologous control in our assay. However, the primers we used target the conserved S1 gene of S1133 strain of avian reovirus, a gene that is different from the counterpart in mammalian reovirus[27]. Therefore, we believe that the RT-PCR is specific for avian reovirus. In summary, we have demonstrated the ability of the RT-PCR to detect of 29 strains/

isolates of avian reoviruses without any detectable cross-reactions with seven other avian viruses and bacteria.

References

[1] ALEXANDER D J. Newcastle disease//PURCHASE H G, ARP L H, DOMERMUTH C H, et al. A laboratory manual for the isolation and identification of avian pathogen. New Bolton Center: American Association of Avian Pathologists, 1989: 114-120.

[2] ENDO-MUNOZ L B. western blot to detect antibody to avian reovirus. Avian Pathol, 1990, 19(3): 477-487.

[3] FAHEY J E, CRAWLEY J F. Studies on chronic respiratory disease of chickens II. Isolation of A Virus. Canadian Journal of Comparative Medicine and Veterinary Science, 1954, 18(1): 13-21.

[4] FREY M L, HANSON R P, ANDRSON D P. A medium for the isolation of avian mycoplasmas. American Journal of Veterinary Research, 1968, 29(11): 2163.

[5] GELB J JR. Infectious bronchitis//PURCHASE H G, ARP L H, DOMERMUTH C H, et al. A laboratory manual for the isolation and identification of avian pathogens. New Bolton Center, Kenneth Squar: American Association of Avian Pathologists, 1989: 124-127.

[6] GERSHOWITZ A, WOOLEY R E. Characterization of two reoviruses isolated from turkeys with infectious enteritis. Avian Dis, 1973, 17(2): 406-414.

[7] DALE R K H, VILLEGAS P, KLEVEN S H. Identification and serological differentiation of several reovirus strains isolated from chickens with suspected malabsorption syndrome. Avian Dis, 1983, 27(1): 246-254.

[8] GLASS S E, NAQI S A, HALL C F, et al. Isolation and characterization of a virus associated with arthritis of chickens. Avian Dis, 1973, 17(2): 415-424.

[9] HORIMOTO T, KAWAOKA Y. Direct reverse transcriptase PCR to determine virulence potential of influenza A viruses in birds. J Clin Microbiol, 1995, 33(3): 748-751.

[10] IDE P R. Avian reovirus antibody assay by indirect immunofluorescence using plastic microculture plates. Canadian Journal of Comparative Medicine, 1982, 46(1): 39-42.

[11] JESTIN V, JESTIN A. Detection of Newcastle disease virus RNA in infected allantoic fluids by in vitro enzymatic amplification (PCR). Archives of Virology, 1991, 118(3-4): 151-161.

[12] KAWAMURA H, TSUBAHARA H. Common antigenicity of avian reoviruses. National Institute of Animal Health quarterly, 1966, 6(4): 187.

[13] TAKEHARA K, KIMURA Y, TANAKA Y, et al. Preparation and characterization of monoclonal antibodies against an avian revirus. Avian Dis, 1987, 31(4): 730-734.

[14] KWON H M, JACKWOOD M W, GELB J JR. Differentiation of infectious bronchitis virus serotypes using polymerase chain reaction and restriction fragment length polymorphism analysis. Avian Dis, 1993, 37(1):194-202.

[15] LI L, GIAMBRONE J J, PANANGALA V S, et al. Production and characterization of monoclonal antibodies against avian reovirus strain S1133. Avian Dis, 1996, 40(2): 349-357.

[16] LEE L H, YU S L, SHIEH H K. Detection of infectious bursal disease virus infection using the polymerase chain reaction. J Virol Methods, 1992, 40(3): 243-253.

[17] MCFERRAN J B, CONNOR T J, MCCRACKEN R M. Isolation of adenoviruses and reoviruses from avian species other than domestic fowl. Avian Dis, 1976, 20(3): 519-524.

[18] MCNULTY M S. Chicken anemia agent // PURCHASE H G, ARP L H, DOMERMUTH C H, et al. A laboratory manual for the isolation and identification of avian pathogen. New Bolton Center, Kenneth Squar: American Association of Avian Pathologists, 1989: 108-109.

[19] NGUYEN A V, KHAN M I, LU Z. Amplification of Salmonella chromosomal DNA using the polymerase chain reaction. Avian Dis, 1994, 38(1): 119-126.

[20] PAGE R K, FLETCHER O J, ROWLAND G N, et al. Malabsorption syndrome in broiler chickens. Avian Dis, 1982, 26(3): 618-624.

[21] PAGE R K, FLETCHER O J, VILLEGAS P. Infectious tenosynovitis in young turkeys. Avian Dis, 1982, 26(4): 924-927.

[22] POULSEN D J, BURTON C R A, O'BRIAN J J, et al. Identification of the infectious laryngotracheitis virus glycoprotein gB gene by the polymerase chain reaction. Virus genes, 1991, 5(4): 335-347.

[23] ROSENBERGER J K, OLSON N O. Reovirus infections//CALNEK B W, BARNES H J, BEARD C W, et al. Diseases of poultry, 9th ed. Ames: Iowa State University Press, 1991: 639-647.

[24] ROSENBERGER J K U O, STERNER F J, BOTTS S, et al. In vitro and in vivo characterization of avian reoviruses. I. Pathogenicity and antigenic relatedness of several avian reovirus isolates. Avian Dis, 1989, 33(3): 535-544.

[25] SAIKI R K, SCKARF S, FALOONA F. Enzymatic amplification of betaglobulin genomic sequences and restriction site analysis for diagnosis of sickle cell anemia. Science, 1985(230): 1350-1354.

[26] SAMBROOK J, FRITSCH E T, MANIATIS T. Molecular cloning: a laboratory manual. New York: Cold Spring Harbor Laboratory Press, 1989.

[27] SHAPOURI M R S, KANE M, LETARTE M, et al. Cloning, sequencing and expression of the S1 gene of avian reovirus. Journal of general virology. J Gen Virol, 1995, 76(6): 1515-1520.

[28] SLAGHT S S, YANG T J, van der HEIDE L, et al. An Enzyme-Linked Immunosorbent Assay (ELISA) for Detecting Chicken Anti-Reovirus Antibody at High Sensitivity. Avian Dis, 1978, 22(4): 802-805.

[29] SOUTHERN E M. Detection of specific sequences among DNA fragments separated by gel electrophoresis. J Mol Biol, 1975: 503-517.

[30] THAM K M, STANISLAWEK W L. Polymerase chain reaction amplification for direct detection of chicken anemia virus DNA in tissues and sera. Avian Dis, 1992, 36(4): 1000-1006.

[31] VAN DER HEIDE L, KALBAC M, BRUSTOLON M, et al. Pathogenicity for chickens of a reovirus isolated from turkeys. Avian Dis, 1980, 24(4): 989-997.

[32] VAN DER HEIDE L, KALBAC M. Infectious tenosynovitis (viral arthritis): characterization of a connecticut viral isolant as a reovirus and evidence of viral egg transmission by reovirus-infected broiler breeders. Avian Dis, 1975, 19(4): 683-688.

Detection of avian adenovirus by polymerase chain reaction

Xie Zhixun, Fadl Amin A, Girshick Theodore, and Khan Mazhar I

Abstract

An avian adenovirus-specific polymerase chain reaction was developed. The origin of primers was from the DNA sequence data of the chicken embryo lethal orphan avian adenovirus virus genome. An avian adenovirus-specific 421 bp DNA product was amplified by these primers from group I of adenovirus containing 12 serotypes and serotypes of adenovirus from group Ⅱ and group Ⅲ. The adenovirus-specific DNA product was also amplified from the 19 field isolates of avian adenoviruses but not from the mammalian adenovirus and other avian pathogenic viruses and bacteria. As little as 1 fg of avian adenovirus DNA was detected by gel electrophoresis and Southern blot analysis.

Keywords

avian, adenovirus, PCR, DNA, primers

Introduction

The large adenovirus family is divided by host range into adenoviruses that infect mammals (the Mastadenoviridae) and adenoviruses that infect avian species (the Aviadenoviridae). The avian adenoviruses are subdivided into three groups. The first includes the conventional group I comprised of 12 serotypes of avian adenoviruses from chickens, turkeys, geese, and other species that share a common group antigen[14, 17, 30]. The second group of avian adenoviruses, known as group Ⅱ, causes infections such as hemorrhagic enteritis in turkeys, marble spleen disease in pheasants, and splenomegaly in chickens. These group Ⅱ viruses share a common group antigen that distinguishes them from the group I adenoviruses[9]. In group Ⅲ, an adenovirus that causes "cgg drop syndrome 76" in laying chickens and a similar virus found to infect ducks only partially share the group I common antigen[18].

The avian adenovirus infections are believed to cause economic losses by increased morbidity, diminished weight gains, dropped egg production, poor egg quality, and poor feed conversion and may be involved in immunosuppression leading to increased incidence of secondary infections[22, 25, 28]. However, recently hydropericardium-hepatitis syndrome in chickens was characterized by high morbidity and mortality[7, 22]. Although the specific etiology of this syndrome has yet to be defined, available evidence suggests the condition is caused by a pathogenic group I adenovirus[7].

Many methods have been developed for the diagnosis of avian adenovirus infections, including virus isolation in cell culture[8, 9], the indirect immunofluorescent assay[1, 2], the virus neutralization test[2], the enzyme-linked immunosorbent assay[4], and the double immunodiffusion[2, 6]. Although these methods are useful for the detection of avian adenoviruses, most of them are laborious and time-consuming. The main problem with any serologic test for adenoviruses is interpretation of the results; antibodies are common in both healthy and diseased birds, and birds are frequently infected with a number of serotypes.

In recent years, the polymerase chain reaction (PCR) method for in vitro amplification of target gene sequences[23] has been applied as a rapid diagnostic tool for the detection of avian viral and bacterial pathogens[12, 13, 15, 16, 20, 21, 27, 29]. This method is not only more rapid but also more sensitive and specific than other diagnostic procedures. In this paper, we describe the development of an avian adenovirus-specific PCR that amplifies a specific DNA region found in all three groups of avian adenoviruses and not in other avian pathogens or mammalian adenoviruses.

Materials and methods

Virus and tissue culture

Table 5-35-2 describes 37 avian adenoviruses, one mammalian adenovirus type 2, and five other avian pathogens used. All serotypes of group Ⅰ, field isolates, and one isolate of avian adenovirus type Ⅲ were propagated in chicken embryo liver cells as described[3]; the turkey hemorrhagic enteritis and marble spleen viruses were propagated in turkey cell line of lymphoblastoid B cells as previously described[19]. The infectious laryngotracheitis virus was grown in primary chicken kidney cell culture as described[10]. The *Mycoplasma gallisepticum* was grown in Frey's medium[11]. The Salmonella and Poicus bacteria were grown in laurial broth (LB)[20].

Table 5-35-2 Viruses and bacteria used

Adenoviruses reference strains	Serotypes/groups	Source
1. Avian adenovirus	Group Ⅰ, type 1	SPAFAS, Inc., Storrs, CT
2. Avian adenovirus	Group Ⅰ, type 2	SPAFAS, Inc., Storrs, CT
3. Avian adenovirus	Group Ⅰ, type 3	SPAFAS, Inc., Storrs, CT
4. Avian adenovirus	Group Ⅰ, type 4	SPAFAS, Inc., Storrs, CT
5. Avian adenovirus	Group Ⅰ, type 5	SPAFAS, Inc., Storrs, CT
6. Avian adenovirus	Group Ⅰ, type 6	SPAFAS, Inc., Storrs, CT
7. Avian adenovirus	Group Ⅰ, type 7	SPAFAS, Inc., Storrs, CT
8. Avian adenovirus	Group Ⅰ, type 8	SPAFAS, Inc., Storrs, CT
9. Avian adenovirus	Group Ⅰ, type 9	SPAFAS, Inc., Storrs, CT
10. Avian adenovirus	Group Ⅰ, type 10	SPAFAS, Inc., Storrs, CT
11. Avian adenovirus	Group Ⅰ, type 11	SPAFAS, Inc., Storrs, CT
12. Avian adenovirus	Group Ⅰ, type 12	SPAFAS, Inc., Storrs, CT
13. Avian adenovirus	Group Ⅱ	SPAFAS, Inc., Storrs, CT
Turkey hemorrhagic enteritis (HE)		
14. Avian adenovirus	Group Ⅱ	Biomune Co., Lenexa, KS
Marble spleen disease virus 1317-IRP19 (5-6-86)		
15. Avian adenovirus	Group Ⅱ	Pennsylvania State University
Marble spleen disease virus AV 2037		
16. Avian adenovirus	Group Ⅲ	SPAFAS, Inc., Storrs, CT

continued

Adenoviruses reference strains	Serotypes/groups	Source
Egg drop syndrome		
17. Mammalian adenovirus type-2 V279		Brookhaven National Lab, Upton, NY
Avian adenovirus field isolates [a]		
18.90-756	EP1	Yolk and allantoic fluid
19.91-768	EP3	Yolk and allantoic fluid
20.92-26	EP1	Yolk and allantoic fluid
21. 92-536	EP1	Yolk and allantoic fluid
22. 92-553	EP1	Yolk and allantoic fluid
23. 92-560	EP1	Yolk and allantoic fluid
24. 92-563	EP1	Yolk and allantoic fluid
25. 92-568	EP1	Yolk and allantoic fluid
26. 92-570	EP1	Yolk and allantoic fluid
27. 92-690	EP1	Yolk and allantoic fluid
28. 94-248	EP1	Yolk and allantoic fluid
29. Indiana	CP1	Chicken embryo liver cell culture
30. Stein	P1	Chicken embryo liver cell culture
31. Tipton	P1	Chicken embryo liver cell culture
32. RT 1329	P1	Chicken embryo liver cell culture
33. RT 1334	P1	Chicken embryo liver cell culture
34. RT 1340	P1	Chicken embryo liver cell culture
35. RT 1345	P1	Chicken embryo liver cell culture
36. RT 1349	P1	Chicken embryo liver cell culture
37. Celo	P1	Chicken embryo liver cell culture
38. RT 1329	P2	Chicken embryo liver cell culture
Other avian pathogen		
39. Salmonella enteritidis		University of Maine
40. *Proteus* spp.		University of Connecticut
41. Chicken anemia virus		SPAFAS, Inc., Storrs, CT
42. Infectious laryngotracheitis virus		University of Delaware
43. *Mycoplasma gallisepticum*-S6		University of California

Note: [a] Avian Diagnostic Lab, Puyallup, WA.

DNA extraction

Adenovirus DNA was extracted from cell culture supernatants as described[20]. Briefly, 600 μL of cell culture supernatants was used and mixed with 60 μL 10% sodium dodecyl sulfate (SDS) and proteinase K (20 μg/mL). The mixture was incubated for 1 hr at 55 ℃ and was extracted twice with an equal volume of phenol : chloroform : isoamyl alcohol (1 : 1 : 24 v/v) (GIBCO BRL, Grand Island, NY). The DNA was precipitated with 1/10 the volume of 3 M sodium acetate and two volumes of absolute ethanol and was incubated overnight at −20 ℃. The precipitated DNA was pelleted by centrifugation at 14 000 × g for 15 min. The pellet was then washed with 1 mL 70% ethanol. After centrifugation at 14 000 × g to recover DNA, the pellet was air dried and resuspended in 50 μL TE (10 mm Tris and 1 mm EDTA) buffer. The concentration of the DNA was determined by spectrophotometry as previously described[24] and stored at −20 ℃ until used.

Oligonucleotide primers

The sequences of 18-mer oligonucleotide MK89 and 19-mer oligonucleotide MK90 (Table 5-35-2), which were used as primers in the PCR, were selected on the basis of the published sequence data of avian adenovirus chicken embryo lethal orphan (CELO) strain[5]. These primers flanked a 421-base DNA sequence in size. These primers were synthesized on a model 380B DNA synthesizer (Applied Biosystem Inc., Foster City, CA) with the assistance of the University of Connecticut Biotechnology Center. The primers were desalted through a Sephadex G-25 column (Pharmacia Inc., Piscataway, NJ). The concentration of the primers was determined by spectrophotometry as previously described[24] and then divided into 20 μL volumes and stored at −20 ℃.

Table 5-35-2 Base sequence of oligonucleotide primers

Primer	Oligonucleotide sequence (5’-3’)
MK 89	CCCTCCCACCGCTTACCA
MK 90	CACGTTGCCCTTATCTTGC

PCR amplification

PCR was carried out as described[23] with reagents from the GeneAmp PCR kit (Perkin Elmer Cetus, Norwalk, CT). A reaction volume of 100 μL of PCR mixture contained 4 μL of 80 mM $MgCl_2$; 8 μL of 10 × PCR buffer; 200 μM each of dATP (deoxyadenosine triphosphate), deoxycytidine triphosphate, deoxyguanosine triphosphate, and deoxythymidine triphosphate; 600 ng each of MK89 and MK90 primers; 0.5 μL (2.5 units) of AmpliTaq DNA polymerase. Then 100 ng of DNA was added to the mixture. The reaction mixture was overlaid with 50 μL of mineral oil. PCR was performed in an automatic DNA thermal cycler (PHC2 Dri-Block, Techne, Princeton, NJ). Initially, the reaction mixture was heated at 94 ℃ for 5 min. Then the PCR was run for 35 cycles at a melting temperature of 94 ℃ for 1 min, annealing temperature of 61 ℃ for 1 min, and extension temperature of 74 ℃ for 1 min. The sample was then heated at 74 ℃ for 10 min for the final extension reaction.

Sensitivity and specificity of avian adenovirus PCR

One hundred nanograms of genomic DNA from each of the 16 different serotypes/groups and 21 field avian adenovirus isolates and DNA from one mammalian adenovirus and five other avian viruses and bacteria were subjected to avian adenovirus-specific PCR. Tenfold dilutions from 100 ng to 10^{-4} g of genomic DNA from avian adenovirus CELO strain virus were also subjected to adenovirus PCR to determine the sensitivity of the test.

Detection of PCR products

Gel electrophoresis and Southern blot hybridization[26] were used to detect PCR products. A volume of 20 μL of PCR products was electrophoretically separated on a 1% agarose gel, stained with ethidium bromide, and exposed to ultraviolet light to determine the presence and size of the amplified DNA product. For Southern blot hybridization, amplified PCR products on agarose gels were transferred to Zeta probe membranes (BioRad, Richmond, CA). Twenty nanograms of PCR-amplified DNA specific for adenovirus was labeled with alpha ^{32}P-dATP (3 000 Ci/ mM) with a DECAprime labeled kit (Ambion Inc., Austin, TX). The membrane was allowed to hybridize with ^{32}P-labeled probe in 1 mM ethylenediaminetetraacetic acid (EDTA), 0.5 M NaH_2PO_4 (pH 7.2), and 7% SDS at 65 ℃ for 18 h. The membrane was washed twice in 1 mM EDTA, 40 mM NaH_2PO_4 (pH 7.2), and 1% SDS. The membrane was allowed to dry and was exposed on X-OMAT AR film (Eastman Kodak Co., Rochester, NY).

Results

One hundred nanograms DNA from 16 different serotypes belonging to three groups and 21 field isolates of avian adenovirus were amplified by 35 cycles of avian adenovirus-specific PCR, and the amplified products consisted of 421 bp, as expected, with the MK89 and MK90 primers. Representative specific avian adenovirus PCR results are shown in Figure 5-35-1 A, B. Southern blot hybridization confirmed that the DNA amplification products seen by gel electrophoresis were indeed those of avian adenovirus DNA as shown in Figure 5-35-1 A, B. The PCR results from three different groups (Ⅰ, Ⅱ, and Ⅲ) of avian adenovirus are shown in Figure 5-35-2 A, B. In clear contrast, the DNA from five avian pathogenic viruses and bacteria and one

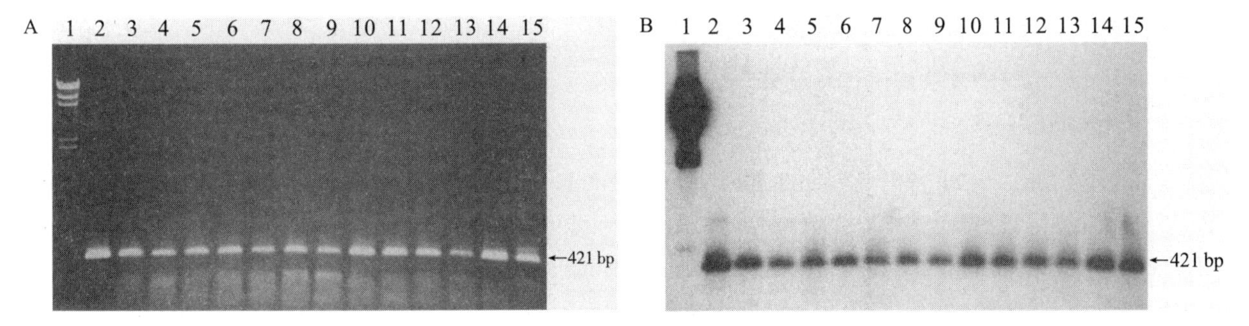

A: Agarose gel electrophoresis of different serotypes/groups of avian adenovirus PCR products; B: Southern blot hybridization of different serotypes/ groups of avian adenovirus PCR products probed with ^{32}Plabeled CELO adenovirus-amplified DNA probe. Lane 1: Molecular marker lambda DNA digested with *Hind* Ⅲ; lane 2: Adenovirus (Ad) type 1; lane 3: Ad type 2; lane 4: Ad type 3; lane 5: Ad type 4; lane 6: Ad type 5; lane 7: Ad type 6; lane 8: Ad type 7; lane 9: Ad type 8; lane 10: Ad type 9; lane 11: Ad type 10; lane 12: Ad type 11; lane 13: Ad type 12; lane 14: Ad-GII HE; lane 15: Ad-GIII EDS. The arrow identifies the 421 bp product of target DNA that was amplified by adenovirus-specific MK89 and MK90 primers.

Figure 5-35-1　Representative specific avian adenovirus PCR results

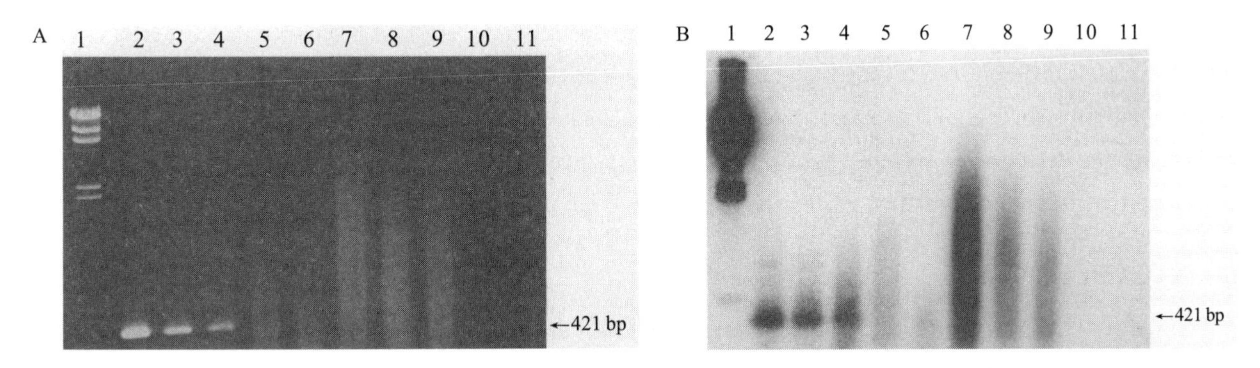

A: Agarose gel electrophoresis of adenovirus-specific DNA products amplified by PCR. B: Southern blot hybridization of adenovirus-specific PCR products probed with ^{32}P-labeled CELO adenovirus-amplified DNA probe. Lane 1: molecular weight marker lambda DNA digested with *Hind* Ⅲ; lane 2: Ad G-I CELO; lane 3: Ad G-Ⅱ HE; lane 4: Ad G-Ⅲ EDS; lane 5: ILTV; lane 6: Salmonella enteritidis; lane 7: chicken anemia virus; lane 8: *Proteus* spp.; lane 9: *Mycoplasma gallisepticum*-S6; lane 10: PCR buffer. The arrow identifies the 421-bp product of target DNA that was amplified by adenovirus-specific MK89 and MK90 primers.

Figure 5-35-2　PCR results from three different groups (Ⅰ, Ⅱ, and Ⅲ) of avian adenovirus

mammalian adenovirus was not amplified under the same PCR conditions (Figure 5-35-2 A). Autoradiographs of Southern blot membrane with PCR products demonstrated that the probe binds to amplified products of the different groups of avian adenovirus but not to those of other viruses and bacteria (Figure 5-35-2 B). The minimum amount detected by avian adenovirus-specific PCR was 1 fg of DNA determined by gel electrophoresis and Southern blot hybridization (Figure 5-35-3 A, B).

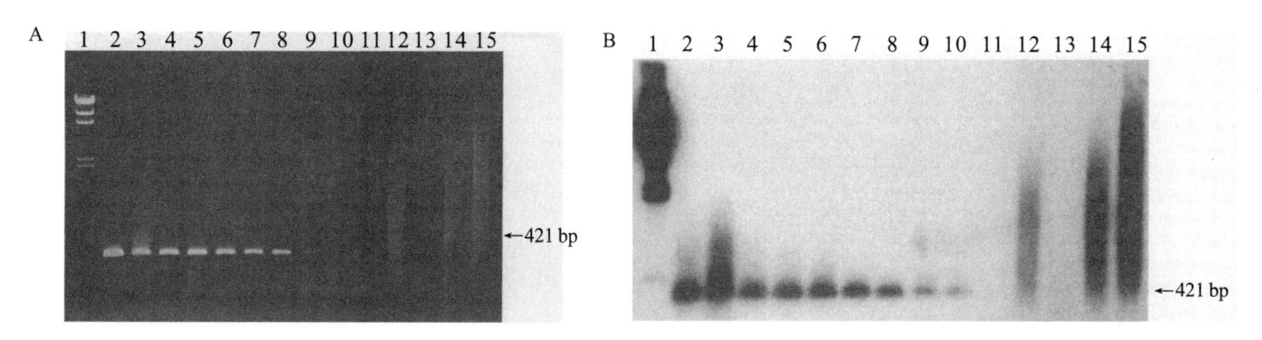

A: Sensitivity of PCR products showing on the agarose gel electrophoresis; B: Southern blot hybridization of PCR products with labeled CELO adenovirus-amplified DNA probe. Lane 1: Molecular weight marker lambda DNA digested with *Hind* Ⅲ ; lane 2: 100 ng; lane 3: 10 ng; lane 4: 1 ng; lane 5: 100 pg; lane 6: 10 pg; lane 7: 1 pg; lane 8: 100 fg; lane 9: 10 fg; lane 10: 1 fg; lane 11: 10^{-1} fg; lane 12: 10^{-2} g; lane 13: 10^{-3} fg; lane 14: 10^{-4} fg; lane 15: PCR buffer. The arrow identifies the 421 bp product of target DNA that was amplified by adenovirus-specific MK89 and MK90 primers.

Figure 5-35-3 Sensitivity of PCR

Discussion

The need for a rapid, sensitive, and specific test for the laboratory diagnosis of viral diseases has long been recognized. The diagnosis of avian adenovirus infections is currently made by virus isolation and detection by various immunologic methods that are laborious and time-consuming. The PCR method described here is able to detect as little as 1 fg of DNA, a level that may be overlooked with the more conventional culture and immunologic methods. The detection of such a small amount of DNA may enable the viral DNA to be amplified directly from clinical samples and may avoid an extra step to grow the virus in the cell culture.

A major advantage of this PCR will be its detection of all three groups of adenoviruses. In addition, the avian adenovirus PCR assay can be used to confirm the identify of the isolations, thus avoiding the need for doing virus neutralization, fluorescent antibody staining, and the double immunodiffusion test.

The avian adenovirus group-specific PCR presented is suggested as a rapid, sensitive, and specific test to demonstrate all different serotypes or groups of avian adenovirus infections and especially might be used for avian adenovirus surveillance of specific-pathogen-free Hocks. In summary, we have demonstrated the utility of PCR for the detection of 16 serotypes of all three groups and 21 field isolates of avian adenoviruses without any cross-reactivity to mammalian adenovirus and other avian pathogens.

References

[1] ADAIR B M, MCFERRAN J B, CALVERT V M. Development of a microtitre fluorescent antibody test for serological detection of adenovirus infection in birds. Avian Pathol, 1980, 9(3): 291-300.

[2] ADAIR B M, TODD D, MCFERRAN J B, et al. Comparative serological studies with egg drop syndrome virus. Avian Pathol, 1986, 15(4): 677-685.

[3] BURKE C N, LUGINBUHL R E, JUNGHERR E L. Avian enteric cytopathogenic viruses. I. Isolation. Avian Dis, 1959, 3(4): 412-419.

[4] CALNEK B W, SHEK W R, MENENDEZ N A, et al. Serological cross-reactivity of avian adenovirus serotypes in an enzyme-linked immunosorbent assay. Avian Dis, 1982, 26(4): 897-906.

[5] CHIOCCA S, KURZBAUER R, SCHAFFNER G, et al. The complete DNA sequence and genomic organization of the avian adenovirus CELO. J Virol, 1996, 70(5): 2939-2949.

[6] COWEN B S. A trivalent antigen for the detection of type I avian adenovirus precipitin. Avian Dis, 1987, 31(2): 351-354.

[7] COWEN B S. Inclusion body hepatitis-anaemia and hydropericardium syndromes: aetiology and control. World's poultry science journal, 1992, 48(3): 247-254.

[8] COWEN B, MITCHELL G B, CALNEK B W. An adenovirus survey of poultry flocks during the growing and laying periods. Avian Dis, 1978, 22(1): 115-121.

[9] DOMERMUTH C H, WESTON C R, COWEN B S, et al. Incidence and distribution of "avian adenovirus group Ⅱ splenomegaly of chickens". Avian Dis, 1980, 24(3): 591-594.

[10] FAHEY K J, BAGUST T J, YORK J J. Laryngotracheitis herpesvirus infection in the chicken: the role of humoral antibody in immunity to a graded challenge infection. Avian Pathol, 1983, 12(4): 505-514.

[11] FREY M L, HANSON R P, RSON D P. A medium for the isolation of avian mycoplasmas. Am J Vet Res, 1968, 29(11): 2163-2171.

[12] HORIMOTO T, KAWAOKA Y. Direct reverse transcriptase PCR to determine virulence potential of influenza A viruses in birds. J Clin Microbiol, 1995, 33(3): 748-751.

[13] JESTIN V, JESTIN A. Detection of Newcastle disease virus RNA in infected allantoic fluids by in vitro enzymatic amplification (PCR). Archives of Virology, 1991, 118(3-4): 151-161.

[14] KAWAMURA H, SHIMIZU F, TSUBAHARA H. Avian adenovirus: its properties and serological classification. Natl Inst Anim Health Q (Tokyo), 1964, 4: 183-193.

[15] KWON H M, JACKWOOD M W, GELB J J. Differentiation of infectious bronchitis virus serotypes using polymerase chain reaction and restriction fragment length polymorphism analysis. Avian Dis, 1993, 37(1): 194-202.

[16] LEE L H, YU S L, SHIEH H K. Detection of infectious bursal disease virus infection using the polymerase chain reaction. J Virol Methods, 1992, 40(3): 243-253.

[17] MCFERRAN J B, ADAIR B, CONNOR T J. Adenoviral antigens (CELO, QBV, GAL). Am J Vet Res, 1975, 36(4 Pt 2): 527-529.

[18] MCFERRAN J B, CONNOR T J, ADAIR B M. Studies on the antigenic relationship between an isolate (127) from the egg drop syndrome 1976 and a fowl adenovirus. Avian Pathol, 1978, 7(4): 629-636.

[19] NAZERIAN K, FADLY A. Propagation of virulent and avirulent turkey hemorrhagic enteritis virus in cell culture. Avian Dis, 1982, 26(4): 816-827.

[20] NGUYEN A V, KHAN M I, LIU Z. Amplification of Salmonella chromosomal DNA using the polymerase chain reaction. Avian Dis, 1994, 38(1): 119-126.

[21] POULSEN D J, BURTON C R, O'BRIAN J J, et al. Identification of the infectious laryngotracheitis virus glycoprotein gB gene by the polymerase chain reaction. Virus Genes, 1991, 5(4): 335-347.

[22] RABBANI M, NAEEM K. In vitro and in vivo evaluation of avian adenovirus isolates from outbreaks of hydropericardium syndrome// PROC. International Symposium on Adenovirus and Reovirus Infections in Poultry. Rauischholzhausen, Germany, 1996: 26-31, 24-27.

[23] SAIKI R K, SCHARF S, FALOONA F, et al. Enzymatic amplification of β-globin genomic sequences and restriction site analysis for diagnosis of sickle cell anemia. Science (American Association for the Advancement of Science), 1985, 230(4732): 1350-1354.

[24] SAMBROOK J, FRITSCH E T, MANIATIS T. Molecular cloning: a laboratory manual. New York: Cold Spring Harbor Laboratory, 1989.

[25] SILK R, SURESH M, NEUMANN U, et al. Pathogenic mechanisms of avian adenovirus type Ⅱ //PROC. International

Symposium on Adenovirus and Reovirus Infections in Poultry. Rauischholzhausen, Germany, 1996: 41-50, 24-27.

[26] SOUTHERN E M. Detection of specific sequences among DNA fragments separated by gel electrophoresis. J Mol Biol, 1975, 98(3): 503-517.

[27] THAM K M, STANISLAWEK W L. Polymerase chain reaction amplification for direct detection of chicken anemia virus DNA in tissues and sera. Avian Dis, 1992, 36(4): 1000-1006.

[28] VAN ECK J H, DAVELAAR F G, HEUVEL-PLESMAN T A, et al. Dropped egg production, soft shelled and shell-less eggs associated with appearance of precipitins to adenovirus in flocks of laying fowls. Avian Pathol, 1976, 5(4): 261-272.

[29] XIE Z, FADL A A, GIRSHICK T, et al. Amplification of avian reovirus RNA using the reverse transcriptase-polymerase chain reaction. Avian Dis, 1997, 41(3): 654-660.

[30] ZSAK L, KISARY J. Characterisation of adenoviruses isolated from geese. Avian Pathol, 1984, 13(2): 253-264.

Reverse transcriptase-polymerase chain reaction to detect avian encephalomyelitis virus

Xie Zhiqin, Khan Mazhar I, Girshick Theodore, and Xie Zhixun

Abstract

A reverse transcriptase-polymerase chain reaction (RT-PCR) was developed and optimized for the detection of avian encephalomyelitis virus (AEV). A pair of primers was prepared based on the VP2 gene of the structural protein Pl region of the AEV genome. An avian encephalomyelitis virus-specific 619-base pair cDNA product was amplified by these primers from five reference/field strains of AEVs but not from 10 other avian pathogenic viruses and bacteria. The RT-PCR assay developed in this study was found to be sensitive and specific with as little as 10 pg of avian encephalomyelitis virus RNA detected using gel electrophoresis. Furthermore, AEV-RT-PCR was able to detect AE virus from chicken embryo brain at 3 days postinoculation as compared with the AE agar gel precipitation test (AGP), which required up to 11 days of incubation in the embryos.

Keywords

avian encephalomyelitis, primers, RT-PCR, VP2

Introduction

Avian encephalomyelitis (AE) was first described in 1932[10]. AE is a viral infection that occurs in young chickens, pheasants, quail, and turkeys[3, 25]. Young chickens infected with this virus can have clinical signs of ataxia, incoordination, paralysis, or rapid tremors of the head and neck, with high morbidity and variable mortality[14, 17]. In adult laying birds, AE infection causes no neurologic signs, but it can cause a slight reduction in egg production[3, 25]. AE virus (AEV) is mainly an egg-transmitted disease, but infection of poultry with AEV by the fecal-oral route is not uncommon[3, 17, 22, 27].

AEV, a member of the family Picornaviridae, contains a single-stranded RNA genome of positive polarity[13, 21]. Its particles are nonenveloped. It possesses a buoyant density of 1.31～1.33 g/mL, a sedimentation coefficient of 148 s, and is 24～32 μm in diameter[2, 7, 13, 26]. It has been now shown that the 7058-base-pair genome sequence has a single, long open reading fragment (ORF) encoding a large polyprotein[13, 26]. Like other picornaviruses, AEV also includes four structural proteins from the P1 region (VP4, VP2, VP3, VP1) and seven nonstructural proteins from the P2 and P3 regions of the genome[13].

Many methods have been developed for the diagnosis of avian encephalomyelitis, including virus isolation by intracerebral inoculation of 1-day-old chicks and yolk-sac inoculation of embryonated chicken eggs[24, 28]. Various serologic methods have been developed, such as hemagglutination, complement fixation, indirect fluorescent-antibody, enzyme-linked immunosorbent assay (ELISA), virus neutralization, and agar-gel-precipitin tests[4, 6, 8, 9, 11, 20]. However, isolation and serological methods are time consuming and labor intensive. Furthermore, serological tests are often hampered by nonspecific reactions or cross-reactions. In recent years, the polymerase chain reaction (PCR) method has been applied as a rapid diagnostic tool for the detection of avian viral and bacterial pathogens[12, 15, 16, 23, 29]. In this study, we have developed and optimized a reverse transcriptase-polymerase chain reaction (RT-PCR) assay using primers targeting to the VP2 gene to detect AEV.

Materials and methods

Virus propagation

All viruses and other bacteria used in this study are listed in Table 5-36-1. AEV were treated with penicillin (10 000 IU/mL) and streptomycin (100 μg/mL). These samples were centrifuged at 1 000 × g for 10 min after incubation at 4 ℃ for 4 h. Supernatants were collected. Aliquots of 0.2 mL of these viral supernatants were inoculated into the allantoic cavity of 6-day-old embryonating specific-pathogen free (SPF) eggs (Charles River SPAFAS, North Franklin, CT). Brains from infected embryos were collected at 11 days postinoculation and resuspended in phosphate-buffered saline (PBS) (10% w/v). Brain tissues were disrupted three times by freezing and thawing and were stored at −70 ℃.

Table 5-36-1 Avian pathogens used for RT-PCR

Species	Strain/serotype	Source
Avian encephalomyelitis	Van Roekel	Charles River SPAFAS, Storrs, CT
	AE 19 (11/77)	University of Connecticut
	AE W1 (5/86)	Washington State University
	AE W2 (5/86)	Washington State University
	AE 1143	Intervet Inc., Millsboro, Delaware
Infectious bronchitis	Mass 41	University of Connecticut Charles River SPAFAS, Storrs, CT
Infectious bursal disease		
Infectious laryngotracheitis	950802	University of Connecticut
Newcastle disease	B1 LaSota	Washington State University
Reovirus	S1133	University of Connecticut
Adenovirus	Group I, type1	Charles River SPAFAS, Storrs, CT
Mycoplasma gallisepticum	S 6 (208)	University of California
Salmonella gallinarum	Field isolate	University of Connecticut
Salmonella enteritidis	Field isolate	University of Connecticut
Escherichia coli	Field isolate	University of Connecticut

Virus semipurification

Semipurified virus was modified as previously described[24]. Brains from six AEV-infected chick embryos were suspended in 200 mL PBS. Brain tissues were treated with 0.5% (w/v) sodium dodecyl sulfate (SDS) and disintegrated by shaking using glass beads for 30 min at 37 ℃ followed by sonication for 20~30 s. Virus suspensions were clarified by centrifugation at 1 000 × g for 30 min at 4 ℃, and supernatants were further centrifuged at 80 000 × g for 3 hr at 4 ℃. The resulting crude virus pellets were resuspended in 1 mL of PBS. These viral preparations were used for extraction of RNA and stored at −20 ℃.

Infectious bronchitis virus (IBV), Newcastle disease virus (NDV), and adenovirus were propagated in the allantoic cavity of 10-day-old SPF embryonated eggs[18]. Infectious laryngotracheitis virus (ILTV) was propagated on the chorioallantoic membrane in 10-day-old SPF embryonated eggs. Infectious bursal disease virus (IBDV) and reovirus were propagated in chicken embryo fibroblast monolayers as described previously[23, 29]. *Mycoplasma gallisepticum* (MG) was grown in Frey medium[5]. *Salmonella* sp.and *Echerichia coli* were grown in Luria-Bertani (LB) broth media[16].

Nucleic acid extraction

RNA extraction from AEV, IBV, NDV, IBDV, and reovirus was carried out according to the Trizol LS manufacturer's protocol. (Trizol LS, Life Technologies, Bethesda, MD). DNA from ILTV, adenovirus, MG, *Salmonella* sp., and *E.coli* were extracted using the phenol : chloroform : isoamyl alcohol (1 : 1 : 24 v/v) and purified (Gibco BRL, Grand Island, NY) method as described by Pang et al.[17]. The concentrations of RNA and DNA were determined by spectrophotometer using the Bio Mate 5 (Thermo Spectronic, Rochester, NY), and nucleic extractions were stored at −20 ℃.

Oligonucleotide primers

Pairs of primers that specifically amplified AEV were designated as MK AE 1 (CTTATGCTGGCCC TGATCGT) and MK AE 2 (TCCCAAATCCACAAACCTAGCC) and were selected on the basis of the published sequence data of AEV [13]. These primers flanked a 619-base pair (bp) DNA sequence containing the VP2 gene. Primers were synthesized at Invitrogen/Gibco (Carlsbad, CA). Primers were aliquoted into 50- μL volumes and stored at −20 ℃.

Reverse transcription (RT) and PCR amplification

The reverse transcription reaction was conducted using Gene Amp PCR kit (Perkin Elmer Cetus, Norwalk, CT). The reaction mixture contained 4 μL of 25 mM $MgCl_2$, 2 μL of 10 × PCR buffer (500 mM KCl, 200 mM Tris·HCl[pH 8.4], 0.5 mg/mL nuclease-free bovine serum albumin), 2 μL of 10 mM each deoxynucleoside triphosphate (dNTP), 1 μL (20 units) of RNase inhibitor, 1 μL (50 pmol) downstream primer (MKAE2), 1 μL (50 units) of Maloney murine leukemia virus, reverse transcriptase, and 50 ng of RNA or DNA. A total volume of 20 μL reactions was obtained by adding sterile diethylpyrocarbonate (DEPC)-treated water. The RT was carried out using a thermal cycler (Model 480, Perkin Elmer Cetus Corporation, Norwalk, CT) set at 42 ℃ for 15 min, 99 ℃ for 5 min, and 5 ℃ for 5 min for one cycle.

For the PCR reaction, 8 μL of 25 mM $MgCl_2$, 8 μL of 10 × PCR buffer (500 mM KCl, 200 mM Tris·HCl (pH 8.4), 0.5 mg/ mL nuclease-free bovine serum albumin), 60 pmol of each primer MK AE 1, and 2.5 units of Ampli Taq DNA polymerase were added to the above RT reaction tubes and a 100- μL total volume was obtained by adding sterile DEPC-treated distilled water. The reaction mixture was denatured at 95 ℃ for 5 min. Then the PCR was run for 35 cycles at 94 ℃ for 1 min with annealing and extension temperature and time of 62 ℃ and 1 min, respectively; the sample was then incubated at 62 ℃ for 10 min for one cycle, with the final extension at 4 ℃.

The specificity and sensitivity of the AE RT-PCR

Fifty nanograms of RNA from each of the five different AE isolates and DNA or RNA from 10 avian viruses and bacteria were subjected to AEV specific PCR. Tenfold dilutions from 100 ng to 0.1 fg of genomic RNA from AEV of Van Rockel strain virus were also subjected to AEV PCR to determine the sensitivity of the PCR test.

Detection-amplified RT-PCR products

Gel electrophoresis[19]was used to detect amplified DNA products. A volume 10 μL of amplified PCR products was subjected to electrophoresis at 80 V in horizontal gels containing 1% agarose (Ultra pure, Bethesda Research Laboratories, Bethesda, MD) with Tris-borate buffer (45 mM Trisborate, 1 mM ethylenediaminetetraacetic acid). The gel was stained with ethidium bromide (0.5 μg/mL), exposed to ultraviolet light to visualize the amplified products, and photographed.

Agar gel precipitation (AGP) test

The AGP test was performed by a standard method[8]. Briefly, 1% agar (Difco) containing 8% NaCl and 0.5% phenol, pH 7.2, was layered onto glass slides to a thickness of approximately 1 mM. A template containing eight peripheral wells surrounding a central well was used. Wells were 2 mM in diameter and 8 mM from the central well. Positive or negative serum (Charles River SPAFAS, Inc.) was place in the central well, and antigens from AE positive and test brain tissue samples were place in the peripheral wells. After 30～60 min at room temperature, wells were filled a second time and slides were incubated in a humid chamber at room temperature overnight. A precipitin line was observed during 12～24 h.

Detection of AEV from chicken embryo organs

Twenty 6-day-old SPF embryonating eggs (Charles River SPAFAS, Inc.) were used. Two tenth of a milliliter of a 10^{-3} dilution of AE van Roekel strain was inoculated into the yolk sac, using a 1 1/2-inch, 22-gauge needle, and incubated. Chicken embryo brains were collected at 3 days postinoculation (PI). Chicken embryo brains, livers, and allantoic fluid samples were also collected at 11 days PI. Brain and liver samples were ground in a tissue grinder with nutrient broth added to make 10% suspensions. Suspensions were freeze-thawed five times, and then samples were centrifuged at 1 000 × g for 10 min. Supernatants were used for RNA extraction and stored at −20 ℃ for future use.

Results

RT-PCR amplification

Analysis by agarose electrophoresis indicated that a DNA fragment of approximately 619 bp was amplified as expected when the RNA extracted from infected AEV chicken brain tissue was subjected to RT-PCR (Figure 5-36-1, lanes 2-6). RNA samples from uninfected chicken brain tissue and RNA or DNA from 10 other avian pathogens was not amplified under the same RT-PCR conditions (Figure 5-36-1, lanes 7-15). The minimum amount detected by the avian encephalomyelitis virus-specific RT-PCR was 10 pg of RNA as determined by gel electrophoresis (Figure 5-36-2, Figure 5-36-3). Negative controls (PBS and PCR reagent) were used to optimize the RT-PCR tests and were included in each RT-PCR run. No spurious PCR amplification reactions were seen. All negative controls were negative (data not shown).

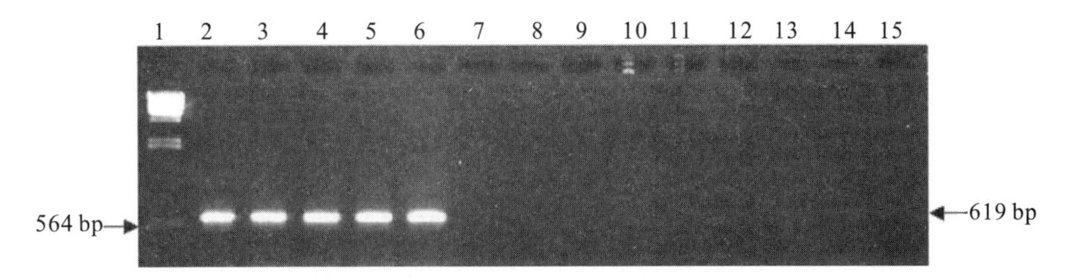

Lane 1: *Hind*Ⅲ-digested lamda DNA size marker; lane 2: AE Van Roekel; lane 3: AE 19; lane 4: AE W1; lane 5: AE W2; lane 6: AE1143; lane 7: IBV (Mass 41) ; lane 8: IBDV; lane 9: ILTV (950802) ; lane10: NDV (B1 LaSota) ; lane 11: Reovirus (S1133) ; lane 12: Adenovirus (group I: type 1) ; lane 13: MG (S6-208) ; lane 14: Salmonella sp. (field isolate) ; lane 15: *E.coli* (field isolate).

Figure 5-36-1 Agarose gel electrophoresis of AEV-specific amplified products by RT-PCR

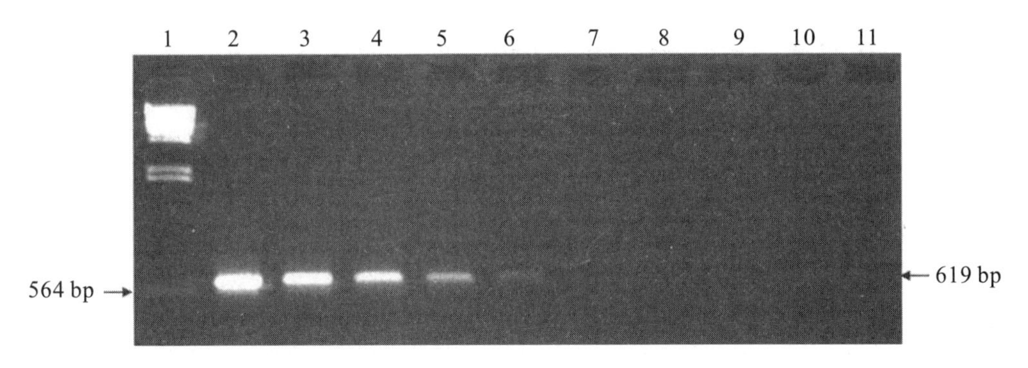

Lane 1: *Hind*Ⅲ-digested lamda DNA size marker; lane 2: 100 ng of AE RNA; lane 3: 10 ng; lane 4: 1 ng; lane 5: 100 pg; lane 6: 10 pg; lane 7: 1 pg; lane 8: 100 fg; lane 9: 10 fg; lane 10: 1 fg; lane 11: 0.1 fg.

Figure 5-36-2 Sensitivity of AE RT-PCR

Detection of AEV from chicken embryo

AEV-specific RT-PCR was able to amplify 619-bp DNA from brain tissues at 3 days PI in embryonating eggs; however, these brain tissue samples were negative for the AGP test. When samples of chicken embryo brains, livers, and allantoic fluid were collected at 11 days PI and tested with AE-RT-PCR, all 10 brain and 10 liver tissue samples were positive. Four out of 10 allantoic fluid samples were positive with the avian encephalomyelitis virus-specific RT-PCR. However, the AGP test was unable to detect AE viral antigen in the allantoic fluid (Table 5-36-2 and Figure 5-36-3).

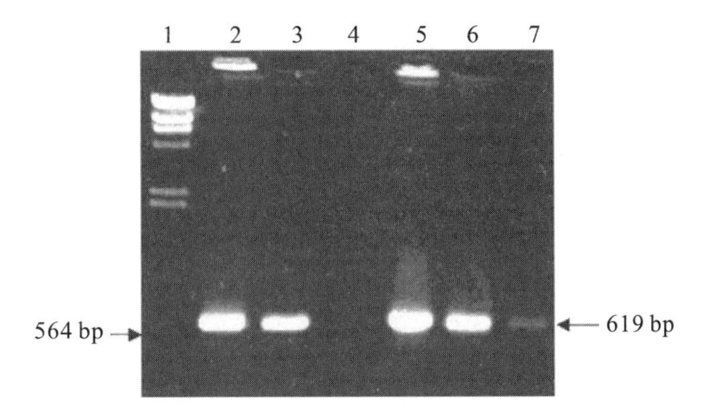

Lane 1: *Hind*Ⅲ-digested lamda DNA size marker; lane 2: Brain 1; lane 3: Liver 1; lane 4: Allantoic fluid 1; lane 5: Brain 2; lane 6: Liver 2; lane 7: Allantoic fluid 2.

Figure 5-36-3 Agarose gel electrophoresis of AE RT-PCR products from chicken embryo tissues infected with AEV

Table 5-36-2 The results of testing AE virus from chicken embryo tissues

Test materials	AGP [a] positive/ test sample	RT-PCR positive/test samples
3 days postinoculation		
Brain tissues	0/10 (0%)	10/10 (100%)
Liver tissues	0/10 (0%)	0/10 (0%)
Allantoic fluid	0/10 (0%)	0/10 (0%)
11 days postinoculation		
Brain tissues	10/10 (100%)	10/10 (100%)
Liver tissues	8/10 (80%)	10/10 (100%)
Allantoic fluid	0/10 (0%)	4/10 (40%)

Note: [a] Used AE van Roekel reference strain antibodies sera (Charles River SPAFAS, Inc.).

Discussion

The primary aim of this study was to develop a RT-PCR method to detect AEV infection in chickens. One pair of primer sets was selected from the VP2 of structural protein genome sequence data of AE. All RNA samples from different AEVs were positive in the AE RT-PCR with all positive products consisting of the expected 619 bp DNA band. In clear contrast, RNA from uninfected chicken brains and RNA or DNA from other avian pathogens was not amplified under the same RT-PCR conditions (Figure 5-36-1, lanes 7-15). AEV RT-PCR results showed as little as 10 pg of RNA from AE viral-infected samples could be amplified. Both the specificity and sensitivity of the AEV RT-PCR indicated that the AEV RT-PCR method could be used for the specific detection of AEV. The AEV RT-PCR showed that a small amount of RNA could be amplified directly from clinical and environmental samples.

Braune and Gentry [1] reported that AEV could be detected in chicken embryonic tissues after 5 days PI using fluorescent-antibody test. However, AEV RT-PCR was able to detect AE RNA at 3 days PI in chicken embryo brain (Table 5-36-2). These results indicated that the AEV RT-PCR method was more sensitive for the detection of AEV from early-infected chicken embryos.

The results of testing chicken embryos infected with AEV revealed that AEV grows very slowly in the embryo allantoic fluid as compared with embryo brain tissues. This may be due to the AEV's ability to propagate in brain cells much better than in the allantoic cavity of embryonating eggs.

In comparison with the AGP test, the RT-PCR method was more effective and specific for detecting AEV. In fact, the AGP test for AEV was affected by various factors, such as antibody quality and different AE antigens. Girshick et al.[6] had shown that the most useful aspect of the AGP test for AEV was the rapidiry, but the sensitivity of the AGP test was its limitation. These results from testing different AEV strains and different tissue samples indicated that the AEV RT-PCR method was rapid and specific.

In summary, a rapid, sensitive, and specific method for the laboratory diagnoses of AE disease is needed. Clinical diagnosis of AEV infection is currently made by virus isolation and detection by various immunologic methods that are laborious and time consuming. In this study, we optimized a RT-PCR method to detect AEV that was not only simple to employ but was also sensitive and specific.

References

[1] BRAUNE M O, GENTRY R F. Avian encephalomyelitis virus. I. Pathogenesis in chicken embryos. Avian Dis, 1971, 15(4): 638-647.

[2] BUTTERFIELD W K, LUGINBUHL R E, HELMBOLDT C F. Characterization of avian encephalomyelitis virus (an avian enterovirus). Avian Dis, 1969, 13(2): 363-378.

[3] CALNEK B W, LUGINBUHL R E, HELMBOLDT C F. Avian encephalomyelitis//CALNCK B W, BARNES H J, BEARD C W, et al. Disease of poultry, 10th ed. Ames, IA: Iowa State University Press, 1997: 571-581.

[4] CHOI W P, MIURA S. Indirect fluorescent antibody technique for the detection of avian encephalomyelitis antibody in chickens. Avian Dis, 1972, 16(4): 949-951.

[5] FREY M L, HANSON R P, ANDRSON D P. A medium for the isolation of avian mycoplasmas. Am J Vet Res, 1968, 29(11): 2163-2171.

[6] GIRSHICK T, CRARY C K. Preparation of an agar-gel-precipitating antigen for avian encephalomyelitis and its use in evaluating the antibody status of poultry. Avian Dis, 1982, 26(4): 798-804.

[7] GOSTING L H, GRINNELL B W, MATSUMOTO M. Physicochemical characteristics of avian encephalomyelitis virus. Vet Microbiol, 1980, 5: 87-91.

[8] HALPIN F B. A Search for a Hemagglutinating Property of the Virus of Infectious Avian Encephalomyelitis. Avian Dis, 1966, 10(4): 513-517.

[9] IKEDA S. Immunodiffusion test in avian encephalomyelitis. II. Detection of precipitating antibody in infected chickens in comparison with neutralizing antibody. Natl Inst Anim Health Q (Tokyo), 1977, 17(3): 88-94.

[10] JONES E E. an encephalomyelitis in the chicken. Science, 1932, 76(1971): 331-332.

[11] LUKERT P D, DAVIS R B. An antigen used in the agar-gel precipitin reaction to detect avian encephalomyelitis virus antibodies. Avian Dis, 1971, 15(4): 935-938.

[12] MANSY M S, ASHOUR M S E, KHAN M I. Development of species-specific polymerase chain reaction (PCR) technique for Proteus mirabilis. Mol Cell Probes, 1999, 13: 133-140.

[13] MARVIL P, KNOWLES N J, MOCKETT A P, et al. Avian encephalomyelitis virus is a picornavirus and is most closely related to hepatitis A virus. J Gen Virol, 1999, 80 (Pt 3): 653-662.

[14] MCNULTY M S, CONNOR T J, MCNEILLY F, et al. Biological characterisation of avian enteroviruses and enterovirus-like viruses. Avian Pathol, 1990, 19(1): 75-87.

[15] NASCIMENTO E R, YAMAMOTO R, KHAN M I. *Mycoplasma gallisepticum* F-vaccine strain-specific polymerase chain reaction. Avian Dis, 1993, 37(1): 203-211.

[16] NGUYEN A V, KHAN M I, LU Z. Amplification of Salmonella chromosomal DNA using the polymerase chain reaction. Avian Dis, 1994, 38(1): 119-126.

[17] OLITSKY P K. experimental studies of the virus of infectious avian encephalomyelitis. J Exp Med, 1939, 70(6): 565-582.

[18] PANG Y S, KHAN M I, WANG H, et al. Multiplex PCR and its application in experimentally infected SPF chickens with respiratory pathogens. Avian Dis, 2002, 46: 691-699.

[19] SAMBROOK J, FRISCH E T, MANIATIS T. Molecular cloning: a laboratory manual. New York: Cold Spring Harbor Laboratory Press, 1989.

[20] SATO G, WATANABE H, MIURA S. An attempt to produce a complement-fixation antigen for the avian encephalomyelitis virus from infected chick embryo brains. Avian Dis, 1969, 13(3): 461-469.

[21] SHAFREN D R, TANNOCK G A. Further evidence that the nucleic acid of avian encephalomyelitis virus consists of RNA. Avian Dis, 1992, 36(4): 1031-1033.

[22] SHAFREN D R, TANNOCK G A, GROVES P J. Antibody responses to avian encephalomyelitis virus vaccines when administered by different routes [chickens; layers]. Australian Veterinary Journal, 1992, 69(11): 272-275.

[23] STRAM Y, MEIR R, MOLAD T, et al. Applications of the polymerase chain reaction to detect infectious bursal disease virus in naturally infected chickens. Avian Dis, 1994, 38(4): 879-884.

[24] TANNOCK G A, SHAFREN D R. A rapid procedure for the purification of avian encephalomyelitis viruses. Avian Dis, 1985, 29(2): 312-321.

[25] TANNOCK G A, SHAFREN D R. Avian encephalomyelitis: a review. Avian Pathol, 1994, 23(4): 603-620.

[26] TODD D, WESTON J H, MAWHINNEY K A, et al. Characterization of the genome of avian encephalomyelitis virus with cloned cDNA fragments. Avian Dis, 1999, 43(2): 219-226.

[27] VAN ROEKEL H, BULLIS K L, CLAEKE M K. Transmission of avian encephalomyelitis. I Am Vet Med Assoc, 1941, 99: 220.

[28] WILS F K, MOULTHROP I M. Propagation of avian encephalomyelitis virus in the chick embryo. Southwest Vet, 1956, 10: 39-42.

[29] XIE Z, FADL A A, GIRSHICK T, et al. Detection of avian adenovirus by polymerase chain reaction. Avian Dis, 1999, 43(1): 98-105.

A polymerase chain reaction assay for detection of virulent and attenuated strains of duck plague virus

Xie Liji, Xie Zhixun, Huang Li, Wang Sheng, Huang Jiaoling, Zhang Yanfang, Zeng Tingting, Luo Sisi

Abstract

Sequence analysis of duck plague virus (DPV) revealed that there was a 528 bp (B fragment) deletion within the UL2 gene of DPV attenuated vaccine strain in comparison with field virulent strains. The finding of gene deletion provides a potential differentiation test between DPV virulent strain and attenuated strain based on their UL2 gene sizes. Thus we developed a polymerase chain reaction (PCR) assay targeting to the DPV UL2 gene for simultaneous detection of DPV virulent strain and attenuated strain, 827 bp for virulent strain and 299 bp for attenuated strain. This newly developed PCR for DPV was highly sensitive and specific. It detected as low as 100 fg of DNA on both DPV virulent and attenuated strains, no same size bands were amplified from other duck viruses including duck paramyxovirus, duck Tembusu virus, duck circovirus, Muscovy duck parvovirus, duck hepatitis virus type I, avian influenza virus and gosling plague virus. Therefore, this PCR assay can be used for the rapid, sensitive and specific detection of DPV virulent and attenuated strains affecting ducks.

Keywords

duck plague virus, virulent strain, attenuated strain, PCR

Duck plague, also known as duck viral enteritis (DVE), is an acute, contagious and lethal disease in ducks, geese, swans and many other waterfowl species[1]. This disease causes a high mortality of up to 100%, and up to 50% of decrease in egg production in domestic and wild waterfowl, resulting in significant economic losses[2-6]. The disease is caused by Anatid herpesvirus 1 or duck plague virus (DPV). Attenuated or nonpathogenic DPV strains are used as live vaccines to provide protection against clinical duck plague in industry settings[7-10]. However, Diao et al. reported that duck plague caused by a DPV virulent strain occurred in duck flocks after immunization with attenuated DPV live vaccines[11].

The current diagnostic methods for DPV, such as ELISA, electron microscopy, PCR, real-time PCR and LAMP[6, 12-15] cannot differentiate DPV virulent strains from attenuated strains. Viral isolation and identification could differentiate virulent from attenuated strain, however, this process can take more than 10 days. Thus, the need for a rapid, sensitive and specific laboratory test that differentiates between the DPV virulent and attenuated strains has long been recognized.

In this present study, we designed specific primers using conserved sequences of DPV UL2 genes, and developed a PCR assay to simultaneously detect and differentiate both DPV virulent and attenuated strains.

The UL2 genes of 10 DPV virulent strains were compared with those of 11 DPV attenuated strains (including all the UL2 gene sequences released in GenBank, Table 5-37-1). According to the UL2 genes sequence analysis a 528 bp (B fragment) deletion within the UL2 gene of DPV attenuated strain (Figure 5-37-1) was observed. The UL2 gene of DPV attenuated strain is 474/477 bp, whereas the UL2 gene of the DPV virulent strains is 1 002 bp. This finding is consistent with previous results reported by Wang et al.[16] and Yang et al.[17] who compared a small number of virulent and attenuated strains.

Table 5-37-1 A list of duck plague virus (DPV) stains used for sequence analysis

Name of virus	strain	Source [a]	PCR results
attenuated strain	VAC	EU082088	477 bp
attenuated strain	Sichuan1	JQ347517	474 bp
attenuated strain	Sichuan2	JQ347518	474 bp
attenuated strain	K	KF487736	474 bp
attenuated strain	AV18		474 bp
attenuated vaccine strain	C-KCE	KF263690	474 bp
attenuated vaccine strain	clone-03	EF449516	474 bp
attenuated vaccine strain	GLBPLC		474 bp
attenuated vaccine strain	SHBPLC		474 bp
attenuated vaccine strain	LYBPLC		474 bp
attenuated vaccine strain	HJBPLC		474 bp
virulent strain	Guangxi 30D	KX925439	1 002 bp
virulent strain	Guangxi 60D	KX925440	1 002 bp
virulent strain	Germany 2085	JF99965	1 002 bp
virulent strain	Chv	EU885419	1 002 bp
virulent strain	CV	JQ673560	1 002 bp
virulent strain	LH2011	KC480262	1 002 bp
virulent strain	LS	JQ248598	1 002 bp
virulent strain	N1	JQ043216	1 002 bp
virulent strain	N2	JQ248596	1 002 bp
virulent strain	N3	JQ248597	1 002 bp

Note: GLBPLC, Guangxi Liyuan Biological Products Limited Company; SHBPLC, Shanghai Haili Biological Products Limited Company; LYBPLC, LiaoningYikang Biological Products Limited Company; HJBPLC, Hayao Jituan Biological Products Limited Company. [a] The blank spaces indicate that there is no GenBank accession number.

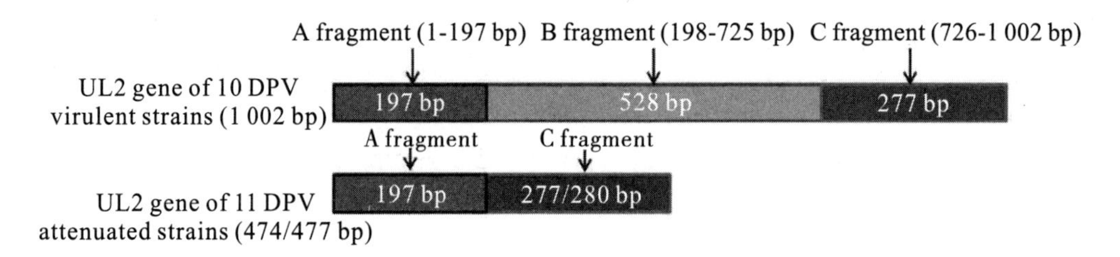

Figure 5-37-1 Schematic representation of the alignment between the UL2 genes of 10DPV virulent strains and 11DPV attenuated strains

Primer design for the DPV PCR assay was based on the published UL2 gene sequence of the DPV strain Chv (GenBank, accession number: EU885419) aligned with sequences of 20 DPV isolates (Table 5-37-1) to identify a conserved region. Primers were designed by using the Primer Premier 5.0 software. The forward primer was situated in the A-fragment, and the reverse primer was situated in the C-fragment of the UL2 gene. The primers are DPlague 299/827-1: 5'-TAGCCATGCCCTTAGTAGGACT-3' and DPlague 299/827-2: 5'-GAACCACTGACGACTACCCTGT-3'. The expected size of the PCR bands were 827 bp for DPV virulent strains and 299 bp for DPV attenuated strains.

Nine DPV strains and seven other viruses affecting ducks were used in the study (Table 5-37-2). Viral DNA or RNA was extracted from 200 μL of each virus sample using the EasyPure Viral DNA/RNAkit (TransGen, Beijing, China) according to the manufacturer's protocol. The DPV PCR was prepared using 2 × PCR Mix (TransGen, Beijing, China). One reaction volume of 50 μL PCR mixture contained 25 μL of 2 × PCR Mix, 0.5 pmol/ μL each of DPlague 299/827-1 and DPlague 299/827-2 primers, 1 μL of the DNA/ RNA, and 22 μL of sterile DEPC water. Initially, the reaction mixture was heated at 94 ℃ for 5 min. Then the PCR was run for 35 cycles at a melting temperature of 94 ℃ for 1 min, annealing temperature of 55 ℃ for 1 min, and extension temperature of 72 ℃ for 1 min. The sample was then heated at 74 ℃ for 10 min for the final extension reaction. Gel electrophoresis was used to detect the amplified PCR products.

Table 5-37-2 Results of duck plague virus (DPV) PCR for detection DPV virulent and attenuated strain, and other duck viruses

Name of virus strain	Source	PCR results
DPV (Yulin2016-30D, virulent strain)	GVRI	+
DPV (Yulin2016-60D, virulent strain)	GVRI	+
DPV (Guangxi01, virulent strain)	GVRI	+
DPV (Guangxi02, virulent strain)	GVRI	+
DPV (AV18, attenuated strain)	CIVDC	+
DPV (attenuated vaccine strain)	GLBPLC	+
DPV (attenuated vaccine strain)	SHBPLC	+
DPV (attenuated vaccine strain)	LYBPLC	+
DPV (attenuated vaccine strain)	HJBPLC	+
Duck paramyxovirus	GVRI	−
Duck Tembusu virus	GVRI	−
Duck Circovirus	GVRI	−
Muscovy duck parvovirus (AV238)	CIVDC	−
Duck hepatitis virus (AV2111)	CIVDC	−
Avian influenza virus subtype H9	GVRI	−
Gosling parvovirus	GVRI	−

Note: GVRI, Guangxi Veterinary Research Institute, China; CIVDC, China Institute of Veterinary Drugs Control, China; GLBPLC, Guangxi Liyuan Biological Products Limited Company; SHBPLC, Shanghai Haili Biological Products Limited Company; LYBPLC, Liaoning Yikang Biological Products Limited Company; HJBPLC, Hayao Jituan Biological Products Limited Company; +, positive. −, negative.

In order to test the analytical sensitivity of the DPV PCR, ten-fold dilutions from 1 ng to 1 fg of genomic DNA from the DPV virulent strain Yulin2016-30D and the attenuated strain AV18 were used. The minimum amount of DNA detected by the DPV-specific PCR was 100 fg as determined by gel electrophoresis (Figure 5-37-2).

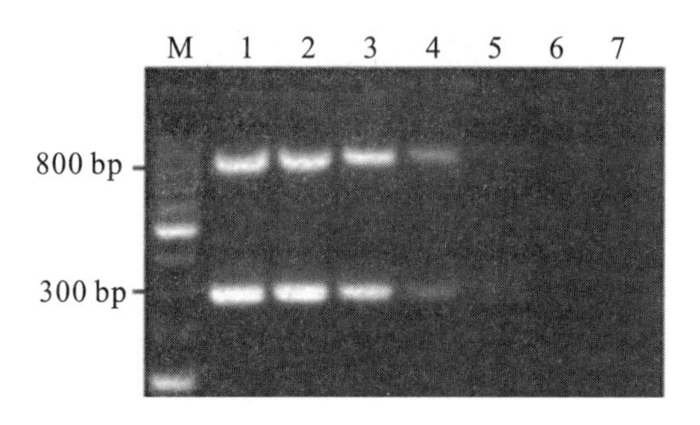

M: 100 bp DNA ladder; 1: 1 ng; 2: 100 pg; 3: 10 pg; 4: 1 pg; 5: 100 fg; 6: 10 fg; 7: 1 fg.

Figure 5-37-2 Sensitive for DPV

In order to test the DPV PCR specificity, DNA from each of the four DPV virulent strains and five DPV attenuated strains and DNA/RNA from seven other avian viruses (Table 5-37-2) was tested by the DPV PCR.

There was no cross-reactivity with the other avian viral pathogens tested, including Duck paramyxovirus, Duck Tembusu virus, Duck cir-covirus, Muscovy duck parvovirus, Duck hepatitis virus type I, Avian influenza virus and Gosling plague virus (Figure 5-37-3). Four DPV virulent strains yielded a positive 827 bp band, and five DPV attenuated strains yielded a positive 299 bp band, after 1.5% agarose gel electrophoresis (Figure 5-37-3). PCR products were cloned into pMD18-T cloning vectors (TaKaRa, Dalian, China) according to the manufacturer's instructions. The constructed recombinant plasmids were identified by PCR and then sequenced by Invitrogen (Guangzhou, China). The sequence analysis confirmed that this developed PCR specifically amplified the 827 bp fragment from the DPV virulent strain and the 299 bp fragment from the DPV attenuated strain.

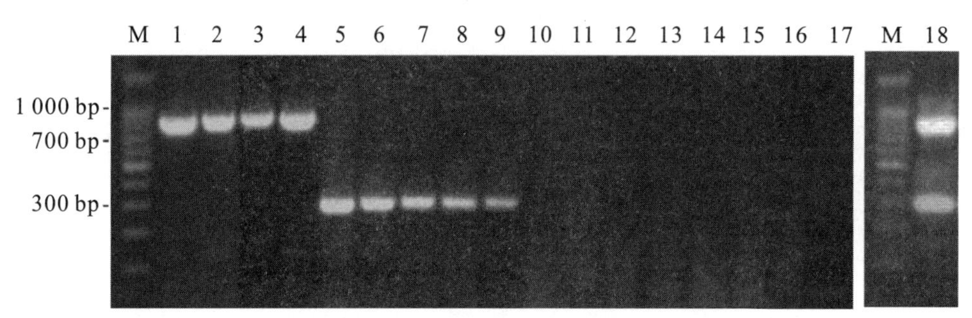

M: 100 bp DNA ladder; samples 1-4: Duck plague virus virulent strains (1=Yulin2016-30D, 2=Yulin2016-60D, 3=DPVGuangxi01, 4=DPVGuangxi02) ; samples 5-9: Five duck plague virus attenuated strains (5=AV18, 6=from SHBPLC, 7=from LYBPLC, 8=from HJBPLC, 9=from GLBPLC) ; 10: Duck paramyxovirus; 11: Duck Tembusu virus; 12: Duck Circovirus; 13: Muscovy duck parvovirus; 14: Duck Hepatitis virus type I; 15: Avian influenza virus; 16: Gosling plague virus; 17: dH_2O; 18: Duck plaguevirus virulent strain (Yulin2016-30D) and duck plague virus attenuated strain (AV18).

Figure 5-37-3 Specific for DPV

This DPV PCR was evaluated for direct detection of DPV from duck tissue specimens. The two livers and two spleen samples from the DPV virulent strainYulin2016-60D infected ducks were tested positive at 827 bp by the DPV PCR. The two liver and two spleen samples from DPV attenuated strain (LYBPLC) vaccinated ducks were tested positive at 299 bp. The liver and spleen samples from DPV negative duck were negative (Figure 5-37-4).

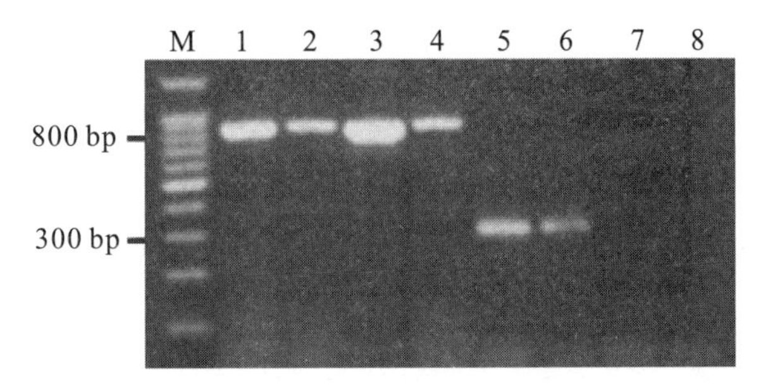

M: 100 bp DNA ladder; 1-2: Liver and spleen samples of duck infected with DPV virulent strain (Yulin2016-60D) ; 3-4: Liver and spleen samples of duck infected with DPV attenuated vaccine strain (from LYBPLC) ; 7-8: DPV negative duck liver and spleen.

Figure 5-37-4 PCR detection of clinical samples

The confirmatory result of a suspected DPV virulent strain and attenuated strain infection was obtained within 4 h. This included the time for DNA extraction, PCR assay and electrophoresis. The viral isolation and identification method used for laboratory differentiation of DPV virulent and attenuated strains are labour-intensive and may lack the sensitivity and speed required to reveal the cause of infection. The DPV PCR assay presented in this study, therefore, can be extremely useful as a fast and sensitive complement to the existing diagnostic methods.

Although we have compared all the UL2 gene sequences available in GenBank to develop the DPV differential PCR, DPV strains could be more variable in domestic ducks or wild waterfowl. To further confirm and ensure the practicality of this method, we will continue to conduct validation tests when more DPV positive samples of virulent and attenuated strains become available.

Polymerase chain reaction (PCR), a highly sensitive and specific test, is considered the best option for detecting virus[18]. The conventional PCR developed in this study successfully differentiate between virulent and attenuate strains. Although real-time PCR has the advantage to do not require agarose gel analysis for the detection of amplification products as conventional PCR, it requires specialized equipment, and false positive results occur frequently due to the high amplification efficiency[19, 20]. Alternatively, LAMP has the advantage of do not require a thermal cycler and/or specialized laboratory, although previous studies indicated that false positive results could occur due to the high amplification efficiency, short target sequences, or contaminations associated with adding the fluorescent dye[21]. Taking the advantages of each method into account and to provide alternative methods for differentiation of DPV virulent from attenuated strain, the development of a differential real-time PCR and LAMP assays are warranted.

References

[1] WILLEMSE M J, CHALMERS W S K, CRONENBERG A M, et al. The gene downstream of the gC homologue in Feline Herpes Virus Type 1 is involved in the expression of virulence. Journal of general virology, 1994, 75(11): 3107-3116.

[2] WALKER J W, PFOW C J, NEWCOMB S S, et al. Status of duck virus enteritis (duck plague) in the United States. Proc Annu Meet U S Anim Health Assoc, 1969, 73: 254-279.

[3] CAMPAGNOLO E R, BANERJEE M, PANIGRAHY B, et al. An outbreak of duck viral enteritis (duck plague) in domestic Muscovy ducks (*Cairina moschata domesticus*) in Illinois. Avian Dis, 2001, 45(2): 522-528.

[4] SANDHU T, SHAWKY S. Duck virus enteritis (duck plague). Dis Poult, 2003, 11: 354-363.

[5] SANDHU T S, METWALLY S A. Diseases of Poultry 12th ed. Singapore: Blackwell, 2008.

[6] JI J, DU L Q, XIE Q M, et al. Rapid diagnosis of duck plagues virus infection by loop-mediated isothermal amplification. Research in Veterinary Science, 2009, 87(1): 53-58.

[7] ISLAM M L, SAMAD M, RAHMAN M, et al. Assessment of immunologic responses in Khaki Cambell Ducks vaccinated against duck plague. International Journal of Poultry Science, 2005(4): 36-38.

[8] LIU S, CHEN S, LI H, et al. Molecular characterization of the herpes simplex virus 1 (HSV-1) homologues, UL25 to UL30, in duck enteritis virus (DEV). Gene, 2007, 401(1): 88-96.

[9] LI Y, HUANG B, MA X, et al. Molecular characterization of the genome of duck enteritis virus. Virology, 2009, 391(2): 151-161.

[10] LIAN B, XU C, CHENG A, et al. Identification and characterization of duck plague virus glycoprotein C gene and gene product. Virol J, 2010, 7: 349.

[11] DIAO Y, LU G, ZHENG Y, et al. Isolation and identification and characteristics of the pathogen of a new duck plague. Chinese Journal of Veterinary Science, 2006, 26(2): 136-139.

[12] PLUMMER P J, ALEFANTIS T, KAPLAN S, et al. Detection of duck enteritis virus by polymerase chain reaction. Avian Dis, 1998, 42(3): 554-564.

[13] KUMAR N V, REDDY Y N, RAO M V S. Enzyme linked immunosorbent assay for detection of duck plague virus. Indian Veterinary Journal, 2004(5): 481-483.

[14] XIE Z, KHAN M I, GIRSHICK T, et al. Reverse transcriptase-polymerase chain reaction to detect avian encephalomyelitis virus. Avian Dis, 2005, 49(2): 227-230.

[15] YANG F, JIA W, YUE H, et al. Development of quantitative real-time polymerase chain reaction for duck enteritis virus DNA. Avian Dis, 2005, 49(3): 397-400.

[16] WANG J, HÖPER D, BEER M, et al. Complete genome sequence of virulent duck enteritis virus (DEV) strain 2085 and comparison with genome sequences of virulent and attenuated DEV strains. Virus Research, 2011, 160(1-2): 316-325.

[17] YANG C, LI Q, LI J, et al. Comparative genomic sequence analysis between a standard challenge strain and a vaccine strain of duck enteritis virus in China. Virus Genes, 2014, 48(2): 296-303.

[18] XIE Z, TANG Y, FAN Q, et al. Rapid detection of group I avian adenoviruses by a Loop-mediated isothermal amplification. Avian Dis, 2011, 55(4): 575-579.

[19] JANOCKO L E, UHMACHER J A, KANT J A. Real time quantitative PCR on the Roche Light Cycler (R): False positive results can occur when using hydrolysis probes with one of two baseline calculation methods. Meeting of the Association For Molecular Pathology, 2005: 696.

[20] NOWROUZIAN F L, ADLERBERTH I, WOLD A E. High frequency of false-positive signals in a real-time PCR-based "Plus/Minus" assay. APMIS: Acta Pathologica, Microbiologica et Immunologica Scandinavica, 2009, 117(1): 68-72.

[21] THEKISOE O MI, BAZIE R S, CORONEL-SERVIAN A M, et al. Stability of Loop-mediated isothermal amplification (LAMP) reagents and its amplification efficiency on crude trypanosome DNA templates. J Vet Med Sci, 2009, 71(4): 471-475.

An attempt to develop a detection method specific for virulent duck plague virus strains based on the UL2 gene B fragment

Xie Liji, Huang Li, Xie Zhixun, Wang Sheng, Huang Jiaoling, Zhang Yanfang, and Zhong Chuande

Abstract

A comparison of the unique long region 2 (UL2) gene sequences between 10 virulent and 11 attenuated duck plague virus (DPV) strains (including all DPV UL2 gene sequences registered in GenBank) showed that the UL2 genes in the attenuated DPV strains had a 528 bp deletion in the B fragment. Primers were designed based on the B fragment sequence of the UL2 gene in an attempt to establish a fluorescence quantitative polymerase chain reaction (PCR) and a conventional PCR detection method that could specifically detect virulent DPV strains (i.e. positive detection for virulent DPV strains and negative detection for attenuated DPV strains). Additionally, PCR products were cloned for sequence analysis. These two methods detected five attenuated DPV strains in addition to the virulent DPV strains. Sequence analysis of the PCR products showed that the amplified products were the B fragments of the UL2 gene. These results indicated that detection methods specific for virulent DPV strains could not be established using primers designed based on the UL2 gene B fragment.

Keywords

duck plague virus, virulent and attenuated strains, UL2 gene, B fragment, detection method

---

Introduction

Duck plague, also known as duck viral enteritis, is an acute, contagious and lethal disease of ducks, geese, swans and many other species of waterfowl within the order Anseriformes[1]. This disease causes high mortality and decreased egg production in domestic and wild waterfowl, resulting in significant economic losses[2-6]. The disease is caused by anatid herpesvirus 1 (duck plague virus, DPV). Attenuated or naturally apathogenic DPV strains are used in live vaccines to provide protection against clinical duck plague in industry settings[7-10]. However, there have been reports that duck plague (caused by a virulent DPV strain) can also break out after immunization with live vaccines containing attenuated DPV strains[11].

The existing diagnostic methods for DPV (such as ELISA, electron microscopy, polymerase chain reaction (PCR), real-time PCR and Loop-mediated isothermal amplification (LAMP)[5, 12-14] cannot differentiate between virulent and attenuated DPV strains. In addition, virus isolation and identification of DPV are time consuming.

The DPV unique long region 2 (UL2) is a homologue of the herpes virus UL2 protein, which encodes a non-essential enzyme with uracil-DNA glycosylase activity[15]. Sequence analysis of the UL2 gene of 10 virulent DPV strains and 11 attenuated DPV strains, including all the UL2 gene sequences released in GenBank, have revealed that there is a 528 bp (B fragment) deletion within the UL2 gene of the attenuated DPV strains. Therefore, based on differences in the UL2 gene between the virulent and attenuated DPV strains, this study aimed to design specific primers based on the UL2 gene-specific B fragment of the virulent strains in an attempt to establish a fluorescence quantitative PCR detection method for virulent DPV strains. However, this study was not successful in achieving its aims. We found that the attenuated DPV strains also contained the UL2 gene-specific B fragment from the virulent DPV strains, and the B fragment of the attenuated DPV strains was separate from the UL2 gene.

Materials and methods

Viruses and DNA/RNA extraction

The DPV strains used in this study are listed in Table 5-138. Genomic DNA were extracted from 200 μL of the viruses listed in Table 5-38-1 using an EasyPure Viral DNA/RNA kit (TransGen, Beijing, China) according to the manufacturer's protocol. DNA samples were immediately stored at –70 ℃ until required.

Table 5-38-1　DPV strains used in the study

Virus strain	Source
DPV (Yulin2016-30D, virulent strain)	GVRI
DPV (Yulin2016-60D, virulent strain)	GVR
DPV (Guangxi01, virulent strain)	GVR
DPV (Guangxi02, virulent strain)	GVRI
DPV (AV18, attenuated strain)	CIVDC
DPV (attenuated vaccine strain)	GLB
DPV (attenuated vaccine strain)	SHB
DPV (attenuated vaccine strain)	LYB
DPV (attenuated vaccine strain)	HJB

GVRI: Guangxi Veterinary Research Institute, China; CIVDC: China Institute of Veterinary Drugs Control, China; GLB: Guangxi Liyuan Biological Co., Ltd.; SHB: Shanghai Hile Bio-Technology, Co., Ltd.; LYB: Liaoning Yikang Biological, Co., Ltd.; HJB: Harbin Pharmaceutical Group Bio-vaccine, Co., Ltd.

Clone and sequence analysis of the UL2 gene from the DPV strains listed in Table 5-38-1

The PCR primers DPV-F: ACGAGGGAGACCCAAAT GAC and DPV-R: TTTATACTGTTCCACAAGGAAG TTG were used to amplify the open reading frame (ORF) of the UL2 gene; the expected PCR amplification size was 474 bp for the attenuated strains and 1 002 bp for the virulent strains. PCR amplification was performed using the conventional method. The DNA samples from the DPV strains listed in Table 5-138 were used as templates. The PCR products were cloned into pMD18-T cloning vectors (TaKaRa Biotechnology, Dalian, China) according to the manufacturer's instructions. The constructed recombinant plasmids were identified by PCR and then sequenced by Invitrogen (Guangzhou, China). The nucleotide sequence analysis was performed using Lasergene software (Version 7.0, Lasergene, USA). Some sequence data were submitted to GenBank.

Analysis of UL2 genes from virulent and attenuated strains

The UL2 genes of 10 virulent DPV strains were compared with those of 11 attenuated DPV strains, including all UL2 gene sequences released in GenBank (Table 5-38-2). The ORF of the UL2 gene of the attenuated DPV strains was 474 bp, but for virulent DPV strains it was 1 002 bp. It was confirmed that there is a 528 bp (B fragment) deletion within the UL2gene of the attenuated DPVstrains, and that the deleted fragment was located at nucleotides 198-725 of the virulent strains (Figure 5-38-1). This finding is consistent with previous results reported by Wang et al.[16] and Yang et al.[6] who compared a small number of virulent and attenuated strains. Therefore, the use of primers designed based on the B fragment sequences of the UL2 gene could theoretically establish a fluorescence quantitative PCR method for the specific detection of virulent DPV strains.

Table 5-38-2 The DPV strains used for sequence analysis

Virulence	Strain name	GenBank accession number[a]	UL2 gene length
Attenuated strain	VAC	EU082088	477 bp
Attenuated strain	Sichuan1	JQ347517	474 bp
Attenuated strain	Sichuan2	JQ347518	474 bp
Attenuated strain	K	KF487736	474 bp
Attenuated strain	AV18		474 bp
Attenuated vaccine strain	C-KCE	KF263690	474 bp
Attenuated vaccine strain	clone-03	EF449516	474 bp
Four attenuated vaccine strains (source: GLB [b], SHB, LYB, HJB)	Unknown		474 bp
Virulent strain	Yulin201630D	KX925439	1 002 bp
Virulent strain	Yulin201660D	KX925440	1 002 bp
Virulent strain	Germany 2085	JF99965	1 002 bp
Virulent strain	Chv	EU885419	1 002 bp
Virulent strain	CV	JQ673560	1 002 bp
Virulent strain	LH2011	KC480262	1 002 bp
Virulent strain	LS	JQ248598	1 002 bp
Virulent strain	N1	JQ043216	1 002 bp
Virulent strain	N2	JQ248596	1 002 bp
Virulent strain	N3	JQ248597	1 002 bp

Note: [a] The blank spaces indicate that no sequence has been released by GenBank to date. [b] GLB, Guangxi Liyuan Biological Co., Ltd.; SHB, Shanghai Hile Bio-Technology, Co., Ltd.; LYB, Liaoning Yikang Biological, Co., Ltd.; HJB, Harbin Pharmaceutical Group Bio-vaccine, Co., Ltd.

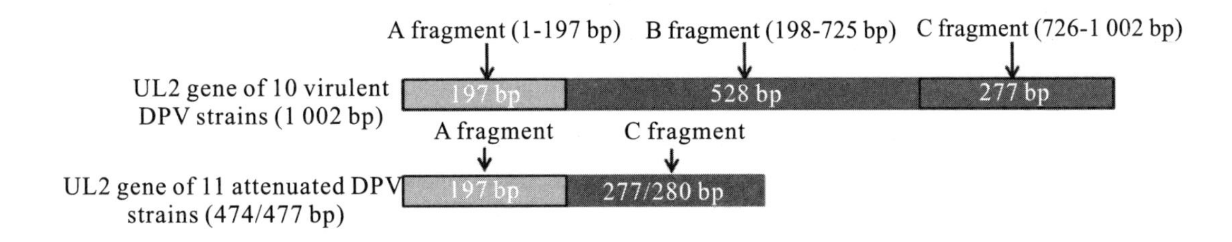

Figure 5-38-1 Schematic diagram of the DPV UL2 gene

Establishment of a quantitative fluorescence PCR method

One pair of specific primers (DPlague virulent F 507-528: TGGGCAAGATCCATATCATCAA; and DPlague virulent R 551-569: CCTCTCCGCACGCTG AAT; expected amplification size 63 bp) and one Taq-Man probe (DPlague virulent T 531-549: FAM-CGG CCAGGCTCACGGGCTC-BHQ1) were designed according to the UL2 gene-specific B fragment sequences of virulent DPV strains using Primer Express 3.0 software. The primers and probe were synthesized by TaKaRa Biotechnology. The DNA samples from four virulent DPV strains (DPVGuangxi01, DPVGuangxi02, Yulin2016-60D and Yulin 2016-30D), and five attenuated DPV strains (AV18 strain and four attenuated DPV vaccine strains) were used as templates.

The fluorescence quantitative PCR mix contained 1 × real-time PCR Premix (Perfect Real Time PCR kit, TaKaRa Biotechnology) ; 0.1 μM primers of DPlague virulent F, DPlague virulent R and probe DPlague virulent T; and 2 μL of template. Distilled H_2O was added to bring the final volume to 20 μL. PCR amplification consisted of an initial step at 95 ℃ for 20 s, followed by 40 cycles at 95 ℃ for 5 s and 60 ℃ for 20 s.

PCR amplification of the UL2 gene B fragment

Because the fluorescence quantitative PCR method established using primers designed based on the B fragment sequences (previously expected to amplify only virulent DPV strains) also detected the attenuated strains, we suspected that the attenuated strains also contained free B fragments. Therefore, we designed conventional PCR primers (DPV F 228-246: TATAGCACAACAACCTCAG and DPV R 700-718: GCCCCTTTCGTACTGTGAG) based on the B fragment sequences to specifically amplify the B fragment (expected size 491 bp). PCR amplification was performed using the conventional method. The DNA samples from four virulent DPV strains (DPVGuangxi01, DPVGuangxi02, Yulin2016-60D and Yulin2016-30D) and five attenuated DPV strains (AV18 and four attenuated DPV vaccine strains) were used as the templates.

After a total of 50 μL of PCR product was subjected to electrophoresis, the PCR product was cloned into pMD18-T cloning vectors (TaKaRa Biotechnology) according to the manufacturer's instructions. The constructed recombinant plasmids were identified by PCR and then sequenced by Invitrogen (Guangzhou, China). The nucleotide sequence analysis was performed using Lasergene sofware (Version 7.0, Madison, WI, USA).

Results

Clone and sequence analysis of the UL2 gene from the DPV strains listed in Table 5-38-1

The PCR amplification results for the ORF of the UL2 gene of the DPV strains listed in Table 5-38-1 are shown in Figure 5-38-2. For the five attenuated DPV strains, the PCR amplification results only showed a 474 bp attenuated strain band (with its correctness also confirmed by sequencing), whereas a 1 002 bp virulent strain band was not observed. The sequence analysis results confirmed that the ORF of the UL2 gene of the attenuated DPV strains (AV18 and the four vaccine strains listed in Table 5-38-1) was 474 bp, but for the virulent DPV strains (DPVGuangxi01, DPVGuangxi02, Yulin2016-60D and Yulin2016-30D, listed in Table 5-38-1), the length was 1 002 bp. Sequence data for the virulent DPV strains Yulin2016-60D and Yulin2016-30D were submitted to GenBank under accession number KX925439 and KX925440, respectively.

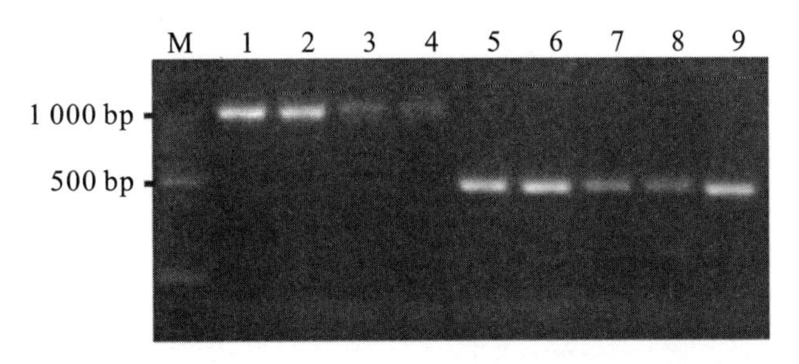

M: DNA marker; 1-4: Virulent strains (DPV Guangxi01 strain, Guangxi02 strain, Yulin2016-60D strain and Yulin2016-30D strain) ; 5-8: Attenuated vaccine strains (Harbin Pharmaceutical Group Bio-vaccine, Co., Ltd., Guangxi Liyuan Biological Co., Ltd., Shanghai Hile Bio-Technology, Co., Ltd.and Liaoning Yikang Biological, Co., Ltd.) ; 9: AV18 strain (attenuated strain).

Figure 5-38-2 PCR amplification of the ORF of the DPV UL2 gene

Fluorescence quantitative PCR detection results

The detection results for the virulent and attenuated DPV strains using fluorescence quantitative PCR established with primers designed based on the UL2 gene B-fragment sequence are shown in Figure 5-58-3.

Theoretically, the fluorescence quantitative PCR method established using primers designed based on the UL2 gene B fragment sequence should have specifically detected the virulent DPV strains (i.e. shown negative detection results for the attenuated strains). However, the actual detection results showed that the established method also detected the attenuated DPV strains (curves 2-6 in Figure 5-38-3). These results indicated that the UL2 gene-specific B fragment of the virulent strains was also present in the attenuated strains and that the B fragment was separate from the UL2 gene in the attenuated strains.

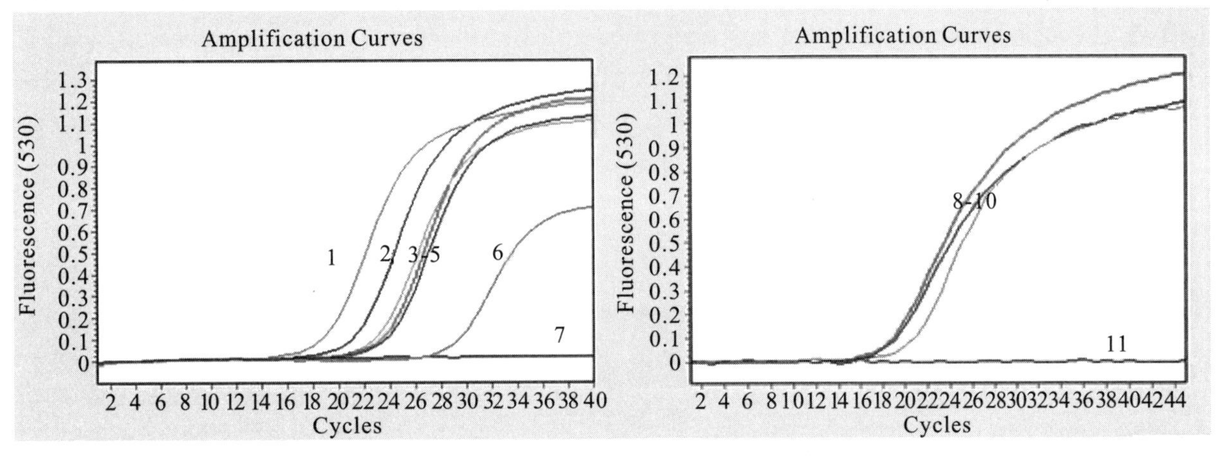

1: Yulin2016-30D strain (virulent) ; 2-5: Attenuated vaccine strains (Harbin Pharmaceutical Group Bio-vaccine Co., Ltd., Guangxi Liyuan Biological Co., Ltd., Shanghai Hile Bio-Technology, Co., Ltd.and Liaoning Yikang Biological, Co., Ltd.) ; 6: AV18 strain (attenuated) ; 7: Normal duck embryo allantoic fluid; 8-10: Virulent strains (DPV Guangxi01strain, Guangxi02 strain and Yulin2016-60D strain) ; 11: Normal duck embryo allantoic fluid.

Figure 5-38-3 Detection of DPV virulent and attenuated strains using fluorescence quantitative PCR

PCR amplification of the UL2 gene B fragment

The results for the virulent and attenuated DPV strains detected using the conventional PCR method established using primers designed based on the UL2 gene B fragment are shown in Figure 5-38-4. This method amplified 491 bp bands consistent with the experimental design based on the four virulent DPV strains (DPVGuangxi01, DPVGuangxi02, Yulin2016-60D and Yulin2016-30D) and five attenuated strains (AV18 and the attenuated vaccine strains listed in Table 5-38-1). The results were in agreement with the fluorescence quantitative PCR results.

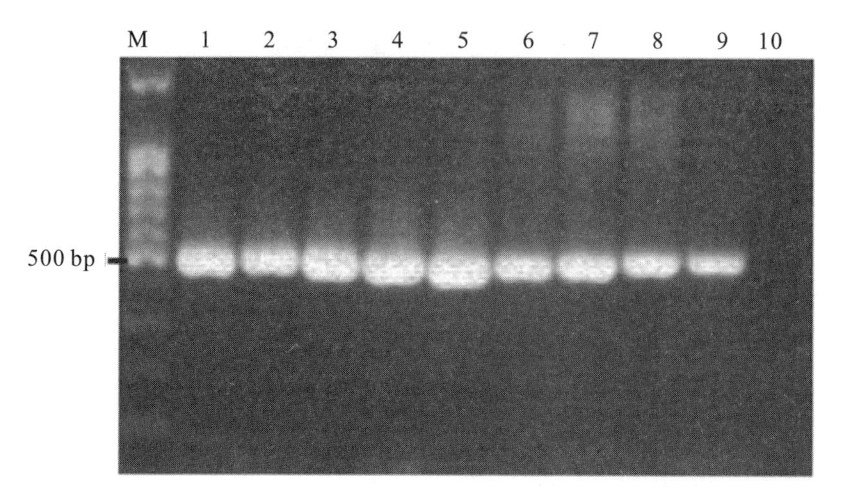

M: DNA marker; 1-4: Virulent strains (DPV Guangxi01strain, Guangxi02 strain, Yulin2016-60D strain and Yulin2016-30D strain) ; 5-9: Attenuated vaccine strains (Harbin Pharmaceutical Group Bio-vaccine, Co., Ltd., Guangxi Liyuan Biological Co., Ltd., Shanghai Hile Bio-Technology, Co., Ltd.and Liaoning Yikang Biological, Co., Ltd.) ; 10: AV18 strain (attenuated strain) ; 10: Normal duck embryo allantoic fluid.

Figure 5-38-4 The PCR-amplified B fragments of the UL2 gene

Sequencing analysis results

The sequencing analysis results confirmed that the amplified PCR products from the virulent and attenuated DPV strains were all 491 bp in size, which is consistent with the anticipated results. The sequences were subjected to Lasergene software analysis (Figure 5-38-5). The results showed that the sequences for the five attenuated strains (AV18 and four attenuated vaccine strains) had consistent homology with the corresponding fragment (i.e. the B fragment of the UL2 gene) of the virulent DPV strains used for primer design.

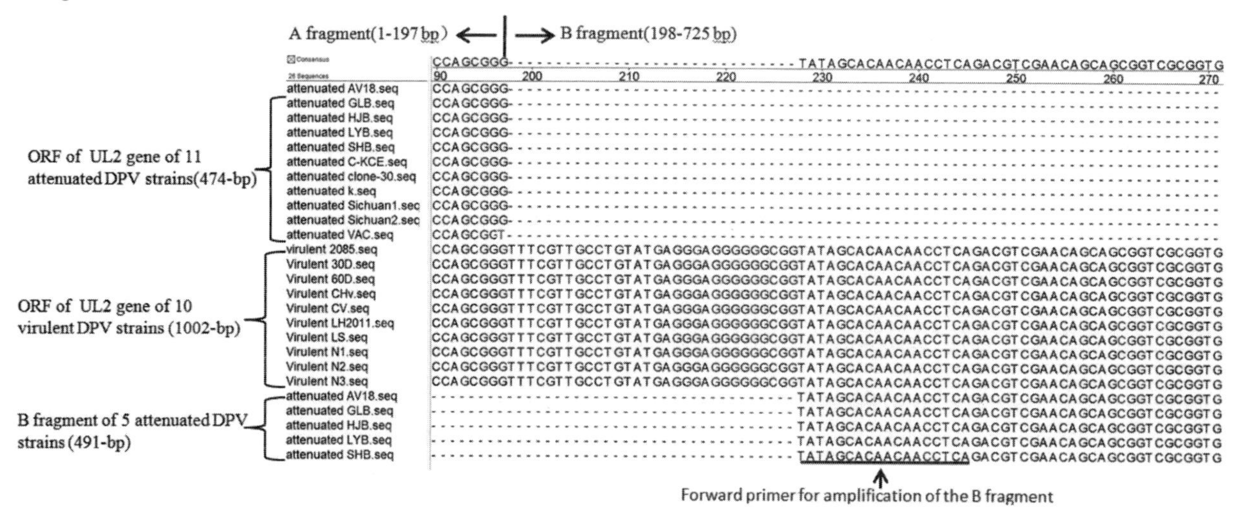

B fragment sequences for the five attenuated strains (AV18 and four attenuated vaccine strains) had consistent homology with the corresponding fragment (i.e.the B fragment of the UL2 gene) of the 10 virulent DPV strains. "-" represent nucleotide deletion.

Figure 5-38-5 Sequence analysis of the B fragment of the UL2 gene by Lasergene software

Discussion

This study compared UL2 gene sequences between 11 attenuated and 10 virulent DPV strains (including all DPV UL2 gene sequences registered in GenBank). The UL2 genes in the attenuated DPV strains all contained a 528 bp deletion in the B fragment. These results were consistent with the results reported by Wang et al.[16] and Yang et al.[6], who compared only a small number of virulent and attenuated strains. Therefore, the fluorescence quantitative PCR method established using primers designed based on the UL2 gene B fragment theoretically should have specifically detected the virulent DPV strains and given negative results for the detection of the attenuated strains. However, the actual results showed that the established fluorescence quantitative PCR method also detected the attenuated DPV strains.

Therefore, we hypothesized that the UL2 gene specific B fragment in the virulent strains was also present in the attenuated DPV strains but was separate from the UL2 gene. To validate this hypothesis, we established a PCR method for the specific amplification of the B fragment. This PCR method amplified the B fragment from five attenuated DPV strains in which the B fragment had been confirmed by sequencing. The results were consistent with the fluorescence quantitative PCR results. We did not discover the insertion of the B fragment into other locations in the DPV genome through analysis of whole gene sequences from the attenuated DPV strains in GenBank. Therefore, we speculated that either the B fragment has been separated from the DPV genome to become a free fragment or that the whole genome sequences of the existing attenuated DPV strains were not comprehensively sequenced. This issue requires further study for clarification.

The results of this study indicated that methods for the specific detection of virulent DPV strains could not be established using primers designed for the UL2 gene B fragment. To establish detection methods for virulent DPV strains, the upstream primer can be designed such that it is based on the UL2 gene A fragment, whereas the downstream primer can be designed such that it is based on the C fragment. Virulent and attenuated DPV strains can be distinguished based on the size of the amplified UL2 gene, because the size of the virulent strains is 528 bp larger than that of the attenuated strains. The results of this study provide very meaningful results to assist others and avoid repeating work that is not feasible.

The five attenuated DPV strains used for fluoescence quantitative PCR and conventional PCR detection included one attenuated strain purchased from the China Institute of Veterinary Drug Control and four commercially available attenuated vaccine strains. The same batch of the attenuated vaccine strains used in the present study was also used for the vaccination of ducks on a farm to confirm the safety of the vaccines and thereby to confirm the absence of contamination by virulent DPV strains. The ducks did not die after immunization with the DPV vaccine, thereby showing that the strains had not been contaminated with virulent DPV. Additionally, our group performed PCR amplification, cloning and sequencing of the UL2 genes from the five attenuated strains used in the present study. The PCR amplification results showed only a 474 bp attenuated strain band (the correctness of which was also confirmed by sequencing), whereas a 1 002 bp virulent strain band was not observed. These results verified that the attenuated strains used in the present study were not contaminated by virulent strains, which ensured the reliability of our results.

Moreover, to avoid laboratory contamination, we asked the Fangchengguang Center for Animal Disease Control and Prevention (FCADC) to replicate our research. All materials were prepared by the FCADC, including four attenuated vaccine strains purchased from four different companies by the FCADC. The primer and all other regents were also purchased by the FCADC. The FCADC's detected results were the same as ours; therefore their results corroborated the reliability of ours.

References

[1] WILLEMSE M J, CHALMERS W S , CRONENBERG A M, et al. The Gene Downstream of the gC Homologue in Feline Herpes Virus Type 1 is Involved in the Expression of Virulence. Journal of General Virology, 1994, 75(11): 3107-3116.

[2] WALKER J W, PFOW C J, NEWCOMB S S, et al. Status of duck virus enteritis (duck plague) in the United States. Proc Annu Meet U S Anim Health Assoc, 1969, 73: 254-279.

[3] CAMPAGNOLO E R, BANERJEE M, PANIGRAHY B, et al. An outbreak of duck viral enteritis (duck plague) in domestic Muscovy ducks (Cairina moschata domesticus) in Illinois. Avian Dis, 2001, 45(2): 522-528.

[4] SANDHU T, SHAWKY S. Duck virus enteritis (duck plague)//SAIF Y M, BARNES H J , GLISSON J R, et al. Diseases of Poultry, 11th ed. Ames: Iowa State Press, 2003: 354-363.

[5] JI J, DU L Q, XIE Q M, et al. Rapid diagnosis of duck plagues virus infection by loop-mediated isothermal amplification. Research in Veterinary Science, 2009, 87(1): 53-58.

[6] YANG C, LI Q, LI J, et al. Comparative genomic sequence analysis between a standard challenge strain and a vaccine strain of duck enteritis virus in China. Virus Genes, 2014, 48(2): 296-303.

[7] ISLAM M A I, SAMAD M A S, RAHMAN M B R, et al. Assessment of immunologic responses in Khaki Cambell Ducks vaccinated against duck plague. International Journal of Poultry Science, 2005(4): 36-38.

[8] LIU S, CHEN S, LI H, et al. Molecular characterization of the herpes simplex virus 1 (HSV-1) homologues, UL25 to UL30, in duck enteritis virus (DEV). Gene, 2007, 401(1): 88-96.

[9] LIAN B, XU C, CHENG A, et al. Identification and characterization of duck plague virus glycoprotein C gene and gene product. Virol J, 2010, 7: 349.

[10] LIAN B, XU C, CHENG A, et al. Identification and characterization of duck plague virus glycoprotein C gene and gene product. Virology Journal, 2010, 7(1):349.

[11] PLUMMER P J, ALEFANTIS T, KAPLAN S, et al. Detection of Duck Enteritis Virus by Polymerase Chain Reaction. Avian Dis, 1998, 42(3): 554-564.

[12] PLUMMER P J, ALEFANTIS T, KAPLAN S, et al. Enzyme linked immunosorbent assay for detection of duck plague virus. Indian Veterinary Journal, 2004(5): 481-483.

[13] YANG F L, JIA W X, YUE H, et al. Development of quantitative real-time polymerase chain reaction for duck enteritis virus DNA. Avian Dis, 2005, 49(3): 397-400.

[14] LI H, LIU S, HAN Z, et al. Comparative analysis of the genes UL1 through UL7 of the duck enteritis virus and other herpesviruses of the subfamily Alphaherpesvirinae. Genet Mol Biol, 2009, 32(1): 121-128.

[15] WANG J, HÖPER D, BEER M, et al. Complete genome sequence of virulent duck enteritis virus (DEV) strain 2085 and comparison with genome sequences of virulent and attenuated DEV strains. Virus research, 2011, 160(1-2): 316-325.

[16] LI Y, HUANG B, MA X, et al. Molecular characterization of the genome of duck enteritis virus. Virology, 2009, 391(2): 151-161.

Part Ⅵ Enzyme-linked Immunosorbent Assay (ELISA)

Recombinant protein-based ELISA for detection and differentiation of antibodies against avian reovirus in vaccinated and non-vaccinated chickens

Xie Zhixun, Qin Chunxiang, Xie Liji, Liu Jiabo, Pang Yaoshan, Deng Xianwen, Xie Zhiqin, and Khan Mazhar I.

Abstract

Two nonstructural genes, σNS and P17, of avian reovirus (ARV) were cloned into the expression plasmid vector pGEX4T-1. Expressed proteins for σNS and P17 of avian reovirus were purified and used as antigens. Three indirect σNS enzyme-linked immunosorbent assays (ELISAs), σNS-ELISA, P17-ELISA and σNS-P17-ELISA were optimized and used as specific tests. Serum samples from reovirus-infected and vaccinated SPF chickens were tested with the three ELISAs and an agar gel precipitin (AGP) method. ELISAs specific for σNS, P17 and σNS-P17 were able to detect specific antibodies for avian reovirus in 88.9%, 61.1%, and 88.9% in infected samples, respectively, whereas the AGP detected 55.6% of the infected samples. The detection rates of ELISA specific antibodies for σNS, P17 and σNS-P17 on sera of vaccinated chickens were 6.7%, 0% and 6.7%. However, in comparison the AGP method detected 60.6% of antibodies in serum samples from vaccinated chickens. The results showed that the use of ELISAs specific for the nonstructural proteins might be able to distinguish between reovirus vaccinated and infected chickens. Further studies are in progress to validate these recombinant protein-based ELISAs under field conditions.

Keywords

ARV, nonstructural proteins, indirect ELSA

Avian reovirus (ARV) is an important cause of diseases in poultry. Reovirus-induced arthritis, chronic respiratory disease, and malabsorption syndrome[1-3] cause significant economic losses. Many laboratory methods have been developed for the detection of antibodies against ARV, including serum neutralization[4], immunodiffusion[5], immunoblot assay[6], and immunofluorescence[7]. Although these methods are useful for the detection of ARV infection, most of them are laborious and time-consuming. On the other hand, enzyme-linked immunosorbant assays (ELISAs) have a high level of sensitivity and reproducibility and allow for automation. They are the method of choice for screening a large number of serum samples. Several ELISAs using whole virus or a recombinant ARV protein have been described[8-15]. None of these ELISAs distinguishes between ARV-vaccinated and infected serum samples.

The nonstructural protein σNS encoded by genome segment S4, is one of the four nonstructural proteins in ARV and contains three epitopes that are highly conserved among virus strains[16-19]. The nonstructural protein p17 that is associated with membranes is conserved in every avian reovirus strain and works as a nuclear targeting protein that shuttles between the nucleus and the cytoplasm. It has been demonstrated that p17 is also capable of causing cell growth retardation[20-22]. ELISAs specific for nonstructural proteins of hepatitis C and influenza viruses have been described that distinguish vaccinated from infected clinical samples[23-25]. Methods for differentiating vaccine and wild type ARV would be useful for diagnosis and for studying the epidemiology of ARV infection in poultry. Recombinant nonstructural protein-based ELISAs were developed and are described below.

The ARV strain S1133 (University of Connecticut, USA) was used for the development of recombinant nonstructural protein-based ELISAs. The virus was propagated in chicken embryo fibroblasts (CEF) at 37 ℃ for 36～48 h[26]. The culture was treated by freezing and thawing three times, and thawed culture fluid was clarified by centrifugation at 10 000 × g for 15 min. Virus titration was done in microtiter plates and the titer was calculated using the Reed and Munch method[27]. Virus with a titer of $10^{6.2}$ $TCID_{50}$/mL was inactivated with 0.1% formaldehyde at 37 ℃ overnight and then emulsified in mineral oil and used as a vaccine.

RNA extractions from ARV strain S1133 were carried out using Trizol, according to the manufacturer's protocol (Trizol, Invitrogen, Carlsbad, CA, USA). To amplify the full-length non-structural encoding genes of the σNS and P17 proteins, purified genomic dsRNA from ARV was used to generate cDNA by reverse transcription and polymerase chain reaction (RT-PCR). The PCR primers XZNS6: 5'-GCGAATTCGCCATGGACAACACCGTGC-3'and XZNS9: 5'-GCCTCGAGCTACGCCATCCTAGCTGG-3'to amplify the 1 107 bp of the σNS gene, and the primers P17-1: 5'-TGAATTCAGCACAATGCAATGGCTCCGC-3'and P17-2: 5'-GCTCGAGTTGGTCAGTCGTTCATA-3'to amplify the 457 bp of the P17 gene were designed and synthesized according to the σNS (U95952) and P17 (AF330703) gene sequences of ARV S1133 available in GenBank. PCR products derived from the σNS and P17-encoding genes of ARV S1133 were digested with enzyme and then ligated into the corresponding sites of a pGEX4T-1 expression vector. Recombinant plasmids were transformed in *Escherichia coli* competent cells (DH5α). Positive colonies pGEX4T-1-σNS and pGEX4T-1-P17 identified by PCR and double digestion were sequenced (TaKaRa, Dalian, China). The procedures for σNS and P17 expression in DH5α *E.coli* have been described previously[28, 29]. Briefly, after induction for 4h with isopropylthin-β-D-galactoside (IPTG) at a final concentration of 0.4 mM in culture medium, pGEX-4T-1-σNS and pGEX-4T-1-P17 were induced. The σNS and P17 protein were isolated and purified as described[28, 29], and the recombinant protein concentration was determined according to the method of Lowry et al.[30].

Purified σNS and P17 fusion proteins were mixed with an equal volume of Laemmeli sample buffer, boiled for 5min, and separated by SDS-PAGE using a 10% gel, in a BioRad miniProtein Ⅱ electrophoresis unit. After electrophoresis, the fusion proteins were transferred to nitrocellulose membranes by the method of Towbin et al.[31]. The blots were incubated with a 1 ∶ 200 dilution of chicken anti-ARV S1133 hyperimmune serum (China Institute of Veterinary Drug Control) for 2 h at room temperature.

To determine the optimum working concentration of ARV antigens in σNS-ELISA. P17-ELISA and σNS-P17-ELISA, checkerboard titrations were carried out between the σNS, P17, σNS-P17 proteins and positive and negative sera (China Institute of Veterinary Drug Control). One hundred microlitres of the purified σNS, P17 and σNS-P17 proteins in 0.05 M carbonate buffer, pH9.6, was coated onto the wells of an ELISA plate and tested with twofold serially diluted ARV positive and negative sera, respectively. Horseradish peroxidase (HRP)-labeled goat anti-chicken IgG conjugate was used at a dilution of 1 ∶ 5 000.

Purified σNS and P17 proteins were used at twofold dilutions (2.33～18.6 μg and 2.87～23 μg separately) in 0.05 M carbonate buffer, pH 9.6.σNS and P17 proteins were mixed and used at a twofold dilution. One hundred microlitre of antigen was added to each well of the ELISA plate. Antigen was coated onto the wells by incubation at 4 ℃ overnight. The wells were washed three times with phosphate-buffered saline (PBS), pH 7.4 and 0.05% Tween 20 washing buffer. Then 50 μL of blocking buffer (PBS, 0.05% Tween 20 and 2.5% bovine serum albumin) were added to each well and incubated for 1 h at 37 ℃ to saturate all unbound sites. Coated plates were washed three times with PBS and 0.05% Tween 20. Chicken sera diluted at twofold dilutions (from 1 ∶ 100 to 1 ∶ 800) using dilution buffer PBS and 0.1% Tween 20, were added at 100 μL/well and incubated for 1 h at 37 ℃. After incubation and three washing cycles as described above, one hundred microlitre per well of a 1 ∶ 5 000 dilution of horseradish peroxidase (HRP)-labeled goat anti-chicken IgG was added, and the plates were incubated for 1 h at 37 ℃ and then washed three times with washing buffer. A volume of 100 μL substrate solution containing 0.04% orthophenylenediamine in 0.05 M phosphate-citrate buffer, pH 5.0 and 0.04% H_2SO_4 was added. The reaction was carried out for 30 min at room temperature. Color development was stopped after 10 min by adding 50 μL of 2.5 M H_2SO_4 to each well. The absorbance was read at 450 nm using a plate reader. Duplicate positive and negative control chicken sera were included in each plate.

Fifty ARV negative serum samples (China Institute of Veterinary Drug Control) were tested at 450 nm to calculate the average (X) and Stdev (SD) of the OD450 nm values. The cut-off value was calculated according to the formula: cut-off value=X+3SD.

Forty-eight 1-week-old specific pathogenic free (SPF) chickens were divided into 3 groups. Fifteen chickens in group I were injected intramuscularly with inactivated vaccine of 10^4 $TCID_{50}$/bird at 2 and 4 weeks old. In group Ⅱ, 18 2-week-old chickens received ARV live virus at 10^4 $TCID_{50}$/bird by eye drop. Fifteen chickens in group Ⅲ were used as negative controls. Serum samples were collected at 7 weeks of age. Thirty-three sera from chickens infected or vaccinated with AVR were tested bv σNS-ELISA. P17-ELISA and σNS-P17-ELISA separately. Using the whole virus antigen of ARV S1133, thirty-three sera from infected chickens and 33 from vaccinated chickens were tested by agar gel precipitation (AGP) as described[6].

For specificity of ELISAs, serum samples from birds with NDV, AIV and IBV were tested by σNS-ELISA, P17-ELISA and σNS-P17-ELISA separately. Statistical analyses of the data were performed by Kruskal-Wallis one-way analysis of variance with the Dunn's post hoc with SigmaStat Version 3.0.

The recombinant plasmids pGEX-4T-1-σNS and pGEX-4T-1-P17 were identified by PCR (data not shown). It was confirmed by DNA sequencing that the acquired recombinant plasmid contained complete σNS and P17 genes, and the pGEX-4T-1-σNS and pGEX-4T-1-P17 recombinant plasmids were successfully constructed. Analysis of the purified protein by SDS-PAGE revealed the expressed σNS and P17 fusion proteins, with an approximate molecular mass of 66.2 kDa and 42.4 kDa, respectively. These were consistent with the expected size of the fusion proteins (Figure 5-39-1 A and B and Figure 5-39-2), and the concentrations were 28 mg/mL and 46 mg/mL, respectively. Conditions for σNS-ELISA. P17-ELISA and σNS-P17-ELISA were standardized by using checkerboard titrations. The results show that, when the concentrations of σNS, P17 and σNS-P17 were 9.3 μg/mL, 11.5 μg/mL and 9.3 μg/mL, respectively, at serum dilutions of 1 ∶ 200, the σNS-ELISA, P17-ELISA and σNS-P17-ELISA gave a clear difference in absorbance values between positive and negative sera. According to 50 ARV negative serum samples, the cut-off value of σNS-ELISA was 0.630, P17-ELISA was 0.560 and σNS-P17-ELISA was 0.640, respectively.

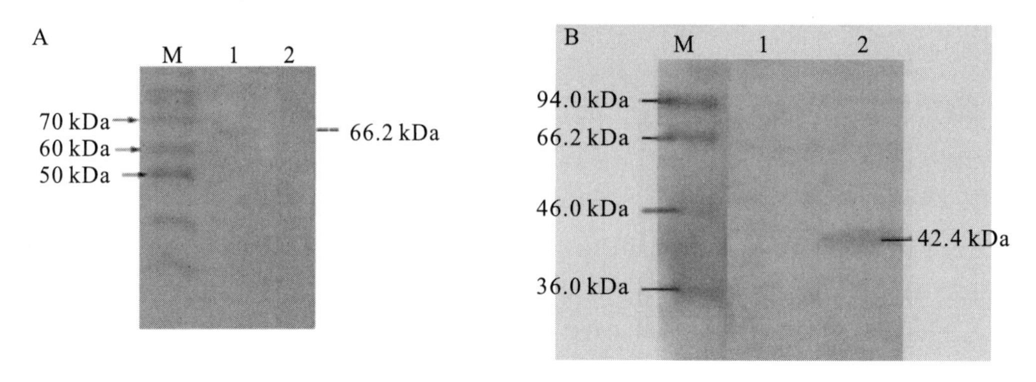

A: Lane M, protein marker; lane 1, purified σNS protein; lane 2, pGEX4T-1 induce at 4 h. B: Lane M, protein marker; lane 1, pGEX4T-1 induce at 4 h; lane 2, purified P17 protein.

Figure 5-39-1 Western blot analysis of ARV positive antiserum specific to the expressed σNS and P17 fusion proteins

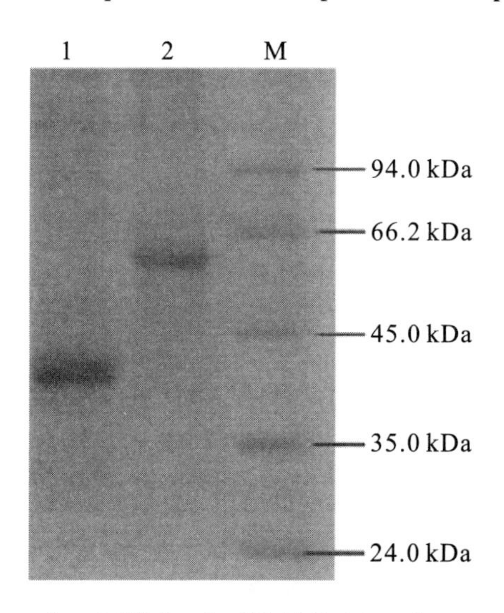

Lane 1: P17; lane 2: σNS; M: Protein marker.

Figure 5-39-2 Purified avian reovirus nonstructural recombinant ARV proteins

σNS-ELISA, P17-ELISA and σNS-P17-ELISA differentiated ARV positive and negative antibodies in experimentally infected and vaccinated chickens (Table 5-39-1 and Table 5-39-2). The σNS-ELISA, P17-ELISA and σNS-P17-ELISA were able to detect 88.9%, 61.1%, and 88.9%, respectively of ARV-infected serum samples, whereas the AGP test detected 55.6%. Detection rates on serum samples from the vaccinated SPF chickens were 6.7%, 0% and 6.7%, respectively for these nonstructural proteins. The AGP method detected 60.6% of samples positive for antibodies from ARV-vaccinated chickens. The σNS-ELISA, P17-ELISA and σNS-P17-ELISA showed no positive reactions with serum samples from birds with NDV, AIV and IBV, which indicated that these indirect ELISAs had good specificity for the detection of ARV antibody.

Table 5-39-1 The OD450 values of individual serum samples

ELISA	σNS	P17	σNS-P17	σNS	P17	σNS-P17	σNS	P17	σNS-P17
Sera collected from chickens infected with ARV	0.781	0.686	0.808	1.167	0.947	1.076	1.395	1.049	1.118
	1.279	0.518	1.057	0.822	0.689	0.842	0.574	0.483	0.612
	0.744	0.871	0.771	0.945	1.013	1.075	1.099	1.158	1.216
	1.086	0.795	1.068	0.674	0.483	0.726	0.899	0.427	0.917

continued

ELISA	σNS	P17	σNS-P17	σNS	P17	σNS-P17	σNS	P17	σNS-P17
Sera collected from chickens infected with ARV	0.606	0.462	0.629	0.752	0.688	0.843	0.720	0.548	0.683
	0.711	0.669	0.779	0.732	0.512	0.820	1.198	0.909	1.105
Sera collected from chickens vaccinated with ARV inactive vaccine	0.431	0.382	0.298	0.428	0.307	0.353	0.306	0.490	0.326
	0.617	0.418	0.437	0.223	0.311	0.341	0.346	0.252	0.389
	0.294	0.521	0.518	0.697	0.491	0.693	0.317	0.385	0.354
	0.379	0.373	0.272	0.582	0.468	0.265	0.407	0.508	0.476
	0.470	0.358	0.365	0.285	0.344	0.525	0.516	0.386	0.622

Table 5-39-2　Detection rates positive for ARV in sera from chickens using ELISA and AGP

Detection method	Infected with ARV	Vaccinated with ARV
σNS-ELISA	88.9% [a]	6.7%
P17-ELISA	61.1%	0%
σNS-P17-ELISA	88.9% [a]	6.7%
AGP	55.6%	60.6%

Note: [a] Detection with σNS-ELISA and σNS-P17-ELISA was significantly ($P \leqslant 0.05$) different from P17-ELISA and AGP groups.

The nonstructural σNS and P17 proteins were expressed successfully and the purified proteins were used as ELISA antigens to detect antibodies against ARV. The purified antigens demonstrated good specificity in ELISAs. These antigens reacted specifically with their respective serum samples, did not cross-react with any heterologous sera and exhibited negative reactivity with antisera against Newcastle disease virus, avian influenza virus and infectious bronchitis virus. Recombinant protein-based ELISA is less sensitive but more specific than the whole virus lysate-based ELISA. A previous report[10] demonstrated that the combined proteins-based ELISA, compared with the single protein-based ELISA or whole virus lysate ELISA, could greatly reduce non-specific binding reactions and had a higher correlation with serum neutralization assays. The current results suggest that σNS-ELISA, P17-ELISA and σNS-P17-ELISA were able to detect antibody in serum samples from the ARV-infected chickens. Using σNS and P17 proteins together as ELISA antigens did not increase the sensitivity of the assay. However, σNS-ELISA, P17-ELISA and σNS-P17-ELISA each was more sensitive than the AGP test (Table 5-39-2).

The nonstructural proteins are expressed in large amounts in virus-infected cells, but they have not been detected in virions[32]. It has been known that ELISAs specific for nonstructural proteins of hepatitis C and equine influenza viruses might be able to distinguish vaccination from infection[23-25]. In our study, positive detection rates for σNS-ELISA, P17-ELISA and σNS-P17-ELISA were high on the serum samples of infected chickens and vaccinated chickens (Table 5-39-2). Thus using σNS and P17 as antigens in ELISA can differentiate between sera from vaccinated and infected chickens. The σNS-ELISA, P17-ELISA and σNS-P17-ELISA were found to be more sensitive than the AGP test (Table 5-39-2). This study showed that σNS-ELISA was more sensitive than the P17-ELISA. A plausible reason may be that the σNS protein carries more broadly specific epitopes than the P17 protein. Further testing on a large numbers of serum samples from field cases will be required to confirm these findings.

Serum samples from chickens infected with ARV were positive with ELISAs between 61.11% and 88.89%, however not all of the samples were positive. This may be because the antibody level was insufficient for detection by ELISA. The positive rate for 15 serum samples from vaccinated chickens was 0~6.7%. Not all the samples were negative. That may be because the vaccines were not purified, or because booster vaccinating makes more nonstructural proteins. Further studies will be carried out to elucidate these phenomena. In conclusion, the σNS-ELISA and σNS-P17-ELISA have advantages over the conventional ELISA, with less non-specific binding reactions and stable sensitivity. These could be used to distinguish vaccinated from infected chickens.

References

[1] FAHEY J E, CRAWLEY J F. Studies on chronic respiratory disease of chickens II isolation of A virus. Canadian Journal of Comparative Medicine and Veterinary Science, 1854, 18(1): 13-21.

[2] HIERONYMUS D R, VILLEGAS P, KLEVEN S H. Identification and serological differentiation of several reovirus strains isolated from chickens with suspected malabsorption syndrome. Avian Dis, 1983, 27(1): 246-254.

[3] MACKENZIE M A, BAINS B S. Tenosynovitis in chickens. Australian Veterinary Journal, 1976, 52(10): 468-470.

[4] WICKRAMASINGHE R, MEANGER J, ENRIQUEZ C E, et al. Avian reovirus proteins associated with neutralization of virus infectivity. Virology, 1993, 194(2), 688–696.

[5] MEANGER J, WICKRAMASINGHE R, ENRIQUEZ C E, et al. Type-specific antigenicity of avian reoviruses. Avian Pathology, 1995, 24(1): 121-134.

[6] IDE P R, DEWITT W. Serological incidence of avian reovirus infection in broiler-breeders and progeny in Nova Scotia. Can Vet J, 1979, 20(12): 348-353.

[7] IDE P R. Avian reovirus antibody assay by indirect immunofluorescence using plastic microculture plates. Canadian Journal of Comparative Medicine and Veterinary Science, 1982, 46(1): 39-42.

[8] SHIEN J H, YIN H S, LEE L H. An enzyme-linked immunosorbent assay for the detection of antibody to avian reovirus by using protein sigma B as the coating antigen. Research in Veterinary Science, 2000, 69(2): 107-112.

[9] LIU H J, GIAMBRONE J J, WU Y H, et al. The use of monoclonal antibody probes for the detection of avian reovirus antigens. Journal of Virological Methods, 2000, 86(2): 115-119.

[10] LIU H J, KUO L C, HU Y C, et al. Development of an ELISA for detection of antibodies to avian reovirus in chickens. Journal of Virological Methods, 2002, 102(1-2): 129-138.

[11] SLAGHT S S, YANG T J, HEIDE L V D, et al. An enzyme-linked immunosorbent assay (ELISA) for detecting chicken anti-reovirus antibody at high sensitivity. Avian Dis, 1978, 22(4): 802-805.

[12] SLAGHT S S, YANG T J, HEIDE L V D. Adaptation of enzyme-linked immunosorbent assay to the avian system. Journal of Clinical Microbiology, 1979, 10(5): 698-702.

[13] ISLAM M R, JONES R C. An enzyme-linked immunosorbent assay for measuring antibody titre against avian reovirus using a single dilution of serum. Avian Pathol, 1988, 17(2): 411-425.

[14] ROBERSON M D, Wilcox G E. Avian reovirus. Vet Bull, 2000, 56: 155-174.

[15] CHEN P N, LIU H J, SHIEN J H, et al. Antibody responses against avian reovirus nonstructural protein sigmaNS in experimentally virus-infected chickens monitored by a monoclonal antibody capture enzyme-linked immunosorbent assay. Research in Veterinary Science, 2004, 76(3): 219-225.

[16] HOU H S, SU Y P, SHIEH H K, et al. Monoclonal antibodies against different epitopes of nonstructural protein sigmaNS of avian reovirus S1133. Virology, 2001, 282(1): 168-175.

[17] HUANG P H, LI Y J, SU Y P, et al. Epitope mapping and functional analysis of sigma A and sigma NS proteins of avian reovirus. Virology, 2005, 332(2): 584-595.

[18] VARELA R, BENAVENTE J. Protein coding assignment of avian reovirus strain S1133. Journal of Virology, 1994, 68(10): 6775-6777.

[19] XIE L J, XIE Z X,Qin C X. Cloning and function analysis of σNS gene of avian reovirus. Chinese Veterinary Science, 2007, 37(4): 291-294.

[20] COSTAS C, MARTÍNEZ-COSTAS J, BODELÓN G, et al. The second open reading frame of the avian reovirus S1 gene encodes a transcription-dependent and CRM1-independent nucleocytoplasmic shuttling protein. Journal of Virology, 2005, 79(4): 2141-2150.

[21] LIU H J, LIN P Y, LEE J W, et al. Retardation of cell growth by avian reovirus p17 through the activation of p53 pathway. Biochemical and Biophysical Research Communications, 2005, 336(2): 709-715.

[22] Qin C X, XIE Z X, XIE L J. Cloning and sequence analysis of the P10 and P17 nonstructural protein genes of avian reovirus. Progress in Veterinary Medicine, 2006, 27(12): 70-74.

[23] INOUE Y, SUZUKI R, MATSUURA Y, et al. Expression of the amino-terminal half of the NS1 region of the hepatitis C virus genome and detection of an antibody to the expressed protein in patients with liver diseases. The Journal of General Virology, 1992, 73 (8): 2151-2154.

[24] BIRCH-MACHIN I, ROWAN A, PICK J, et al. Expression of the nonstructural protein NS1 of equine influenza A virus: detection of anti-NS1 antibody in post infection equine sera. Journal of Virological Methods, 1997, 65(2): 255-263.

[25] OZAKI H, SUGIURA T, SUGITA S, et al. Detection of antibodies to the nonstructural protein (NS1) of influenza A virus allows distinction between vaccinated and infected horses. Veterinary Microbiology, 2001, 82(2): 111-119.

[26] WU W Y, SHIEN J H, LEE L H, et al. Analysis of the double-stranded RNA genome segments among avian reovirus field isolates. Journal of Virological Methods, 1994, 48(1): 119-122.

[27] Reed L J, Munch, H. A simple method of estimating fifty percent end-points. American Journal of Epidemiology, 1938, 27(3):493-497.

[28] XIE L J, XIE Z X,LIU J B, et al. Prokaryotic expression and purification of σNS protein of avian reovirus. Chinese Journal of Veterinary Science, 2008, 38(4): 375-378.

[29] XIE Z X, Qin C X, XIE L J. Expression of P17 protein gene of avian reovirus in Escherichia coli and establishment of an ELISA assay for detection of the virus. Chinese Journal of Veterinary Science, 2007, 37(9): 777-782.

[30] LOWRY O H, ROSEBROUGH N J, FARR A L, et al. Protein measurement with the Folin phenol reagent. The Journal of Biological Chemistry, 1951, 193(1): 265-275.

[31] TOWBIN H, STAEHELIN T, GORDON J. Electrophoretic transfer of proteins from polyacrylamide gels to nitrocellulose sheets: procedure and some applications. Proceedings of the National Academy of Sciences of the United States of America, 1979, 76(9): 4350-4354.

[32] KRUG R F, ETKIND P R, ETKIND P R. Cytoplasmic and nuclear virus-specific proteins in influenza virus-infected MDCK cells. Virology, 1973, 56(1): 334-348.

Detection of antibodies specific to the non-structural proteins of fowl adenoviruses in infected chickens but not in vaccinated chickens

Xie Zhixun, Luo Sisi, Fan Qing, Xie Liji, Liu Jiabo, Xie Zhiqin, Pang Yaoshan, Deng Xianwen, and Wang Xiuqing

Abstract

Antibodies specific to the non-structural proteins of viruses are detected in virus-infected animals and show promise as a reliable diagnostic marker for virus infections. We examined the potential use of two non-structural proteins of fowl adenovirus (FAdV)-based, 100 K and 33 K, enzyme-linked immunosorbent assays (ELISAs) in the diagnosis of FAdVs. We cloned and expressed the 100 K and 33 K non-structural protein genes of the FAdVs in the pGEX-4T-1 plasmid vector. Purified 100 K and 33 K proteins alone or in combination were used as antigens in ELISAs. Antibodies specific to the 100 K and 33 K non-structural proteins were detected in chickens experimentally infected with FAdVs, but not in chickens vaccinated with inactivated FAdVs. In contrast, the agar gel precipitation (AGP) test detected FAdV-specific antibodies in 70.3% of the vaccinated chickens, suggesting that the non-structural protein-based ELISA could be used in the differential diagnosis of infected and vaccinated chickens. To further validate the 100 K and 33 K-based ELISA (100 K-33 K-ELISA) method, we compared its sensitivity and specificity with that of a whole virus-based ELISA and an AGP test in detecting FAdV-specific antibodies in 350 field samples. The results showed that the 100 K-33 K-ELISA exhibited a higher sensitivity than the AGP test and a comparable sensitivity and specificity to the whole virus ELISA. Overall, the 100 K-33 K-ELISA method is sensitive, specific and can be used to distinguish an acute FAdV infection from an inactivated virus-based vaccination response.

Keywords

fowl adenoviruses, look non-structural protein, 33K non-structural protein, ELISA

Introduction

Fowl adenoviruses (FAdVs) belong to the genus *Aviadenovirus* of the family Adenoviridae. FAdVs cause a wide range of diseases including respiratory infection, inclusion body hepatitis, infectious hydropericardium, Angara disease and hydropericardium syndrome[1-4]. FAdVs encode 10 major structural proteins (pⅡ, pⅢ, pⅣ, pⅢa, pⅧ, pⅤ, pⅦ, pⅩ, μ, and TP) and 11 non-structural proteins (E1A, E1B, ADP[E3], E4, pol, DBP[E2A], pⅣaⅡ, 52/55 K, EP, 100 K, and 33 K). The 100 K non-structural protein is necessary for the efficient translation of viral late mRNA species[5]. The 33 K non-structural protein plays a role in capsid assembly[6]. The 100 K and 33 K proteins are both highly immunogenic and antisera generated against them have been used to study their cellular localization[7].

A polymerase chain reaction (PCR) and a whole virus-based enzyme-linked immunosorbent assay (ELISA) are used in the diagnosis of FAdVs[8, 9]. The ELISA is highly sensitive compared with the traditional agar gel precipitation (AGP) method. A previous study has demonstrated that a recombinant protein-based ELISA greatly reduces non-specific binding reactions and appears to have a higher correlation with serum neutralization titres than the conventional whole virus-based ELISA[10]. Additionally, a non-structural protein-based ELISA has been used successfully to distinguish vaccinated from infected animals for a number of viruses including foot-and-mouth disease virus[11], avian reovirus[12] and influenza A virus[13]. To our knowledge, a recombinant protein-based ELISA for FAdVs has not so far been reported. In this paper, we investigated the potential use of two non-structural proteins, 100 K and 33 K, in the development of a recombinant protein-based ELISA for FAdVs. Since inactivated FAdV vaccines have been used in some countries, including China, the successful development of a non-structural protein-based ELISA would not only distinguish an acute infection from a vaccine-induced response, but could also be used as an alternative to a whole virus-based ELISA in clinical diagnosis. Compared with AGP and whole virus ELISA, a recombinant protein-based ELISA should be safer and exhibit higher sensitivity and specificity.

Materials and methods

Viruses

The FAdV strain CELO obtained from the China Institute of Veterinary Drug Control, Beijing, China, was used in this study. The virus was grown on chicken embryo liver cells prepared from 14-day-old specific pathogen free (SPF) embryos (Merial Inc., Beijing, China) as described previously[14]. At 48 h after infection, cells and supernatant were harvested and subjected to three cycles of freezing and thawing. Culture supernatant was collected after centrifugation at 10 000 × g for 15 min. The virus titre was determined in 96-well microtitre plates as described previously[15].

Serum samples

SPF chicken embryos were purchased from Merial Inc. and used for the production of SPF chickens in our laboratory. The 110 SPF chickens were randomly divided into three groups. One group of 30, 4-week-old chickens was inoculated intranasally with $\log_{10}$ 5.5 median tissue culture infectious doses ($TCID_{50}$) /mL CELO virus using a nebulizer, every 2 weeks for a total of three times. Blood was collected 2 weeks after the third inoculation. Another group of 30, 2-week-old SPF chickens was vaccinated with $\log_{10}$ 5.5 $TCID_{50}$/mL inactivated FAdV intramuscularly every 2 weeks for a total of three times. Blood samples were collected 2 weeks after the third vaccination. FAdV was inactivated with 0.1% formaldehyde at 37 ℃ overnight and then emulsified in mineral oil composed of equal volumes of Span-80 (GOODWAY, Shanghai, China) and 10# white oil (SINOPEC, Hangzhou, China). The efficacy of the inactivated vaccine has been reported previously[16, 17]. Fifty SPF chickens were used as negative controls. Blood samples were collected at 10 weeks of age. Animal protocols used in this study were approved by the Institutional Animal Care and Use Committee (Approval # A001-2010).

Control antisera

Serum samples known to be positive for infectious bursal disease virus, Newcastle disease virus, egg drop syndrome and avian influenza virus were stored at the Guangxi Veterinary Research Institute.

Cloning of the 100 K and 33 K genes into the pGEX-4T-1 plasmid vector

DNA extraction from CELO virus was performed as described previously[9]. For the 100 K gene, the forward primer (5'-CGGAATTCTCATCGAAAATGGCAGACAAGAT-3') and the reverse primer (5'-TT GAATTCCCCGAGACGCCATTTAGGTA-3') were designed and used to amplify the 1 062 base pairs of 100 K-encoding gene. Primers for 33 K (forward, 5'-CCGGAATTCTCTCTCCCTCATTTCT-3'; reverse, 5'-CCGGAATTCCTTGATAGCATCGCTCTTC-3') were used to amplify the 288 base pairs of 33 K-encoding gene (accession number U46933.1). PCR amplification was performed with the PCR master mix kit (Tiangen, Beijing, China) by using 1 μL (20 ng) DNA template and 50 pmol each primer in a 25 μL reaction volume by following the manufacturer's protocol with the following cycling times and temperatures: 94 ℃ for 5 min and 35 cycles of 94 ℃ for 1 min, 55 ℃ for 1 min and 72 ℃ for 1 min. The PCR product was digested with *ECoR* I, and then ligated into *ECoR* I-digested pGEX-4T-1 (GE Healthcare, Piscataway, NJ, USA). The ligation mixture was transformed into *Escherichia coli* BL21 competent cells. Plasmids containing 100 K or 33 K genes were identified by PCR and confirmed by DNA sequencing (TaKaRa, Dalian, China).

Expression and purification of recombinant 100 K-GST and 33 K-GST fusion proteins

Recombinant bacteria were grown in 5 mL Luria broth (LB) medium containing 100 μg/mL ampicillin overnight at 37 ℃. The culture was then added to 1 000 mL LB medium containing ampicillin (100 μg/mL) and cultured for approximately 3 h in an orbital shaker at 37 ℃. When the optical density at 600 nm value of the culture reached 0.55 to 0.65, 1 mM isopropyl *β*-D-1-thiogalactopyranoside (Sigma, St Louis, MO, USA) was added to induce the expression of cloned 100 K and 33 K genes. After incubation for 5 h at 30 ℃, the recombinant glutathione S-transferase (GST) fusion proteins and GST protein were purified using the following procedure. First, the concentrated cell pellets were resuspended in 10 mL phosphate-buffered saline (PBS) supplemented with lysozyme to a final concentration of 100 mg/mL. The mixture was incubated at room temperature for 15 min and sonicated on ice until the solution was no longer viscous. The mixture was then clarified by centrifugation at 10 000 × g for 15 min. The supernatant containing the solubilized protein was collected for subsequent binding to GST-agarose resin (GST binding resin; Glutathione Sepharose™, 4B, lot 0074574011; GE Healthcare) according to the manufacturer's instructions. The protein concentration was determined according to the Lowry method[18]. The purified 100K-GST and 33 K-GST proteins were diluted 1 ∶ 1 in glycerol and stored at –70 ℃ for use as antigens in the ELISA.

To further confirm the expression of the recombinant GST fusion proteins, western blotting was performed using serum collected from either FAdV-infected chickens or vaccinated chickens. Briefly, the membrane was washed three times in PBS containing 0.05% Tween 20 (PBST) (20 sec/ wash) and then incubated with goat anti-chicken IgG-horseradish peroxidase conjugate (KPL Inc., Gaithersburg, MD, USA) for 30 min at room temperature. Following three washes with PBST at room temperature, the membrane was incubated with the DAB Color Development Kit (Tiangen).

Indirect 100 K-ELISA, 33 K-ELISA, and 100 K-33 K-ELISA

To determine the optimal concentrations of recombinant proteins for use in the ELISA, checkerboard titrations were carried out for the 100 K-GST and 33 K-GST fusion protein alone and in combination using known FAdV-positive and FAdV-negative sera. Purified 100 K and 33 K proteins were used at a series of dilutions ranging from 10.8 to 1.4 μg/mL and 12 to 2 μg/mL, respectively, in 0.05 M carbonate buffer (pH 9.6). To increase the specificity and sensitivity of the ELISA, different combinations of 100 K-GST and 33 K-GST fusion proteins were tested. One hundred microlitres of antigen was added to the ELISA plate and incubated at

4 ℃ overnight. The wells were washed three times with PBST, and then blocked with 50 μL blocking buffer (PBS, 0.05% Tween 20, and 2.5% bovine serum) for 1 h at 37 ℃ to saturate all unbound sites. Coated plates were then washed three times with PBST. Chicken sera were diluted in a series of dilutions ranging from 1 ∶ 50 to 1 ∶ 400 using 1% bovine serum albumin, then added at 100 μL/well and incubated for 1 h at 37 ℃. After incubation, the plates were washed three times. One hundred microlitres per well of a 1 ∶ 3 000 dilution of horseradish peroxidase-labelled goat anti-chicken IgG (KPL, Inc.) was added, and the plates were incubated for 1 h at 37 ℃ and then washed three times. The colour was developed using the chromogen/substrate mixture TMB/H_2O_2 (Sigma). After 10 min, the reaction was stopped by the addition of 2.0 M H_2SO_4.

The absorbance was read at 450 nm using a plate reader. Duplicate positive and negative control chicken sera were included on each plate. In addition, 5 μg/mL purified GST protein was used as a coating antigen and served as a negative control. For the whole virus ELISA, CELO virus was purified by the sucrose density gradient centrifugation method and diluted 10-fold for use as the coating antigen.

To determine the cut-off value, 50 known negative serum samples and 30 known positive serum samples were tested using the 100 K-ELISA, 33 K-ELISA and 100 K-33 K-ELISA to calculate the average and standard deviation (SD) of the optical density at 450 nm values. The cut-off value was calculated using the formula: cut-off values =±3SD.

Serum samples from commercial chicken flocks

Three hundred and fifty serum samples were collected from Guangxi. Two hundred sera were randomly collected from chicken farms that were experiencing FAdV infections in some counties of the province during 2010. One hundred and fifty chickens that were known to be FAdV-negative were vaccinated with an inactivated commercial FAdV vaccine (Xinxing, Nanning, China). Blood samples were collected 6 weeks after vaccination. These serum samples came from different chicken farms. The sera were tested for FAdV by the 100 K-ELISA, 33 K-ELISA, 100 K-33 K-ELISA, GST-ELISA, AGP and whole virus ELISA developed in our laboratory. The detection rate of each method was determined.

Results

Expression and purification of 100 K-GST and 33 K-GST fusion proteins

Analysis of the purified proteins by SDS-PAGE revealed the expression of 100 K-GST and 33 K-GST fusion proteins with an approximate molecular mass of 64.5 kDa and 36.7 kDa respectively (Figure 5-40-1 A). The concentrations of the 100 K-GST and 33 K-GST fusion proteins were 1.08 mg/mL and 0.6 mg/mL, respectively. The blurred band below the 33 K-GST protein in Figure 5-40-1 A was thought to be a degraded form of 33 K-GST protein, possibly due to the ultrasonic homogenization procedure used. The degraded smaller protein was not detected in the Western blotting analysis (Figure 5-40-1 B, C), suggesting that the truncated 33K-GST protein did not react with FAdV antiserum. Thus it was not expected to affect the ELISA.

To further confirm the expression of the expected proteins, western blotting analysis was used. As shown in Figure 5-40-1 B, C, the purified proteins reacted strongly with FAdV-positive serum, but not with FAdV-negative serum (data not shown).

Optimization of 100 K-ELISA, 33 K-ELISA and 100 K-33 K-ELISA

Checkerboard titrations were used to determine the optimal conditions for the 100 K-ELISA, 33 K-ELISA

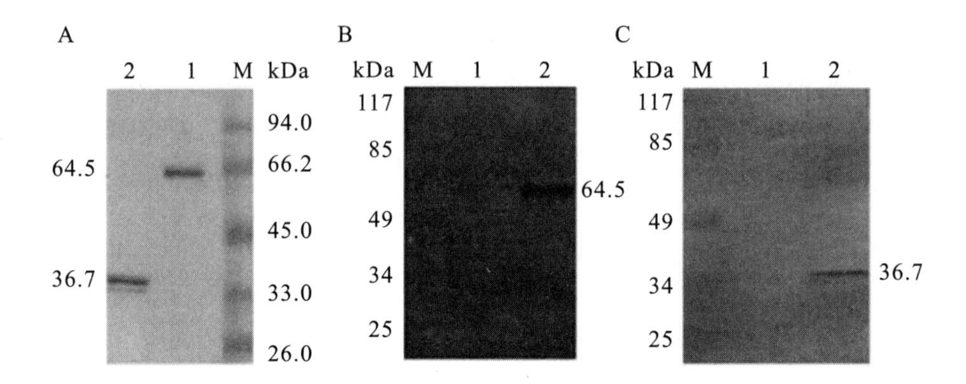

A : Coomassie brilliant blue staining of the purified 100 K-GST and 33 K-GST fusion proteins; lane M: protein marker standards; lane 1: 100 K-GST fusion protein; lane 2: 33 K-GST fusion protein.
B: Western blotting analysis of the 100 K-GST fusion proteins using known FAdV-positive sera; lane M: protein marker standards; lane 1: pGEX-4T-I induced 5 h; lane 2: pGEX-4T-1 100K-GST fusion protein induced 5 h.
C: Western blotting analysis of the 33 K-GST fusion proteins using known FAdV-positive sera; lane M: protein marker standard; lane 1: pGEX-4T-I induced 5 h; lane 2: PGEX-4T-1 33K-GST fusion protein induced 5 h.

Figure 5-40-1 SDS-PAGE and western blotting analysis of purified 100 K-GST and 33 K-GST fusion proteins

and 100 K-33 K-ELISA. Representative results are shown in Table 5-40-1. The optimal concentrations for 100 K, 33 K and 100 K-33 K fusion proteins were 5.4 μg/mL, 6 μg/mL, and 5.4+6 μg/mL, respectively. The serum dilutions of 1 ∶ 100 gave a clear difference in absorbance values between the positive and negative sera.

Table 5-40-1 Optimal concentrations of recombinant proteins and of serum dilutions used in ELISAs as determined by checkerboard titration assays

Type of ELISA	Concentration of /(μg/mL)	Protein	Serum dilution							
			1 ∶ 50		1 ∶ 100		1 ∶ 200		1 ∶ 400	
			+ serum[a]	− serum[b]	+ serum	− serum	+ serum	− serum	+ serum	− serum
100 K-ELISA	10.8		1.235[c]	0.234	1.203	0.215	0.872	0.164	0.611	0.133
	5.4		1.019	0.173	0.904	0.140	0.726	0.131	0.529	0.110
	2.7		0.832	0.156	0.742	0.137	0.689	0.129	0.443	0.115
	1.4		0.591	0.137	0.451	0.120	0.374	0.110	0.244	0.098
33 K-ELISA	12		0.977	0.195	0.903	0.182	0.753	0.159	0.536	0.132
	6		0.853	0.164	0.821	0.153	0.620	0.137	0.428	0.121
	4		0.760	0.159	0.712	0.148	0.584	0.123	0.391	0.115
	2		0.607	0.148	0.531	0.121	0.356	0.119	0.278	0.110
100 K-33 K-ELISA	10.8 (100 K) +12 (33 K)		1.635	0.306	1.428	0.275	1.278	0.238	0.843	0.195
	5.4 (100 K) +6 (33 K)		1.445	0.208	1.233	0.172	0.974	0.153	0.723	0.124
	2.7 (100 K) +4 (33 K)		1.124	0.185	0.993	0.152	0.874	0.137	0.607	0.118
	1.4 (100 K) +2 (33 K)		0.905	0.153	0.818	0.137	0.662	0.112	0.415	0.102
GST-ELISA[d]	6 (GST)		0.275	0.089	0.153	0.066	0.088	0.052	0.067	0.040

Note: [a] FAdV-positive serum. [b] FAdV-negative serum. [c] The values shown in the columns are the optical absorbances. [d] Purified GST protein was used as a coating antigen.

Determination of the cut-off values

The average and the SD of 50 known negative sera collected from 10-week-old SPF chickens were used to calculate the cut-off values using the formula: cut-off values =±3SD. The cut-off values for the 100 K-ELISA, 33 K-ELISA and 100 K-33 K-ELISA were 100 K-ELISA > 0.345, 33 K-ELISA > 0.316, and 100 K-33 K-ELISA > 0.342.

Detection of FAdV-specific antibodies with non-structural protein-based ELISAs and AGP

The 100 K-ELISA, 33 K-ELISA and 100 K-33 K-ELISA were used to detect FAdV-specific antibodies in sera collected from the experimentally infected and vaccinated chickens. All three ELISAs were able to detect antibodies in all the sera from chickens experimentally infected with FAdV, whereas the AGP test only detected antibodies in 62.1% of the serum samples from the experimentally infected chickens. Interestingly, the 100 K-ELISA, 33 K-ELISA and 100 K-33 K-ELISA failed to detect any antibodies in the sera collected from experimentally vaccinated chickens, whereas by the AGP method antibodies were detected in 70.3% of the chickens experimentally vaccinated with FAdV. The results suggest that the 100 K and 33 K non-structural protein-based ELISA could differentiate the response to FAdV vaccination from that following an acute infection.

Validation of the non-structural protein-based ELISA for use to test serum samples from the field

A total of 350 field serum samples were collected from different farms that were either experiencing FAdV infection or were vaccinated using inactivated FAdV. The presence of FAdV-specific antibodies in these samples was examined by AGP, the whole virus ELISA developed in our laboratory and the non-structural protein-based ELISAs. The results are summarized in Table 5-40-2. The non-structural protein-based ELISA only detected antibodies in samples collected from infected farms, but not from vaccinated farms, further confirming the specificity of these ELISAs in distinguishing the response to infection from that following vaccination. The detection rates for the 100 K-ELISA, 33 K-ELISA and 100 K-33 K-ELISA were 35.5% (71/200), 32.5% (65/200) and 38% (76/200) respectively, while the detection rates for the whole virus ELISA and the AGP were 40.5% (81/200) and 26.5% (53/200) respectively. The 100 K-33 K-ELISA showed the highest sensitivity among the three non-structural protein-based ELISAs, but this was a little lower than that found using the whole virus ELISA. The optical density values for these sera ranged from 0.45 to 1.8. This may reflect the extent of virus activity in the chickens and the timing of sample collection. Moreover, the detection rate for the GST-ELISA was 0%, indicating that the results obtained with the recombinant protein-base ELISAs were not influenced by the GST-target protein. The 100 K-ELISA, 33 K-ELISA and 100 K-33 K-ELISA showed no cross-reaction with known positive serum samples against infectious bursal disease virus, Newcastle disease virus, egg drop syndrome or avian influenza virus (data not shown), which indicated that these indirect ELISAs are specific for the detection of FAdV antibodies.

Table 5-40-2 Comparison of different methods for detecting FAdV-specific antibodies in field samples

Origin of serum samples	Location of chicken farm	Number of samples	100 K-ELISA-positive samples	33 K-ELISA-positive samples	100 K+33 K-ELISA-positive samples	AGP-positive samples	Whole virus ELISA-positive samples
Chickens with suspected FAdV infection	Jvdong	50	21	20	21	18	23
	Jingxi	50	18	16	21	14	21
	Liangfeng	50	10	9	10	6	11
	Longan	50	22	20	24	15	26
	Total number of samples	200	71	65	76	53	81
	positive/%		35.5	32.5	38	26.5	40.5
Chickens vaccinated with inactivated FAdV vaccine	Jvdong	43	0	0	0	31	40
	Jingxi	36	0	0	0	26	33
	Liangfeng	32	0	0	0	21	32
	Longan	39	0	0	0	24	37
	Total number of samples	150	0	0	0	102	142
	positive/%		0	0	0	68	94.7

Discussion

Several methods have been developed to detect antibodies against FAdVs, including AGP[19], virus neutralization[20], and ELISA[8]. Because the ELISA technique has a high level of sensitivity and reproducibility and allows for automation, it is the method of choice for screening a large number of serum samples.

Whole virus lysate-based ELISAs have been widely used to detect the virus-specific antibody response. However, non-specific reactions are more likely to occur using the whole virus lysate-based ELISA as has been shown for reoviruses[21]. Furthermore, the use of whole virus lysates as the coating antigen leads to the danger of spreading live virus particles. Previous reports have demonstrated that using two proteins as the coating antigens in ELISAs used for detecting antibodies greatly reduces non-specific binding reactions and gives a higher correlation with serum neutralization assays than does the conventional whole virus-based ELISA[10, 21]. Non-structural proteins are normally expressed abundantly in virus-infected cells, but they are not detected in the virions[22]. It has also been reported for influenza virus and reovirus that a non-structural protein-based ELISA might be able to distinguish the vaccination response from that following an acute viral infection[12, 13].

In this study, we cloned two non-structural protein genes (100 K and 33 K) of CELO FAdV into a pGEX-4T-1 vector to generate recombinant GST fusion proteins. The potential use of these two GST fusion proteins in capturing antibodies specific to the two non-structural proteins was evaluated in experimentally infected and in vaccinated chickens. The results showed that the 100 K-ELISA, 33 K-ELISA and 100 K-33 K-ELISA were able to differentiate experimentally infected chickens from vaccinated ones, while the conventional whole virus-based ELISA and the AGP test could not distinguish an acute infection response from that following vaccination. All 30 sera collected from experimentally infected chickens were found to be positive by these ELISAs. In contrast, none of the experimentally vaccinated chickens were positive for FAdV. Overall, the 100 K-ELISA, 33 K-ELISA and 100 K-33 K-ELISA have advantages over the conventional whole virus-based ELISA[8, 19], with less non-specific binding reactions and high sensitivity, and could be used as a differential diagnostic tool to distinguish infected chickens from birds given inactivated virus vaccines. The availability of such diagnostic tests will allow researchers to gather accurate epidemiological data on FAdVs, which in turn will greatly facilitate the regional elimination of FAdVs.

Formaldehyde can effectively inactivate a number of viruses including poliovirus, human enterovirus 71, Japanese encephalitis virus, foot-and-mouth disease virus, enteroviruses, Ross River virus, and human immunodeficiency virus[15, 23-27]. In addition, formaldehyde has been shown to be suitable for commercial vaccine production due to its low cost. In our previous study, we compared the effect of formaldehyde and β-propriolactone for the inactivation of FAdV. Treatment of FAdV with 0.1% formaldehyde or with 0.05% β-propriolactone effectively inactivated virus activity[16, 17]. Although inactivation with formaldehyde can have a protein altering effect, the data reported here show that the ELISAs we developed were not negatively influenced by the inactivation procedure; this finding is important, because all of the current FAdV vaccines used in China are inactivated by formaldehyde treatment.

We used the 100 K-ELISA, 33 K-ELISA and 100 K-33 K-ELISA to detect FAdV-specific antibodies from 350 field samples collected from Guangxi in China. A good correlation in sensitivity and specificity was observed between the three ELISAs and also with the whole virus-based ELISA. Furthermore, these non-structural protein-based ELISAs, but not the AGP test or the whole virus-based ELISA, were capable of distinguishing infected chickens from vaccinated ones, supporting the data we obtained from the experimentally infected and vaccinated chickens. Overall, the results reported here suggest that these non-structural protein-based ELISAs can be used in screening for the prevalence of FAdV in chickens and in the differential diagnosis of acute infections and responses to vaccination. Since there are many different serotypes of FAdV[19, 28], it remains to be determined whether these non-structural protein-based ELISAs can be used in the detection of all of the different serotypes.

References

[1] WINTERFIELD R W, FADLY A M, GALLINA A M. Adenovirus infection and disease. I. Some characteristics of an isolate from chickens in Indiana. Avian Dis, 1973, 17(2): 334-342.

[2] DHILLON A S, WINTERFIELD R W. Pathogenicity of various adenovirus serotypes in the presence of Escherichia coli in chickens. Avian Dis, 1984, 28(1): 147-153.

[3] MAZAHERI A, PRUSAS C, VOSS M, et al. Some strains of serotype 4 fowl adenoviruses cause inclusion body hepatitis and hydropericardium syndrome in chickens. Avian Pathol, 1998, 27(3): 269-276.

[4] NAKAMURA K, MASE M, YAMAGUCHI S, et al. Pathologic study of specific-pathogen-free chicks and hens inoculated with adenovirus isolated from hydropericardium syndrome. Avian Dis, 1999, 43(3): 414-423.

[5] HAYES B W, TELLING G C, MYAT M M, et al. The adenovirus L4 100-kilodalton protein is necessary for efficient translation of viral late mRNA species. J Virol, 1990, 64(6): 2732-2742.

[6] MORIN N, BOULANGER P. Morphogenesis of human adenovirus type 2: sequence of entry of proteins into previral and viral particles. Virology, 1984, 136(1): 153-167.

[7] GAMBKE C, DEPPERT W. Late nonstructural 100, 000- and 33, 000-dalton proteins of adenovirus type 2. I. Subcellular localization during the course of infection. J Virol, 1981, 40(2): 585-593.

[8] DAWSON G J, ORSI L N, YATES V J, et al. An enzyme-linked immunosorbent assay for detection of antibodies to avian adenovirus and avian adenovirus-associated virus in chickens. Avian Dis, 1980, 24(2): 393-402.

[9] XIE Z, FADL A A, GIRSHICK T, et al. Detection of avian adenovirus by polymerase chain reaction. Avian Dis, 1999, 43(1): 98-105.

[10] LIU H J, KUO L C, HU Y C, et al. Development of an ELISA for detection of antibodies to avian reovirus in chickens. J Virol Methods, 2002, 102(1-2): 129-138.

[11] LU Z, CAO Y, GUO J, et al. Development and validation of a 3ABC indirect ELISA for differentiation of foot-and-mouth disease virus infected from vaccinated animals. Vet Microbiol, 2007, 125(1-2): 157-169.

[12] XIE Z, QIN C, XIE L, et al. Recombinant protein-based ELISA for detection and differentiation of antibodies against avian reovirus in vaccinated and non-vaccinated chickens. J Virol Methods, 2010, 165(1): 108-111.

[13] OZAKI H, SUGIURA T, SUGITA S, et al. Detection of antibodies to the nonstructural protein (NS1) of influenza A virus allows distinction between vaccinated and infected horses. Vet Microbiol, 2001, 82(2): 111-119.

[14] MEULEMANS G, BOSCHMANS M, BERG T P, et al. Polymerase chain reaction combined with restriction enzyme analysis for detection and differentiation of fowl adenoviruses. Avian Pathol, 2001, 30(6): 655-660.

[15] CHEN C W, LEE Y P, WANG Y F, et al. Formaldehyde-inactivated human enterovirus 71 vaccine is compatible for co-immunization with a commercial pentavalent vaccine. Vaccine, 2011, 29(15): 2772-2776.

[16] WEI Y, XIE Z, Fan Q, et al. Development of an inactivated vaccine against infectious fowl adenovirus Group I serum type 1 strain. Progress in Veterinary Medicine, 2011a, 32: 14-19.

[17] WEI Y, XIE Z, Fan Q, et al. Development of bivalent inactivated oil emulsion vaccine for fowl adenovirus group-I. Journal of Southern Agriculture, 2011b, 42: 1550-1554.

[18] LOWRY O H, ROSEBROUGH N J, FARR A L, et al. Protein measurement with the Folin phenol reagent. J Biol Chem, 1951, 193(1): 265-275.

[19] COWEN B S. A trivalent antigen for the detection of type I avian adenovirus precipitin. Avian Dis, 1987, 31(2): 351-354.

[20] CALNEK B W, SHEK W R, MENENDEZ N A, et al. Serological cross-reactivity of avian adenovirus serotypes in an enzyme-linked immunosorbent assay. Avian Dis, 1982, 26(4): 897-906.

[21] ISLAM M R, JONES R C. An enzyme-linked immunosorbent assay for measuring antibody titre against avian reovirus using a single dilution of serum. Avian Pathol, 1988, 17(2): 411-425.

[22] KRUG R M, ETKIND P R. Cytoplasmic and nuclear virus-specific proteins in influenza virus-infected MDCK cells. Virology, 1973, 56(1): 334-348.

[23] MARTIN J, CROSSLAND G, WOOD D J, et al. Characterization of formaldehyde-inactivated poliovirus preparations made from live-attenuated strains. J Gen Virol, 2003, 84(Pt 7): 1781-1788.

[24] APPAIAHGARI M B, VRATI S. Immunogenicity and protective efficacy in mice of a formaldehyde-inactivated Indian strain of Japanese encephalitis virus grown in Vero cells. Vaccine, 2004, 22(27-28): 3669-3675.

[25] BARTELING S J, CASSIM N I. Very fast (and safe) inactivation of foot-and-mouth disease virus and enteroviruses by a combination of binary ethyleneimine and formaldehyde. Dev Biol (Basel), 2004, 119: 449-455.

[26] POON B, SAFRIT J T, MCCLURE H, et al. Induction of humoral immune responses following vaccination with envelope-containing, formaldehyde-treated, thermally inactivated human immunodeficiency virus type 1. J Virol, 2005, 79(8): 4927-4935.

[27] KISTNER O, BARRETT N, BRUHMANN A, et al. The preclinical testing of a formaldehyde inactivated Ross River virus vaccine designed for use in humans. Vaccine, 2007, 25(25):4845-4852.

[28] MCFERRAN J B, SMYTH J A. Avian adenoviruses. Rev Sci Tech, 2000, 19(2): 589-601.

Production and identification of monoclonal antibodies and development of a sandwich ELISA for detection of the H3-subtype avian influenza virus antigen

Luo Sisi, Deng Xianwen, Xie Zhixun, Huang Jiaoling, Zhang Minxiu, Li Meng, Xie Liji,
Li Dan, Fan Qing, Wang Sheng, Zeng Tingting, Zhang Yanfang, and Xie Zhiqin

Abstract

The H3 subtype of avian influenza virus (AIV) is widespread in avian species and is frequently isolated in surveillance projects; thus, we have developed a more effective diagnostic approach of a monoclonal antibody (mAb)-based sandwich ELISA for the H3 AIV detection. First, we have produced the essential reagent of mAb against AIV H3 strains with the development of an mAb-Mouse immunization with a purified H3-subtype AIV strain and cell fusion to generate hybridoma cells. These cells were screened with hemagglutination inhibition (HI) tests, and optimal cells were subcloned. We chose a hybridoma cell line that steadily secreted a specific H3-subtype AIV mAb, designated 9F12, that belongs to the IgG1 subclass and has a K-type light chain. 9F12 was shown to bind specifically to the H3-subtype AIV antigen by both immunofluorescence assay and Western blot analysis. Finally, a 9F12-based sandwich ELISA was successfully developed and used to specifically test for this antigen. The sandwich ELISA conditions were optimized, and the specificity and sensitivity were validated. The results for clinical sample detection were consistent with viral isolation. Consequently, the 9F12-based sandwich ELISA is a specific, sensitive, robust, rapid and versatile diagnostic tool for H3-subtype AIV and provides a promising strategy for effective influenza virus prevention and control.

Keywords

H3 subtype, influenza, monoclonal antibody, ELISA

Introduction

Influenza A virus, a member of the genus Orthomyxovirus, is classified into different H and N subtypes according to differences in two major antigen genes, hemagglutinin (HA) and neuraminidase (NA). The primary reservoirs of influenza A virus are waterfowl and shorebirds[1]. The H1-H16 and N1-N9 subtypes have been identified in poultry and wild birds, while H17-H18 and N10-N11 have recently been discovered in bats[2, 3]. The H3 influenza A virus is an important subtype that has distinct significance for public health. This subtype has a wide range of hosts, infects mainly avian species and spreads directly to humans across interspecies barriers, circulating in swine, canines, equines, and felines and causing sporadic outbreaks in sea mammals[4]. The H3 influenza A virus has different infection characteristics in different hosts. In poultry, H3-subtype avian influenza virus (AIV), the most ubiquitous HA subtype, is frequently isolated from live bird markets (LBMs) in China[5]. In humans, the pandemic virus A/Hong Kong/1968 (H3N2) became endemic after the first year post emergence, and it has since caused yearly seasonal epidemics[4]. In swine, the H3N2, H1N1 and H1N2 subtypes are largely responsible for annual outbreaks of swine influenza[6]. In addition, the avian-origin H3N2 canine influenza virus emerged in dogs in China or Republic of Korea in approximately 2005 and has since become an enzootic virus in Southeast Asia and the United States, causing occasional epizootics[7]. The H3N8 equine influenza virus was first isolated from horses in Miami, Florida, in 1963[8]. Currently, it is the only subtype affecting equine populations worldwide and has serious implications for public health[9, 10]. In addition, an H3N8 influenza virus carrying mammalian adaptation mutations has been isolated from seals[11].

The H3 influenza A virus plays unique roles in different hosts. Avian species are primary reservoirs of influenza A virus and can harbor viruses with significant pandemic potential. AIV is a pathogen of economic significance to the poultry industry, and the H3-subtype AIV plays an important role in the emergence of zoonotic infections. Some avian-origin H3N2 viruses transmitted to dogs in Republic of Korea cause acute respiratory disease[12]. H3N2-subtype AIV originating in ducks has been found to acquire the potential to infect humans after multiple infections in a pig population[13]. Reassortment may occur between H3N2 and other influenza virus subtypes, and an H3N2 isolate containing genes from H7N3 and H7N7 has been found to show the highest sequence homology to the H7N9 virus[14]. H3N2-subtype AIV poses a clear threat to human health, and ongoing surveillance of H3N2-subtype AIV in birds is warranted[15]; thus, we think it is critical to develop an antibody and diagnostic reagent for H3-subtype AIV. In our study, we produced a monoclonal antibody (mAb) against H3-subtype AIV and developed a sandwich enzyme-linked immunosorbent assay (ELISA) method for the detection of this subtype.

The traditional diagnostic method for H3-subtype AIV involves viral isolation and identification. These techniques are time consuming and tedious, require biosafety facilities and exhibit some limitations in practical application. Currently, detection methods for H3-subtype AIV are urgently needed as technical reserves, and mAbs and sandwich ELISA methods based on mAbs against H3-subtype AIV have rarely been reported. mAbs have a single epitope, display high specificity and have been widely adopted for medical examination. ELISA is highly suitable for epidemiological analyses involving a large number of samples, and sandwich ELISA methods based on mAbs have been developed for other pathogens[16-18]. This study had two purposes. First, we sought to produce an mAb against H3-subtype AIV in hybridoma cells and characterize its reactivity. Second, we sought to develop a sandwich ELISA method based on the mAb that could be used as an effective tool for the diagnosis of H3-subtype AIV.

Materials and methods

Cells, viruses and sera

The Madin-Darby canine kidney (MDCK) and mouse myeloma (SP2/0) cell lines were purchased from the China Center for Type Culture Collection (CCTCC). The two cell lines were maintained in Dulbecco's modified Eagle medium (DMEM) (Gibco, USA) supplemented with 10% fetal bovine serum (FBS) (Gibco, USA) and penicillin (100 U/mL)-streptomycin (100 μg/mL). The H3-subtype AIV strains and other viruses used in this study are shown in Table 5-41-1. Viral stocks were grown in 10-day-old specific pathogen-free (SPF) embryonated eggs or in MDCK cells, as stated in the text. Polyclonal sera were obtained from H3-positive and H3-negative SPF chickens.

Table 5-41-1　Origins of the H3-subtype AIV isolates used in this study

Number	Strain name	Source	Whether hemagglutination was inhibited by the mAb 9F12 using HI test	This sandwich ELISA assay
1	A/Duck/Guangxi/015D2/2009 (H3N2)	A	Y	+
2	A/Chicken/Guangxi/015C10/2009 (H3N2)	A	Y	+
3	A/Goose/Guangxi/020G/2009 (H3N8)	A	Y	+
4	A/Pigeon/Guangxi/020P/2009 (H3N6)	A	Y	+
5	A/Duck/Guangxi/057D6/2010 (H3N2)	A	Y	+
6	A/Chicken/Guangxi/073C2/2010 (H3N2)	A	Y	+
7	A/Duck/Guangxi/112D4/2012 (H3N2)	A	Y	+
8	A/Chicken/Guangxi/125C8/2012 (H3N2)	A	Y	+
9	A/Pigeon/Guangxi/128P9/2012 (H3N2)	A	Y	+
10	A/Chicken/Guangxi/135C10/2013 (H3N2)	A	Y	+
11	A/Duck/Guangxi/135D20/2013 (H3N2)	A	Y	+
12	A/Goose/Guangxi/139G20/2013 (H3N2)	A	Y	+
13	A/Chicken/Guangxi/165C7/2014 (H3N2)	A	Y	+
14	A/Duck/Guangxi/175D12/2014 (H3N6)	A	Y	+
15	A/Chicken/Guangxi/252C6/2016 (H3N2)	A	Y	+
16	A/Duck/Guangxi/272D18/2016 (H3N2)	A	Y	+
17	A/Chicken/Guangxi/284C1/2017 (H3N2)	A	Y	+
18	A/Pigeon/Guangxi/286P45/2017 (H3N2)	A	Y	+
19	A/Pigeon/Guangxi/288P43/2017 (H3N6)	A	Y	+
20	A/Goose/Guangxi/318G39/2018 (H3N2)	A	Y	+
21	A/Chicken/Guangxi/117B3/2018 (H3N2)	A	Y	+
22	A/Duck/Guangxi/030D/2009 (H1N1)	A	N	−
23	A/Duck/HK/77/76 (H2N3)	B	N	−
24	A/Duck/Guangxi/125D17/2012 (H4N2)	A	N	−
25	A/Duck/Guangxi/1/04 (H5N1)	A	N	−
26	A/Duck/Guangxi/GXd-5/2010 (H6N1)	A	N	−
27	A/Chicken/NY/273874/03 (H7N2)	C	N	−
28	A/Turkey/Ontario/6118/68 (H8N4)	B	N	−
29	A/Chinese Francolin/Guangxi/020B7/2010 (H9N2)	A	N	−
30	A/Duck/HK/876/80 (H10N3)	B	N	−
31	A/Duck/PA/2099/12 (H11N9)	C	N	−
32	A/Duck/HK/862/80 (H12N5)	B	N	−
33	A/Gull/Md/704/77 (H13N5)	B	N	−
34	A/Mallard/Astrakhan/263/82 (H14N5)	D	N	−
35	A/Shearwater/Western Australia/2576/79 (H15N9)	D	N	−
36	A/Shorebird/Delaware/168/06 (H16N3)	D	N	−
37	NDV F48	E	N	−
38	ARV S1133	E	U	−
39	EDSV	E	N	−
40	FAdV4-GX001	A	U	−
41	IBV M41	E	U	−

Note: Y, yes; N, no; U, the HI test was unavailable for the strain; A, Guangxi Veterinary Research Institute, China; B, University of Hong Kong, China; C, University of Pennsylvania, USA; D, University of Connecticut, USA; E, China Institute of Veterinary Drug Control, China.

Production of the mAb

The strain A/Duck/Guangxi/112D4/2012 (H3N2) was chosen as an immunogen for preparation of the H3 mAb. The impurities were removed via low-speed centrifugation at 4 500 r/min for 10 min, and then the viral supernatant was further purified by ultracentrifugation at 32 000 r/min for 90 min. The viral particle pellet was resuspended using phosphate-buffered saline (PBS) at 4 ℃ overnight and finely purified by sucrose density gradient ultracentrifugation through a 10% /50% sucrose cushion. To inactivate the H3 virus, β-propiolactone was added, and the virus was incubated at 4 ℃ for 16~24 h. Successful inactivation was confirmed by a lack of growth under chick embryo propagation. Six-week-old female SPF BALB/c mice were inoculated subcutaneously at multiple sites with an emulsion of the inactive H3 virus with an equal amount of Freund's complete adjuvant. At 21 days and 35 days after the first injection, the second and third immunizations, respectively, were performed with the same antigen and Freund's incomplete adjuvant. At 49 days after the first injection, the antibody titers against H3-subtype AIV in the mice were detected by hemagglutination inhibition (HI) test. BALB/c mice with high antibody titers were selected, and immunity was boosted by intraperitoneal inoculation of the antigen without adjuvant 3 days before cell fusion.

Spleen cells were harvested from the immunized mice. SP2/0 cells and spleen cells were fused at a ratio of 1 : 5 with polyethylene glycol (PEG) 4 000. The treated cells were suspended in HAT (RPMI 1 640 medium containing 20% FBS, 100 mg/mL streptomycin, 100 U/mL penicillin, 100 mM hypoxanthine, 16 mM thymidine, and 400 mM aminopterin) and plated into 96-well tissue culture plates at a density of 1×10^5 cells per well in 200 μL of medium. After cultivation at 37 ℃ in 5% CO_2 for 9 days, 12 days, and 15 days, the medium was assayed for H3-subtype AIV-specific antibodies by HI test. Positive hybridoma cells with high antibody titers were subcloned three times. The broad-spectrum activity and specificity of the hybridoma cells were assessed with purified AIV (H1-H16 subtypes), Newcastle disease virus (NDV), and egg drop syndrome virus (EDSV) by HI test. Finally, we chose a high-titer antibody and specific hybridoma cells for subsequent research. The selected hybridoma cells were inoculated into BALB/c mice, and ascetic fluid was purified by saturated ammonium sulfate (SAS) precipitation as described previously[19]. The mAb isotypes were determined using a Mouse MonoAb-ID Kit according to the manufacturer's instructions (Sigma, USA).

Western blot analysis

To investigate the reactivity of the mAb 9F12 with H3-subtype AIV antigen, Western blot was performed according to the kit manufacturer's instructions (ProteinSimple Wes, part#PS-T001). In brief, the reagents were first prepared, the H3 virus was diluted with HeLa Lysate Control, and 4 parts of diluted lysate were combined with 1 part of 5 × Fluorescent Master Mix in a microcentrifuge tube. The mixture was then heated at 95 ℃ for 5 min to denature the sample. Second, the sample and reagents were pipetted into the specified plate; the mAb acted as the primary antibody, and a horseradish peroxidase (HRP)-labeled goat anti-mouse antibody (Sigma, USA) acted as the secondary antibody. Finally, Western blot was completed with a fully automated instrument (ProteinSimple Wes, ProteinSimple Corp. 2014).

Immunofluorescence assay (IFA)

MDCK cells cultured in 48-well plates were inoculated with the strain 112D4 (H3) at a multiplicity of infection (MOI) of 0.1 and incubated at 37 ℃ in 5% CO_2 for 1 h. In addition, MDCK cells were inoculated with PBS as mock-infected controls. Following infection, viral growth medium consisting of fresh DMEM with 0.3% bovine serum albumin (BSA), 1 μg/mL tolylsulfonyl phenylalanyl chloromethyl ketone (TPCK)-treated trypsin and penicillin (100 U/mL)-streptomycin (100 μg/mL) was added, and the cells were further incubated at 37 ℃ in 5% CO_2 for 48 h. The infected cells were fixed with ice-cold methanol for 20 min at room temperature. After washing the cells three times with PBS, blocking buffer was added, and the cells were incubated at 37 ℃ for 30 min. The cells were then gently washed with PBS, the hybridoma culture supernatant or diluted murine ascetic fluid was added, and the cells were incubated at 37 ℃ for 1 h. The cells were again washed, and FITC-conjugated goat anti-mouse IgG (Sigma, USA) was added at a dilution of 1 ∶ 1 000. The cells were incubated for 1 h at 37 ℃, washed again and observed by fluorescence microscopy.

Development of a sandwich ELISA for H3-subtype AIV antigen detection

A sandwich ELISA for the H3-subtype AIV antigen was developed using H3-positive polyclonal serum as the coating antibody, the mAb 9F12 as the capture antibody and HRP-conjugated goat anti-mouse IgG as the detection antibody. The basic protocol was composed of the following steps: coating of the ELISA plate with H3 polyclonal serum, three washes with PBS with 0.05% Tween (PBST), addition of blocking buffer with 5% skim milk, incubation, three washes with PBST, addition of the antigen sample, incubation, three washes with PBST, addition of HRP-conjugated goat anti-mouse IgG, incubation, three washes with PBST, addition of 0.24 mg/mL 3, 3', 5, 5'-tetramethyl benzidine and 0.003% H_2O_2 (TMB) substrate, and termination of the reaction with H_2SO_4. The working concentrations and incubation times for the polyclonal antibody, antigen, mAb, HRP-conjugated goat anti-mouse IgG, and other components of this ELISA were optimized though the use of checkerboard titrations. The optimal conditions were determined by evaluating the optical density (OD) values at 450 nm ($OD_{450\ nm}$) and the positive/negative (P/N) ratios of the samples.

Determination of the cutoff value for the sandwich ELISA

To determine the cutoff value for the sandwich ELISA, 60 H3-free samples, including chicken cloacal swab samples and tissue samples, were tested with the developed sandwich ELISA, and the average (X) and standard deviation (SD) of the $OD_{450\ nm}$ values of the 60 samples were calculated. X + 3SD was the cutoff value. A sample with an $OD_{450\ nm}$ greater than or equal to the cutoff value was considered positive, and a sample with a value less than the cutoff value was considered negative.

Validation of the specificity and sensitivity of the sandwich ELISA

H3 subtype isolates and isolates of other AIV subtypes (H1-H2, H4-H16), NDV, avian reovirus (ARV), EDSV, fowl adenovirus 4 (FAdV4) and infectious bronchitis virus (IBV) (Table 5-41-1) were analyzed by sandwich ELISA to evaluate the specificity and cross-reactivity of the method.

To validate the sensitivity of the method, 10, 100, 10^3, 10^4, 10^5, 10^6, 10^7, and 10^8 dilutions of the H3 virus (HA titer 2^8) were created, and these dilutions were tested with the developed sandwich ELISA.

Clinical samples

A total of 180 swab samples (oral pharyngeal and cloacal swabs from the same bird were pooled as a single sample) were obtained as part of the AIV surveillance program in LBMs in Nanning, the capital of Guangxi, from 2018 to 2019. The swab solutions were tested with the developed H3-subtype AIV sandwich ELISA method. In parallel, 10-day-old SPF embryonated chicken eggs were inoculated with the swab solutions for viral isolation, and allantoic fluid was collected and analyzed by HI test, also amplified with conventional RT-PCR followed by HA and NA amplicon sequencing.

Statistical analysis

All data are presented as the means with SDs from three independent experiments and were visualized in GraphPad Prism 5.0 software (GraphPad Software, CA, USA).

Results

Characterization of the mAb

Two hybridoma cells secreted antibodies against H3-subtype AIV that were isolated and screened by the HI test. After three cycles of subcloning, 1 mAb, designated 9F12, had broad-spectrum reactivity to H3-subtype AIV strains and was chosen for subsequent analyses. The mAb 9F12 reacted with the H3 isolates tested but lacked cross-reactivity with other AIV subtypes, NDV and EDSV, as assessed by the HI analysis (Table 5-41-1). The hemagglutination of H3-subtype AIV was inhibited by hybridoma supernatants and ascites, and the inhibition test titers were $2^6\sim2^8$ and $2^{16}\sim2^{18}$, respectively. The heavy-chain subclass of 9F12 was determined to be IgG1, and the light-chain type was K. Nine H3 isolates were randomly chosen for identification of the reactivity of the mAb 9F12 by Western blot assay, and the results suggested that the H3 mAb could react with all nine H3 isolates, which showed specific bands at 66 kDa; in contrast, the AIV-H9, NDV and mock reactions had no band (Figure 5-41-1). IFA using the mAb 9F12 as the primary antibody revealed strong green fluorescence in infected cells (Figure 5-41-2 A) ; mock-infected cells were stained in parallel, but no positive signal was observed (Figure 5-41-2 B). The mAb 9F12 was thus used to develop a sandwich ELISA.

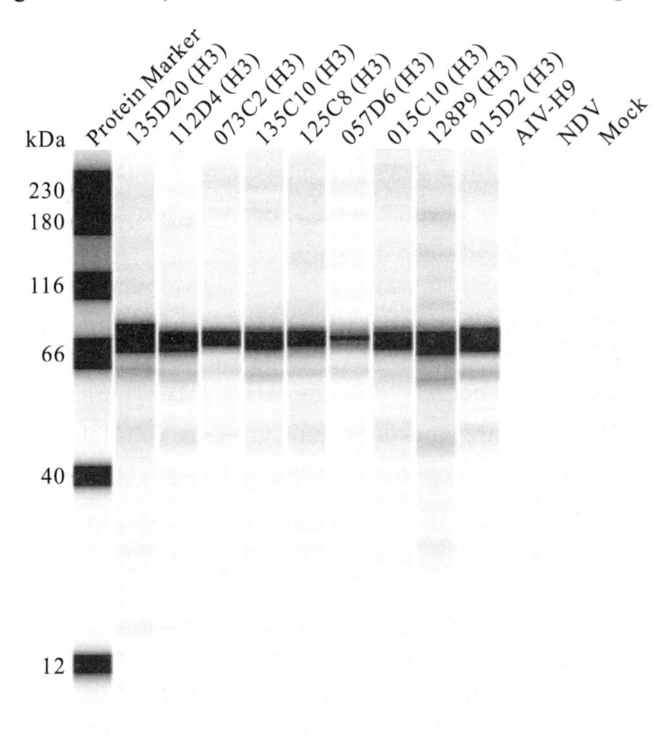

Figure 5-41-1 The reactivities of different H3 isolates with the mAb 9F12 were determined in a Western blot assay

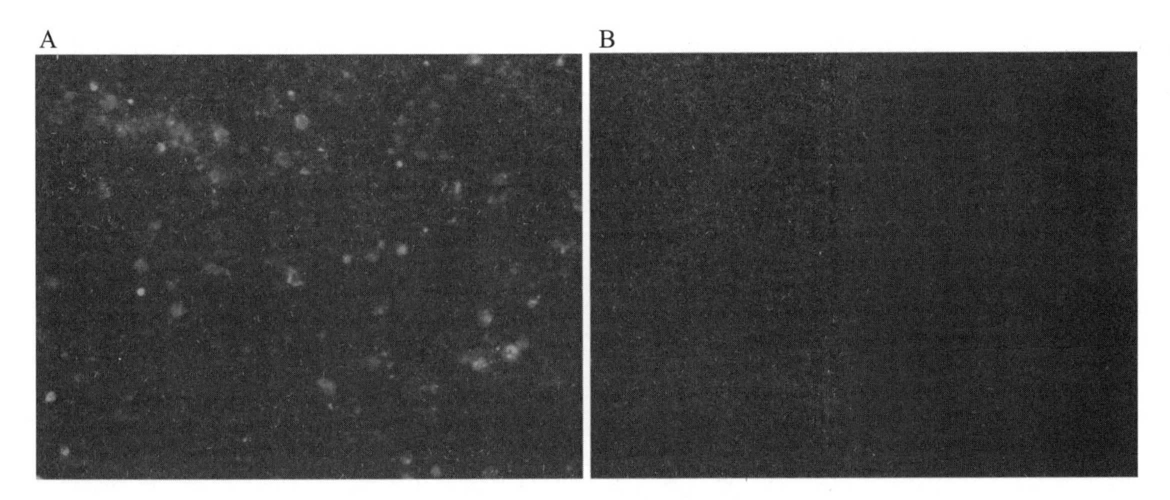

A: H3-subtype AIV-infected MDCK cells; B: Mock-infected MDCK cells. H3-subtype AIV-infected or mock-infected MDCK cells were fixed with ice-cold methanol and then incubated with the mAb 9F12. After incubation with a FITC-conjugated goat anti-mouse IgG secondary antibody, the cells were visualized by fluorescence microscopy.

Figure 5-41-2　IFA of the mAb 9F12

Protocol of the sandwich ELISA for H3-subtype AIV antigen detection

First, 96-well polystyrene plates were precoated with H3-positive polyclonal serum (1 : 160 dilution) and incubated at 37 ℃ for 1 h and at 4 ℃ overnight. The plates were then blocked with 5% skim milk at 37 ℃ for 1 h. Antigen samples were distributed into each well, and the plates were incubated at 37 ℃ for 1 h. The mAb 9F12 was then diluted 1 : 100 in 1% BSA and added to the plates, which were again incubated at 37 ℃ for 1 h. HRP-conjugated goat anti-mouse IgG was diluted to 1 : 1 000 with 1% BSA and added to the plates before incubation at 37 ℃ for 1 h. After each incubation and before adding a new reagent, the plate wells were washed three times using PBST to remove unbound reagent. The plates were thoroughly washed, and 100 μL of a TMB substrate solution was added. Following incubation at room temperature in the dark for 10 min, the chromogenic reaction was quenched with 50 μL of 0.5 M H_2SO_4, and the $OD_{450\ nm}$ values were measured using an ELISA plate reader.

Cutoff value for the sandwich ELISA

To determine the cutoff value for the sandwich ELISA, 60 H3-free samples, including chicken cloacal swab samples and tissue samples, were tested. The average (X) $OD_{450\ nm}$ was 0.129 7, the SD was 0.041 5, and the X + 3SD was 0.254 2. For an $OD_{450\ nm} \geqslant 0.254\ 2$, the sample was determined to be positive; for an $OD_{450\ nm} < 0.254\ 2$, the sample was determined to be negative.

Specificity and sensitivity of the sandwich ELISA

H3 isolates were subjected to this ELISA, and the results were all positive (Table 5-41-1 and Figure 5-41-3 A). H1-H2, H4-H16, NDV, ARV, EDS, FAdV-4 and IBV were also assayed by the sandwich ELISA, and the $OD_{450\ nm}$ values were all less than the cutoff value; thus, the samples were considered negative (Table 5-41-1 and Figure 5-41-3 B). Each sample was assayed three times under the same conditions on different days, and the replicate analyses produced very similar results.

H3-subtype AIV (HA: 2^8) was diluted 1 : 10~1 : 10^5, and testing revealed that the $OD_{450\ nm}$ value was greater than the cutoff value. When the H3 virus was diluted 1 : 10^6, the $OD_{450\ nm}$ value fluctuated around the cutoff value. When the dilution was 1 : 10^7, the $OD_{450\ nm}$ value was less than the cutoff value, thus the sample was reported as negative (Figure 5-41-3 C). Each sample was assayed three times under the same conditions on different days, and the replicate analyses produced very similar results.

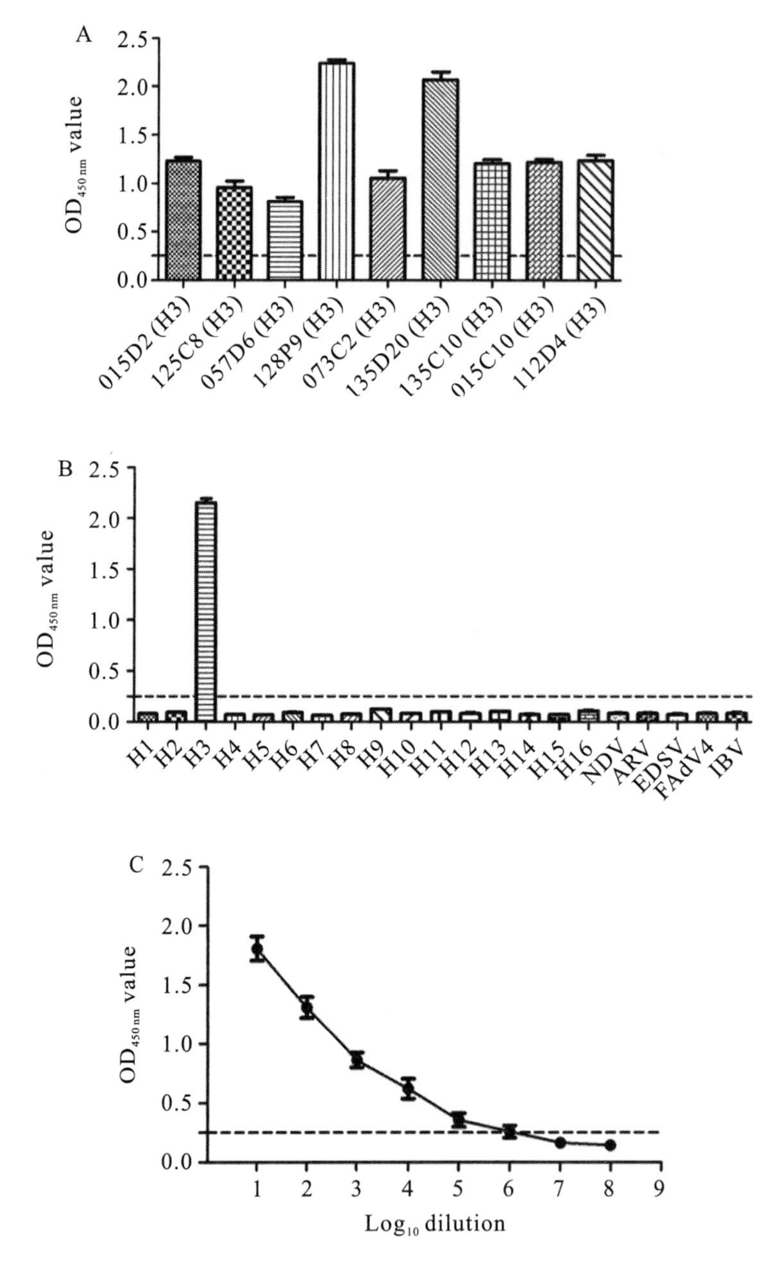

A: Different H3 isolates were tested and shown to be positive; B: Other HA subtypes of AIV, NDV, ARV, EDSV, FAdV-4 and IBV were tested and shown to be negative; C: Sensitivity of the sandwich ELISA for H3 virus detection. Dotted line indicates the cutoff value.

Figure 5-41-3 Specificity and sensitivity of the sandwich ELISA

Detection in clinical samples

Among 180 clinical samples, twenty-three samples were determined to be positive for H3-subtype AIV by the developed sandwich ELISA method. The same 23 samples were also determined to be positive by viral isolation, HI testing and amplicon sequencing (Table 5-41-2). The remaining samples were all negative. Therefore, the results of the two methods were coincident.

Table 5-41-2 Results obtained from clinical samples using this sandwich ELISA and isolation and identification

Host	Sandwich ELISA	Isolation and identification
Duck	H3 (12)	H3N2 (10), H3N6 (1), H3N8 (1)
Chicken	H3 (7)	H3N2 (7)
Goose	H3 (3)	H3N2 (2), H3N6 (1)
Pigeon	H3 (1)	H3N2 (1)
Total	H3 (23)	H3N2 (20), H3N6 (2), H3N8 (1)

Discussion

H3-subtype AIV is a major subtype threatening both human and animal health. It is important to enhance surveillance for H3-subtype AIV, and a simple, rapid, specific, sensitive and effective detection method suitable for mass detection is urgently needed. Some molecular methods for the detection of H3-subtype AIV have been developed. Thus far, several reports have described the diagnosis of H3-subtype AIV using RT-LAMP, real-time RT-PCR and multiplex RT-PCR[20, 21, 22]. However, serological detection methods for H3-subtype AIV have rarely been reported. Routine monitoring of H3-subtype AIV frequently involves large numbers of samples; notably, the developed ELISA method, which is performed with a 96-well plate, can be used to simultaneously analyze 94 samples in addition to the positive and negative control samples.

In recent years, the use of gene expression or virus like particles to produce antigens by genetic recombination has yielded antigens with increased purity and concentrations and has enabled long-lasting humoral immunity and cross-protection to be achieved. In our study, direct inoculation of mice with inactive virus kept the epitopes intact and preserved the natural spatial structure of the antigen, thus yielding an antigen that may be superior to those produced by gene expression. It is very likely that a protein does not fully recover its original stereoscopic conformation during the process of protein renaturation. Thus, we attempted to use inactive whole virus to prepare and produce the H3 mAb. Whole viruses have widely been used as immunogens for the production of mAbs to develop ELISAs for the diagnosis of disease[16, 17, 18]. The HA glycoprotein, the major surface protein of influenza A virus, plays a critical role in viral infection and is a primary target of neutralizing antibodies[23]. In our study, the mAb was screened by HI assay; thus, the mAb 9F12 was determined to recognize the HA of H3-subtype AIV. The Western blot and IFA results revealed that the mAb 9F12 specifically bound to the H3 virus with effective reactivity. In an analysis of 180 clinical samples, the detection results of this sandwich ELISA were consistent with those of gold standard methods for isolation and identification, but further analysis of more clinical samples is needed to validate this method and extend its use to the routine diagnostic epidemiological detection of H3-subtype AIV infection.

BALB/c mice were inoculated with an inactive local epidemic strain of H3-subtype AIV, and hybridoma cells that could steadily secrete a high titer of antibody specific for H3-subtype AIV was obtained through a series of cell fusion screening, HI testing and subcloning steps. The mAb obtained from the resulting hybridoma cells had good specificity and sensitivity. The successful production of this mAb raised three key points. (1) For immunized mice, a higher antibody titer is associated with a higher success rate. If the antibody titer is low, the levels of corresponding specific antibodies in spleen cells are reduced, leading to a lack of a specific and high-titer mAb in the process of cell fusion. (2) PEG was used for cell fusion and was added within 1～2 min. In addition, attention was paid to the formation of particles. Once particles appear, the remaining PEG must be quickly added. The time of PEG addition must be controlled to avoid excessive fusion.

(3) Myeloma cell activity must be ensured because cell morphology is not fully intact, and poor refraction will affect cell fusion, resulting in cell death. In this study, the preparation of the mAb laid a foundation for the establishment of a sandwich ELISA for the detection of H3-subtype AIV.

The sandwich ELISA included mainly the following five components: a polyclonal antibody, a sample, an mAb, a conjugated HRP antibody, and a substrate. The ELISA plate was coated with polyclonal antibody serum, and the antigen sample was added. If the sample had H3-subtype AIV, it combined with the polyclonal antibody to form an antigen-antibody complex. After this step, the wells were rinsed with PBST to remove unbound components. The H3 mAb was then added and reacted with the antigen in the complex, binding to the ELISA plate. An HRP-labeled goat anti-mouse antibody was added to interact with the mAb. The results were determined qualitatively by a color change and quantitatively by the reaction of the TMB substrate, as measured by an ELISA reader. This ELISA allowed samples to first react with the polyclonal antibody and then react with the mAb. The mAb recognized a single antigen epitope. The specificity of the mAb was better than that of the polyclonal antibody, but the binding range was not as wide as that of the polyclonal antibody. The initial combination of clinical samples with polyclonal antibodies for preliminary screening and the subsequent reaction of the samples with the mAb increased the detection rate and prevented false negatives.

The hybridoma cell producing H3-subtype AIV prepared in this study can steadily secrete homogeneous and high-titer antibodies and provide a very useful mAb for further research on H3-subtype AIV. In addition, the sandwich ELISA method is time saving, convenient, highly specific, sensitive and repeatable. Furthermore, the method can produce results without a specific instrument or the need for professional technical staff; it is performed according to a simple protocol and provides an effective method for the diagnosis of H3-subtype AIV.

References

[1] OSHANSKY C M, WONG S S, JEEVAN T, et al. Seasonal influenza vaccination is the strongest correlate of cross-reactive antibody responses in migratory bird handlers. MBio, 2014, 5(6): e2107.

[2] TONG S, LI Y, RIVAILLER P, et al. A distinct lineage of influenza A virus from bats. Proceedings of the National Academy of Sciences of the United States of America, 2012, 109(11): 4269-4274.

[3] TONG S, ZHU X, LI Y, et al. New world bats harbor diverse influenza A viruses. PLOS Pathogens, 2013, 9(10):e1003657.

[4] BAILEY E, LONG L P, ZHAO N, et al. Antigenic characterization of H3 subtypes of avian influenza A viruses from North America. Avian Disease, 2016, 60(1): 346-353.

[5] LUO S, XIE Z, XIE Z, et al. Surveillance of live poultry markets for low pathogenic avian influenza viruses in Guangxi Province, southern China, from 2012-2015. Scientific Reports, 2017, 7(1): 17577.

[6] WANG Z, YU J, THOMAS M, et al. Pre-exposure with influenza A virus A/WSN/1933(H1N1) resulted in viral shedding reduction from pigs challenged with either swine H1N1 or H3N2 virus. Veterinary Microbiology, 2019, 228: 26-31.

[7] HE W, LI G, ZHU H, et al. Emergence and adaptation of H3N2 canine influenza virus from avian influenza virus: An overlooked role of dogs in interspecies transmission. Transboundary and Emerging Diseases, 2019, 66(2): 842-851.

[8] KITCHEN R H, KEHLER W H, HENTHORNE J C. The 1963 equine influenza epizootic. Journal of the American Veterinary Medical Association, 1963, 143: 1108-1110.

[9] ALVES B E, WOODWARD A, RASH A, et al. Characterisation of the epidemic strain of H3N8 equine influenza virus responsible for outbreaks in South America in 2012. Virology Journal, 2016, 13: 45.

[10] SREENIVASAN C C, JANDHYALA S S, LUO S, et al. Phylogenetic analysis and characterization of a sporadic isolate of equine influenza A H3N8 from an unvaccinated horse in 2015. Viruses, 2018, 10(1): 31.

[11] SOLÓRZANO A, FONI E, CÓRDOBA L, et al. Cross-species infectivity of H3N8 influenza virus in an experimental

infection in swine. Journal of Virology, 2015, 89(22): 11190-11202.

[12] SONG D, KANG B, LEE C, et al. Transmission of avian influenza virus (H3N2) to dogs. Emerging Infectious Diseases, 2008, 14(5): 741-746.

[13] SHICHINOHE S, OKAMATSU M, SAKODA Y, et al. Selection of H3 avian influenza viruses with SAα2,6Gal receptor specificity in pigs. Virology, 2013, 444(1-2): 404-408.

[14] LI C, YU M, LIU L, et al. Characterization of a novel H3N2 influenza virus isolated from domestic ducks in China. Virus Genes, 2016, 52(4): 568-572.

[15] GUAN L, SHI J, KONG X, et al. H3N2 avian influenza viruses detected in live poultry markets in China bind to human-type receptors and transmit in guinea pigs and ferrets. Emerging Microbes Infections, 2019, 8(1): 1280-1290.

[16] CHEN L, RUAN F, SUN Y, et al. Establishment of sandwich ELISA for detecting the H7 subtype influenza A virus. Journal of Medical Virology, 2019, 91(6): 1168-1171.

[17] HUANG L, WEI Y, XIA D, et al. A broad spectrum monoclonal antibody against porcine circovirus type 2 for antigen and antibody detection. Applied Microbiology and Biotechnology, 2019, 103(8): 3453-3464.

[18] ZHANG L, LI Z, JIN H, et al. Development and application of a monoclonal antibody-based blocking ELISA for detection of antibodies to Tembusu virus in multiple poultry species. BMC Veterinary Research, 2018, 14(1): 201.

[19] DARCY E, LEONARD P, FITZGERALD J, et al. Purification of antibodies using affinity chromatography. Methods in Molecular Biology, 2011, 681: 369-382.

[20] PENG Y, XIE Z, LIU J, et al. Visual detection of H3 subtype avian influenza viruses by reverse transcription loop-mediated isothermal amplification assay. Virology Journal, 2011, 8: 337.

[21] TANG Q, WANG J, BAO J, et al. A multiplex RT-PCR assay for detection and differentiation of avian H3, H5, and H9 subtype influenza viruses and Newcastle disease viruses. Journal of Virological Methods, 2012, 181(2): 164-169.

[22] TENG Q, SHEN W, YAN D, et al. Development of a TaqMan MGB RT-PCR for the rapid detection of H3 subtype avian influenza virus circulating in China. Journal of Virological Methods, 2015, 217: 64-69.

[23] ITO M, YAMAYOSHI S, MURAKAMI K, et al. Characterization of mouse monoclonal antibodies against the HA of A (H7N9) influenza virus. Viruses, 2019, 11(2): 149.

Development of a double-antibody sandwich ELISA based on a monoclonal antibody against the viral NS1 protein for the detection of chicken parvovirus

Zhang Minxiu, Liao Jianqi, Xie Zhixun, Zhang Yanfang, Luo Sisi, Li Meng, Xie Liji, Fan Qing, Zeng Tingting, Huang Jiaoling, and Wang Sheng

Abstract

Chicken parvovirus (ChPV) infection can cause runting-stunting syndrome (RSS) in chickens. There is currently no commercially available vaccine for controlling ChPV, and ChPV infection in chickens is widespread globally. The rapid detection of ChPV is crucial for promptly capturing epidemiological data on ChPV. Two monoclonal antibodies (mAbs), 1B12 and 2B2, against the ChPV NS1 protein were generated. A double-antibody sandwich enzyme-linked immunosorbent assay (DAS-ELISA) was developed for detecting ChPV based on the mAb 1B12 and an anti-chicken polyclonal antibody against the ChPV NS1 protein. The detection limit for the ChPV recombinant pET32a-NS1 protein was approximately 31.2 ng/mL. A total of 192 throat and cloaca swab samples were analyzed for ChPV by the established DAS-ELISA and nested PCR methods. The concordance rate between the DAS-ELISA and the nested PCR method was 89.1%. The DAS-ELISA can detect the ChPV antigen without any cross-reaction with FAdV-4, FAdV-1, NDV, AIV, MS, CIAV, aMPV, EDSV, IBV, or AGV2. The method also has high repeatability, with a coefficient of variation (CV) of less than 5%. These findings indicate that the DAS-ELISA exhibits high accuracy, good sensitivity, and specificity, making it suitable for viral detection, field surveillance, and epidemiological studies.

Keywords

chicken parvovirus, NS1 protein, monoclonal antibodies, DAS-ELISA

Introduction

Chicken parvovirus (ChPV) is a nonenveloped, single-stranded DNA virus that belongs to the genus *Aveparvovirus* within the subfamily Parvovirinae of the family Parvoviridae[1]. ChPV was first discovered in the feces of young chickens with runting-stunting syndrome (RSS) in Hungary in 1984[2]. ChPV can cause watery diarrhea and growth retardation in broiler chicks[3, 4]. The size of the ChPV genome is approximately 5 kb, and it contains three open reading frames (ORFs) that encode four proteins: two structural proteins (VP1 and VP2) and two nonstructural proteins (NS1 and NP1)[5]. The VP1 protein is composed of 675 amino acids and may play an essential role in the process of ChPV entering cells and eventually releasing viruses. The VP2 protein is a capsid protein composed of 536 amino acids and is associated with functions such as DNA replication and virus packaging[6]. The NP1 protein contains approximately 101 amino acids. However, the function of the NP1 protein in ChPV is unclear. Previous studies have shown that the NP1 protein is a nonstructural protein necessary for the efficient replication of viral DNA and control of capsid protein expression in human Boca virus, which also belongs to the same family, Parvoviridae[7]. NS1 is the most important nonstructural protein and consists of 694 amino acids. This protein is a nuclear phosphoprotein that is primarily located in the cell nucleus and is involved in viral replication and assembly[8].

The pathological characteristics and clinical symptoms of ChPV-infected chickens are similar to those of chickens infected by chicken astrovirus (CAstV), avian rotavirus (AvRV), and picornavirus[9-11], leading to difficulties in differential diagnosis based on clinical features. Therefore, it is necessary to develop a detection method for identifying ChPV. The NS1 gene is highly conserved among chicken parvoviruses and is often used as a target gene for the detection of ChPV nucleic acids[12, 13]. Currently, polymerase chain reaction (PCR) and real-time PCR (RT-qPCR) are two commonly used methods for determining the presence of ChPV[12, 13]. However, serological assays for detecting ChPV are rare. Enzyme-linked immunosorbent assays (ELISAs) are simple and cost-efficient serological assays that do not require viral DNA or RNA extraction. ELISA-based methods have been developed for pathogen detection[14-16].

RSS is an enteric disease in young poultry characterized by clinical symptoms such as diarrhea, depression, decreased weight gain, and growth delay, causing significant economic losses in the poultry industry[17]. The etiological agents that cause RSS are complex. The occurrence of RSS in poultry has been described as possibly being related to infection by one or more poultry enteroviruses, including poultry parvovirus, CAstV, AvRV, picornavirus, avian reovirus (ARV), and infectious bronchitis virus (IBV)[9-11, 18-20]. ChPV has been detected in chickens with RSS in a few countries, such as India, Brazil, Republic of Korea, Poland, and China[21-25]. Zsak et al.[4] and Nuñez et al.[26] showed that SPF chicks infected with ChPV exhibit obvious clinical symptoms of RSS. In addition, ChPV infections are prevalent in healthy chickens[27]. Currently, there is no vaccine available to prevent or control ChPV infections, so it is essential to detect the virus to evaluate the impact of ChPV infections.

In this study, two monoclonal antibodies (mAbs) targeting the NS1 protein of ChPV were generated, and a double-antibody sandwich ELISA (DAS-ELISA) was used to detect ChPV based on a mAb and polyclonal antibody. The established DAS-ELISA was sensitive and specific for detecting ChPV infection, providing a new tool for ChPV surveillance.

Materials and methods

Cells, clinical samples and viruses

SP2/0 myeloma cells and chicken liver cancer cells (LMHs) were preserved by the Guangxi Veterinary Research Institute (Guangxi, China) ; 50 ChPV-negative throat and cloacal swab samples were collected from specific pathogen-free (SPF) chickens and used to determine the cut-off value; and 192 throat and cloacal swab samples were collected from chickens in live poultry markets in Guangxi, China.

Newcastle disease virus (NDV), *Mycoplasma gallisepticum* (MS), fowl adenovirus serotype 4 (FAdV-4), fowl adenovirus serotype 1 (FAdV-1), chicken infectious anemia virus (CIAV), chicken infectious bronchitis virus (IBV), avian metapneumonia virus (aMPV), H9N2 subtype avian influenza virus (AIV), avian egg drop syndrome virus (EDSV), and an avian circovirus 2 (AGV2)-positive throat and cloacal swab sample were preserved in the laboratory and used for testing the specificity of the DAS-ELISA. The details of these pathogens are shown in Table 5-42-1.

Table 5-42-1 Details of pathogens used for testing the specificity of the DAS-ELISA

Pathogens	Strain names	Source	Accession number	Year	Cultures/Tissues	Mean $OD_{450\ nm}$ values
NDV	GX6/02	GVRI		2002	Chicken embryo allantoic fluid	0.032
NDV	Duck/China/ Guangxi19/2011	GVRI	KC920893	2011	Chicken embryo allantoic fluid	0.036
NDV	Chicken/Guangxi/ B14/2021	GVRI		2021	Chicken embryo allantoic fluid	0.039
MS	PMS156	UC			MS liquid culture medium	0.067
FAdV-1	GX201802	GVRI	MZ322953	2018	LMH cells cultures	0.06
FAdV-1	GX201803	GVRI	MZ322954	2018	LMH cells cultures	0.051
FAdV-4	GX2018-07	GVRI	MN577983	2018	LMH cells cultures	0.028
FAdV-4	GX2017-06	GVRI	MN577982	2017	LMH cells cultures	0.031
FAdV-4	GX2019-09	GVRI	MN577985	2019	LMH cells cultures	0.032
CIAV	GXC060821	GVRI	JX964755	2006	MDCC-MSB1 cell cultures	0.029
IBV	GXIB/02	GVRI		2002	Chicken embryo allantoic fluid	0.043
aMPV	MN-10	UC			Chicken embryo allantoic fluid	0.050
AIV	A/Chicken/Guangxi/ LZ066C/2020 (H9N2)	GVRI		2020	Chicken embryo allantoic fluid	0.100
AIV	A/Chicken/Guangxi/ C1228/2015 (H9N2)	GVRI	KX185890	2015	Chicken embryo allantoic fluid	0.067
AIV	A/Chicken/Guangxi/ C227/2015 (H9N2)	GVRI	KX130848	2015	Chicken embryo allantoic fluid	0.058
EDSV	GEV	GVRI		1995	Chicken embryo allantoic fluid	0.045
AGV2	AGV2-GX19010	GVRI	MW404236	2019	Throat and cloaca swab sample	0.017
ChPV infectious clone	GX-CH-PV-21	GVRI	MG602511	2016	LMH cells cultures	0.895

Note: GVRI, Guangxi Veterinary Research Institute, China; UC, University of Connecticut, USA.

Full-length infectious plasmid of ChPV, recombinant pET32a-NS1 protein, and polyclonal antibodies

The full-length infectious plasmid of ChPV (pBluescript Ⅱ SK (+)-ChPV), which contains the whole ORF of the ChPV strain GX-CH-PV-21 (GenBank: MG602511), was transfected into LMH cells (to obtain the ChPV infectious clone) for use in the specific DAS-ELISA experiment. LMH cells infected with the pBluescript Ⅱ SK (+)-ChPV plasmid were also subjected to immunofluorescence analysis (IFA) and Western blotting (WB). The purified recombinant pET32a-NS1 protein (104 kDa) containing the whole NS1 protein from strain GX-CH-PV-21 was preserved at −80 ℃ before use for animal immunity and WB analysis with mAbs. The purified polyclonal antibodies (1.5 mg/mL) were preserved at −80 ℃ and produced by immunizing SPF chickens with recombinant pET32a-NS1 protein[28].

Production and identification of mAbs against the ChPV-NS1 protein

Eight-week-old BALB/c mice were immunized four times with purified recombinant pET32a-NS1 protein. The mice were subcutaneously immunized with 100 μg/mouse recombinant pET32a-NS1 protein emulsified with complete Freund's Adjuvant (Sigma-Aldrich, St. Louis, MO, USA) at the first immunization. The second and third immunizations were performed with the same dose of recombinant pET32a-NS1 protein emulsified with incomplete Freund's adjuvant at 21 and 35 days after the first immunization. After 49 days, the antibody titers against the recombinant pET32a-NS1 protein in the immunized mice were determined via indirect ELISA. Mice with high antibody titers were immunized with 1.0 mL of recombinant pET32a-NS1 protein solution (400 μg/mL) without adjuvant via intraperitoneal injection at 56 days (the fourth immunization). On the third day after the fourth immunization, the splenocytes of the mice were harvested. The fusion of splenocytes with SP2/0 myeloma cells was performed according to methods previously described by Luo et al.[29]. Positive hybridoma cells with high antibody titers against the recombinant pET32a-NS1 protein were screened by indirect ELISA. The positive hybridoma cells were subcloned three times, and the antibody titers of the hybridoma supernatants of each subclone were determined using an indirect ELISA. Finally, positive hybridoma cells with high antibody titers were selected and injected into the abdominal cavities of the mice. The ascitic fluid secreted after the injection of positive hybridoma cells was harvested and purified according to methods previously described by Wang et al.[14]. The mAb isotypes were determined using a commercial mouse mAb isotyping kit (Sigma-Aldrich, St. Louis, MO, USA).

The steps of the indirect ELISA mentioned above were as follows: (1) 96-well microtiter plates were coated with 100 μL/well recombinant pET32a-NS1 protein at a concentration of 5 μg/mL in phosphate-buffered saline (PBS) and then incubated at 37 ℃ for 1 h. (2) The plates were subsequently blocked with blocking buffer (5% skim milk powder in PBS) at 37 ℃ for 1 h. (3) The plates were washed three times with PBS containing 0.1% Tween-20 (PBST). (4) Serum samples from immunized mice or the supernatant of hybridoma cells were added to the wells of the plates. Serum samples from nonimmunized mice were used as negative controls. The plates were incubated at 37 ℃ for 1 h and then washed again. (5) HRP-labelled goat anti-mouse IgG (Beyotime Biotechnology Co., Ltd., Shanghai, China) (100 μL/well) at a dilution of 1 ∶ 2 000 was added to each well, and the plates were incubated at 37 ℃ for 45 min. (6) After a washing step, 3, 3', 5, 5'-tetramethylbenzidine (TMB) solution (100 μL/well) was added, and the plates were incubated in the dark at room temperature for 10 min. (7) Then, 100 μL/well sulfuric acid (2 M) was added, and the absorbance at 450 nm was measured immediately using an ELISA plate reader (Shanghai Kehua Bio-Engineering Co., Ltd., Shanghai, China). When the $OD_{450\ nm}$ of the sample/$OD_{450\ nm}$ value of the negative control was greater than 2.1, the sample was considered positive.

WB analysis and IFA

WB analysis and IFA were performed to determine the reactivity and specificity of the mAbs against the recombinant pET32a-NS1 protein and the NS1 protein expressed in the LMH cells. (1) LMH cells were transfected with the plasmid pBluescript Ⅱ SK (+)-ChPV. After 3 days, the LMH cells were collected and lysed for SDS-PAGE. The NS1 protein expressed in LMH cells was approximately 79 kDa in length. (2) In addition, the purified recombinant pET32a-NS1 protein (104 kDa) was subjected to SDS-PAGE. (3) Then, the proteins in the gel were transferred onto a polyvinylidene fluoride (PVDF) membrane. The membrane was blocked with 5% skim milk overnight at 4 ℃. The membrane was subsequently washed three times with PBST. (4) Primary antibodies (mAbs) (1 ∶ 1 000 dilution) against the NS1 protein harvested as described in

Section *Full-length infectious plasmid of ChPV, recombinant pET32a-NS1 protein and polyclonal antibodies* were added to the membrane and incubated at 37 ℃ for 1 h. (5) The membrane was washed three times. Then, alkaline phosphatase (AP)-labelled goat anti-mouse IgG (1 ∶ 2 000 dilution) was used as a secondary antibody and was added, and the membrane was incubated at 37 ℃ for 1 h. (6) The membrane was washed and then stained with a commercial BCIP/NBT alkaline phosphatase color development kit (Beyotime Biotechnology Co., Ltd., Shanghai, China).

LMH cells cultured in 6-well plates were transfected with the plasmid pBluescript Ⅱ SK (+)-ChPV. LMH cells were fixed with 4% paraformaldehyde three days after transfection. The mAbs (1 ∶ 500 dilution) were incubated with LMH cells at 37 ℃ for 1 h. A FITC-labelled goat anti-mouse IgG antibody (1 ∶ 500) was added after the cells were washed three times with PBST, and the plate was incubated at 37 ℃ for 1 h. The LMH cells were subsequently washed again and observed via fluorescence microscopy.

Selection of the capture antibody

The mAbs were used as the capture antibodies and were coated on a 96-well microtiter plate to determine the optimal capture antibody by the DAS-ELISA. For this purpose, 1 ∶ 1 000 dilutions of 2.0 mg/mL stocks of purified mAbs were coated (100 μL/well) on a 96-well microtiter plate. The recombinant pET32a-NS1 protein in PBS (5 μg/mL) and the cell suspension of the ChPV infectious clone were used as the sandwich antigen, and the polyclonal antibody against the NS1 protein from the SPF chickens was used as the detection antibody. PBS and a cell suspension of negative LMH cells were used as negative controls. Each sample was tested in triplicate. The absorbance at 450 nm of the mixture in the wells of the plate was measured. If the $OD_{450\ nm}$ value of the sample/$OD_{450\ nm}$ value of the negative control was greater than 2.1, the sample was considered positive.

Development of an NS1-DAS-ELISA for ChPV detection

The purified mAb and polyclonal antibody were used as the capture and detection antibodies, respectively. The optimal concentrations of the mAbs and polyclonal antibodies were determined via checkerboard titration. The steps of NS1-DAS-ELISA were as follows: (1) 1 ∶ 1 000, 1 ∶ 2 000, 1 ∶ 4 000, 1 ∶ 8 000, 1 ∶ 16 000, and 1 ∶ 32 000 dilutions of 2.2 mg/mL stocks of mAbs were coated (100 μL/well) on the 96-well microtiter plate at 4 ℃ for 18 h. The plate was subsequently blocked with blocking buffer (5% skim milk powder in PBS) at 37 ℃ for 45 min. (2) Then, the plate was washed three times with PBST, after which, 100 μL/well recombinant pET32a-NS1 protein in PBS (5 μg/mL) was added. (3) According to the checkerboard titration, 1 ∶ 1 000, 1 ∶ 2 000, 1 ∶ 4 000, 1 ∶ 8 000, 1 ∶ 16 000, 1 ∶ 32 000, 1 ∶ 64 000, and 1 ∶ 128 000 dilutions of 1.5 mg/mL stocks of polyclonal antibodies were added to each well after washing with PBST, and the plate was then incubated at 37 ℃ for 45 min. (4) The plate was washed again. HRP-labelled goat anti-chicken IgG (Beyotime Biotechnology Co., Ltd., Shanghai, China) (100 μL/well) at a dilution of 1 ∶ 2 000 was added, and the mixture was incubated at 37 ℃ for 45 min. (5) Then, the wells were washed with PBST, and TMB solution (100 μL/well) was added. The mixture was incubated in the dark for 11 min at 25 ℃. The color reaction was stopped with 100 μL of sulfuric acid (2 M, 100 μL/well), after which the absorbance at 450 nm was measured. (6) The ideal concentrations of the capture antibody and detection antibody were determined according to the above 5 steps. Then, the concentration of HRP-labelled goat anti-chicken IgG (diluted 1 ∶ 500, 1 ∶ 1 000, 1 ∶ 2 000, 1 ∶ 4 000 and 1 ∶ 8 000) was further optimized.

NS1-DAS-ELISA positive and negative cut-off values

A total of 50 ChPV-negative throat and cloacal swab samples were collected from SPF chickens. These samples were analyzed by NS1-DAS-ELISA, and the $OD_{450\ nm}$ values of 50 samples were obtained. The critical value was x+3SD (where "x" represents the mean $OD_{450\ nm}$ value of 50 samples, and "SD" represents the standard deviation).

Specificity and sensitivity of the NS1-DAS-ELISA

The specificity of the NS1-DAS-ELISA was tested using suspensions of LMH cells transfected with pBluescript Ⅱ SK (+)-ChPV, NDV, MS, FAdV-1, FAdV-4, CIAV, IBV, aMPV, H9N2 subtype AIV, EDSV, and AGV2. Each pathogen was tested in triplicate.

Due to the lack of available ChPV isolates and the low viral titer of the ChPV infectious clones in the LMH cells, the recombinant pET32a-NS1 protein was used for sensitivity analysis. The recombinant pET32a-NS1 protein was diluted to concentrations of 1 000, 500, 250, 125, 62.5, 31.2, 15.6, 7.8, and 0 ng/mL with PBS. The sensitivity of the NS1-DAS-ELISA was evaluated with different recombinant pET32a-NS1 protein concentrations (100 μL/well), and each concentration of the recombinant pET32a-NS1 protein was tested in triplicate.

Repeatability analysis of the NS1-DAS-ELISA

The repeatability of the NS1-DAS-ELISA was tested using different concentrations of the recombinant pET32a-NS1 protein (at dilutions of 250 ng/mL, 500 ng/mL, 750 ng/mL, and 1 000 ng/mL), and the ChPV-negative sample from SPF chickens was used as a negative control. The intra-and interassay repeatability tests were performed using the same batch of the ELISA plates or three different batches of the ELISA plates, respectively. The same concentration of the recombinant pET32a-NS1 protein and the negative control samples were tested in triplicate. The means of the $OD_{450\ nm}$ values, standard deviations, and percent coefficients of variation (% CVs) were calculated with SPSS software version 22.0 (IBM SPSS Inc., Chicago, IL, USA).

Comparison of the NS1-DAS-ELISA and nested PCR methods

The established NS1-DAS-ELISA and a nested PCR method[30] were used to analyze 192 throat and cloacal swab samples. The primers used for nested PCR were designed based on the NS1 gene. For nested PCR, the primer pair used in the first round of amplification were 661F (5'-GGTACAAGATATGCTAGATTT-3') and 1073R (5'-CGGATGGCTAAATTATCATCT-3'). In the second round, the primer pair used was 718F (5'-CCATCGCAGGAATTAACTCCAG-3') and 1043R (5'-GTGTCAACATCTCCATGTATTG-3'). The first and second rounds of the nested PCR amplification procedure were as follows: 95 ℃ for 3 min; 30 cycles of 95 ℃ for 1 min, 53 ℃ for 30 s, and 72 ℃ for 1 min; and 72 ℃ for 5 min. The detection results of the two methods were comparatively analyzed.

Results

Production and identification of mAbs against the ChPV-NS1 protein

Two positive hybridoma cell lines capable of secreting mAbs against the NS1 protein, namely, 1B12 and 2B2, were obtained. The titers of the mAbs (1B12 and 2B2) in the mouse ascites fluid were 1 : 1.6×10^7 and

1 : 4.1×10^6, respectively. The isotyping assay showed that 1B12 and 2B2 are IgG1 with κ light chains (Table 5-42-2).

Table 5-42-2 $OD_{450\ nm}$ values of the isotyping assay for mAbs (1B12 and 2B2)

mAbs	IgG1	IgG2a	IgG2b	IgG2c	IgG3	IgM	Kappa	Lambda
1B12	1.023	0.112	0.088	0.089	0.093	0.130	0.521	0.076
2B2	0.744	0.098	0.201	0.176	0.101	0.128	0.634	0.099

WB analysis and IFA of the mAbs

The binding abilities of mAbs 1B12 and 2B2 were verified using WB and IFA. The NS1 proteins used in the WB analysis were the recombinant pET32a-NS1 protein (104 kDa) expressed in *Escherichia coli* and NS1 protein (79 kDa) expressed in LMH cells (Figure 5-42-1 A). IFA was performed to detect NS1 protein expression in LMH cells transfected with the ChPV infectious plasmid pBluescript Ⅱ SK (+)-ChPV (Figure 5-42-1 B).

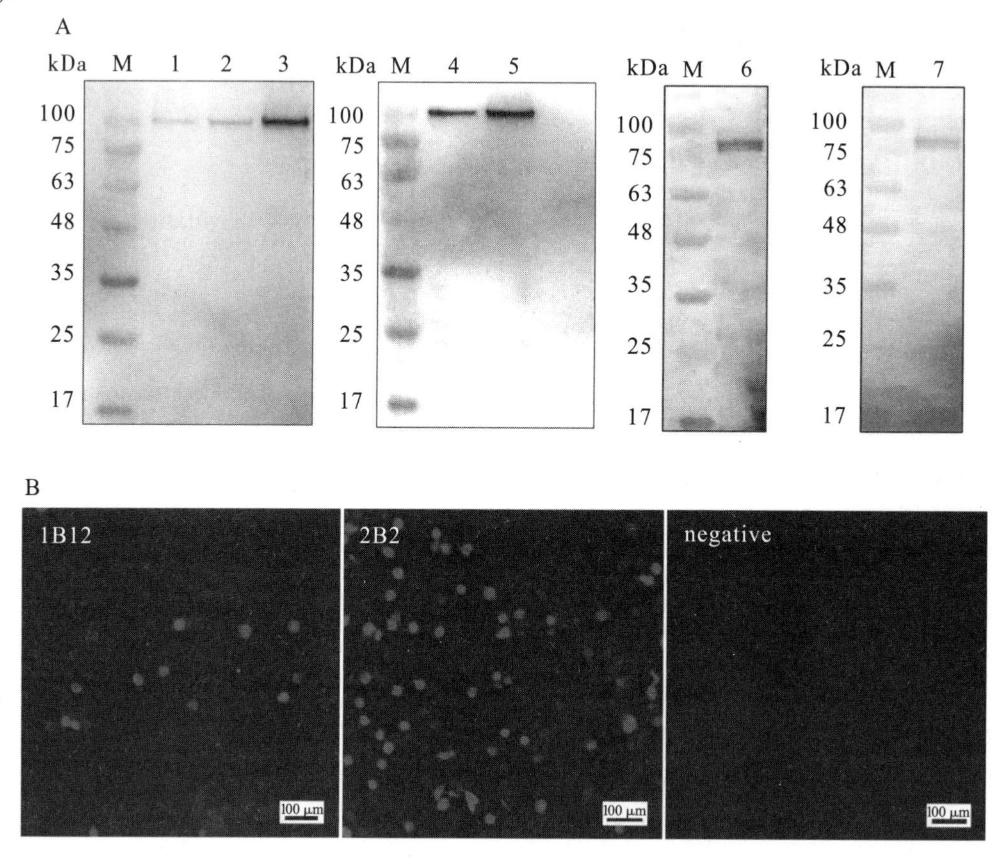

A: mAb 1B12 (lane 1, lane 2, and lane 3) reacted with the 104 kDa recombinant pET32a-NS1 protein, mAb 2B2 (lane 4 and lane 5) reacted with the 104 kDa recombinant pET32a-NS1 protein, the mAbs 1B12 and 2B2 (lane 6 and lane 7) reacted with the 79 kDa NS1 protein expressed in LMH cells. B: IFA was performed on LMH cells transfected with the ChPV infectious plasmid pBluescript II SK(+)-ChPV, Scale bars, 100 μm (color figure in appendix).

Figure 5-42-1 The reactivity of the monoclonal antibodies

Selection of the capture antibody

The mAbs 1B12 and 2B2 were used as the capture antibodies in the DAS-ELISA. The DAS-ELISA was performed to determine the optimal capture antibody. The results are shown in Table 5-42-3. The mAb 1B12 had a higher $OD_{450\ nm}$ value than the mAb 2B2 (Table 5-42-3), and the titer of the mAb 1B12 was also greater than that of the mAb 2B2 (see Section *Production and identification of mAbs against the ChPV-NS1 protein*). Therefore, the mAb 1B12 was chosen as the capture antibody for the development of the NS1-DAS-ELISA.

Table 5-42-3　The $OD_{450\ nm}$ values of 1B12 and 2B2 used as the capture antibodies in the DAS-ELISA

mAbs	Mean $OD_{450\ nm}$ values			
	Recombinant pET32a-NS1 protein	PBS control	ChPV infectious clone	Negative LMH cells
1B12	1.025	0.056	0.806	0.069
2B2	0.648	0.073	0.450	0.102

Development of the NS1-DAS-ELISA

The optimal concentrations of the capture antibody, detection antibody, and HRP-labelled goat anti-chicken IgG were as follows: a 1 ∶ 2 000 dilution of a 2.2 mg/mL stock of mAbs was applied to the ELISA plate, and the mixture was incubated at 4 ℃ for 18 h. Polyclonal antibodies (1 ∶ 4 000 dilutions of 1.5 mg/mL stocks) were added after washing the clinical samples. The HRP-labelled goat anti-chicken IgG (100 μL/well) was diluted to a concentration of 1 ∶ 2 000.

Cut-off values for the NS1-DAS-ELISA

The $OD_{450\ nm}$ values of 50 clinically negative ChPV samples were determined by the optimal NS1-DAS-ELISA protocol to evaluate the cut-off value of the assay. The mean (x) $OD_{450\ nm}$ of the 50 samples was 0.088, the standard deviation (SD) was 0.017, and the critical value was 0.139 according to the formula x + 3SD. If the $OD_{450\ nm}$ of the sample was greater than or equal to 0.139, the sample was considered a ChPV-positive sample. A value less than 0.139 was considered to indicate a negative sample.

Specificity of the NS1-DAS-ELISA

The specificities of the supernatants of LMH cells transfected with the plasmid pBlue-script Ⅱ SK (+)–ChPV (ChPV infectious clone), NDV, MS, FAdV-1, FAdV-4, CIAV, IBV, aMPV, H9N2 subtype AIV, or EDSV and an AGV2-positive sample were tested via the NS1-DAS-ELISA. PBST was used as a negative control. As shown in Table 5-41-1, $OD_{450\ nm}$ values less than 0.139 for ten pathogens (NDV, MS, FAdV-1, FAdV-4, CIAV, IBV, aMPV, H9N2 subtype AIV, EDSV, and AGV2) were considered to indicate a negative sample (Table 5-41-1), which suggested that there was no cross-reactivity with NDV, MS, FAdV-1, FAdV-4, CIAV, IBV, aMPV, H9N2 subtype AIV, EDSV, or AGV2 with the NS1-DAS-ELISA method. The NS1-DAS-ELISA detected only the ChPV-positive supernatant ($OD_{450\ nm}$=0.895) (Table 5-41-1), indicating that the NS1-DAS-ELISA was highly specific for ChPV detection.

Sensitivity of the NS1-DAS-ELISA

The sensitivity of the NS1-DAS-ELISA was evaluated with different recombinant pET32a-NS1 protein concentrations (at dilutions of 1 000, 500, 250, 125, 62.5, 31.2, 15.6, 7.8, and 0 ng/mL). The limit of detection was 31.2 ng/mL for the recombinant pET32a-NS1 protein (Figure 5-42-2).

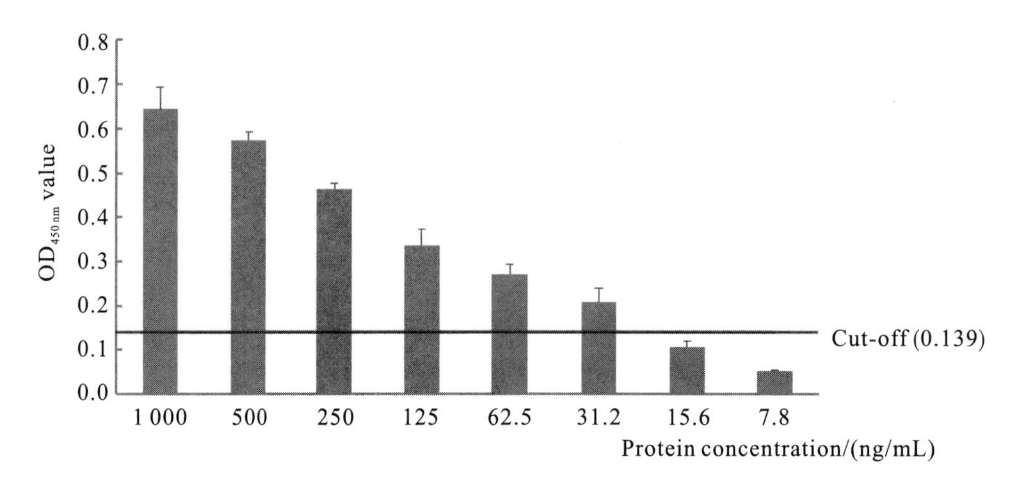

Figure 5-42-2 Determination of the limit of detection of the NS1-DAS-ELISA using the recombinant pET32a-NS1 protein antigen

Repeatability analysis of the NS1-DAS-ELISA

To evaluate the repeatability of the NS1-DAS-ELISA, the $OD_{450\ nm}$ was measured for different concentrations of the recombinant pET32a-NS1 protein (at dilutions of 250 ng/mL, 500 ng/mL, 750 ng/mL, and 1 000 ng/mL) using the NS1-DAS-ELISA. The CVs of the intra-and interassay data were less than 5%, which indicated that the NS1-DAS-ELISA has good repeatability (Table 5-42-4).

Table 5-42-4 Intra-and interassay repeatability of the NS1-DAS-ELISA

Concentrations of recombinant pET32a-NS1 protein	Intra-assay			Inter-assay		
	Mean $OD_{450\ nm}$	SD	CV	Mean $OD_{450\ nm}$	SD	CV
250 ng/mL	0.50	0.018	3.5%	0.511	0.017	3.3%
500 ng/mL	0.51	0.023	4.6%	0.492	0.012	2.4%
750 ng/mL	0.70	0.027	3.8%	0.643	0.032	4.9%
1 000 ng/mL	0.77	0.018	2.3%	0.671	0.032	4.7%
0 ng/mL	0.06	0.002	3.4%	0.069	0.003	4.9%

Comparison of the NS1-DAS-ELISA and nested PCR methods

A total of 192 clinical samples were tested for ChPV by the established NS1-DAS-ELISA and a nested PCR method. A total of 141 samples were ChPV positive according to NS1-DAS-ELISA. Thirty samples were negative and 162 samples were positive according to the nested PCR method (Table 5-42-5). The coincidence rate between the NS1-DAS-ELISA and the nested PCR method for ChPV positivity was 87.0% (141/162). The total coincidence rate of the two detection methods was 89.1% (((141 + 30) /192) × 100%).

Table 5-42-5 Comparison of the NS1-DAS-ELISA and nested PCR for the detection of ChPV in clinical samples

		Nested PCR		
		Positive	Negative	Total
NS-DAS-ELISA	Positive	141	0	141
	Negative	21	30	51
	Total	162	30	192

Discussion

Minute virus of mice (MVM), goose parvovirus (GPV), and porcine parvovirus (PPV) belong to the family Parvoviridae. The NS1 protein is a multifunctional protein in MVM, GPV, and PPV and is associated with the replication of viral DNA, the induction of host cell apoptosis, and the induction of inflammatory reactions[31-34]. The NS1 protein is a nonstructural protein of ChPV that is highly conserved among ChPV strains[13]. A report by Nuñez et al.[35]showed that ChPV could be isolated through chicken embryo inoculation. We attempted to isolate ChPV in our laboratory according to the method described by Nuñez et al.[35]. In addition, we used different cell lines for ChPV isolation, but all the methods used failed. An appropriate cell line for the isolation of ChPV remains to be identified; therefore, research on the function of the NS1 protein in ChPV is very limited. Although the function of the NS1 protein in ChPV is unclear, the preparation of mAbs against the NS1 protein is a prerequisite for understanding protein function and for detecting ChPV. The mAbs against the NS1 protein of canine parvovirus (CPV) and MVM have been used in ELISA, Western blot, and IFA analyses[36, 37]. The mAbs against the NS1 protein of CPV were prepared and used to analyze the distribution of the NS1 protein in cells during the infection phase[36]. A mAb against the NS1 protein was used in IFA and WB analysis by Larsen et al. to determine how MVM localizes to cellular sites of DNA damage[37]. The above data indicate that the development of mAbs against viral proteins is crucial for comprehending viral protein functions and creating specific detection tools for viruses. Two mAbs (1B12 and 2B2) against the NS1 protein of ChPV were successfully produced in this study. The titers of 1B12 and 2B2 were 1.6×10^7 and 4.1×10^6, respectively. The Western blot and IFA analyses indicated that 1B12 and 2B2 could react specifically with the recombinant pET32a-NS1 protein expressed in prokaryotes, as well as the NS1 protein expressed in LMH cells. This study represents a phased achievement, and the prepared mAbs can be used for Western blot and IFA analyses in the future to determine the function of the NS1 protein of ChPV.

Currently, ChPV infection is widespread among commercial chickens, including healthy chickens. ChPV detection has been conducted mainly via nucleic acid testing. However, nucleic acid extraction and PCR-based methods require specialized instruments and strong technical expertise for detection and are not suitable for the rapid analysis of numerous samples on farms. ELISAs are fast and convenient assays that do not involve complex processing of analyzed samples. In addition, ELISAs do not require expensive instruments, are low cost and are easy to perform on farms[38]. In this study, a DAS-ELISA based on a mAb against the ChPV NS1 protein was developed for ChPV detection. The mAb 1B12, which had a high antibody titer, was selected as the capture antibody, and the optimal concentrations of the mAb and polyclonal antibody were determined. The mAb 1B12 only reacted with ChPV according to the results of the NS1-DAS-ELISA, and no cross-reactivity with other pathogens, such as NDV, MS, FAdV-1, FAdV-4, CIAV, IBV, aMPV, H9N2 subtype AIV, EDSV, or AGV2, was observed. These findings indicate that the established NS1-DAS-ELISA has a high specificity for the detection of ChPV. The limit of detection was 31.2 ng/mL for the recombinant pET32a-NS1 protein based on the optimal conditions for the DAS-ELISA. The NS1-DAS-ELISA can be applied to analyze numerous clinical samples and is the first DAS-ELISA for the detection of the ChPV antigen.

A total of 192 samples were analyzed by the NS1-DAS-ELISA and nested PCR method in this study. The results for 21 samples were inconsistent between the two methods; these samples were ChPV-positive according to nested PCR but negative according to the NS1-DAS-ELISA. It is possible that the low concentrations of ChPV in the samples were not detected by the NS1-DAS-ELISA. In addition, there were two limitations associated with the NS1-DAS-ELISA. No ChPV isolates were available for the DAS-ELISA sensitivity test in this study, and the recombinant pET32a-NS1 protein was used as a standard in the sensitivity test, which may lead to bias in the analysis of clinical infection samples by the NS1-DAS-ELISA. Another drawback is that it was unknown whether the NS1-DAS-ELISA could also detect turkey parvovirus (TuPV) because of the high homology of the NS1 protein between TuPV and ChPV.

For the initial establishment of the NS1-DAS-ELISA, chicken polyclonal antibodies and mAbs were used as capture antibodies and detection antibodies, respectively. The $OD_{450\ nm}$ of the ChPV positive control was less than 0.2, suggesting that the specific binding site between the NS1 protein and the mAb was blocked. This could be due to the preferential binding of polyclonal antibodies to these specific sites, preventing the mAb from binding to the NS1 protein. Therefore, the mAbs and polyclonal antibodies were exchanged for capture antibodies and detection antibodies, respectively, in subsequent experiments, and significant increases in the $OD_{450\ nm}$ and P/N values of the ChPV-positive control were observed.

In summary, two specific mAbs (1B12 and 2B2) against the NS1 protein of ChPV were screened, and 1B12 was used as the capture antibody for developing an NS1-DAS-ELISA to detect ChPV. This NS1-DAS-ELISA offers a simple and low-cost tool with good specificity and sensitivity for the diagnosis of ChPV infection in chickens.

References

[1] COTMORE S F, AGBANDJE-MCKENNA M, CHIORINI J A, et al. The family Parvoviridae. Arch Virol, 2014, 159(5): 1239-1247.

[2] KISARY J, NAGY B, BITAY Z. Presence of parvoviruses in the intestine of chickens showing stunting syndrome. Avian Pathol, 1984, 13(2): 339-343.

[3] KISARY J. Experimental infection of chicken embryos and day-old chickens with parvovirus of chicken origin. Avian Pathol, 1985, 14(1): 1-7.

[4] ZSAK L, CHA R M, DAY J M. Chicken parvovirus-induced runting-stunting syndrome in young broilers. Avian Dis, 2013, 57(1): 123-127.

[5] DAY J M, ZSAK L. Determination and analysis of the full-length chicken parvovirus genome. Virology, 2010, 399(1): 59-64.

[6] KOO B S, LEE H R, JEON E O, et al. Genetic characterization of three novel chicken parvovirus strains based on analysis of their coding sequences. Avian Pathol, 2015, 44(1): 28-34.

[7] HAO S, ZHANG J, CHEN Z, et al. Alternative polyadenylation of human bocavirus at its 3' end is regulated by multiple elements and affects capsid expression. J Virol, 2017, 97(3): e02026-16 .

[8] IHALAINEN T O, NISKANEN E A, JYLHAVA J, et al. Dynamics and interactions of parvoviral NS1 protein in the nucleus. Cell Microbiol, 2007, 9(8): 1946-1959.

[9] KANG K I, LINNEMANN E, ICARD A H, et al. Chicken astrovirus as an aetiological agent of runting-stunting syndrome in broiler chickens. J Gen Virol, 2018, 99(4): 512-524.

[10] KANG K I, EL-GAZZAR M, SELLERS H S, et al. Investigation into the aetiology of runting and stunting syndrome in chickens. Avian Pathol, 2012, 41(1): 41-50.

[11] DE OLIVEIRA L B, STANTON J B, ZHANG J, et al. Runting and stunting syndrome in broiler chickens: histopathology and association with a novel picornavirus. Vet Pathol, 2021, 58(1): 123-135.

[12] NUNEZ L F, SANTANDER-PARRA S H, CHAIBLE L, et al. Development of a sensitive real-time fast-qPCR based on SYBR® Green for detection and quantification of chicken parvovirus (ChPV). Vet Sci, 2018, 5(3): 69.

[13] ZSAK L, STROTHER K O, DAY J M. Development of a polymerase chain reaction procedure for detection of chicken and turkey parvoviruses. Avian Dis, 2009, 53(1): 83-88.

[14] WANG W, LI J, FAN B, et al. Development of a novel double antibody sandwich ELISA for quantitative detection of

porcine deltacoronavirus antigen. Viruses, 2021, 13(12): 2403.
[15] SHRIVASTAVA N, KUMAR J S, YADAV P, et al. Development of double antibody sandwich ELISA as potential diagnostic tool for rapid detection of Crimean-Congo hemorrhagic fever virus. Sci Rep, 2021, 11(1): 14699.
[16] GU K, SONG Z, ZHOU C, et al. Development of nanobody-horseradish peroxidase-based sandwich ELISA to detect Salmonella Enteritidis in milk and in vivo colonization in chicken. J Nanobiotechnology, 2022, 20(1): 167.
[17] BARNES H J, GUY J S, VAILLANCOURT J P. Poult enteritis complex. Rev Sci Tech, 2000, 19(2): 565-588.
[18] KAPGATE S S, KUMANAN K, VIJAYARANI K, et al. Avian parvovirus: classification, phylogeny, pathogenesis and diagnosis. Avian Pathol, 2018, 47(6): 536-545.
[19] DEVANEY R, TRUDGETT J, TRUDGETT A, et al. A metagenomic comparison of endemic viruses from broiler chickens with runting-stunting syndrome and from normal birds. Avian Pathol, 2016, 45(6): 616-629.
[20] PANTIN-JACKWOOD M J, DAY J M, JACKWOOD M W, et al. Enteric viruses detected by molecular methods in commercial chicken and turkey flocks in the United States between 2005 and 2006. Avian Dis, 2008, 52(2): 235-544.
[21] PRADEEP M, REDDY M R, KANNAKI T R. Molecular identification and characterization of chicken parvovirus from indian chicken and association with runting and stunting syndrome. Indian J Anim Res, 2020, 54(12): 1517-1524.
[22] DE LA TORRE D, NUNEZ L F N, PUGA-TORRES B, et al. Molecular diagnostic of chicken parvovirus (ChPV) affecting broiler flocks in ecuador. Braz J Poult Sci, 2018, 20: 643-650.
[23] KOO B S, LEE H R, JEON E O, et al. Molecular survey of enteric viruses in commercial chicken farms in Korea with a history of enteritis. Poult Sci, 2013, 92(11): 2876-2885.
[24] DOMANSKA-BLICHARZ K, JACUKOWICZ A, LISOWSKA A, et al. Genetic characterization of parvoviruses circulating in turkey and chicken flocks in Poland. Arch Virol, 2012, 157(12): 2425-2430.
[25] CHEN L, CHEN L, WANG X, et al. Detection and molecular characterization of enteric viruses in poultry flocks in Hebei province, China. Animals (Basel), 2022, 12(20): 2873.
[26] NUNEZ L F N, SANTANDER-PARRA S H, DE LA TORRE D I, et al. Molecular characterization and pathogenicity of chicken parvovirus (ChPV) in specific pathogen-free chicks infected experimentally. Pathogens, 2020, 9(8): 606.
[27] ZHANG Y, FENG B, XIE Z, et al. Epidemiological surveillance of parvoviruses in commercial chicken and Turkey farms in Guangxi, Southern China, during 2014-2019. Front Vet Sci, 2020, 7: 561371.
[28] LIAO J, XIE Z, ZHANG M, et al. Prokaryotic expression of chicken parvovirus NS1 and VP2 proteins and preparation of polyclonal antibodies. Southwest China Journal of Agricultural Sciences, 2023, 36(2): 8.
[29] LUO S, DENG X, XIE Z, et al. Production and identification of monoclonal antibodies and development of a sandwich ELISA for detection of the H3-subtype avian influenza virus antigen. AMB Express, 2020, 10(1): 49.
[30] CARRATALA A, RUSINOL M, HUNDESA A, et al. A novel tool for specific detection and quantification of chicken/turkey parvoviruses to trace poultry fecal contamination in the environment. Appl Environ Microbiol, 2012, 78 (20): 7496-7499.
[31] JINDAL H K, YONG C B, WILSON G M, et al. Mutations in the NTP-binding motif of minute virus of mice (MVM) NS-1 protein uncouple ATPase and DNA helicase functions. J Biol Chem, 1994, 269(5): 3283-3289.
[32] NUESCH J P, ROMMELAERE J. A viral adaptor protein modulating casein kinase II activity induces cytopathic effects in permissive cells. Proc Natl Acad Sci U S A, 2007, 104(30): 12482-12487.
[33] YAN Y Q, JIN L B, WANG Y, et al. Goose parvovirus and the protein NS1 induce apoptosis through the AIF-mitochondrial pathway in goose embryo fibroblasts. Res Vet Sci, 2021, 137: 68-76.
[34] JIN X, YUAN Y, ZHANG C, et al. Porcine parvovirus nonstructural protein NS1 activates NF-κB and it involves TLR2 signaling pathway. J Vet Sci, 2020, 21(3): e50.
[35] NUNEZ L F, SANTANDER PARRA S H, METTIFOGO E, et al. Isolation and molecular characterisation of chicken parvovirus from Brazilian flocks with enteric disorders. Br Poult Sci, 2015, 56 (1): 39-47.
[36] WANG X, ZHANG J, HUO S, et al. Development of a monoclonal antibody against canine parvovirus NS1 protein and investigation of NS1 dynamics and localization in CPV-infected cells. Protein Expr Purif, 2020, 174: 105682.

[37] LARSEN C I S, MAJUMDER K. The autonomous parvovirus minute virus of mice localizes to cellular sites of DNA damage using ATR signaling. Viruses, 2023, 15(6): 1243.
[38] TEN HAAF A, KOHL J, PSCHERER S, et al. Development of a monoclonal sandwich ELISA for direct detection of bluetongue virus 8 in infected animals. J Virol Methods, 2017, 243: 172-176.

Chapter Six
Studies on Vaccines and Antibodies

Efficacy of a fowl adenovirus 4 (FAdV-4) live vaccine candidate based on the wild-type low-virulence strain GX2019-014 against newly emerging highly pathogenic FAdV-4 strains

Xie Zhixun, Wei You, Xie Zhiqin, Deng Xianwen, Wu Aiqiong, Li Xiaofeng, Xie Liji, Luo Sisi

Abstract

Circulating widely in China, hydropericardium-hepatitis syndrome (HHS) caused by fowl adenovirus serotype 4 (FAdV-4) places a significant economic burden on the poultry farming industry. GX2019-014 is a novel wild-type low-virulence strain of FAdV-4 that is characterized by excellent genetic stability. The results of animal experiment indicate that specific pathogen-free (SPF) chickens and local broiler breeds are protected against virulent FAdV-4 following immunization. Oral administration is recommended for immunization. Inoculation with GX2019-014 at 3×10^4 $TCID_{50}$ did not induce viremia, caused no pathological organ damage, did not affect the feed conversion ratio, and demonstrated excellent safety. Chickens aged 5 to 28 days exhibited strong immune responses, with protection beginning on the 3rd day post-inoculation and 100% resistance to lethal doses of FAdV-4 by the 7th day. Additionally, 80% of the noninoculated chickens exhibited significant protection. Overall, GX2019-014 was evaluated as a candidate low-virulence vaccine for both safety and efficacy. These findings provide a scientific foundation for future vaccine design and development to provide new tools for HHS prevention and control.

Keywords

fowl adenovirus serotype 4 (FAdV-4), wild-type low-virulence strain, live vaccine, immunogenicity, hydropericardium hepatitis syndrome, GX2019-014

Introduction

Fowl adenovirus serotype 4 (FAdV-4) belongs to the Adenoviridae family, *Aviadenovirus* genus, and *Aviadenovirus* C species[1, 2]. It is a nonenveloped, linear, double-stranded DNA virus with a genome length of 43～46 kb. FAdV-4 has been confirmed as the pathogen causing hydropericardium hepatitis syndrome (HHS)[3]. HHS was first reported in the Ankara region of Pakistan in 1987, and outbreaks have subsequently occurred in Pakistan, India, Japan, South Korea, and other regions[4-8]. Since 2015, sporadic cases of HHS have emerged in China, quickly spreading on a large scale in major poultry-raising provinces such as Shandong, Henan, Yunnan, Jilin, Guangdong, and Guangxi[9-11]. It has become one of the major emerging diseases in young chickens in recent years. HHS results in mortality and growth impairment in both broiler and layer chickens. The clinical mortality rate can range from 20% to 80%[12]. Additionally, it damages the immune organs of surviving chickens, leading to decreased flock immunity and reduced feed efficiency[13], thereby causing substantial economic losses to the poultry industry. Consequently, disease prevention and control measures have become pivotal focuses in both poultry farming and clinical prevention in recent years.

Vaccine immunization is currently the most proactive, effective, and operationally feasible method to combat HHS caused by FAdV-4. In China, there are currently five commercially available combination inactivated vaccines targeting new genotypes of highly pathogenic FAdV-4, as well as a single-component inactivated vaccine[14]. These vaccines serve as powerful tools for HHS prevention and control, effectively curbing outbreaks of the disease. Subunit vaccines and gene-edited live vaccines are also current research hotspots. Schachner et al.[15] developed subunit vaccines using the major structural proteins of FAdV-4, with studies indicating that fiber-2 provides the best immune protection, which is consistent with the findings of Wang et al.[16]. Xie et al.[17] and Zhang et al.[18] used gene editing techniques to obtained attenuated strains of FAdV-4, which could provide protection against highly pathogenic strains of FAdV-4.

We isolated a naturally attenuated FAdV-4 strain, named GX2019-014 (GenBank: MW439040; CCTCC NO: V202353), during an epidemiological survey of emerging viral infectious diseases in Guangxi[19]. When administered via intramuscular injection to 4-week-old specific pathogen-free (SPF) chickens at a dosage of 10^6 $TCID_{50}$, this strain did not induce any clinical symptoms or mortality. Whole-genome analysis revealed that GX2019-014 occupies an evolutionary position between highly pathogenic and nonpathogenic FAdV-4 strains. In particular, at positions 219AA and 380AA of the fiber-2 protein, GX2019-014 shares identical sequences with known nonpathogenic strains (B1-7, KR5, and ON1). Analysis via RDP4 and SimPlot software excluded the possibility that GX2019-014 is a recombinant strain between pathogenic and nonpathogenic strains. There is market demand for a variety of adenovirus vaccines with different formulations and immune combinations to optimize immunization programs for commercial poultry and reduce immunization costs. Therefore, we are evaluating GX2019-014 as a candidate strain for an attenuated vaccine to assess its protective efficacy against highly pathogenic FAdV-4-induced HHS, aiming to provide a robust tool for preventing and controlling FAdV-4 infections, with significant market value.

Materials and methods

Virus and experimental animals

The FAdV-4 strains GX2017-005 (GenBank number: MN577981) and GX2019-014 were isolated and stored in our laboratory. SPF chickens were hatched from embryos purchased from Beijing Boehringer Ingelheim Vital Biotechnology Co., Ltd. (Beijing, China), and reared in negative-pressure SPF isolators until they reached 4 weeks of age for use in animal experiments. Field trials were conducted with 2～3-week-old local breed chicks at a large-scale broiler farm in Guangxi Province, China. Animal experiments were conducted in accordance with regulations and approved by the Ethics Committee of the Guangxi Institute of Animal Science and Veterinary Medicine (approval number: 2020c0601).

Determination of the maximum safe dose and minimum protective dose

Four-week-old SPF chickens were divided into 2 groups, each receiving various doses of GX2019-014 via intramuscular injection or oral administration, with 10 chickens per dose. The control group received an equivalent volume of PBS via the same routes. The disease onset time, time of death, and postmortem examination findings of surviving chickens were recorded up to 14 days post-inoculation (dpi). A dose inducing no deaths or lesions was designated a safe dose, and the highest such dose was considered the maximum safe dose.

After immunization, SPF chickens in the safe dose group were challenged with a lethal dose of virulent FAdV-4 (10^5 $TCID_{50}$ of the GX2017-005) on day 14. A challenge control group was established. Disease onset, time of death, and postmortem examination findings of surviving chickens were recorded up to 14 days post-challenge (dpc). A group with no deaths or characteristic lesions was designated a completely protected group, with the lowest effective dose considered the minimum protective dose. The therapeutic index (TI) was calculated as follows: maximum safe dose/minimum protective dose.

Effects of immunization with Gx2019-014 on SPF chickens

3×10^4 $TCID_{50}$ dose of GX2019-014 was applied as the immunization dose. Forty 4-week-old SPF chickens were inoculated via intramuscular injection or oral administration with the immunization dose, and a blank control group was established.

At 1, 3, 5, 7, 9, 11, 14, and 17 dpi, oral and cloacal swabs were collected from the birds for real-time quantitative PCR (RT-PCR) assessment of virus shedding. Serum samples collected at 1, 3, 5, 7, 9, 11, and 14 dpi were used for RT-PCR detection of viremia. Blood samples collected at 1, 3, 5, 7, and 9 dpi were used for the quantification of five liver function indicators: total bilirubin (TBil), alanine aminotransferase (ALT), aspartate aminotransferase (AST), glutamyl transferase (GGT), and alkaline phosphatase (ALP).

At 3 dpi and 7 dpi, 3 immunized birds and 3 control birds were euthanized. Tissues, including the heart, liver, spleen, lungs, kidneys, bursa of Fabricius, and thymus, were collected for histopathological examination via H&E staining to observe pathological changes.

Serum samples were collected at weeks 1, 2, 3, 4, 5, 6, 7, 8, 10, and 12 and every subsequent week up to 60 weeks post-immunization (wpv). The total antibody concentration was determined via a commercial fowl adenovirus group Ⅰ antibody ELISA kit (Product Code: CK132, Biochek, Holland). To quantify neutralizing antibodies, serum samples were serially diluted and incubated with 10^3 $TCID_{50}$ of GX2017-005 for 1 hour at 37 ℃. The mixture was then inoculated onto LMH cells, and an immunofluorescence assay (IFA) was used to assess cell infection and measure neutralizing antibody titers.

At 1, 5, 10, and 15 days of age, 20 SPF chickens per group received GX2019-014 via intramuscular injection or oral administration, with 10 chickens per administration route. A blank control group was included. Disease onset, time of death, and postmortem examination findings of surviving chickens were recorded up to 14 dpi to assess GX2019-014 safety across different ages.

Twenty 4-week-old local-breed roosters and hens were selected and immunized with GX2019-014 via intramuscular injection or oral administration. The control groups of roosters and hens received equivalent doses of PBS via the same routes. The body weights of both the roosters and the hens were measured weekly for 7 wpv to assess the impact of immunization on chicken growth.

Determination of immune challenge protection efficacy

At 4 weeks of age, SPF chickens were divided into 2 groups of 70 chickens each and immunized with the GX2019-014 strain via intramuscular injection or oral administration. At 1, 3, 5, 7, 9, 11, and 14 dpi, 10 chickens from each group were challenged with 10^5 $TCID_{50}$ of GX2017-005. A challenge control group was included. Observations continued until 21 dpc, when disease onset and the time of death were recorded, and postmortem examinations were conducted on surviving chickens to determine the earliest onset of immune protection.

At 1, 5, 10, and 15 days of age, 20 SPF chickens each were divided into 2 groups and immunized with GX2019-014 via intramuscular injection or oral administration, with 10 chickens per age group per route. At 14 dpi, chickens were challenged with 10^5 $TCID_{50}$ of GX2017-005, with a challenge control group included.

Observations continued until 14 dpi, followed by postmortem examinations of surviving chickens to evaluate the protective efficacy against FAdV-4 infection across different ages.

At 4 weeks of age, SPF chickens were immunized with GX2019-014 via intramuscular injection and then challenged with 10^5 $TCID_{50}$ of GX2017-005 at 14 dpi. At 1, 3, 5, 7, 10, 14, and 21 dpc, three immunized and three control chickens were euthanized, and DNA was extracted from the heart, liver, spleen, lung, kidney, bursa of Fabricius, thymus, and muscle tissues for RT-PCR to quantify the tissue viral load. Liver samples were also collected for histopathological examination via H&E staining to observe pathological changes.

Ten SPF chickens and ten chickens immunized with GX2019-014 were cohoused in isolation units. On day 14 of cohabitation, all chickens were challenged with 10^5 $TCID_{50}$ of GX2017-005, with a challenge control group included. Observations continued until 14 dpc, followed by postmortem examinations of surviving chickens to evaluate the level of horizontal immune protection provided by GX2019-014.

Determination of genetic stability

GX2019-014 was serially passaged to the 20th generation in LMH cells for in vitro passaging (OPV). Furthermore, GX2019-014 was serially passaged to the 5th generation in SPF chickens for in vivo passaging (IPV). These passaged samples were subjected to whole-genome sequencing to assess genetic stability.

For safety testing, the samples were diluted to 10^8 $TCID_{50}$/0.25 mL and administered via intramuscular injection to 4-week-old SPF chickens (10 chickens per sample).

The sample was diluted to 3×10^4 $TCID_{50}$/0.25 mL and administered intramuscularly to 10 SPF chickens per sample at 4 weeks of age. At 14 days after inoculation, the chickens were challenged with 10^5 $TCID_{50}$ of GX2017-005 to assess its immune protection efficacy.

Field experiment

GX2019-014 was administered to batches of 2～3-week-old local-breed chickens via intramuscular injection and oral routes at immunization doses, with three batches per administration route (1000-3000 birds per batch). Observations continued for 14 dpi, and postmortem examinations were conducted on chickens that died during this period.

After 14 days, 20 chickens were randomly selected from each batch and challenged with 10^5 $TCID_{50}$ of GX2017-005 to assess immune protection efficacy. A separate group served as the challenge control to observe the effectiveness of the immunization.

Results

Maximum safe dose and minimum protective dose of GX2019-014

Four-week-old SPF chickens were inoculated with different doses of the GX2019-014 strain. The results indicate that the maximum safe dose via intramuscular injection is 10^6 $TCID_{50}$, the minimum protective dose is 4×10^3 $TCID_{50}$, and the TI value is 250. For oral administration, the maximum safe dose is 10^8 $TCID_{50}$, the minimum protective dose is 10^3 $TCID_{50}$, and the TI value is 100 000 (Table 6-1-1 and Table 6-1-2).

Table 6-1-1 Determination of the maximum safe dose of the GX2019-014 strain

Group	Dose/mL	Virus titer $TCID_{50}$/0.25 mL	Intramuscular injection		Oral administration	
			Onset/Total	Death/Total	Onset/Total	Death/Total
1	0.25	10^9	10/10	10/10	2/10	2/10
2	0.25	10^8	10/10	10/10	0/10	0/10
3	0.25	5×10^7	10/10	10/10	0/10	0/10
4	0.25	10^7	10/10	8/10	0/10	0/10
5	0.25	5×10^6	4/10	2/10	0/10	0/10
6	0.25	10^6	0/10	0/10	0/10	0/10
7	0.25	10^5	0/10	0/10	0/10	0/10
8	0.25	10^4	0/10	0/10	0/10	0/10
Control	0.25	PBS	0/10	0/10	0/10	0/10

Table 6-1-2 Determination of the minimum protective dose of the GX2019-014 strain

Groups	Dose/mL	Virus titer $TCID_{50}$/0.25 mL	Intramuscular injection		Oral administration	
			Onset/Total	Death/Total	Onset/Total	Death/Total
1	0.25	10^5	0/10	0/10	0/10	0/10
2	0.25	2×10^4	0/10	0/10	0/10	0/10
3	0.25	10^4	0/10	0/10	0/10	0/10
4	0.25	8×10^3	0/10	0/10	0/10	0/10
5	0.25	6×10^3	0/10	0/10	0/10	0/10
6	0.25	4×10^3	0/10	0/10	0/10	0/10
7	0.25	2×10^3	1/10	1/10	0/10	0/10
8	0.25	10^3	3/10	3/10	0/10	0/10
9	0.25	5×10^2	10/10	10/10	1/10	1/10
Control	0.25	PBS	0/10	0/10	10/10	10/10

Effects of GX2019-014 Immunization on SPF Chickens

Viral shedding after immunization

After vaccination with the recommended immunization dose of GX2019-014 via both the intramuscular injection and oral administration routes, viral shedding was detectable in cloacal and oral samples. From 5 to 14 dpi, cloacal shedding was significantly greater than oral shedding ($P<0.01$), peaking at 7 dpi before gradually declining. In the oral administration group, cloacal shedding was detected in 100% of the chickens at 1 dpi, whereas in the intramuscular administration group, shedding started to be detectable at 5 dpi, indicating earlier onset of cloacal shedding following oral immunization than following intramuscular immunization (Figure 6-1-1).

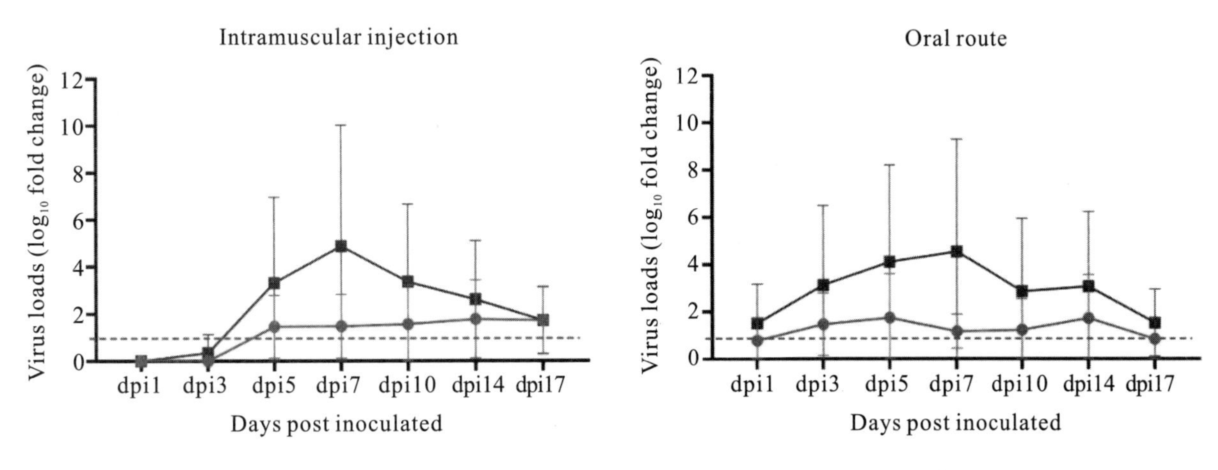

Figure 6-1-1 Detection of viral shedding after immunization with GX2019-014 in SPF chickens

Viremia after immunization

SPF chickens vaccinated with the immunization dose of GX2019-014 did not have detectable levels of FAdV-4 virus in blood samples collected at 1, 3, 5, 7, 9, 11, and 14 dpi. These findings suggests that immunization does not induce viremia in these animals.

Effects of immunization on liver function indicators

SPF chickens immunized with GX2019-014 via intramuscular injection or the oral route presented no significant increase in the levels of the five main liver function indicators—AST, T-Bil, γ-GT, ALT, and ALP—at 1, 3, 5, 7, and 9 dpi compared with those in the blank control group, indicating that there was no effect on the liver function of the chickens (Table 6-1-3).

Table 6-1-3 Detection of liver function indicators after immunization

AST/ (U/L)			
	Control	I.M	P.O
1dpi	276.76±70.71	271.86±29.94	262.69±13.3
3dpi		253.17±19.76	231.44±23.95[a]
5dpi		228.89±14.00[a]	210.90±8.00[a]
7dpi		277.07±29.04	234.99±19.78[a]
9dpi		274.16±38.12	263.85±58.93
ALT /(U/L)			
	Control	I.M	P.O
1dpi	2.02±0.59	2.12±1.10	2.11±0.34
3dpi		1.49±0.31	1.46±0.65
5dpi		1.99±0.58	1.15±0.32[a]
7dpi		1.79±0.50	1.36±0.30
9dpi		2.70±0.34	2.50±0.59
TBil/(μmol/L)			
	Control	I.M	P.O
1dpi	16.6±2.75	15.67±2.04	17.12±1.69
3dpi		15.35±1.94	16.34±3.02
5dpi		13.52±1.48	14.23±2.94
7dpi		11.20±2.83[a]	13.52±2.62
9dpi		11.83±1.28[b]	12.11±2.89[a]

continued

γ-GT /(U/L)			
	Control	I.M	P.O
1dpi	26.54±2.84	26.11±2.46	25.83±3.23
3dpi		23.49±1.45	23.69±3.29
5dpi		24.11±2.74	24.73±1.57
7dpi		24.01±3.52	24.95±2.51
9dpi		26.19±2.96	24.27±2.21
ALP (U/L)			
	Control	I.M	P.O
1dpi	5 606.36±1 066.10	3 231.86±1 159.23 [b]	4 175.33±2 001.65
3dpi		4 058.15±2 402.58	4 024.80±3 655.94
5dpi		3 859.46±1 510.51	3 431.69±1 295.00 [a]
7dpi		3 250.39±1 058.77 [b]	3 432.57±1 325.16 [a]
9dpi		2 054.25±1 402.39 [b]	3 089.35±1 954.90 [a]

Note: [a] $0.01 < P < 0.05$; [b] $P < 0.01$.

Effects of immunization on organ tissues

SPF chickens immunized with GX2019-014 via intramuscular injection or the oral route presented no pathological changes in the heart, liver, spleen, lungs, kidneys, bursa of Fabricius, or thymus tissues at 3 or 7 dpi. Those findings suggest that the immunization did not cause pathological damage to organ tissues (Figure 6-1-2 and Figure 6-1-3).

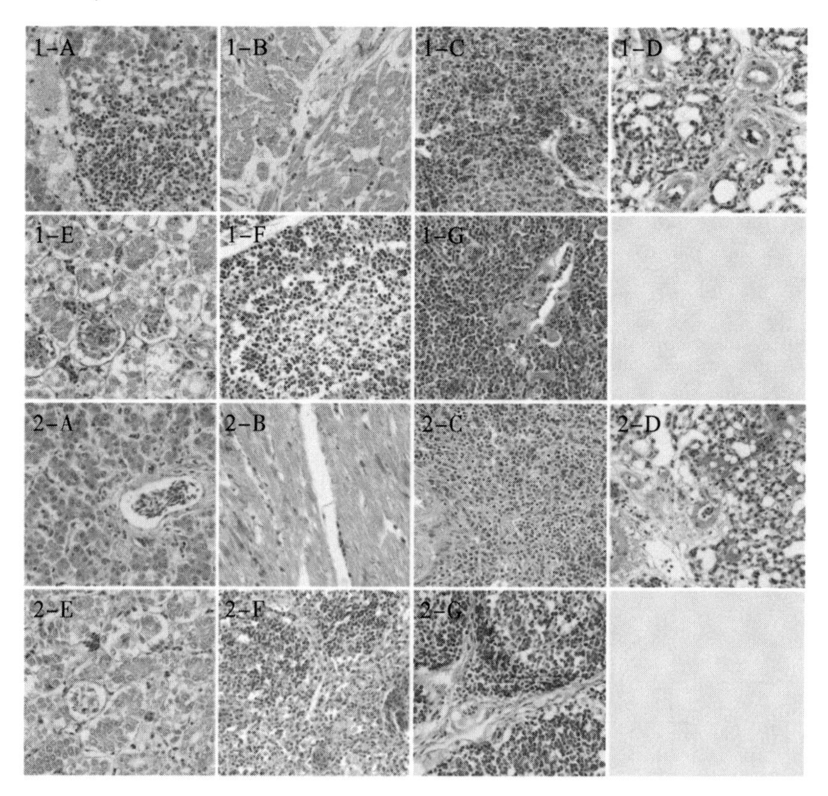

1-A–1-G: Three days post-immunization: liver, heart, spleen, lungs, kidneys, bursa of Fabricius, and thymus; 2-A–2-G: Seven days post-immunization: liver, heart, spleen, lungs, kidneys, bursa of Fabricius, and thymus (color figure in appendix).

Figure 6-1-2 Pathological tissue changes in the organs of SPF chickens after immunization with GX2019-014 via intramuscular injection

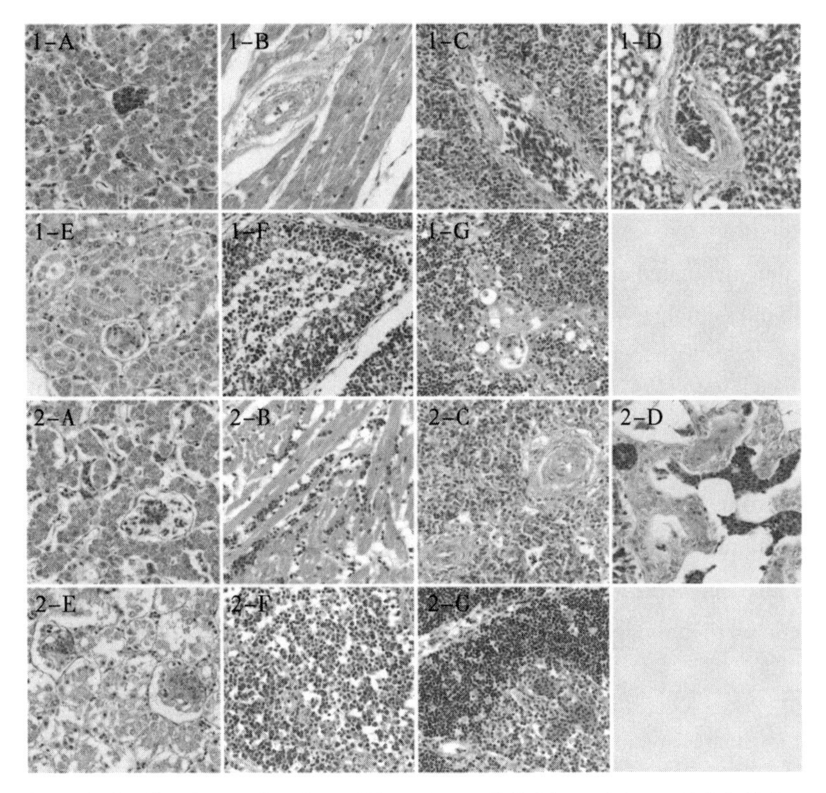

1-A–1-G: Three days post-immunization: liver, heart, spleen, lungs, kidneys, bursa of Fabricius, and thymus; 2-A–2-G: Seven days post-immunization: liver, heart, spleen, lungs, kidneys, bursa of Fabricius, and thymus (color figure in appendix).

Figure 6-1-3 Pathological tissue changes in the organs of SPF chickens after immunization with GX2019-014 via oral administration

Total antibodies and neutralizing antibodies after immunization

The mean ELISA S/P ratios and neutralizing antibody results obtained during the study are shown in Figure 6-1-4. For both the muscle injection group and the oral group, the mean ELISA S/P values for total antibodies were above the positive threshold specified in the kit instructions at 6 wpv and 4 wpv, respectively. These values plateaued from 8 to 28 wpv before gradually declining. Elevated levels of neutralizing antibodies were detected in both experimental groups at 7 dpi ($log_2 4$ and $log_2 5$, respcctively), peaking between 24 and 26 wpv. Both groups maintained high levels of neutralizing antibodies until 60 wpv ($log_2 7$ and $log_2 8$, respectively), as illustrated in Figure 6-1-4.

Safety of GX2019-014 in chickens of different ages

Those results indicate that the immunization dose of GX2019-014 does not cause illness or death in SPF chickens at 1, 5, 10, or 15 days of age, indicating that it is safe for chicks at these early ages (Table 6-1-4).

Table 6-1-4 Safety of GX2019-014 in chickens of different ages

Age at vaccination·	Intramuscular injection		Oral administration		Blank control	
	Onset/Total	Death/Total	Onset/Total	Death/Total	Onset/Total	Death/Total
1 days old	0/10	0/10	0/10	0/10	0/10	0/10
5 days old	0/10	0/10	0/10	0/10	0/10	0/10
10 days old	0/10	0/10	0/10	0/10	0/10	0/10
15 days old	0/10	0/10	0/10	0/10	0/10	0/10

Effects of the immunization dose on chicken body weight

For the 4-week-old local chickens vaccinated with GX2019-014, the intramuscular administration group presented significantly greater weight gain than did the control group at 1 week but lower weight gain at 4 weeks. Compared with the control group, the oral administration group presented greater weight gain at 1 and 4 weeks but lower weight gain at 2 and 7 weeks. Overall, the average weekly weight gain from 1 to 7 weeks was similar between the vaccinated and control groups, indicating that the vaccine does not affect weight gain or the feed conversion ratio (Figure 6-1-5).

Assessment of Protective Efficacy

Determination of the onset of immunoprotection

In the intramuscular injection group, protection started at 3 dpi (4/10 chickens protected) and increased over time, reaching 100% protection against the lethal dose of virulent FAdV-4 by day 9. The oral administration group also showed protection starting from day 3 (2/10 chickens protected) and achieved full protection by day 7, earlier than the intramuscular administration group (Table 6-1-5).

Table 6-1-5 Determination of the onset of immunoprotection

Challenge time	Intramuscular injection		Oral administration		Challenge control	
	Onset/Total	Death/Total	Onset/Total	Death/Total	Onset/Total	Death/Total
1 dpi	10/10	10/10	10/10	10/10	10/10	10/10
3 dpi	10/10	6/10	10/10	8/10	10/10	10/10
5 dpi	7/10	4/10	3/10	2/10	10/10	10/10
7 dpi	3/10	2/10	0/10	0/10	10/10	10/10
9 dpi	0/10	0/10	0/10	0/10	10/10	10/10
11 dpi	0/10	0/10	0/10	0/10	10/10	10/10
14 dpi	0/10	0/10	0/10	0/10	10/10	10/10

Protective efficacy of immunization in chickens at various ages

The results revealed that 10% of the 1-day-old chickens did not achieve complete protection, regardless of whether they were vaccinated via the intramuscular or oral route. However, 100% protection was observed in chickens that were 5, 10, and 15 days old (Table 6-1-6).

Table 6-1-6 Protective efficacy of immunization in chickens at various ages

Age at vaccination	Intramuscular injection		Oral administration		Challenge control	
	Onset/Total	Death/Total	Onset/Total	Onset/Total	Death/Total	Onset/Total
1 days old	1/10	0/10	1/10	1/10	10/10	10/10
5 days old	0/10	0/10	0/10	0/10	10/10	10/10
10 days old	0/10	0/10	0/10	0/10	10/10	10/10
15 days old	0/10	0/10	0/10	0/10	10/10	10/10

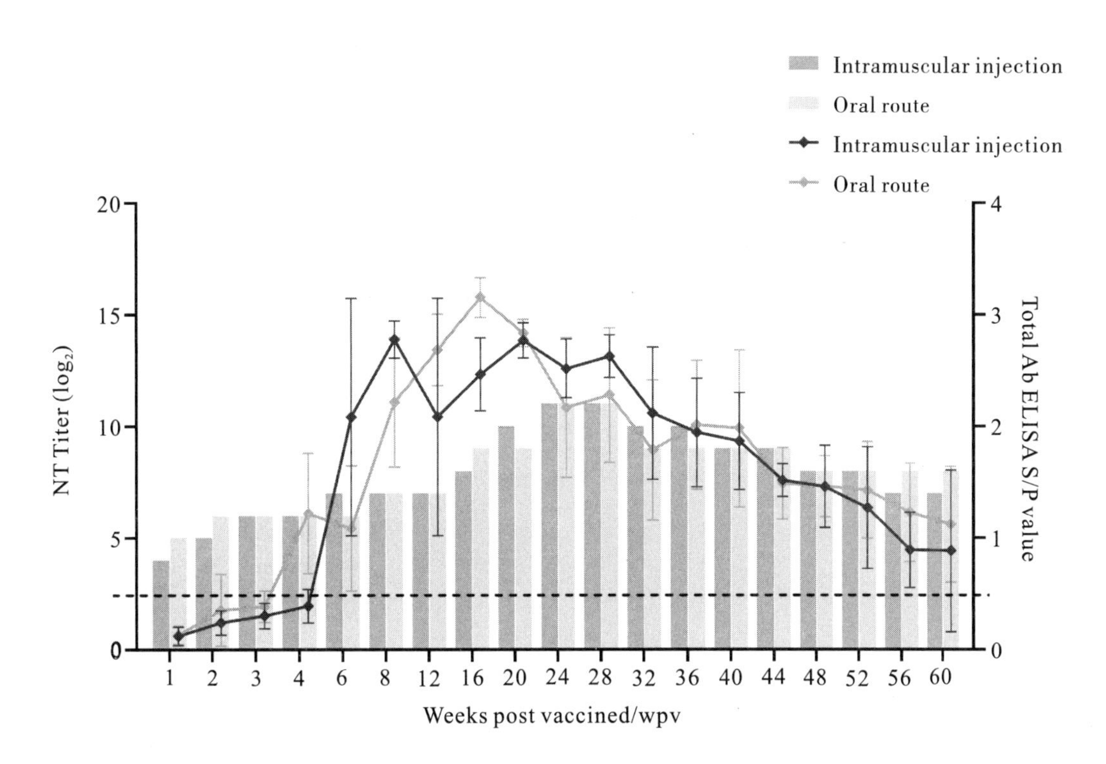

Figure 6-1-4 Quantification of total antibodies and neutralizing antibodies after immunization

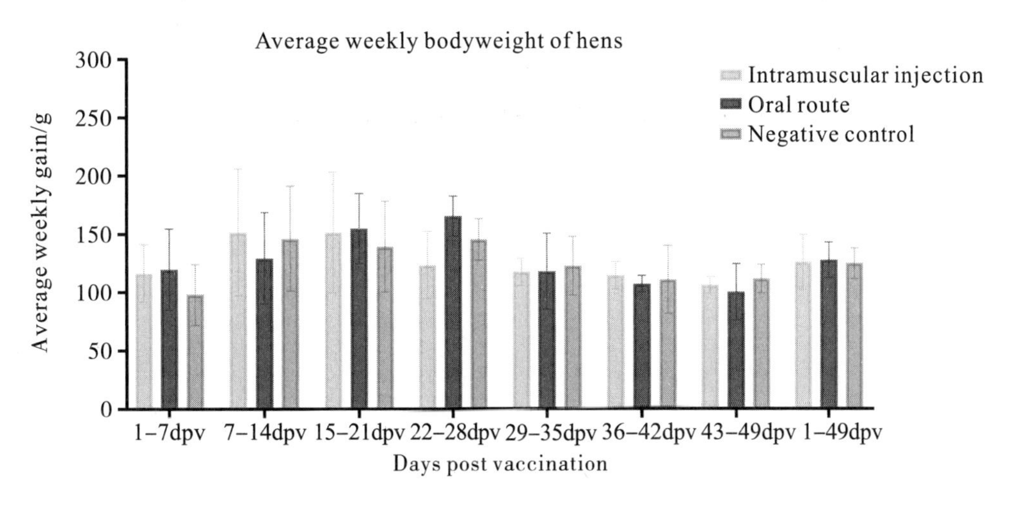

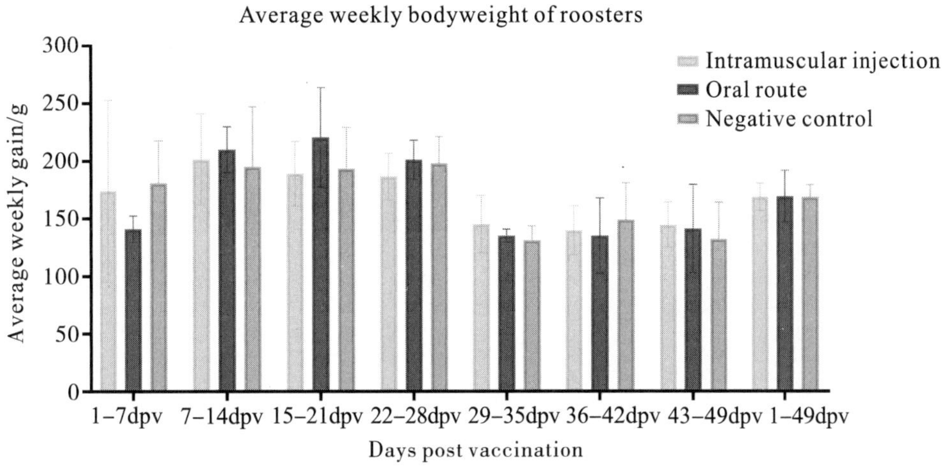

Figure 6-1-5 Average weekly body weight after immunization

Protection efficacy in different organs

At 1 dpc, low levels of virus (less than $10^{3.4}$ copies) were detected in the muscles and lungs; at 3 dpc, in the thymus and lungs; and at 5 dpc, in the lungs, kidneys, liver, heart, spleen, and bursa of Fabricius. No virus was detected in any organ after 7 dpc. Histopathological examination of liver tissues from 1 to 21 dpc revealed no pathological changes, and no pathological damage was observed in the target organs of FAdV-4 infection.

Histopathological examination of liver tissues from 1 to 21 dpc revealed no pathological changes, and no damage was observed in the target organs of FAdV-4 infection (Figure 6-1-6 and Figure 6-1-7).

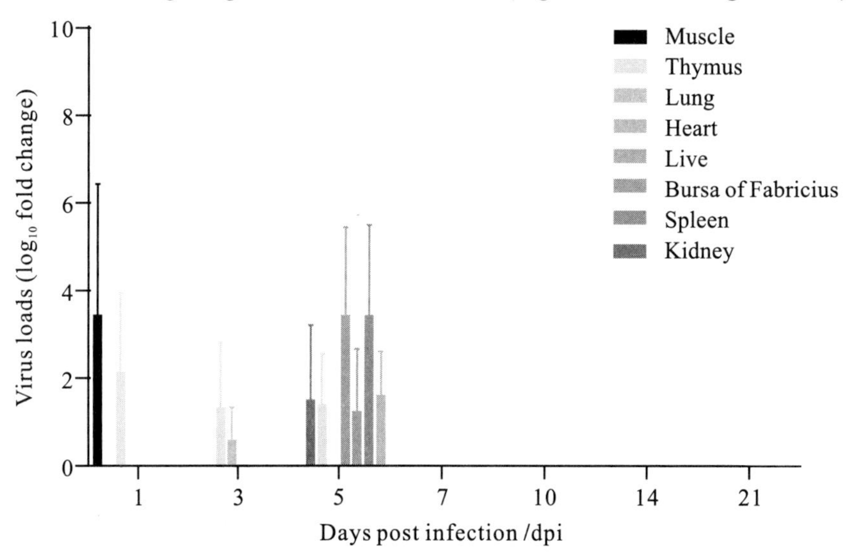

Figure 6-1-6 Viral load in various organs after challenge

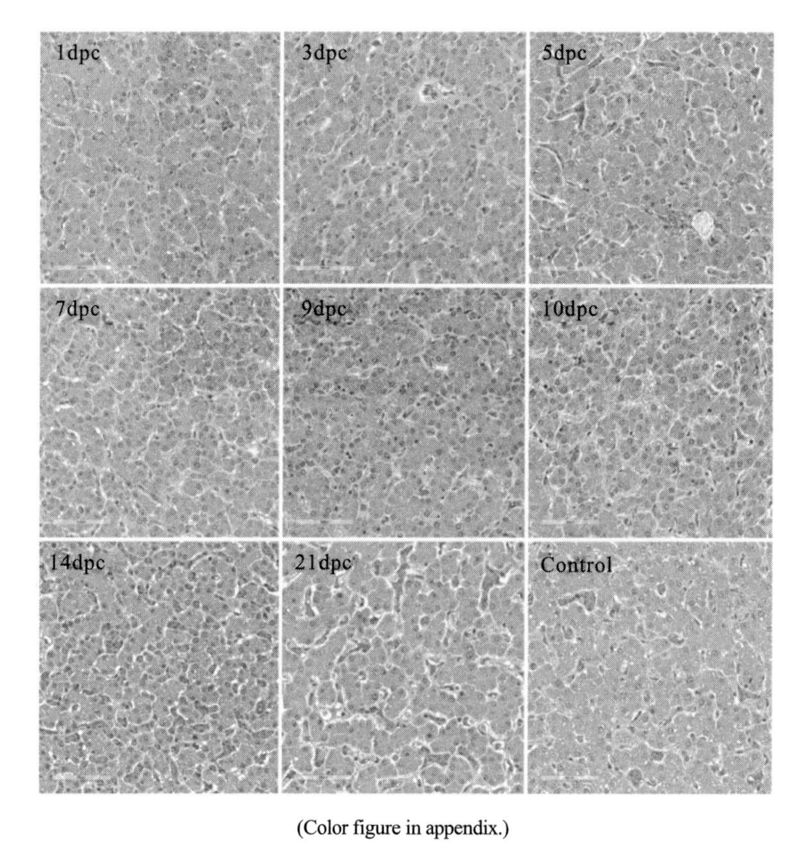

(Color figure in appendix.)

Figure 6-1-7 Histopathological examination of the liver after challenge

Effectiveness of horizontal transmission immunization

On the 14th day after cohousing, the chickens were challenged with a lethal dose of FAdV-4. The horizontal immunization group had a morbidity rate of 60% and a mortality rate of 20%, with 80% of the chickens achieving protection, whereas the intramuscular injection group exhibited 100% protection (Figure 6-1-8).

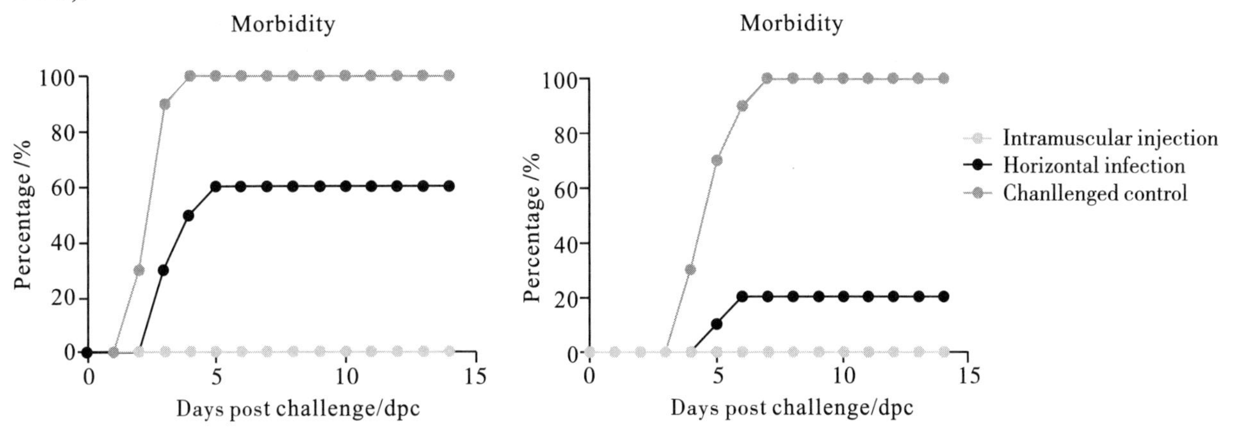

Figure 6-1-8 Effectiveness of horizontal transmission immunization

Determination of Genetic Stability

After 20 passages in vitro or 5 passages in vivo, the whole-genome sequence of GX2019-014 remained unchanged. After 4-week-old SPF chickens were inoculated with a 10^6 $TCID_{50}$ of the passaged viruses (OPV F20 or IPV F5), no typical clinical symptoms of HHS or mortality were observed, indicating that the virulence did not increase. The recommended immunization doses of OPV F20 and IPV F5 both provided protection against lethal doses of FAdV-4, with no change in the efficacy of immunoprotection (Table 6-1-7).

Table 6-1-7 Genetic Stability of GX2019-014

Passaging method	Passage number	Genomic homogeneity	Safety test		Immunoprotective test		
			Vaccination	Survival rate	Vaccination	challenge	Protection rate
	F0		GX2019-014 10^6 $TCID_{50}$	100% (10/10)	GX2019-014 3×10^4 $TCID_{50}$	GX2017-005 10^5 $TCID_{50}$	100% (10/10)
In vitro	OPV F20	100%		100% (10/10)			100% (10/10)
In vivo	IPV F5	100%		100% (10/10)			100% (10/10)

Field experiment

GX2019-014 was administered via intramuscular injection and oral administration to 2～3 week-old broilers of three local breeds at three different-scale farms. The postimmunization survival rates ranged from 99.5% to 99.74% and 99.54% to 99.8%, which were comparable to those of the unvaccinated flocks. Necropsies of the deceased chickens did not reveal typical HHS lesions, such as pericardial effusion, swelling and jaundice of the liver. Twenty chickens from each group were randomly selected and challenged with a lethal dose of FAdV-4, resulting in a 100% protection rate (Table 6-1-8).

Table 6-1-8 Field experiment test of GX2019-014

<table>
<tr><th>No.</th><th>Group</th><th>Number of chickens</th><th>Age of chickens</th><th colspan="2">Chicken breed</th><th colspan="2">Farm</th></tr>
<tr><td>1</td><td>I.M group 1</td><td>1 865</td><td>16</td><td colspan="2">Local breed A</td><td colspan="2">Field A</td></tr>
<tr><td>2</td><td>I.M group 2</td><td>2 890</td><td>18</td><td colspan="2">Local breed B</td><td colspan="2">Field B</td></tr>
<tr><td>3</td><td>I.M group 3</td><td>2 426</td><td>19</td><td colspan="2">Local breed C</td><td colspan="2">Field C</td></tr>
<tr><td>4</td><td>P.O group 1</td><td>1 756</td><td>16</td><td colspan="2">Local breed A</td><td colspan="2">Field A</td></tr>
<tr><td>5</td><td>P.O group 2</td><td>3 004</td><td>18</td><td colspan="2">Local breed B</td><td colspan="2">Field B</td></tr>
<tr><td>6</td><td>P.O group 3</td><td>2 218</td><td>19</td><td colspan="2">Local breed C</td><td colspan="2">Field C</td></tr>
<tr><th rowspan="2">No.</th><th rowspan="2">Vaccination</th><th rowspan="2">Survival rate</th><th rowspan="2">FAdV-4 lesions on necropsy</th><th rowspan="2">Challenge</th><th colspan="3">Protection rate after challenge</th></tr>
<tr><th colspan="2">Vaccinated group</th><th>Challenge control</th></tr>
<tr><td>1</td><td rowspan="6">GX2019-014 3×10^{4} $TCID_{50}$</td><td>99.68% (1 859/1 865)</td><td>0% (0/6)</td><td rowspan="6">GX2017-005 10^{5} $TCID_{50}$</td><td colspan="2">100% (20/20)</td><td>0% (0/10)</td></tr>
<tr><td>2</td><td>99.74% (3 182/3 190)</td><td>0% (0/8)</td><td colspan="2">100% (20/20)</td><td>0% (0/10)</td></tr>
<tr><td>3</td><td>99.50% (2 414/2 426)</td><td>0% (0/12)</td><td colspan="2">100% (20/20)</td><td>0% (0/10)</td></tr>
<tr><td>4</td><td>99.54% (1 745/1 756)</td><td>0% (0/11)</td><td colspan="2">100% (20/20)</td><td>0% (0/10)</td></tr>
<tr><td>5</td><td>99.8% (2 990/3 004)</td><td>0% (0/14)</td><td colspan="2">100% (20/20)</td><td>0% (0/10)</td></tr>
<tr><td>6</td><td>99.59% (2 209/2 218)</td><td>0% (0/9)</td><td colspan="2">100% (20/20)</td><td>0% (0/10)</td></tr>
</table>

Discussion

Since 2015, outbreaks of the novel highly pathogenic FAdV-4 strain causing HHS have extensively spread across poultry farms in East China, resulting in significant economic losses to the poultry farming industry[10, 11, 20, 21]. During the HHS epidemic period from 2017 to 2020, our group isolated and purified 19 strains of FAdV-4 (named as GX2017-001 to GX2020-019), and performed comprehensive genome sequencing and pathogenicity assessments[10, 19]. Previous studies identified strain GX2019-014 as having molecular characteristics typical of novel FAdV-4 strains, including a 1966 bp deletion, however, phylogenetic analysis placed it closer to nonpathogenic strains KR5, ON1, and B1-7. Strain GX2019-014 did not exhibit the molecular features observed in the novel virulent FAdV-4 strain, such as a 10-base insertion in the tandem repeat sequence (TR-B), 3-base insertions in the coding regions of ORF2, ORF22, and 52/55K proteins, and a 3-base deletion resulting in the loss of one histidine residue at the C-terminus of fiber-1. The amino acids at positions 219 and 380 on the fiber-2 protein are identical to those found in nonpathogenic strains (B1-7, KR5, and ON1). Analysis via RDP4 and SimPlot software excluded the possibility that strain GX2019-014 derived from recombination between pathogenic and nonpathogenic strains. When administered intramuscularly to 4-week-old SPF chickens at a dose of 10^{6} $TCID_{50}$, GX2019-014 did not induce any clinical symptoms or mortality. This strain is the first reported wild-type novel FAdV-4 strain with low virulence in China[19].

The therapeutic index (TI) is a measure used in pharmacology to evaluate the safety and effectiveness of a drug. It is defined as the ratio of the dose that produces toxicity to the dose that produces a therapeutic effect. The therapeutic index helps determine the relative safety of a drug and is an important consideration in drug development and clinical use.[22, 23] We determined the maximum safe dose and minimum effective dose for strain GX2019-014 administered via intramuscular injection and oral administration. The maximum safe dose for the oral administration route was 10^8 $TCID_{50}$ per bird, with a TI value of 100,000, whereas for the intramuscular administration route, the maximum safe dose was 10^6 $TCID_{50}$, with a TI value of 250. Therefore, oral administration is recommended as the preferred immunization route because of its greater safety, whereas intramuscular injection is considered an alternative option.

We set the recommended immunization dose and evaluated the effects of GX2019-014 immunization on SPF chickens. Within 14 dpi, FAdV-4 was not detected in the blood of vaccinated chickens, indicating that immunization does not cause viremia. The levels of five key liver function markers—AST, T-Bil, γ-GT, ALT, and ALP—were not significantly greater than those in the control group at 9 dpi, suggesting that immunization does not affect liver function. Histopathological examination revealed no pathological damage to major organs such as the heart, liver, spleen, lungs, kidneys, bursa of Fabricius, or thymus in vaccinated chickens. Additionally, there was no significant difference in weight gain between the immunized and control groups ($P>0.05$), indicating that immunization does not negatively impact feed conversion efficiency. No typical symptoms or mortality related to FAdV-4 infection were observed in chickens vaccinated at 1, 5, 10, or 15 days of age. Therefore, we conclude that the recommended immunization dose of GX2019-014 is safe for chickens.

Four-week-old SPF chickens, after immunization and subsequent challenge with a lethal dose of virulent FAdV-4, exhibited the following results. for the chickens in both the intramuscular and oral administration groups were protected beginning at 3 dpc. Initially, 20% of chickens in the intramuscular administration group and 40% in the oral administration group were protected. Over time, the level of protection increased, reaching 100% by 9 dpc in the intramuscular administration group and by 7 dpc in the oral administration group. This result is consistent with the timing of the production of neutralizing antibodies induced by GX2019-014. At 7 dpi, the average neutralizing antibody titers in the intramuscular and oral administration groups reached $\log_2 4$ and $\log_2 5$, respectively. As time progressed, the titers gradually increased, peaking between 24 and 26 wpv. The high titers were maintained for an extended period, with $\log_2 7$ and $\log_2 8$ still measured at 60 wpv. Therefore, we believe that a single vaccination of GX2019-014 provides sufficient long-lasting immunoprotection for chickens. However, when a commercial ELISA kit was used to quantify total antibodies against group I FAdV (coated with inactivated whole virus), the S/P values for the intramuscular and oral administration groups were clearly higher than the positive cutoff value (0.5) at 6 wpv and 4 wpv, respectively, which does not align with the observed timing of actual protection. Similarly, Schachner et al. immunized SPF chickens with a fiber-2 protein subunit vaccine and achieved 96% protection against virulent FAdV-4. However, no antibodies were detected via a commercial ELISA kit for anti-group I FAdV antibodies before challenge[15]. Therefore, we believe that neutralizing antibody tests are more suitable than whole virus-coated ELISA kits for evaluating the protective efficacy of FAdV-4 live vaccines. Additionally, research on subunit vaccines for FAdV-4[15,16, 24], including those based on fiber-2, fiber-1, penton base, and hexon proteins, has demonstrated that protein immunization can effectively protect against these challenges. However, these studies also revealed a general lack of neutralizing activity in the serum. This finding suggests that neutralization is likely due to antibody binding to multiple adenoviral capsid components rather than a single antigen. Moreover, protection against virulent FAdV-4 may not depend solely on neutralizing antibodies, indicating the need for further investigation into the role of cellular immunity in protection.

Upon challenge with virulent FAdV-4, immunization of chickens at 5 to 20 days of age demonstrated excellent protective efficacy. No pathological damage was detected in the liver, the target organ for FAdV-4, and no virus was detected in any organ at 7 dpc. However, chickens immunized at 1 day of age did not achieve complete protection. This is likely because secondary immune organs in neonatal chickens are not fully developed. Structures such as the periarteriolar lymphoid sheath (PALS) and the ellipsoid with its surrounding macrophage ring in the spleen mature later, and the immune system is not fully functional[25]. As a result, the vaccine cannot effectively stimulate B-cell responses that lead to antibody production.

Additionally, chickens vaccinated via both intramuscular injection and the oral administration route excrete some GX2019-014 virus into the environment. Therefore, we evaluated the protective efficacy of horizontal transmission of this vaccine. The results revealed that 80% of the chickens that naturally acquired GX2019-014 from the environment achieved protection against virulent FAdV-4, which is highly important for herd immunity.

In summary, existing FAdV-4 vaccines on the market are limited to inactivated formulations, underscoring the need to develop new vaccine types to improve immunization strategies. We conducted a thorough evaluation of the safety and immunogenicity of the FAdV-4 natural attenuated strain GX2019-014 as a candidate live vaccine. We hope this strain will offer a new and effective tool for the prevention and control of HHS.

References

[1] BENKŐ M, KOKI A, NIKLAS A, et al. ICTV virus taxonomy profile: adenoviridae. journal of general virology, 2022.

[2] CHEN S, LIN F, JIANG B, et al. Isolation and characterization of a novel strain of duck aviadenovirus B from Muscovy ducklings with acute hepatitis in China. Transboundary and Emerging Diseases, 2022, 69(5): 2769-2778.

[3] ABE T, NAKAMURA K, TOJO H, et al. Histology, immunohistochemistry, and ultrastructure of hydropericardium syndrome in adult broiler breeders and broiler chicks. Avian Dis, 1998, 42(3): 606-607.

[4] ANJUM A D, SABRI M A, IQBAL Z. Hydropericarditis syndrome in broiler chickens in Pakistan. The Veterinary Record, 1989, 124(10): 247-248.

[5] ASRANI R K, GUPTA V K, SHARMA S K, et al. Hydropericardium-hepatopathy syndrome in Asian poultry. The Veterinary Record, 1997, 141(11): 271-273.

[6] ONO M, OKUDA Y, YAZAWA S, et al. Outbreaks of adenoviral gizzard erosion in slaughtered broiler chickens in Japan. The Veterinary Record, 2003, 153(25): 775-779.

[7] CHOI K S, KYE S J, KIM J Y, et al. Epidemiological investigation of outbreaks of fowl adenovirus infection in commercial chickens in Korea. Poultry Science, 2012, 91(10): 2502-2506.

[8] DAHIYA S, SRIVASTAVA R N, HESS M, et al. Fowl adenovirus serotype 4 associated with outbreaks of infectious hydropericardium in Haryana, India. Avian Dis, 2002, 46(1): 230-233.

[9] NIU Y, SUN Q, ZHANG G, et al. Epidemiological investigation of outbreaks of fowl adenovirus infections in commercial chickens in China. Transboundary and Emerging Diseases, 2018, 65(1): e121-e126.

[10] RASHID F, XIE Z, ZHANG L, et al. Genetic characterization of fowl aviadenovirus 4 isolates from Guangxi, China, during 2017-2019. Poultry Science, 2020, 99(9): 4166-4173.

[11] CHEN L, YIN L, ZHOU Q, et al. Epidemiological investigation of fowl adenovirus infections in poultry in China during 2015-2018. BMC Veterinary Research, 2019, 15(1): 271.

[12] MO K K, LYU C F, CAO S S, et al. Pathogenicity of an FAdV-4 isolate to chickens and its genomic analysis. J Zhejiang Univ Sci B, 2019, 20(9): 740-752.

[13] WEI Y., DENG X W, XIE Z X, et al. Pathological observation of outcome of hydropericardium hepatitis syndrome. China Poultry, 2023, 45(1): 45-50.

[14] CHINA INSTITUTE OF VETERINARY DRUG CONTROL. National veterinary drug basic database.[2024-07-05]. http://

vdts.ivdc.org.cn:8081/cx/#.

[15] SCHACHNER A, MAREK A, JASKULSKA B, et al. Recombinant FAdV-4 fiber-2 protein protects chickens against hepatitis–hydropericardium syndrome (HHS). Vaccine, 2014, 32(9): 1086-1092.

[16] WANG X, TANG Q, CHU Z, et al. Immune protection efficacy of FAdV-4 surface proteins fiber-1, fiber-2, hexon and penton base. Virus Res, 2018, 245:1-6.

[17] XIE Q, WANG W, KAN Q, et al. FAdV-4 without fiber-2 is a highly attenuated and protective vaccine candidate. Microbiology Spectrum, 2022, 10(1): e0143621.

[18] ZHANG Y, PAN Q, GUO R, et al. Immunogenicity of novel live vaccine based on an artificial RHN20 strain against emerging fowl adenovirus 4. Viruses, 2021, 13(11): 2153.

[19] WEI Y, XIE Z Z, XIE Z Z, et al. Differences in the pathogenicity and molecular characteristics of fowl adenovirus serotype 4 epidemic strains in Guangxi, southern China. Frontiers in Microbiology, 2024, 15: 1428958.

[20] KAI W, HAIWEI S, YUNZHANG L, et al. Characterization and pathogenicity of fowl adenovirus serotype 4 isolated from eastern China. BMC Veterinary Research, 2019, 15(1): 373.

[21] GUAN R, TIAN Y, HAN X, et al. Complete genome sequence and pathogenicity of fowl adenovirus serotype 4 involved in hydropericardium syndrome in southwest China. Microb Pathog, 2018, 117: 290-298.

[22] DALEY-YATES P T. Inhaled corticosteroids: potency, dose equivalence and therapeutic index. Br J Clin Pharmacol, 2015, 80(3): 372-380.

[23] GERBER H P, GANGWAR S, BETTS A. Therapeutic index improvement of antibody-drug conjugates. MAbs, 2023, 15(1): 2230618.

[24] WANG W, LIU Q, LI T, et al. fiber-1, not fiber-2, directly mediates the infection of the pathogenic serotype 4 fowl adenovirus via its shaft and knob domains. Journal of Virology, 2020, 94(17): e00954-20.

[25] MAST J, GODDEERIS B M. Development of immunocompetence of broiler chickens. Vet Immunol Immunopathol, 1999, 70(3-4): 245-256.

Immune response to oil-emulsion vaccines with single or mixed antigens of Newcastle disease, avian influenza, and infectious bronchitis

Xie Zhixun, and Stone Henry D

Abstract

Inactivated Newcastle disease virus (NDV), avian influenza virus (AIV), and infectious bronchitis virus (IBV) antigens were evaluated for immunological efficacy in monovalent and polyvalent vaccines. Vaccinated broilers were bled for hemagglutination-inhibition (HI) tests at 1- or 2-week intervals. Half of the chickens were challenged with the Largo isolate of velogenic viscerotropic (VV) NDV at 8 weeks post-vaccination, and the remainder were challenged with the Massachusetts 41 strain IBV at 9 weeks post-vaccination.

Newcastle disease HI titers were reduced significantly ($P<0.05$) from those of monovalent control vaccine groups when IBV antigen was emulsified in mixtures with low (1～3 x) < concentrated NDV or NDV and AIV antigens. Avian influenza HI titers were significant ($P<0.05$) lower than those of the control monovalent groups when highly concentrated NDV was part of the polyvalent vaccine. Infectious bronchitis HI titers were higher than those of control monovalent groups in 13 of 15 vaccine groups when IBV antigen was in polyvalent formulations.

VV NDV challenge killed all non-NDV vaccinates and induced increased HI titers in NDV vaccinates but no morbidity or mortality.

Sixty of 80 IBV vaccinates experienced a fourfold or greater HI titer increase following challenge. All non-IBV vaccinates seroconverted at 1 week post-challenge.

Keywords

Newcastle disease, avian influenza, infections bronchitis, oil-emulsion vaccines, immune response

Introduction

Highly virulent strains of Newcastle disease virus (NDV) affect both domestic and wild birds[2, 7, 20] and have been a major problem in the poultry industry throughout the world for many years. During the past decade, avian influenza has been recognized as a disease of considerable economic importance to the industry[15]. Infectious bronchitis causes serious economic losses in laying hens and young chicks world-wide[23].

During recent years, injectable oil-emulsion inactivated vaccines against Newcastle disease (ND) and infectious bronchitis (IB) (monovalent or polyvalent) have been extensively used[1, 6, 13, 17, 18, 28, 30]. Inactivated emulsion vaccine for avian influenza (AI) and its combination with different subtypes of avian influenza virus (AIV) resulting in polyvalent vaccines have been described[3, 10, 11, 16, 27]. These monovalent or polyvalent vaccines have been used by the poultry industry for protection against NDV, AIV, and IBV. The purpose of the present study was to investigate possible deleterious effects on efficacy due to mixing different disease antigens in multivalent vaccines.

Materials and method

Experimental chickens

Specific-pathogen-free white rock broilers obtained from the same hatch at the Southeast Poultry Research Laboratory supply flocks were used for all vaccine groups. They were known to be free of Newcastle disease, avian influenza, infectious bronchitis, mycoplasma, Marek's disease, and avian encephalomyelitis. They were held in disease-containment buildings with appropriate diet and husbandry.

Antigen preparation

The LaSota strain of NDV, avian influenza A/Turkey/Wisconsin/68 (H5N9), and the Mass 41 strain of IBV were propagated as described[24, 28, 29]. The allantoic fluid virus harvests were clarified by centrifugation at 300 × g for 20 min at 4 ℃. Before inactivation, the NDV suspension had an infectivity titer of $10^{9.1}$ mean embryo lethal dose (ELD_{50}) /0.1 mL and a hemagglutinin titer of 1 : 1 024. The AIV suspension had an infectivity titer of $10^{8.9}$ ELD_{50}/0.1 mL and a hemagglutinin titer of 1 : 1 024. The IBV suspension had an infectivity titer of 1 075 mean embryo infectious dose (ELD_{50}) /0.1 mL. The ND, AI, and IB viruses were inactivated with 0.1% beta-propiolactone for 4 h at room temperature while stirring. Inactivation was confirmed by embryo inoculation. Antigens to be concentrated were clarified by centrifugation at 1 200 × g for 45 min at 4 ℃. The pellet was discarded, and the supernatant fluid was then centrifuged at 186 009 × g for 75 min at 4 ℃. The resulting virus-containing pellet was resuspended in its supernatant allantoic fluid to achieve a 100 × concentration. All antigens were stored at −75 ℃. The concentrated antigens were thawed and sonicated 1 min per 10 mL continuously at a setting number of 4 with a small tip (Branson Cell disrupter 200, Branson Power Co., Danbury, Conn.). The antigen was then diluted with its supernate to accommodate the volume and quantity of antigen needed for the experimental design.

Preparation of vaccines

Water-in-oil-emulsion vaccines were prepared as described[27] with 10% of oil phase as surfactant. Surfactant contained 10 parts Arlacel 80 and one part Tween 80. A ratio of oil-to-aqueous-phase of 4 : 1 was used for all vaccines. All vaccines were made from the same batch of oil phase, and each antigen was from a single antigen pool. Completed vaccines were stored at 4 ℃ until used.

Challenge

Chickens challenged with the Largo strain of viscerotropic velogenic (VV) NDV received $10^{7.3}$ ELD_{50} in 0.1 mL intratracheally. Chickens challenged with IBV (Mass 41) received $10^{4.7}$ ELD_{50} in 0.1 mL intratracheally. No chickens were challenged with AIV.

Challenge assessment

The response to challenge with NDV was measured by serologic evidence of infection, morbidity, and mortality. Morbidity was defined as clinical signs of disease, such as diarrhea, labored breathing, depression, ataxia, torticollis, paralysis, involuntary movements, and coma during 14 days after challenge. Infection with IBV was assumed if there was a fourfold or greater increase in HI titer at 1 week post-challenge.

Serology

All serum samples were tested individually for the presence of HI antibodies by standard procedures in microtiter plates using eight hemagglutination (HA) units for NDV[8], eight HA units for AI[14], and four HA units for IBV[24].

Analysis of HI results

Statistics were applied to the HI results of the 8-week cumulative means with analysis of variance and Duncan's multiple range test[12]. For each mean, n=5. The use of the term significant in this paper has a statistical meaning ($P<0.05$).

Experimental design

Three different types of vaccines were used: (1) Control monovalent vaccines (type M): a single antigen was emulsified to prepare the vaccine; (2) Polyvalent vaccines: vaccines were made by manually mixing equal volumes of type M monovalent vaccines of NDV, AIV, or IBV after they were emulsified (type AFT); (3) Polyvalent vaccines: equal volumes of specified antigens were mixed before emulsification (type BEF).

Four-week-old chickens were vaccinated subcutaneously in the mid-dorsal neck region, with 10 birds in each vaccine group. All chickens were bled at 2, 3, 4, 6, and 8 weeks post-vaccination for NDV, AIV, and IBV HI serology. Half of all vaccine groups were challenged at 8 weeks post-vaccination with vvNDV. The remaining birds were challenged with IBV at 9 weeks post-vaccination. All challenged chickens were bled for serology at the time of challenge, at 2 weeks following challenge with vvNDV, and at 1 week after challenge with IBV.

Doses of each vaccine were varied to administer the same quantity of the antigen being evaluated regardless of the formulation.

Results

Table 6-2-1 shows NDV HI results from groups vaccinated with monovalent or type AFT and type BEF polyvalent vaccines. HI titers of type BEF polyvalent vaccine groups containing IBV antigen (Nos.6, 7, 9, and 10) with low (1 × ~3 ×) NDV and AIV antigen concentration were significantly lower than their control type M monovalent vaccine group (No.1). Type AFT polyvalent vaccine group (Nos.2, 3, and 4) HI titers were not significantly different from their control type M monovalent vaccine group (No.1). Titers induced by highly concentrated (25 × ~37.5 ×) NDV and AIV antigen type AFT and type BEF polyvalent vaccines (Nos.12-17) were not significantly different from their monovalent control vaccine group (No.11). In general, cumulative mean titers of all polyvalent vaccine groups trended lower as more antigens were combined in the vaccines. Titers between the two control vaccine groups (Nos.1 and 11) were not significantly different.

Table 6-2-1 Newcastle disease geometric mean hemagglutination-inhibition titers of chickens after vaccination with monovalent or polyvalent vaccines made by combined Newcastle disease, avian influenza, and infectious bronchitis antigens before or after emulsification

Vaccine group No. and antigen	Vaccine type [A]	Antigen concentration (×) [B]	Dose/mL	8-week cumulative HI GMT [C]
Low concentrations of NDV, AIV antigen				
1. NDV control	M	1	0.5	429 [a]
2. NDV/AIV	AFT	1/1	1.0	339 [a]
3. NDV/IBV	AFT	1/25	1.0	295 [a]
4. NDV/AIV/IBV	AFT	1/1/25	1.5	269 [a]
5. NDV/AIV	BFF	1/1	1.0	314 [a]
6. NDV/IBV	BEF	1/25	1.0	281 [a]
7. NDV/AIV/IBV	BEF	1/1/25	1.5	183 [a]
8. NDV/AIV	BEF	2/2	0.5	317 [ac]
9. NDV/IBV	BEF	2/50	0.5	210 [bc]
10. NDV/AIV/IBV	BEF	3/3/75	0.5	171 [b]
High concentrations of NDV, AIV antigen				
11. NDV control	M	12.5	0.5	576 [a]
12. NDV/AIV	AFT	25/25	0.5	536 [a]
13. NDV/IBV	AFT	25/50	0.5	474 [a]
14. NDV/AIV/IBV	AFT	37.5/37.5/75	0.5	368 [a]
15. NDV/AIV	BEF	25/25	0.5	458 [a]
16. NDV/IBV	BEF	25/50	0.5	439 [a]
17. NDV/AIV/IBV	BEF	37.5/37.5/75	0.5	434 [a]

Note: [A] M, monovalent vaccines; AFT, polyvalent vaccines from antigens combined after emulsification; BEF, polyvalent vaccines from antigens combined before emulsification. [B] 1 × concentration, original antigen concentration in allantoic fluid. [C] Cumulative means determined from geometric mean titers (GMT) at 2, 3, 4, 6, and 8 weeks after vaccination. Vaccine titers followed by superscripts different from those of the control vaccines are significantly different from the control ($P<0.05$). For each mean, n=5.

Table 6-2-2 shows AIV-HI results from groups vaccinated with monovalent or type AFT and type BEF polyvalent vaccines. Groups vaccinated with low concentrations (1 × ~3 ×) of NDV and AIV antigen of type M, AFT, or BEF vaccine groups had no significant differences in HI titers. HI titers of groups were significantly lower than their controls (No.22) when highly concentrated (25 × ~37.5 × conc.) NDV and AIV antigen of type AFT and BEF vaccines contained NDV antigen (Nos.12, 14, 15, and 17). The control monovalent vaccine with concentrated (12.5 ×) antigen (No.22) induced significantly higher HI titers than its counterpart with 1 × antigen concentration (No.18).

Table 6-2-2 Avian influenza geometric mean hemagglutination-inhibition titers of chickens after vaccination with monovalent or polyvalent vaccines made by combined Newcastle disease, avian influenza, and infectious bronchitis antigens before or after emulsification

Vaccine group No. and antigen	Vaccine type [A]	Antigen concentration (×) [B]	Dose/mL	8-week cumulative HI GMT [C]
Low concentrations of NDV, AIV antigen				
18. AIV control	M	1	0.5	841 [a]
2. AIV/NDV	AFT	1/1	1.0	757 [a]
19. AIV/IBV	AFT	1/25	1.0	833 [a]
4. AIV/NDV/IBV	AFT	1/1/25	1.5	848 [a]
5. AIV/NDV	BEF	1/1	1.0	843 [a]
20. AIV/IBV	BEF	1/25	1.0	1 026 [a]
7. AIV/NDV/IBV	BEF	1/1/25	1.5	860 [a]
8. AIV/NDV	BEF	2/2	0.5	1 585 [a]
21. AIV/IBV	BEF	2/50	0.5	1 695 [a]
10. AIV/NDV/IBV	BEF	3/3/75	0.5	942 [a]
High concentrations of NDV, AIV antigen				
22. AI control M	M	12.5	0.5	2 055 [b]
12. AIV/NDV	AFT	25/25	0.5	1 127 [ac]
23. AIV/IBV	AFT	25/50	0.5	1 560 [bc]
14. AIV/NDV/IBV	AFT	37.5/37.5/75	0.5	993 [a]
15. AIV/NDV	BEF	25/25	0.5	920 [a]
24. AIV/IBV	BEF	25/50	0.5	153 [b]
17. AIV/NDV/IBV	BEF	37.5/37.5/75	0.5	985 [a]

Note: [A] M, monovalent vaccines; AFT, polyvalent vaccines from antigens combined after emulsification; BEF, polyvalent vaccines from antigens combined before emulsification. [B] 1 × concentration, original antigen concentration in allantoic fluid. [C] Cumulative means determined from geometric mean titers (GMT) at 2, 3, 4, 6, and 8 weeks after vaccination. Vaccine titers followed by superscripts different from those of the control vaccines are significantly different from the control ($P<0.05$). For each mean, $n=5$.

Table 6-2-3 shows IBV-HI results of type M, AFT, and BEF IBV vaccines. No significant differences in HI titers were found between vaccine groups. Thirteen of 15 polyvalent vaccines did, however, induce higher HI titers than did the monovalent control vaccine (No.25).

Table 6-2-3 Infectious bronchitis geometric mean hemagglutination-inhibition titers of chickens after vaccination with monovalent or polyvalent vaccines made by combined Newcastle disease, avian influenza, and infectious bronchitis antigens before or after emulsification

Vaccine group No. and antigen	Vaccine Type [A]	Antigen concentration (×) [B]	Dose/mL	8-week cumulative HI GMT [C]
Low concentrations of NDV, AIV antigen				
25. IBV control	M	25	0.5	989 [a]
3. IBV/NDV	AFT	25/1	1.0	1 298 [a]
19. IBV/AIV	AFT	25/1	1.0	1 248 [a]
4. IBV/NDV/AIV	AFT	25/1/1	1.5	1 489 [a]
6. IBV/NDV	BEF	25/1	1.0	1 696 [a]

continued

Vaccine group No. and antigen	Vaccine Type [A]	Antigen concentration (×) [B]	Dose/mL	8-week cumulative HI GMT [C]
20. IBV/AIV	BEF	25/1	1.0	1 536[a]
7. IBV/NDV/AIV	BEF	25/1/1	1.5	1 453[a]
9. IBV/NDV	BEF	50/2	0.5	1 260[a]
21. IBV/AIV	BEF	50/2	0.5	1 395[a]
10. IBV/NDV/AIV	BEF	75/3/3	0.5	1 235[a]
High concentrations of NDV, AI antigen				
13. IBV/NDV	AFT	50/25	0.5	2 050[a]
23. IBV/AIV	AFT	50/25	0.5	1 256[a]
14. IBV/NDV/AIV	AFT	75/37.5/37.5	0.5	877[a]
16. IBV/NDV	BEF	50/25	0.5	1 103[a]
24. IBV/AIV	BEF	50/25	0.5	1 040[a]
17. IBV/NDV/AIV	BEF	75/37.5/37.5	0.5	850[a]

Note: [A] M, monovalent vaccines; AFT, polyvalent vaccines from antigens combined after emulsification; BEF, polyvalent vaccines from antigens combined before emulsification. [B] 1 × concentration, original antigen concentration in allantoic fluid. [C] Cumulative means determined from geometric means at 2, 3, 4, 6, and 8 weeks after vaccination. Vaccine titers followed by superscripts different from those of the control vaccine are significantly different from the control ($P < 0.05$). For each mean, $n = 5$.

Table 6-2-4 shows challenge results from VV NDV exposure. All groups experienced a high (two-fold or greater) secondary serologic response following challenge. All non-NDV vaccinates died, whereas none of the vaccinates had clinical ND symptoms or died.

Table 6-2-4 Response of white rock chickens challenged with viscerotropic velogenic Newcastle disease virus following vaccination with monovalent and polyvalent oil-emulsion vaccines [A]

Vaccine group No. and antigen	Vaccine type [B]	Antigen concentration (×)[C]	Dose /mL	HI GMT [D]		No. dead/ total	No. morbid/ total
				At challenge	2 weeks post-challenge		
NDV vaccinates							
1. NDV control	M		0.5	320	970	0/5	0/5
2. NDV/AIV	AFT	1/1	1.0	299	1 140	0/5	0/5
3. NDV/IBV	AFT	1/25	1.0	279	1 120	0/5	0/5
4. NDV/AIV/IBV	AFT	1/1/25	1.5	279	905	0/5	0/5
5. NDV/AIV	BEF	1/1	1.0	211	844	0/5	0/5
6. NDV/IBV	BEF	1/25	1.0	243	485	0/5	0/5
7. NDV/AIV/IBV	BEF	1/1/25	1.5	149	788	0/5	0/5
8. NDV/AIV	BEF	2/2	0.5	211	1 280	0/5	0/5
9. NDV/IBV	BEF	2/50	0.5	130	970	0/5	0/5
10. NDV/AI/IBV	BEF	3/3/75	0.5	92	2 941	0/5	0/5
11. NDV control	M	12.5	0.5	343	1 689	0/5	0/5
12. NDV/AIV	AFT	25/25	0.5	243	970	0/5	0/5
13. NDV/IBV	AFT	25/50	0.5	243	1 280	0/5	0/5

continued

Vaccine group No. and antigen	Vaccine type [B]	Antigen concentration (×)[C]	Dose /mL	HI GMT[D]		No. dead/ total	No. morbid/ total
				At challenge	2 weeks post-challenge		
14. NDV/AIV/IBV	AFT	37.5/37.5/75	0.5	197	1689	0/5	0/5
15. NDV/AIV	BEF	25/25	0.5	149	2 229	0/5	0/5
16. NDV/IBV	BEF	25/50	0.5	211	1 689	0/5	0/5
17. NDV/AIV/IBV	BEF	37.5/37.5/75	0.5	211	1 470	0/5	0/5
Non-NDV vaccinates							
18. AIV	M	1	0.5	0			5/5
22. AIV	M	12.5	0.5	0			5/5
25. IBV	M	25	0.5	0			5/5
19. AIV/IBV	AFT	1/25	1.0	0			5/5
20. AIV/IBV	BEF	1/25	1.0	0			5/5
21. AIV/IBV	BEF	2/50	0.5	0			5/5
23. AIV/IBV	AFT	25/50	0.5	0			5/5
24. AIV/IBV	BEF	25/50	0.5	0			5/5

Note: [A] Challenge was by intratracheal instillation of 0.1 mL of $10^{7.3}$ ELD_{50} of VVNDV, Largo strain. [B] M, monovalent vaccines; AFT, polyvalent vaccines from antigens combined after emulsification; BEF, polyvalent vaccines from antigens combined before emulsification. [C] 1 × concentration, original antigen concentration in allantoic fluid. [D] Reciprocal of geometric mean hemagglutination-inhibition titer based on total of five birds per challenge group.

Table 6-2-5 shows results from challenge with IBV. All controls seroconverted at 1 week post-challenge. All vaccinated groups had an HI secondary response of twofold or more. Sixty of 80 chickens from the different IBV-vaccinated groups responded with a fourfold or greater increase in HI titer.

Table 6-2-5 Response of white rock chickens challenged with Mass 41 infectious bronchitis virus following vaccination with monovalent and polyvalent oil-emulsion vaccines[A]

Vaccine group No. and antigen	Vaccine type [B]	Antigen concentration (×)[C]	Dose/mL	HI GMT[D]		No. with 4 × HI/ total challenged
				At challenge	1 week post-challenge	
IBV vaccinates						
25. IBV control	M	25	0.5	788	3 380	5/5
3. IBV/NDV	AFT	25/1	1.0	1 280	5 120	4/5
19. IBV/AIV	AFT	25/1	1.0	1 280	2 229	0/5
4. IBV/NDV/AIV	AFT	25/1/1	1.5	1 114	3 378	3/5
6. IBV/NDV	BEF	25/1	1.0	1 280	2 560	1/5
20. IBV/AIV	BEF	25/1	1.0	733	2 560	3/5
7. IBV/NDV/AIV	BEF	25/1/1	1.5	733	3 380	5/5
9. IBV/NDV	BEF	50/2	0.5	733	2 229	3/5
21. IBV/AIV	BEF	50/2	0.5	1 470	5 120	3/5
10. IBV/NDV/AIV	BEF	75/3/3	0.5	557	3 880	5/5

continued

Vaccine group No. and antigen	Vaccine type [B]	Antigen concentration (×)[C]	Dose/mL	HI GMT [D]		No. with 4 × HI/ total challenged
				At challenge	1 week post-challenge	
13. IBV/NDV	AFT	50/25	0.5	640	2 941	4/5
23. IBV/AIV	AFT	50/25	0.5	557	5 120	5/5
14. IBV/NDV/AIV	AFT	75/37.5/37.5	0.5	844	5 120	5/5
16. IBV/NDV	BEF	50/25	0.5	905	4 457	4/5
24. IBV/AIV	BEF	50/25	0.5	1 280	5 120	5/5
17. IBV/NDV/AIV	BEF	75/37.5/37.5	0.5	394	5 120	5/5
Non-IBV Vaccinates						
1. NDV	M	1	0.5	0	226	5/5
11. NDV	M	12.5	0.5	0	279	5/5
18. AIV	M	1	0.5	0	160	5/5
22. AIV	M	12.5	0.5	0	279	5/5
2. NDV/AIV	AFT	1/1	1.0	0	226	5/5
12. NDV/AIV	AFT	25/25	1.0	0	226	5/5
5. NDV/AIV	BEF	1/1	1.0	0	98	5/5
8. NDV/AIV	BEF	2/2	0.5	0	279	5/5
15. NDV/AIV	BEF	25/25	0.5	0	226	5/5

Note: [A] Challenge was by intratracheal instillation 0.1 mL of $10^{4.7}$ ELD_{50} of Mass 41 IBV. [B] M, monovalent vaccines; AFT, polyvalent vaccines from antigens combined after emulsification; BEF, polyvalent vaccines from antigens combined before emulsification. [C] 1 × concentration, original antigen concentration in allantoic fluid. [D] Reciprocal of geometric mean hemagglutination-inhibition titers based on a total five birds per challenge group.

Discussion

It has been shown previously that viable IBV strains can interfere with NDV growth in chickens, chicken embryos, and cell cultures, as measured by NDV hemagglutination activity[4, 5, 19, 21, 26]. Live IBV vaccine has been reported to interfere with the B1 strain of live NDV vaccine; i.e., it produced significantly less immunity and less protection against ND challenge[29]. Hitchner et al.[22] reported a reduction in NDV HI titer when IBV and NDV were administered in a combined vaccine. The main criticism of live combined NDV/IB vaccines has been that the IB component of the vaccine may interfere with the ability of the chicken to respond to the NDV component[9, 26, 29]. However, similar interference has not been demonstrated with inactivated NDV/IBV, NDV/AIV, or NDV/AIV/IBV combined vaccines.

Table 6-1-1 shows that unconcentrated vaccines and low-concentration NDV/AIV vaccines of antigen type BEF (Nos.6, 7, 9, and 10) with IBV antigen (antigens mixed before emulsification) induced significantly lower NDV HI titers than those conferred by NDV control monovalent vaccine (No.1). The interference phenomenon was less when the NDV and AI antigen concentrations were increased in type BEF vaccines (Nos.16 and 17). All type AFT vaccine groups had HI titers that were not significantly different from their control vaccine groups.

The interference phenomenon is apparently related to the direct mixing of unemulsified NDV and IBV antigen, because mixing of the monovalent vaccines of the same antigen mass (Nos. 3 and 4) did not cause a significant lowering of NDV-HI response. The ratio of the two antigens may somehow determine when the interference occurs, because vaccines with concentrated NDV antigen combined with IBV antigen apparently compensated for or did not have the interference and conferred HI titers equivalent to those conferred by concentrated NDV/AIV vaccines (Nos.12 and 15). IBV virus particles characteristically have bald areas on the surface of a large proportion of their particles. Surface charges at NDV HA sites and IBV bald areas could possibly differ enough to allow electrostatic reaction and thus obscure the antigenic surfaces of NDV. The data suggest that monovalent NDV vaccines are most efficacious when used alone and that polyvalent vaccines perform better when made up of combined monovalent vaccines rather than with mixed antigens that are then emulsified.

Work by others has clearly demonstrated the effectiveness of inactivated avian influenza monovalent oil-emulsion vaccine and multisubtype polyvalent AIV oil-emulsion vaccines in chickens[3, 10, 11, 16, 27].

Table 6-2-2 shows that AIV-HI titers of vaccinated groups was lowered significantly compared with their monovalent control (No.22) only when AIV and NDV antigen were concentrated 25 × and 37.5 × in both type AFT and type BEF vaccines (Nos.12, 14, 15, and 17). That the HI reduction effect occurred also in type AFT vaccines suggests that a direct physical interaction between AIV and NDV antigens was not the cause, because gently mixed monovalent vaccines as prepared in type AFT would isolate the two antigens in separate emulsified antigen particles. In addition, all type BEF vaccines of low antigen concentration with AIV and NDV (No.5, 7, 8, 10, 20, and 21) conferred higher titers than the control monovalent vaccine (No.18). These results suggest that the lowering of HI titers was directly related to raised NDV and AIV concentration of antigens in the vaccines (No.12, 14, 15, and 17). At present, the mechanism of this phenomenon is unclear. In general, efficacy of highly concentrated AIV polyvalent vaccine was compromised by highly concentrated NDV antigen, and these combinations should be avoided when maximum efficacy is required. Because there were only 10 chickens in each vaccine group, it was decided that the AIV challenge would be omitted so that there would be five birds for each of the NDV and IBV challenge groups. Others have shown that HI antibody levels are a reliable indication of immunity to AIV[10, 27].

IBV combined with NDV or other antigen in polyvalent vaccines has been reported for many years[17, 25, 29, 30]. Otsuki and Iritani[25] reported that trivalent vaccine containing inactivated NDV and IBV and *Haemophilus gallinarum* induced a better immunological response to IBV than a vaccine containing inactivated IBV only. Gough et al.[17] reported that NDV/IB inactivated bivalent vaccines induced an enhanced immune response to IBV.

A similar phenomenon is shown in Table 6-2-3. Thirteen of 15 polyvalent IBV vaccines containing NDV or AIV or NDV/AIV antigen induced a higher HI response to IB than a vaccine containing equivalent inactivated IBV antigen mass. The data suggest that IBV antigen is highly compatible in the polyvalent vaccines described.

Table 6-2-4 shows that challenge with VV NDV did not cause morbidity or mortality in any of vaccinated chickens. Also, all chickens that received non-NDV vaccine died. Therefore, the challenge was adequate to prove vaccine efficacy. These results were expected, as all vaccinated groups had protective HI titer levels (1 : 92 to 1 : 343) at the time of challenge. The high secondary response indicated that most groups were infected by challenge virus. Vaccine groups given highly concentrated vaccines responded with higher secondary HI antibody levels overall, reflecting the greater degree of sensitization from increased level of vaccine antigen. These antibody levels still did not prevent infection by the challenge virus. The group (No. 10) with the lowest HI titer at challenge responded with the highest secondary titer (2941), which suggests the greatest extent of infection by the challenge virus. The data suggest that: (1) the polyvalent vaccines did not compromise the NDV portion of the vaccines; and (2) the high concentration of the NDV antigen did not improve vaccine efficacy sufficient to warrant its use.

Results of challenge shown in Table 6-2-5 reflect an apparent enhanced HI response but also an interfering effect on IBV in polyvalent vaccines. The nine polyvalent vaccine groups with low concentration (1 × ~3 ×) AI or NDV comprised 27 of 45 birds without a 4 × or greater increase in HI titer following challenge. The vaccine groups with highly concentrated (25 × ~37.5 ×) AIV or NDV antigens comprised only two of 30 birds without a 4 × or greater increase in HI titer following challenge. These results suggest an interfering effect on vaccine efficacy from highly concentrated AIV and NDV. Also, the six polyvalent vaccines with highly concentrated NDV and AIV antigen conferred lower HI titers at time of challenge than did their counterparts with low concentrations of NDV and AIV polyvalent vaccines. The pre-challenge HI data supports the post-challenge HI response.

The results of this limited study should serve as the basis for exercising caution when considering the preparation of multi antigen inactivated oil-emulsion vaccines. Because of different effects due to the relative concentrations and type of antigens, each individual combination should be evaluated for effects of enhancement or interference.

References

[1] ALLAN W H. Newcastle disease control. London: Agriculture, 1972.

[2] ALLAN W H. Newcastle disease vaccines, their production and use. Rome: Food and Agriculture Organization of The United Nations Press, 1978.

[3] ALLAN W H, MADELEY C R, KENDAL A P. Studies with avian influenza A viruses: cross protection experiments in chickens. The Journal of General Virology, 1971, 12(2): 79-84.

[4] BEARD C W. Infectious bronchitis virus interference with Newcastle disease virus in monolayers of chicken kidney cells. Avian Dis, 1967, 11(3): 399-406.

[5] BEARD C W. An interference type of serological test for infectious bronchitis virus using Newcastle disease virus. Avian Dis, 1968, 12(4): 658-665.

[6] BEARD C W, BRUGH M. Immunity to Newcastle disease. American Journal of Veterinary Research, 1975, 36(4 Pt 2): 509-512.

[7] BEARD C W, HANSON P R. Newcastle disease//HOFSTAD M S, BARNES H J, CALNEK B W, et al. Diseases of poultry, 8th ed. Iowa: Iowa State University Press, 1984.

[8] BEARD C W, WILKES W J. A simple and rapid microtest procedure for determining Newcastle hemagglutination-inhibition (HI) antibody titers. Proceedings, annual meeting of the United States Animal Health Association, 1973(77): 569-600.

[9] BENGELSDORFF H J. Protective effect of vaccination and the antibody production after combined administration of Newcastle disease and infectious bronchitis vaccines. Blue Book for the Veterinary Profession, 1972, 22: 144-151.

[10] BRUGH M, Stone H D. Immunization of chickens against influenza with Hemagglutinin-specific (H5) emulsion vaccine. Avian Dis, 1986, 47, 283-292.

[11] BRUGH M, BEARD C W, STONE H D. Immunization of chickens and turkeys against avian influenza with monovalent and polyvalent oil emulsion vaccines. American Journal of Veterinary Research, 1979, 40(2): 165-169.

[12] BRUNING J L, KINTZ B L. Computational handbook of statistics. Glenview: Scott, Foresman and Co. Press, 1968.

[13] CORIA M F, HOFSTAD M S. Immune response in chickens to infectious bronchitis virus, strain 33. I. Response to beta-propiolactone-inactivated virus. Avian Dis, 1971, 15(4): 688-695.

[14] DOWDLE W R. Influenza virus//ROSE N H, FRIEDMAN F. Manual of clinical immunology. Washington, D C: American

Society for Microbiology Press, 1976.

[15] EASTERDAY B C, BEARD C W. Avian influenza// HOFSTAD M S, BARNES H J, CALNEK B W, et al. Diseases of poultry, 8th ed. Iowa: Iowa State University Press, 1984.

[16] GOUGH R E, ALLAN W H, KNIGHT D J, et al. Further studies on the adjuvant effect of an interferon inducer (BRL 5907) on Newcastle disease and avian influenza inactivated vaccines. Research in veterinary science, 1975, 19(2): 185-188.

[17] GOUGH R E, ALLAN W H, NEDELCIU D. Immune response to monovalent and bivalent Newcastle disease and infectious bronchitis inactivated vaccines. Avian Pathol, 1977, 6(2): 131-142.

[18] GOUGH R E, WYETH P J, BRACEWELL C D. Immune responses of breeding chickens to trivalent oil emulsion vaccines: responses to infectious bronchitis. The Veterinary Record, 1981, 108(5): 99-101.

[19] HANSON L E, WHITE F H, ALBERTS J O. Interference between Newcastle disease and infectious bronchitis viruses. American Journal of Veterinary Research, 1956, 17(63): 294-298.

[20] HANSON R P, SPALATIN J, JACOBSON G S. The viscerotropic pathotype of Newcastle disease virus. Avian Dis, 1973, 17(2): 354-361.

[21] HIDALGO H, GALLARDO R, VIVAR J, et al. Identification of field isolates of infectious bronchitis virus by interference with the LaSota of strain of Newcastle disease virus. Avian Dis, 1985, 29(2): 335-340.

[22] HITCHNER S B, WHITE P G, LOZANO E A. Screening tests on modified strains of infectious bronchitis virus. ASL Research Representative. Madison: American Scientific Laboratories Inc Press, 1955.

[23] HOFSTAD M S. Avian infectious bronchitis// HOFSTAD M S, BARNES H J, CALNEK B W, et al. Diseases of poultry, 8th ed. Iowa: Iowa State University Press, 1984.

[24] KING D J, HOPKINS S R. Evaluation of the hemagglutination-inhibition test for measuring the response of chickens to avian infectious bronchitis virus vaccination. Avian Dis, 1983, 27(1): 100-112.

[25] OTSUKI K, IRITANI Y. Preparation and immunological response to a new mixed vaccine composed of inactivated Newcastle disease virus, inactivated infectious bronchitis virus, and inactivated Hemophilus gallinarum. Avian Dis, 1974, 18(3): 297-304.

[26] RAGGI L G, LEE G G. Infectious bronchitis virus interference with growth of Newcastle disease virus. I. Study of interference in chicken embryos. Avian Dis, 1963, 7(1): 106-122.

[27] STONE H D. Efficacy of avian influenza oil-emulsion vaccines in chickens of various ages. Avian Dis, 1987, 31(3): 483-490.

[28] STONE H D, BRUGH M, HOPKINS S R, et al. Preparation of inactivated oil-emulsion vaccines with avian viral or Mycoplasma antigens. Avian Dis, 1978, 22(4): 666-674.

[29] THORNTON D H, MUSKETT J C. Effect of infectious bronchitis vaccination on the performance of live Newcastle disease vaccine. The Veterinary Record, 1975, 96(21): 467-468.

[30] UCHINUNO Y, YAMADA S, FUJIKAWA H, et al. Newcastle disease-infectious bronchitis inactivated combined vaccines. Journal of the Japan Veterinary Medical Association, 1973, 26: 124-128.

Phylogeny and pathogenicity of subtype XIIb NDVs from francolins in southwestern China and effective protection by an inactivated vaccine

Zeng Tingting, Xie Liji, Xie Zhixun, Huang Jiaoling, Xie Zhiqin, Huang Qinghong, Luo Sisi, Wang Sheng, Li Meng, Hua Jun, Zhang Yanfang, and Zhang Minxiu

Abstract

Most genotype XII Newcastle disease viruses (NDVs) were isolated from poultry, chickens, or geese, with the exception of one subtype, XIIa NDV, which was isolated from a peacock. Here, two subtype XIIb NDVs, francolin/China/GX01/2017 and francolin/China/GX02/2017 (GX01 and GX02 hereafter), were isolated from francolins, which are resident birds in southern China. GX01 and GX02 were characterized as velogenic NDVs. Based on the weaker pathogenicity of these viruses in chickens, the amino acid sequences of seven proteins from genotype XII NDVs were compared, which revealed 17, 40, 15, 7, 32, 25, and 31 variations in the NP, P, M, F, HN, L, and V proteins, respectively, some of which could be responsible for this decreased pathogenicity. Epidemiological and phylogenetic analyses suggest that subtype XIIb NDVs have multiple transmission chains, and that resident birds may be involved in this process as intermediate hosts in which viruses keep evolving. Because of the increased pathogenicity of subtype XIIb NDVs, the protective efficacy of GX01 as an inactivated vaccine was evaluated and compared with that of two commercial inactivated vaccines in chickens. The results showed that the subtype XIIb NDVs could be candidate genotype-matched vaccine strains against genotype XII NDVs.

Keywords

XIIb NDVs, phylogeny, pathogenicity, genotype-match vaccine

Introduction

Newcastle disease viruses (NDVs), often referred to as avian paramyxoviruses 1 (APMV-1), which have been classified to the genus Avian *orthoavulavirus* 1 (AOAV-1) within the family Paramyxoviridae and cause diseases in a variety of domestic and wild birds worldwide[1]. Virulent strains, which are highly contagious and pathogenic, are viruses that have an intracerebral pathogenicity index (ICPI) of at least 0.7 (2.0 is the maximum) or a fusion cleavage site with several basic amino acids and phenylalanine at position 117[2]. Currently, NDVs are divided into two classes, class I with subtypes 1.1.1, 1.1.2, and 1.2, and class II with at least 20 genotypes (I to XXI) in class II according to new naming criteria[3]. Epidemiological data have revealed that genotypes VI and VII NDVs are the predominant epidemic strains in China and cause substantial poultry industry financial losses[4-7]. Genotype XII NDVs were isolated and characterized in 2010 in Guangdong Province, China[8].

Genotype XII NDVs isolated in China cluster in subtype XIIb[8], whereas subtype XIIa NDVs were isolated in South America, and subtype XIId NDVs were isolated in Vietnam[9-11]. Subtype XIIa and XIId NDVs have all been isolated from birds with typical ND manifestations, and the fusion cleavage site or the ICPI value indicated that these strains belonged to velogenic NDVs and caused lethality in chickens in experiments[10]. However, no outbreak associated with this genotype has been reported in China, and subtype ⅫⅡb NDVs have been isolated from apparently healthy geese since 2010. In this study, two subtype ⅫⅡb NDV strains were isolated from apparently healthy francolins. Although the fusion cleavage site or the ICPI value indicated that all subtype ⅫⅡb NDVs were velogenic NDVs, subtype ⅫⅡb NDVs caused ND manifestations but not death in chickens in both previous studies[8] and our work. Due to the differences in pathogenicity, the variations in amino acids of seven proteins (NP, P, M, F, HN, L, and V) were examined. The protection efficiency of inactivated vaccines of genotype ⅫⅡ NDVs with commercial vaccines were evaluated.

Materials and methods

Virus isolation and pathogenic evaluation

The NDVs francolin/China/GX01/2017 and francolin/China/GX02/ 2017 (abbreviated GX01 and GX02, respectively, below) were isolated from oropharyngeal and cloacal swabs of apparently healthy francolins in the suburban region of Nanning city, Guangxi, southwestern China, during a surveillance project in 2017. Oropharyngeal and cloacal suspend oropharyngeal and cloacal swab samples in 1 mL of phosphate-buffered saline (PBS) containing antibiotics (penicillin, 1 000 U/mL; streptomycin, 100 μg/mL; amphotericin B, 0.25 μg/mL). Virus isolation was carried out using 9-day-old SPF chicken embryos (Beijing Merial Vital Laboratory Animal Technology Co., Ltd., China) and then incubated at 37 ℃.

Allantois fluids were harvested and identified by the hemagglutination (HA)-inhibition (HI) test following the OIE protocol[2]. Briefly, the HA titers of allantois fluids were determined with 0.5% chicken red blood cells (RBCs) then were generated against 4 hemagglutination units of standardized antigen (4U antigen). Antisera of NDV, avian influenza virus (AIV) subtypes H1-H16, and egg drop syndrome virus (EDSV) were serially double diluted and mixed with 4U of antigen. RBCs were added to assess the HI titers. Viruses were identified when HA was blocked by the corresponding antiserum in the HI assay.

After four plaque purifications in DF-1 cells, the viruses were propagated in 9-day-old SPF chicken embryos and stored at −80 ℃ until use. The virulence of the NDVs was evaluated by determining the mean death time (MDT) in 10-day-old SPF chicken embryos and intracerebral pathogenicity test (ICPI) in 1-day-old SPF chicks following the OIE protocol[2].

Genome sequencing

Viral genomic RNA was extracted with a viral RNA/DNA extraction kit. Complete genomes were amplified by one-step RT-PCR with the High Fidelity One Step RT-PCR Kit and previously described primers, [12]according to the kit's protocol. RT-PCR fragments were amplified, ligated into a pMD18-T vector then transformed into *Escherichia coli* DH5α. All reagents and kits were purchased from TaKaRa (TaKaRa, Dalian, China). The positive clones were sequenced by BGI (BGI, Shenzhen, China). The two genomes were assembled by using the SeqMan program (DNASTAR, Inc. Madison, USA). The GenBank accession numbers of GX01 and GX02 are MZ306226 and MZ306225, respectively.

Genome analysis

The deduced amino acid sequences of all protein-encoding genes were compared with those of the other genotype Ⅻ NDVs by the MegAlign program (DNASTAR, Inc.). To analyze the evolutionary relationship, a maximum likelihood tree was constructed with a pilot tree dataset using full-length F gene sequences[3]. The maximum likelihood tree was constructed via the CIPRES Science Gateway website[13] based on a general time-reversible (GTR) model with the combination GTR+Γ+I and 1 000 bootstrap replicates. The mean evolutionary distances between groups were calculated by MEGA v. X software based on the number of base substitutions per site from averaging over all sequence pairs between groups utilizing the maximum composite likelihood model with rate variation among sites with a gamma distribution (shape parameter 1). Trees were visualized using FigTree v1.4.2.

Estimated protection efficiency of an inactivated vaccine of genotype Ⅻ NDVs compared with commercial vaccines

Inactive vaccine generation

The protective efficacy of the commercial vaccines in China, LaSota and A-Ⅶ[14], were evaluated in a previous study[8]. In this study, GX01 was used as an inactivated vaccine for comparison with the protective efficacy of LaSota and A-Ⅶ. GX01 was propagated in 9-day-old SPF chicken embryos, and allantois fluids were harvested. The allantois fluids were filtered and concentrated by sucrose gradient centrifugation to a 10 $\log_2$ HA titer. Viral suspensions were inactivated by incubation with 0.02% β-propiolactone (Sigma-Aldrich, Darmstadt, Germany) at 4 ℃ for 24 hand then mixed with Tween 80 at a 9 : 1 ratio to constitute the aqueous phase. Lipids mixed with Span 80 at a 25 : 1 ratio constituted the oil phase. The aqueous phase was mixed with the oil phase at a 1 : 4 ratio to generate the inactive vaccine. PBS was used to replace GX01 to generate an oil-emulsion vaccine as a control. All reagents used to generate the oil-emulsion vaccine were purchased from Sigma (Sigma-Aldrich, Darmstadt, Germany).

Animal experiments

SPF chickens were divided into four groups and immunized with the inactivated vaccines GX01, LaSota, and A-Ⅶ or PBS. Each chicken was immunized twice, at two weeks old and four weeks old, with a volume of 0.2 mL. The antibody titers were monitored every week using the corresponding antigen by HI assay as described in Section *Virus isolation and pathogenic evaluation*. Five weeks after the first immunization, eleven SPF chickens with antibody titers reaching 10 $\log_2$ were selected for challenge with 10^6 EID_{50} in a volume 0.2 mL of GX01 through intraocular and intranasal routes, and six SPF chickens in the PBS-immunized group were challenged with 0.2 mL PBS through the same routes.

Detection of viral shedding

All birds were observed daily for clinical symptoms for 14 days. To detect viral shedding, oropharyngeal and cloacal swabs were collected at 1, 3, 5, 7, 9, 11, and 13 days post infection (dpi). The swab samples were suspended in 1 mL of PBS containing antibiotics as described before, and then viral titers of swab samples were determined by a 50% egg infectious dose (EID_{50}) assay and calculated by the Reed and Muench's method.

Pathological observation and analysis

In addition, three SPF chickens in each group were randomly selected and dissected at 3 dpi for observation, and tissue lesions of the liver, spleen, kidney, lung, bursa, pancreas, glandular stomach, cecum tonsil, and brain were collected and fixed in a 10% formalin solution to prepare pathological sections. The collected tissues and organs were fixed in a 10% neutral formalin solution for more than 24 h, dehydrated and cleared with an alcohol series. After paraffin sealing, the tissues were embedded, and 4～10 μm-thick sections were prepared. After HE staining, the sections were dehydrated and sealed, and pathological changes in the sections were observed by microscopy (Nikon, Tokyo, Japan) and image capture.

Statistical analysis

Data were analyzed by using Prism (v.9.4.0) using the ROUT test (GraphPad Software Inc., USA). The D'Agostino-Pearson normality test was used to estimate whether the values in each group were derived from according with a Gaussian distribution. One-way ANOVA statistical analysis was used for multiple comparisons of viral titers in oropharyngeal and cloacal swab samples from each group with the same sampling point based on the normality distribution. The statistical significance threshold was set at a P value of <0.05.

Results

Isolation and characterization of NDVs

The NDVs GX01 and GX02 were HI positive with NDV-specific antiserum but negative with antisera for avian influenza viruses (AIVs) of subtypes H1-H16 and egg drop syndrome virus (EDSV) [15]. The MDT values of GX01 and GX02 were 58 h and 57 h, and the ICPI values were 1.613 and 1.65, respectively. According to the OIE protocol, GX01 and GX02 were characterized as velogenic NDVs.

Genome sequencing and phylogenetic analyses with a pilot tree

The lengths of both GX01 and GX02 were 15 192 nt, composing six genes in the order 3'-NP-P-M-F-HN-L-5'. Both strains had leader and trailer sequences comprising 55 and 114 nt, respectively. The length of the intergenic region of the NP-P, P-M, and M-F genes was 1 nt and that of the F-HN and HN-L genes was 31 and 47 nt, respectively.

A maximum-likelihood tree was constructed based on the full length of the F gene with the pilot tree plus all genotype Ⅻ NDVs as reported previously (Figure 6-3-1). The distance between GX01 and GX02 was 0.000 6. These two strains were clustered with subtype Ⅻb NDVs isolated in China, and the distance from other subtype Ⅻb NDVs was 0.017 4～0.022 0. The average genetic distance between Chinese strains (subtype Ⅻb) and South American (subtype Ⅻa) strains was 0.090 9, and it was 0.069 1 between Chinese strains (subtype Ⅻb) and Vietnam strains (subtype Ⅻd).

Analysis of six protein-encoding genes

The open reading frame (ORF) of the F gene of these two strains was 1 662 nt in length, encoding 553 amino acids (aa). The deduced amino acid at the cleavage site of the protein F was 112RRQKR ↓ F117, which is the determinant of velogenic NDVs. There were 7 deduced amino acid variations between subtype Ⅻb and Ⅻa, Ⅻd NDVs, with 1 variation in the signal peptide, 3 in heptad repeat regions, 1 in the cytoplasmic

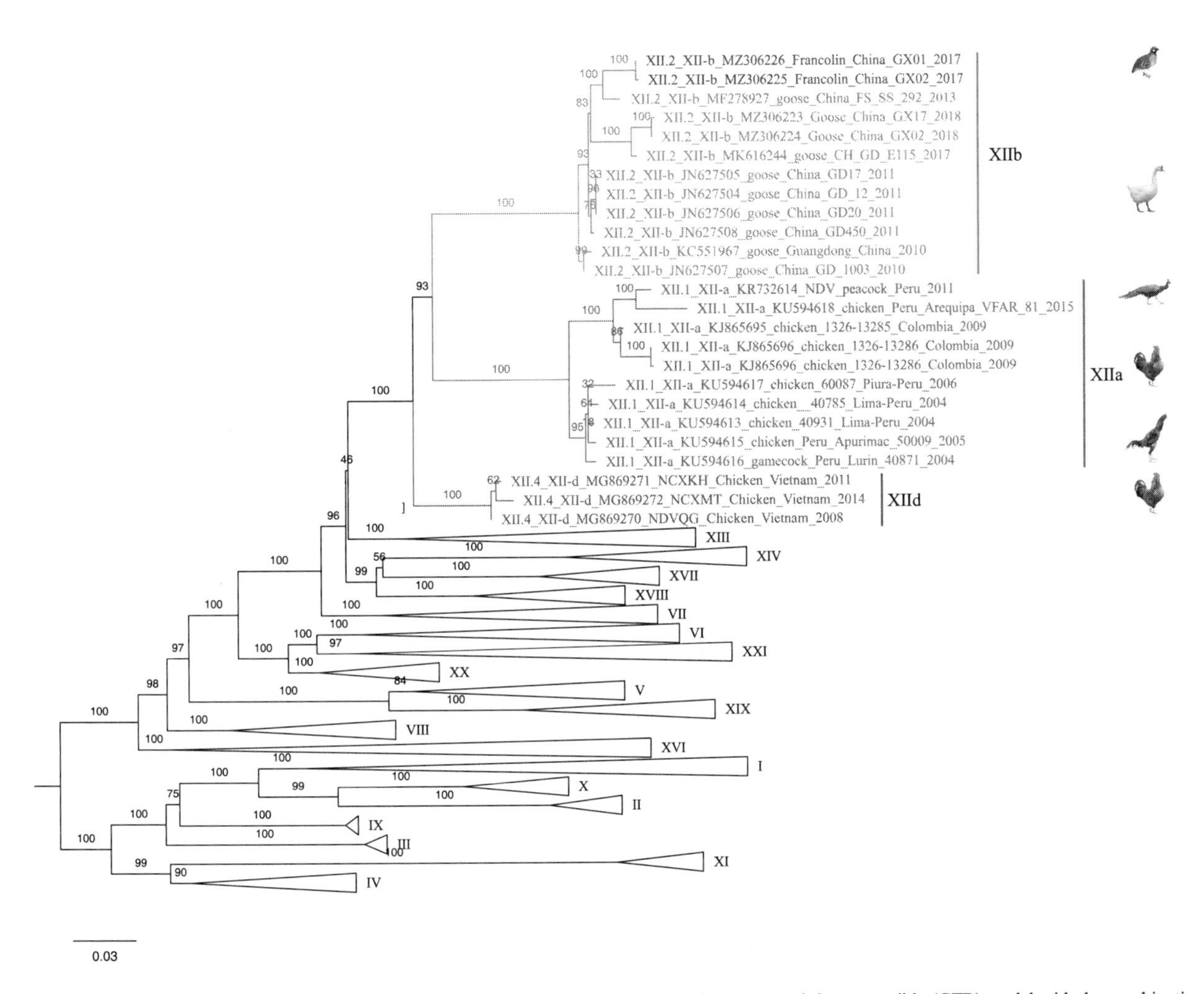

The maximum likelihood tree was constructed via the CIPRES science gateway based on a general time-reversible (GTR) model with the combination GTR + Γ + I and 1 000 bootstrap replicates. The strains of subtype XIIa are shown in green, subtype XIId is shown in blue, GX01 and GX02 are shown in purple, and the other genotype II subtype XIIb NDVs are shown in orange. The other NDVs of class II were collapsed for better illustration.

Figure 6-3-1 Maximum likelihood tree of the full-length F gene with the class II pilot tree plus all genotype XII NDVs reported previously

tail, and 2 in other regions[16]. Six N-linked glycosylation sites[17] were conserved among the strains and were analyzed.

The ORFs of the HN gene of these two strains were 1 716 nt in length, encoding 571 amino acids (aa). HN protein sequences extracted from four subtype XIIb NDVs from Guangxi, two from Guangdong Province, China, and two subtype XIIa NDVs were analyzed since only F gene information is available in GenBank for the remaining genotype XII NDVs. Comparing the HN protein sequences of genotype XII NDVs showed 5, 5, 2, and 20 variations in the cytoplasmic tail[18], transmembrane domain[19], stalk region, and globular head[20, 21], respectively. Ten residues of the first sialic acid binding site[22] and eight conserved residues of the second sialic acid binding site[23] were conserved among genotype XII NDVs. Six N-linked glycosylation sites (119, 341, 433, 481, 508, and 538) were predicted in most NDVs; all subtype XIIb NDVs lost site 538, NDV/peacock/Peru/2011 lost sites 508 and 538, and chicken/Peru/1918-03/603/2008 lost site 508.

Regarding the remaining proteins, the P gene encoded 395 amino acids (aa) and encoded a V protein with 239 amino acids after insertion of one G residue at the conserved editing site (UUUUUCCC, genome sense). There were 40 and 31 amino acid variations in the proteins P and V between subtype XIIb and XIIa NDVs, respectively. The NP, M, and L genes encoded 489, 364, and 2 204 amino acids (aa) and shared 17, 15, and 25 amino acid variations between subtype XIIb and XIIa NDVs, respectively.

Upon comparing seven neutralizing epitopes of protein F located at residues 72, 74, 75, 78, 79, 157 to 171, and 343[24] with those of the vaccine strains LaSota and A-VII[25], subtype XIIb NDVs had a substitution of D to N at residue 170 as well as in the HR1 region, whereas the remaining neutralizing epitope residues were conserved with LaSota and A-VII. As there were seven overlapping antigenic sites of protein HN (sites 1 : 345; site 2 : 513, 514, 521, and 569; site 3: 236, 287, and 321; site 4 : 332, 333, and 356; site 12 : 494, and 516; site 14 : 347, 350, and 353; site 23: 193, 194, and 201) [26, 27] with those of vaccine strains, two residues (347 D and 521 N) differed from those in A-VII and LaSota (347 E and 521 S).

In addition, 17 variations were found between francolin strains and other genotype XII NDVs: 4 variations in protein F, 3 in protein HN, 1 in protein NP, 1 in protein L, 4 in protein P, and 4 in protein V. Among these variations, 2 in transmembrane domain and 1 in globular head of protein HN. Four variations in the amino acid compositions of the proteins were found between GX01 and GX02 genomes; these amino acids are residue 474 (I in GX01 and M in GX02) of protein F, residue 171 (G in GX01 and S in GX02) of protein HN, and residues 717 (V in GX01 and I in GX02), and 1603 (I in GX01 and T in GX02) of protein L.

Comparison of the inactivated genotype XIINDV vaccine with commercial vaccines

The antibody titers of each chicken was monitored every week after immunization and reached 9 $\log_2$-10 $\log_2$ at four weeks after the first immunization, whereas the titers of the PBS-immunized group remained negative (Figure 6-3-2). Challenge with GX01 was carried out five weeks after the first immunization. Chickens in the GX01-, LaSota-, and A-VII-immunized groups did not show any obvious clinical symptoms throughout the 14-day observation timeframe, and chickens in the PBS-immunized group showed slight depression, ruffled feathers, diarrhea, and rhinorrhea between 5 and 10 days post infection (dpi). Examination of histopathologic slides from 9 organs at 3 dpi showed that the bronchiolar mucosa was edematous, with a large number of infiltrated inflammatory cells and some detached epithelial cells, and the lumen of the lung

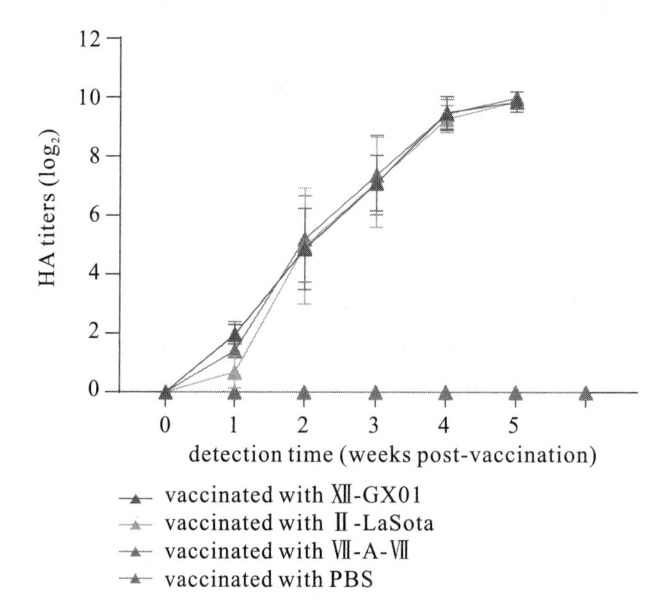

The data are presented as the means and SDs. Error bars indicate the standard deviations of measurements of 29 blood samples (n = 29) (color figure in appendix).

Figure 6-3-2 Antibody titers in the four groups inoculated with different vaccines

was filled with inflammatory exudate in the PBS-immunized group (Figure 6-3-3 D). A small number of inflammatory cells had infiltrated the bronchioles, and inflammatory exudate was observed in the GX01/LaSota/A-Ⅶ-immunized group (Figure 6-3-3 A to C). The normal lung of a chicken inoculated with PBS is shown in Figure 6-3-3 E. Cecal tonsil hemorrhage was observed in the PBS-immunized group (Figure 6-3-4 D) but not the GX01/LaSota/A-Ⅶ-immunized group (Figure 6-3-4 A to C). A normal cecal tonsil from a chicken inoculated with PBS is shown in Figure 6-3-4 E. The rest of the organs in each group showed no apparent lesions.

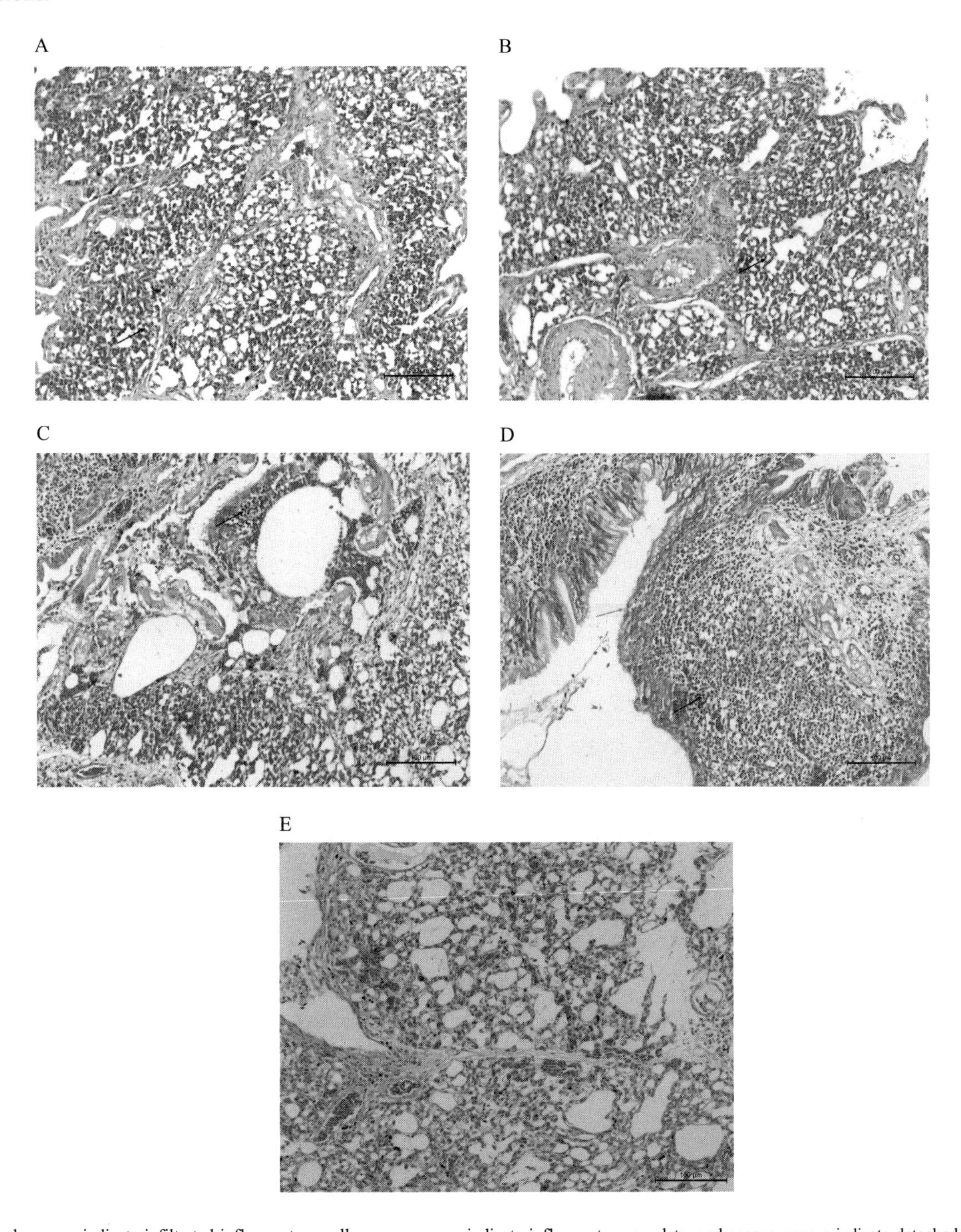

Black arrows indicate infiltrated inflammatory cells, green arrows indicate inflammatory exudate, and orange arrows indicate detached epithelial cells (color figure in appendix).

Figure 6-3-3 Pathological observation of the lungs

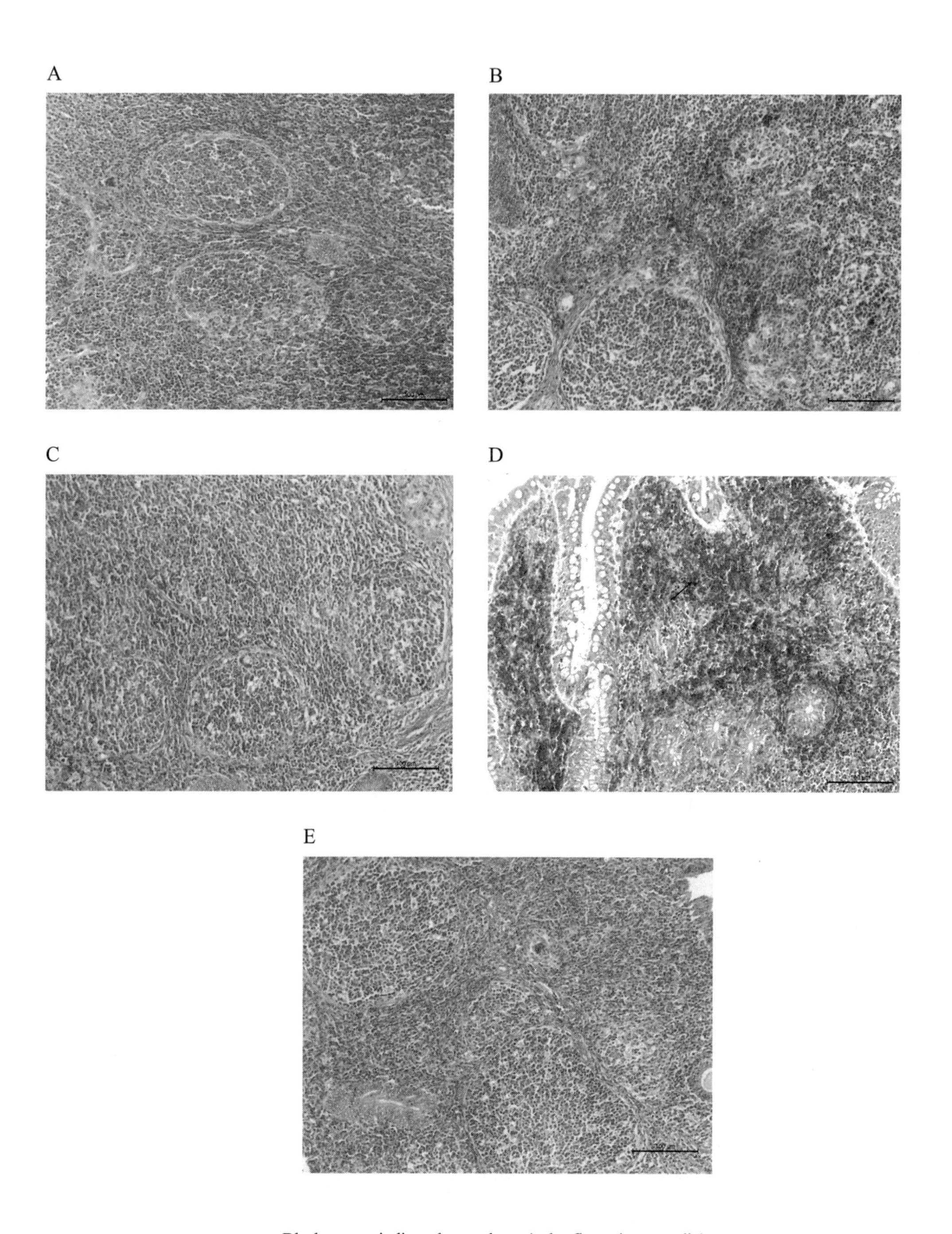

Black arrows indicate hemorrhage (color figure in appendix).

Figure 6-3-4 Pathological observation of the cecal tonsil

Oropharyngeal and cloacal swabs were collected for the EID_{50} assay to evaluate viral shedding at 1, 3, 5, 7, 9, 11, 13, and 15 dpi. In the GX01-immunized group, viral shedding was detected at 1 and 3 dpi (7/8 and 8/8) via the oropharyngeal route, with titers ranging from 1.681 to 2.833 $log_{10}EID_{50}$/0.1 mL, and via the cloacal route (3/8 and 2/8), with titers ranging from 1.168 to 1.5 $log_{10}EID_{50}$/0.1 mL (Figures 6-3-5 A and 6-3-6 A). In the LaSota-immunized group, virus titers varied from 1.681 to 3.5 $log_{10}EID_{50}$/0.1 mL at 1, 3, and 5 dpi (6/8, 7/8, and 5/8) in the oropharyngeal route and 1.167 to 1.833 $log_{10}EID_{50}$/0.1 mL at 1, 3, 5, and 7 dpi (6/8, 1/8, 2/8, and 2/8) in the cloacal route (Figure 6-3-5 B and Figure 6-3-6 B). In the A-Ⅶ-immunized group, viral shedding was detected at 1, 3, and 5 dpi (6/8, 8/8, and 4/8) via the oropharyngeal route, with titers ranging from 1.681 to 3.5 $log_{10}EID_{50}$, and at 1 and 3 dpi via the cloacal (4/8 and 3/8) route, with titers ranging from 1.375 to 1.833 $log_{10}EID_{50}$ (Figure 6-3-5 C and Figure 6-3-6 C). In the PBS-immunized group, viral shedding was detected from 1 to 11 dpi, in the oropharyngeal route (8/8 from 1 to 9 dpi and 6/8 at 11 dpi), titers varied from 1.167 to 4.833 $log_{10}EID_{50}$, and in the cloacal route (7/8, 8/8, 8/8, 8/8, 4/8, and 3/8) with 1.167 to 4.5 $log_{10}EID_{50}$ (Figure 6-3-5 D and Figure 6-3-6 D).

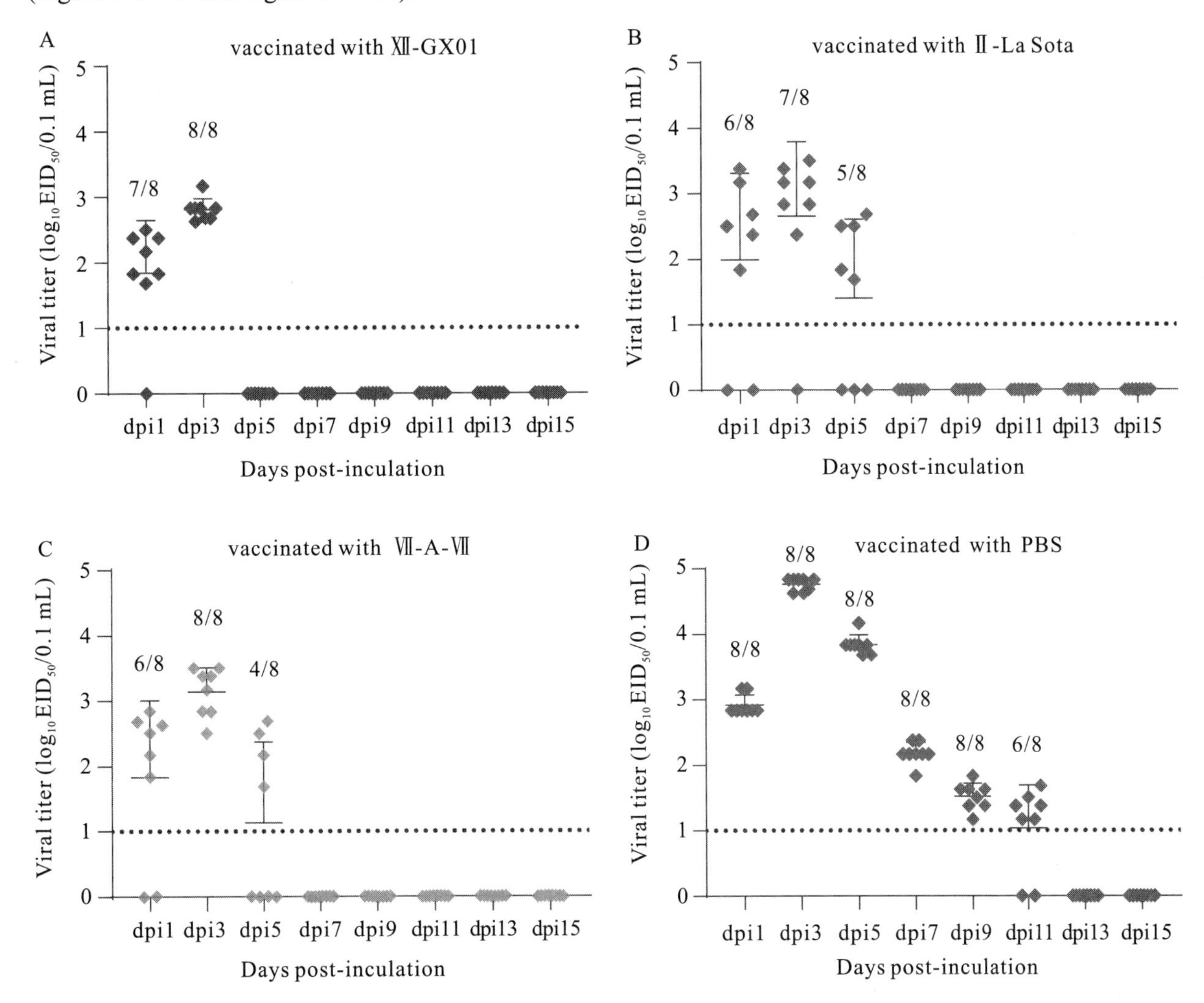

The data are presented as the individual values, means, and SDs. Above each time point, the ratio of swabs was displaying positive recovery to the number of all tested swabs in each group at the sampling time points. The error bars indicate the standard deviations of measurements for 8 swabs ($n = 8$). The dashed lines represent the lower limits of detection.

Figure 6-3-5　Viral titers cloacal swabs after inoculating with GX01

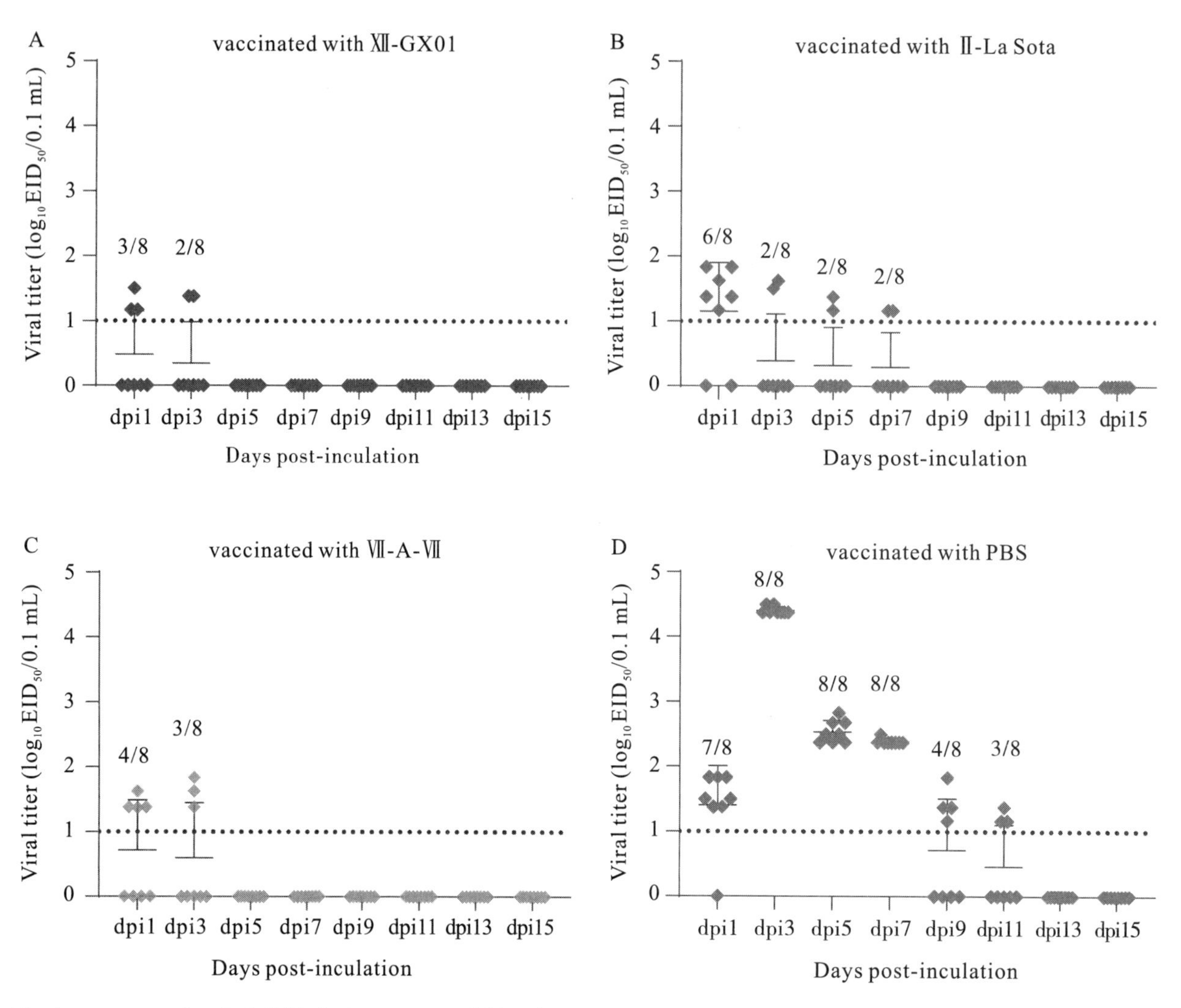

The data are presented as the individual values, means, and SDs. Above each time point, the ratio of swabs was displaying positive recovery to the number of all tested swabs in each group at the sampling time points. The error bars indicate the standard deviations of measurements for 8 swabs (n = 8). The dashed lines represent the lower limits of detection.

Figure 6-3-6 Viral titers in oropharyngeal swabs after inoculating with GX01

Viral shedding titers in both routes in the GX01 immunized group were significantly lower than PBS-immunized group but not significantly different between the LaSota-immunized and A-Ⅶ-immunized groups at 1 dpi. The viral shedding titers of the three vaccine groups in both routes were significantly lower than PBS-immunized group at 3, 5, and 7 dpi, and there were no significant differences among them (Table 6-3-1).

Table 6-3-1 Statistical analysis of viral shedding in different vaccine-immunized groups

Days post inoculation	Group					
	Vaccinated with XII-GX01/ vaccinated with II-LaSota	Vaccinated with XII-GX01/ vaccinated with VII-A-VII	Vaccinated with XII-GX01/ nonvaccinated	Vaccinated with II-LaSota/ vaccinated with VII-A-VII	Vaccinated with II-LaSota/ nonvaccinated	Vaccinated with VII-A-VII/ nonvaccinated
Cloacal						
1	n.s.	n.s.	*	n.s.	n.s.	n.s.
3	n.s.	n.s.	***	n.s.	***	***
5	n.s.	n.s.	***	n.s.	***	***

continued

Days post inoculation	Group					
	Vaccinated with XII-GX01/ vaccinated with II-LaSota	Vaccinated with XII-GX01/ vaccinated with VII-A-VII	Vaccinated with XII-GX01/ nonvaccinated	Vaccinated with II-LaSota/ vaccinated with VII-A-VII	Vaccinated with II-LaSota/ nonvaccinated	Vaccinated with VII-A-VII/ nonvaccinated
7	n.s.	n.s.	***	n.s.	***	***
Oropharyngeal						
1	n.s.	n.s.	*	n.s.	n.s.	n.s.
3	n.s.	n.s.	***	n.s.	***	***
5	n.s.	n.s.	***	n.s.	***	***
7	n.s.	n.s.	***	n.s.	***	***

Note: n.s., $P > 0.05$, not significantly different. * $0.01 < P < 0.05$, **$0.001 < P < 0.01$, and ***$P < 0.001$.

Discussion

Subtype XIIa NDVs were isolated and characterized in 2004 in Peru[10], and they caused severe clinical manifestations and up to 100% mortality in gamecocks. Then, infection spread to vaccinated layers (with a 1.8% mortality rate, minor respiratory difficulties, and decreased egg production) and broilers (with 4%~12% mortality, respiratory difficulties, and depression). A strain from a peacock was isolated in 2011 and caused 90% mortality of a peacock fock, and killed 100% of nonvaccinated SPF chickens after inoculation with 2×10^2 PFU[9]. On the other hand, subtype XIId NDVs have been isolated and characterized in Vietnam since 2008 and have caused typical clinical manifestations and death in infected chickens. Subtype XIIb NDVs of China have been isolated in apparently healthy geese since 2010[8, 28]. The MDT, ICPI and cleavage site of protein F indicated that all genotype XII NDVs were velogenic, but subtype XIIb caused only clinical manifestations with 0% mortality after inoculation with 10^6 EID_{50} of strain E115 via intraocular and intranasal routes in 6-week-old SPF chickens, and artificially infected geese exhibited no clinical manifestations[8]. In this study, GX01 and GX02 were isolated from apparently healthy francolins, and two other strains, goose/China/GX02/2018 and goose/China/GX17/2018, were isolated from apparently healthy geese. The MDT, ICPI, and cleavage site of protein F of these four strains indicated that they are velogenic NDVs. Seven-week-old SPF chickens were challenged with 10^6 EID_{50} of GX01 through intraocular and intranasal routes, which caused slight depression, ruffled feathers, diarrhea, and rhinorrhea without death. It is interesting that these velogenic NDVs from China were not lethal to SPF chickens and showed significant pathogenicity differences from subtype XIIa and XIId NDVs. Examination of histopathologic slides from 9 organs at 3 dpi showed that there was an inflammatory response in the lungs, hemorrhage in the cecal tonsil in PBS-immunized group, yet the rest of the organs in each group showed no apparent lesions. It is demonstrated that the lesion were milder than other velogenic NDVs[29].

All seven proteins and noncoding regions contributed to the virulence of NDV, whereas proteins F, HN, and V majorly contribute to virulence according to a compilation of the data[29]. Although the deduced amino acids of the cleavage site in protein F were characterized as a major molecular determinant of NDV virulence[30, 31], some other functional regions in protein F may also influence NDV pathogenicity, as fusogenic ability is responsible for virulence. For example, nonconserved and conserved mutations in positions "a" and "d" of HR1, nonconserved and conserved mutations in position "a" of HR2 and conserved mutations in position "a" of HR3 could inhibit fusion activity, except residue 289 mutation[32-36]. Compared to those of the other genotype XII NDVs, there were three variations in heptad repeats (residues 170 in HR1, 496 in HR2, and 270 in HR2) in positions "f" "c" and "b" respectively. The role of these three nonconserved variations remains unknown, although a single mutation in the "b" position did not significantly disrupt the fusogenic ability of the F protein[33]. Signal peptides are located at the N-terminus of the F protein, marking the protein secretion pathway and the protein target location. It plays a role in controlling the rate of protein secretion, determining the protein folding state, affecting downstream transmembrane behavior, N-terminal glycosylation, and nuclear localization signal functionality and viral infectivity[37]; thus, the impact of nonconserved variation at residue 28 is unknown. The cytoplasmic tail is located at residues 523~553, which are the C-terminal residues of protein F. The mutation M553A resulted in increased (32% higher) fusion indices and viral titers in DF-1cells in a previous study[38]; therefore, whether the variation A553V from subtypes XIIa and XIId to XIIb NDVs would impact pathogenicity and replication also needs to be further studied.

HN is a multifunctional protein and consists of four regions: the cytoplasmic tail, transmembrane domain, stalk regions, and globular head. The cytoplasmic tail is critical for replication, and the species-specific phenotypes based on only the homologous cytoplasmic tails of F and HN contained in chimeric NDVs could be recovered[39]. The mutation S6A in the cytoplasmic tail can affect cell fusion and colocalization of protein HN and M in cells[18]. On the other hand, two subtype XIIa NDVs had the nonconserved substitution S6D, which impaired viral assembly and replication in a chimeric genotype XII NDV vaccine[40], while the role in four subtype XIIb NDVs with the conserved variation S6N remains unknown. Variations at residue 2 may not have any impact, as a previous report described that deleting the first 2 amino acids did not affect the biological characteristics of NDV[18]. The HR motif in the transmembrane domain of protein HN has three conserved leucine residues (residues 30, 37, and 44 in position "a" in the heptad-repeat), and mutation of these residues impairs the attachment and fusion promotion activity of the protein HN[19]. The stalk region of the protein HN interacts with homologous F proteins to mediate the fusion process and is critical to replication[39, 41-43]. Mutations in the intervening region of the stalk region (89 to 95) or in two HRs of the stalk region (74, 81, 88, 90, 96, 97, 102, 103, 110, or combination mutation) variously decrease fusion promotion and HN-F interaction[41, 44]. Yan et al.[43] characterized L110 and R116 as playing a key role in determining the difference in virulence between two strains with 99.9% similarity. Five variations in the transmembrane domain and two variations in the stalk region were found between subtype XIIa and XIIb NDVs, and none of them were included in the residues that have been studied before. Two sialic acid-binding sites located in the globular head that play roles in binding to sialic acid receptors, removing sialic acid from newly synthesized viral coat proteins to prevent the aggregation of new viruses, and activating fusion by the protein F[22, 23, 45]. Mutation of these sialic acid-binding sites impaired NA and/or HAd activities. These two conserved residues of sialic acid-binding sites were conserved among all analyzed genotype XII NDVs in this study. In addition to two sialic acid-binding sites, mutations in residues 158, 160, 216, 220, 224, 232, 280, 289, 332, 336, 384, 495, 518, 536, and 557 reduced the fusion promotion activity of NDV[46]. Variations in residues 232 and 280 were found in this study. Only four (119, 341, 433, and 481) of six potential N-linked glycosylation sites used for carbohydrate addition were involved in cell attachment and fusion promotion activities of the protein in a previous report[47]; thus, variation of sites 508 and 538 may not influence the bioactivities of HN.

Protein V contributes to virus virulence via its IFN antagonistic activity[48]. One report on LaSota and Beaudette C showed that variation at residues 144, 153, 161, and 234 contributed to IFN antagonistic activity, and residues 30, 41, 46, 65, and 82 at the N-terminal region may also contribute to IFN antagonistic activity by affecting the structure of the protein, making the C-terminus less accessible to interacting host proteins to affect the structure of the protein[49]. In this study, residue 161 of subtype ⅫI b NDVs was an S, consistent with that of LaSota, whereas it was a P in subtype Ⅻa NDVs, consistent with Beaudette C. Residue 65 of subtype Ⅻb NDVs was an N, as also observed in LaSota, whereas it was an S in subtype Ⅻa NDVs, but this variation did not change the properties of the amino acids.

By exchanging genes between velogenic strain Herts and lentogenic strain AV324A, it was observed that the correlation between virulence and the efficiency of viral replication in the viral replication complex (NP, P, and L) of NDVs[50]. Nuclear localization of protein M could also impact the virulence of NDVs through inhibiting host cell transcription, and promoting viral replication by affecting viral RNA synthesis and transcription[51]. Residues 40, 17, 25, and 15 variations were found in the P, NP, L, and M proteins.

In this study, numerous variations were found in six proteins between subtype Ⅻa and Ⅻb NDVs. Some of them have been studied in cells to determine their roles in previous studies, yet their role in animals still needs to be researched. Considering the pathogenicity differences in chickens via the natural route, some regions must play a role in this phenomenon, which remains unknown, and the variations among genotype Ⅻ NDVs would be suitable materials.

Genotype Ⅻ NDVs have been reported in chickens, gamecocks, geese, and peacocks, while they were isolated from wild birds for the first time in this study, as subtype Ⅻb NDVs were previously isolated only in geese. Poultry-derived NDVs have been proven to spill over to wild birds[52], especially migratory birds, and spillover events from wild birds to poultry also exist[53]. Francolins are common resident birds in southern China that live around poultry farms where the poultry are fed on the open ground, which is a popular breeding method in southern China. Unlike migratory birds, resident birds often stay in the poultry farm, e.g., eating poultry feed and getting in close contact with poultry. It would facilitate infectious disease spread to each other. Subtype Ⅻb NDVs have only been found in Guangdong and Guangxi to date, and it seems that resident birds play a more major role than migratory birds in spreading subtype Ⅻb NDVs over short distances to form an endemic situation. According to the phylogenetic tree, two francolins-derived NDVs were divided into a branch with an NDV isolated in Guangdong in 2013, whereas two goose-derived NDVs in Guangxi in 2018 with E115 in Guangdong in 2017 were divided into another branch, and none of the NDVs in this genotype were monitored between 2014 and 2017 under the continuous surveillance program in poultry, indicating that subtype Ⅻb NDVs have multiple transmission chains and that resident birds maybe involved in this branch as intermediate hosts in which the virus also continues to evolve. In summary, it reminds us that infectious disease surveillance should be conducted in both migratory and resident birds, and subtype Ⅻd NDVs in Vietnam could be a potential risk to poultry in China because Guangxi shares along border with Vietnam.

The protective efficacy of subtype Ⅻb NDVs as inactivated vaccines was evaluated in chickens in this study. However, according to the statistical analysis results for viral shedding, there were no significant differences among the three vaccine-treated groups at 3, 5, and 7 dpi for both routes, and viral shedding could not be detected in the GX01-immunized group from 5 dpi for both routes, whereas it could be detected at 5 dpi in the LaSota and the A-Ⅶ-immunized group for the oropharyngeal route, and was detected at 5 dpi in the LaSota-immunized group for the cloacal route. Tis demonstrated that the GX01 vaccine could eliminate viral shedding faster. These results were similar to those showing the protective efficacy of a subtype Ⅻa chimeric vaccine against the homologous virus to a certain degree[40].

The chickens with the HI titer at 10 $\log_2$ were chosen for the inoculation with GX01 to eliminate the effects due to inconsistent antibody levels, and the results support the theory that a genotype-matched vaccine with similar antigen epitopes was more efficient than the serum antibody enrichment as an immunologic mechanism that still needs to be elucidated, with two major possibilities suggested by previous studies: local immunity such as mucosal immunity was increased and more specific with similar antigen epitopes in vaccine; cellular immunity was increased to clear the infection faster. Based on the result, subtype XIIb NDVs could be a candidate vaccine strain against subtype XIId NDVs on account of the potential risk of spreading to China and increased pathogenicity of subtype XIIb NDVs.

In conclusion, two francolin-derived subtype XIIb NDV strains, GX01 and GX02 were isolated and characterized as velogenic NDVs in Guangxi, southwestern China. Based on the weaker pathogenicity in chickens, 17, 40, 15, 7, 32, 25, and 31 amino acid variations were found in proteins NP, P, M, F, HN, L, and V, and some of these variations may be responsible for this phenomenon. Based on epidemiology and phylogenetic analyses, we were reminded that infectious disease surveillance should be conducted in both migratory and resident birds. Because of the potential risk of subtype XIId NDVs spreading to China and the increased pathogenicity of subtype XIIb NDVs, the protective efficacy of GX01 as an inactivated vaccine was evaluated with two commercial inactivated vaccines in chickens. The results showed that subtype XIIb NDVs could be candidate genotype-matched vaccine strains against genotype XII NDVs.

References

[1] LEFKOWITZ E J, DEMPSEY D M, CURTIS H R, et al. Virus taxonomy: the database of the International Committee on Taxonomy of Viruses (ICTV). Nucleic Acids Research, 2018(01): 708-717.

[2] OIE. Manual of diagnostic tests and vaccines for terrestrial animals. Paris: World Organisation for Animal Health, 2018.

[3] DIMITROV K M, ABOLNIK C, AFONSO C L, et al. Updated unified phylogenetic classification system and revised nomenclature for Newcastle disease virus. Infection, Genetics Evolution, 2019, 74: 103917.

[4] GUO H, LIU X, HAN Z, et al. Phylogenetic analysis and comparison of eight strains of pigeon paramyxovirus type 1 (PPMV-1) isolated in China between 2010 and 2012. Archives of Virology, 2013, 158(6): 1121-1131.

[5] HE Y, LU B, DIMITROV K M, et al. Complete genome sequencing, molecular epidemiological, and pathogenicity analysis of pigeon paramyxoviruses type 1 isolated in Guangxi, China during 2012-2018. Viruses, 2020, 12(4): 336.

[6] LIU X F, WAN H Q, NI X X, et al. Pathotypical and genotypical characterization of strains of Newcastle disease virus isolated from outbreaks in chicken and goose flocks in some regions of China during 1985-2001. Archives of Virology, 2003, 148(7): 1387-1403.

[7] QIN Z M, TAN L T, XU H Y, et al. Pathotypical characterization and molecular epidemiology of Newcastle disease virus isolates from different hosts in China from 1996 to 2005. Journal of Clinical Microbiology, 2008, 46(2): 601-611.

[8] XIANG B, CHEN R, LIANG J, et al. Phylogeny, pathogenicity and transmissibility of a genotype XII Newcastle disease virus in chicken and goose. Transboundary and Emerging Diseases, 2020, 67(1): 159-170.

[9] CHUMBE A, IZQUIERDO-LARA R, TATAJE-LAVANDA L, et al. Characterization and sequencing of a genotype XII Newcastle disease virus isolated from a peacock (Pavo cristatus) in Peru. Genome Announcements, 2015, 3(4): e00792-00715.

[10] CHUMBE A, IZQUIERDO-LARA R, TATAJE L, et al. Pathotyping and phylogenetic characterization of Newcastle disease viruses isolated in Peru: defining two novel subgenotypes within genotype XII. Avian Dis, 2017, 61(1): 16-24.

[11] LE X T K, DOAN H T T, LE T H. Molecular analysis of Newcastle disease virus isolates reveals a novel XIId subgenotype in Vietnam. Archives of Virology, 2018, 163(11): 3125-3130.

[12] DIEL D G, SUSTA L, CARDENAS GARCIA S, et al. Complete genome and clinicopathological characterization of a virulent Newcastle disease virus isolate from South America. Journal of Clinical Microbiology, 2012, 50(2): 378-387.

[13] MILLER M A, PFEIFFER W T, SCHWARTZ T. Creating the CIPRES science gateway for inference of large phylogenetic trees//Gateway Computing Environments Workshop (GCE). New Orleans, LA: IEEE, 2010: 1-8.

[14] HU S, MA H, WU Y, et al. A vaccine candidate of attenuated genotype VII Newcastle disease virus generated by reverse genetics. Vaccine, 2009, 27(6): 904-910.

[15] LUO S, XIE Z, XIE Z, et al. Surveillance of live poultry markets for low pathogenic avian influenza viruses in Guangxi Province, southern China, from 2012-2015. Scientific Reports, 2017, 7(1): 1-9.

[16] ZHU J, LI P, WU T, et al. Design and analysis of post-fusion 6-helix bundle of heptad repeat regions from Newcastle disease virus F protein. Protein Engineering, 2003(5): 373-379.

[17] SAMAL S, KHATTAR S K, KUMAR S, et al. Coordinate deletion of N-Glycans from the heptad repeats of the fusion F protein of Newcastle disease virus yields a hyperfusogenic virus with increased replication, Virulence, and Immunogenicity. Journal of Virology, 2012, 86(5): 2501-2511.

[18] KIM S H, YAN Y, SAMAL S K. Role of the cytoplasmic tail amino acid sequences of Newcastle disease virus hemagglutinin-neuraminidase protein in virion incorporation, cell fusion, and pathogenicity. Journal of Virology, 2009, 83(19): 10250-10255.

[19] MCGINNES L, SERGEL T, MORRISON T. Mutations in the transmembrane domain of the HN protein of Newcastle disease virus affect the structure and activity of the protein. Virology, 1993, 196(1): 101-110.

[20] CRENNELL S, TAKIMOTO T, PORTNER A, et al. Crystal structure of the multifunctional paramyxovirus hemagglutinin-neuraminidase. Nature Structural Biology, 2000, 7(11): 1068-1074.

[21] PING Y, SWANSON K A, LESER G P, et al. Structure of the Newcastle disease virus hemagglutinin-neuraminidase (HN) ectodomain reveals a four-helix bundle stalk. Proceedings of the National Academy of Sciences, 2011, 108(36): 14920-14925.

[22] HELEN C, TORU T, RUPERT R, et al. Probing the sialic acid binding site of the hemagglutinin-neuraminidase of Newcastle disease virus: identification of key amino acids involved in cell binding, catalysis, and fusion. Journal of Virology, 2002, 76(4): 1816.

[23] ZAITSEV V, ITZSTEIN M V, GROVES D, et al. Second sialic acid binding site in Newcastle disease virus hemagglutinin-neuraminidase: implications for fusion. Journal of Virology, 2004, 78(7): 3733-3741.

[24] UMALI D V, ITO H, SUZUKI T, et al. Molecular epidemiology of Newcastle disease virus isolates from vaccinated commercial poultry farms in non-epidemic areas of Japan. Virology Journal, 2013, 10(1): 330.

[25] HU Z, HU S, MENG C, et al. Generation of a genotype VII Newcastle disease virus vaccine candidate with high yield in embryonated chicken eggs. Avian Dis, 2011, 55(3): 391-397.

[26] IORIO R M, GLICKMAN R L, SHEEHAN J P. Inhibition of fusion by neutralizing monoclonal antibodies to the haemagglutinin-neuraminidase glycoprotein of Newcastle disease virus. Journal of General Virology, 1992, 73(5): 1167-1176.

[27] IORIO R M, SYDDALL R J, SHEEHAN J P, et al. Neutralization map of the hemagglutinin-neuraminidase glycoprotein of Newcastle disease virus: domains recognized by monoclonal antibodies that prevent receptor recognition. Journal of Virology, 1991, 65(9): 4999-5006.

[28] LIU H, LÜ Y, WANG J, et al. Distribution and characterization of emerged genotype XII Newcastle disease viruses in China. China Animal Health Inspection, 2017, 34(9): 1-3.

[29] DORTMANS J C, KOCH G, ROTTIER P J, et al. Virulence of Newcastle disease virus: what is known so far?. Veterinary Research, 2011, 42(1): 122-122.

[30] PANDA A, HUANG Z, ELANKUMARAN S, et al. Role of fusion protein cleavage site in the virulence of Newcastle disease virus. Microbial Pathogenesis, 2004, 36(1): 1-10.

[31] PEETERS B P, DE LEEUW O S, KOCH G, et al. Rescue of Newcastle disease virus from cloned cDNA: evidence that cleavability of the fusion protein is a major determinant for virulence. Journal of Virology, 1999, 73(6): 5001-5009.

[32] MCGINNES L W, SERGEL T, CHEN H, et al. Mutational analysis of the membrane proximal heptad repeat of the Newcastle disease virus fusion protein. Virology, 2001, 289(2): 343-352.

[33] REITTER J N, SERGEL T, MORRISON T G. Mutational analysis of the leucine zipper motif in the Newcastle disease virus

fusion protein. Journal of Virology, 1995, 69(10): 5995-6004.
[34] SERGEL-GERMANO T, MCQUAIN C, MORRISON T. Mutations in the fusion peptide and heptad repeat regions of the Newcastle disease virus fusion protein block fusion. Journal of Virology, 1994, 68(11): 7654-7658.
[35] SERGEL T A, MCGINNES L W, MORRISON T G. A single amino acid change in the Newcastle disease virus fusion protein alters the requirement for HN protein in fusion. Journal of Virology, 2000, 74(11): 5101-5107.
[36] SERGEL T A, MCGINNES L W, MORRISON T G. Mutations in the fusion peptide and adjacent heptad repeat inhibit folding or activity of the Newcastle disease virus fusion protein. Journal of Virology, 2001, 75(17): 7934-7943.
[37] HAJAR, OWJI, NAVID, et al. A comprehensive review of signal peptides: Structure, roles, and applications. European Journal of Cell Biology, 2018, 97(6): 422-441.
[38] SAMA L, KHATTAR S K, PALDURAI A, et al. Mutations in the cytoplasmic domain of the Newcastle disease virus fusion protein confer hyperfusogenic phenotypes modulating viral replication and pathogenicity. Journal of Virology, 2013, 87(18): 10083-10093.
[39] KIM S H, SUBBIAH M, SAMUEL A S, et al. Roles of the fusion and hemagglutinin-neuraminidase proteins in replication, tropism, and pathogenicity of avian paramyxoviruses. Journal of Virology, 2011, 85(17): 8582-8596.
[40] IZQUIERDO-LARA R, CHUMBE A, CALDERÓN K, et al. Genotype-matched Newcastle disease virus vaccine confers improved protection against genotype XII challenge: The importance of cytoplasmic tails in viral replication and vaccine design. PLOS ONE, 2019, 14(11): e0209539.
[41] MELANSON V R, IORIO R M. Amino acid substitutions in the F-Specific domain in the stalk of the Newcastle disease virus HN protein modulate fusion and interfere with its interaction with the F protein. Journal of Virology, 2004, 78(23): 13053-13061.
[42] MELANSON V R, IORIO R M. Addition of N-Glycans in the stalk of the Newcastle disease virus HN protein blocks Its interaction with the F protein and prevents fusion. Journal of Virology, 2006, 80(2): 623-633.
[43] YAN C, LIU H, JIA Y, et al. Screening and mechanistic study of key sites of the hemagglutinin-neuraminidase protein related to the virulence of Newcastle disease virus. Poultry Science, 2020, 99(7): 3374-3384.
[44] STONE-HULSLANDER J, MORRISON T G. Mutational analysis of heptad repeats in the membrane-proximal region of Newcastle disease virus HN protein. Journal of Virology, 1999, 73(5): 3630-3637.
[45] POROTTO M, SALAH Z, DEVITO I, et al. The second receptor binding site of the globular head of the Newcastle disease virus hemagglutinin-neuraminidase activates the stalk of multiple paramyxovirus receptor binding proteins to trigger fusion. Journal of virology, 2012, 86(10): 5730-5741.
[46] TAKIMOTO T, TAYLOR G L, CONNARIS H C, et al. Role of the hemagglutinin-neuraminidase protein in the mechanism of paramyxovirus-cell membrane fusion. Journal of Virology, 2002, 76(24): 13028-13033.
[47] MCGINNES L W, MORRISON T G. The role of individual oligosaccharide chains in the activities of the HN glycoprotein of Newcastle disease virus-science direct. Virology, 1995, 212(2): 398-410.
[48] HUANG Z, KRISHNAMURTHY S, PANDA A, et al. Newcastle disease virus V protein is associated with viral pathogenesis and functions as an alpha interferon antagonist. Journal of Virology, 2003, 77(16): 8676-8685.
[49] ALAMARES J G, ELANKUMARAN S, SAMAL S K, et al. The interferon antagonistic activities of the V proteins from two strains of Newcastle disease virus correlate with their known virulence properties. Virus Research, 2010, 147(1): 153-157.
[50] DORTMANS J, ROTTIER P, KOCH G, et al. The viral replication complex is associated with the virulence of Newcastle disease virus. Journal of Virology, 2010, 84(19): 10113-10120.
[51] DUAN Z, DENG S, JI X, et al. Nuclear localization of Newcastle disease virus matrix protein promotes virus replication by affecting viral RNA synthesis and transcription and inhibiting host cell transcription. Veterinary Research, 2019, 50(1): 1-19.
[52] XIANG B, HAN L, GAO P, et al. Spillover of Newcastle disease viruses from poultry to wild birds in Guangdong province, southern China. Infection, Genetics Evolution, 2017, 55: 199-204.
[53] BROWN V R, BEVINS S N. A review of virulent Newcastle disease viruses in the United States and the role of wild birds in viral persistence and spread. Veterinary Research, 2017, 48(1): 1-15.

Expression and identification of the avian reovirus σC gene in transgenic tobacco

Wang Sheng, Xie Zhixun, Huang Li, Xie Liji, Deng Xianwen, Xie Zhiqin, Liu Jiabo, Luo Sisi, Zeng Tingting, and Chen Zhongwei

Abstract

In this study, the expression of the avian reovirus σC gene in tobacco was examined and the reactogenicity of the resulting expression products was analyzed. A pair of primers was designed according to the major antigen region of the σC gene derived from GenBank, and a plant vector pBI121-σC that constitutively expresses the σC gene was constructed. The pBI121-σC vector was then transformed into Agrobacterium strain EHA105. After the tobacco plants were transformed via *Agrobacterium tumefaciens*, kanamycin-resistant plants were obtained by screening. The resistant plants were first analyzed by PCR, and real-time fluorescence quantitative PCR was used to further estimate the copy number of σC genes in the positive control plants. We successfully obtained regenerated transgenic σC transgenic tobacco plants. Western blot analysis was used to examine σC protein expression in transgenic tobacco. Recombinant σC protein has good reactivity with anti-ARV-positive serum. Our study provides a basis for further analysis of plants used in bioreactor development and for the production of σC oral vaccines.

Keywords

avian reovirus, σC gene, plant expression vector, tobacco

Introduction

Avian reovirus (ARV), which belongs to the genus *Orthoreovirus* in the Reoviridae family, is an important poultry pathogen[1]. ARV infection causes numerous avian diseases, including viral arthritis, chronic respiratory disease and malnutrition, the incidence of which has been increasing in recent years[2]. ARV infection in animal industries has caused considerable economic losses, and an effective strategy is needed to control the spread of ARV[3]. Different strategies, such as vaccination, are available to control the spread of ARV[4-5]. However, inactivated vaccines are generally not effective enough and do not provide adequate immune protection. Although modified live vaccines can provide decent protection, virulence can be reversed. Therefore, developing an effective ARV subunit vaccine has become a research topic of increasing concern[6-9].

The ARV genome is composed of double-stranded RNA, which contains 10 segments, including large (L), medium (M), and small (S) segments, and the outer layer is coated by a bilayer of capsid protein. Among them, the σC protein encoded by the third ORF of the S1 gene is the smallest protein in the ARV capsid protein and carries the surface antigen of the ARV-specific neutralization reaction; this antigen can stimulate the body to produce neutralizing antibodies and is related to the adsorption and proliferation of virions[10-11]. Therefore, σC is the primary target of a protective immune response against ARV infection. The σC protein has been successfully expressed in a variety of expression systems, such as *Escherichia coli* and baculovirus systems[12]. Edible vaccines based on plant bioreactors have a number of advantages over traditional vaccines. Recently, the surface antigens of multiple pathogens, including hGH, HPV, HIV and hGLP1, have been successfully expressed in plants[13-16]. These successful examples provide a reference for the development of an edible vaccine against pathogen infection in poultry.

In this study, tobacco was selected as the transformation object, and the main antigenic region of the σC protein encoded by the avian reovirus S1 gene was constructed on the plant expression vector pBI121, which was subsequently transformed via the Agrobacterium-mediated method. The low-copy transformed tobacco was screened via PCR amplification and real-time PCR, and the reactogenicity of the recombinant protein expressed via Western blotting was analyzed. The aim of this study was to provide a theoretical basis and technical support for the research and development of oral vaccines to produce avian reolarized plants by using transgenic plants as bioreactors.

Materials and methods

Equipment and materials

The plasmids pMD18-σC and ARV-positive serum were developed in our laboratory in an earlier study. The pBI121 plasmid, *Agrobacterium rhizogenes* EHA105 strains and tobacco plants (*Nicotiana tabacum* L., K346) were kindly provided by Dr. Luocong (Guangxi University, China). Taq DNA polymerase, SYBR® Premix Ex Taq™ Ⅱ, restriction enzymes and T4 ligase were purchased from TaKaRa Biotechnology Co. Anti-rabbit anti-HRP/HRP was purchased from KPL Company.

Plasmid construction

The σC gene was amplified from the ARV S1133 strain, which contains the complete gene, using the forward primer (5'-TTGGATCCATGGCGGGTCTCAATCCATC-3', *BamH*I site underlined) and the reverse primer (5'-TTCCCGGGGGTGTCGATGCCGGTACGC-3', *Sma*I site underlined). The PCR products were digested with *BamH*I and *Sma*I and inserted into the plant expression vector pBI121, which was subsequently digested with the same enzymes to create the recombinant plasmid pBI121-σC. The obtained recombinant vector was subsequently introduced into the *A. rhizogenes* EHA105 strain via heat shock. We examined the positive clones via PCR screening and evaluated the purified plasmid via restriction endonuclease analysis.

Transformation procedure and genetic analysis

Tobacco (*N. tabacum* L.) petiole explants were cut into 0.5～0.7 cm sections and grown for 2 days on Murashige and Skoog (MS) media. Then, the leaf discs were incubated with *Agrobacteria tumefaciens* EHA 105 carrying the binary vector pBI121-σC with suction for 8～10 min. The explants were cocultivated for 3 days in the dark on MS media and then washed with sterile water containing 350 μg cefotaxime/mL. After washing, the explants were maintained in MS media containing 350 μg of cefotaxime/mL, 350 μg of kanamycin/mL, 0.5 μg of IAA/mL and 1 μg of 6-BA/mL for selection, and fresh media were supplied every 2 weeks. The explants were differentiated into adventitious buds, and these adventitious buds were kept in MS media supplemented with 350 μg of cefotaxime/mL, 350 μg of kanamycin/mL, or 0.5 μg of IAA/mL for root induction. When the explants regenerated into complete tobacco plantlets, the plantlets were transferred to pots and cultured in a greenhouse. At maturity, the transgenic plants were identified via PCR analysis with the forward primer 5'-ATGGCGGGTCTCAATCCATC-3' and the reverse primer 5'-GGTGTCGATGCCGGTACGC-3'.

Real-time PCR analysis

High-throughput and sensitive SYBR Green I real-time fluorescence quantitative PCR was used to further estimate the number of copies of the σC gene in the positive control plants, and the endogenous RNR2 gene in tobacco was used as a reference gene. The copy number was quantified using the relative double standard curve quantification method. With a series of dilutions, the standard curves of the cycle threshold (CT) relative to the log of each initial template copy of the σC and RNR2 genes were obtained. The transgenic copy number was determined by comparing the initial template copy number of the σC gene with that of RNR2.

Western blot analysis

To further demonstrate the reactogenicity of the σC protein expressed in transgenic tobacco leaves, the σC recombinant protein was detected by Western blot analysis. Total protein was extracted from the transgenic tobacco plants as test samples and from wild-type tobacco plants as a negative control. The protein samples were resuspended in loading buffer, analyzed and separated via 12% SDS-PAGE, and then transferred to a nitrocellulose membrane. Western blotting was carried out using ARV-positive serum (1 ∶ 300 dilution) as the primary antibody and HRP-conjugated rabbit anti-chicken IgG (1 ∶ 200 dilution) as the secondary antibody. The immune signals were identified using DAB.

Results

Gene construct and agrobacterium transformation

The correct insertion of the construct carrying the sequence for the σC gene into pBI121 was identified by restriction endonuclease analysis (Figure 6-4-1). DNA sequencing confirmed that the recombinant plasmid pBI121-σC contains a complete σC gene, a correct insertion site and the expected open reading frame (ORF). The σC gene was expressed in the plant nuclear vectors using the CaMV 35S promoter, the NOS terminator and a kanamycin resistance marker (NPT Ⅱ). The introduction of pBI121-σC into the *A. tumefaciens* strain EHA105 was confirmed by PCR.

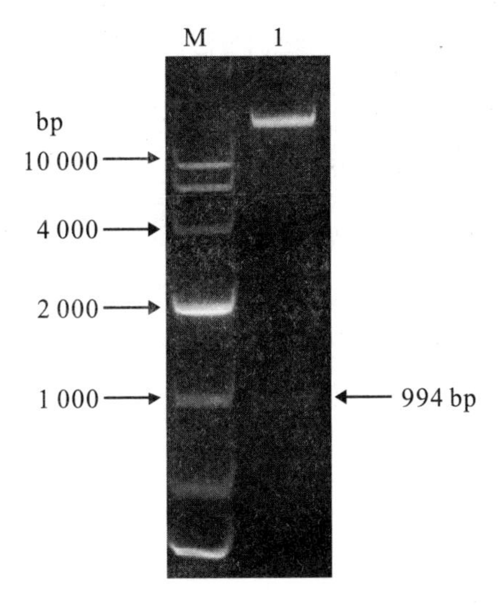

1: pBI121-σC digested by two enzymes of *BamH*I and *Sma*I; M: DL 10 000 Marker.

Figure 6-4-1 Restriction enzyme digestion of pBI121-σC

Detection of transgenic tobacco plants

After transformation of the tobacco plants via *A. tumefaciens* strain EHA105, the resistant plants were selected with kanamycin. MS media supplemented with 350 mg of cefotaxime/L inhibited the growth of Agrobacteria, and 100 mg of kanamycin/L provided enough selective pressure for growth inhibition. Transgenic tobacco plants were successfully obtained (Figure 6-4-2). The resistant plants were first analyzed via PCR (Figure 6-4-3), and eight resistant plants harboring the σC gene were obtained via PCR screening. We successfully generated transgenic tobacco plants that expressed σC recombinants.

A, B: Induction differentiation; C, D: Induce root; E: Transplantation.

Figure 6-4-2 The regeneration of transgenic tobacco expressing ARV σC

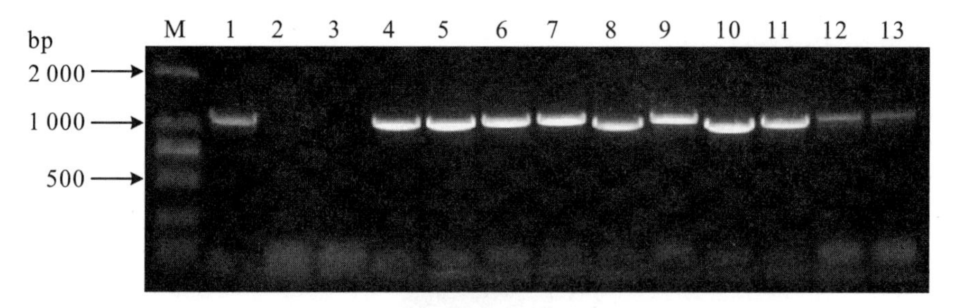

1: Positive control; 2: Negative control; 3: Water; 4-13: Transgenic plant; M: DNA marker.

Figure 6-4-3 PCR assay of σC transgenic plants

σC copy numbers in transgenic tobacco

To quantify the transcription level of the σC gene, we used real-time fluorescence quantitative PCR. The endogenous RNR2 gene was added as the spike reference for the target σC gene. With a series of dilutions, standard curves of the CT relative to the log of each initial template copy of the σC and RNR2 genes were constructed. The standard curve of σC was $y=-3.536\,6x+1.826\,5$, and that of RNR2 was $y=-3.535\,2x+15.143$. The correlation coefficients were 0.998 4 and 0.999 8, respectively, and the samples were accurately analyzed quantitatively (Figure 6-4-4). The transgenic copy number was determined by comparing the initial template copy number of σC with that of RNR2. Among the five putative transgenic lines, the copy number was 4, and the negative control had zero copies.

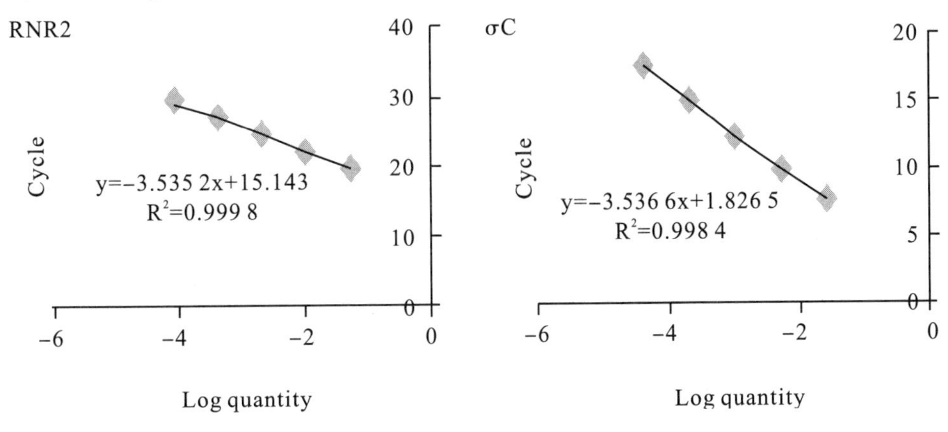

Figure 6-4-4 The standard curves of RNR2 and σC gene

σC protein expression in transgenic tobacco

Western blotting was used to determine the expression of the σC protein in the transgenic tobacco plants. A specific band at 61.6 ku corresponding to the σC protein was detected by Western blot in the transgenic plants (Figure 6-4-5) but not in the negative controls. This finding revealed that the σC protein was highly expressed in the transgenic tobacco plants and it was the corresponding reactogenicity.

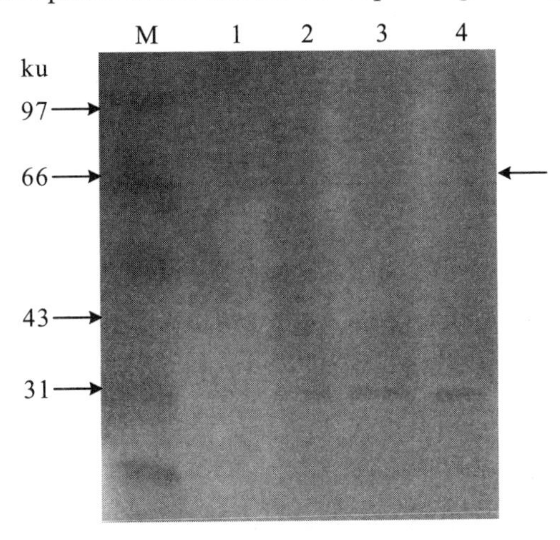

1: Positive control; 2-3: Transgentic tabacco; M: Protein marker.

Figure 6-4-5 Western blotting detection results of ARV σC gene from the transgenic tobacco plant

Discussion

ARV infection has caused serious economic losses in the poultry industry. Live attenuated and inactivated vaccines from strain S1133 are available for use against ARV infection[17-18]. However, some problems, such as safety risks associated with pathogen contamination and vaccine immunity, still exist. The ARV epidemic has made the development of effective and inexpensive vaccines that can protect poultry from ARV infection especially necessary[19]. The protein σC, encoded by the S1 segment, has many functions, including providing protection against the ARV[20]. With the development of molecular biology, plants have offered a new platform for the production of heterologous proteins. Plants that are used as bioreactors and exploited to produce heterologous proteins possess many advantages over other production systems and are thus becoming more attractive[21]. These plants present advantages such as low cost, easy storage and transportation, safety and convenient incubation. Vaccines produced in plants and delivered to animals orally can induce mucosal, cellular and humoral immunity [22-25]. In 2006, the US Department of Agriculture approved the first plant-based vaccine [26]. At present, plants used as a platform to produce vaccines combine food and genetic engineering and are becoming a popular research topic in the field of plant genetic engineering [27, 28].

Tobacco, as a model plant, has been widely used in these types of bioreactor studies. Antigen proteins expressed in plants have been increasingly reported to stimulate humoral and mucosal immune responses when they are provided as food for animals and humans [29]. In the present study, we successfully obtained the ARV σC protein in transgenic tobacco through *A. rhizogenes*. The σC protein was correctly expressed and accumulated in the leaves of the transgenic tobacco plants. Western blot analysis indicated that the σC protein could react with polyclonal antiserum against ARV with good antigenicity. However, there are still several problems associated with these methods. For example, we need to increase the expression levels and stabilize the expression of exogenous genes in plants while enhancing immunogenicity. This study provides a new approach for using a transgenic plant system to express a vaccine candidate.

Conclusions

In summary, the σC protein was effectively expressed in the transgenic tobacco plants. These research findings provide a basis for further analysis of the use of plants as bioreactors in the development and production of σC oral vaccines.

References

[1] VAN DER HEIDE L. The history of avian reovirus. Avian diseases, 2000, 44(3): 638-641.

[2] FANG Q, ZHU Z Y. Progress in avian reovirus structure and function reovirus. Chinese Journal of Virology, 2003, 19(4): 381-384. (in Chinese)

[3] BENAVENTE J, MARTINEZ-COSTAS J. Avian reovirus: structure and biology. Virus Res, 2007, 123(2): 105-119.

[4] WANG Q, WU B, LI G, et al. Effects of duck infected by Muscovy duck reovirus on humoral immunity function. Scientia Agricultura Sinica, 2010, 43(2): 424-429. (in Chinese)

[5] SHMULEVITZ M, YAMEEN Z, DAWE S, et al. Sequential partially overlapping gene arrangement in the tricistronic S1 genome segments of avian reovirus and nelson bay reovirus: implications for translation initiation. Journal of Virology, 2002, 76(2): 609-618.

[6] LIAO M, XIE Z, LIU J, et al. Isolation and identification of avian reovirus. China Poultry, 2002, (22)1: 12-14. (in Chinese)

[7] CHEN W, WU Y, CHEN T, et al. Proteomics analysis of the DF-1 chicken fibroblasts infected with avian reovirus strain S1133: e92154. PLOS ONE, 2014, 9(3): e92154.

[8] HELLAL K Y, BOUROGAA H, GRIBAA L, et al. Molecular characterization of avian reovirus isolates in Tunisia. Virol J, 2013, 10(12): 2-10.

[9] TENG L, XIE Z, XIE L, et al. Sequencing and phylogenetic analysis of an avian reovirus genome. Virus Genes, 2014, 48(2): 381-386.

[10] XIE Z, XIE Z, LIU J, et al. Expression of σC gene in recombined avian reovirus S1133 and its immunogenicity. Journal of Southern Agriculture, 2012, 43(8): 1223-1226. (in Chinese)

[11] SUN M, QIN L, GAO Y, et al. Expression and identification of σC gene of avain reovirus by baculovirus expression system.

Chinese Journal of Virology, 2011, 27(4): 353-357. (in Chinese)
[12]XIE Z, XIE Z, LIU J, et al. Preparation and identification of monoclonal antibody of proteinσC from avain reovirus. Acta Agriculturae Boreali-sinica, 2013, 28(1): 135-139. (in Chinese)
[13] STAUB J M, GARCIA B, GRAVES J, et al. High-yield production of a human therapeutic protein in tobacco chloroplasts. Nat Biotechnol, 2000, 18(3): 333-338.
[14] BIEMELT S, SONNEWALD U, GALMBACHER P, et al. Production of human papillomavirus Type 16 virus-like particles in transgenic plants. Journal of Virology, 2003, 77(17): 9211-9220.
[15] MEYERS A, CHAKAUYA E, SHEPHARD E, et al. Expression of HIV-1 antigens in plants as potential subunit vaccines. BMC Biotechnol, 2008, 8: 1-53.
[16] LIANG H, CHEN F, WANG Y, et al. Expression of human glucagon-like peptide 1 in transgenic tobaccos. Biotechnology Bulletin, 2013, 1: 139-143. (in Chinese)
[17] MIAOMIAO Y, GUANGCHENG W, XIAOHONG P, et al. A method suitable for extracting genomic DNA from animal and plant-modified CTAB method. Hunan Agricultural Science & Technology Newsletter : HASTN, 2008, 9(2): 39-41.
[18] WOOD G W, NICHOLAS R A, HEBERT C N, et al. Serological comparisons of avian reoviruses. J Comp Pathol, 1980, 90(1): 29-38.
[19] MEANGER J, WICKRAMASINGHE R, ENRIQUEZ C E, et al. Association between the sigma C protein of avian reovirus and virus-induced fusion of cells. Arch Virol, 1999, 144(1): 193-197.
[20] LIU H, KUO L, HU Y, et al. Development of an ELISA for detection of antibodies to avian reovirus in chickens. J Virol Methods, 2002, 102(1-2): 129-138.
[21] XIE Z, QIN Y, XIE L, et al. Development of an indirect ELISA based on antigen of recombinant σC protein of avian reovirus. Animal Husbandry and Veterinary Medicine, 2007, 30(11): 3-7. (in Chinese)
[22] SMITH G, WALMSLEY A, POLKINGHORNE I. Plant-derived immunocontraceptive vaccines. Reprod Fertil Dev, 1997, 9(1): 85-89.
[23] LI H, CHYE M. Chapter 19: Use of GFP to investigate expression of plant-derived vaccines. Methods in molecular biology (Clifton, N.J.), 2009, 515: 275-285.
[24] FU L, MIAO Y, LO S W, et al. Production and characterization of soluble human lysosomal enzyme α-iduronidase with high activity from culture media of transgenic tobacco BY-2 cells. Plant science (Limerick), 2009, 177(6): 668-675.
[25] JUNG S, KIM S, BAE H, et al. Expression of thermostable bacterial β-glucosidase (BglB) in transgenic tobacco plants. Bioresource Technology, 2010, 101(18): 7144-7150.
[26] CHIA M, HSIAO S, CHAN H, et al. Immunogenicity of recombinant GP5 protein of porcine reproductive and respiratory syndrome virus expressed in tobacco plant. Veterinary Immunology and Immunopathology, 2010, 135(3): 234-242.
[27] DUS SANTOS M J, WIGDOROVITZ A, TRONO K, et al. A novel methodology to develop a foot and mouth disease virus (FMDV) peptide-based vaccine in transgenic plants. Vaccine, 2002, 20(7): 1141-1147.
[28] CHIA M Y, HSIAO S H, CHAN H T, et al. Immunogenicity of recombinant GP5 protein of porcine reproductive and respiratory syndrome virus expressed in tobacco plant. Vet Immunol Immunopathol, 2010, 135(3-4): 234-242.
[29] ZHOU J, WU J, CHENG L, et al. Expression of Immunogenic S1 Glycoprotein of Infectious Bronchitis Virus in Transgenic Potatoes. Journal of Virology, 2003, 77(16): 9090-9093.

Expression of the E2 gene of Bovine viral diarrhea virus in tobacco

Wang Sheng, Xie Zhixun, Fan Qing, Xie Liji, Deng Xianwen, Huang Li, Xie Zhiqin, Liu Jiabo, Pang Yaoshan, Luo Sisi, and Chen Zhongwei

Abstract

A pair of primers was designed to study the expression of the E2 gene of bovine viral diarrhea virus in plants, and these primers were based on the major antigenic region of the E2 gene using the sequence found in the GenBank database. A plant constitutive expression plasmid that contained pBI121-E2, which possessed the E2 gene, was constructed and transformed into tobacco plants via *Agrobacterium tumefaciens*. Colonies containing this construct were obtained by kanamycin negative selection. A single copy of the E2 gene was detected in five tobacco plants by PCR amplification and real-time PCR analysis. Western blot analysis revealed that the E2 protein was specifically recognized by anti-BVDV-positive serum. Our study provides a foundation for further investigations into the use of plants as bioreactors for the development and production of E2 oral vaccines.

Keywords

bovine viral diarrhea virus, E2 gene, tobacco, oral vaccine

Introduction

Bovine viral diarrhea virus (BVDV) is a positive-stranded RNA *pestivirus* of the family Flaviviridae. BVDV can infect a wide range of hosts, including cattle, sheep and pigs, and has a high degree of infectivity; after infection, BVDV manifests as viral diarrhea, acute and chronic mucosal diseases, persistent infection and immunosuppression, etc., and can cause embryonic malformations, miscarriage or stillbirth in sick animals[1-3]. At present, the disease is widespread worldwide and has caused great damage to the aquaculture industry. However, safe and effective vaccines have not been reported for the prevention and treatment of BVDV at home or abroad, so obtaining effective vaccines for the prevention and treatment of this disease is important.

The BVDV genome consists of approximately 12 300 nucleotides and encodes at least 11 proteins[4]. Among them, glycoprotein E2 is the main component of virions. The E2 protein is composed of 374 amino acids and has a molecular weight of 42 kDa; this protein is less conserved and is divided into two variant regions, V1 and V2. The protein is the main cause of impaired protection against BVDV vaccines and persistent infection. The E2 protein is a structural protein that comprises the capsid and capsule of BVDV; this protein is involved in the adhesion and invasion process of the virus, mediates the immune neutralization response, and can induce the production of neutralizing antibodies[5]. Therefore, research on E2 gene vaccines has become a new direction for the prevention and treatment of BVDV infection. The E2 protein has been successfully expressed in *Escherichia coli* and in the baculovirus system. Tobacco is widely used in the production of antibodies, medicinal proteins, and plant oral vaccines due to the maturity of tobacco, its perfected transformation system and the relatively short cycle needed to obtain transgenic plants [6], and a variety of medicinal proteins or vaccines have been successfully expressed in tobacco[7-10].

In this study, the main antigenic region of the E2 protein-encoding gene of bovine viral diarrhea virus was constructed from the plant expression vector pBI121, which was subsequently transformed via the Agrobacterium-mediated method. In addition, low-copy-transformed tobacco was screened via PCR amplification and real-time PCR, and the reactogenicity of the recombinant protein expressed by the transgenic plants was analyzed. The aim of this study was to provide a theoretical basis and technical support for the production of plant oral vaccine for bovine viral diarrhea using transgenic plants as a bioreactor.

Materials and methods

Reagents, bacteria, and plasmids

The vector pBI121, *Agrobacterium rhizogenes* EHA105 strain, and tobacco (*Nicotiana tabacum*, K346) were kindly provided by Dr. Luo Cong (Guangxi University, China). Bovine viral diarrhea virus (BVDV) strain CV24 and BVDV positive serum were provided by the China Institute of Veterinary Drugs Control.

Plasmid construction

The BVDV E2 open reading frame was amplified from strain CV24, which contains the complete gene, using a forward primer (5'-GGTCTAGACCTGAATACTCATATGCCATAGC-3', *Xba*I site underlined) and a reverse primer (5'-TTCCCGGGGGTCAGTGACCTCCAGGTC-3', *Sma*I site underlined). The PCR products were digested with *Xba*I and *Sma*I and inserted into the plant binary expression vector pBI121, which was digested with the same enzymes, to create the recombinant plasmid pBI121-E2. The construct was validated by PCR, restriction endonuclease mapping and sequencing. This recombinant vector was subsequently introduced into the *A. rhizogenes* EHA105 strain by heat shock. The recombinant strain was obtained by PCR analysis.

Plant transformation and genetic analysis

Tobacco (*N. tabacum* L.) were cut into 0.5～0.7 cm sections and incubated with *A. tumefaciens* EHA 105 carrying pBI121-E2 with suction for 8～10 min. The explants were cocultivated for 3 days in the dark on Murashige and Skoog (MS) media and then washed with sterile water containing 350 μg cefotaxime/mL. After washing, the explants were grown using MS medium containing 350 μg cefotaxime/mL, 350 μg kanamycin/mL, 0.5 μg IAA/mL and 1 μg 6-BA/mL for selection, with fresh medium supplied every 2 weeks. When the explants differentiated into adventitious buds, we transferred the buds to MS media supplemented with 350 μg cefotaxime/mL, 350 μg kanamycin/mL, or 0.5 μg IAA/mL for root induction. When the explants regenerated into complete tobacco plantlets, they were transplanted to pots and grown in greenhouses. At the mature stage, the transgenic plants were identified via PCR analysis with the forward primer 5'-GGTCTAGA CCTGAATACTCATATGCCATAGC-3' and the reverse primer 5'-TTCCCGGGGGTCAGTGACCTCCAGG TC-3'.

Real-time PCR analysis

High-throughput and sensitive SYBR Green I real-time fluorescence quantitative PCR was used to further estimate the copy number of the E2 gene in the positive control plants. The endogenous RNR2 gene in tobacco was used as a reference gene. We quantified the relative density of the double standard curve. Using serial dilutions, the standard curves of the cycle threshold (CT) relative to the log of each initial template copy of the E2 and RNR2 genes were obtained. The transgenic copy number was obtained by comparing the initial template copy number of the E2 gene with that of RNR2.

Western blot analysis

To demonstrate the reactogenicity of the glycoprotein E2 expressed in transgenic leaves, the recombinant protein was detected by western blot analysis. The total protein of the tobacco plants was extracted from the low-copy number samples of transgenic tobacco and analyzed via 12% sodium dodecyl sulfate-polyacrylamide gel electrophoresis (SDS-PAGE). The separated protein was blotted onto a nitrocellulose membrane. Western blotting was performed with BVDV-positive bovine serum as the primary antibody at a 1 ∶ 300 dilution, and HRP-conjugated rabbit anti-bovine IgG was used as the secondary antibody at a dilution of 1 ∶ 200. Total protein extracted from wild-type tobacco was used as a negative control. Immune complexes were detected after incubation with a DAB kit.

Results

Genetic analysis of transformed tobacco

The insertion of the construct carrying the E2 gene into pBI121 was confirmed by PCR (Figure 6-5-1), restriction endonuclease mapping and sequencing. The introduction of pBI121-E2 into *A. tumefaciens* strain EHA105 was confirmed by PCR. Transgenic tobacco plants were successfully obtained (Figure 6-5-2). To confirm the integration of the E2 gene into the transgenic tobacco plants, DNA extracted from the leaves of the transgenic tobacco plants was tested via PCR: a 1 004 bp band was detected in most groups, except for the negative control group (Figure 6-5-3). Thus, we successfully obtained transgenic tobacco plants that expressed recombinant E2.

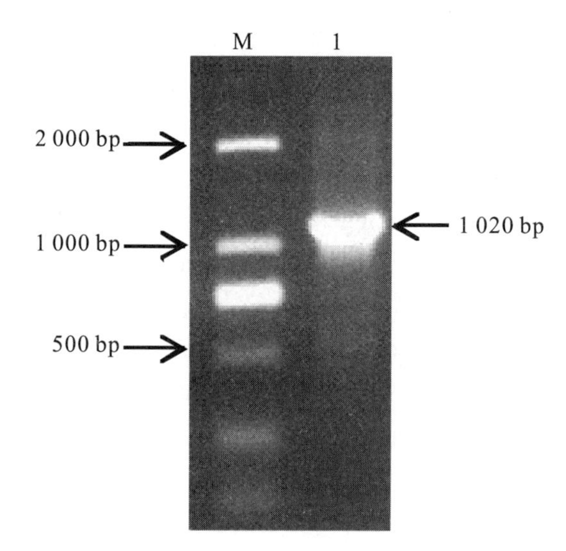

Lane 1: PCR product of E2 gene; M: DNA maker.

Figure 6-5-1 PCR product of E2 gene from BVDV

A, B: Induction differentiation; C, D: Induce root; E: Transplantation. After transformation of tobacco plants via Agrobacterium tumefactions, the resistant plants were selected with kanamycin.

Figure 6-5-2 The regeneration of transgenic tabacco expressing E2

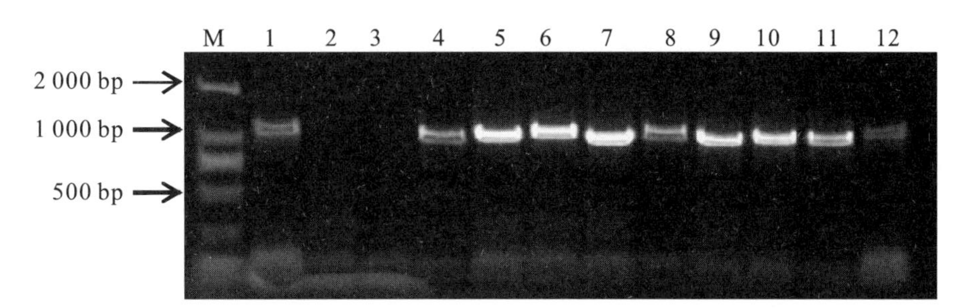

Lane 1: Positive control; Lane 2-3: Negative control; Lanes 4-12: Transgentic plant, a specific band of 1 020 bp was detected.

Figure 6-5-3 PCR assay of E2 transgenic plants

E2 copy numbers in transgenic tobacco

We used real-time fluorescence quantitative PCR to quantify the transcription level of the E2 gene. The endogenous RNR2 gene was added as the spike reference for the target E2 gene. The standard curves of the cycle (CT) threshold relative to the log of each initial template copy of the E2 and RNR2 genes were obtained using the serial dilution technique (the standard curves of E2: $y=-3.529\ 5x+14.797$; RNR2: $y=-3.882\ 9x+2.756\ 7$). The correlation coefficients were 0.999 2 and 0.999 5, respectively, suggesting that we could accurately analyze the samples quantitatively (Figure 6-5-4). The transgenic copy number was obtained by comparing the initial template copy number of E2 with that of RNR2. Among the five putative transgenic lines, a copy number of 3 was obtained, whereas the negative control had zero copies.

E2 protein expression in transgenic tobacco

Western blotting was used to determine the expression of the glycoprotein E2 in the transgenic tobacco plants. Western blot analysis revealed a specific band at 70 kDa corresponding to the glycoprotein E2 (Figure 6-5-5), the negative control was not detected. This finding revealed that the highly expressed glycoprotein E2 in transgenic tobacco was reactive to the anti-BVDV antiserum.

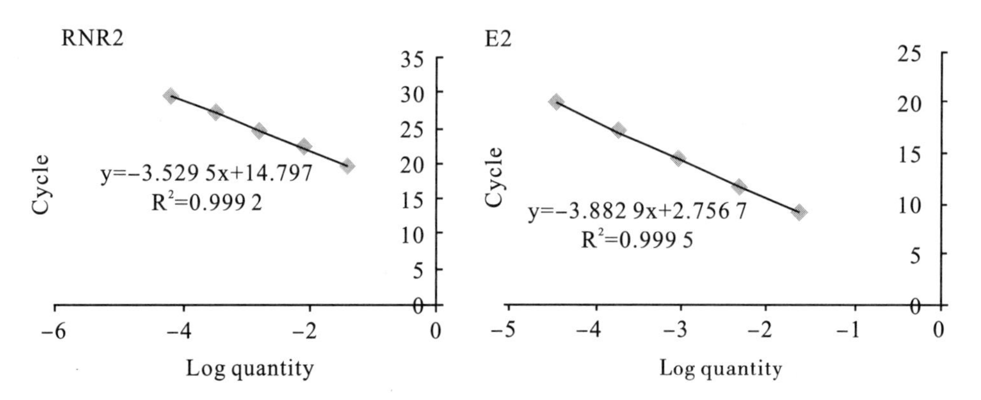

The standard curves of E2 y= −3.529 5x+14.797, the standard curves of RNR2 y=−3.882 9x+2.756 7. The correlation coefficients were 0.999 2 and 0.999 5, respectively.

Figure 6-5-4 The standard curve of RNR2 and E2 gene

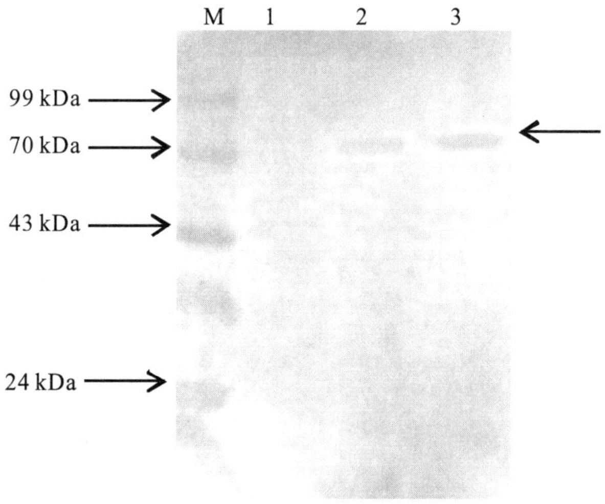

M 1 2 3

99 kDa

70 kDa

43 kDa

24 kDa

Western blot analysis showing immune activity of the E2 protein expressing in transgenic tobacco. In transgenic group (lanes 2, 3) a specific band of 70 kDa was detected. Lane 1: Negative control (proteins extracted from wild-type tobacco).

Figure 6-5-5 Purification results of E2 transgenic tabacco

Discussion

Bovine viral diarrhea causes severe mucosal lesions and clinical manifestations, thus causing economic losses to the cattle industry worldwide. Statistics suggest that BVDV infection has a significant negative effect on dairy operations[11]. Due to the severity of the BVDV epidemic, it is necessary to develop inexpensive and effective vaccines to protect animals from BVDV infection. The glycoprotein E2 has many functions, such as facilitating viral absorption and cell entry, increasing antibody yields with high neutralizing activity, and protecting against challenge with the BVDV virus[12-13].

Currently, with the development of molecular biology and plant gene engineering, plants serve as new targets for the production and delivery of antigenic proteins for use in plant-derived oral vaccines[14-18]. Compared with typical vaccines, transgenic plants are utilized as vaccine products because of their many advantages, including easy storage and transport, low cost, rapid scalability, safety and convenient incubation. Vaccines produced from transgenic plants can maintain native immunogenicity[19]. Oral vaccines can stimulate strong humoral systemic and mucosal antigen-specific antibody responses.

Tobacco is a model plant for molecular biology research and has been widely used in the study of bioreactors[20]. However, there are no immediate reports of the use of tobacco as a bioreactor for BVDV vaccine production. In this study, we successfully obtained transgenic tobacco plants expressing the glycoprotein E2 of BVDV through the use of *A. rhizogenes*. The protein E2 was expressed in transgenic tobacco leaves. Western blot analysis indicated that the glycoprotein E2 could react with polyclonal antiserum against BVDV, with good antigenicity. This study offers a new approach for the use of transgenic plant systems to investigate candidate vaccines.

Currently, plants are used as a platform for producing vaccines. The function of plants in food and medicine has become a popular research topic in the study of plant genetic engineering. In general, the stable expression level of exogenous genes in plants should be enhanced while also improving their immunogenicity.

Conclusions

In summary, the E2 protein was effectively expressed in the transgenic tobacco plants. Western blot analysis revealed that the E2 protein was specifically recognized by anti-BVDV-positive serum. These findings provide a basis for further investigations into the use of plants as bioreactors for the development and production of E2 oral vaccines.

References

[1] JONES L R, WEBER E L. Application of single-strand conformation polymorphism to the study of bovine viral diarrhea virus isolates. Diagn Invest, 2001, 13(1): 50-56.

[2] BROCK K V. The persistence of bovine viral diarrhea virus. Biologicals, 2003, 31(2): 133-135.

[3] LIANG R, VAN DEN HURK J V, ZHENG C, et al. Immunization with plasmid DNA encoding a truncated, secreted form of the bovine viral diarrhea virus E2 protein elicits strong humoral and cellular immune responses. Vaccine, 2005, 23(45): 5252-5262.

[4] BRUSHKE C J, VAN OIRSHOT J T, VAN RIJN P A. An experimental multivalent bovine virus diarrhea virus E2 subunit vaccine and two experimental conventionally inactivated vaccines induce partial fetal protection in sheep. Vaccine, 1999, 17(15-16): 1983-1991.

[5] FAN Q, XIE Z, LIU J, et al. Expression and E2 gene of bovine virus diarrhea in insect cells. Progress in Veterinary Medicine, 2013, 34(4): 9-12. (in Chinese)

[6] ZENG F, ZHANG Y, ZHANG M, et al. Cloning and optimizing expression of E2 gene of bovine viral diarrhea-mucosal disease virus in BCG. Acta Veterinaria et Zootechnica Sinica, 2014, 45(1): 94-100. (in Chinese)

[7] BIEMELT S, SONNEWALD U, GALMBACHER P, et al. Production of human papillomavirus type 16 virus-like particles in transgenic plants. J Virol, 2003, 77(17): 9211-9220.

[8] FAN Q, XIE Z, XIE L, et al. A reverse transcription loop-mediated isothermal amplification method for rapid detection of bovine viral diarrhea virus. J Virol Methods, 2012, 186(1-2): 43-48.

[9] LIANG H, CHEN F, WANG Y et al. Expression of Human Glucagon-like Peptide 1 in Transgenic Tobaccos. BIOTECHNOLOGY BULLETIN, 2013, 1: 139-143. (in Chinese)

[10] BIEMELT S, SONNEWALD U, GALMBACHER P, et al. Production of human papillomavirus type 16 virus-like particles in transgenic plants. J Virol, 2003, 77(17): 9211-9220.

[11] QIU C, GUO H, CHENG S, et al. Serlogical monitoring of bovine viral diarrhea/mucosal disease in buffalos at the partial regions in Anhui, Jiangsu and Guangxi. Chinese Journal of Preventive Veterinary Medicine, 2000, 22(6): 453-454. (in Chinese)

[12] FAN Q, XIE Z, LIU J, et al. Establishment of real-time flourescent quantitative PCR for detection of bovine viral diarrhea virus. Progress in Veterinary Medicine, 2010, 31(10): 10-14. (in Chinese)

[13] FAN Q, XIE Z, XIE L, et al. A reverse transcription loop-mediated isothermal amplification method for rapid detection of bovine viral diarrhea virus. Journal of Virological Methods, 2012, 186(1-2): 43-48.

[14] ZHANG B, YANG Y, NI W, et al. Studies on avian influenza virus Matrix 2 Transgenic tobacco. Jiangsu J of Agr Sci, 2010, 26(1): 51-54. (in Chinese)

[15] LI H, CHYE M. Use of GFP to investigate expression of plant-derived vaccines. Methods in Molecular Biology, 2009, 515: 275-285.

[16] FU L, MIAO Y, LO S, et al. Production and characterization of soluble human lysosomal enzyme α-iduronidase with high activity from culture media of transgenic tobacco BY-2 Cells. Plant Science, 2009, 177(6): 668-675.

[17] JUNG S, KIM S, BAE H, et al. Expression of thermo stable bacterial β-glucosidase(BgIB) intransgenic tobacco plants. Bioresource Technology, 2010, 101(18): 7144-7150. (in Chinese)

[18] CHIA M, HSIAO S, CHAN H, et al. Immunogenicity of recombinant GP5 protein of porcine reproductive and respiratory syndrome virus expressed in tobacco plant. Veterinary Immunology and Immunopathology, 2010, 135(3-4): 234-242.

[19] WANG Y, ZHAO S, CHEN F, XIAO G. Estimation of the copy number of exogenous gene in transgenic rice by real-time fluorescence quantitative PCR. Life Science Research, 2007, 11(4): 301-305. (in Chinese)

[20] DU X, HE Z, CHEN L. Recent progress on plant bioreactor expressing pharmaceutical proteins. China Biotechnology, 2008, 28(9): 135-143. (in Chinese)

Immunological tests in the fields on attenuated B_{26}-T_{1200} vaccines of *Pasteurella multocida*

Xie Zhixun, Liu Jiabo, Mo Zhaolan, and Pang Yaoshan

Abstract

Five batches of the attenuated B_{26}-T_{1200} vaccines of *Pasteurella multocida* (*P. multocida*) were inoculated into 30 000 000 chickens. Among them, 62 437 vaccinated chickens were monitored in the fields. Our experimental data showed that the tested attenuated B_{26}-T_{1200} vaccines were safe for various types of chickens. Only 3.8% of the vaccinated chickens displayed transient anorexia and lethargy. However, the chickens that were transiently affected recovered after two days of inoculation, and behaved normally. Death induced by a direct known cause among the vaccinated chickens was not observed. One-hundred and forty days after vaccination, the chickens vaccinated by each batch of the vaccines were challenged with wild-type bacteria, respectively. Among the chickens that were challenged, approximately 80% of the vaccinated chickens were protected, and no *P. multocida* was observed among the vaccinated chickens after vaccination of the live attenuated vaccines.

Keywords

B_{26}-T_{1200}, vaccine pasteurella multocida, immunological

---

Introduction

Our research group had previously reported on the immunogenicity, safety, induction period and duration of immunity for the attenuated B_{26}-T_{1200} live vaccine of *P. multocida*. To further evaluate on the practical use and effect of the attenuated vaccine, we conducted a series of experiments testing on the immune effect of the vaccines in the fields. Our experimental data are reported in the following sections.

Materials and Methods

Experimental vaccines

Five batches (No. 8903, 9001, 9002, 9003 and 9004) of lyophilized, attenuated B_{26}-T_{1200} vaccines of *P. multocida* were manufactured by the Guangxi Biological Drugs Factory according to specialized production procedures for the vaccines.

Experimental chicken

Flocks of chickens over two-month old were collected from different regions. The types of chickens used in the experiments include: Shiqiza, Yisha, B6 and Zhusi.

A highly virulent *P. multocida* strain

The C_{48-1} was a highly virulent *P. multocida* strain provided by China Veterinary Drug Surveillance Institute.

Experimental designs

Five batches of the attenuated B_{26}-T_{1200} vaccines were tested in the fields with more than 300 000 chickens from the poultry farms in different regions in Guangxi, including counties (such as Yongning, Lingshan, and Yuling) and cities (such as Qinzhou, Nanning, Liuzhou and Beihai). More than 10 000 immunized chickens were monitored by experienced workers after vaccination of the attenuated vaccines. Data were collected within 15 days of vaccination, including chickens with general responses (transient lethargy, anorexia, and those that were recovered and became normal in two days after inoculation), chickens with severe responses (e.g., paralysis), and the number of dead chickens. Autopsies were performed on all the inoculated chickens that died within 15 days after inoculation. Isolation of *P. multocida* from the dead chickens was conducted according to Rhoads protocol and procedures. If the isolated bacteria are *P. multocida*, then the same type of chickens used for field-testing will be utilized for further testing on bacterial virulence. The isolated bacteria were cultured, and 1 mL of cultured bacteria containing 3×10^9 CFU/mL live bacteria was inoculated into each chicken, respectively. The chickens were monitored for 15 days after inoculation of bacteria to observe clinical signs.

Each batch of live vaccine was used to vaccinate the naïve chickens, which were monitored continuously for 140～180 days. After 140 days of vaccination, the vaccinated chickens were evaluated for appearance of *P. multocida*. Four chickens were randomly chosen from the vaccinated ones. In addition to four vaccine-naïve control chickens, the four selected vaccinated chickens were also challenged with a highly virulent bacterial strain, respectively. The protection rate was calculated at the end of experiments.

Results

Clinical manifestations of the vaccinated chickens

Five batches of experimental vaccines were used for vaccination of more than 30 000 000 chickens. Only 62 437 chickens were monitored (Table 6-6-1). Among them 2 112 chickens possess general responses including those who had transient lethargy and anorexia. These 2 112 chickens constitute 3.3% of the 64 437 chickens that were monitored. Within the 15-day observation period, four vaccinated chickens died of *P. multocida* (0.006%). One hundred and forty-five chickens died of other diseases such as *E.* coli disease, Marek's disease, and staphylococcal infection.

Table 6-6-1 Field testing results on the attenuated B_{26}-T_{1200} vaccine of *P. multocida*

Vaccine batch	Number of vaccinated chickens observed	Number of chickens with general responses	Number of chickens died of *P. multocida*	Number of chickens died of other pathogens
8903	11 500	630	4	42
9001	13 419	237	0	31
9002	13 070	407	0	22
9003	12 366	248	0	32
9004	12 082	590	0	18
Total	64 437	2 112	4	145

Protection rates of vaccinated chickens

Table 6-6-2 shows the protection rates of the vaccinated chickens. Chickens in the fields were vaccinated. One hundred and forty days thereafter, the vaccinated chickens were challenged with the highly virulent bacterial strain C_{48-1}. The protection rates for the five batches of the attenuated vaccines were: 75%, 75%, 75%, 75% and 100%, respectively. The total protection rate is 80%. In contrast, in the control group all the chickens that were challenged died (10/10, 100% mortality). Interestingly, *P. multocida* was not detected in the vaccinated chickens within 140 days of inoculation.

Table 6-6-2 Results of field challenging on chickens that had been vaccinated for 140 days

Vaccine batch	Chicken farms	Types of chickens	Number of protected chickens/No. of chickens that were challenged	Number of chickens died in the controls
8903	School of Agricultural Machinery	Shiqiza	3/4	2/2
9001	Institute of Agricultural Machinery	Shiqiza	3/4	2/2
9002	Food Company	Shiqiza	3/4	2/2
9003	Sanqiao	Shiqiza	4/4	2/2
9004	Xinan	Shiqiza	3/4	2/2
Total			16/20	10/10

Summary and Discussion

Our field testing results clearly demonstrated that the attenuated B_{26}-T_{1200} vaccines of *P. multocida* are safe for chickens. Only 3.38% of the inoculated chickens had transient responses. We isolated four strains of *P. multocida* from four dead chickens, respectively. When the healthy chickens were inoculated intramuscularly, respectively, with 3×10^9 CFU/mL live bacteria of each strain, the inoculated bacteria resulted in death of chickens. We reasoned that the death of chickens inoculated with the isolated bacteria was not due directly to the effect of the attenuated vaccines. We proposed that it was likely that these four chickens that were inoculated with the isolated bacteria might have been infected with wild-type bacteria or these four chickens might be healthy carriers of the bacteria. Under the testing conditions mentioned above, when the infected chickens were inoculated with a live attenuated vaccine, the immune systems in the chickens might be weakened or compromised, resulting in *P. multocida*. We had previously passaged the live attenuated vaccine strain used in this study consecutively for five generations in the chickens. Each healthy chicken was inoculated with 3×10^9 CFU/mL live bacteria that were passaged, and no abnormal reactions were observed. In addition, no reversion-to-virulence phenomenon was noticed for the attenuated vaccines tested in our study.

Chickens in the fields were first vaccinated with the live attenuated vaccines for 140 days. Thereafter, the vaccinated chickens were challenged with a highly virulent bacterial strain. Our results demonstrated that 80% of the vaccinated chickens were protected. These results were consistent with those obtained from our previous study in our laboratory. In that study, a protection rate of 75% was observed in the vaccinated chickens after five-month inoculation of vaccines.

Selection and cultivation of the attenuated B_{26}-T_{1200} vaccine strain of *Pasteurella multocida*

Zuo Wanshun, Ning Zhenhua, and Xie Zhixun

Abstract

The *Pasteurella multocida* (*P. multocida*) B_{26}-T_{1200} is an attenuated strain obtained from a locally isolated B_{26} virulent strain of duck by consecutive passages for 1 200 generations on 0.1% hemagglutinin-containing Martin culture medium. The attenuated B_{26}-T_{1200} strain did not show virulent reversion after continued passage for 30 generations on slant blood culture medium or with continued passage in susceptible chickens. Chickens immunized with 3×10^7 live bacteria were protected from challenge with the wild-type virulent bacterial strain. The attenuated B_{26}-T_{1200} strain displayed a good safety profile and potent immunogenicity.

Keywords

B_{26}-T_{1200}, vaccine, pasteurella multocida safety

Introduction

Pasteurellosis is a contiguous and highly infectious disease caused by *P. multocida* in domestic and wild poultry. Since 1979 we have been conducting research on the attenuated strain of *P. multocida*. Here we report that an attenuated *P. multocida* strain B_{26}-T_{1200} was obtained by passage of a regional, highly virulent strain B_{26} for 1 200 generations. We further evaluated the safety, stability and immunogenicity of the B_{26}-T_{1200} strain. Our experimental data are presented in the following sections:

Experimental materials

Culture media used for consecutive passage

Martin-liquid culture medium containing 0.1% hemagglutinin and Martinagar plate containing 4% sheep serum.

Bacterial strains

Local wild-type virulent strain: These bacterial strains were isolated from the liver of ducks, which died of acute pasteurellosis. The three isolated strains were named B_{25}, B_{26} and B_{27} virulent strains derived from ducks, respectively. The capsular type was identified as A type, and the fluorescence of bacterial colony was classified as Fo type.

Standard highly virulent strain: *P. multocida* $C_{48\text{-}1}$, a bacterial strain provided by China Veterinary Drug Surveillance Institute.

Experimental chickens

Three-month old susceptible Leghorn chickens were provided by the Chicken Farms owned by the Guangxi Veterinary Research Institute.

Methods and Results

Determination of virulence and immunogenicity of local virulent bacterial strain

Testing on virulence

The virulent bacterial strains B_{25}, B4 and B_{27} were inoculated, respectively, into Martin-liquid culture medium containing 0.1% hemagglutinin, and cultured at 37 ℃ for 24 h. By using Martin culture medium, the bacterial culture was divided into five groups as follows: 2 colonies/mL, 4 colonies/mL, 6 colonies/mL, 8 colonies/mL, and 10 colonies/mL, respectively. Two chickens in each group were tested. Individual chicken in each group was inoculated intramuscularly with 1 mL of diluted bacterial culture. The inoculated chickens were monitored for 14 days and mortality rate was recorded. Our experimental data showed that the B_{26} strain is the most virulence one among the three virulent strains tested, because both of the two chickens inoculated with two bacterial colonies died. In contrast, chickens inoculated with two colonies of either bacterial strain B_{25} or B_{27} did not die or only a portion of chickens died.

Testing on immunogenicity

According to the method described in "China Guideline for Veterinary Biological Drugs" , three vaccines containing aluminum adjuvant were prepared from bacterial strains B_{25}, B_{26} and B_{27}, respectively. The chickens were divided into three groups and each group contains four chickens. The chickens in each group were injected subcutaneously with equal doses of vaccines. The chickens in the control group were challenged with the lethal dose of highly virulent strain C_{48-1}. Protection rates were recorded two weeks after challenges. Our results showed that protection rate for the B_{26} aluminum vaccine was 75% (3/4), whereas those of B_{25} and B_{27} aluminum vaccines were 50% (2/4). All the chickens in the control group died two weeks after challenge with *P. multocida* C_{48-1}. Identical experiments were repeated twice and reproducible results were obtained.

Artificial attenuation of the highly virulent strain

The B_{26}, a highly virulent strain with potent immunogenicity, was inoculated into Martin-liquid culture medium containing 0.1% hemagglutinin. The culture was incubated at 37 ℃ for 24 h and passaged consecutively. During the passage, temperature in the incubator was gradually increased to 45 ℃. After consecutive passages of 20～30 generations, colonies that will be further passaged were selected based on fluorescence standard and morphology of the bacterial colonies. The selected colonies were passaged continuously to 1 200 generations. Alterations in virulence and immunogenicity of the passaged bacterial strains are as follows.

Alteration on virulence of the attenuated bacterial strains

When the bacterial strain B_{26} was passaged over 401 generations, virulence of the passaged bacterial strains started to decrease. Upon passage on 501 generations, four sensitive chickens were injected with

7.1×10^4 CFU/mL live bacteria, respectively. All the four injected chickens survived, indicating that the virulence of the passaged live bacteria was substantially reduced. Upon passage on 701 generations, four chickens were inoculated with 1.2×10^6 CFU/mL live bacteria, respectively, and all the four inoculated chickens survived, and the virulence of the passaged live bacteria decreased over 600 000 fold. Upon passage for 1 000 generations, two chickens were injected with 2.6×10^9 CFU/mL live bacteria, respectively, and both chickens survived. Upon passage for 1 200 generations, five chickens were injected with 9.3×10^9 CFU/mL live bacteria, respectively, and all the five chickens survived, demonstrating that the virulence of the passaged live bacteria was reduced by 4.65×10^9 fold relative to the B_{26} primary bacteria.

Alteration on immunogenicity in the attenuated bacterial strains

Upon passage of the B_{26} for 600～1 000 generations, 28 chickens were immunized via chest muscle with 1.2×10^6～5.2×10^8 CFU/mL live bacteria, respectively. Fourteen days later, the immunized chickens were challenged with a highly virulent strain *P. multocida* $C_{48\text{-}1}$. Our results showed that 27 out of 28 chickens were protected, and the protection rate is 96.7%. All the sixteen chickens in the control group died (16/16, 100% mortality). Upon passage for 1 200 generations, each chicken was inoculated with 3×10^7 CFU/mL live bacteria. Fourteen days after immunization, all the 30 immunized chickens as well as chickens in the control group were challenged with wild-type virulent bacteria. Our results demonstrated that the 30 immunized chickens were all protected, whereas all the chickens in the control group died.

Virulence reversion test on the attenuated B_{26}-T_{1200} strain

The attenuated B_{26}-T_{1200} bacterial strain was passaged consecutively on slant fresh blood Martin-agar culture medium for 30 generations or in susceptible chickens for five generations. Ten sensitive chickens were injected intramuscularly with 1 mL of bacterial culture containing 3×10^9 CFU/mL live bacteria. All the injected chickens were monitored for 15 days. Our results showed that all the injected chickens survived without abnormal reactions.

Summary

By performing consecutive passages in Martin culture medium , an attenuated bacterial strain B_{26}-T_{1200} is obtained from a highly virulent strain B_{26} isolated in Guangxi, China. The B_{26} strain is highly virulent and has a good immunogenicity.

Our results demonstrated that chickens immunized with 3×10^7 CFU/mL colonies of live, attenuated avian *P. multocida* were protected from challenge with a highly virulent bacterial strain. That is, the live attenuated avian *P. multocida* strain of B_{26}-T_{1200} has a good immunogenicity.

When the attenuated avian *P. multocida* strain of B_{26}-T_{1200} was passaged on slant blood culture medium for 30 generations or passaged in susceptible chickens for five generations, the phenomenon of virulence reversion has not been observed. Therefore, the B_{26}-T_{1200} strain has a good safety profile as well as a stable phenotype.

Determination of safety range and immune dosage in chickens for the attenuated B_{26}-T_{1200} vaccine of *Pasteurella multocida*

Xie Zhixun, Liu Jiabo, Mo Zhaolan, and Pang Yaoshan

Abstract

The attenuated B_{26}-T_{1200} vaccine of *Pasteurella multocida* (*P. multocida*) was diluted by normal saline containing 20% aluminum gel. Different doses of the diluted vaccine were inoculated subcutaneously into chickens. Two weeks later, the inoculated chickens were challenged by virulent wild-type bacteria. Our results showed that for chickens that were inoculated with 4.1×10^5 CFU/mL live bacteria, one out of five chickens survived challenge with the lethal dose of live virulent bacteria. For chickens inoculated with 8.3×10^5 CFU/mL live bacteria, two out of five chickens survived the lethal challenge. For the nine groups of chickens that were inoculated with live bacteria ranging from 1.3×10^6 to 3×10^7 CFU/mL, respectively, the number of the protected chickens was consistent, ranging from 4/5 to 5/5. Chickens inoculated with higher doses of live bacteria were monitored for 15 days. Our results showed that for the eight groups of chickens inoculated with the live bacteria ranging from 2.6×10^9 to 9.3×10^9 CFU/mL, both of the inoculated chickens in each group survived without any abnormality. For the chicken groups inoculated with 1×10^{10}, 1.09×10^{10} and 1.5×10^{10} CFU/mL live bacteria, respectively, one half of the inoculated chickens survived, whereas the other half died. In contrast, for the chicken group inoculated with 1.71×10^{10} CFU/mL live bacteria, both inoculated chickens died.

Keywords

B_{26}-T_{1200} vaccine, pasteurella multocida, immune dosage safety

Introduction

The attenuated B_{26}-T_{1200} vaccine strain of *P. multocida* was obtained via consecutive passages on cultures of local wild-type virulent bacterial strain in our research institute. In recent years, a variety of poultry with a total number of more than 2×10^7 CFU/mL had been inoculated with the lyophilized B_{26}-T_{1200} vaccine, and a good immunization effect was observed. We systemically evaluated safety profile and immune dosage in chickens for the attenuated B_{26}-T_{1200} vaccine. Our experimental data will be presented in the following sections.

Materials and methods

A highly virulent bacterial strain

The *Pasteurella multocida* C_{48-1} was a highly virulent bacterial strain provided by China Veterinary Drug Surveillance Institute.

Experimental chickens

Three-month old susceptible Leghorn cocks were hatched in the same batch in a laboratory at the Guangxi Veterinary Research Institute.

Lyophilized B_{26}-T_{1200} attenuated vaccine of *P. multocida*

This vaccine was manufactured according to specialized production procedures for lyophilized biological products.

Testing on immune dosage

Lyophilized vaccines were diluted with normal saline containing 20% aluminum gel. Different doses of the diluted vaccines were injected subcutaneously into the necks of chickens. Two weeks later, the vaccinated and the control group chickens were challenged, respectively, with the lethal dose of a highly virulent bacterial strain $C_{48\text{-}1}$. All the challenged chickens were monitored for 15 days, and the protection rates were recorded.

Determination on safety profiles

Lyophilized vaccines were diluted with normal saline containing 20% aluminum gel. For chickens in the experimental groups, vaccines containing various numbers of live bacteria were injected subcutaneously into the necks of chickens, respectively. For chickens in the control group, only normal saline containing 20% aluminum gel was injected subcutaneously into the necks of chickens. Chickens inoculated with various amounts of vaccines or with normal saline containing 20% aluminum gel were monitored for 15 days, and the mortality rates were recorded.

Results

Testing on immune dosage

The experimental results are shown in Table 6-8-1.

Table 6-8-1 Tests on immune dosage of the B_{26}-T_{1200} attenuated vaccine of *P. multocida*

Test	Live bacteria inoculated (1×10^4)	Death number/ Test number	Death number in the control group
1	3 000	0/5	2/2
	2 030	1/5	2/2
	800	0/5	2/2
	400	1/5	2/2
	83	3/5	2/2
	41	4/5	2/2
2	2 800	0/5	2/2
	2 000	1/5	2/2
	800	1/5	2/2
	637	0/5	2/2
	130	1/5	2/2

Determination on safety profiles

The experimental results are shown in Table 6-8-2.

Table 6-8-2 Tests on safety range of the attenuated B_{26}-T_{1200} live vaccine of *P. multocida*

Test	Live bacteria inoculated (1×10^4)	Number of death / Number of tests
1	39	0/2
	60	0/2
	70.2	0/2
	93	0/2
	109	1/2
	171	2/2
Control group injected with normal saline containing 20% aluminum gel		0/2
2	26	0/2
	48	0/2
	63	0/2
	75	0/2
	100	12
	150	1/2
Control group injected with normal saline containing 20% aluminum gel		0/2

Summary and Discussion

As shown in the results for determination on immune dosage, when each chicken was immunized with 4.1×10^5 CFU/mL live bacteria, respectively, 20% of the vaccinated chickens were protected (One out of five chickens survived). With the elevated immune doses, the protection rates also increase accordingly. When each chicken was immunized with 8.3×10^5 CFU/mL live bacteria, two out of five chickens survived the challenge from the highly virulent bacteria. In addition, when 1.3×10^6 CFU/mL live bacteria or even more bacteria were used for vaccination, the vaccinated chickens in each experimental group had good protection rates ranging from 80% to 100%. Therefore, we consider that the minimum immune dosage for the attenuated B_{26}-T_{1200} vaccine was 1.3×10^6 CFU/mL live bacteria. These results clearly demonstrated that the B_{26}-T_{1200} vaccine of *P. multocida* has a good immunogenicity.

We tested the safety profiles of higher doses of the vaccine, and we found that for the eight groups of chickens inoculated with 2.6×10^9, 3.9×10^9, 4.8×10^9, 6.0×10^9, 6.3×10^9, 7.0×10^9, 7.5×10^9 and 9.3×10^9CFU/mL, respectively, all the vaccinated chickens in each group survived (2/2, 100% survival) without any abnormality. When chickens were inoculated with 1.0×10^{10} and 1.5×10^{10} CFU/mL live bacteria, respectively, half of the inoculated chickens died (1/2, 50% survival). When chickens were inoculated with 1.7×10^{10} CFU/mL live bacteria, all the inoculated chickens died (2/2, 100% death). These results demonstrated that the maximum safety dosage for the lyophilized B_{26}-T_{1200} attenuated vaccine of *P. multocida* was 9.3×10^9 CFU/mL live bacteria.

Our results demonstrated that for the attenuated B_{26}-T_{1200} vaccine of *P. multocida*, the minimum immune dosage was 1.3×10^6 live bacteria, whereas the maximum safety dosage was 9.3×10^9 live bacteria. Considering the fact that the difference between the minimum immune dosage and the maximum safety dosage was 7 154 fold, we consider that the safety profile of the B_{26}-T_{1200} vaccine is satisfactory and the vaccine is suitable for practical application.

Induction of immune responses and duration of immunity in chickens inoculated with attenuated B_{26}-T_{1200} vaccine of *Pasteurella multocida*

Xie Zhixun, Liu Jiabo, Mo Zhaolan, Pang Yaoshan, Xie Zhiqin, and Xu Zhenfong

Abstract

Chickens that were vaccinated with the attenuated B_{26}-T_{1200} vaccine of *Pasteurella multocida* (*P. multocida*) 8 h later acquired immunity (One out of five chickens was protected from getting disease). Two to three days thereafter, three out of five chickens were protected. Four to five days later, the majority of chickens acquired robust immunity (4/5～5/5 of chickens were protected). Total protection rate after vaccination in two, three, four and five months all together, was 83.3% (80/96). Average protection rate for the six batches of vaccines after five months of vaccination was 75% (18/24).

Keywords

B_{26}-T_{1200}, vaccine pasteurella multocida, immune responses, immunity

---

Introduction

The attenuated B_{26}-T_{1200} vaccines of *P. multocida* were tested in the laboratory, and the results showed that the B_{26}-T_{1200} is an attenuated vaccine with a good safety profile and potent immunogenicity. We conducted a series of experiments to further explore when the immunity was induced and how long the established immunity could last. The observed experimental data are presented in the following sections.

Materials and Methods

Experimental vaccines

Six batches of the attenuated B_{26}-T_{1200} vaccines of *P. multocida* were produced. The first three batches of vaccines (9001, 9002 and 9003) were manufactured by the Guangxi Biological Drugs Factory according to specialized production procedures for the vaccines. The rest of three batches of vaccines (9101, 9102 and 9103) were produced by the Guangxi Veterinary Research Institute.

Experimental chickens

Three-month old susceptible Leghorn cocks were hatched in the same batch by a laboratory at the Guangxi Veterinary Research Institute.

A highly virulent bacterial strain

C_{48-1} was a highly virulent *P. multocida* strain provided by China Veterinary Drug Surveillance Institute.

Testing on induction of immunity

Lyophilized vaccines were diluted with normal saline containing 20% aluminum gel according to labels on the vials. The diluted vaccines were injected subcutaneously into the necks with one dose (3×10^7 live bacteria/dose). After inoculation of vaccines for eight hours, and on days 1, 2, 3, 4 and 5, respectively, the controls and the vaccinated chickens were challenged, respectively, by a lethal dose of the highly virulent *P. multocida* strain $C_{48\text{-}1}$.

Testing on duration of immunity

Lyophilized vaccines were diluted with normal saline containing 20% aluminum gel according to labels on the vials. The diluted vaccines were injected subcutaneously into the necks with one dose (3×10^7 CFU/mL live bacteria/dose). After inoculation of vaccines for 2, 3, 4, 5 and 6 months, respectively, four immunized chickens were chosen from each batch of vaccine-inoculated chicken groups. The vaccinated and control chickens were challenged, respectively, by a lethal dose of the highly virulent *P. multocida* strain $C_{48\text{-}1}$.

Results

Testing on induction period of immunity

As shown in Table 6-9-1, chickens vaccinated with the attenuated B_{26}-T_{1200} vaccines generated immunity at 8 h after inoculation. Four days after vaccination, the vaccinated chickens acquired potent immunity.

Table 6-9-1 Tests on induction of immunity by the attenuated B_{26}-T_{1200} vaccines of *P. multocida*

Test	Vaccine Batch	Time of challenge after immunization	Immunized chickens after challenge Chickens died/Chickens challenged	Chickens died in the control group
1	910	5 d	0/5	2/2
		4 d	1/5	2/2
		3 d	2/5	2/2
		2 d	2/5	2/2
		1 d	3/5	2/2
		8 h	4/5	2/2
2	910	5 d	1/5	2/2
		4 d	0/5	2/2
		3 d	2/5	2/2
		2 d	2/5	2/2
		1 d	3/5	2/2
		8 h	3/5	2/2

Testing on duration of established immunity

As shown in Table 6-9-2, at the fourth month of vaccination, the protection rates for the six batches of vaccines varied from 75% to 100%, and the total protection rate was 79.2%. At the fifth month of vaccination, the protection rates varied from 50% to 100%, and the total protection rate was 75%. At the sixth month of vaccination, the protection rates for the six batches of vaccines varied from 25% to 75%, and the total protection rate was 45.8%. In contrast, all the chickens in the control groups that were challenged by the highly virulent *P. multocida* strain died in the experiments.

Table 6-9-2 Tests on induction of immunity by the attenuated B_{26}-T_{1200} vaccines of *P. multocida*

Test	Vaccine batch	Time of challenge and the number of chickens protected				
		2 months	3 months	4 months	5 months	6 months
1	9001	4/4	3/4	4/4	4/4	2/4
	9002	3/4	4/4	3/4	2/4	1/4
	9003	4/4	4/4	3/4	3/4	1/4
2	9101	3/4	4/4	3/4	2/4	2/4
	9102	3/4	3/4	3/4	3/4	2/4
	9103	4/4	4/4	3/4	4/4	3/4
Total protection rate	4～8 ℃	21/24	22/24	19/24	18/24	11/24
Mortality in the control group	4～8 ℃	12/12	12/12	12/12	12/12	12/12

Summary and Discussion

8 h after vaccination with the attenuated B_{26}-T_{1200} vaccine of *P. multocida*, the inoculated chickens acquired immunity, and 20% of the vaccinated chickens were protected. Two days after vaccination, 60% the vaccinated chickens were protected. On day 4 or 5, the inoculated chickens could generate potent immunity, and 80%～100% of the vaccinated chickens resisted challenge from the highly virulent *P. multocida* strain. Therefore, in order to control epidemic of *P. multocida*, it would be very beneficial of conducting emergency vaccination in flocks of chickens where outbreak and spreading of *P. multocida* did happen.

Duration of immunity induced by the attenuated B_{26}-T_{1200} vaccines of *P. multocida* could last for 4～5 months, and the total protection rate within 1～5 months was 82.3%. Considering the fact that at present some attenuated vaccines of *P. multocida* have an established immunity for only 2～3 months, we consider that the less virulent B_{26}-T_{1200} vaccine strain of *P. multocida* is an attenuated vaccine with a satisfactory safety profile and a relatively longer duration of immunity.

Tests on stability of B_{26}-T_{1200} attenuated vaccine strain of *Pasteurella multocida*

Xie Zhixun, Liu Jiabo, Mo Zhaolan, Pang Yaoshan, Xie Zhiqin, and Xu Zhenfong

Abstract

Ten batches of the attenuated B_{26}-T_{1200} vaccine of *Pasteurella multocida* (*P. multocida*) were stored at 4～8 ℃ for one year. The results showed that the viability of the bacteria was 76.7%～93.3%, and that the average viability of the bacteria is 86.4%. The vaccines were diluted according to the doses on the vials, and the diluted vaccines were subsequently used to immunize chickens. The protection rates for the ten batches of vaccines varied from 75% to 100%, and the total protection rate was 87.5% (35/40). We evaluated the stability and immune efficacy of lyophilized B_{26}-T_{1200} attenuated vaccines of *P. multocida* after the vaccines had been stored at 4～8 ℃ for one year. Our experimental data are presented in the following sections.

Keywords

B_{26}-T_{1200}, vaccine, pasteurella multocida, stability

Materials and methods

Experimental vaccines

Ten batches of attenuated B_{26}-T_{1200} vaccines of *P. multocida* were produced and used in this study. The first five batches of vaccines (8903, 9001, 9002, 9003 and 9004) were manufactured by the Guangxi Biological Drugs Factory, and the rest of five batches of vaccines (9101, 9102, 9103, 9104 and 9105) were produced by the Guangxi Veterinary Research Institute. All the vaccines were stored at 4～8 ℃ for one year.

Experimental chickens

Three-month old cocks were hatched in the same batch by a farm owned by the Guangxi Veterinary Research Institute.

Highly virulent bacterial strain

C_{48-1}, a highly virulent *P. multocida* strain, was provided by China Veterinary Drug Surveillance Institute.

Immunization method

Bacterial vaccines stored at 4～8 ℃ for one year were diluted with normal saline containing 20% aluminum gel according to labels on the vials.

The diluted vaccines were injected into the necks subcutaneously for one dose (0.5 mL/ chicken). Martin-agar plates containing 4% sheep serum were used to calculate actual live bacterial colonies for one dose of bacterial vaccine stored at 4～8 ℃ for one year.

Method used for challenging chickens

Two weeks after immunization, the immunized and control chickens were challenged, respectively, with lethal doses of a highly virulent bacterial strain $C_{48\text{-}1}$. After challenging with wild-type bacteria, the chickens had been monitored for 14 days. Mortality rate and protection rate were recorded.

Results

As shown in Table 6-10-1, different batches of lyophilized B_{26}-T_{1200} vaccines of *P. multocida* had been stored at 4～8 ℃ for one year. The viability of the ten batches of vaccines varied from 76.6% to 93.3% (2.3×10^7～2.8×10^7 CFU/mL live bacteria), and the average viability of bacteria was 86.4%. The protection rate for the ten batches of vaccines varied from 75% to 100%, and the total protection rate was 87.5% (35/40).

Table 6-10-1 Stability tests on the B_{26}-T_{1200} attenuated vaccines of *P. multocida*

Vaccine batch	Storage temperature	Number of live bacteria / (CFU/mL) per dose (rate/%)	Number of protected chickens or vaccines stored for one year	Chickens died in control group her pathogens
8903	4～8 ℃	2.65×10^7 (88.3)	3/4	2/2
9001	4～8 ℃	2.3×10^7 (76.7)	4/4	2/2
9002	4～8 ℃	2.3×10^7 (76.7)	3/4	2/2
9003	4～8 ℃	2.8×10^7 (93.3)	4/4	2/2
9004	4～8 ℃	2.56×10^7 (85.3)	3/4	2/2
9101	4～8 ℃	2.7×10^7 (90.0)	4/4	2/2
9102	4～8 ℃	2.63×10^7 (87.7)	3/4	2/2
9103	4～8 ℃	2.8×10^7 (9.3.3)	4/4	2/2
9104	4～8 ℃	2.53×10^7 (84.3)	3/4	2/2
9105	4～8 ℃	2.65×10^7 (88.3)	4/4	2/2
Total		86.4	35/40	20/20

Summary and discussion

Ten batches of lyophilized B_{26}-T_{1200} vaccines of *P. multocida* had been stored at 4～8 ℃ for one year. The viability of these batches of vaccines varied from 76.6% to 93.3% (2.3×10^7～2.8×10^7 CFU/mL live bacteria), and the average viability of bacteria was 86.4% (2.592×10^7 CFU/mL live bacteria). We had previously reported that the minimum immune dose of vaccine was 1.3×10^6 CFU/mL live bacteria. Considering the fact that after one-year storage, there were still 17.7～21.5 minimum immune doses of live attenuated bacteria, it is clear that the B_{26}-T_{1200} vaccines stored at 4～8 ℃ for one year are still potent in inducing immunity, and can be applied in practice.

After storage of the lyophilized B_{26}-T_{1200} attenuated vaccines of *P. multocida* at 4～8 ℃ for one year, the protection rates of the ten batches of vaccines varied from 75% to 100%, and the total protection rate was 87.5% (35/40). In the present study, our results demonstrated that the B_{26}-T_{1200} attenuated vaccines retained good immunogenicity after being stored at 4～8 ℃ for one year. In addition, when the stored vaccines were used to inoculate chickens, the chickens were able to establish potent immune responses, and were protected from subsequent challenge with a highly virulent bacterial strain, suggesting that after one-year storage, the efficacy of B_{26}-T_{1200} vaccines was not substantially altered.

Recombinant hemagglutinin protein from H9N2 avian influenza virus exerts good immune effects in mice

Li Xiaofeng, Xie Zhixun, Wei You, Li Meng, Zhang Minxiu, Luo Sisi, Xie Liji

Abstract

The H9N2 subtype of avian influenza virus (AIV) causes enormous economic losses and poses a significant threat to public health; the development of vaccines against avian influenza is ongoing. To study the immunogenicity of hemagglutinin (HA) protein, we constructed a recombinant pET-32a-HA plasmid, induced HA protein expression with isopropyl β-D-1-thiogalactopyranoside (IPTG), verified it by SDS–PAGE and Western blotting, and determined the sensitivity of the recombinant protein to acid and heat. Subsequently, mice were immunized with the purified HA protein, and the immunization effect was evaluated according to the hemagglutination inhibition (HI) titer, serum IgG antibody titer, and cytokine secretion level of the mice. The results showed that the molecular weight of the HA protein was approximately 84 kDa, and the protein existed in both soluble and insoluble forms; in addition, the HA protein exhibited good acid and thermal stability, the HI antibody titer reached 6 log2–8 log2, and the IgG-binding antibody titer was 1:1000 000. Moreover, the levels of IL-2, IL-4, and IL-5 in the immunized mouse spleen cells were significantly increased compared with those in the control group. However, the levels of IL-1β, IL-6, IL-13, IFN-γ, IL-18, TNF-α, and GM-CSF were decreased in the immunized group. The recombinant HA protein utilized in this study exhibited good stability and exerted beneficial immune effects, providing a theoretical basis for further research on influenza vaccines.

Keywords

AIVs, recombinant, HA protein, stability, immune effects, cytokines

Introduction

Avian influenza viruses (AIVs) are members of the Orthomyxoviridae family and can infect a wide variety of hosts, including birds, pigs, dogs, horses, bats, and humans [1-4]. According to differences in virion envelope surface hemagglutinin (HA) proteins and neuraminidase (NA) antigens, AIVs can harbor H1-H16 and N1-N9[5], which are combined into a variety of virus subtypes, all of which endanger the development of the poultry industry. In China, the H9N2 AIV was first reported in Guangdong Province in 1992; subsequently, the disease spread to other provinces, where it became an endemic strain. Due to genetic evolution, the pathogenicity and transmissibility of the virus have shown increasing trends annually, thus posing serious threats to the poultry industry. In addition, H9N2 can not only directly infect humans but also recombine with other subtypes of viruses (H5N1, H7N9, H10N8, and H5N6) to produce variant subtypes of viruses that threaten human health[6-10].

In general, local farming and trade patterns provide a means for the spread of H9N2 epidemics. Traditional small-scale free-range and polyculture systems, which account for a portion of national poultry production, have low vaccination coverage and biosecurity awareness, and freshly killed poultry meat is preferred over frozen meat in live poultry markets (LPMs). These methods have significantly contributed to a vast gene pool of AIVs, as evidenced by the increased prevalence of viruses, including multiple HA/NA subtypes[11].

Currently, H9N2 AIV is widespread in China and has become a stable strain on commercial flocks[12]. However, antigenic variation or drift persists, putting pressure on the compatibility of seasonal vaccines. Mutations in the HA protein loci (R164Q, T150A, and I220T) promote virus replication in both poultry and mammals. The molecular mechanism of the antigenic transformation of avian influenza A subtype H9N2 has been elucidated, providing an important reference for the selection of vaccine candidates [13]. Since 2013, H9N2 circulating in Chinese poultry has undergone frequent mutations at HA residue 193, and viruses carrying asparagine (N) have been replaced by viruses carrying alanine (A), aspartic acid (D), glutamic acid (E), glycine (G), and serine (S). These mutations greatly alter the antigenicity of H9N2 and have profound impacts on its replication and transmission in chickens [14]. Moreover, the H9N2 AIV is divided into two antigenic clades from the original H9.4.2.5 lineage, with new clades emerging after 2015 and old clades reappearing in other regions in 2018. Hemagglutinin antigenic drift has led to the emergence of new clades, and vaccine immunity is a challenge [15]. A survey has shown that H9N2 is the predominant AIV subtype in wholesale and retail markets, mainly in the forms of G57 and three new genotypes (NG164, NG165, and NG166), and the emergence of NG165 genotypes, which are more suitable for transmission among poultry and mammals, poses a threat to public health [16]. In one case, H9N2 influenza viruses isolated from chickens, humans, and pet cats were found to have a high degree of homology, suggesting that H9N2 AIV may coexist in poultry and mammals, thus threatening both human and animal health[17]. Although the mortality rate of H9N2-infected flocks is usually no more than 20%, an H9N2 infection usually leads to respiratory and egg drop symptoms as well as severe secondary infections with other respiratory diseases, thus affecting poultry productivity [18]. The continuous antigenic mutation of the H9N2 subtype of avian influenza virus has challenged the prevention and control effects of vaccines, and research and development into immune mechanisms and new vaccines are also urgently needed.

Inactivated vaccines are currently the main vaccines available for influenza viruses, and they rely on embryo production. Following immunization, humoral immunity predominates, and the abilities to induce effective mucosal and cellular immunity are lacking. The effectiveness of a vaccine is limited by differences in genetic variants and subtypes of the virus. Therefore, the development of new, safe, and highly effective avian influenza vaccine candidates is critical, and new DIVA vaccines against H9N2 have emerged, including recombinant live virus vector vaccines, subunit vaccines, DNA vaccines, and virus-like particle vaccines. Subunit vaccines are generally developed based on the extraction of AIV immunogenic proteins rather than on the introduction of viral particles.

As the major viral membrane protein, HA is the primary protective antigen used in current influenza vaccines due to its immunodominant induction of high neutralizing antibody titers. Studies have shown that the combination of HA and NA mRNAs from different influenza subtypes can induce potent antibody and cellular immune responses in immunized non-human primates [19]. The HA protein of AIVs has rich epitopes and exhibits strong immunogenicity; therefore, this protein stimulates the host to produce neutralizing antibodies and has become the main target protein in the development of influenza vaccines. The insertion of HA gene fragments into different viral vectors can result in a strong resistance to host virus infections [20-22]. Some studies have also explored the preparation of HA stems and glycosylation sites in different groups of influenza viruses for use in vaccines to improve cross-antibiotic protection [20-22]. Research has shown that H5 COBRA

HA vaccines induce protective antibodies against two historical H5Nx and currently circulating H5N1, H5N6, and H5N8 clade influenza viruses [24]. Deng et al. [25] expressed the HA of the H1 and H3 strains and the M2e sequences common to humans, pigs, and poultry to form a bilayer nanovaccine, which achieved good immune effects in immunized mice. The abovementioned studies showed that the HA protein exerts a good immune effect, but the cytokines produced via cellular immunity remain to be explored.

In the present study, we cloned the H9N2 HA protein and constructed plasmid pET-32a-HA. The acid resistance and thermal stability of the HA protein were determined after purification. The HA protein was prepared for the immunization of BALB/c mice, and the immune effect was evaluated based on the resulting antibody and cytokine levels. Our results suggest that vaccination with the HA protein induces an immune response, increases or decreases cytokine expression, and prepares the host for protection without significant side effects, which have important implications for the development of new antiviral vaccines.

Materials and methods

Ethics statement

The animal experiments and sample collection were conducted in accordance with the guidelines of protocol #2019C0409, issued by the Animal Ethics Committee of the Guangxi Veterinary Research Institute.

Virus and mouse strains

The A/chicken/Guangxi/CWM/2019 (H9N2) virus strain used in this study was isolated from the livers of diseased chickens in Nanning, Guangxi, and stored at −70℃ at the Guangxi Veterinary Research Institute. Our laboratory studied this strain and found that it has potential use for a vaccine. Six- to eight-week-old female BALB/c mice (Guangdong Animal Center, Foshan, China) were maintained in cages with independent air supply.

HA gene encoding and plasmid construction

Based on the HA sequence of the A/chicken/Guangxi/CWM/2019 (H9N2) strain, we amplified the full-length coding sequences (CDS) of the HA gene, and the amino acids from positions 1 to 560 encoded by the full-length CDS were expressed, referring to the program (https://web.expasy.org/translate/, accessed on 24 July 2024). The primers (HA-U and HA-D) were designed (Table 6-11-1) and synthesized by Guangzhou Ruibo Biotech (Guangzhou, China). The HA gene was cloned from cDNA obtained from reverse transcription of RNA extracted from the virus and used as a template, and the PCR products were separated by 1.5% agarose gel electrophoresis; then, the PCR product was recovered and inserted into the pMD18-T plasmid (TaKaRa Bio, Dalian, China, Code No. 6011) and transferred into competent DH5α cells, where the plasmid pMD18-T-HA was constructed. The pET-32a (+) plasmids and pMD18-T-HA plasmids were digested with BamH Ⅰ (TaKaRa Bio, Code No. 1010A) and Not Ⅰ (TaKaRa Bio, Code No. 1166A) at 37 ℃ for 4 h; then, the target fragments were recovered and ligated overnight at 16 ℃. Finally, the ligated product was transfected into competent BL21 cells, which were verified by sequencing by Ruibo Biotech Company (Guangzhou, China). Since the pET-32a (+) vector carries two His tags, when the HA gene fragment with the start codon and the stop codon was inserted into the plasmid, the HA protein was expressed and fused with one of the His tags on the vector.

Table 6-11-1 Primers used in this study

Primer name	Sequence (5'-3')	Amplified sequence length
HA-U	ATAGGATCCATGGAGACAGTATCACTAATAACTA	1 683 bp
HA-D	ATAAGAATGCGGCCGCTTATATACAAATGTTGCATCTG	

SDS-PAGE and Western blotting

HA protein expression was induced with 1.0 mmol/L isopropyl β -D-1-thiogalactopyranoside (IPTG, Beyotime Biotechnology, Beijing, China, Code No. I1020) for 6 h. The bacteria were centrifuged at 4,000 rpm for 5 min and then washed three times with 1 × PBS (Solarbio, Beijing, China, Code No. P1020). The cells were then lysed with RIPA lysis buffer (Solarbio, Code No. R0010) on ice for 30 min and subjected to ultrasonication at 50 Hz until thoroughly lysed. The cells were boiled for 10 min and then incubated on ice for 10 min. The suspension (total protein), supernatant (soluble protein), and precipitate (insoluble protein) were collected after centrifugation for SDS-PAGE (Solarbio, Code No. PG01010). The recombinant proteins were stained with Coomassie brilliant blue (Beyotime Biotechnology, Code No. P1300). If the HA protein was present in the supernatant, it was considered a soluble protein; otherwise, it was considered an insoluble protein. After SDS-PAGE separation, the proteins were transferred to PVDF membranes, and the membranes were incubated with Western blot blocking solution for 4 h. The membranes were incubated with a diluted anti-His mouse monoclonal antibody (Invitrogen, Carlsbad, CA, USA, Code No.26183). After the primary antibody was removed, the membranes were washed three times (10 min each time) with 1 × PBST buffer (Solarbio, Code No. 1033), incubated with HRP-labeled goat anti-mouse IgG (Invitrogen, Code No. C31430100) for 1 h, and washed four times with 1 × PBST buffer. Photographs were taken after color development using a 3,3'-diaminobenzidine (DAB) horseradish peroxidase color development kit (Beyotime Biotechnology, Code No. P0202).

Purification HA protein

After HA protein expression for 6 h, the bacteria were centrifuged at 8 000 r/min for 5 min and then washed three times with 1 × PBS, lysed with lysis buffer (CW BIO, Beijing, China, Code No. CW0894S) on ice for 30 min, and then lysed via ultrasonication at 50 Hz until the cells were thoroughly lysed. The cells were then centrifuged at 10 000 r/min for 10 min to collect the supernatant, and the supernatant was mixed with binding buffer (CW BIO, Code No. CW0894S). The mixture was loaded onto a Ni-Agarose resin column (CW BIO, Code No. CW0894S), and the columns were mixed on a horizontal shaker for 30 min. The protein was completely bound to the resin. After elution of impurities with 15 volumes of binding buffer, the HA protein was eluted with different concentrations (100, 120, 160, 170, 180, 190, 200, 210 mM) of imidazole elution buffer (CW BIO, Code No. CW0894S), and the flow-through solution was collected. After purification, the protein was dialyzed with PBS to completely dialyze the imidazole of the eluate. SDS-PAGE and Western blot were also performed to verify that the obtained proteins was HA protein. Aliquots of purified HA protein were stored at −70 ℃ .

Hemagglutination assay of HA protein

To test the effect of the HA purified protein (antigen) on red blood cells (antibodies), the purified HA protein was tested with 1% chicken red blood cells (prepared and provided by the Guangxi Key Laboratory of Veterinary Biotechnology) to determine whether it caused red blood cell aggregation. Normal saline (25 μL, Solarbio, Code No. IN9000) was added to wells 1～12 of the coagulation plate, HA protein (25 μL) was added to the first well and mixed well, and 25 μL of the mixture from the first well was added to the second well. The 10th well was diluted in this manner. That is, the purified HA protein solution was diluted 1:2, 1:4, 1:8, 1:16, 1:32, 1:64, 1:128, 1:256, 1:512, and 1:1 024, where the last two wells were for red blood cell control, and the other row of wells were for saline control. Finally, 25 μL of 1% red blood cells was added to wells 1–12, which were gently vortexed and mixed. The results were observed after manual mixing and incubation at 37 ℃ for 30 min. The experiment was repeated 3 times.

HA protein acid sensitivity

The HA purified protein (200 μg/mL, 100 μL) was mixed with equal volumes of 100 mmol/L (pH4.0, pH5.0) acetate buffer (Yuanye Bio-Technology, Shanghai, China, Code No. R26128, R26131), pH 6.0 phosphate buffer (Yuanye Bio-Technology, Code No. 26268), and pH 7.0 neutral phosphate buffer (Yuanye Bio-Technology, Code No. 26273) and incubated at 37 ℃ for 10 min, after which the change in stability was determined using a hemagglutination assay.

HA protein thermal sensitivity

The HA purified protein (100 μg/mL, 100 μL) was incubated in a water bath maintained at 56 ℃ for durations of 0, 5, 10, 15, 30, or 60 min, after which its stability was determined via a hemagglutination assay. The HA protein was incubated at 56 ℃ for 5 min, after which the coagulation stability decreased by 2 unit titers, indicating thermal instability. The protein was then incubated at 56 ℃ for 30 min. If the hemagglutination titer decreased by 2 unit titers, the protein was considered thermostable. Otherwise, it was considered to be moderately thermally stable, exhibiting properties between those of thermostable and moderately thermally stable proteins.

Animal experiments and safety evaluation

All animal studies were conducted in accordance with the recommendations of the Guide for the Care and Use of Laboratory Animals of the Ministry of Science and Technology of the People's Republic of China as well as with the institutional guidelines of the Animal Ethics Committee of Guangxi Veterinary Research Institute. All animal experiments and sample collections were conducted in accordance with the guidance set forth in protocol #2019C0409, which was issued by the Animal Ethics Committee of the Guangxi Veterinary Research Institute.

Six- to eight-week-old female BALB/c mice were randomly assigned to the two following groups (n=10) for our study: the HA protein-immunized group and the Freund's adjuvant-immunized group (mock). The HA protein (100 μg/mL) was thoroughly mixed with complete Freund's adjuvant (Sigma-Aldrich, Shanghai, China, Code No. 1003150981) and emulsified to a uniform consistency. Intraperitoneal solution at a dose of 0.5 mL was applied on a 14-day schedule. For the second immunization, the complete adjuvant was replaced with incomplete Freund's adjuvant (Sigma-Aldrich, Code No. 1003212978), and the immunization methods were identical. The Freund's adjuvant-immunized group was injected with an equal volume of adjuvant and PBS. The mental state of the mice was monitored on a daily basis. Following each immunization, the mice were weighed, and their deaths were recorded over a 14-day period. The results are presented as the mean body weight of five mice.

Hemagglutination inhibition antibody assays

Following the administration of a booster immunization to the mice at 14 days, blood was obtained retro-orbitally, and sera were treated with receptor-destroying enzyme (RDE, Denka Seiken, Japan, Code No. 340016) at 56 ℃ for 30 min. Serial dilutions of the mouse sera (25 μL) were mixed with 4 units of H9N2 viral fluid and incubated at 37 ℃ for 30 min. Fresh chicken red blood cells were shaken, and 50 μL of the cell suspension was pipetted into each well. The mixture was thoroughly mixed and allowed to stand at room temperature for 30 min to observe the results. The test was conducted using serum from each mouse.

IgG antibody assay

Mouse serum was collected as described above. The purified HA protein (10 μg/mL) was diluted with 1 × ELISA coating buffer (Solarbio, Code No. C1055), which was used as the antigen, and added to a 96-well ELISA plate (100 μL/well). The plate was stored at 4 ℃ overnight and then washed 3 times with 1×PBST buffer. The plate was blocked with 1% bovine serum albumin (Solarbio, Code No. 7940) at 37 ℃ for 1 h. The plate was washed as described above; mouse serum was diluted with PBS at dilutions of 1 : 10 000, 1 : 20 000, 1 : 30 000, 1 : 40 000, 1 : 50 000, 1 : 60 000, 1 : 70 000, 1 : 80 000, 1 : 90 000, 1 : 100 000, 1 : 150 000, and 1 : 200 000 (each concentration was replicated in 3 wells), and negative mouse serum or PBS was used as the control. The plate was washed 3 times with 1×PBST, incubated with goat anti-mouse IgG (Proteintech, Wuhan, China, Code No. SA00001-1) at 37 ℃ for 1 h, and washed 3 times with 1×PBST. The OD450 was measured using a VICTOR® Nivo ™ Multimode Plate Reader (PerkinElmer) after color development using chromogenic kit (Solarbio, Code No. PR1200) and stop solution (Solarbio, Code No. C1058), and the maximum dilution of the positive group OD value (P)/negative group OD value (N) >2 represented the mouse serum IgG antibody titer.

Mouse splenocyte supernatant multicytokine assay

Following the removal of the spleens from the sacrificed mice and thorough homogenization, the cells were incubated at room temperature for five minutes and then resuspended in red blood cell lysis buffer (Solarbio, Code No. R1010). Lysis was terminated with Dulbecco's modified Eagle's medium (DMEM; Gibco, Code No. C11885500BT). The cells were washed 3 times with PBS, resuspended in supplemented DMEM, supplemented with 5% fetal bovine serum (FBS; Gibco, Grand Island, NY, USA, Code No. 10099141C1), distributed into 6-well plates (1.0×10^6 cells/well), and cultured at 37 ℃ in 5% CO_2 and 95% humidity. Subsequently, the cells in the HA protein-immunized group were divided into two groups, designated as HA+ and HA–, while cells from the mock group were divided into two groups, designated as mock+ and mock–. The HA+ and mock+ groups were subjected to stimulation with the HA protein, whereas the cells in the HA– and mock– groups were not stimulated. After 72 h, the cell supernatant was collected and sent to the Shanghai Leids Biotechnology Company for cytokine detection. The experiment was repeated three times, and results were calculated as follows, IL-2 (HA)=mean (HA+) – mean(HA–), IL-2 (mock)=mean (mock+) – mean (mock–).

Statistical analysis

The statistical analyses were conducted using t-tests with GraphPad Prism 8.0 software. Asterisks indicate statistical significance as follows: * indicates $P<0.05$; ** indicates $P<0.01$.

Results

HA protein was successfully expressed in *Escherichia coli* cells

The HA gene was amplified using cDNA as the PCR template, and cDNA was reverse transcribed from RNA extracted from viruses. The sequence analysis revealed that the size of the HA gene was 1,683 bp (Figure 6-11-1 A). The SDS-PAGE results showed that the HA protein was present in both the supernatant and precipitate, indicating that it existed in both soluble and insoluble protein forms (Figure 6-11-1 B). The size of the HA protein with His-tag was approximately 84 kDa; meanwhile, using an anti-His mouse monoclonal antibody to verify the HA protein, a band of approximately 84 kDa was confirmed for the recombinant plasmid pET-32a-HA, but the corresponding band was not observed for the empty vector (Figure 6-11-1 C). When the HA protein mixture was purified, the SDS-PAGE results showed that the 180 mM imidazole eluate eluted best (Figure 6-11-1 D) and highly concentrated and pure proteins can be obtained. Meanwhile, the Western blot result showed that the obtained purified protein was the HA protein (Figure 6-11-1 E).

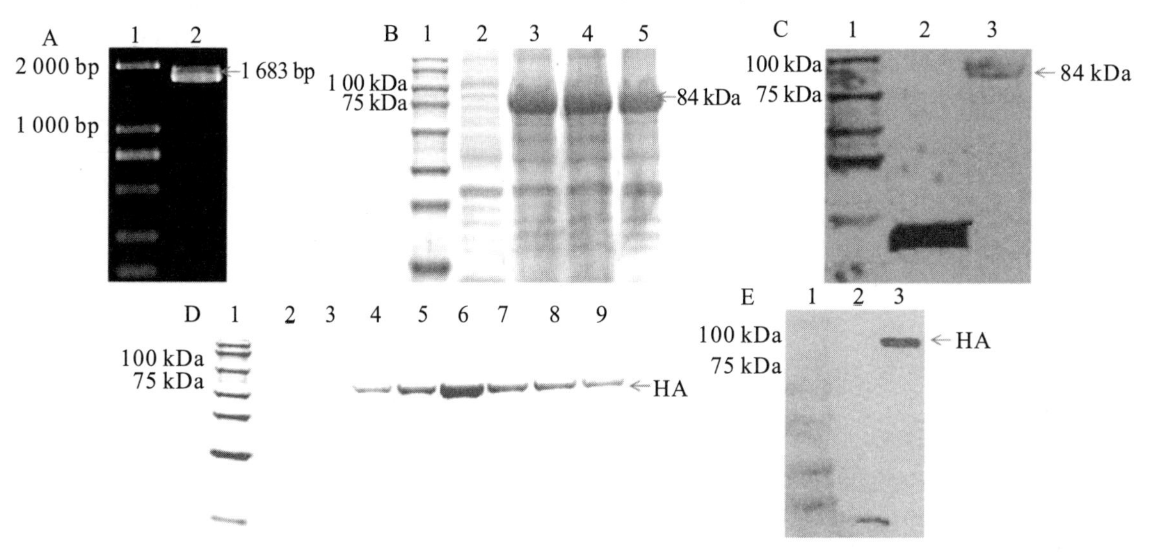

A: The PCR product of the H9 HA gene. Lane 1: DL 2, 000 DNA marker; lane 2: PCR amplification products. The red arrow indicates the amplified HA fragment, the size of which is 1,683 bp. B: The SDS-PAGE and solubility analysis of the recombinant protein. Lane 1: protein marker; lane 2: protein obtained from the empty pET-32a plasmid; lane 3: protein obtained from the pET-32a-HA plasmid (total protein); lane 4: protein from the supernatant (soluble protein); lane 5: protein from the precipitation (insoluble protein). The red arrow indicates the recombinant HA protein, which is 84 kDa, and the His-tag protein is included. C: The identification of recombinant HA via Western blotting. Lane 1: protein marker; lane 2: protein obtained from the pET-32a empty plasmid; lane 3: protein obtained from the pET-32a-HA plasmid. The red arrow indicates the recombinant HA protein, which is 84 kDa, and the His-tag protein is included. D: The SDS-PAGE of HA purified protein. Lane 1: protein marker; lane 2–9: Proteins eluted with 100, 120, 160, 170, 180, 190, 200, and 210 mM imidazole eluent; 180 mM imidazole eluted best. The red arrow indicates the HA purified protein. E: The identification of HA purified protein via Western blotting. Lane 1: protein marker; lane 2: protein obtained from the pET-32a empty plasmid; lane 3: His tag HA purified protein. The red arrow indicates the HA purified protein.

Figure 6-11-1 Analysis of the HA gene and recombinant HA protein

HA protein has good acid and thermal stability

The results show that the HA protein can cause hemagglutination on red blood cells at a dilution of 1 : 128 (Figure 6-11-2 A). The hemagglutination titer of the HA protein was 7 log2. No hemagglutination of red blood cells occurred in the red blood cells (well C) and saline control group. When the purified HA protein was treated with different pH buffer solutions and different temperatures, the same method was used to experiment and determine the results.

The acid sensitivity of the HA protein was measured using hemagglutination titers after incubation with solutions at different pH values. First, the HA protein was incubated with neutral phosphate buffer (pH 7.0), acetate buffer (pH 5.0), and acetate buffer (pH 4.0), and the hemagglutination titers were 7 log2, 6 log2, and 4.7 log2, respectively (Figure 6-11-2 A). The stability of the HA protein was good in the neutral solution, and the stability decreased rapidly in a more acidic environment. The thermal stability of the HA protein changed with increasing incubation time. The HA proteins were incubated at 56 ℃ for 5 min or 15 min, and the titer decreased from 6 log2 to 5 log2, indicating that the HA protein had good stability in this temperature range. When the HA protein was incubated for 30 min or 60 min, the titers decreased to 4.7 log2 and 4.3 log2, respectively (Figure 6-11-2 B), representing a decrease in the titer of 2 units from the preincubation hemagglutination titer and indicating that the recombinant HA protein had good stability; these results show that the stability of the HA protein decreased with increasing incubation time.

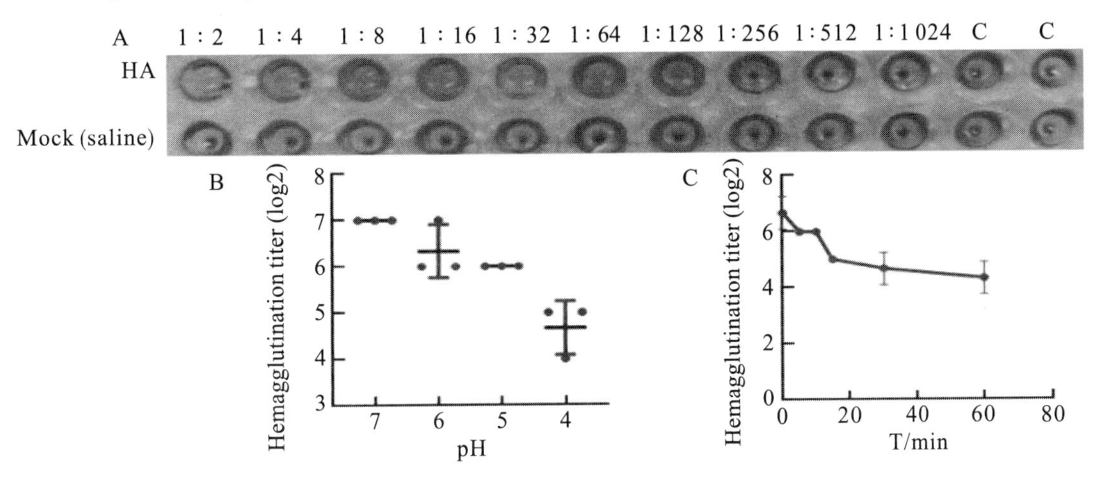

A: The result of HA protein hemagglutination. HA protein hemagglutination at 1:2, 1:4, 1:8, 1:16, 1:32, 1:64, and 1:128 dilutions of red blood cells. The hemagglutination titer of the HA protein was 7 log2. There is no hemagglutination of red blood cells that occurred in the red blood cells (well C) and saline control group. B: The acid stability results. The HA protein was incubated with a neutral phosphate buffer (pH 7.0), phosphate buffer (pH 6.0), and acetate buffer (pH 4.0 or 5.0), and the hemagglutination titers were 7 log2, 6.3 log2, 6 log2, and 4.7 log2, respectively. C: The result of thermal stability. The HA protein was incubated in a 56 ℃ water bath for 0, 5, 10, 15, 30, or 60 min, and the hemagglutination titers were 6.7 log2, 6 log2, 6 log2, 5 log2, 4.7 log2, and 4.3 log2, respectively. These hemagglutination assay results were determined through log2 calculations.

Figure 6-11-2 Acid stability and thermal stability results.

The HA protein is safe in mice

The safety of immunizing mice with the recombinant HA protein was evaluated according to the weight change and survival rate. After immunization with HA recombinant protein, the change in the body weight of the mice was greater than that of the control group; the mice showed a decrease in body weight and then returned to their original body weight at either the first (HA-1) or second (HA-2) immunization, and they showed an increase in body weight over time compared with those in the control group (Figure 6-11-3 A). All mice survived after the first (Mock-1) and second (Mock-2) immunizations (Figure 6-11-3 B), suggesting that the purified HA protein is relatively safe for use in mice.

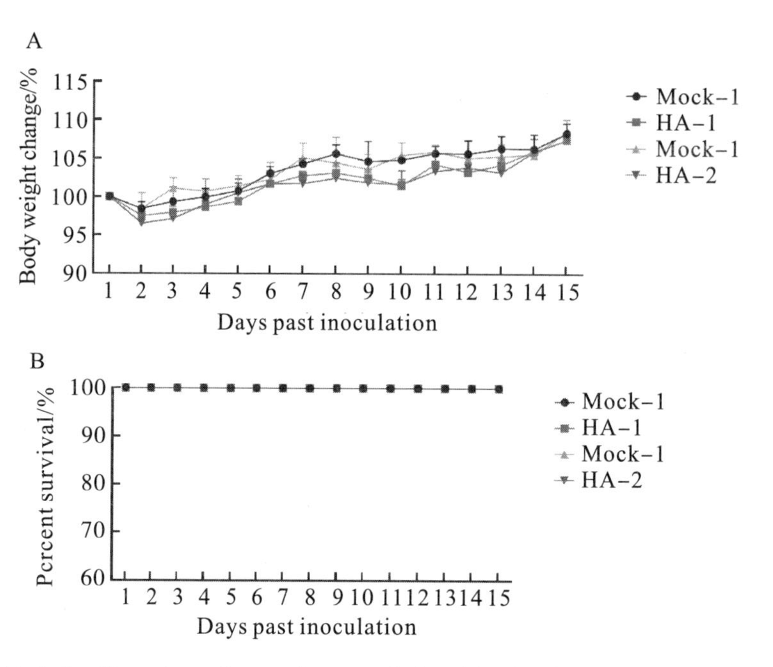

A: Changes in the body weight of mice. The results are shown as the average body weight of 5 mice. Mock-1, HA-1 indicates a change in body weight after the first immunization; Mock-2, HA-2. B: Percent survival (%). There were no deaths after the first and second immunizations. Mock-1, HA-1 indicates the first immunization, Mock-2, HA-2 indicates the second immunization.

Figure 6-11-3 The safety of the recombinant HA protein in mice

High levels of hemagglutinin-inhibiting antibodies and IgG antibodies

The immunogenicity of the HA protein was determined by measuring antigen-specific humoral immune responses after boost immunizations. An HI antibody titer assay was performed with serum from immunized mice. After 2 weeks of immunization, specific antibodies against the H9 subtype of AIV were produced in the serum of the immunized mice (Figure 6-11-4 A). The HI antibody titers of HA protein-immunized mice reached 6 log2–8 log2, and the antibody titers of the control group were negative.

To determine the level of IgG antibody produced in mice after immunization, the HA protein was used as the antigen, and mouse serum was used as the antibody. The ELISA results showed that the IgG titer produced by the immunized mice was 1:1 000 000 (Figure 6-11-4 B), which was significantly different from that of the control group ($P<0.01$). The results indicated that HA protein-immunized mice produced specific IgG-binding antibodies.

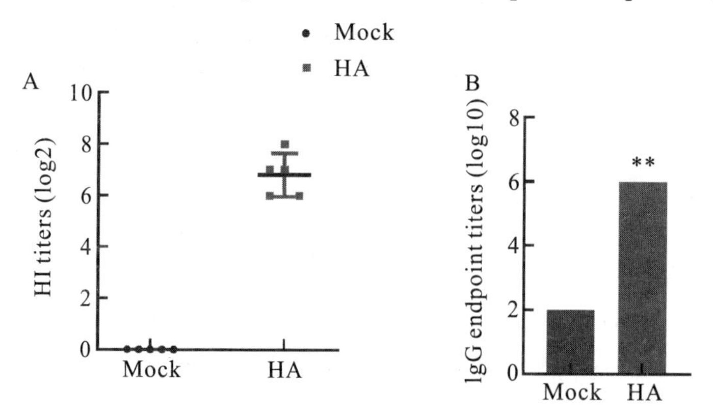

A : The results of the HI antibody. The serum of the control group mice did not show anticoagulation in the HI experiment, and the HI antibody titer was negative. The HI antibody titer of HA protein-immunized mice reached 6 log2–8 log2. B: The IgG antibody results. The results showed that HA protein-immunized mice produced specific IgG-binding antibodies. Asterisks indicate significant differences (** indicates $P<0.01$).

Figure 6-11-4 The results of the HI antibody and IgG antibody

Effects of multiple cytokines on mouse spleen cell culture

The above results suggest that the recombinant HA protein produces antibodies after immunization in mice; however, it is unclear whether cytokines are involved in the immune response. To explore the changes in cytokine levels, we collected mouse splenocyte supernatants for cytokine assays. Cytokine concentrations in the mouse splenocyte culture supernatant were determined using Luminex. The levels of IFN-γ, IL-2, IL-4, IL-5, TNF-α, IL-1β, IL-6, IL-13, IL-18, and GM-CSF were measured (Figure 6-11-5). Compared with those in the control group, the levels of IL-2 (Figure 6-11-5 A), IL-4 (Figure 6-11-5 B), and IL-5 (Figure 6-11-5 C) in the cells of the immunized group were significantly increased ($P<0.01$); among them, the levels of IL-2 increased the most, and the average level in the control group was 5.15 pg/mL, which increased to 60.48 pg/mL after immunization. However, the levels of IL-1β, IL-13, IL-18, IFN-γ, and GM-CSF were significantly decreased ($P<0.01$). In particular, the average level of IL-18 in the immune group (7 191.07 pg/mL) was significantly lower than that in the control group (21 596.6 pg/mL). In addition, the levels of IL-6 and TNF-α were lower in the immunized group than in the control group, but the difference was not significant ($P>0.05$). These results suggest that the HA protein is involved in antiviral immunity, increasing the levels of IL-2, IL-4, and IL-5 or decreasing the levels of other cytokines.

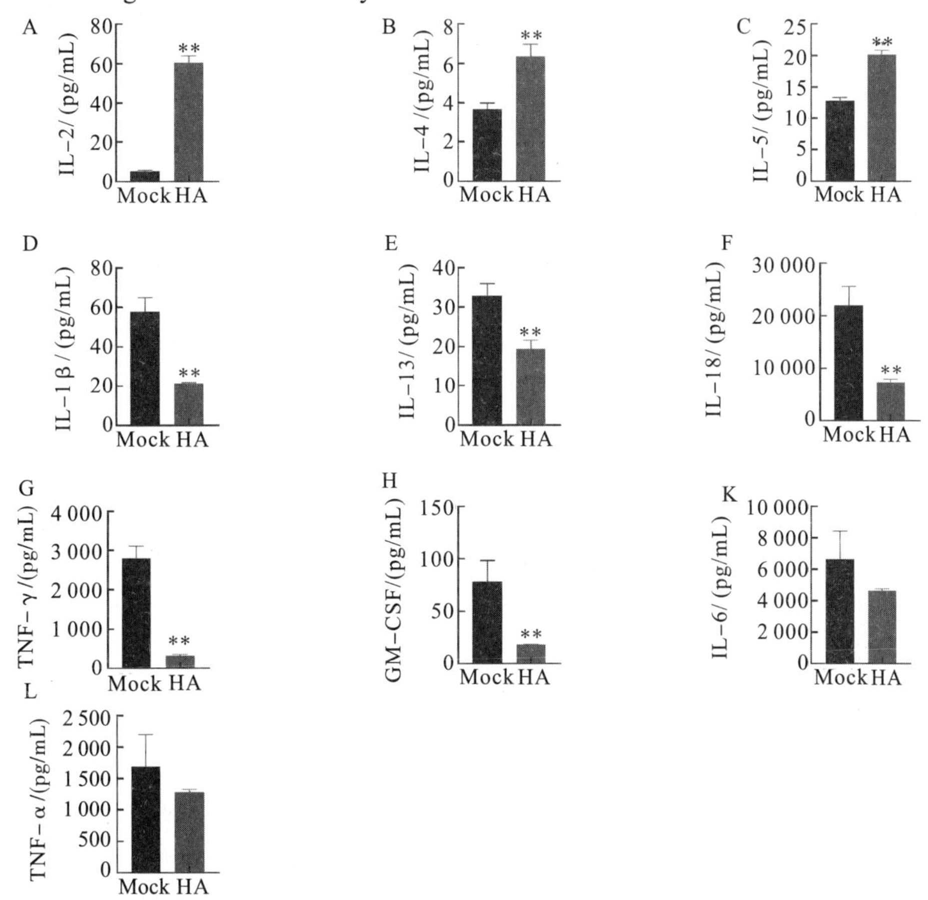

A: IL-2 levels in both groups were 5.15 pg/mL and 60.48 pg/mL, respectively; B: IL-4 levels in both groups were 3.64 pg/mL and 6.36 pg/mL, respectively; C: IL-5 levels in both groups were 12.74 pg/mL and 20.10 pg/mL, respectively; D: IL-1β levels in both groups were 57.84 pg/mL and 21.25 pg/mL, respectively; E: IL-13 levels in both groups were 32.92 pg/mL and 19.37 pg/mL, respectively; F: IL-18 levels in both groups were 21 956.6 pg/mL and 7 191.07 pg/mL, respectively; G: IFN-γ levels in both groups were 2 794.4 pg/mL and 318.7 pg/mL, respectively; H: GM-CSF levels in both groups were 78.08 pg/mL and 18.35 pg/mL, respectively; K: IL-6 levels in both groups were 6 633.1 pg/mL and 4 621 pg/mL, respectively; L: TNF-α levels in both groups were 1 688.03 pg/mL and 1 286.08 pg/mL, respectively. Asterisks indicate significant differences (** indicates $P<0.01$).

Figure 6-11-5 Luminex assay results for mouse splenocyte culture supernatants.

Discussion

The HA protein is the major antigenic protein on the surface of AIV and is the antigen of choice for subunit avian influenza vaccines. Recombinant antigenic proteins are produced by genetic engineering, in which genes with multiple antigenic sites are cloned into expression vectors and protein expression is induced to yield a single sexual antigen, a well-established technique that produces antigens with a favorable safety profile for both humans and animals.

In this study, the HA protein was expressed by prokaryotes and existed in both soluble and insoluble protein forms. The HA protein was incubated in a buffer at pH values of 4.0, 5.0, 6.0, or 7.0 for 10 min, and the stability of the HA protein decreased with decreasing pH. In addition, when the HA protein was mixed with a buffer (pH 6.0) and acidic buffer (pH 4.0), the hemagglutination titer of the protein began to decrease by approximately 2 units. The reason for this difference may be that the prokaryotically expressed proteins are unstable under acidic conditions due to inadequate structural folding and may degrade. In addition, the thermal stability measurements showed that the titer of the HA protein changed slightly after incubation at 56 ℃ for 5 min to 15 min, suggesting that the stability of the HA protein was strong. After incubation at 56 ℃ for 30 min, the titer decreased to 4.7 log2, indicating that the thermal stability of the HA protein was good. These results indicate that the recombinant HA protein has good thermal stability; therefore, it may also have good thermal stability when prepared as a vaccine.

Subunit vaccines induce humoral and cellular immune responses following the immunization of the host. In the present study, after the mice were immunized with the HA protein, their weight first decreased and then increased, and all mice survived, indicating that the recombinant HA protein is safe for mice. After two weeks, the HI antibody level reached 6 log2–8 log2, and the IgG antibody titer was 1:1 000 000, indicating that the recombinant HA protein induced a good immune response in mice and that this antibody can promote the neutralization of the virus. T cell-mediated cellular immunity is now a popular area of vaccine research, including as a direction in the production of general influenza vaccines [26-29]. The influenza vaccine induces a cellular immune response that can provide broad-spectrum protection; multiple immune responses contribute to the prevention of an influenza infection. Among these responses, antibodies alone provide adequate protection against an infection, and T cell-mediated responses appear to play an important role in recovery [30]. After the mice were immunized with HA in our study, the concentrations of IL-2, IL-4, and IL-5 in the supernatant were significantly higher than those in the mock group, and the levels of IL-2 and IL-4 were consistent with the results reported by Feifei Xiong et al. [31]. IL-2, IL-4, and IL-5, which are mainly secreted by cells, stimulate innate immunity and modulate host immune responses [32, 33] ; therefore, the secretion of these cytokines indicates that the immunization of mice elicits a strong cellular response.

Compared with those in the control group, the levels of IL-1 β, IL-13, IFN-γ, IL-18, and GM-CSF were significantly reduced in the cells of the immunized group ($P<0.01$), and the levels of IL-6 and TNF-α were reduced, but the differences were not significant ($P>0.05$).These results suggest that compared with the control mice injected with a mixture of Freund's adjuvant and PBS, in the mice immunized with the HA protein and Freund's adjuvant, the treatment causes an increase or decrease in cytokine levels and protects the host through different pathways [31, 34, 35]; thus, the immunization results showed that HA had a good immune effect. The production of inflammatory factors has dual effects; at a certain concentration, inflammatory factors can inhibit viral replication, but excessive accumulation can lead to host inflammatory damage and acute death. The cytokines IL-1 β and IL-18 induce an inflammatory response after an influenza virus infection and recruit other inflammatory cells to infected tissues to clear the virus. Levels of IL-1 β and IL-18 were significantly lower in the immune group than in the control group ($P<0.01$). In the early stage of a viral infection, IL-1 β and IL-18 promote the activity of CD8+ T cells and induce antibody secretion, thus playing protective roles; however, when overproduced, these cytokines destroy benign tissues [36]. IL-1 β and IL-18 bind to receptors and induce NF-κ β-dependent inflammation [35]. IL-18 can mediate IFN- γ production via T and NK cells[37]. IFN- γ is a primary antiviral agent, but when mass production leads to adverse outcomes, IFN- γ has been shown to play an important role in acute lung injury caused by an H1N1 virus infection [38]. Some studies have shown that when the host is infected with H1N1, H3N2, or H7N9, excessive IL-1 β can exacerbate the condition and cause serious consequences. In both the early and late stages of H1N1 or H3N2 infections, treatment with targeted anti-IL-1 β antibodies can alleviate lung inflammation and favorably improve survival [36, 39]. In our study, IL-1 β levels decreased after the immunization of mice with HA, indicating that recombinant HA can alleviate lung inflammation in mice by reducing the secretion of inflammatory cytokines. IL-1 β and IL-18 play complex roles in the influenza virus-induced cytokine storm. These cytokines not only play important roles themselves but also regulate the production of TNF- α and IL-6. Compared to the control group, the levels of IL-5, IFN- γ, and GM-CSF in the immunized group were significantly lower ($P<0.05$), while the levels of IL-6 and TNF- α were not significantly different ($P>0.05$). IL-6 is an inflammatory marker, and clinical studies have shown that excess IL-6 in people with influenza is associated with adverse effects [36, 39]. Compared to those in WT mice, IL-6 levels in SOCS3-/- mice infected with influenza virus are significantly lower and return to normal levels, preventing the production of cytokine storms [43]. IL-6 plays an important role in the cytokine storm caused by influenza viruses and represents a novel target for immunotherapeutic strategies. TNF- α is a typical proinflammatory factor located at the center of the cytokine storm . Like IFN, TNF- α can disrupt the endothelial barrier, causing pulmonary edema and tissue damage [45]. Compared to WT mice, mice without TNFR are more resistant to lethal H5N1, have an average survival time of two days, and exhibit lower levels of cytokines, including IFN- γ and interleukins, in their lungs [46, 47]. Therefore, controlling TNF- α levels to reduce cytokine storms while inhibiting viral replication is a potential strategy through which to reduce pathologic damage in the lungs. Both inflammatory factors can protect the host from a viral infection; conversely, they can also exacerbate inflammatory damage to the host. The results described above indicate that the recombinant HA protein can not only produce IgG antibodies but also promote the secretion of cytokines and reduce the accumulation of some inflammatory factors, which may involve different methods of protecting the host.

Conclusions

In conclusion, the H9N2 recombinant HA protein induces cellular and humoral immune responses, results that lay the foundation for an in-depth study of the function of the HA protein in immunity and vaccination; these findings also provide a reference for the effective use of viral proteins to inhibit viral replication as well as insights for the further research and development of influenza vaccines.

References

[1] PARRISH C R, KAWAOKA Y. The origins of new pandemic viruses: the acquisition of new host ranges by canine parvovirus and influenza A viruses. Annu Rev Microbiol, 2005, 59: 553-586.

[2] STEEL J, LOWEN A C, MUBAREKA S, et al. Transmission of influenza virus in a mammalian host is increased by PB2 amino acids 627K or 627E/701N. PLOS Pathog, 2009, 5(1): e1000252.

[3] GONZALEZ G, MARSHALL J F, MORRELL J, et al. Infection and pathogenesis of canine, equine, and human influenza viruses in canine tracheas. J Virol, 2014, 88(16): 9208-9219.

[4] BAILEY E S, CHOI J Y, FIELDHOUSE J K, et al. The continual threat of influenza virus infections at the human–animal interface. Evolution, Medicine, and Public Health, 2018, 2018(1): 192-198.

[5] LISTED N A. A revision of the system of nomenclature for influenza viruses: a WHO memorandum. Bulletin of the World Health Organization, 1980, 58(4): 585-591.

[6] GUAN Y, SHORTRIDGE K F, KRAUSS S, et al. H9N2 influenza viruses possessing H5N1-like internal genomes continue to circulate in poultry in southeastern China. J Virol, 2000, 74(20): 9372-9380.

[7] GU M, CHEN H, LI Q, et al. Enzootic genotype S of H9N2 avian influenza viruses donates internal genes to emerging zoonotic influenza viruses in China. Vet Microbiol, 2014, 174(3-4): 309-315.

[8] RAHIMIRAD S, ALIZADEH A, ALIZADEH E, et al. The avian influenza H9N2 at avian-human interface: A possible risk for the future pandemics. J Res Med Sci, 2016, 21: 51.

[9] Shen Y, Ke C, Li Q, et al. Novel reassortant avian influenza A(H5N6) viruses in humans, Guangdong, China, 2015. Emerging infectious diseases, 2016, 22(8): 1507-1509.

[10] ZHANG Z, LI R, JIANG L, et al. The complexity of human infected AIV H5N6 isolated from China. BMC Infect Dis, 2016, 16(1): 600.

[11] CHEN L J, LIN X D, TIAN J H, et al. Diversity, evolution and population dynamics of avian influenza viruses circulating in the live poultry markets in China. Virology, 2017, 505: 33-41.

[12] LIU Y, LAI H, LI L, et al. Endemic variation of H9N2 avian influenza virus in China. Avian Dis, 2016, 60(4): 817-825.

[13] ZHANG J, WANG X, CHEN Y, et al. Mutational antigenic landscape of prevailing H9N2 influenza virus hemagglutinin spectrum. Cell reports (Cambridge), 2023, 42(11): 113409.

[14] WAN Z, ZHAO Z, SANG J, Et al. Amino acid variation at hemagglutinin position 193 impacts the properties of H9N2 avian influenza virus. Journal of Virology, 2023, 97(2): e0137922.

[15] ZHANG N, QUAN K, CHEN Z, et al. The emergence of new antigen branches of H9N2 avian influenza virus in China due to antigenic drift on hemagglutinin through antibody escape at immunodominant sites. Emerging Microbes & Infections, 2023, 12(2): 2246582.

[16] WANG Z, LI H, LI Y, et al. Mixed selling of different poultry species facilitates emergence of public-health-threating avian influenza viruses. Emerging Microbes & Infections, 2023, 12(1): 2214255.

[17] YANG J, YAN J, ZHANG C, et al. Genetic, biological and epidemiological study on a cluster of H9N2 avian influenza virus infections among chickens, a pet cat, and humans at a backyard farm in Guangxi, China. Emerging Microbes & Infections, 2023, 12(1): 2143828.

[18] YI PENG SUN J L. H9N2 influenza virus in China: a cause of concern. Protein Cell, 2015,1(6):18-25.

[19] CHIVUKULA S, PLITNIK T, TIBBITTS T, et al. Development of multivalent mRNA vaccine candidates for seasonal or pandemic influenza. npj Vaccines, 2021, 6(1): 153.

[20] LI Z, ZAISER S A, SHANG P, et al. A chimeric influenza hemagglutinin delivered by parainfluenza virus 5 vector induces broadly protective immunity against genetically divergent influenza a H1 viruses in swine. Veterinary microbiology, 2020, 250: 108859.

[21] KAUGARS K, DARDICK J, DE OLIVEIRA A P, et al. A recombinant herpes virus expressing influenza hemagglutinin confers protection and induces antibody-dependent cellular cytotoxicity. Proc Natl Acad Sci U S A, 2021, 118(34): e2110714118.

[22] KIM Y, ZHENG X, ESCHKE K, et al. MCMV-based vaccine vectors expressing full-length viral proteins provide long-term humoral immune protection upon a single-shot vaccination. Cell Mol Immunol, 2022, 19(2): 234-244.

[23] BOYOGLU-BARNUM S, HUTCHINSON G B, BOYINGTON J C, et al. Glycan repositioning of influenza hemagglutinin stem facilitates the elicitation of protective cross-group antibody responses. Nat Commun, 2020, 11(1): 791.

[24] NUNEZ I A, HUANG Y, ROSS T M. Next-generation computationally designed influenza hemagglutinin vaccines protect against H5Nx virus infections. Pathogens, 2021, 10(11): 1359.

[25] DENG L, MOHAN T, CHANG T Z, et al. Double-layered protein nanoparticles induce broad protection against divergent influenza A viruses. Nature Communications, 2018, 9(1): 359.

[26] KORENKOV D, ISAKOVA-SIVAK I, RUDENKO L. Basics of CD8 T-cell immune responses after influenza infection and vaccination with inactivated or live attenuated influenza vaccine. Expert review of vaccines, 2018, 17(11): 977-987.

[27] LEE S Y, KANG J O, CHANG J. Nucleoprotein vaccine induces cross-protective cytotoxic T lymphocytes against both lineages of influenza B virus. Clin Exp Vaccine Res, 2019, 8(1): 54-63.

[28] NELSON S A, DILEEPAN T, RASLEY A, et al. Intranasal nanoparticle vaccination elicits a persistent, polyfunctional CD4 T Cell response in the murine lung specific for a highly conserved influenza virus antigen that is sufficient to mediate protection from influenza virus challenge. Journal of Virology, 2021, 95(16): e84121.

[29] NOISUMDAENG P, ROYTRAKUL T, PRASERTSOPON J, et al. T cell mediated immunity against influenza H5N1 nucleoprotein, matrix and hemagglutinin derived epitopes in H5N1 survivors and non-H5N1 subjects. Peer J, 2021,9:e11021.

[30] COMPANS R W, COOPER M, KOPROWSKI H. The Systemic and mucosal immune response of humans to influenza A virus //Germany: Springer Berlin / Heidelberg, 1989: 107-116.

[31] XIONG F, ZHANG C, SHANG B, et al. An mRNA-based broad-spectrum vaccine candidate confers cross-protection against heterosubtypic influenza A viruses. Emerg Microbes Infect, 2023, 12(2): 2256422.

[32] HERNANDEZ R, PÕDER J, LAPORTE K M, et al. Engineering IL-2 for immunotherapy of autoimmunity and cancer. Nature reviews. Immunology, 2022, 22(10): 614-628.

[33] PELAIA G, VATRELLA A, BUSCETI M T, et al. Role of biologics in severe eosinophilic asthma- focus on reslizumab. Ther Clin Risk Manag, 2016, 12: 1075-1082.

[34] KONG X, LU X, WANG S, et al. Type I interferon/STAT1 signaling regulates UBE2M-mediated antiviral innate immunity in a negative feedback manner. Cell reports (Cambridge), 2023, 42(1): 112002.

[35] PALOMO J, DIETRICH D, MARTIN P, et al. The interleukin (IL)-1 cytokine family—Balance between agonists and antagonists in inflammatory diseases. Cytokine, 2015, 76(1): 25-37.

[36] TATE M D, ONG J, DOWLING J K, et al. Reassessing the role of the NLRP3 inflammasome during pathogenic influenza A virus infection via temporal inhibition. Sci Rep, 2016, 6: 27912.

[37] NOVICK D, KIM S, KAPLANSKI G, et al. Interleukin-18, more than a Th1 cytokine. Semin Immunol, 2013, 25(6): 439-448.

[38] LIU B, BAO L, WANG L, et al. Anti-IFN-gamma therapy alleviates acute lung injury induced by severe influenza A (H1N1) pdm09 infection in mice. J Microbiol Immunol Infect, 2021, 54(3): 396-403.

[39] BAWAZEER A O, ROSLI S, HARPUR C M, et al. Interleukin-1 β exacerbates disease and is a potential therapeutic target to reduce pulmonary inflammation during severe influenza A virus infection. Immunology and Cell Biology, 2021, 99(7): 737-748.

[40] KAISER L, FRITZ R S, STRAUS S E, et al. Symptom pathogenesis during acute influenza: interleukin-6 and other cytokine responses. J Med Virol, 2001, 64(3): 262-268.

[41] HAGAU N, SLAVCOVICI A, GONGANAU D N, et al. Clinical aspects and cytokine response in severe H1N1 influenza A virus infection. Crit Care, 2010, 14(6): R203.

[42] OSHANSKY C M, GARTLAND A J, WONG S S, et al. Mucosal immune responses predict clinical outcomes during influenza infection independently of age and viral load. Am J Respir Crit Care Med, 2014, 189(4): 449-462.

[43] LIU S, YAN R, CHEN B, et al. Influenza virus-induced robust expression of SOCS3 contributes to excessive production of IL-6. Front Immunol, 2019, 10: 1843.

[44] PANDEY P, KARUPIAH G. Targeting tumour necrosis factor to ameliorate viral pneumonia. FEBS J, 2022, 289(4):

883-900.
[45] WITTEKINDT O H. Tight junctions in pulmonary epithelia during lung inflammation. Pflugers Arch, 2017, 469(1): 135-147.
[46] PERRONE L A, SZRETTER K J, KATZ J M, et al. Mice lacking both TNF and IL-1 receptors exhibit reduced lung inflammation and delay in onset of death following infection with a highly virulent H5N1 virus. J Infect Dis, 2010, 202(8): 1161-1170.
[47] SZRETTER K J, GANGAPPA S, LU X, et al. Role of host cytokine responses in the pathogenesis of avian H5N1 influenza viruses in mice. J Virol, 2007, 81(6): 2736-2744.

Chapter Seven Studies on Livestock and Poultry Genetic Resources

Molecular characterization of the Cenxi classical three-buff chicken (*Gallus gallus domesticus*) based on mitochondrial DNA

Xie Zhixun, Zhang Yanfang, Deng Xianwen, Xie Zhiqin, Liu Jiabo, Huang Li, Huang Jiaoling, Zeng Tingting, and Wang Sheng

Abstract

The complete mitochondrial genome sequence of the Cenxi classical three-buff chicken was measured by PCR-based methods and analyzed in detail. Our research findings reveal that the entire mitochondrial genome of the Cenxi classical three-buff chicken is a circular molecule consisting of 16 786 bp (GenBank accession number: KM433666). The contents of A, T, C, and G in the mitochondrial genome were found to be 30.27%, 23.75%, 32.49% and 13.49%, respectively. The complete mitochondrial genome of the Cenxi classical three-buff chicken contains a typical structure, including 13 protein-coding genes, 2 rRNA genes, 22 tRNA genes and 1 control region (D-loop region). This complete mitochondrial genome sequence provides essential information in understanding phylogenetic relationships among Gallus gallus domesticus mitochondrial genomes.

Keywords

Cenxi classical three-buff chicken, genome organization, mitochondrial genome

---

The Cenxi classical three-buff chicken is a unique high-quality breed of native chicken with a long history. The Cenxi three-buff chicken was named for its yellow feathers, yellow mouth and yellow feet.Villages in the Nuotong townships regions in Cenxi district of Guangxi are home origins in raising the native three-buff chicken, the body size of the Cenxi classical three-buff chicken is small (1.5 kg in average).The body shape is elegant, the body muscles are very tender and the body bones are quiet thin.Thus, the Cenxi three-buff chicken is very active, especially in pecking and running. The Cenxi three-buff chicken is the only absolutely pure native breed chicken without any foreign genealogical genes. In the 1980s Cenxi three-buff chicken was documented the title "national quality three-buff chicken" by China State ministry of Agriculture.It is important and of high priority to preserve genetic resources of native species for urgent conservation and restoration of the species [1]. The most advanced techniques in studying mitochondrial genome provide a unique tool to identify genetically pure breed for the purpose of preservation[2].

In this report, we describe the Cenxi classical three-buff chicken's genomic DNA extracted from its liver tissue using the EasyPure Genomic DNA Kit (TransGen, Beijing, China). The complete mitochondrial genome was amplified by 22 pairs of primers designed according to the sequence of the Gallus gallus (GenBank accession number: AP003322).The gel electrophoresis PCR products were purified by using AxyPrep™ DNA Gel Extraction Kit (Hangzhou, China) and then were submitted to Invitrogen (Guangzhou, China) for genomic sequencing. The DNA sequence was analyzed using DNAStar 7.1 software (Madison, WI, USA) and the base composition and distribution of the mitochondrial DNA (mtDNA) sequence were analyzed using tRNA Scan-

SE1.21 software [3] and DOGMA software [4], respectively.

Our research findings have completed the entire mitochondrial genome sequence of Cenxi classical three-buff chicken which is 16 786 bp containing 13 protein-coding genes, 22 tRNA genes, 2 rRNA genes and 1 control region (D-loop region), which are similar to those of other avian species in gene arrangement and composition [5]. The overall nucleotide composition of A, T, C and G were found to be 30.27%, 23.75%, 32.49% and 13.49%, respectively. Most of the genes are encoded on the H-strand (Table 7-1-1) and they are similar in structure to the typical mitochondrial genome of vertebrates[6]. The Cenxi classical three-buff chicken's entire mitochondrial genome sequence has been deposited in GenBank recently (GenBank accession number: KM433666).

The Cenxi classical three-buff chicken mitochondrial genes not encoded on the H-strand include 1 protein-coding gene (ND6) and 8 tRNA genes ($tRNA^{Gln}$, $tRNA^{Ala}$, $tRNA^{Asn}$, $tRNA^{Cys}$, $tRNA^{Tyr}$, $tRNA^{Ser}$, $tRNA^{Pro}$ and $tRNA^{Glu}$), which are similar to typical avian mtDNAs[7, 8]. The initiation codon of the protein-coding genes is ATG, except for COX1, which show a GTG initiation codon (Table 1). These genes have three different types of termination codons: typel, TAA for ND1, COX2, ATPase8, ATPase6, ND3, ND4L, ND5, Cytb and ND6; type 2, AGG for COX1; type 3, incomplete termination codon "T--" for ND2, COX3 and ND4, which is thought to be completed by the polyadenylation of the mRNAs after cleavage of their primary transcripts[9] and is found in most vertebrates[10-12].

Table 7-1-1 MtDNA genome organization of the Cenxi classical three-buff chicken

Gene (element)	Position (from-to) site	Chain		Space (+) overlap (–)	Codons		
		Length	(H/L)		Start	Stop	Anti
D-loop	1-1 232	1 232	H	0			
$tRNA^{Phe}$	1 233-1 301	69	H	0			GAA
12S rRNA	1 302-2 277	976	H	0			
$tRNA^{Val\ (GUM)}$	2 278-2 350	73	H	0			GAC
16S rRNA	2 351-3 973	1 623	H	0			
$tRNA^{Leu\ (YUH)}$	3 974-4 047	74	H	0			GAG
ND1	4 057-5 031	975	H	+9	ATG	TAA	
$tRNA^{Ile\ (AUM)}$	5 032-5 103	72	H	0			GAT
$tRNA^{Gln\ (CAR)}$	5 109-5 179	71	L	+5			TTG
$tRNA^{Met}$	5 179-5 247	69	H	-1			CAT
ND2	5 248-6 286	1 039	H	0	ATG	T--	
$tRNA^{Trp}$	6 287-6 362	76	H	0			CCA
$tRNA^{Ala\ (GCH)}$	6 369-6 437	69	L	+6			AGC
$tRNA^{Asn\ (AAY)}$	6 441-6 513	73	L	+3			ATT
$tRNA^{Cys\ (UGY)}$	6 515-6 580	66	L	+1			GCA
$tRNA^{Tyr}$	6 580-6 650	71	L	–1			GTA
COX1	6 652-8 202	1 551	H	+1	GTG	AGG	

continued

Gene (element)	Position (from-to) site	Chain		Space (+) overlap (–)	Codons		
		Length	(H/L)		Start	Stop	Anti
tRNA$^{Ser (UCD)}$	8 194-8 268	75	L	–9			TGA
tRNA^{Asp}	8 271-8 339	69	H	+2			GTC
COX2	8 341-9 024	684	H	+1	ATG	TAA	
tRNA^{Lys}	9 026-9 093	68	H	+1			CTT
ATPase8	9 095-9 259	165	H	+1	ATG	TAA	
ATPase6	9 250-9 933	684	H	-10	ATG	TAA	
COX3	9 933-10 716	784	H	-1	ATG	T--	
tRNA^{Gly}	10 718-10 785	69	H	+1			ACC
ND3	10 786-11 137	352	H	0	ATG	TAA	
tRNA^{Arg}	11 139-11 206	68	H	+1			TCT
ND4L	11 207-11 503	297	H	0	ATG	TAA	
ND4	11 497-12 874	1 378	H	-7	ATG	T--	
tRNA^{His}	12 875-12 943	69	H	0			ATG
tRNA$^{Ser (AGY)}$	12 945-13 009	65	H	+1			GCT
tRNA$^{Len (YUN)}$	13 011-13 081	71	H	+1			AAG
ND5	13 082-14 899	1 818	H	0	ATG	TAA	
Cyt b	14 904-16 046	1 143	H	+4	ATG	TAA	
tRNA$^{Thr (ACH)}$	16 050-16 118	69	H	+3			TGA
tRNA$^{Pro (Ccw)}$	16 119-16 188	68	L	0			TGG
ND6	16 195-16 716	552	L	+6	ATG	TAA	
tRNA^{Glu}	16 719-16 786	68	L	+2			TTC

Note: "T--"represent incomplete stop codons. Positive numbers correspond to the nucleotides separating adjacent genes. Negative numbers indicate overlapping nucleotides.

Among all gene elements (Table 7-1-1), 7 overlaps and/or 18 spaces were found in the length between 1 and 10 bp.The lengths of the 12S rRNA and 16S rRNA genes are 976 bp and 1 623 bp, respectively, they are located between the tRNA^{Leu} and tRNA^{Phe} genes and separated by the tRNA^{Val} gene. Twenty-two deduced tRNA genes were found to be distributed in rRNA and protein-coding genes, ranging from 65 to 76 bp in size. The D-loop is a non-coding control region located at 1 232 bp position between the tRNA^{Glu} and tRNA^{Phe} genes, where is rich in A and T. This region is most likely to show sequence variations[13]. The D-loop contains regulatory elements that control mtDNA replication and transcription.

References

[1] NIU D, FU Y, LUO J, et al. The origin and genetic diversity of Chinese native chicken breeds. Biochem Genet, 2002, 40(5-6): 163-174.

[2] SHEN Y Y, LIANG L, SUN Y B, et al. A mitogenomic perspective on the ancient, rapid radiation in the Galliformes with an emphasis on the Phasianidae. BMC Evol Biol, 2010, 10: 132-141.

[3] LOWE T M, EDDY S R. tRNAscan-SE: a program for improved detection of transfer RNA genes in genomic sequence. Nucleic Acids Res, 1997, 25(5): 955-964.

[4] WYMAN S K, JANSEN R K, BOORE J L. Automatic annotation of organellar genomes with DOGMA. Bioinformatics, 2004, 20(17): 3252-3255.

[5] XIE Z, ZHANG Y, XIE L, et al. Sequence and gene organization of the Xilin small partridge duck (Anseriformes, Anatidae, *Anas*) mitochondrial DNA. Mitochondrial DNA A DNA Mapp Seq Anal. 2016, 27(3): 1579-1580.

[6] BOORE JL. Animal mitochondria genomes. Nucleic Acids Res, 1999, 27: 1767-1780.

[7] SNYDER J C, MACKANESS C A, SOPHER M R, et al. The complete mitochondrial genome sequence of the Canada goose (*Branta canadensis*). Mitochondrial DNA, 2015, 26(5): 672-673.

[8] XIE Z, ZHANG Y, XIE L, et al. The complete mitochondrial genome of the Jingxi duck (*Anas platyrhynchos*). Mitochondrial DNA A DNA Mapp Seq Anal, 2016, 27(2): 809-810.

[9] ANDERSON S, BANKIER A T, BARRELL B G, et al. Sequence and organization of the human mitochondrial genome. Nature, 1981, 290(5806): 457-465.

[10] MINDELL D P, SORENSON M D, DIMCHEFF D E. An extra nucleotide is not translated in mitochondrial ND3 of some birds and turtles. Mol Biol Evol, 1998, 15(11): 1568-1571.

[11] OJALA D, MONTOYA J, ATTARDI G. tRNA punctuation model of RNA processing in human mitochondria. Nature. 1981, 290(5806): 470-474.

[12] ZHANG Y, XIE Z, XIE L, et al. Genetic characterization of the Longsheng duck (*Anas platyrhynchos*) based on the mitochondrial DNA. Mitochondrial DNA A DNA Mapp Seq Anal, 2016, 27(2): 1146-1147.

[13] ZHANG Y, XIE Z, XIE L, et al. Analysis of the Rongshui Xiang duck (*Anseriformes*, Anatidae, Anas) mitochondrial DNA. Mitochondrial DNA A DNA Mapp Seq Anal, 2016, 27(3): 1867-1868.

Complete sequence determination and analysis of the Guangxi Phasianus colchicus (Galliformes, Phasianidae) mitochondrial genome

Zhang Yanfang, Xie Zhixun, Xie Zhiqin, Deng Xianwen, Xie Liji, Fan Qing, Luo Sisi, Liu Jiabo, Huang Li, Huang Jiaoling, Zeng Tingting, and Wang Sheng

Abstract

The complete mitochondrial genome sequence of the Guangxi *Phasianus colchicus* was measured by PCR-based methods and analyzed in detail. Our research findings reveal that the entire mitochondrial genome of *P.colchicus* is a circular molecule of length 16 687 bp (GenBank accession number: KT364526). The contents of A, T, C, and G in the mitochondrial genome were found to be 30.64%, 25.29%, 30.81% and 13.26%, respectively. The complete mitochondrial genome of the *P.colchicus* is a typical structure, including 13 protein-coding genes, 2 rRNA genes, 22 tRNA genes, and 1 control region (D-loop region). This complete mitochondrial genome sequence provides essential information in understanding phylogenetic relationships among Galliformes mitochondrial genomes.

Keywords

genome organization, mitochondrial genome, *Phasianus colchicus*, phasianidae

---

The Galliformes is a well-known and widely distributed order in Aves, containing more than 290 species within some 82 genera[1]. *Phasianus colchicus* (Ring-necked Pheasant) is a genus of bird in the family Phasiandae (Galliformes: Phasiandae) and widely distributed in the world, including 31 subspecies[2]. Many *P.colchicus* have beautiful ornamentations and they play an important role in hunting and entertainment. The *P.colchicus* in our report is one of the subspecies from Guangxi, China, which was classified as Least Concerned (LC) in the International Union for the Conservation of Nature and Natural Resources (IUCN) Red list (IUCN 2009). Many studies showed that mitochondrial DNA sequences are suitable markers to infer genetic diversity and phylogeny[3-5]. It is very important and of high priority to preserves genetic resources of native species for urgent conservation and restoration of the species[6].

P. colchicus specimen was collected from Nanning, Guangxi, China. In this report, we describe the mitochondrial genome of *P.colchicus* genomic DNA extracted from its muscle tissue using the EasyPure Genomic DNA Kit (TransGen, Beijing, China). The complete mitochondrial genome was amplified by 24 pairs of primers designed according to the sequence of the *P.colchicus* (GenBank accession number: JF739859 and FJ752430). The gel electrophoresis PCR products were purified by using AxyPrep™ DNA Gel Extraction Kit (Axygen, Hangzhou, China) and then were submitted to Invitrogen (Guangzhou, China) for genomic sequencing. The DNA sequence was analyzed using DNAStar 7.1 software (Madison, WI, USA) and the base composition and distribution of the mitochondrial DNA (mtDNA) sequence were analyzed using tRNA Scan-SE1.21 software [7] and DOGMA software [8], respectively.

Our research findings have completed the entire mitochondrial genome sequence of *P.colchicus* which is 16 687 bp containing 13 protein-coding genes, 22 tRNA genes, 2 rRNA genes, and 1 control region (D-loop region), which are similar to those of other avian species in gene arrangement and composition[9]. The overall nucleotide composition of A, T, C, and G were found to be 30.64%, 25.29%, 30.81% and 13.26%, respectively. The *P.colchicus*'s entire mitochondrial genome sequence has been deposited in GenBank recently (GenBank accession number: KT364526). Mitochondrial genome analysis revealed that the D-loop is a non-coding control region located at 1 147 bp position between the tRNA^{Glu} and tRNA^{Phe} genes, which is longer than the reported one (1 144 bp, FJ752430). Twenty-two deduced tRNA genes were found to be distributed in rRNA and protein-coding genes, ranging from 65 to 78 bp in size. The lengths of the 12S rRNA and 16S rRNA genes are 966 bp and 1 618 bp, respectively, they are located between the tRNA^{Leu} and tRNA^{Phe} genes and separated by the tRNA^{Val} gene.

The *P. colchicus* mitochondrial genes are not encoded on the H-strand and include 1 protein-coding gene (ND6) and 8 tRNA genes (tRNA^{Gln}, tRN^{Ala}, tRNA^{Asn}, tRNA^{Cys}, tRNA^{Tyr}, tRNA$^{Ser\ (UCM)}$, tRNA^{Pro} and tRNA^{Glu}), which are similar to typical avian mtDNAs [10, 11]. The initiation codon of the protein-coding genes is ATG, except for COX1, which shows a GTG initiation codon (Table 7-2-1). These genes have 4 different types of termination codons: type 1, TAA for ND1, COX2, ATPase8, ATPase6, ND3, ND4L and ND5; type 2, AGG for COX1; type3, TAG for Cytb and ND6; type 4, incomplete termination codon "T--" for ND2, COX3 and ND4, which is thought to be completed by the polyadenylation of the mRNAs after cleavage of their primary transcripts[12] and is found in most vertebrates[13].

Table 7-2-1 Mt DNA genome organization of the *Phasianus colchicus*

Gene (element)	Position (from-to) site	Chain		Space (+) overlap (–)	Codons		
		Length	(H/L)		Start	Stop	Anti
D-loop	1-1 147	1 147	H	0			
tRNA^{Phe}	1 148-1 214	67	H	0			GAA
12S rRNA	1 215-2 180	966	H	0			
tRNA^{Val}	2 181-2 255	75	H	0			GAC
16S rRNA	2 256-3 873	1 618	H	0			
tRNA$^{Leu\ (UUH)}$	3 873-3 946	74	H	–1			GAG
ND1	3 958-4 932	975	H	+11	ATG	TAA	
tRNA^{Ile}	4 933-5 004	72	H	0			GAT
tRNA^{Gln}	5 011-5 081	71	L	+6			TTG
tRNA^{Met}	5 081-5 149	69	H	–1			CAT
ND2	5 150-6 188	1 039	H	0	ATG	T--	
tRNA^{Trp}	6 189-6 266	78	H	+2			CCA
tRNA^{Ala}	6 269-6 337	69	L	+2			AGC
tRNA^{Asn}	6 341-6 413	73	L	+3			ATT
tRNA^{Cys}	6 416-6 482	67	L	+2			GCA
tRNA^{Tyr}	6 482-6 552	71	L	–1			GTA
COX1	6 552-8 104	1 551	H	+1	GTG	AGG	
tRNA$^{Ser\ (UCN)}$	8 096-8 170	75	L	–9			TGA
tRNA^{Asp}	8 173-8 242	69	H	+2			GTC

continued

Gene (element)	Position (from-to) site	Chain		Space (+) overlap (−)	Codons		
		Length	(H/L)		Start	Stop	Anti
COX2	8 243-8 926	684	H	+1	ATG	TAA	
tRNA^{Lys}	8 928-8 995	68	H	+1			CTT
ATPase8	8 997-9 161	165	H	+1	ATG	TAA	
ATPase6	9 152-9 835	684	H	−10	ATG	TAA	
COX3	9 835-10 618	784	H	−1	ATG	T--	
tRNA^{Gly}	10 620-10 687	68	H	+1			ACC
ND3	10 688-11 039	352	H	0	ATG	TAA	
tRNA^{Arg}	11 041-11 109	69	H	+1			TCT
ND4L	11 110-11 406	297	H	0	ATG	TAA	
ND4	11 400-12 777	1 378	H	−7	ATG	T--	
tRNA^{His}	12 778-12 846	69	H	0			ATG
tRNA$^{Ser (AGY)}$	12 848-12 912	65	H	+1			GCT
tRNA$^{Len (CUN)}$	12 914-12 984	71	H	+1			AAG
ND5	12 985-14 802	1 818	H	0	ATG	TAA	
Cyt b	14 807-15 949	1 143	H	+4	ATG	TAG	
tRNA^{Thr}	15 951-16 020	69	H	+1			TGA
tRNA^{Pro}	16 022-16 090	69	L	+2			TGG
ND6	16 097-16 618	522	L	+6	ATG	TAG	
tRNA^{Glu}	16 620-16 687	68	L	+1			TTC

Note: "T--" represent incomplete stop codons; +Numbers correspond to the nucleotides separating adjacent genes; negative numbers indicate overlapping nudeotides.

For the mitochondrial sequence obtained in this study, the DNA data of 13 protein coding genes of *P. colchicus* and 10 other Phasianidaes used to build the neighbor-joining phylogenetic tree were downloaded from GenBank. The NJ tree method was performed using MEGA 4.1 (MEGA Inc., Englewood, NJ) [14] with 1 000 bootstrap replicates (Figure 7-2-1). The data presented here will be very useful for studying the evolutionary relationships and genetic diversity of *P.colchicus*.

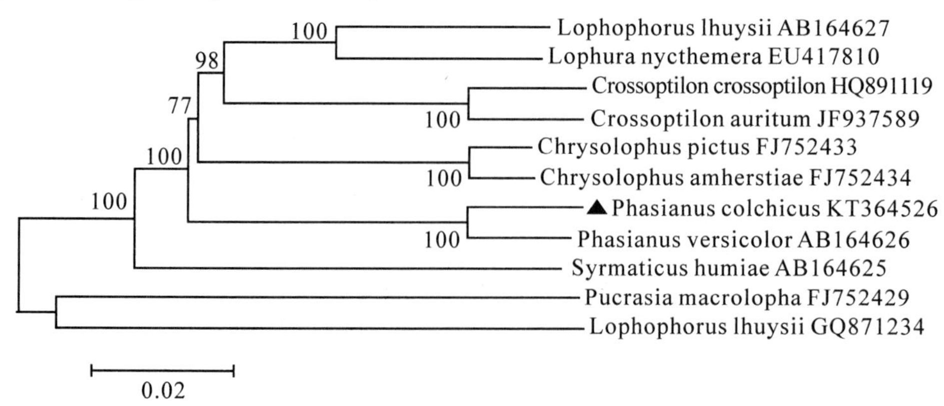

Figure 7-2-1 A neighbor-joining (NJ) tree of 11 species from Phasianidaes was constructed based on the dataset of 13 concatenated mitochondrial protein coding genes

References

[1] GILL F, DONSKER D. IOC World Bird Names (v3.1). (2012-07-21) [2015-08-01]. http://www.worldbirdnames.org.

[2] JOHNSGARD P A. The pheasants of the world-2nd Edition. Washington: Smithsonian Institution Press, 1999.

[3] XIE Z, ZHANG Y, DENG X, et al. Molecular characterization of the Cenxi classical three-buff chicken (*Gallus gallus domesticus*) based on mitochondrial DNA. Mitochondrial DNA A DNA Mapp Seq Anal, 2016, 27(6): 3968-3970.

[4] ZHANG Y, XIE Z, XIE L, et al. Genetic characterization of the Longsheng duck (*Anas platyrhynchos*) based on the mitochondrial DNA. Mitochondrial DNA A DNA Mapp Seq Anal, 2016, 27(2): 1146-1147.

[5] ZHANG Y, XIE Z, DENG X, et al. Mitochondrial genome of the Luchuan pig (*Sus scrofa*). Mitochondrial DNA A DNA Mapp Seq Anal, 2016, 27(6): 4139-4141.

[6] ZHANG Y, XIE Z, XIE L, et al. Analysis of the Rongshui Xiang duck (Anseriformes, Anatidae, Anas) mitochondrial DNA. Mitochondrial DNA A DNA Mapp Seq Anal, 2016, 27(3): 1867-1868.

[7] LOWE T M, EDDY S R. tRNAscan-SE: a program for improved detection of transfer RNA genes in genomic sequence. Nucleic Acids Res, 1997, 25(5): 955-964.

[8] WYMAN S K, JANSEN R K, BOORE J L. Automatic annotation of organellar genomes with DOGMA. Bioinformatics, 2004, 20(17): 3252-3255.

[9] XIE Z, ZHANG Y, XIE L, et al. Sequence and gene organization of the Xilin small partridge duck (Anseriformes, Anatidae, *Anas*) mitochondrial DNA. Mitochondrial DNA A DNA Mapp Seq Anal, 2016, 27(3): 1579-1580.

[10] SNYDER J C, MACKANESS C A, SOPHER M R, et al. The complete mitochondrial genome sequence of the Canada goose (*Branta canadensis*). Mitochondrial DNA, 2015, 26(5): 672-673.

[11] XIE Z, ZHANG Y, XIE L, et al. The complete mitochondrial genome of the Jingxi duck (*Anas platyrhynchos*). Mitochondrial DNA A DNA Mapp Seq Anal, 2016, 27(2): 809-810.

[12] ANDERSON S, BANKIER A T, BARRELL B G, et al. Sequence and organization of the human mitochondrial genome. Nature, 1981, 290(5806): 457-465.

[13] MINDELL D P, SORENSON M D, DIMCHEFF D E. An extra nucleotide is not translated in mitochondrial ND3 of some birds and turtles. Mol Biol Evol, 1998, 15(11): 1568-1571.

[14] TAMURA K, DUDLEY J, NEI M, et al. MEGA4: molecular evolutionary genetics analysis (MEGA) software version 4.0. Mol Biol Evol, 2007, 24(8): 1596-1599.

Complete sequence cloning and bioinformatics analysis of the chukar partridge (*Alectoris chukar*, Aves, Galliformes) in Guangxi based on mitochondrial genome

Zhang Yanfang, Xie Zhixun, Deng Xianwen, Xie Zhiqin, Xie Liji, Fan Qing, Luo Sisi,
Liu Jiabo, Huang Li, Huang Jiaoling, Zeng Tingting, and Wang Sheng

Abstract

The objective of this study was to obtain the complete mitochondrial DNA sequence of chukar partridge, and to provide reference data for protection and utilization of these resources of chukar partridge. The complete mitochondrial genome sequence of the China chukar partridge was measured by PCR-based methods and analysed in detail. Our research findings reveal that the entire mitochondrial genome of the chukar partridge is a circular molecule consisting of 16 688 bp (GenBank accession number: KY829450). The contents of A, T, C, and G in the mitochondrial genome were found to be 30.44%, 24.43%, 31.57%, and 13.56%, respectively. The complete mitochondrial genome of the chukar partridge has a typical structure, including 13 protein-coding genes, two rRNA genes, 22 tRNA genes, and one control region (D-loop region). This complete mitochondrial genome sequence provides essential information in understanding phylogenetic relationships among Galliformes mitochondrial genomes.

Keywords

mitochondrial genome, chukar partridge, *Alectoris chukar*, genome organization

---

The chukar partridge (*Alectoris chukar*, Aves, Galliformes) has a very wide distribution, ranging from east Balkans and the adjacent Mediterranean islands to central Asia up to northeastern China[1-3]. The chukar is a polytypic species, with 16 reported subspecies in the world[4] and six described subspecies in China[5]. The Chukar partridge in our report, is one of the subspecies from Guangxi, China, was classified as Least Concerned (LC) in the International Union for the Conservation of Nature and Natural Resources (IUCN) Red list (IUCN 2012). Many studies showed that mitochondrial (mt) DNA sequences are suitable markers to infer genetic diversity and phylogeny[6-9].

Chukar partridge specimen was collected from Nanning, Guangxi, China. The specimens were kept in the laboratory at −80 ℃ under the accession serial no.60813984. In this report, we describe the mt genome of Chukar partridge genomic DNA extracted from its muscle tissue using the EasyPure Genomic DNA Kit (TransGen, Beijing, China). DNA was stored at −20 ℃. The DNA sequence was analysed using DNAStar 7.1 software (Madison, WI, USA) and the base composition and distribution of the mtDNA sequence were analysed using tRNA ScanSE1.21 (Santa Cruz, CA) [10] and DOGMA software (Austin, TX) [11], respectively.

Whole mtDNA sequence of Chukar partridge has a circular genome of 16 688 bp containing 13 protein-coding genes, 22 tRNA genes, two rRNA genes, and one control region, which are similar to those of other avian species in gene arrangement and composition[12]. A, T, C, and G in the mt genome were found to be 30.44%, 24.43%, 31.57%, and 13.56%, respectively. The entire mt genome sequence has been deposited in GenBank recently (GenBank accession number: KY829450). MtDNA analysis revealed that the D-loop is a non-coding control region located at 1 154 bp position between the $tRNA^{Glu}$ and $tRNA^{Phe}$ genes. Twenty-two deduced tRNA genes were found to be distributed in rRNA and protein-coding genes, ranging from 66 to 76 bp in size. The lengths of the 12S rRNA and 16S rRNA genes are 966 bp and 1 617 bp, respectively.

The mt genes do not encode on the H-strand including 1 protein-coding gene (ND6) and eight tRNA genes, which are similar to typical avian mtDNAs[13]. The initiation codon of the protein-coding genes is ATG, except for COX1, which shows a GTG initiation codon. These genes have four different types of termination codons and are found in most vertebrates[14].

With respect to the mtDNA obtained in this study, the DNA data of 13 protein coding genes of chukar partridge and 11 other Galliformes were used to build the neighbour-joining phylogenetic tree. The NJ tree method was performed using MEGA 6.0 (Chemnitz, Germany) with 1 000 bootstrap replicates (Figure 7-3-1).

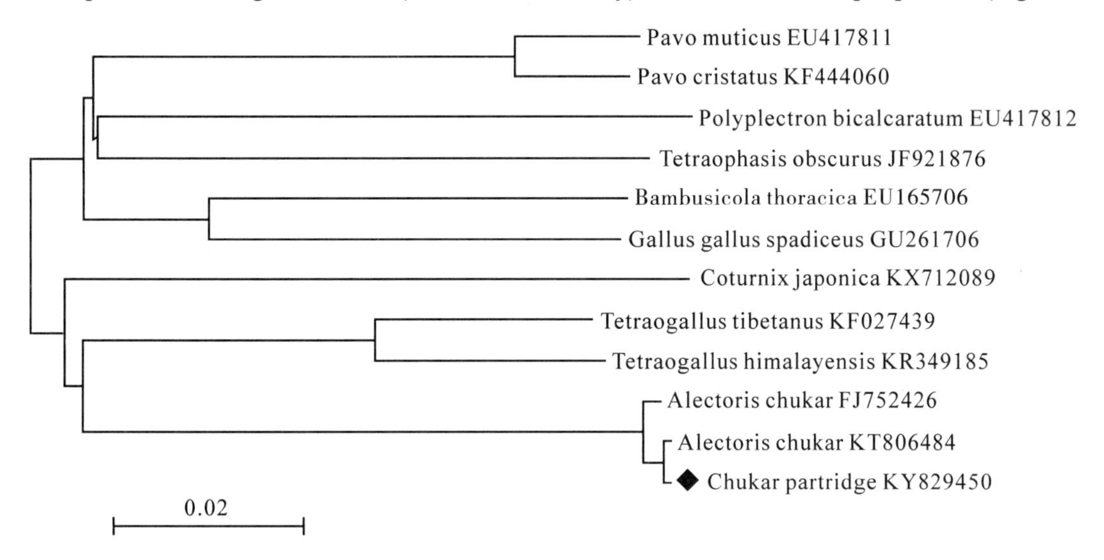

Figure 7-3-1 A neighbour-joining (NJ) tree of 12 species from Galliformes was constructed based on the dataset of 13 concatenated mitochondrial protein coding genes

References

[1] CHENG T H. Fauna sinica: Aves. Beijing: Science Press, 1978, 4: 61-66. (in Chinese)

[2] AEBISCHER N. Alectoris chukar//HAGENMEIJER W J M, BLAIR M J. The EBCC atlas of European breeding birds: their distribution and abundance. London: T&AD Poyser, 1997: 206-211.

[3] MCGOWAN P J K, KIRWAN G M, MADGE S. Pheasants, partridges, and grouse: a guide to the pheasants, partridges, quails, grouse, guineafowl, buttonquails, and sandgrouse of the world. Princeton, New Jersey: Princeton University Press, 2002: 488.

[4] SONG S, Liu N F. A review of the researches on Alectoris partridge. Acta Ecol Sin, 2013, 33: 4215-4225. (in Chinese)

[5] ZHENG G M. A checklist on the classification and distribution of the birds of China. Beijing: Science Press, 2005.

[6] XIE Z, ZHANG Y, XIE L, et al. Sequence and gene organization of the Xilin small partridge duck (Anseriformes, Anatidae, *Anas*) mitochondrial DNA. Mitochondrial DNA A DNA Mapp Seq Anal, 2016, 27(3): 1579-1580.

[7] ZHANG Y, XIE Z, XIE L, et al. Genetic characterization of the Longsheng duck (*Anas platyrhynchos*) based on the mitochondrial DNA. Mitochondrial DNA A DNA Mapp Seq Anal, 2016, 27(2): 1146-1147.

[8] ZHANG Y, XIE Z, XIE L, et al. Analysis of the Rongshui Xiang duck (Anseriformes, Anatidae, *Anas*) mitochondrial DNA.

Mitochondrial DNA A DNA Mapp Seq Anal, 2016, 27(3): 1867-1868.

[9] ZHANG Y, XIE Z, XIE Z, et al. Complete sequence determination and analysis of the Guangxi Phasianus colchicus (Galliformes: Phasianidae) mitochondrial genome. Mitochondrial DNA A DNA Mapp Seq Anal, 2016, 27(6): 4600-4602.

[10] LOWE T M, EDDY S R. tRNAscan-SE: a program for improved detection of transfer RNA genes in genomic sequence. Nucleic Acids Res, 1997, 25(5): 955-964.

[11] WYMAN S K, JANSEN R K, BOORE J L. Automatic annotation of organellar genomes with DOGMA. Bioinformatics, 2004, 20(17): 3252-3255.

[12] XIE Z, ZHANG Y, XIE L, et al. The complete mitochondrial genome of the Jingxi duck (*Anas platyrhynchos*). Mitochondrial DNA A DNA Mapp Seq Anal, 2016, 27(2): 809-810.

[13] XIE Z, ZHANG Y, DENG X, et al. Molecular characterization of the Cenxi classical three-buff chicken (*Gallus gallus domesticus*) based on mitochondrial DNA. Mitochondrial DNA A DNA Mapp Seq Anal, 2016, 27(6): 3968-3970.

[14] MINDELL D P, SORENSON M D, DIMCHEFF D E. An extra nucleotide is not translated in mitochondrial ND3 of some birds and turtles. Mol Biol Evol, 1998, 15(11): 1568-1571.

The complete mitochondrial genome of the Jingxi duck (*Anas platyrhynchos*)

Xie Zhixun, Zhang Yanfang, Xie Liji, Liu Jiabo, Deng Xianwen, Xie Zhiqin, Fan Qing, and Luo Sisi

Abstract

The entire mitochondrial genome of Jingxi duck from China was 16 603bp in length, and has been analyzed for gene locations, length, start codons and stop codons. With the base composition of 29.20% for A, 22.19% for T, 32.82% for C, 15.79% for G, so the percentage of A and T (51.39%) was slightly higher than those of G and C. The Jingxi duck mitochondrial genome contained two ribosomal RNA genes, 13 protein-coding genes, 22 transfer RNA genes and one non-coding control region (D-loop region). The arrangement of these genes was the same as most birds. The complete mitochondrial genome sequence of the Jingxi duck will be useful for phylogenetics, and provide an important data set for further study on the germplasm resources.

Keywords

Anas platyrhynchos, Jingxi duck, mitochondrial genome

Jingxi duck was the native breed of Guangxi in China. As an important species of Chinese domestic duck, the complete mitochondrial genome sequence was not known. Mitochondrial genome is one of the good candidates for the purpose[1]. Total genomic DNA of Jingxi duck was extracted from the liver tissues using the EasyPure Genomic DNA Kit (TransGen, Beijing, China) according to the instruction manuals. PCR products of the gel electrophoresis were purified by AxyPrep™ DNA Gel Extraction Kit (Axygen, Union City, CA) and sequenced by Invitrogen (Guangzhou, China). DNA sequence was analyzed using DNAStar 7.1 software (Madison, WI, USA).

The base composition and distribution of mitochondrial DNA (mtDNA) sequence was analyzed using tRNA Scan-SE1.21 software [2] and DOGMA software[3]. The complete mitochondrial genome has been submitted to GenBank (accession number: KJ689447).

The total length of the mitogenome was 16 603 bp, with the base composition of 29.20% for A, 22.19% for T, 32.82% for C, 15.79% for G and an A-T rich (51.39%) feature in the Jingxi duck. It comprises two ribosomal RNA genes (rRNA), 13 protein-coding genes, 22 transfer RNA genes (tRNA) and one displacement loop region (D-loop). The arrangement of these genes was the same as most birds genes[4, 5]. Besides the COX1, COX2 and ND5 start codon was GTG and the rest of the protein-coding genes were ATG (Table 7-4-1). These tRNA genes had four types of stop codons including AGG for ND1 and COX1, TAG for ND2 and ND6, incomplete stop codon “T” for COX3 and ND4. The latter case has an as shown in most in length vertebrates[6, 7], which is thought to be completed by polyadenlyation of the mRNAs after cleavage of their primary transcripts[8], TAA for the others.

Table 7-4-1 Organization of the mitochondrial genome of the Jingxi duck

Gene (element)	Position (from-to) site	Length	Space (+) Overlap (−)	Start codon	Stop codon	Anti-codon
D-loop	1-1 048	1 048				GAA
tRNA^{Phe}	1 049-1 118	70				
12S rRNA	1 119-2 103	985				
tRNA^{Val}	2 104-2 174	71				TAC
16S rRNA	2 175-3 776	1 602				
tRNA$^{Leu\,(UUR)}$	3 777-3 850	74	+4			TAA
ND1	3 855-4 832	978	−2	ATG	AGG	
tRNA^{Ile}	4 831-4 902	72	+7			GAT
tRNA^{Gln*}	4 910-4 980	71	−1			TTG
tRNA^{Met}	4 980-5 048	69				CAT
ND2	5 049-6 089	1 041	−2	ATG	TAG	
tRNA^{Trp}	6 088-6 163	76	+3			TCA
tRNA^{Ala*}	6 167-6 235	69	+2			TGC
tRNA^{Asn*}	6 238-6 310	73				GTT
tRNA^{Cys*}	6 311-6 376	66	−1			GCA
tRNA^{Tyr*}	6 376-6 447	72	+1			GTA
COX1	6449-7999	1 551	−9	GTG	AGG	
tRNA$^{Ser\,(UCN)\,*}$	7 991-8 063	73	+2			TGA
tRNA^{Asp}	8 066-8 134	69	+1			GTC
COX2	8 136-8 822	687	+1	GTG	TAA	
tRNA^{Lys}	8 824-8 892	69	+1			TTT
ATPase8	8 894-9 061	168	−10	ATG	TAA	
ATPase6	9 052-9 735	684	−1	ATG	TAA	
COX3	9735-10 518	784		ATG	T--	
tRNA^{Gly}	10 519-10 587	69				TCC
ND3	10 588-10 939	352	+1	ATG	TAA	
tRNA^{Arg}	10 941-11 010	70				TCG
ND4L	11 011-11 307	297	−7	ATG	TAA	
*ND*4	11 301-12 678	1378		ATG	T--	
tRNA^{His}	12 679-12 747	69				GTG
tRNA$^{Ser\,(AGY)}$	12 748-12 813	66	−1			GCT
tRNA$^{Leu\,(CUN)}$	12 813-12 883	71				TAG
ND5	12 884-14 707	1 824	−1	GTG	TAA	
Cyt b	14 707-15 849	1 143	+2	ATG	TAA	
tRNA^{Thr}	15 852-15 920	69	+10			TGT
tRNA^{Pro*}	15 931-16 000	70	+10			TGG
ND6*	16 011-16 532	522		ATG	TAG	
tRNA^{Glu*}	16 533-16 603	71				TTC

Note: *Means coded on complement (L) strand.

All the mitogenome genes were encoded on the H strand except for one protein-coding gene (ND6) and eight tRNA genes ($tRNA^{Gln}$, $tRNA^{Ala}$, $tRNA^{Asn}$, $tRNA^{Cys}$, $tRNA^{Tyr}$, $tRNA^{Ser}$, $tRNA^{Pro}$ and $tRNA^{Glu}$). In all these genes, 13 spaces and 10 overlaps in the length of 1～10 bp were found. The lengths of 12S rRNA and the 16S rRNA were 985 bp and 1 602 bp. They were located between the $tRNA^{Leu}$ and $tRNA^{Phe}$ genes and separated by the $tRNA^{Val}$ gene, and the result was the same as those found in mammals. Deduced 22 tRNA genes were distributed in rRNA and protein-coding genes, ranging from 66 to 74 bp in size. D-loop had a length of 1 048 bp as a non-coding control region, between $tRNA^{Glu}$ and $tRNA^{Phe}$ genes. It is rich in A and T and most likely to occur in sequence variation. D-loop contains regulatory elements, which control mtDNA replication and transcription.

References

[1] SHEN Y Y, LIANG L, SUN Y B, et al. A mitogenomic perspective on the ancient, rapid radiation in the Galliformes with an emphasis on the Phasianidae. BMC Evol Biol, 2010, 10: 132.

[2] LOWE T M, EDDY S R. tRNAscan-SE: a program for improved detection of transfer RNA genes in genomic sequence. Nucleic Acids Res, 1997, 25(5): 955-964.

[3] WYMAN S K, JANSEN R K, BOORE J L. Automatic annotation of organellar genomes with DOGMA. Bioinformatics, 2004, 20(17): 3252-3255.

[4] TU J, SI F, WU Q, et al. The complete mitochondrial genome of the Muscovy duck, *Cairina moschata* (Anseriformes, Anatidae, *Cairina*). Mitochondrial DNA, 2014, 25(2): 102-103.

[5] SNYDER J C, MACKANESS C A, SOPHER M R, et al. The complete mitochondrial genome sequence of the Canada goose (*Branta canadensis*). Mitochondrial DNA, 2015, 26(5): 672-673.

[6] MINDELL D P, SORENSON M D, DIMCHEFF D E. An extra nucleotide is not translated in mitochondrial ND3 of some birds and turtles. Mol Biol Evol, 1998, 15(11): 1568-1571.

[7] OJALA D, MONTOYA J, ATTARDI G. tRNA punctuation model of RNA processing in human mitochondria. Nature, 1981, 290(5806): 470-474.

[8] ANDERSON S, BANKIER A T, BARRELL B G, et al. Sequence and organization of the human mitochondrial genome. Nature, 1981, 290(5806): 457-465.

Genetic characterization of the Longsheng duck (*Anas platyrhynchos*) based on the mitochondrial DNA

Zhang Yanfang, Xie Zhixun, Xie Liji, Tan Wei, Liu Jiabo, Deng Xianwen, Xie Zhiqin , and Luo Sisi

Abstract

The complete mitochondrial genome sequence of the Longsheng duck was measured by PCR-based methods. Our research findings revealed that the entire mitochondrial genome of the Longsheng duck was 16 603 bp (GenBank accession number: KJ739616). The contents of A, T, C, and G in the mitochondrial genome were 29.22%, 22.21%, 32.79% and 15.77%, respectively, which were similar to the majority of most avian species. The complete mitochondrial genome of the Longsheng duck contains 13 protein-coding genes, 2 rRNA genes, 22 tRNA genes, and one control region. Components of the Longsheng duck's mitochondrial genome were similar to those of other *Anas platyrhynchos* in gene arrangement and composition. The complete mitochondrial genome of the Longsheng duck should provide essential information for understanding phylogenetic relationships of duck mitochondrial genome.

Keywords

Anas platyrhynchos, longsheng duck, mitochondrial genome

As one of the three black duck species at Longsheng county of Guangxi, China, the Longsheng duck is featured with black feet and black feathers. The feathers on head, feet, and wings are shining as blackish green color like jadeite gloss when seen at certain angle. The native duck's name is also called "Yang Dong duck", raised by the Miao nationality, and mainly distributed in the Madi and Weijiang townships regions. The Longsheng duck is good at gregariousness and suitable for free-ranging in paddy fields and its egg shell is green. As a rare genetic resource of the native duck breed, the Longsheng not only has the unique shape and appearance, but also has extremely high ornamental and tender meat. It is important to preserve genetic resources of native species for urgent conservation and restoration of the species[1, 2]. Mitochondrial genome is one of the unique genetic characters suitable for the purpose[3]. In this research, we report the complete mitochondrial genome sequence of the Longsheng duck.

The Longsheng duck's genomic DNA was extracted from its liver using the EasyPure Genomic DNA Kit (Tiangen, Beijing, China). The complete mitochondrial genome was amplified by 15 pairs of primers, which were designed according to the sequence of Beijing duck (GenBank accession number: EU755252). PCR products from gel electrophoresis were purified by using AxyPrep^{TM} DNA Gel Extraction Kit (Axygen, Union city, CA) and then submitted to Invitrogen (Guangzhou, China) for sequencing.

The Longsheng duck's entire mitochondrial genome sequence has been deposited to the GenBank (GenBank accession number: KJ739616). The complete mitochondrial genome of Longsheng duck was 16 603 bp, which had 13 protein-coding genes, 22 tRNA genes, 2 rRNA genes, and one control region. Most genes were encoded on the H-strand (Table 7-5-1), and are similar in structure to the typical mitochondrial genome of vertebrates[4].

Table 7-5-1 The mtDNA genome organization of the Longsheng duck

Gene (element)	Position (from-to) Chain Codon						Space (+) Overlap (–)
	Site	Length	(H/L)	Start	Stop	Anti	
D-loop	1-1 048	1 048	H				0
tRNA^{Phe}	1 049-1 118	70	H			GAA	0
12S rRNA	1 119-2 103	985	H				0
tRNA^{Val}	2 104-2 174	71	H			TAC	0
16S rRNA	2 175-3 776	1 602	H				0
tRNA$^{Leu(UUR)}$	3 777-3 850	74	H			TAA	+4
ND1	3 855-4 832	978	H	ATG	AGG		–2
tRNA^{Ile}	4 831-4 902	72	H			GAT	+7
tRNA^{Gln}	4 910-4 980	71	L			TTG	–1
tRNA^{Met}	4 980-5 048	69	H			CAT	0
ND2	5 049-6 089	1 041	H	ATG	TAG		–2
tRNA^{Trp}	6 088-6 163	76	H			TCA	+3
tRNA^{Ala}	6 167-6 235	69	L			TGC	+2
tRNA^{Asn}	6 238-6 310	73	L			GTT	0
tRNA^{Cys}	6 311-6 376	66	L			GCA	–1
tRNA^{Try}	6 376-6 447	72	L			GTA	+1
COX1	6 449-7 999	1 551	H	GTG	AGG		–9
RNA$^{Ser(UCN)}$	7 991-8 063	73	L			TGA	+2
tRNA^{Asp}	8 066-8 134	69	H			GTC	+1
COX2	8 136-8 822	687	H	GTG	TAA		+1
tRNA^{Lys}	8 824-8 892	69	H			TTT	+1
ATPase8	8 894-9 061	168	H	ATG	TAA		–10
ATPase6	9 052-9 735	684	H	ATG	TAA		–1
COX3	9 735-10 518	784	H	ATG	T--		0
tRNA^{Gly}	10 519-10 587	69	H			TCC	0
ND3	10 588-10 939	352	H	ATG	TAA		+1
tRNA^{Arg}	10 941-11 010	70	H			TCG	0
ND4L	11 011-11 307	297	H	ATG	TAA		–7
ND4	11 301-12 678	1 378	H	ATG	T--		0
tRNA^{His}	12 679-12 747	69	H			GTG	0
tRNA$^{Ser(AGY)}$	12 748-12 813	66	H			GCT	–1
tRNA$^{Leu(CUN)}$	12 813-12 883	71	H			TAG	0
ND5	12 884-14 707	1 824	H	GTG	TAA		–1
Cyt b	14 707-15 849	1 143	H	ATG	TAA		+2
tRNA^{Thr}	15 852-15 920	69	H			TGT	+10

continued

Gene (element)	Position (from-to) Chain Codon						Space (+) Overlap (–)
	Site	Length	(H/L)	Start	Stop	Anti	
tRNA^{Pro}	15 931-16 000	70	L			TGG	+10
ND6	16 011-16 532	522	L	ATG	TAG		0
tRNA^{Glu}	16 533-16 603	71	L			TTC	0

Genes were not encoded on the H-strand include *ND*6 and eight tRNA genes (tRNA^{Pro}, tRNA^{Gln}, tRNA^{Ala}, tRNA^{Asn}, tRNA^{Cys}, tRNA^{Tyr}, tRNA^{Ser} and tRNA^{Glu}), which were similar to typical bird mtDNAs[5, 6]. The overall nucleotide compositions of the A, T, C, and G were 29.22, 22.21, 32.79 and 15.77%, respectively. The start codon of the protein-coding genes was ATG except for COX1, COX2 and ND5 which had a GTG start codon (Table 7-5-1). These tRNA genes had 4 types of stop codons: type1: AGG for ND1 and COX1; type 2: TAG for ND2 and ND6; type 3: incomplete stop codon "T" for COX3 and ND4; type 4: TAA for the others. Type 3 was shown in most vertebrates[7, 8], which was thought to be completed by polyadenlyation of the mRNAs after cleavage of their primary transcripts[9].

Among all gene elements (Table 7-5-1), 13 spaces and/or 10 overlaps were found in the length between 1～10 bp. The lengths of 12S rRNA and the 16S rRNA genes were 985 bp and 1 602 bp, respectively, which were located between the tRNA^{Leu} and tRNA^{Phe} genes and separated by the tRNA^{Val} gene. Deduced 22 tRNA genes were distributed in rRNA and protein-coding genes, ranging from 66 to 74 bp in size. The D-loop located between tRNA^{Glu} and tRNA^{Phe} genes, was 1 048 bp as a non-coding control region. It was rich in A and T and most likely to occur as sequence variations[10]. D-loop contains regulatory elements, which control mtDNA replication and transcription.

References

[1] MOISEYEVA I G, ROMANOV M N, NIKIFOROV A A, et al. Evolutionary relationships of red jungle fowl and chicken breeds. Genet Sel Evol, 2003, 35(4): 403-423.

[2] NIU D, FU Y, LUO J, et al. The origin and genetic diversity of Chinese native chicken breeds. Biochem Genet, 2002, 40(5-6): 163-174.

[3] SHEN Y Y, LIANG L, SUN Y B, et al. A mitogenomic perspective on the ancient, rapid radiation in the Galliformes with an emphasis on the Phasianidae. BMC Evol Biol, 2010, 10: 132.

[4] BOORE J L. Animal mitochondrial genomes. Nucleic Acids Res, 1999, 27(8): 1767-1780.

[5] SNYDER J C, MACKANESS C A, SOPHER M R, et al. The complete mitochondrial genome sequence of the Canada goose (*Branta canadensis*). Mitochondrial DNA, 2015, 26(5): 672-673.

[6] XIE Z, ZHANG Y, XIE L, et al. The complete mitochondrial genome of the Jingxi duck (*Anas platyrhynchos*). Mitochondrial DNA A DNA Mapp Seq Anal, 2016, 27(2): 809-810.

[7] MINDELL D P, SORENSON M D, DIMCHEFF D E. An extra nucleotide is not translated in mitochondrial ND3 of some birds and turtles. Mol Biol Evol, 1998, 15(11): 1568-1571.

[8] OJALA D, MONTOYA J, ATTARDI G. tRNA punctuation model of RNA processing in human mitochondria. Nature, 1981, 290(5806): 470-474.

[9] ANDERSON S, BANKIER A T, BARRELL B G, et al. Sequence and organization of the human mitochondrial genome. Nature, 1981, 290(5806): 457-465.

[10] JIANG F, MIAO Y, LIANG W, et al. The complete mitochondrial genomes of the whistling duck (*Dendrocygna javanica*) and black swan (*Cygnus atratus*): dating evolutionary divergence in Galloanserae. Mol Biol Rep, 2010, 37(6): 3001-3015.

Sequence and gene organization of the Xilin small partridge duck (Anseriformes, Anatidae, *Anas*) mitochondrial DNA

Xie Zhixun, Zhang Yanfang, Xie Liji, Deng Xianwen, Xie Zhiqin, Huang Li, Huang Jiaoling, and Zeng Tingting

Abstract

The complete mitochondrial genome sequence of the Xilin small partridge duck was measured by PCR-based methods. Our research findings revealed that the entire mitochondrial genome of the Xilin small partridge duck was 16 604 bp (GenBank accession number: KJ833586). The contents of A, T, C, and G in the mitochondrial genome were 29.20%, 22.19%, 32.82%and 15.79%, respectively, which were similar to the majority of most avian species. The complete mitochondrial genome of the Xilin small partridge duck contains 13 protein-coding genes, 2 rRNA genes, 22 tRNA genes, and 1 control region. Components of the Xilin small partridge duck's mitochondrial genome were similar to those of other Anatidae in gene arrangement and composition. The complete mitochondrial genome of the Xilin small partridge duck should provide essential information for understanding phylogenetic relationships of duck mitochondrial genome.

Keywords

genome organization, mitochondrial genome, Xilin small partridge duck

---

The Xilin small partridge duck (Anseriformes, Anatidae, *Ana*) was well-known inside and outside in the district of Guangxi. As the name suggests, the female Xilin small partridge duck is featured with spots feathers. The female duck is always with spotted feather (yellow-brown or black) and the male duck's color is deeper, which is honored as "Guangxi first duck". With small mouth and short neck, the Xilin small partridge duck is small and exquisite, which is good at gregariousness and suitable for free-ranging in paddy fields. The Xilin small partridge duck was be listed in the list of Chinese livestocks or poultry genetic resources as native duck in 2006, and then the Xilin small partridge duck race storage area was officially selected and the field was established by the Guangxi aquatic animal husbandry and veterinary in 2010. The original germplasm breeding base of Xilin small partridge duck was built in 2011. It is important to preserve genetic resources of native species for urgent conservation and restoration of the species[1]. Mitochondrial genome is one of the unique genetic characters suitable for the purpose[2]. In this research, we report the complete mitochondrial genome sequence of the Xilin small partridge duck.

The Xilin small partridge duck's genomic DNA was extracted from its liver using the EasyPure Genomic DNA Kit (Tiangen, Beijing, China). The complete mitochondrial genome was amplified by 15 pairs of primers, which were designed according to the sequence of Beijing duck (GenBank accession number: EU755252). PCR products from gel electrophoresis were purified by using AxyPrep™ DNA Gel Extraction Kit (Axygen, Union city,CA) and then submitted to Invitrogen (Guangzhou, China) for sequencing. DNA sequence was analyzed using DNAStar 7.1 software (Madison, WI, USA). The base composition and distribution of mitochondrial DNA (mtDNA) sequence was analyzed using tRNA Scan-SE1.21 software [3] and DOGMA software [4].

The complete mitochondrial genome of Xilin small partridge duck was 16 604 bp, which contains 13 protein-coding genes, 22 tRNA genes, 2 rRNA genes, and 1 control region. The Xilin small partridge duck's entire mitochondrial genome sequence has been deposited to the GenBank (GenBank accession number: KJ739616). Most genes were encoded on the H-strand (Table 7-6-1), and are similar in structure to the typical mitochondrial genome of vertebrates[5]. Genes were not encoded on the H-strand include ND6 and eight tRNA genes ($tRNA^{Pro}$, $tRNA^{Gln}$, $tRNA^{Ala}$, $tRNA^{Asn}$, $tRNA^{Cys}$, $tRNA^{Tyr}$, $tRNA^{Ser}$ and $tRNA^{Glu}$), which were similar to typical bird mtDNAs[6, 7].

Table 7-6-1 The mtDNA genome organization of the Xilin small partridge duck

Gene (element)	Position (from-to) site	Chain		Codon			Space (+) overlap (–)
		Size	(H/L)	Star	Stop	Anti	
D-loop	1-1 049	1 049	H				0
$tRNA^{Phe}$	1 050-1 119	70	H			GAA	0
12S rRNA	1 120-2 104	985	H				0
$tRNA^{Val}$	2 105-2 175	71	H			TAC	0
16S rRNA	2 176-3 777	1 602	H				0
$tRNA^{Lue(UUR)}$	3 778-3 851	74	H			TAA	+4
ND1	3 856-4 833	978	H	ATG	AGG		-2
$tRNA^{Ile}$	4 832-4 903	72	H			GAT	+7
$tRNA^{Gln}$	4 911-4 981	71	L			TTG	–1
$tRNA^{Met}$	4 981-5 049	69	H			CAT	0
ND2	5 050-6 090	1 041	H	ATG	TAG		–2
$tRNA^{Trp}$	6 089-6 164	76	H			TCA	+3
$tRNA^{Ala}$	6 168-6 236	69	L			TGC	+2
$tRNA^{Asn}$	6 239-6 311	73	L			GTT	0
$tRNA^{Cys}$	6 312-6 377	66	L			GCA	–1
$tRNA^{Tyr}$	6 377-6 448	72	L			GTA	+1
COX1	6 450-8 000	1 551	H	GTG	AGG		–9
$tRNA^{Ser(UCN)}$	7 992-8 064	73	L			TGA	+2
$tRNA^{Asp}$	8 067-8 135	69	H			GTC	+1
COX2	8 137-8 823	687	H	GTG	TAA		+1
$tRNA^{Lys}$	8 825-8 893	69	H			TTT	+1
ATPase8	8 895-9 062	168	H	ATG	TAA		–10
ATPase6	9 053-9 736	684	H	ATG	TAA		–1
COX3	9 736-10 519	784	H	ATG	T--		0
$tRNA^{Gly}$	10 520-10 588	69	H			TCC	0
ND3	10 589-10 940	352	H	ATG	TAA		+1
$tRNA^{Arg}$	10 942-11 011	70	H			TCG	0

continued

Gene (element)	Position (from-to) site	Chain		Codon			Space (+) overlap (–)
		Size	(H/L)	Star	Stop	Anti	
ND4L	11 012-11 308	297	H	ATG	TAA		–7
ND4	11 302-12 679	1 378	H	ATG	T--		0
$tRNA^{His}$	12 680-12 748	69	H			GTG	0
$tRNA^{Ser(AGY)}$	12 749-12 814	66	H			GCT	–1
$tRNA^{Leu(CUN)}$	12 815-12 884	71	H			TAG	0
ND5	12 885-14 708	1 824	H	GTG	TAA		–1
Cyt b	14 708-15 850	1 143	H	ATG	TAA		+2
$tRNA^{Thr}$	15 853-15 921	69	H			TGT	+10
$tRNA^{Pro}$	15 932-16 001	70	L			TGG	+10
ND6	16 012-16 533	522	L	ATG	TAG		0
$tRNA^{Glu}$	16 534-16 604	71	L			TTC	0

The overall nucleotide compositions of the A, T, C, and G were 29.20%, 22.19%, 32.82% and 15.79%, respectively. The start codon of the protein-coding genes was ATG except for COX1, COX2 and ND5 which had a GTG start codon (Table 7-6-1). These tRNA genes had four types of stop codons: type 1: AGG for ND1 and COX1; type 2: TAG for ND2 and ND6; type 3: incomplete stop codon "T" for COX3 and ND4; type 4: TAA for the others. Type 3 was shown in most vertebrates[8, 9], which was thought to be completed by polyadenlyation of the mRNAs after cleavage of their primary transcripts[10].

Among all gene elements (Table 7-6-1), 13 spaces and/or 10 overlaps were found in the length between 1～10 bp. The lengths of 12S rRNA and the 16S rRNA genes were 985 bp and 1 602 bp, respectively, which were located between the $tRNA^{Leu}$ and $tRNA^{Phe}$ genes and separated by the $tRNA^{Val}$ gene. Deduced 22 tRNA genes were distributed in rRNA and protein-coding genes, ranging from 66 to 74 bp in size. The D-loop located between $tRNA^{Glu}$ and $tRNA^{Phe}$ genes, was 1 049 bp as a non-coding control region. It was rich in A and T and most likely to occur as sequence variations[11]. D-loop contains regulatory elements, which control mtDNA replication and transcription.

References

[1] NIU D, FU Y, LUO J, et al. The origin and genetic diversity of Chinese native chicken breeds. Biochem Genet, 2002, 40(5-6): 163-174.

[2] SHEN Y Y, LIANG L, SUN Y B, et al. A mitogenomic perspective on the ancient, rapid radiation in the Galliformes with an emphasis on the Phasianidae. BMC Evol Biol, 2010, 10: 132.

[3] LOWE T M, EDDY S R. tRNAscan-SE: a program for improved detection of transfer RNA genes in genomic sequence. Nucleic Acids Res, 1997, 25(5): 955-964.

[4] WYMAN S K, JANSEN R K, BOORE J L. Automatic annotation of organellar genomes with DOGMA. Bioinformatics, 2004, 20(17): 3252-3255.

[5] BOORE J L. Animal mitochondrial genomes. Nucleic Acids Res, 1999, 27(8): 1767-1780.

[6] SNYDER J C, MACKANESS C A, SOPHER M R, et al. The complete mitochondrial genome sequence of the Canada goose (*Branta canadensis*). Mitochondrial DNA, 2015, 26(5): 672-673.

[7] XIE Z, ZHANG Y, XIE L, et al. The complete mitochondrial genome of the Jingxi duck (*Anas platyrhynchos*). Mitochondrial DNA A DNA Mapp Seq Anal, 2016, 27(2): 809-810.

[8] MINDELL D P, SORENSON M D, DIMCHEFF D E. An extra nucleotide is not translated in mitochondrial ND3 of some birds and turtles. Mol Biol Evol, 1998, 15(11): 1568-1571.

[9] OJALA D, MONTOYA J, ATTARDI G. tRNA punctuation model of RNA processing in human mitochondria. Nature, 1981, 290(5806): 470-474.

[10] ANDERSON S, BANKIER A T, BARRELL B G, et al. Sequence and organization of the human mitochondrial genome. Nature, 1981, 290(5806): 457-465.

[11] ZHANG Y, XIE Z, XIE L, et al. Genetic characterization of the Longsheng duck (*Anas platyrhynchos*) based on the mitochondrial DNA. Mitochondrial DNA A DNA Mapp Seq Anal, 2016, 27(2): 1146-1147.

Analysis of the Rongshui Xiang duck (*Anseriformes, Anatidae, Anas*) mitochondrial DNA

Zhang Yanfang, Xie Zhixun, Xie Liji, Deng Xianwen,
Xie Zhiqin, Huang Li, Huang Jiaoling, and Zeng Tingting

Abstract

The complete mitochondrial genome sequence of the Rongshui Xiang duck was measured by PCR-based methods. Our research findings reveal that the entire mitochondrial genome of the Rongshui Xiang duck is a circular molecule consisting of 16 605 bp (GenBank accession number: KJ833587). The contents of A, T, C, and G in the mitochondrial genome were found to be 29.20%, 22.20%, 32.82% and 15.79%, respectively, similar to the majority of avian species. The complete mitochondrial genome of the Rongshui Xiang duck contains 13 protein-coding genes, 2 rRNA genes, 22 tRNA genes, and 1 control region. The characteristics of the mitochondrial genome were analyzed in detail. Our complete mitochondrial genome sequence should provide essential information for understanding phylogenetic relationships among duck mitochondrial genomes.

Keywords

genome organization, mitochondrial genome, Rongshui Xiang duck

---

The Rongshui Xiang duck originates from Rongshui Miao Autonomous County (Guangxi, China) and it was recognized as a Guangxi local poultry breed in 2006. Because the meat has a fresh aroma, savory flavor and appealing texture, it is therefore called Rongshui Xiang duck. Drakes have the notable characteristics of head and wing feathers that are of a metallic emerald-green color, with a white circle on the neck and black and white tail feathers. The female duck's head and wing feathers are speckled white, the neck's feathers are white, and the ventral feathers are white and light grey. To preserve the pure Rongshui Xiang duck breed, five breed preservation areas have been established since 1999. As a rare genetic resource of the native duck breed, the Rongshui Xiang duck is not only unique in shape and appearance, but also has extremely high ornamental value and tender meat. Indeed, it is important to preserve genetic resources of native species for urgent conservation and restoration of the species[1], and the mitochondrial genome is one of the unique genetic characters suitable for this purpose[2]. Here, we report the complete mitochondrial genome sequence of the Rongshui Xiang duck.

The Rongshui Xiang duck's genomic DNA was extracted from its liver tissue using the EasyPure Genomic DNA Kit (Tiangen, Beijing, China). The complete mitochondrial genome was amplified by 15 pairs of primers designed according to the sequence of the Beijing duck (GenBank accession number: EU755252). PCR products separated by gel electrophoresis were purified by using AxyPrep™ DNA Gel Extraction Kit (Hangzhou, China) and then submitted to Invitrogen (Guangzhou, China) for sequencing. The DNA sequence was analyzed using DNAStar 7.1 software (Madison, WI, USA) and the base composition and distribution of the mitochondrial DNA (mtDNA) sequence was analysed using tRNA Scan-SE1.21 software [3]and DOGMA software[4].

The Rongshui Xiang duck's entire mitochondrial genome sequence has been deposited in GenBank (GenBank accession number: KJ833587). The complete mitochondrial genome of the Rongshui Xiang duck was 16 605 bp, which contains 13 protein-coding genes, 22 tRNA genes, 2 rRNA genes, and 1 control region and was similar to that of other Anatidae in gene arrangement and composition[5]. Most of the genes were encoded on the H-strand (Table 7-7-1) and were similar in structure to the typical mitochondrial genome of vertebrates[6].

Table 7-7-1 MtDNA genome organization of the Rongshui Xiang duck

Gene (element)	Position (from-to) site	Chain		Codons			Space (+) overlap (–)
		Size	(H/L)	Start	Stop	Anti	
D-loop	1-1 050	1 050	H				0
tRNA*Phe*	1 051-1 120	70	H			GAA	0
12S rRNA	1 121-2 105	985	H				0
tRNA*Val*	2 106-2 176	71	H			TAC	0
16S rRNA	2 177-3 778	1 602	H				0
tRNA*Lue (UUR)*	3 779-3 852	74	H			TAA	+4
ND1	3 857-4 834	978	H	ATG	AGG		–2
tRNA*Ile*	4 833-4 904	72	H			GAT	+7
tRNA*Gln*	4 913-4 982	71	L			TTG	–1
tRNA*Met*	4 982-5 050	69	H			CAT	0
ND2	5 051-6 091	1 041	H	ATG	TAG		–2
tRNA*Trp*	6 090-6 165	76	H			TCA	+3
tRNA*Ala*	6 169-6 237	69	L			TGC	+2
tRNA*Asn*	6 240-6 312	73	L			GTT	0
tRNA*Cys*	6 313-6 378	66	L			GCA	–1
tRNA*Tyr*	6 378-6 449	72	L			GTA	+1
COX1	6 451-8 001	1 551	H	GTG	AGG		–9
tRNA*Ser (UCN)*	7 993-8 065	73	L			TGA	+2
tRNA*Asp*	8 068-8 136	69	H			GTC	+1
COX2	8 138-8 824	687	H	GTG	TAA		+1
tRNA*Lys*	8 826-8 894	69	H			TTT	+1
ATPase8	8 896-9 063	168	H	ATG	TAA		–10
ATPase6	9 054-9 737	684	H	ATG	TAA		–1
COX3	9 737-10 520	784	H	ATG	T--		0
tRNA*Gly*	10 521-10 589	69	H			TCC	0
ND3	10 591-10 941	352	H	ATG	TAA		+1
tRNA*Arg*	10 943-11 012	70	H			TCG	0
ND4L	11 013-11 309	297	H	ATG	TAA		–7
ND4	11 303-12 680	1 378	H	ATG	T--		0
tRNA*His*	12 681-12 749	69	H			GTG	0

continued

Gene (element)	Position (from-to) site	Chain		Codons			Space (+) overlap (–)
		Size	(H/L)	Start	Stop	Anti	
tRNA$^{Ser\,(AGY)}$	12 751-12 815	66	H			GCT	–1
tRNA$^{Leu\,(CUN)}$	12 815-12 885	71	H			TAG	0
ND5	12 886-14 709	1 824	H	GTG	TAA		–1
Cyt b	14 709-15 851	1 143	H	ATG	TAA		+2
tRNA^{Thr}	15 854-15 922	69	H			TGT	+10
tRNA^{Pro}	15 933-16 002	70	L			TGG	+10
ND6	16 013-16 534	522	L	ATG	TAG		0
tRNA^{Glu}	16 535-16 605	71	L			TTC	0

The genes that were not encoded on the H-strand include ND6 and eight tRNA genes (tRNA^{Pro}, tRNA^{Gln}, tRNA^{Ala}, tRNA^{Asn}, tRNA^{Cys}, tRNA^{Tyr}, tRNA^{Ser}, and tRNA^{Glu}), similar to typical bird mtDNAs[7, 8]. The overall nucleotide composition of A, T, C, and G were found to be 29.22%, 22.21%, 32.79% and 15.77%, respectively. The start codon of the protein-coding genes is ATG, except for COX1, COX2, and ND5, which show a GTG start codon (Table 7-7-1). These tRNA genes had 4 types of stop codons: typel, AGG for ND2 and COX1; type 2, TAG for ND2 and ND6; type 3, incomplete stop codon "T" for COX3 and ND4; type 4, TAA for the others. Type 3 is found in most vertebrates[9, 10], and is thought to be completed by the polyadenylation of the mRNAs after cleavage of their primary transcripts[11].

Among all gene elements (Table 7-7-1), 13 spaces and/or 10 overlaps were found in the length between 1～10 bp. The lengths of the 12S rRNA and 16S rRNA genes are 985 bp and 1 602 bp, respectively, and are located between the tRNA^{Leu} and tRNA^{Phe} genes and separated by the tRNA^{Val} gene. Twenty-two deduced tRNA genes were found to be distributed in rRNA and protein-coding genes, ranging from 66 to 74 bp in size. The D-loop as a non-coding control region, located between the tRNA^{Glu} and tRNA^{Phe} genes is 1 050 bp and is rich in A and T. This region is most likely to show sequence variations[12]. The D-loop contains regulatory elements that control mtDNA replication and transcription.

References

[1] NIU D, FU Y, LUO J, et al. The origin and genetic diversity of Chinese native chicken breeds. Biochem Genet, 2002, 40(5-6): 163-174.

[2] SHEN Y Y, LIANG L, SUN Y B, et al. A mitogenomic perspective on the ancient, rapid radiation in the Galliformes with an emphasis on the Phasianidae. BMC Evol Biol, 2010, 10: 132.

[3] LOWE T M, EDDY S R. tRNAscan-SE: a program for improved detection of transfer RNA genes in genomic sequence. Nucleic Acids Res, 1997, 25(5): 955-964.

[4] WYMAN S K, JANSEN R K, BOORE J L. Automatic annotation of organellar genomes with DOGMA. Bioinformatics, 2004, 20(17): 3252-3255.

[5] XIE Z, ZHANG Y, XIE L, et al. Sequence and gene organization of the Xilin small partridge duck (Anseriformes, Anatidae, *Anas*) mitochondrial DNA. Mitochondrial DNA A DNA Mapp Seq Anal, 2016, 27(3): 1579-1580.

[6] BOORE J L. Animal mitochondrial genomes. Nucleic Acids Res, 1999, 27(8): 1767-1780.

[7] SNYDER J C, MACKANESS C A, SOPHER M R, et al. The complete mitochondrial genome sequence of the Canada goose

(*Branta canadensis*). Mitochondrial DNA, 2015, 26(5): 672-673.

[8] XIE Z, ZHANG Y, XIE L, et al. The complete mitochondrial genome of the Jingxi duck (*Anas platyrhynchos*). Mitochondrial DNA A DNA Mapp Seq Anal, 2016, 27(2): 809-810.

[9] MINDELL D P, SORENSON M D, DIMCHEFF D E. An extra nucleotide is not translated in mitochondrial ND3 of some birds and turtles. Mol Biol Evol. 1998, 15(11): 1568-1571.

[10] OJALA D, MONTOYA J, ATTARDI G. tRNA punctuation model of RNA processing in human mitochondria. Nature, 1981, 290(5806): 470-474.

[11] ANDERSON S, BANKIER A T, BARRELL B G, et al. Sequence and organization of the human mitochondrial genome. Nature, 1981, 290(5806): 457-465.

[12] ZHANG Y, XIE Z, XIE L, et al. Genetic characterization of the Longsheng duck (*Anas platyrhynchos*) based on the mitochondrial DNA. Mitochondrial DNA A DNA Mapp Seq Anal, 2016, 27(2): 1146-1147.

Mitochondrial genome of the Luchuan pig (*Sus scrofa*)

Zhang Yanfang, Xie Zhixun, Deng Xianwen, Xie Zhiqin, Liu Jiabo,
Xie Liji , Luo Sisi, Huang Li, Huang Jiaoling, Zeng Tingting, and Wang Sheng

Abstract

The complete mitochondrial genome sequence of the Luchuan pig was measured by PCR-based methods, primer-walking sequencing, and fragment cloning. The entire mitochondrial genome of the Luchuan pig was identified as a circular molecule consisting of 16 730 bp (GenBank accession number: KP126954). The contents of A, T, C, and G in the mitochondrial genome were found to be 34.65%, 25.80%, 26.19%, and 13.36%, respectively. The complete mitochondrial genome of the Luchuan pig contains a typical structure, including 13 protein-coding genes, two rRNA genes, 22 tRNA genes, and one control region (D-loop region). This complete mitochondrial genome sequence provides essential information in understanding phylogenetic relationships among *Sus scrofa* domestic mitochondrial genomes.

Keywords

genome organization, Luchuan pig, mitochondrial genome

Luchuan pig is derived from the Luchuan county of Yulin city in the south eastern area of Guangxi in China. The Luchuan pig is one of the kinds of eight famous local varieties in China, and belongs to the national class protected varieties[1]. The Luchuan pig has some advantages, such as good meat quality, rich nutrition, special growth property, and reproduction performance. Thus, Luchuan pig holds an important position in the market, which is deeply loved by the consumers. At present, with the gradually increasing demand for the number of pork from Luchuan pig, the biological technology and the theory in animal reproduction are needed to apply in the Luchuan pig industry. In addition, the research is developed in reproductive area of Luchuan pig for further improving the breeding performance of Luchuan pig. The excellent resource characteristics should be enhanced, and the protection measurement for Luchuan pig is needed to further do well, for continuously maintaining the predominant status of the Luchuan pig in the national class protected varieties. The most advanced techniques in studying mitochondrial genome provide a unique tool to identify genetically pure bread for the purpose of preservation [2].

In this report, we describe the organization Luchuan pig's mitochondrial genomic DNA extracted from its blood using the EasyPure blood Genomic DNA Kit (TransGen, Beijing, China). The complete mitochondrial genome was amplified by 24 pairs of primers designed according to the sequence of the *Sus scrofa* (GenBank accession number: KF472178). The gel electrophoresis PCR products were purified by using the AxyPrep™ DNA Gel Extraction Kit (Axygen, Hangzhou, China) and submitted to Invitrogen (Guangzhou, China) for genomic sequencing. The DNA sequence was analyzed using the DNAStar 7.1 software (DNASTAR, Madison, WI, USA) and the base composition and the distribution of the mitochondrial DNA (mtDNA) sequence were analyzed using the tRNA Scan-SE1.21 software [3] and DOGMA software [4], respectively.

Our research findings have completed the entire mitochondrial genome sequence of Luchuan pig which is 16 730 bp containing 13 protein-coding genes, 22 tRNA genes, two rRNA genes, and one control region (D-loop region), which are similar to those of other avian species in gene arrangement and composition[5]. The overall nucleotide composition of A, T, C, and G were found to be 34.65%, 25.80%, 26.19%, and 13.36%, respectively. Most of the genes are encoded on the H-strand (Table 7-8-1) and they are similar in structure to the typical mitochondrial genome of vertebrates[6]. The Luchuan pig's entire mitochondrial genome sequence has been deposited in GenBank recently (GenBank accession number: KP126954).

Table 7-8-1 MtDNA genome organization of the Luchuan pig

Gene (element)	Position (from-to) site	Chain		Codon			Space (+) overlap (−)
		Size	(H/L)	Start	Stop	Anti	
D-loop	1-1 294	1 294	H				0
tRNA^{Phe}	1 295-1 364	70	H			GAA	0
12S rRNA	1 365-2 325	961	H				0
tRNA^{Val}	2 326-2 393	68	H			AAC	0
16S rRNA	2 394-3 963	1 570	H				0
tRNA$^{Lue\ (YUH)}$	3 964-4 038	75	H			GAG	0
ND1	4 041-4 997	957	H	ATG	TAG		+2
tRNA$^{Ile\ (AUH)}$	4 996-5 064	69	H			GAT	-2
tRNA$^{Gln\ (CAR)}$	5 062-5 134	73	L			TTG	-3
tRNA^{Met}	5 136-5 205	70	H			CAT	+1
ND2	5 206-6 249	1 044	H	ATA	TAG		0
tRNA^{Trp}	6 248-6 315	68	H			CCA	-2
tRNA^{Ala}	6 322-6 389	68	L			AGC	+6
tRNA^{Asn}	6 391-6 465	75	L			GTT	+1
tRNA^{Cys}	6 498-6 563	66	L			GCA	+32
tRNA^{Tyr}	6 563-6 628	66	L			GTA	-1
COX1	6 630-8 174	1 545	H	ATG	TAA		+1
tRNA$^{Ser\ (UCN)}$	8 187-8 248	71	L			AGA	+3
tRNA^{Asp}	8 254-8 321	68	H			ATC	+5
COX2	8 322-9 009	688	H	ATG	T--		0
tRNA^{Lys}	9 010-9 076	67	H			CTT	0
ATPase8	9 078-9 281	204	H	ATG	TAA		+1
ATPase6	9 239-9 919	681	H	ATG	TAA		−43
COX3	9 919-10 702	784	H	ATG	T--		−1
tRNA^{Gly}	10 703-10 771	69	H			GCC	0
ND3	10 772-11 117	346	H	ATA	T--		0
tRNA^{Arg}	11 119-11 187	69	H			TCT	+1
ND4L	11 188-11 484	297	H	GTG	TAA		0
ND4	11 478-12 855	1 378	H	ATG	T--		−7

continued

Gene (element)	Position (from-to) site	Chain		Codon			Space (+) overlap (−)
		Size	(H/L)	Start	Stop	Anti	
tRNA^{His}	12 856-12 924	69	H			ATG	0
tRNA$^{Ser\,(AGY)}$	12 925-12 983	59	H			TGA	0
tRNA$^{Leu\,(CUN)}$	12 984-13 053	70	H			TAA	0
ND5	13 054-14 874	1 821	H	ATA	TAA		0
Cyt b	14 858-15 385	528	L	ATG	TAA		−17
tRNA^{Thr}	15 386-15 454	69	L			TTC	0
tRNA^{Pro}	15 459-16 598	1 140	H	ATG	AGA		+4
ND6	16 599-16 666	68	H			GGT	0
tRNA^{Glu}	16 599-16 730	64	L			TGG	0

Note: "T--" represents incomplete stop codons.

The Luchuan pig mitochondrial genes that are not encoded on the H-strand include one protein-coding gene (ND6) and eight tRNA genes (tRNA$^{Gln\,(CAR)}$, tRNA^{Ala}, tRNA$^{Asn\,(AAY)}$, tRNA^{Cys} tRNA^{Tyr}, tRNA$^{Sen\,(KCH)}$, tRNA^{Glu}, and tRNA$^{Pro\,(CCN)}$), which are similar to typical vertebrate mtDNAs[7, 8]. Of the 13 protein-codon genes for Luchuan pig mitochondria, nine subunits require ATG as the initiation codon, while ND2, ND3, and ND5 gene utilize ATA, ND4L gene use GTG (Table 7-8-1). These tRNA genes had four types of stop codons: type 1: AGA for Cyt b; type 2: TAG for ND1 and ND2; type 3: incomplete stop codon "T" for COX2, COX3, ND3, and ND4; type 4: TAA for the others. Type 3 is known to be found in most vertebrates[9-12], was probably completed by polyadenlyation of the mRNAs after cleavage of their primary transcripts[13].

Positive numbers correspond to the nucleotides separating adjacent genes. Negative numbers indicate overlapping nucleotides.

Among all the gene elements (Table 7-8-1), 11 overlaps and/or eight spaces were found in the length between 1 and 43 bp. The lengths of the 12S rRNA and 16S rRNA genes are 961 bp and 1 570 bp, respectively, they are located between the tRNA^{Phe} and tRNA$^{Leu\,(YUH)}$ genes and separated by the tRNA^{Val} gene. In addition, 22 deduced tRNA genes were found to be distributed in rRNA and protein-coding genes, ranging from 59 to 75 bp in size. The D-loop is a non-coding control region located at 1 294 bp position between the tRNA^{Glu} and tRNA^{Phe} genes, where is rich in A and T. However, the length of the control region can potentially fluctuate among individuals due to variation in copy number of the tandem repeat 5'-TACACGTGCG-3'. This region is most likely to show sequence variations. The D-loop contains regulatory elements that control mtDNA replication and transcription.

References

[1] HUANG Y, JIANG Y, LIANG W, et al. Developmental status on national-class protected species Luchuan pig. Heilongjiang Animal Sci Veterinary Med, 2013, (4): 43-45. (in Chinese)

[2] SHEN Y Y, LIANG L, SUN Y B, et al. A mitogenomic perspective on the ancient, rapid radiation in the Galliformes with an emphasis on the Phasianidae. BMC Evol Biol, 2010, 10: 132.

[3] LOWE T M, EDDY S R. tRNAscan-SE: a program for improved detection of transfer RNA genes in genomic sequence. Nucleic Acids Res, 1997, 25(5): 955-964.

[4] WYMAN S K, JANSEN R K, BOORE J L. Automatic annotation of organellar genomes with DOGMA. Bioinformatics, 2004, 20(17): 3252-3255.

[5] XIE Z, ZHANG Y, XIE L, et al. Sequence and gene organization of the Xilin small partridge duck (Anseriformes, Anatidae, *Anas*) mitochondrial DNA. Mitochondrial DNA A DNA Mapp Seq Anal, 2016, 27(3): 1579-1580.

[6] BOORE J L. Animal mitochondrial genomes. Nucleic Acids Res, 1999, 27(8): 1767-1780.

[7] ARNASON U, ADEGOKE J A, BODIN K, et al. Mammalian mitogenomic relationships and the root of the eutherian tree. Proc Natl Acad Sci USA, 2002, 99(12): 8151-8156.

[8] XIE Z, ZHANG Y, XIE L, et al. The complete mitochondrial genome of the Jingxi duck (*Anas platyrhynchos*). Mitochondrial DNA A DNA Mapp Seq Anal, 2016, 27(2): 809-810.

[9] MINDELL D P, SORENSON M D, DIMCHEFF D E. An extra nucleotide is not translated in mitochondrial ND3 of some birds and turtles. Mol Biol Evol, 1998, 15(11): 1568-1571.

[10] OJALA D, MONTOYA J, ATTARDI G. tRNA punctuation model of RNA processing in human mitochondria. Nature, 1981, 290(5806): 470-474.

[11] ZHANG Y, XIE Z, XIE L, et al. Genetic characterization of the Longsheng duck (*Anas platyrhynchos*) based on the mitochondrial DNA. Mitochondrial DNA A DNA Mapp Seq Anal, 2016, 27(2): 1146-1147.

[12] ZHANG Y, XIE Z, XIE L, et al. Analysis of the Rongshui Xiang duck (Anseriformes, Anatidae, *Anas*) mitochondrial DNA. Mitochondrial DNA A DNA Mapp Seq Anal, 2016, 27(3): 1867-1868.

[13] ANDERSON S, BANKIER A T, BARRELL B G, et al. Sequence and organization of the human mitochondrial genome. Nature, 1981, 290(5806): 457-465.

Appendix

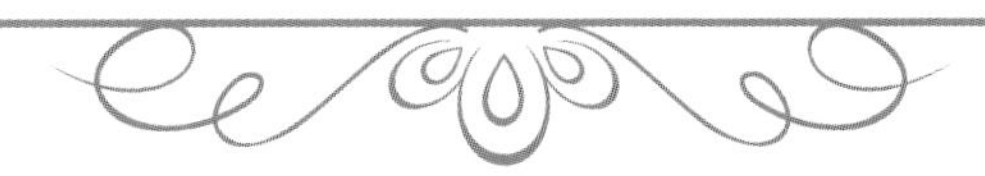

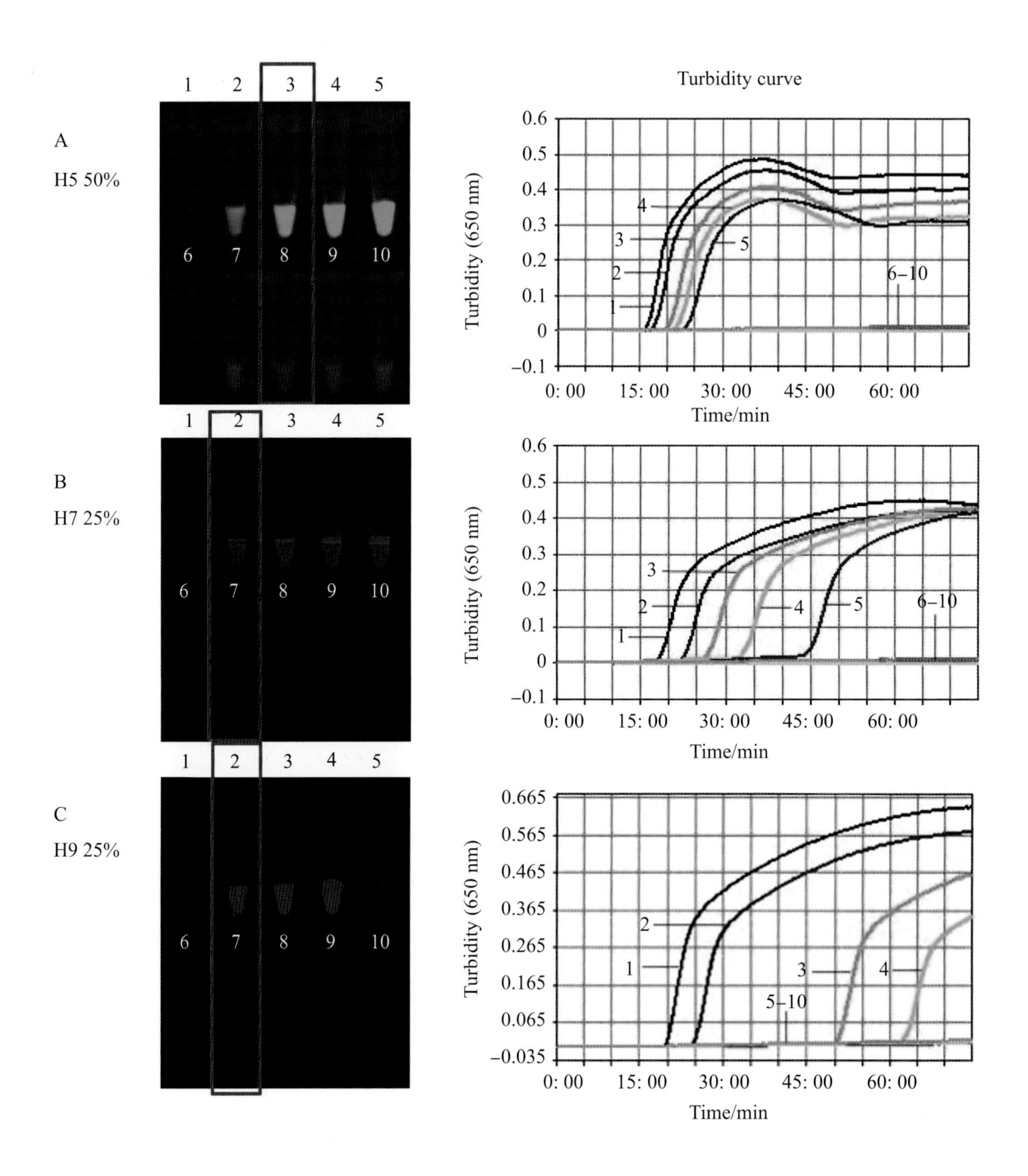

A: The optimal proportion of H5-FIP-FD among the total H5-FIP primers in the 520 nm channel was 50% at a H5-FIP-FD concentration of 0.133 μmol/L and an unlabelled H5-FIP concentration of 0.134 μmol/L and an initiation time of 20 min; B: The optimal proportion of H7 in the 670 nm channel was 25% with 0.067 μmol/L H7-FIP-FD, 0.2 μmol/L unlabelled H7-FIP, and an initiation time of 23 min; C: The optimal proportion of H9 in the 570 nm channel was 25% with 0.067 μmol/L H9-FIP-FD, 0.2 μmol/L unlabelled H9-FIP, and an initiation time of 25 min. When the H9 proportion was 100%, only 0.267 μmol/L H9-FIP-FD was used with no H9-FIP, and the reaction was completely suppressed. 1: FD-FIP composite probe among the total FIP = 0; 2: 25%; 3: 50%; 4: 75%; and 5: 100%; 6, 7, 8, 9, and 10 are the negative controls for the corresponding proportions.

Figure 5-10-3 Optimal proportion of FIP-FD among the total FIP primers for each subtype

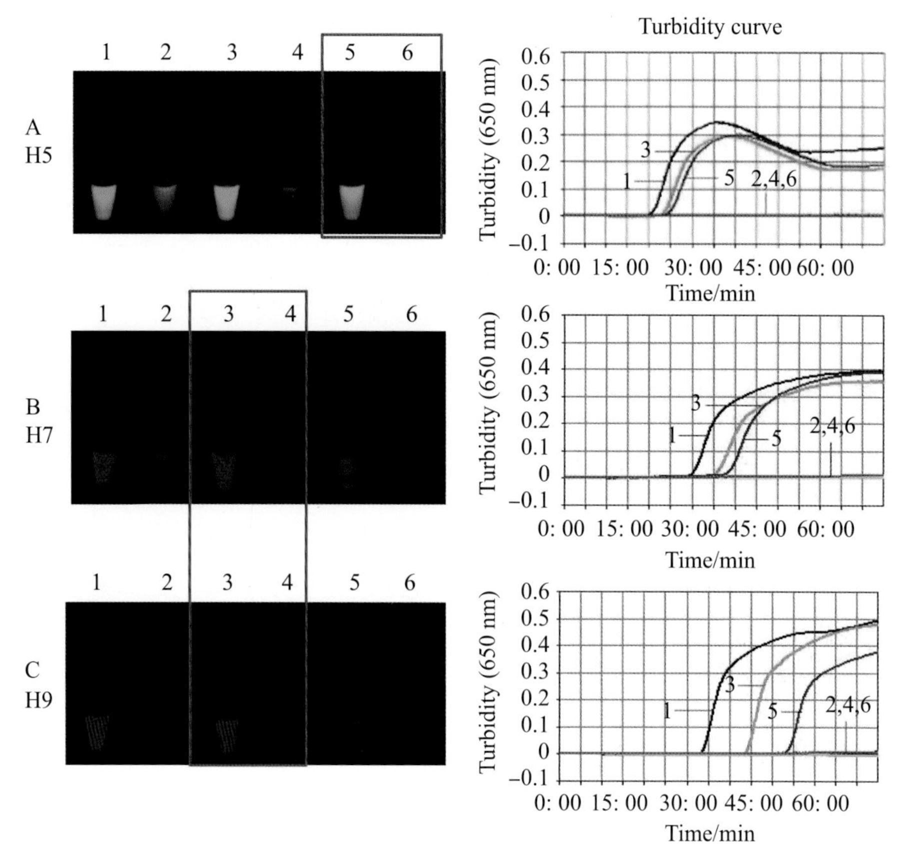

A: When only 0.134 μmol/L H5-FIP-Q was used instead of unlabelled H5-FIP in the H5 primer set, the reaction initiation time was 24 min, the background fluorescence was the lowest, and the difference between the negative and positive samples was the most pronounced; B: In the optimal reaction, the concentrations of the H7 primer group were 0.1 μmol/L unlabelled H7-FIP and 0.1 μmol/L H7-FIP-Q, the initiation time was 35 min, and the difference between the negative and positive samples was the most pronounced; C: In the optima reaction, the concentration of the H9 primer set was the same as that of the H7 primer set with a 43 min initiation time, and the difference between the negative and positive samples was the most pronounced. 1: Unlabelled FIP only (no FIP-Q was present); 2: The negative control of 1; 3: FIP-Q: unlabelled FIP = 1:1; 4: The negative control of 3; 5: FIP-Q only (no unlabelled FIP was present); and 6: The negative control of 5.

Figure 5-10-4 Optimal concentrations of FIP-Q and unlabelled FIP for each subtype

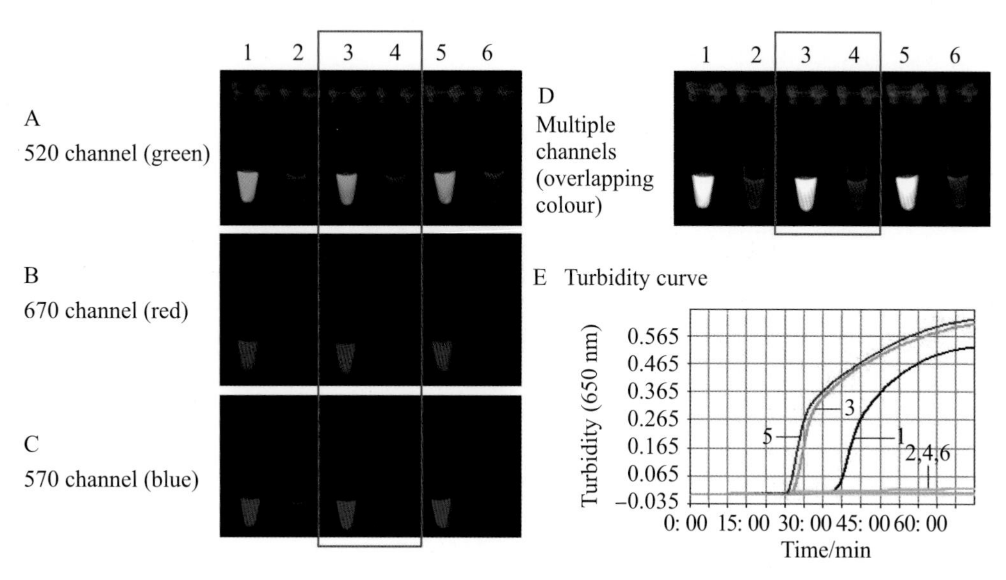

A: 520 nm channel; B: 670 nm channel; C: 570 nm channel; D: Multiple channels; E: Turbidity curve. 1: The ratio of H9 outer primer: inner primer : loop primer=1:8:2; 2: The negative control of 1; 3: H9 primer set with a ratio of 1 : 16 : 4; 4: The negative control of 3; 5: H9 primer set with a ratio of 1 : 24 : 6; 6: The negative control of 5.

Figure 5-10-5 Optimal amount of the H9 primers for the TLAMP reaction

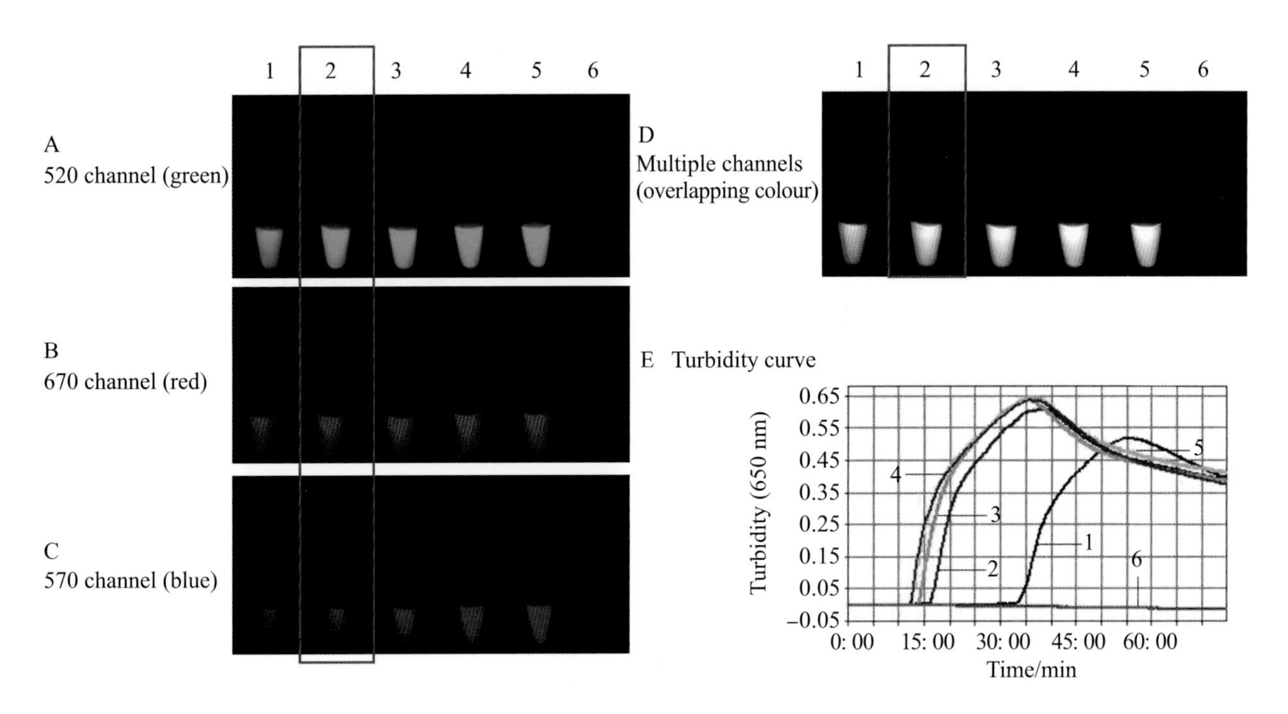

A: 520 nm channel; B: 670 nm channel; C: 570 nm channel; D: Multiple channels; E: Turbidity curve. 1: When no DNA polymerase was added, the reaction was slow, and 33 minutes was needed to initiate the reaction; 2: Each reaction with 2 U Bst 2.0 had an initial reaction time of 16 minutes; 3: 4 U Bst 2.0 was added over 14 minutes; 4: 6 U Bst 2.0 was added over 12 minutes; 5: 8 U Bst 2.0 was added over 12 minutes, and the effect of adding 8 U Bst 2.0 and 6 U; 6: The negative control.

Figure 5-10-6 Optimal amount of Bst 2.0 WarmStart DNA polymerase in the TLAMP reaction

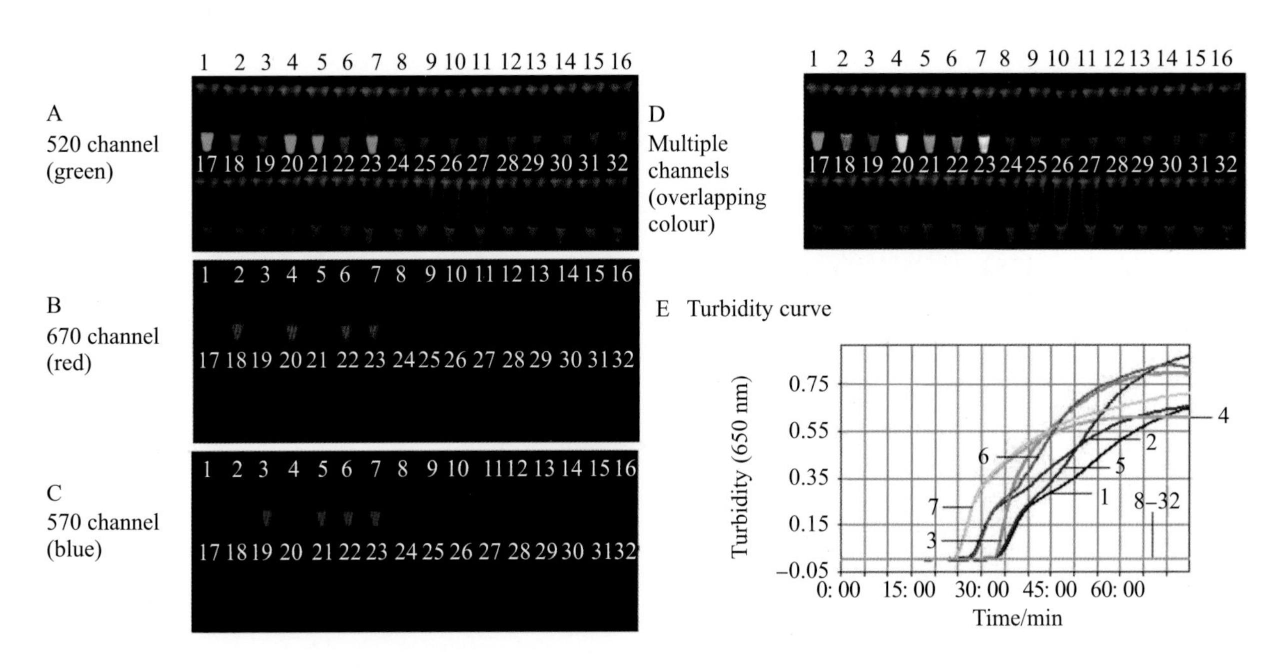

A: 520 nm channel; B: 670 nm channel; C: 570 nm channel; D: Multiple channels; E: Turbidity curve. 1: H5; 2: H7; 3: H9; 4: H5+H7; 5: H5+H9; 6: H7+H9; 7: H5+H7+H9; 8: H1; 9: H2; 10: H3; 11: H4; 12: H6; 13: H8; 14: H10; 15: H11; 16: H12; 17: H13: 18: H14; 19: H15; 20: H16; 21: NDV; 22: IBV; 23: ARV; 24: ILTV; 25: CIAV; 26: FAdV-4; 27: ChPV; 28: MDV; and 29-32: Negative control.

Figure 5-10-7 Specificity of the TLAMP assay

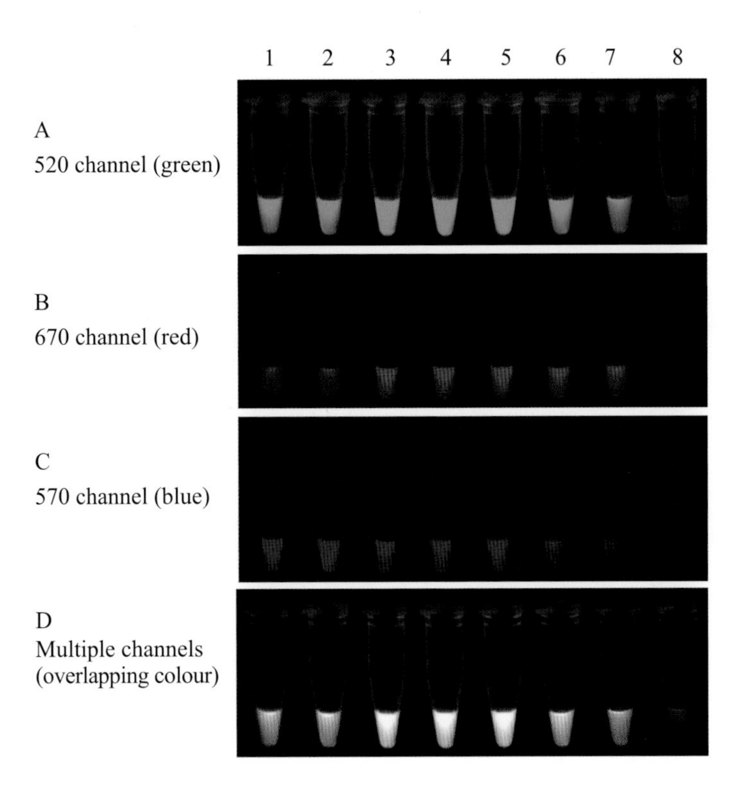

A: 520 nm channel; B: 670 nm channel; C: 570 nm channel; D: Multiple channels; E: Turbidity curve. 1: Sample 1, H5 (10^2 copies/μL) +H7 (10^2 copies/μL) +H9 (10^8 copies/μL); 2: Sample 2, H5 (10^4 copies/μL) +H7 (10^4 copies/μL) +H9 (10^7 copies/μL); 3: Sample 3, H5 (10^4 copies/μL) +H7 (10^5 copies/μL) +H9 (10^6 copies/μL); 4: Sample 4, H5 (10^5 copies/ μ L) +H7 (10^3 copies/ μ L) +H9 (10^5 copies/ μ L); 5: Sample 5, H5 (10^6 copies/μL) +H7 (10^8 copies/μL) +H9 (10^4 copies/μL); 6: Sample 6, H5 (10^4 copies/μL) +H7 (10^7 copies/μL) +H9 (10^3 copies/μL); 7: Sample 7, H5 (10^2 copies/μL) +H7 (10^8 copies/μL) +H9 (10^2 copies/μL); 8: Sample 8, negative control.

Figure 5-10-8 The results of the TLAMP assay

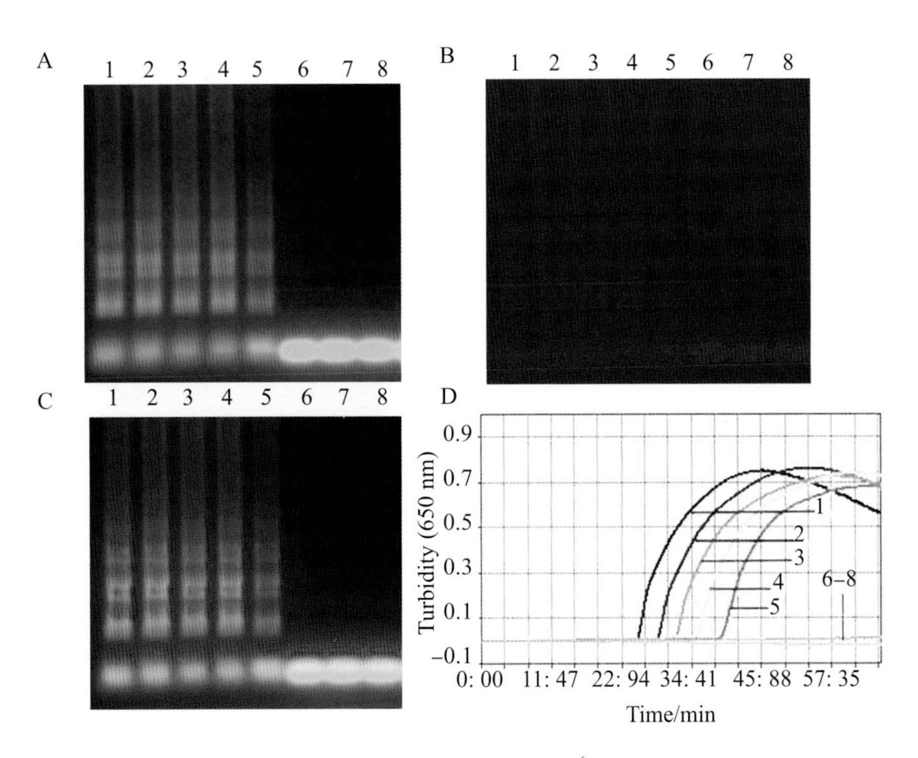

A: 520 nm channel; B: 670 nm channel; C: duplex channel; D: the turbidity curve. 1: 10^6 copies/μL (standards of equivalent pMD18T-MB and pMD18-T-BHV-1); 2: 10^5 copies/μL; 3: 10^4 copies/μL; 4: 10^3 copies/μL; 5: 10^2 copies/μL; 6: 10^1 copies/μL; 7: 1 copy/μL; 8: Negative control.

Figure 5-16-3 Sensitivity results of DLAMP for BHV-1 and MB

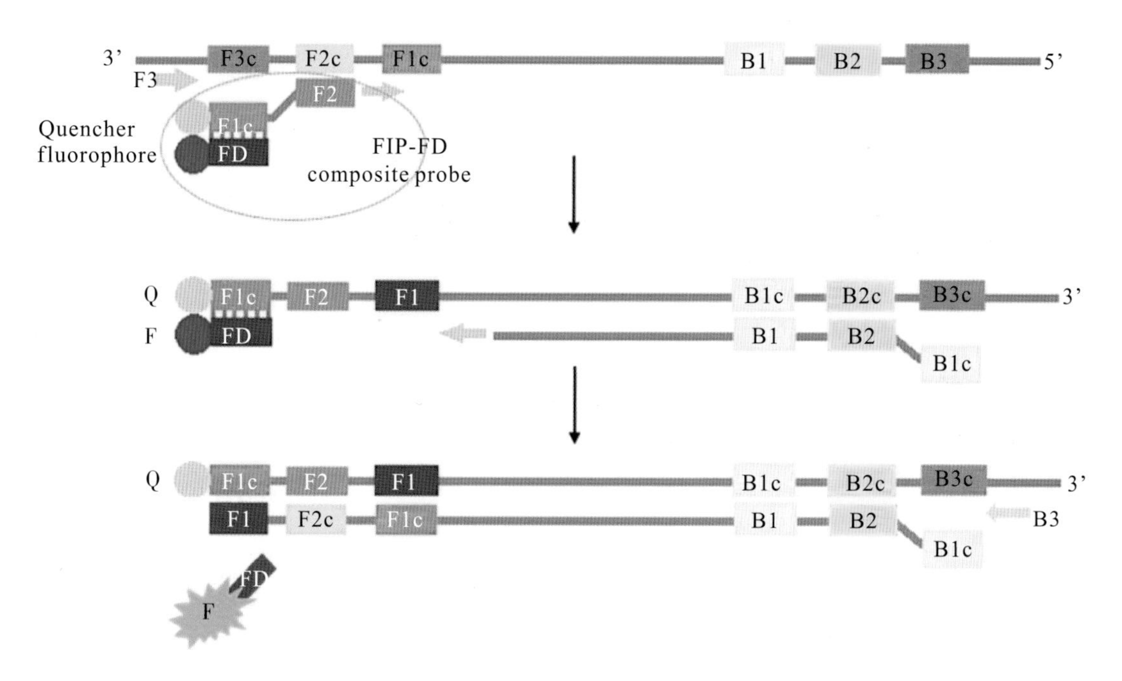

Figure 5-17-1 Schematic presentation of the mechanism of mLAMP

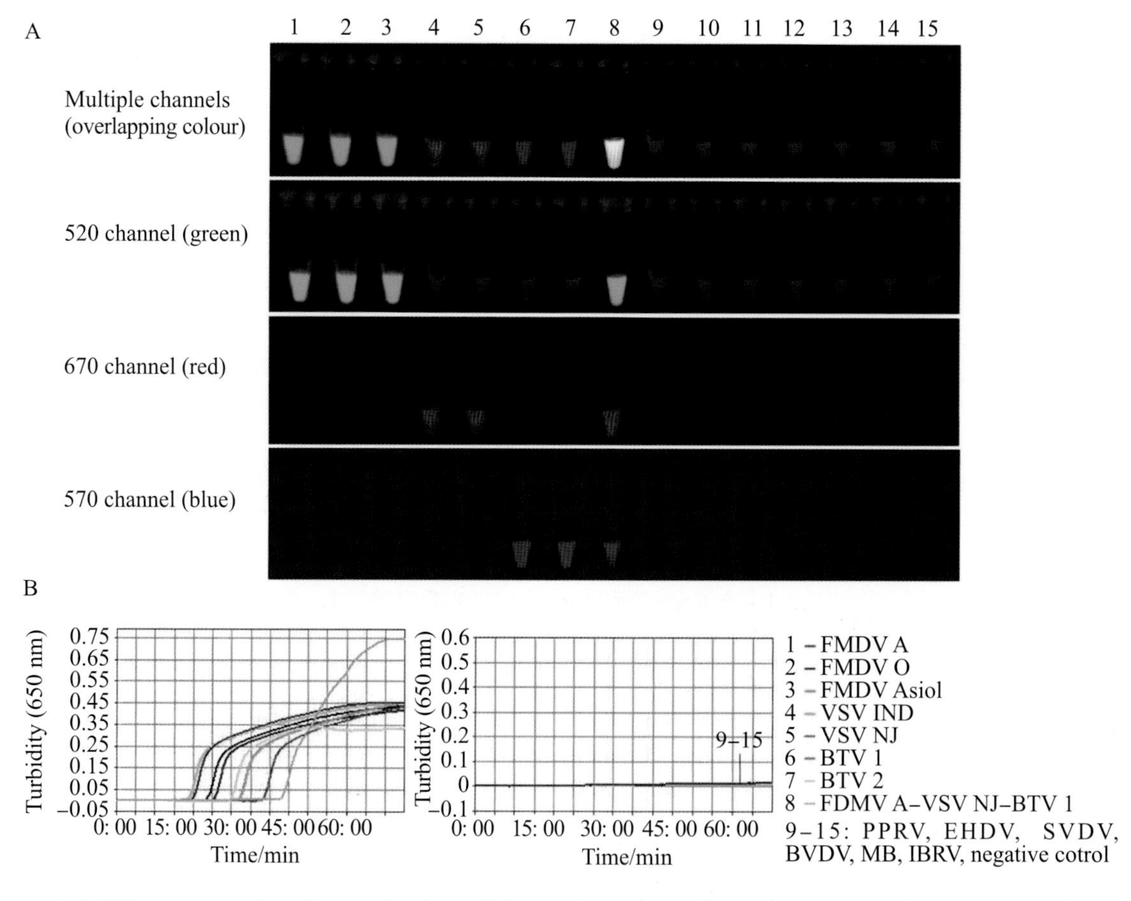

A: The fluorescent mLAMP products were imaged separately using multiple channels; B: The turbidity curve was generated by a real-time turbidimeter to interpret the process of amplification. Green fluorescence (FAM) indicates FMDV-positive amplification, red fluorescence (Cy5) indicates VSV-positive amplification, and blue fluorescence (Cy3) indicates BTV-positive amplification. Overlapping fluorescence indicates multiple positive amplifications. 1: FMDV A; 2: FMDV O; 3: FMDV Asia l; 4: VSV IND; 5: VSV ND; 6: BTV 1; 7: BTV 2; 8: FMDV A+VSV IND+BTV; 9-15: PPRV, EHDV, SVDV, BVDV, MB, IBRV, negative control.

Figure 5-17-6 Specificity of mLAMP

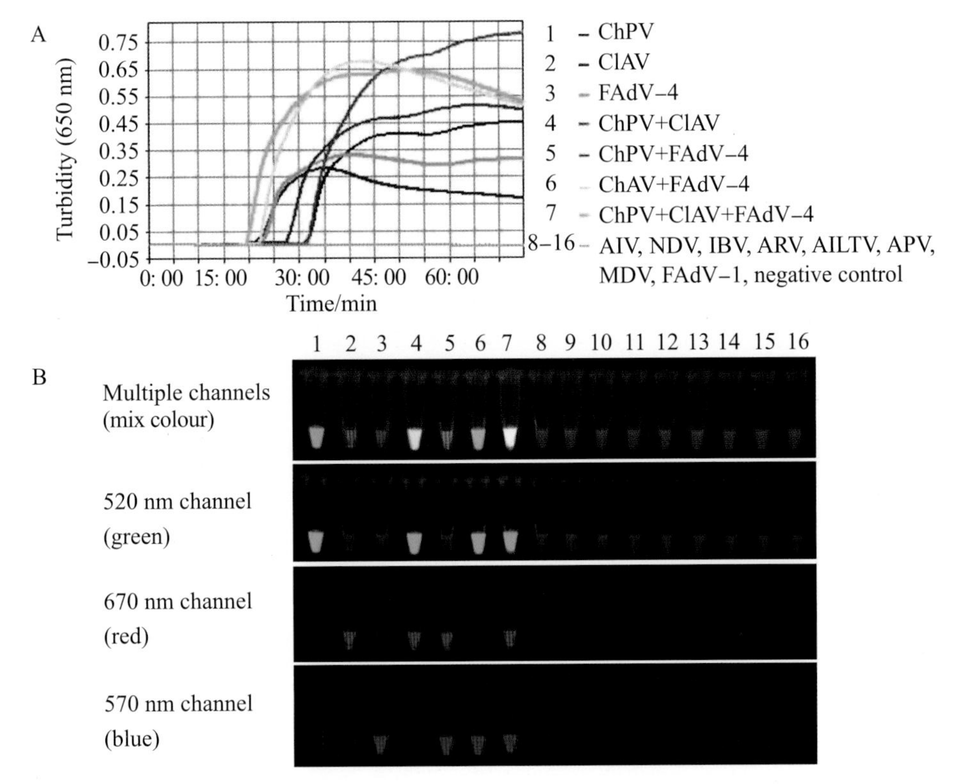

A: The turbidity curve; B: The fluorescent mLAMP products were imaged separately using multiple channels, the single reaction exhibits a single fluorescence signal in the corresponding channel.

Figure 5-18-1 Specificity of mLAMP

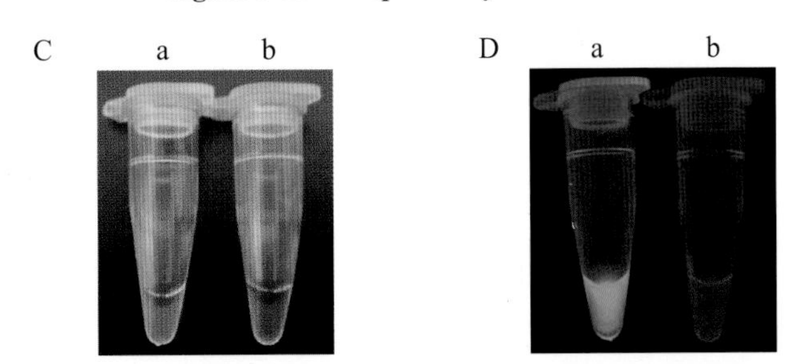

C:The result of RT-LAMP assay under daylight; D: The result of RT-LAMP assay under UV light. a: Positive sample (green); b: Negative sample.

Figure 5-20-2 (C and D) Detection of H3 subtype AIVs by RT-LAMP assay

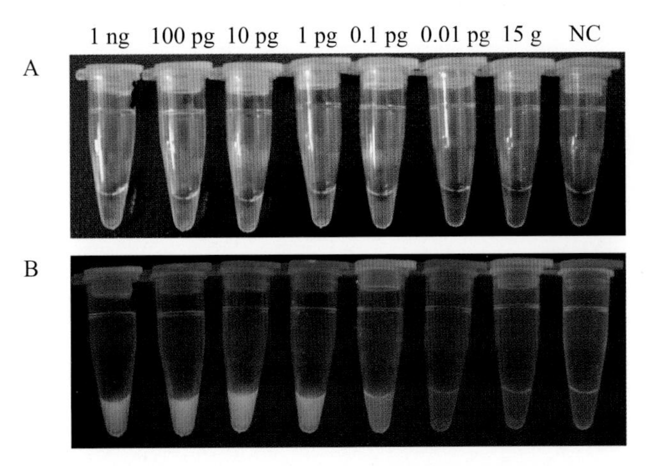

A: The result of sensitivity test for RT-LAMP assay under daylight; B: The result of sensitivity test for RT-LAMP assay under UV light.

Figure 5-20-3 (A and B) Comparative sensitivity tests between RT-LAMP and RT-PCR assays for detection of H3 subtype AIVs

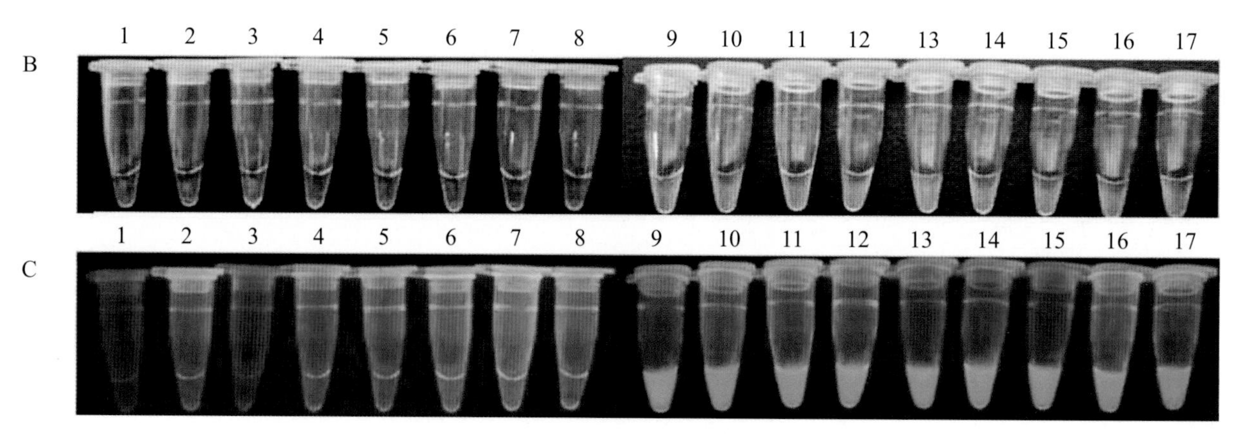

B: Visualization of the RT-LAMP products under daylight; C: Visualization of the RT-LAMP products under UV light. 1: H5 subtype avian influenza virus; 2: H9 subtype avian influenza virus; 3: Newcastle disease virus; 4: Infectious bronchitis virus; 5: Infectious laryngotracheitis virus; 6: Adenovirus; 7: *M. gallisepticum*; 8: *S. enteritidis*; 9: S1133 ARV; 10: 1733 ARV; 11: 203-M3 ARV; 12: C78 ARV; 13: R1 ARV; 14: R2 ARV; 15: ARV Iso1; 16: ARV Iso2; 17: ARV Iso3.

Figure 5-22-2(B and C)　Specificity of the RT-LAMP assay

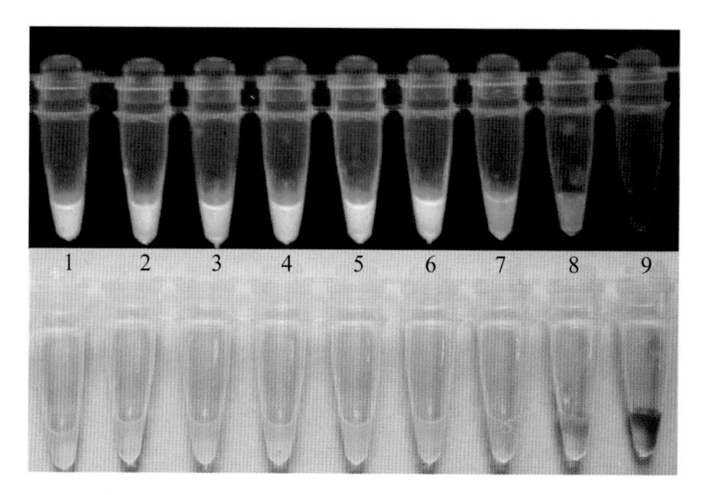

RT-LAMP product. 1: 3.32×10^7 copies/tube; 2: 3.32×10^6 copies/tube; 3: 3.32×10^5 copies/tube; 4: 3.32×10^4 copies/tube; 5: 3.32×10^3 copies/tube; 6: 3.32×10^2 copies/tube; 7: 3.32×10^1 copies/tube; 8: 3.32×10^0 copies/tube; 9: 3.32×10^{-1} copies/tube.

Figure 5-26-2 (A)　Sensitivity of RT-LAMP-and real-time RT-PCR

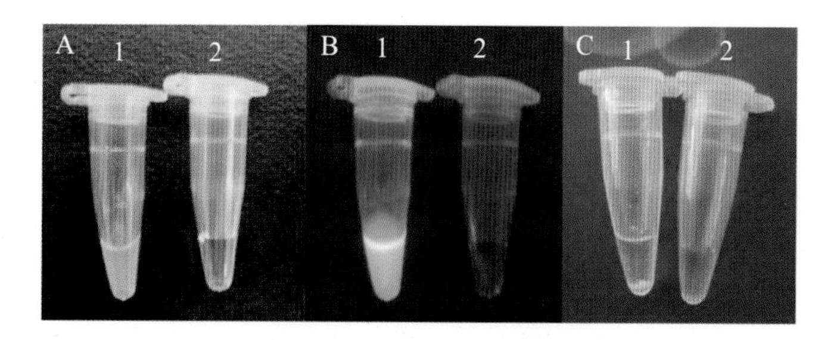

A: Fluorescent dye added seen without ultraviolet light; B: Fluorescent dye added seen with ultraviolet light; C: By turbidity with white sediment. 1: Positive control sample; 2: Negative control sample.

Figure 5-26-3　Detection of BRV RT-LAMP product

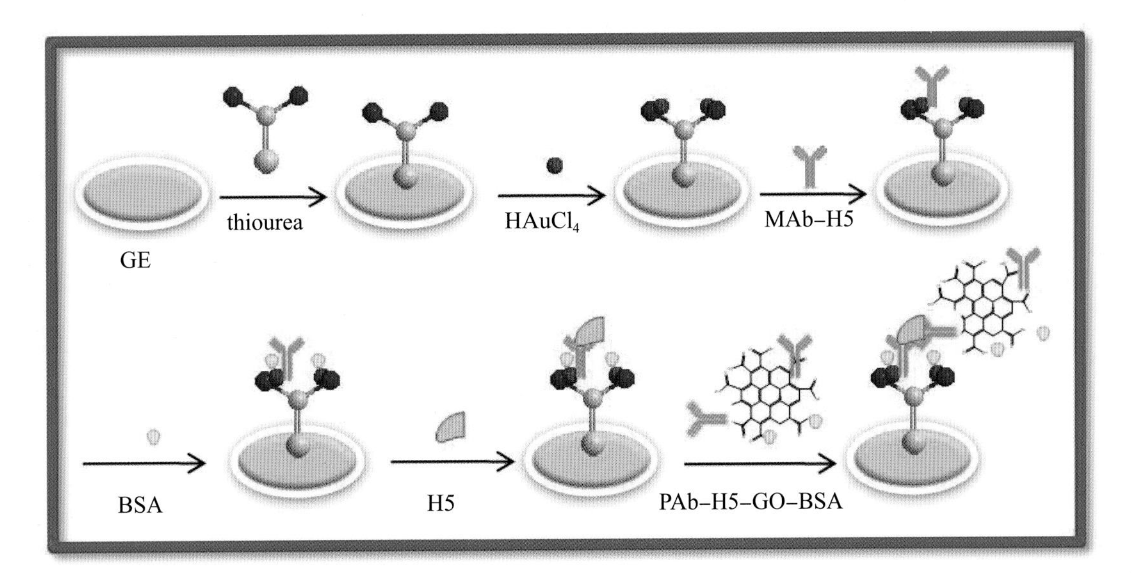

Figure 5-27-1 Immunosensor fabrication process

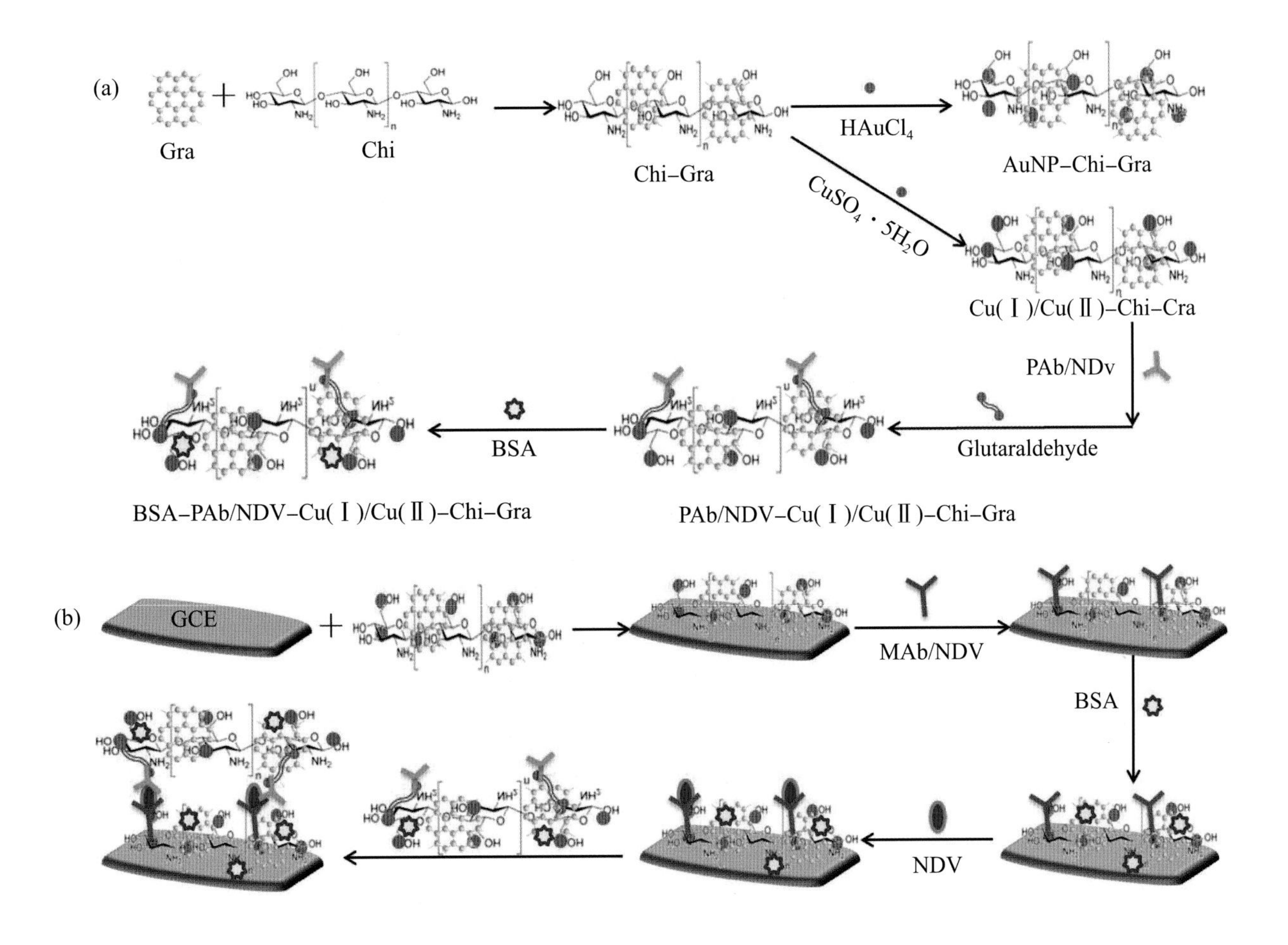

Figure 5-29-1 Preparation procedures of AuNP-Chi-Gra, Cu(I)/Cu(II)-Chi-Gra and the immunosensor

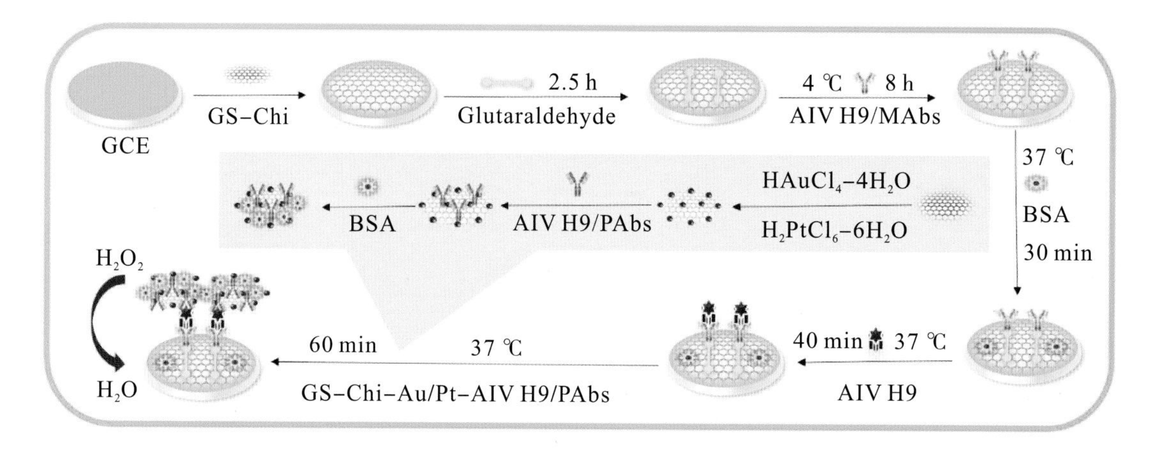

Figure 5-31-1 Schematic illustrating the electrochemical immunosensor

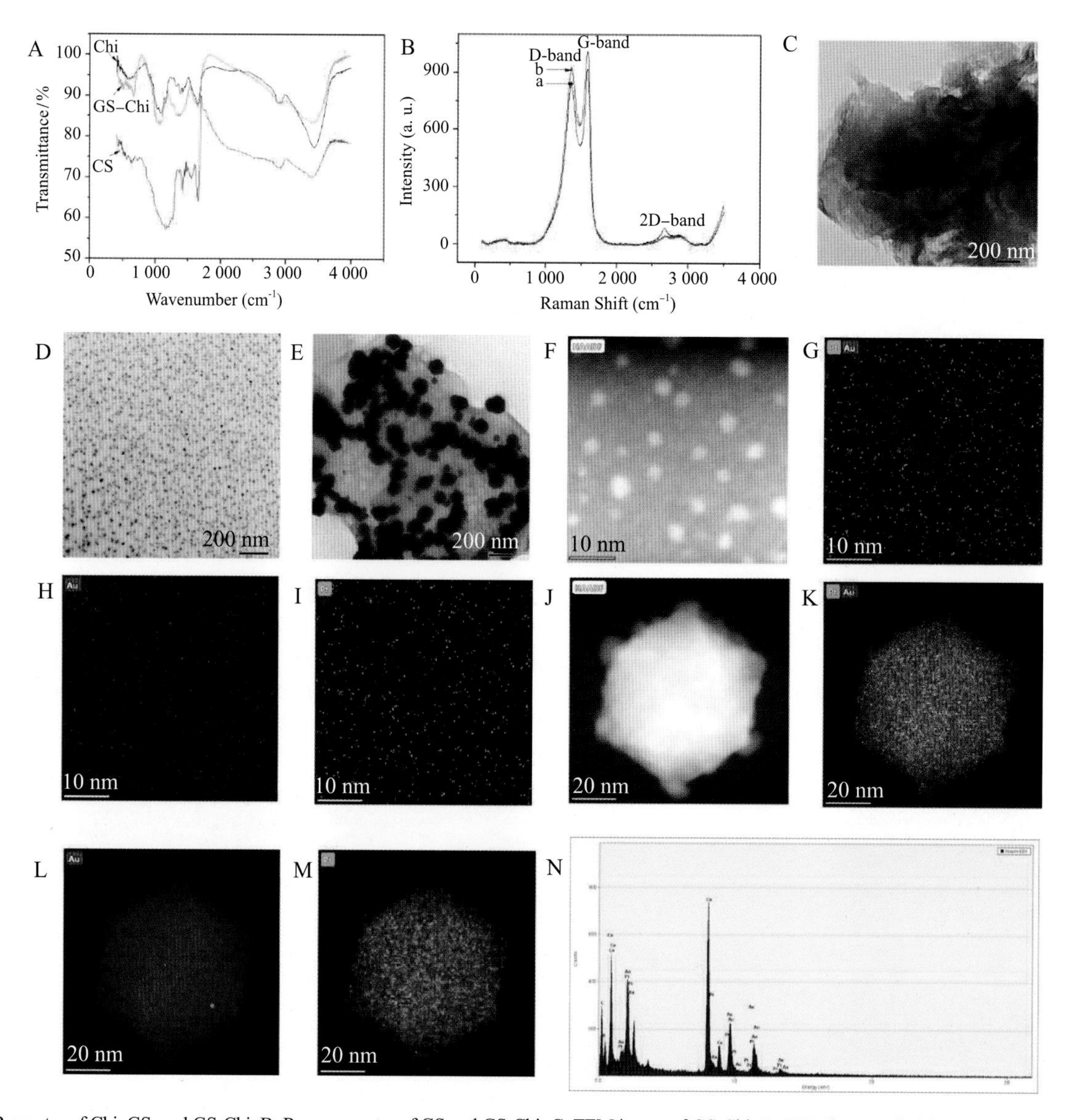

A: FT-R spectra of Chi, GS, and GS-Chi; B: Raman spectra of GS and GS-Chi; C: TEM image of GS-Chi; D: TEM image of Chi-Au/Pt; E: TEM image of GS-Chi-Au/Pt; F: HRTEM image of Chi-Au/Pt; G: Overlay on the image of Chi-Au/Pt; H: Au elemental mapping image of Chi-Au/Pt; I: Pt elemental mapping image of Chi-Au/Pt; J: GS-Chi-Au/Pt; K: Overlay on an image of GS-Chi-Au/Pt; L: Au elemental mapping image of GS-Chi-Au/Pt; M: Pt elemental mapping image of GS-Chi-Au/Pt; N: EDS elemental analysis of GS-Chi-Au/Pt.

Figure 5-31-2 Characterization of the related nanocomposites

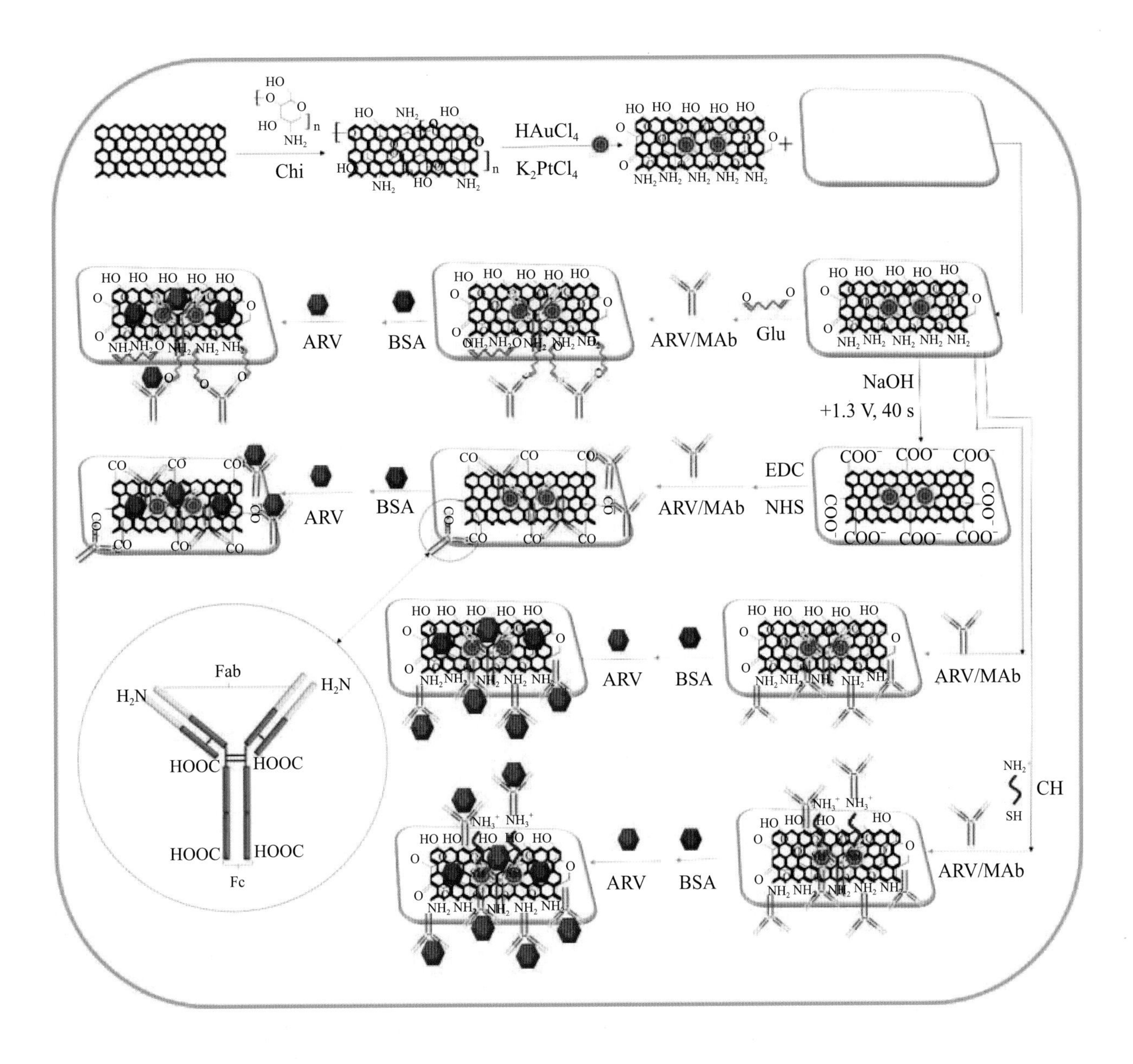

Figure 5-32-1　Schematic illustration of the electrochemical immunosensors

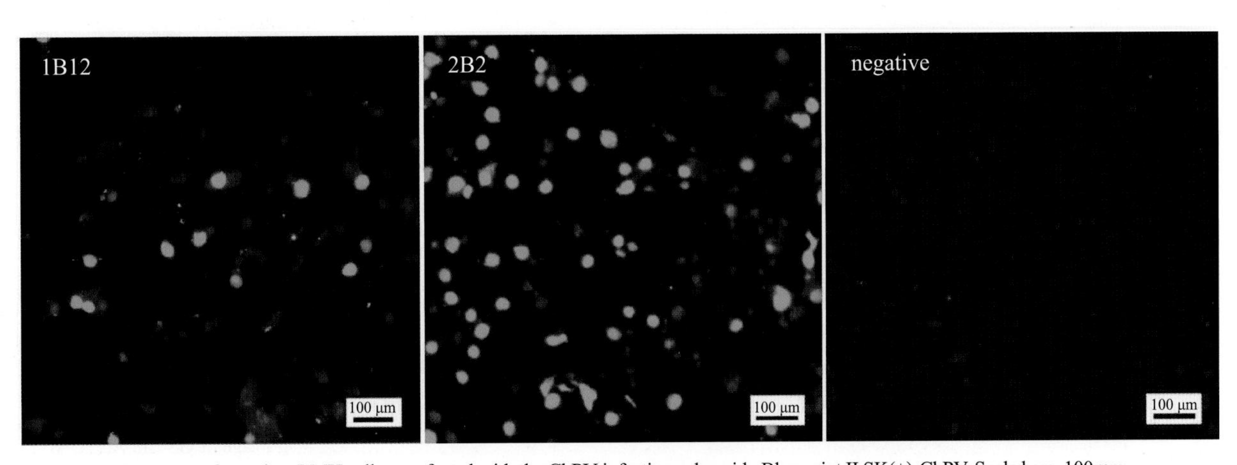

IFA was performed on LMH cells transfected with the ChPV infectious plasmid pBluescript II SK(+)-ChPV, Scale bars, 100 μm.

Figure 5-42-1(B)　The reactivity of the monoclonal antibodies

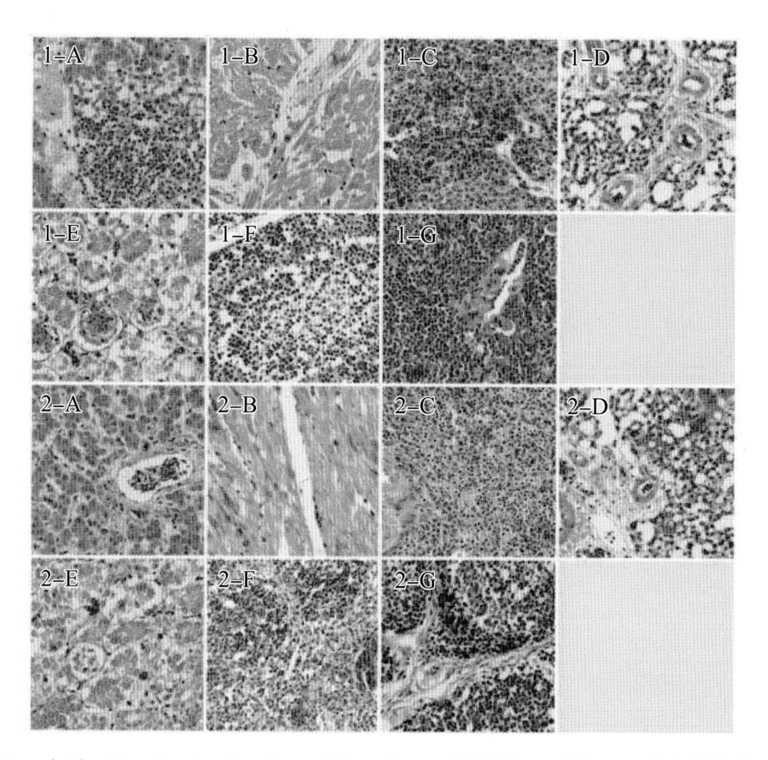

1-A–1-G: Three days post-immunization: liver, heart, spleen, lungs, kidneys, bursa of Fabricius, and thymus; 2-A–2-G: Seven days post-immunization: liver, heart, spleen, lungs, kidneys, bursa of Fabricius, and thymus.

Figure 6-1-2 Pathological tissue changes in the organs of SPF chickens after immunization with GX2019-014 via intramuscular injection

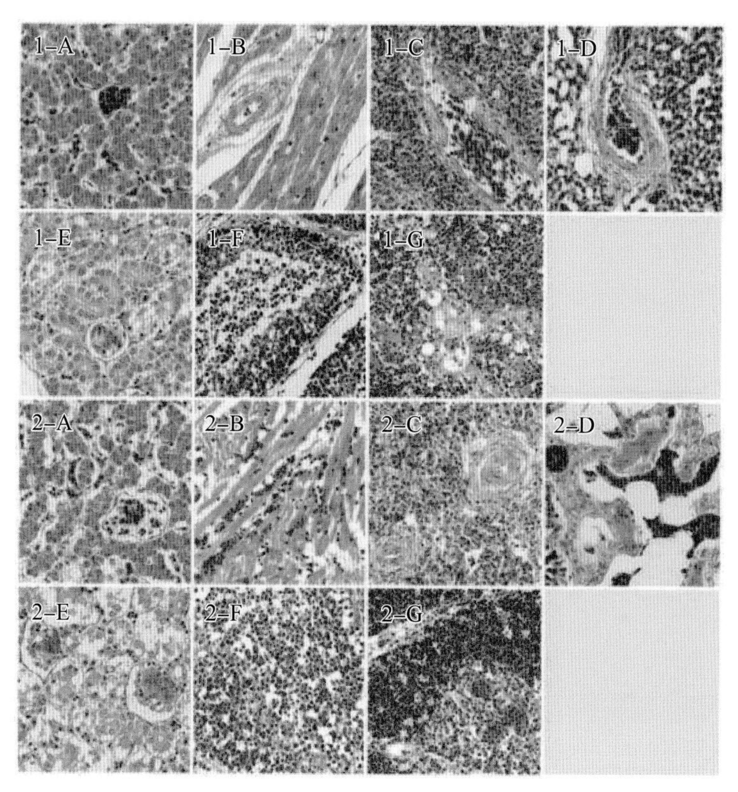

1-A–1-G: Three days post-immunization: liver, heart, spleen, lungs, kidneys, bursa of Fabricius, and thymus; 2-A–2-G: Seven days post-immunization: liver, heart, spleen, lungs, kidneys, bursa of Fabricius, and thymus.

Figure 6-1-3 Pathological tissue changes in the organs of SPF chickens after immunization with GX2019-014 via oral administration

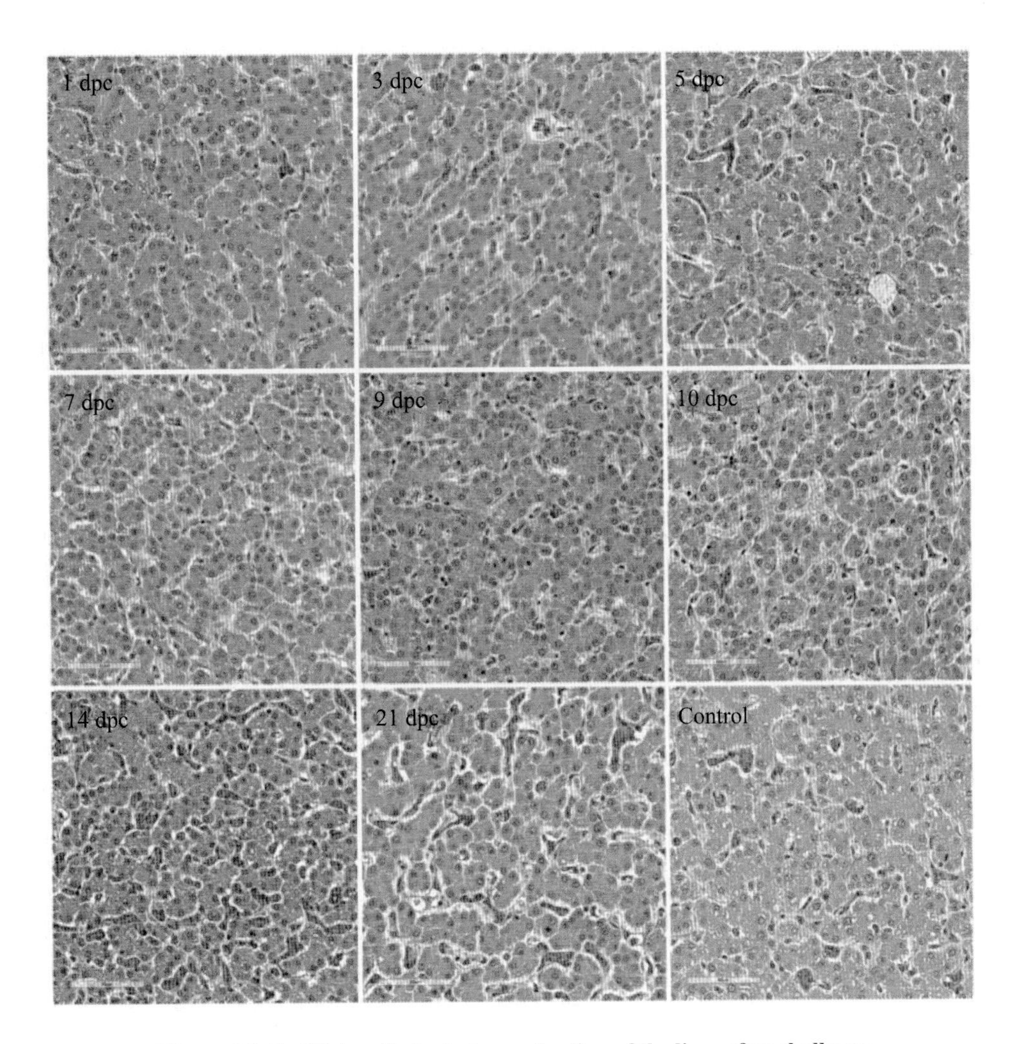

Figure 6-1-7 Histopathological examination of the liver after challenge

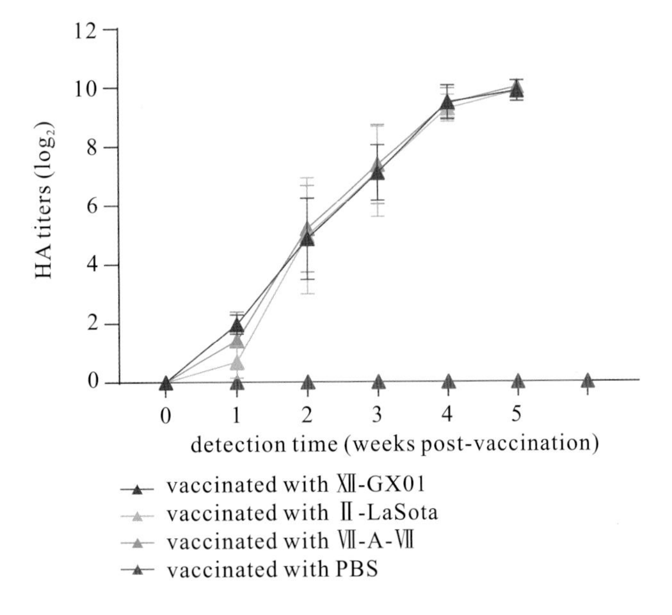

The data are presented as the means and SDs. Error bars indicate the standard deviations of measurements of 29 blood samples (n=29)

Figure 6-3-2 Antibody titers in the four groups inoculated with different vaccines

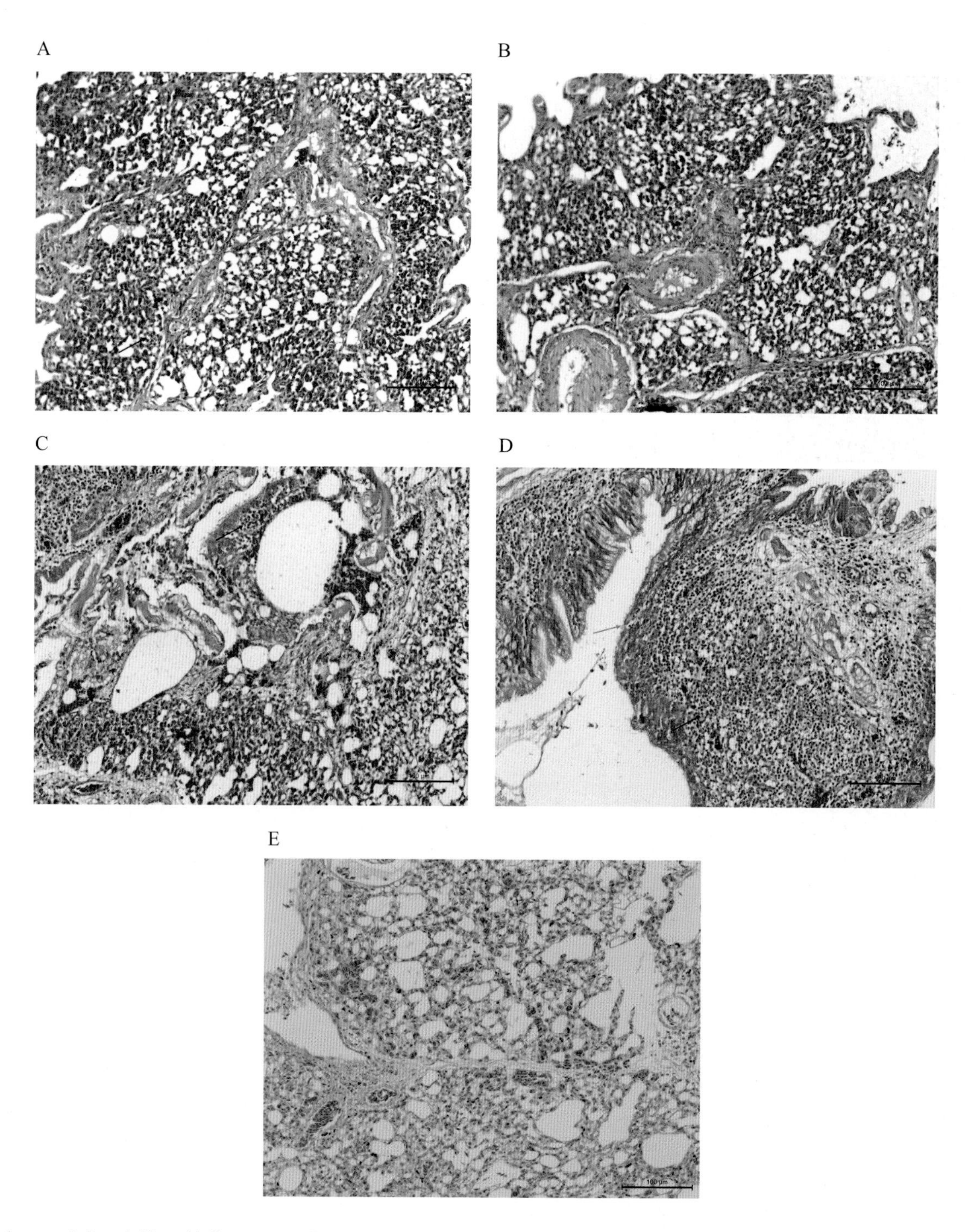

Black arrows indicate infiltrated inflammatory cells, green arrows indicate inflammatory exudate, and orange arrows indicate detached epithelial cells.

Figure 6-3-3 Pathological observation of the lungs

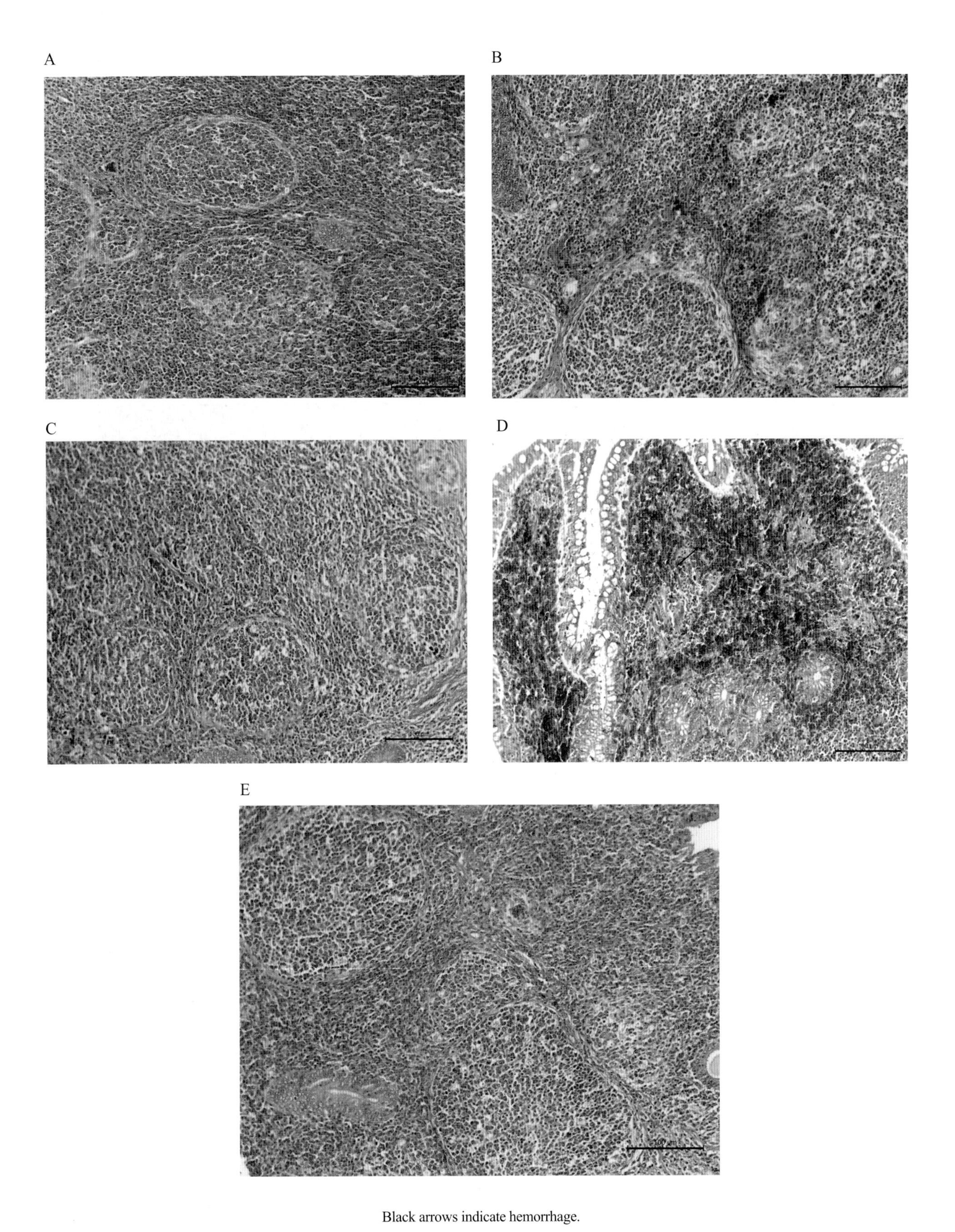

Black arrows indicate hemorrhage.

Figure 6-3-4 Pathological observation of the cecal tonsil